AR

ANNUAL REVIEW OF MICROBIOLOGY

ANNUAL REVIEW OF MICROBIOLOGY

VOLUME 41, 1987

L. NICHOLAS ORNSTON, *Editor*
Yale University

ALBERT BALOWS, *Associate Editor*
Centers for Disease Control, Atlanta

PAUL BAUMANN, *Associate Editor*
University of California, Davis

ANNUAL REVIEWS INC. 4139 EL CAMINO WAY P.O. BOX 10139 PALO ALTO, CALIFORNIA 94303-0897

ANNUAL REVIEWS INC.
Palo Alto, California, USA

International Standard Serial Number: 0066–4227
International Standard Book Number: 0–8243–1141-8
Library of Congress Catalog Card Number: 49-432

Typesetting by Kachina Typesetting Inc., Tempe, Arizona; John Olson, President
Typesetting coordinator, Janis Hoffman

PRINTED AND BOUND IN THE UNITED STATES OF AMERICA

PREFACE

Probably every child spends some time wondering why adults are so crazy. The question is usually answered in early adulthood with the realization that our parents' revenge is that we become them. Several decades ago microbiology spawned molecular biology. Brash and relentlessly reductionist in its youth, molecular biology seemed on the verge of solving every biological problem in sight. Biological problems did not vanish as molecular biology matured, and now the science is accused by those who do not understand it of being merely a collection of techniques. This is nonsense. Good biology—observation, discernment of pattern, testing of generalization—can be practiced quite effectively at the molecular level. This is appreciated full well by microbiologists, who depend on biochemistry and molecular biology in order to understand microorganisms. It is gratifying that molecular biologists increasingly appreciate the extraordinary diversity of microorganisms as a source of good biological questions. Microbiology also demands a sense of physiology, genetics, population biology, ecology, and evolution, so the field expands the minds of its practitioners. All biologists, indeed all scientists, are welcome in microbiology, a place where visitors tend to stay.

The mind-expanding experience of planning this volume of the *Annual Review of Microbiology* was shared by the editorial committee members named at the front of the volume. We thank Sharon Long for providing valued contributions as a guest. With the planning of Volume 41 we welcomed E. Peter Greenberg as a new member of the editorial committee. With regret we note the retirement of Stanley Falkow and Francesco Parenti from the editorial committee after the planning of this volume. Their stimulating ideas and thoughtful advice have been greatly appreciated. The volume was brought to fruition on schedule under the graceful guidance of Production Editor Andrea Perlis. Whatever contributions emerged from my own office were made possible by the admirable organizational talents of Susan Voigt and Kim Williams, on whom my dependence is absolute.

L. NICHOLAS ORNSTON
EDITOR

ANNUAL REVIEWS INC. is a nonprofit scientific publisher established to promote the advancement of the sciences. Beginning in 1932 with the *Annual Review of Biochemistry*, the Company has pursued as its principal function the publication of high quality, reasonably priced *Annual Review* volumes. The volumes are organized by Editors and Editorial Committees who invite qualified authors to contribute critical articles reviewing significant developments within each major discipline. The Editor-in-Chief invites those interested in serving as future Editorial Committee members to communicate directly with him. Annual Reviews Inc. is administered by a Board of Directors, whose members serve without compensation.

Annual Review of Microbiology
Volume 41, 1987

CONTENTS

(*continued*)

OTHER REVIEWS OF INTEREST TO MICROBIOLOGISTS

From the *Annual Review of Biochemistry,* Volume 56 (1987)

Complexes of Sequential Metabolic Enzymes, Paul A. Srere
Enzymes of General Recombination, Michael M. Cox and I. R. Lehman
Transfer RNA Modification, Glenn R. Björk, Johanna U. Ericson, Claes E. D. Gustafsson, Tord G. Hagervall, Yvonne H. Jönsson, and P. Mikael Wikström
The Structure and Function of the Hemagglutinin Membrane Glycoprotein of Influenza Virus, Don C. Wiley and John J. Skehel
DNA Mismatch Correction, Paul Modrich
The Molecular Biology of the Hepatitis B Viruses, Don Ganem and Harold E. Varmus
Cyclic AMP and Other Signals Controlling Cell Development and Differentiation in Dictyostelium, Günther Gerisch
Protein Glycosylation in Yeast, M. A. Kukuruzinska, M. L. E. Bergh, and B. J. Jackson

From the *Annual Review of Biophysics and Biophysical Chemistry,* Volume 16 (1987)

Polar Lipids of Thermophilic Prokaryotic Organisms: Chemical and Physical Structure, Vittorio Luzzati, Agata Gambacorta, Mario DeRosa, and Annette Gulik

From the *Annual Review of Cell Biology,* Volume 3 (1987)

Replication of Plasmids Derived from Bovine Papilloma Virus Type 1 and Epstein-Barr Virus in Cells in Culture, Joan Mecsas and Bill Sugden

From the *Annual Review of Genetics,* Volume 21 (1987)

The Genetics of Active Transport in Bacteria, Howard A. Shuman
Regulatory Proteins in Yeast, Leonard Guarente
Behavioral Genetics of Paramecium, Yoshiro Saimi and Ching Kung
Mechanism and Control of Homologous Recombination in Escherichia coli, Gerald R. Smith
DNA Methylation in Escherichia coli, M. G. Marinus
The Essential Role of Recombination in Phage T4 Growth, Gisela Mosig

From the *Annual Review of Medicine,* Volume 38 (1987)

Live Attenuated Varicella Vaccine, Anne A. Gershon
Chronic Epstein-Barr Virus Infection, James F. Jones and Stephen E. Straus

(*continued*)

From the *Annual Review of Phytopathology,* Volume 25 (1987)

The Changing Scene in Plant Virology, R. E. F. Matthews

Ecology and Pathogenicity of Anastomosis and Intraspecific Groups of Rhizoctonia solani *Kühn,* Akira Ogoshi

Current Status and Future Prospects of Research on Bacterial Blight of Rice, T. W. Mew

Molecular Genetics of Pathogenesis by Soft-Rot Erwinias, A. Kotoujansky

Beetle Transmission of Plant Viruses, J. P. Fulton, R. C. Gergerich, and H. A. Scott

Structural and Chemical Changes Among the Rust Fungi During Appressorium Development, H. C. Hoch and R. C. Staples

Molecular Markers for Genetic Analysis of Phytopathogenic Fungi, R. W. Michelmore and S. H. Hulbert

Modeling the Long-Range Transport of Plant Pathogens in the Atmosphere, Jerry M. Davis

Screening for Fungicides, M. C. Shephard

Rhizobium—*The Refined Parasite of Legumes,* Michael A. Djordjevic, Dean W. Gabriel, and Barry G. Rolfe

Interactions of Deleterious and Beneficial Rhizosphere Microorganisms and the Effect of Cropping Practices, Bob Schippers, Albert W. Bakker, and Peter A. H. M. Bakker

Epidemiology of Aflatoxin Formation by Aspergillus flavus, Urban L. Diener, Richard J. Cole, T. H. Sanders, Gary A. Payne, Louise S. Lee, and Maren A. Klich

Fungal Endophytes of Grasses, M. R. Siegel, G. C. M. Latch, and M. C. Johnson

From the *Annual Review of Plant Physiology,* Volume 38 (1987)

Plant Virus-Host Interactions, Milton Zaitlin and Roger Hull

Stanley Dagley

Ann. Rev. Microbiol. 1987. 41:1–23

LESSONS FROM BIODEGRADATION

Stanley Dagley

Department of Biochemistry, University of Minnesota, College of Biological Sciences, St. Paul, Minnesota 55108

CONTENTS

PHYSICAL CHEMISTRY

It is perhaps unusual to be able to pinpoint to within two hours the start of fifty years' interest in a research area. However, I well recall the sunlit morning of early summer in 1937 when I called upon C. N. Hinshelwood in his rooms at Trinity College, Oxford, to find out whether he would be willing to let me join his group to do research in chemical kinetics. Yes, he was willing, but on condition that I helped him in a new venture: a study of the kinetics of growth of certain bacteria. My initial astonishment quickly gave place to dismay: As a chemist I knew almost nothing about bacteria and, what was more important, I had no experience in handling them. However, "Hinsh" was persuasive. He pointed out that if the sign of the exponential term is changed from negative to positive, the rate of increase of a dividing bacterial population is governed by an equation similar to the equation that describes the course of a unimolecular gas reaction such as I had expected to study. Moreover, nothing was known at that time about the dependence of growth rate upon nutrient concentration, or about the factors that determined the point at which growth stopped. My curiosity about bacteria was aroused, but the starting point for the investigation was most exceptional: chemical kinetics.

0066-4227/87/1001-0001$02.00

The development of this approach was later received in biological circles with less than universal enthusiasm; this was sometimes (but not always) for good reasons, which I shall mention. It is therefore appropriate for me to make clear at the outset that Sir Cyril Hinshelwood was the most erudite person I have ever known, and also one of the kindest. As his obituary in *Nature* (65) states, added to his fluency in numerous European languages was a competence in Russian and Chinese, with a little Arabic. He was an accomplished painter in oils, and he often contributed papers to the Oxford Dante Society. He was in turn president of the Chemical Society, the Faraday Society, the Royal Society of London, and the British Classical Association. From my own experience I can say that, in a life so full, he would usually answer any letter seeking advice or help, by return of post.

Compared with the controversial later publications from Hinshelwood's laboratory, our first three papers on bacterial growth (28–30) made little impression. However, two of them dealt with topics that became more significant later. The first paper (28) showed that for a well-buffered medium with other nutrients in excess, the size of the bacterial crop was a linear function of the amount of carbon source (glucose) provided for growth. This relationship held for a range of concentrations over which there was no systematic change in growth rate. Incidentally, the construction of just one good growth curve using the method then available, hemocytometer counts, required patience. Since it had generally been believed in the 1930s that bacterial cultures ceased to grow when the cells had filled all the "available biological space" (a concept difficult to quantify), it was gratifying to be able to demonstrate a simple linear relationship between crop and nutrient concentration. This relationship proved to be remarkably reproducible so long as the nutrient was limiting. At higher concentrations, accumulation of toxic products, including development of an adverse pH, could limit growth. Twenty years later, when the role of ATP in biochemistry was appreciated, Bauchop & Elsden (6) measured amounts of bacteria and yeast grown under anaerobic conditions and showed that crop sizes were directly proportional to the ATP that could be derived from the catabolism of the carbon source. Dawes (39) has recently discussed the value of the concept of molar growth yield (*Y*).

The work described in the second (29) of the above-mentioned papers was a follow-up of an observation made three years earlier: When mineral salts media were thoroughly freed from CO_2, the growth of several species of bacteria was delayed indefinitely (46). We measured growth rates of *Aerobacter aerogenes* and expressed them as a function of the CO_2 content of the airstream that was aspirated continuously through the growing cultures (29). Growth increased steadily to 0.15% CO_2 (the content of fresh air was about 0.03%) and had significantly decreased by the time the CO_2 content reached 5%. The proposal that CO_2 is an obligatory metabolite for a heterotroph was not readily accepted in the 1930s; when I presented a paper one evening to an Oxford scientific club, I was given a hard time by a more senior bacteriologist who had no doubt that my findings could be accounted for by inadequate experimentation. Neither of us was aware that Wood & Werkman (85) had recently presented a paper in Minneapolis to the North Central Branch of the Society of American Bacteriologists, as the ASM was then designated, in which they proposed that CO_2 was metabolized by such typically heterotrophic organisms as the propionic acid bacteria. These authors established the new metabolic status of CO_2 when, in an outstanding display of painstaking, accurate chemical analysis, they showed that an equimolar relationship existed between CO_2 fixed and succinic acid formed by these bacteria (86).

We are now sufficiently far removed from the dominant interests of the 1930s and 1940s to be able to appreciate the interplay between scientific creativity and the formulation of the theories of that period. The study of metabolism provides examples. This is an area where a good theory is one that makes sense of a wide range of current observations, but an even better theory may not be all-embracing. Satisfaction does not then generate complacency; new research is stimulated and a modified theory emerges to accommodate a wider range of new findings. During the period under discussion, for example, the Krebs cycle had been formulated and had then immediately been rejected for publication by *Nature;* indeed, the cycle did not assume its present form until the discovery of acetyl coenzyme A, and its scope was not fully appreciated until its central place in metabolism, where it is related both to biosynthesis and to terminal oxidation, was recognized.

In Hinshelwood's laboratory, after I left, many experiments were performed to test his kinetic model of the bacterial cell based upon the "principle of total integration" and a related concept termed the "network theorem" (40). Differential equations with variables in common were developed to describe the dependence of bacterial growth upon the operation of closed cycles of mutually dependent interactions. This approach rightly emphasized that of the multitude of components that constitute a bacterial cell, none is formed or functions in isolation from the rest. The cell achieves a steady state of mutual dependence: The proportions of the various cell constituents settle down to constant values, and when the biochemical environment of the cell is altered these proportions modulate to new values that are consistent with optimum growth. This kinetic adjustment is revealed by mathematical analysis (40). Given the initial assumptions, which are reasonable, the system has its own built-in mechanism for regulation. The model is valuable for bacterial physiologists, although the mathematical approach is more congenial to chemical engineers than to most biologists.

With the wisdom of hindsight, it now appears that the model was not as fruitful as it promised to be simply because it initially succeeded in explaining regulation entirely in terms of concepts that were then current. Further hypotheses were not needed. But they might well have been needed had the initial research been extended. I had used only glucose as carbon source when I demonstrated the linear dependence of bacterial crop upon nutrient concentration. Quite independently, Monod (64) had started to measure kinetics of growth of *Escherichia coli* or *Bacillus subtilis* in media containing mixed sugars in limiting amounts, and had used the linear function as an analytical tool. "Diauxie" curves, with two growth phases, were often obtained; but of greater interest was the observation that the cells did not always use first the particular sugar on which the inoculum had been grown. Glucose present in the mixture could give rise to catabolite repression and so be consumed first; the second sugar would then be metabolized only after readaptation during a lag phase. At the time, such results were unexpected; new concepts consequently emerged and further experiments were suggested, notably on enzymes such as β-galactosidase that are involved in metabolism of specific sugars. The initial growth experiments may have been simple, but their eventual elaboration, notably by the Paris school, had a tremendous influence on general concepts in the field of biochemical regulation. In particular, the emerging concept of allosterism revealed that the control of enzyme rates was often a matter of specific responses to effectors rather than a general consequence of adjustment of kinetic parameters. But these wonderful new insights did not please everyone, and it was after I moved to the United States that I heard the new excitement in enzymology described as "allohysteria."

It has been said that a university exists to find and communicate the truth. That is, research and teaching ought to be aspects of the same enterprise. In my lifetime this ideal has become much more difficult to realize as large universities have become more like small corporations. The type of market research most easily comprehended by university administrators consists of ranking departments by quizzes answered by faculty in other institutions. These rankings largely reflect the amount of attention that the local research, for various reasons, has attracted elsewhere. Further, departments in the biological sciences are often financially dependent upon grants from government and other sources, awarded to support faculty research. To young faculty, therefore, the voice of authority is loud and clear on how priorities must be ordered if their careers are to be extended beyond a probationary five-year period: Publish or perish. But despite these tensions the close relationship of teaching and research emerges clearly from the topics dealt with by successful textbooks of any particular decade.

I have examined elsewhere the evolution of biochemical texts (16). Only those that are best fitted to prepare the young for the research of the next few years survive, and then in their time these texts also become extinct. In the 1930s it seemed to be a matter of common sense that the area of greatest relevance for future developments in biochemistry would be colloid chemistry. One of the best, most widely adopted textbooks of the period was R. A. Gortner's *Outlines of Biochemistry;* one third of the second edition was devoted to colloids (47). As with bacterial kinetics, the physicochemical treatment provided a useful general framework, but profound modification and change of emphasis occurred in later texts such as the excellent book of Fruton & Simmonds (45). These changes were called for as attention was focused increasingly upon events at the molecular level, and they reflect how knowledge is extended, as well as communicated. For example, adsorption at a surface, enzymatic reaction rates, and bacterial growth rates can each be described by a relationship of the same mathematical form, which gives a rectangular hyperbola when plotted. These relationships are, respectively, the Langmuir isotherm, the Michaelis-Menten equation, and the Monod equation. They predict how the fraction of available surface filled, the enzymatic reaction rate, and the specific bacterial growth rate, respectively, increase with concentration to a saturating or maximal value. Since the mathematical relationships are similar in each case, some early derivations of the dependence of enzymatic reaction rate on substrate concentration were based on the same assumptions as those for the Langmuir isotherm; namely, that substrate molecules are adsorbed and partially saturate the enzyme surface, so the reaction rate is proportional to the surface density of the adsorbed molecules. By contrast, the alternative, modern treatment proved more fruitful because it focused attention on organic chemistry and reaction mechanisms. Enzymology opened up to more precise investigation when urease and pepsin were crystallized. Enzymes were elevated from their rather ill-defined status as colloids to that of organic compounds that undergo stoichiometric and stereospecific reactions with substrates. However, it was a branch of physical chemistry, thermodynamics, that supplied the study of metabolism with its most fruitful and enduring generalization.

The classical essay of Lipmann (60), published in 1941, provided the concepts that were needed to describe how solar energy is harnessed and ultimately put to use to sustain all life, the universal carrier of harnessed energy being ATP. The term "energy-rich" was introduced for ATP and was later extended to other metabolites whose standard free energies of hydrolysis are relatively high. Although it is fortunately much less true of science than of religion that original insights can become distorted or enfeebled when developed by the uninspired, I could nevertheless cite college texts, even from the 1960s and 1970s, that provide bizarre examples of misinstruction of the young concerning thermodynamics when applied to metabolism. The basic errors, although there were others, were the following: First, the free energy released in a reaction was treated as though it had originally resided exclusively in one bond of one reactant; second, it was considered a fixed quantity that did not depend in any way upon the prevailing physiological conditions. Misconceptions were sometimes abetted by imaginative drawings of mechanical devices such as water wheels and pulleys, or of streams of electrons cascading over jutting rocks of cytochrome. But these aberrations can be corrected, and they do not detract from the power of the original concepts. The physical chemist G. N. Lewis (59), writing only fifteen years before Lipmann, observed, "Living organisms alone seem to breast the great stream of apparently irreversible processes. These processes tear down, living things build up. When the rest of the world seems to move towards a dead level of uniformity, the living organism is evolving new substances and more intricate forms." Lewis consequently wrote of living matter as cheating at the game of entropy, and earlier, Lord Kelvin had cautiously excluded the operations of "animate agencies" when he formulated the Second Law of Thermodynamics. As in research, so in teaching: A general overview eventually requires elaboration from other sources for detail. For metabolites such as ATP and esters of coenzyme A, explanations of energy richness in terms of structure and reaction mechanisms have been available for some time in the textbooks of Walsh (81) and Metzler (63), both of which I reread with pleasure.

Such reflections upon the relevance of physicochemical concepts for biochemistry became part of my teaching assignment when I was appointed to the faculty of the Department of Biochemistry at the University of Leeds forty years ago. A recent environmental problem has reminded me of yet another physical generalization that I used to discuss with students at that time, but which is not universally appreciated even to this day. Consider, first, an alcohol possessing a straight, short carbon chain, which exerts mild toxicity against some type of cell. As the chain is lengthened, toxicity increases exponentially, and measurements have demonstrated that bacteriocidal (80) or bacteriostatic (30) action more than doubles as each carbon atom is added to the chain. The reason for this is readily appreciated when the distribution of alcohol molecules between phases, i.e. between aqueous phase and either cell surface or a nonpolar internal cellular phase, is considered. If the reasonable assumption is made that the energy needed to pull a carbon atom out of the phase and into the water is about the same for any carbon in the chain, simple thermodynamic treatment establishes that the coefficient of distribution will increase exponentially as each carbon is added (30). More complex problems attend biodegradation of alkylphenol ethoxylates in use as nonionic surfactants, but the principles involved are the same. Nonylphenol ethoxylates possess a branched chain of nine carbons that seek to pull the benzene ring, to which they are attached, into cellular nonaqueous phases. The hydrophilic polyethyleneglycol chain substituent has the opposite effect, and a balance between the two factors determines the effective concentration of the nonylphenol ethoxylate inside the cell or at its surface. Now it happens that ethylene glycol residues can be removed from one of the substituent side chains by elimination reactions that are catalyzed quite readily by bacterial enzymes, whereas the other (branched alkyl) chain of the phenol is not attacked. What is left, therefore, is an alkyl-substituted phenol, considerably more toxic than the original pollutant for the physicochemical reasons given. This is a case where partial biodegradability is undesirable; I have given other examples in a review (15). Nonylphenols and also nonylphenol ethoxylates with only one and two oxyethylene groups have been identified in Swiss sewage treatment plants, and the nonylphenols have been shown to be highly toxic to aquatic life (76). These are doubtless residues that remain from incomplete biodegradation of nonylphenol polyethoxylate surfactants. The toxicity of the residues is an important observation that currently deserves emphasis, but it would not have surprised some investigators fifty years ago.

I suspect that when they look back over careers in predominantly experimental sciences such as microbiology or biochemistry, most older research workers find that the progress they actually made contrasts strongly with what was predicted in successful grant applications. The latter were logical deductions based upon acceptable observations and so predicted linear, uninterrupted progress. The former more often resembled Brownian movement arising from collisions with the unexpected and were stimulated rather infrequently by equally random hunches. It is now clear to me how I came to abandon physicochemical generalizations and models for the nitty-gritty of bacterial metabolism. I did not set myself new goals, to use a fashionable phrase. On the contrary, measurements of bacterial growth kinetics began to focus upon lag phases; and these were largely abolished when certain biochemicals, usually Krebs cycle metabolites or their related amino acids, were added to cultures (27). This was then an area of great biochemical interest because the role of the cycle in aerobic metabolism was keenly debated in the late 1940s. Today, students are astonished, and this can be a useful part of their education, to learn that one could pack an auditorium at a Federation meeting by announcing that such a debate was to be held. For what excellent reasons (at that time) were competent scientists convinced that the complete cycle did not function in aerobes? And may not some present concepts also be incomplete or even mistaken? Students have a great many generally accepted concepts to learn and so many examinations to take that necessitate one correct reply to fit each question. It is therefore difficult to cultivate an attitude of mind desirable in a research worker: namely, the attitude that discoveries are not made when they can be precisely predicted, and that research may be most promising when results conflict with expectations. While it may be difficult for many science students, at first, to get used to the idea that discoveries are made because understanding is not complete, for some sections of the lay population this outlook is totally incomprehensible, so that theories of evolution, for example, are thought to be entirely wrong because they are not completely right.

A shift of my interests to the Krebs cycle, at a time when its operation in some aerobic organisms was questioned, generated an abiding interest in the various criteria that ought to be satisfied before any metabolic pathway can be regarded as established. These criteria were discussed in a book (31) and in reviews (17, 23), and they are referred to later in this article. In addition, the attention of my research group was attracted, not surprisingly, to the enzymology of citric acid, and also to the biochemical problems raised by the need to replenish the cycle with intermediates when these were used for the biosynthesis of cellular constituents in organisms grown with C_2 compounds as sole sources of carbon. But before my complete switch to metabolic interests took place, one physiological legacy from physical biochemistry remained. The Department of Biochemistry at Leeds was one of the first departments in England to acquire a Beckman model E analytical ultracentrifuge. On its arrival it was used to examine bacterial cell extracts, whose ribosomes are responsible for prominent features in schlieren patterns furnished by the instrument. These cellular components were called ribonucleoprotein particles (8) until 1957, when several groups working in the area met in Boston for one of the first symposia organized by the newly founded Biophysical Society. They agreed to call these particles by their present name: ribosomes. We also investigated ribosomelike particles present in certain "relaxed" mutants of *Escherichia coli* (37) and in cells treated with chloramphenicol (34, 38); but as my own attention was increasingly monopolized by metabolic research, investigations of ribosomes were taken over by colleagues (62, 78).

CITRIC ACID AND VARIOUS SEMIALDEHYDES

In the early 1950s several groups, including those of Gunsalus, Ajl, Dawes and myself, began to investigate citrate oxaloacetate-lyase (bacterial citrate lyase; EC 4.1.3.6). The enzyme is confined to prokaryotes, and these initial investigators soon established that it does not involve coenzyme A. In this respect it contrasts with two other aldol-type cleavage enzymes: the ubiquitous citrate synthase (EC 4.1.3.7), present in both eukaryotes and prokaryotes, and ATP citrate lyase (EC 4.1.3.8), which is generally confined to the cytosol of eukaryotes, although it does occur in *Chlorobium limicola* (3, 51). Bacterial citrate lyase is elaborated by *Klebsiella aerogenes, Streptococcus diacetilactis, Clostridium sphenoides,* and other organisms, and it enables them to use citrate as a source of carbon in the absence of air. *Rhodopseudomonas gelatinosa* grows phototropically with citrate, rapidly degrading the substrate by means of citrate lyase so that acetate, with malate formed from oxaloacetate, accumulates in the medium. Growth continues at the expense of these compounds after all the citrate has been decomposed. References to the distribution of the enzyme and to investigations of its structure and mode of action by the research groups of Eggerer, Gottschalk, Srere, and others are given in References 17 and 74. Dagley & Dawes (25) showed that citrate lyase is induced in *K. aerogenes* when conditions are anaerobic, whereas its synthesis is repressed by oxygen; citrate is then metabolized by the reactions of the Krebs cycle. They also partially purified the enzyme and showed (26) that it is autoinactivated during the course of the reaction, leaving some citrate unconsumed. The frequency of autoinactivation, and so the extent of the reaction, depends upon the nature of the divalent ion (e.g. Mg^{2+}, Zn^{2+}) required for enzyme activity. Autoinactivation was one puzzling feature of the enzyme: It was ascribed at first to the inhibitory action of one of its highly reactive products, oxaloacetate; but Singh & Srere (reviewed in 74) showed that this was not the cause. The second unusual feature was the large size of the enzyme, contrasted with the small free-energy change involved in citrate cleavage; ultracentrifuge runs with crude extracts of citrate-adapted cells indicated that the enzyme, of molecular weight of about 5×10^5, was present in amounts that gave a single peak clearly separated from those of other soluble cellular constituents. We had the impression that a sledgehammer was being used to crack a nut.

Subsequent work in other laboratories has explained some of these characteristics (74), but the enzyme remains a remarkable one. First, autoinactivation is explained on the basis of the enzyme mechanism. When active, the enzyme is in an acetylated form that reacts with citrate; the citryl-*S*-enzyme then undergoes an aldol-type cleavage to release oxaloacetate and regenerate the acetyl-*S*-enzyme:

$$E\text{-}S\text{-acetyl} + \text{citrate} \rightleftharpoons E\text{-}S\text{-citryl} + \text{acetate}$$

$$E\text{-}S\text{-citryl} \rightleftharpoons E\text{-}S\text{-acetyl} + \text{oxaloacetate}$$

The sum of these equations gives that for the overall reaction catalyzed by citrate lyase:

$$\text{Citrate} \rightleftharpoons \text{oxaloacetate} + \text{acetate}$$

Transfer of acyl to water during catalytic turnover gives rise to an inactive deacetyl enzyme. The enzyme can be reactivated by chemical treatment with acetic anhydride, and alkaline hydrolysis releases phosphopantothenate (74). This is the group that is acetylated in the active enzyme; it is part of the structure of an acyl carrier protein of 10–11 kd (42). Various organisms contain a separate enzyme that uses ATP and acetate to reactivate the deacetyl enzyme. In an elegant investigation, Rokita et al (71) used a novel fluorinated analog of citrate, 3-fluoro-3-deoxycitrate, to study interaction with each one of the three citrate-cleaving enzymes (EC 4.1.3.6, 4.1.3.7, and 4.1.3.8). The fate of the analog was uniquely related to the mechanism of the particular enzyme used. In the case of bacterial citrate lyase, the unusual sequence of events during catalysis accounts for the observation that 3-fluoro-3-deoxycitrate is a mechanism-based inhibitor that acts at two points during turnover of the acetyl enzyme (71).

The second exceptional feature of citrate lyase (a very large enzyme catalyzing an apparently simple cleavage) is explained by the fact that its mechanism involves three types of proteins: the acyl carrier protein (ACP) and two others catalyzing the separate steps of the above equations. Mainly because ACP is a small unit in a large holoenzyme, the subunit composition was extremely difficult to determine; at various times it has engaged the attention of several investigating groups. It now appears that, except for that of *Escherichia coli,* bacterial citrate lyase holoenzymes contain six of each of the three types of subunits (43, 72). Citrate metabolism by *E. coli* is of particular interest, since typical strains do not grow with citrate as the carbon source in mineral salts media and are thus readily distinguished from *K. aerogenes*. Since *E. coli* operates the Krebs cycle under aerobic conditions, its failure to grow may be ascribed to lack of a system to transport citrate into the cells. However, growth occurs with citrate, even in anaerobic conditions, when amino acids and glucose are added. Citrate lyase is then elaborated and enzyme activity continues to increase until the added glucose is used up, when the activity begins to decline; activity is regained upon further addition of glucose (13). I also observed another difference between *K. aerogenes* and *E. coli:* The former organism, when grown with citrate, develops a powerful oxaloacetate decarboxylase, furnishing pyruvate, whereas the latter does not. However, *E. coli* does convert the oxaloacetate formed from citrate into malate and hence into fumarate, for which it has a membrane-bound fumarate reductase that can generate ATP (7). It appears, then, that glucose in anaerobic citrate metabolism of *E. coli* functions as a source of reducing power for the energy-generating conversion of fumarate to succinate (61). Finally, Nilekani & SivaRaman (66) have observed another unusual feature of the *E. coli* enzyme: Its acyl transferase and acyl lyase subunits are the same size as those in other organisms, and there are six of each in the haloenzyme molecule; but there is only one ACP polypeptide, and it is about six times larger than that of the *K. aerogenes* enzyme.

I have already mentioned how studies of citrate fostered an interest in the need for what has been called the anaplerotic (55) sequence, which is required to replenish Krebs cycle intermediates when bacteria grow with acetate as carbon source. This requirement was strikingly illustrated by our experiments (32) with an organism later classified as a strain of *Acinetobacter* (44). After growth at the expense of acetate or a fatty acid yielding acetate by β-oxidation, washed cell suspensions formed α-ketoglutarate when they were aerated solely with acetate; the amounts formed were more than sufficient for isolation and characterization by the analytical methods then available. Thus, 0.5 g of purified α-ketoglutarate 2,4-dinitrophenylhydrazone was isolated after aeration of cells for 3 hr in 2 liters of 0.04 M acetate (32). This conversion did not occur when the cells were grown and harvested from other carbon sources. Moreover, the required enzymes were not induced if the cells were simply allowed to oxidize acetate; growth on this substrate was essential. The problem of how α-ketoglutarate and other growth intermediates are biosynthesized solely from acetate was solved when Kornberg and his coworkers (56, 57) proposed the glyoxylate pathway. Soon afterwards, my own group (11) obtained cell extracts of this *Acinetobacter* that, after growth with octanoate or nonanoate, oxidized various fatty acids. The consumption of oxygen gave acetate or acetate plus propionate from fatty acids of appropriate chain length. Cell extracts contained the inducible enzymes involved in the newly proposed pathway (11).

These developments prompted another question: By what reactions are the requirements of growing cultures satisfied when glyoxylate itself is the only source of carbon? This is essentially the circumstance when either glycine, glycollate, or oxalate is the growth substrate, since they provide glyoxylate by deamination, oxidation, or, in the case of oxalate, reduction of a coenzyme A derivative. My own group investigated glycine (10, 36) and those of Kornberg and Quayle studied glycollate (55) and oxalate (69), respectively. From these investigations the glycerate anaplerotic sequence was constructed. This sequence involves formation of tartronate semialdehyde by the enzyme tartronate semialdehyde synthase, formerly known as glyoxylate carboligase (49, 58). The semialdehyde is then reduced to glycerate (48), which enters central metabolic processes upon phosphorylation. Tartronate semialdehyde was one of a group of five metabolites we encountered that have certain structural features in common; the first was glyoxylate and the others were the semialdehydes of succinate, α-ketoglutarate, and α-hydroxymuconate.

Our experiences with these metabolites suggest to me that research is stimulated in diverse ways; there is no general recipe for progress. Thus, when the glycerate anaplerotic sequence had been elucidated, tartronate semialdehyde immediately took its place in metabolism. However, as explained, the need for some such sequence to exist was recognized at the outset, although the details could only be guessed; and the same was true for the glyoxylate pathway. In both sequences, some of the enzymes that now found a place had been known for some time and had, so to speak, been in search of a physiological function; these were isocitrate lyase, malate synthase, and glyoxylate carboligase. Likewise, there was no surprise when succinate semialdehyde was found to be a catabolite of homoprotocatechuate (73), since the biochemical characteristics of the pathway were already anticipated. But in the case of α-ketoglutarate semialdehyde we were not seeking to fit a compound into a metabolic slot; we bumped into it in the course of tracing an enzyme-catalyzed route leading to a destination unknown at the time. We were prepared beforehand only in the sense that we had by then gained experience in detecting and characterizing semialdehydes.

Our recognition of the metabolic status of the carboxy derivative of α-hydroxymuconate semialdehyde was entirely fortuitous. Moreover, it led to the discovery of *meta*-fission pathways for catechols. Patel and I (33) isolated a white, crystalline dibasic acid as the product that remained when cell extracts of a *p*-cresol–grown pseudomonad oxidized protocatechuate. The only catabolic sequence known for catechols at that time was *ortho*-fission, which had been brilliantly elucidated by the groups of Stanier, Hayaishi, and Evans. Since they had demonstrated the formation of lactones, we thought at first that we had isolated such a compound, perhaps a new one, whose ring could not be opened by our extracts. However, to our surprise and initial embarrassment, elemental analysis showed that nitrogen was present in the molecule. Our colleague S. Trippett, then a faculty member of the Leeds Department of Chemistry, examined the crystals and made a preliminary identification: 2,4-lutidinic acid (pyridine 2,4-dicarboxylate). It appeared, therefore, that we had unwittingly achieved the improbable feat of converting the benzene nucleus into the pyridine nucleus at 25°C and pH 7. Trippett went to a good deal of trouble over identification, and he was not helped by finding that the authentic lutidinic acid received from his commercial suppliers did not contain a trace of this compound. Moreover, in order to establish the precise point at which the benzene ring had opened we had to know for certain which isomer had been isolated. Three of the isomers of pyridine dicarboxylic acid have close melting points, so we converted samples to their methyl esters, which exhibit a wider range (14). One is reminded how much easier characterization has become in less than three decades. When we realized that cell extracts contained sufficient ammonium ions to permit spontaneous cyclization with incorporation of nitrogen, we routinely submitted them to dialysis and were able to show formation of a metabolite that had the properties of a semialdehyde. Since we had a good idea of its chemical structure from the 2,4-lutidinic acid work, a metabolic pathway was suggested as a basis for future research, thereby reversing the order of the previous approaches to the problems that I have described.

Trudgill & I (35) synthesized authentic α-ketoglutarate semialdehyde to facilitate characterization of biologically formed material. The compound is the product of two distinct bacterial dehydratases whose catalytic mechanisms involve Schiff base formation between enzyme and substrate. Through these and other reactions (17) all five carbons of L-arabinose (67, 82) or D-xylose (35) can enter the Krebs cycle as α-ketoglutarate, which is formed from its semialdehyde by NAD-dependent dehydrogenation. This last reaction is also used during catabolism of hexaric acids; thus Jeffcoat et al (52, 53) showed that D-glucarate undergoes two successive dehydrations to form the semialdehyde. The second of these dehydrations is unusual because the same enzyme catalyzes removal of water with simultaneous decarboxylation of the substrate, D-4-deoxy-5-ketoglucarate. We purified and characterized the enzyme and accounted for its mode of action by a mechanism involving Schiff base formation between the substrate and a lysine residue at the active site (53). Our publication of these two papers (52, 53) involved my only advantageous personal experience associated with the use of computers that I can recall. A certain textbook was published in the mid 1970s and still serves as a valuable literature resource. According to the author's preface it was assembled after a sophisticated computer search had been completed. Now it happened that during our investigations the *Biochemical Journal* allowed the trivial name 2-oxoglutarate (α-ketoglutarate) semialdehyde at first, whereas they insisted upon the more precise designation 2,5-dioxovalerate in later publications. In consequence, the textbook now gives us the undeserved credit of having discovered two separate metabolic pathways during our investigation of the involvement of α-ketoglutarate semialdehyde in D-glucarate catabolism.

AROMATIC CATABOLISM AND THE CARBON CYCLE

When I started work on aromatic catabolism my colleagues were inclined to concede that while the general significance of this area was peripheral, it was not without a certain fascination. Thus, it would be a bored organic chemist whose attention was not held by a demonstration of benzene rings being opened smoothly in neutral aqueous solution when shaken with oxygen gas at room temperature in the presence of a bacterial protein. But at that time, the importance of biochemical research was measured by its ultimate relevance for mammalian processes; and human enzymes, compared with those of bacteria, open the benzene rings of relatively few metabolites, these being derived from dietary aromatic amino acids. It is true, as I have written elsewhere (20), that much of our knowledge of mammalian biochemistry was obtained conveniently, but indirectly, from *E. coli,* so that by the 1960s this organism had attained the status of honorary mammal. However, it was not believed that *E. coli* was able to catabolize aromatics until the publications of

Cooper & Skinner (12) and Burlingame & Chapman (9) made their appearance. In the 1950s, many microbiologists regarded phenolics as disinfectants rather than metabolites.

The importance of aromatic catabolism is now more generally appreciated. Next to glucosyl residues, the benzene ring is the most widely distributed unit of chemical structure in the biosphere, and to a large extent the continuous operation of the carbon cycle depends upon its rapid fission by microorganisms. The physiological significance of each one of the vast array of aromatic compounds synthesized by plants is by no means clear, but lignin, a plant aromatic biopolymer more abundant than protein, has certainly played an essential part in evolution by allowing plants to stand erect and compete effectively for solar energy. The evolution of living organisms, from simple forms to those of high complexity, required the establishment of a continuous flow of energy through the biosphere; this was promoted by accumulation of dioxygen, which provides an atmosphere in which carbon dioxide is the most stable compound of carbon. Higher organisms were able to evolve in the energy gradient between photosynthesized biochemicals and carbon dioxide. A carbon cycle is a prerequisite for evolution, for within the cycle organisms can reproduce and undergo selection: They die, and dead material is recycled, largely through the reactions of microbial catabolism.

Recognition of the fact that attack by dioxygen accomplishes rapid turnover of otherwise inert materal, often aromatic, does not imply that we can ignore anaerobic processes, which also make an important contribution to the operation of the carbon cycle. Thus aromatic acids, formed when lignin is degraded, are converted anaerobically into methane by microbial consortia (87). Also, reductive dechlorinations are important steps that are catalyzed by anaerobes when they remove certain chemical pollutants from the environment (77). However, lignin itself is only degraded to any significant extent when aerobic conditions are maintained, while the slow anaerobic degradation of some aromatic hydrocarbons, in contrast to their rapid oxidation, has only recently been detected. Moreover, the techniques required to study anaerobic aromatic reactions are demanding, and growth is slow. Aerobic investigations are easier and our knowledge of aerobic processes is, in consequence, extensive. Since reviews are available (15, 17, 21, 22), I refer here only to some generalizations, and to a few aspects that seem to me to be of particular interest.

Before a benzene ring can be opened by a dioxygenase it must carry two hydroxyl groups. There are exceptions to this generalization, but they are quite rare. Biochemical features of interest in this process are as follows. *Ortho*-fission dioxygenases for catechols are red and contain tightly bound ferric iron; *meta*-fission enzymes are colorless and require ferrous iron, which is usually more easily removed (50). Recently, my colleagues E. Münck, J.

Lipscomb, and others have been able to suggest different mechanisms for the two types of dioxygenases based upon spectroscopic and kinetic observations. Thus the iron of the red *ortho*-fission enzyme, protocatechuate 3,4-dioxygenase, remains in the ferric form throughout the catalytic cycle (70) and appears to bind the aromatic substrate in a chelate complex (83). These researchers proposed that dioxygen directly attacks this substrate in the iron complex. By contrast, the ferrous iron of protocatechuate 4,5-dioxygenase, rather than the substrate itself, appears to be the site of initial attack by dioxygen (1).

The study of the aromatic monooxygenases that prepare the benzene nucleus for fission has also made significant contributions to general knowledge of the enzymology of dioxygen. These enzymes are flavoproteins, and many detailed kinetic investigations, notably those by V. Massey's group at Ann Arbor, have shown that the FAD is first reduced by NAD(P)H; NAD^+ then leaves the active site before O_2 gains access and forms a hydroperoxide with the reduced FAD. These features are illustrated by a mechanism recently proposed for phenol hydroxylase (4). 4-Hydroxybenzoate 3-hydroxylase is another monooxygenase that has been particularly well characterized. It is a dimer; each subunit contains 394 amino acid residues and binds one FAD covalently. Sites for binding NADPH and 4-hydroxybenzoate (which causes a conformational change when bound) have been identified, and a channel through which O_2 reaches the active site has been revealed by X-ray analysis (84). In the case of aromatic hydrocarbons, two hydroxyl groups have to be inserted simultaneously by a dioxygenase to prepare for ring-opening. Most of the enzymology and also the structures of reaction intermediates were established in the laboratory of D. T. Gibson, who was a pioneer investigator of the *meta*-fission pathway for catechols (24). The outstanding investigations of his group have been reviewed recently (22). This group recognized *cis*-dihydrodiols as reaction intermediates in the bacterial catabolism of hydrocarbons, whereas in mammalian and fungal metabolism *trans*-diols and epoxides are formed. The absolute configurations of several *cis*-dihydrodiols were determined, convenient methods for their isolation were developed, and their enzymology was elucidated.

If we consider more general aspects of aerobic catabolism, it is not surprising that the vast turnover of aromatics in the biosphere has resulted, during the course of millennia, in the use of just a few well-worn channels of catabolism for those compounds most commonly encountered. Complex structures, after various degradative modifications, give rise to a limited number of dihydric phenols that serve as ring-fission substrates for dioxygenases. Less than a dozen separate and distinct pathways direct the flow of carbon, through these phenols, into the Krebs cycle (22). Pathways are kept separate by the operation of two factors: substrate specificity and specificity of enzyme derepression. For example, gentisic and homogentisic acids are two of the dihydric phenols. Their structures differ only by one methylene group, but the dioxygenase for the first of these acids does not attack the second, and the subsequent enzymes for one pathway do not significantly metabolize the homologous substrates for the other. Two aldolases function on the *meta*-fission pathways for two homologs, protocatechuate and homoprotocatechuate. The aldolase substrates are, respectively, 4-carboxy-4-hydroxy-2-ketoadipate and 4-hydroxy-2-ketopimelate. The enzymes are similar proteins, both being approximately spherical and composed of six similar subunits, but neither enzyme attacks the other's substrate (18). A point of historical interest is that the first of these two aldolase substrates is the product expected from the condensation of pyruvate with oxaloacetate; it was considered, at one time, as a possible precursor of citric acid in what is now the Krebs cycle. Some 35 years later a function was found for this tricarboxylic keto acid as a metabolite in aromatic catabolism.

The specificity of enzyme derepression is illustrated by the following example. If a particular bacterial strain is found to oxidize 2,5-dihydroxybenzoate (gentisate) after it has grown with benzoate as source of carbon, it does not typically oxidize other feasible ring-fission substrates such as 3,4-dihydroxybenzoate (protocatechuate) or catechol until it is grown on another appropriate substrate. Conversely, another strain might derepress and use either catechol or protocatechuate enzymes for metabolizing benzoate, but gentisate would not be oxidized. This behavior is the basis of "simultaneous adaptation," an experimental technique by which pathways of aromatic catabolism have often been elucidated from measurements of oxygen consumption by cell suspensions (23, 75).

Investigations of degradative routes for aromatics stimulate thought about the general criteria that ought to be applied before a pathway is accepted. Chapman and I (23) pointed out that when a compound can be readily isolated from cultures, it ought to be shown that kinetics of formation and disappearance are those expected for an intermediate; the easier the isolation, the more one should entertain the possibility of side-product formation. Results from the use of inhibitors to block pathways should also be interpreted cautiously (23). The characteristics of purified enzymes of the pathway are relevant since they form a part of its complete description. However, since a living cell is much more than a mere bag of enzymes, one must return at some point to experiments with intact cells. The pathway proposed must be compatible with their behavior, and this is where simultaneous adaptation is valuable. Stanier (75) laid down clearly the limitations of the method, and disregard of his rules has resulted in some worthless publications. In our investigations of *Trichosporon cutaneum* it soon became evident that the technique was not applicable. This yeast is as versatile in aromatic catabolism as any pseudomonad, yet it lacks ring-fission dioxygenases for protocatechuate, homoprotocatechuate, and gentisate while possessing the ability to oxidize all three compounds completely when grown on the appropriate substrate (2). It compensates for its limitations by hydroxylating the benzene ring three times where appropriate, instead of twice, to give a suitable ring-fission substrate. But broad specificity of enzyme-induction was the particular characteristic that ruled out the use of simultaneous adaptation; cells grown on benzoate could oxidize benzoate readily, as well as catechol, protocatechuate, and 2-, 3-, and 4-hydroxybenzoates. In consequence, checks on the proposed pathways relied heavily upon establishment, for cell extracts, of stoichiometry involving substrates and their catabolites, O_2 consumption, and NADH/NAD involved in the various hydroxylations of the pathways (2).

CHEMICAL POLLUTANTS

I have described elsewhere (19) how I came to isolate *Pseudomonas U* in 1964 from a highly polluted stream that flowed through the Boneyard Creek below my residence in Urbana, Illinois. This organism found extensive use when *meta*-fission reactions for catechol degradation were elucidated. I returned to Leeds, and after two years there I was appointed professor of biochemistry at the University of Minnesota in St. Paul. This was an unusual appointment in the US for an overseas scientist, insofar as teaching experience was as important as, and perhaps more important than, my research activities. A new College of Biological Sciences was being set up; biochemistry was a core subject in the undergraduate curriculum, and I had extensive experience in all aspects of teaching at this level, including the initiation of untried programs. Through my teaching interests, I maintained contacts with colleagues in England. I became American editor of *Biochemical Education* and for several years also wrote articles for this journal, which has its editorial office at the University of Leeds. I continued to interact with former students and colleagues who came over for postdoctoral research or spent sabbaticals in St. Paul; and similar exchanges were arranged in England or Wales. I also renewed an association with I. C. Gunsalus, which influenced the trend of my thoughts about microbial catabolism. In 1970 a committee was set up by the National Academy of Sciences, under the chairmanship of Gunsalus, to consider the impact of agricultural practices upon environmental quality. As part of my duties as a committee member I helped to organize a symposium (64a) which focused the attention of biochemically oriented microbiologists upon the relevance of their research for environmental problems (15). The volume of publications in this area has grown enormously in the intervening 15 years, and to help accommodate the trend, the words "*and Environmental*" were added to the title of the journal *Applied Microbiology* at the end of 1975.

The relevance of academic microbiology for problems of chemical pollution has been questioned or ignored by persons in two widely separated categories: those who are only too familiar with the complexities of natural environments and those who are ignorant of biochemical microbiology. The latter feel that we should be able to construct a sort of periodic table that relates in simple fashion such properties as bond strength and atomic radii to biodegradability, whereas the former quite rightly say that biochemists oversimplify complex systems by selecting for study, for convenience, bacterial athletes that grow overnight upon single carbon sources possessing simple chemical structures. It would be salutary for both groups to consider how they would attempt to rid contaminated soil of a chlorophenol using biological methods. As soon as the matter is given thought it becomes apparent that information about the best conditions for dechlorination (77), choices of catabolic sequences for aromatics, and mechanisms of the participating enzymes are all relevant considerations. For initial attack the structure of the substrate is important, but so are the structures of its catabolites, which may, indeed, prove more of a nuisance than the original compound (15).

Cometabolism is a concept that has been much debated (20). There is no doubt that in nature compounds are often degraded by combined efforts, so that when an organism fails at one stage of degradation another takes over. At a time before cometabolism had even been mentioned we found that a pseudomonad grown with *p*-cresol could also oxidize certain xylenols to the stage of substituted benzoates; these failed to serve as carbon sources (33). There was no evidence that the organism derived any benefit from this activity. In other cases two or more organisms are able to pool biochemical resources for mutual benefit. Although perhaps this situation should not be termed cometabolism, I believe that many people will continue to call it so. The most serious oversimplification we can make is to assume that because a particular organism catalyzes a certain reaction in the laboratory to produce or remove a noxious chemical, this is the only or even the most important way that the result is achieved in nature. In nature the same job may in principle be done in several different ways by diverse microbial species. Just as it is advisable to return to intact cells to confirm a pathway with single cultures when enzymology has been completed, so environmental checks should complement research with isolated strains. This may be done by radiotracer studies, provided one is sure that measured $^{14}CO_2$ originates from the labeled pollutant; this has not always been the case.

The fates of chemical pollutants cannot be reliably predicted without knowing what types of catabolic reactions are available. This is one justification for the traditional approach of using single cultures and simple carbon sources. Another reason is that this approach was necessary for work with plasmids whose role in disseminating catabolic expertise among pseudomonads is now recognized. In bacteria, the types of reactions used in the *meta*-fission and gentisate pathways (15, 17) are more suited to degrading substituted aromatics than are those of *ortho*-fission. Thus, extracts of catechol-adapted *P. putida* readily and extensively degrade methyl-substituted catechols. Incidentally, this is not at variance with the generalization that degradative routes for commonly encountered natural diphenols are kept separate by stringent substrate specificity. Substrate analogs, provided they are not the common metabolites on these pathways, are quite frequently open to attack by *meta*-fission. When extracts degrade 3-methylcatechol, the methyl substituent is released as acetic acid by hydrolysis early in the sequence. This type of reaction liberates a phenyl or benzyl group, as benzoate or phenylacetate, respectively, and so enhances catabolic potential. However, this does not mean that the hydrolase and the other enzymes involved are identical as protein molecules in each competent organism. By contrast, bacteria that use *ortho*-fission are handicapped for degrading even 4-methylcatechol because they accumulate a methyl-substituted lactone that they cannot metabolize further. A methyl substituent is no obstacle to the soil yeast *T. cutaneum,* although it uses *ortho*-fission for 4-methylcatechol. An isomeric methyl-lactone is produced, which is degraded without difficulty to acetyl CoA and pyruvate and so used for growth (68).

For some substrates the lactonizing step assists degradation. This is the case when a chlorine substituent is eliminated by lactonization that follows *ortho*-fission of 3,5-dichlorocatechol, a catabolite of (2,4-dichlorophenoxy)acetic acid. On the other hand, *meta*-cleavage can prove to be a handicap. Thus, the ring-fission products formed from 3-halocatechols by catechol 2,3-dioxygenase are acyl halides; they are suicide inhibitors that acylate the enzyme (5). Degradation is consequently blocked. However, the discovery of a new metabolite, namely the lactone 2-pyrone-4,6-dicarboxylic acid, which is formed during degradation of protocatechuate, indicates how halides might be eliminated by *meta*-fission without damage to enzymes. When protocatechuic acids substituted with a halogen at C-5 are attacked by the 4,5-dioxygenase, it appears that the pyrone ring is formed with halide ion expulsion faster than the enzyme can be acylated (54). A similar reaction takes place when the substituent is a methyl group, and methanol is expelled. Recent whole-cell NMR work in my laboratory (79) strongly suggests that in these cases the pyrone is formed by spontaneous, rapid nonenzymic cyclization of the ring-fission product. This raises the question of the extent to which nonenzymic reactions play a part in the catabolism of compounds such as pollutants that are not produced enzymically in nature. An organism's success in degrading a pollutant may depend upon whether or not a catabolite is formed that has a chemical structure that permits degradation to proceed, e.g. a structure that can cyclize spontaneously, expelling a halide, or one in which a bond of uncommon occurrence in nature, say a C-P bond, is easily broken by virtue of some unusual structural feature of the catabolite. If a compound is totally foreign to the microbial world, it may not always be profitable to seek an enzyme-catalyzed reaction that happens to degrade it accidentally. However, C-P bond cleavage by bacteria has recently been studied (80a); a cell-free system has not yet been obtained.

Literature Cited

1. Anciero, D. M., Lipscomb, J. D. 1986. Binding of ^{17}O-labeled substrate and inhibitors to protocatechuate 4,5-dioxygenase-nitrosyl complex. *J. Biol. Chem.* 261:2170–78
2. Anderson, J. J., Dagley, S. 1980. Catabolism of aromatic acids in *Trichosporon cutaneum*. *J. Bacteriol.* 141: 534–43
3. Antranikian, G., Herzberg, C., Gottschalk, G. 1982. Characterization of ATP citrate lyase from *Chlorobium limicola*. *J. Bacteriol.* 152:1284–87
4. Ballou, D. P. 1982. Flavoprotein monooxygenases. In *Flavins and Flavoproteins*, ed. V. Massey, C. H. Williams, pp. 301–10. New York: Elsevier. 890 pp.
5. Bartels, I., Knackmuss, H.-J., Reineke, W. 1984. Suicide inactivation of catechol 2,3-dioxygenase from *Pseudomonas putida* mt-2 by halocatechols. *Appl. Environ. Microbiol.* 47:500–5
6. Bauchop, T., Elsden, S. R. 1960. The growth of microorganisms in relation to their energy supply. *J. Gen. Microbiol.* 23:457–69
7. Bernhard, T., Gottschalk, G. 1978. Cell yield of *Escherichia coli* during anaerobic growth on fumarate and molecular hydrogen. *Arch. Microbiol.* 116:235–38
8. Bowen, T. J., Dagley, S., Sykes, J. 1959. A ribonucleoprotein component of *E. coli*. *Biochem. J.* 72:419–25
9. Burlingame, R., Chapman, P. J. 1983. Catabolism of phenylpropionic acid and its 3-hydroxy derivative by *Escherichia coli*. *J. Bacteriol.* 155:113–21
10. Callely, A. G., Dagley, S. 1959. Metabolism of glycine by a pseudomonad. *Nature* 183:1793–94
11. Callely, A. G., Dagley, S., Hodgson, B. 1958. Oxidation of fatty acids by cell-free extracts of a vibrio. *Biochem. J.* 69:173–81
12. Cooper, R. A., Skinner, M. A. 1980. Catabolism of 3- and 4-hydroxyphenylacetate by the 3,4-dihydroxyphenylacetate pathway in *Escherichia coli*. *J. Bacteriol.* 143:302–6
13. Dagley, S. 1954. Dissimilation of citric acid by *A. aerogenes* and *E. coli*. *J. Gen. Microbiol.* 11:218–27
14. Dagley, S. 1967. The microbial metabolism of phenolics. In *Soil Biochemistry*, ed. A. D. McLaren, G. H. Peterson, pp. 290–317. New York: Dekker. 509 pp.
15. Dagley, S. 1975. A biochemical approach to some problems of environmental pollution. *Essays Biochem.* 11:81–138
16. Dagley, S. 1975. A review assay for the textbook *Biochemistry* by L. Stryer. *Biochem. Ed.* 3:44–46
17. Dagley, S. 1978. Pathways for the utilization of organic growth substrates. In *The Bacteria*, Vol. 6, ed. L. N. Ornston, J. R. Sokatch, pp. 305–88. New York: Academic
18. Dagley, S. 1982. 4-Hydroxy-2-ketopimelate aldolase. *Methods Enzymol.* 90:277–80
19. Dagley, S. 1982. Our microbial world. In *Experiences in Biochemical Perception*, ed. L. N. Ornston, S. G. Sligar, pp. 45–55. New York: Academic. 381 pp.
20. Dagley, S. 1984. Introduction. In *Microbial Degradation of Organic Compounds*, ed. D. T. Gibson, pp. 1–10. New York: Dekker. 535 pp.
21. Dagley, S. 1985. Microbial metabolism of aromatic compounds. In *Comprehensive Biotechnology*, Vol. 1, ed. M. Moo-Young, pp. 483–505. Oxford: Pergamon
22. Dagley, S. 1986. Biochemistry of aromatic hydrocarbon degradation by pseudomonads. In *The Bacteria*, Vol. 10, ed. J. R. Sokatch, pp. 527–55. New York: Academic. 617 pp.
23. Dagley, S., Chapman, P. J. 1971. Evaluation of methods used to determine metabolic pathways. *Methods Microbiol.* 6A:217–68
24. Dagley, S., Chapman, P. J., Gibson, D. T., Wood, J. M. 1964. Degradation of the benzene nucleus by bacteria. *Nature* 202:775–78

25. Dagley, S., Dawes, E. A. 1953. Citric acid metabolism of *Aerobacter aerogenes*. *J. Bacteriol.* 66:259–65
26. Dagley, S., Dawes, E. A. 1955. Citridesmolase, its properties and mode of action. *Biochim. Biophys. Acta* 17:177–84
27. Dagley, S., Dawes, E. A., Morrison, G. A. 1950. Factors influencing the early phases of growth of *Aerobacter aerogenes*. *J. Gen. Microbiol.* 4:437–47
28. Dagley, S., Hinshelwood, C. N. 1938. Dependence of growth of *Bact. lactis aerogenes* on concentration of medium. *J. Chem. Soc.* 1938:1930–36
29. Dagley, S., Hinshelwood, C. N. 1938. Quantitative dependence of growth rate of *Bact. lactis aerogenes* on the carbon dioxide content of the gas atmosphere. *J. Chem. Soc.* 1938:1936–42
30. Dagley, S., Hinshelwood, C. N. 1938. Influence of alcohols on the growth of *Bact. lactis aerogenes*. *J. Chem. Soc.* 1938:1942–48
31. Dagley, S., Nicholson, D. E. 1970. *An Introduction to Metabolic Pathways*. Oxford: Blackwell. 343 pp.
32. Dagley, S., Patel, M. D. 1955. Excretion of α-ketoglutarate by a vibrio. *Biochim. Biophys. Acta* 16:418–23
33. Dagley, S., Patel, M. D. 1957. Oxidation of *p*-cresol and related compounds by a *Pseudomonas*. *Biochem. J.* 66:227–33
34. Dagley, S., Sykes, J. 1959. Bacterial RNP synthesized in the presence of chloramphenicol. *Biochem. J.* 74:11P
35. Dagley, S., Trudgill, P. W. 1965. The metabolism of galactarate, D-glucarate and various pentoses by species of *Pseudomonas*. *Biochem. J.* 95:48–58
36. Dagley, S., Trudgill, P. W., Callely, A. G. 1961. Synthesis of cell constituents from glycine by a *Pseudomonas*. *Biochem. J.* 81:623–31
37. Dagley, S., Turnock, G., Wild, D. G. 1963. The accumulation of RNA by a mutant of *Escherichia coli*. *Biochem. J.* 88:555–66
38. Dagley, S., White, A. E., Wild, D. G. 1962. The effect of various drugs and inorganic ions on bacterial ribonucleoprotein. *J. Gen. Microbiol.* 29:59–63
39. Dawes, E. A. 1986. *Microbial Energetics*. New York: Chapman & Hall. 187 pp.
40. Dean, A. C. R., Hinshelwood, C. N. 1966. *Growth, Function and Regulation in Bacterial Cells*. Oxford: Clarendon. 439 pp.
41. Deleted in proof
42. Dimroth, P., Dittmar, W., Walther, G., Eggerer, H. 1973. The acyl-carrier protein of citrate lyase. *Eur. J. Biochem.* 37:305–15
43. Dimroth, P., Eggerer, H. 1975. Evaluation of the protein components of citrate lyase from *Klebsiella aerogenes*. *Eur. J. Biochem.* 53:227–35
44. Fewson, C. A. 1967. The identity of the gram-negative bacterium NCIB 8250 ('Vibrio 01'). *J. Gen. Microbiol.* 48:107–10
45. Fruton, J. S., Simmonds, S. 1963. *General Biochemistry*. New York: Wiley. 1077 pp. 2nd ed.
46. Gladstone, G. P., Fildes, P., Richardson, G. M. 1935. Carbon dioxide as an essential growth factor in bacteria. *Br. J. Exp. Pathol.* 16:335–48
47. Gortner, R. A. 1938. *Outlines of Biochemistry*. New York: Wiley. 1017 pp. 2nd ed.
48. Gotto, A. M., Kornberg, H. L. 1961. The metabolism of C_2 compounds in microorganisms. 7. Preparation and properties of crystalline tartronic semialdehyde reductase. *Biochem. J.* 81:273–84
49. Gupta, N. K., Vennesland, B. 1964. Glyoxylate carboligase of *Escherichia coli:* a flavoprotein. *J. Biol. Chem.* 239:3787–89
50. Hayaishi, O. 1966. Crystalline oxygenases of pseudomonads. *Bacteriol. Rev.* 30:720–31
51. Ivanovski, R. N., Sintsov, N. W., Kondratieva, E. N. 1980. ATP-linked citrate lyase activity in the green sulfur bacterium *Chlorobium limicola* forma *thiosulfatophilium*. *Arch. Microbiol.* 128:239–41
52. Jeffcoat, R., Hassall, H., Dagley, S. 1969. The metabolism of D-glucarate by *Pseudomonas acidovorans*. *Biochem. J.* 115:969–76
53. Jeffcoat, R., Hassall, H., Dagley, S. 1969. Purification and properties of D-4-deoxy-5-oxoglucarate hydro-lyase (decarboxylating). *Biochem. J.* 115:977–83
54. Kersten, P. J., Chapman, P. J., Dagley, S. 1985. Enzymatic release of halogens or methanol from some substituted protocatechuic acids. *J. Bacteriol.* 162:693–97
55. Kornberg, H. L. 1966. Anaplerotic sequences and their role in metabolism. *Essays Biochem.* 2:1–31
56. Kornberg, H. L., Krebs, H. A. 1957. Synthesis of cell constituents from C_2 units by a modified tricarboxylic acid cycle. *Nature* 179:988–91
57. Kornberg, H. L., Madsen, N. B. 1958. The metabolism of C_2 compounds in

microorganisms. 3. Synthesis of malate from acetate via the glyoxylate cycle. *Biochem. J.* 68:549–57

58. Krakow, G., Barkulis, S. S. 1956. Conversion of glyoxylate to hydroxypyruvate by extracts of *Escherichia coli*. *Biochim. Biophys. Acta* 21:593–94
59. Lewis, G. N. 1926. *The Anatomy of Science*. New Haven: Yale Univ. Press. 221 pp.
60. Lipmann, F. 1941. Metabolic generation and utilization of phosphate bond energy. *Adv. Enzymol.* 1:99–162
61. Lutgens, M., Gottschalk, G. 1980. Why a co-substrate is required for anaerobic growth of *Escherichia coli* on citrate. *J. Gen. Microbiol.* 119:63–70
62. Markey, F., Wild, D. G. 1975. A 30S precursor of 30S ribosomes in a mutant of *Escherichia coli*. *Biochem. J.* 151:463–65
63. Metzler, D. E. 1977. *Biochemistry. The Chemical Reactions of Living Cells*. New York: Academic. 1129 pp.
64. Monod, J. 1942. *La Croissance des Cultures Bacteriennes*. Paris: Herman & Cie. 210 pp.

64a. National Academy of Sciences. 1971. *Degradation of Synthetic Organic Molecules in the Biosphere*. Washington, DC: Natl. Acad. Sci. 350 pp.

65. *Nature* 1967. Obituary of Sir Cyril Hinshelwood. 216:832–33
66. Nilekani, S., SivaRaman, C. 1983. Purification and properties of citrate lyase from *Escherichia coli*. *Biochemistry* 22:4657–63
67. Portsmouth, D., Stoolmiller, A. C., Abeles, R. H. 1967. Studies on the mechanism of action of 2-keto-3-deoxy-L-arabonate dehydratase. *J. Biol. Chem.* 242:2751–59
68. Powlowski, J. B., Dagley, S. 1985. β-Ketoadipate pathway in *Trichosporon cutaneum* modified for methyl-substituted metabolites. *J. Bacteriol.* 163:1126–35
69. Quayle, J. R., Keech, D. B., Taylor, G. A. 1961. Carbon assimilation by *Pseudomonas oxalaticus* (OX1). 4. Metabolism of oxalate in cell-free extracts of the organism grown on oxalate. *Biochem. J.* 78:225–36
70. Que, L. Jr., Lipscomb, J. D., Zimmermann, R., Münck, E., Orme-Johnson, N. R., et al. 1976. Mossbauer and EPR spectroscopy on protocatechuate 3,4-dioxygenase from *Pseudomonas aeruginosa*. *Biochim. Biophys. Acta* 452:320–44
71. Rokita, S. E., Srere, P. A., Walsh, C. T. 1982. 3-Fluoro-2-deoxycitrate: a probe for mechanistic study of citrate-utilizing enzymes. *Biochemistry* 16: 3765–74
72. Singh, H., Srere, P. A., Klapper, D. G., Capra, J. D. 1976. Subunit and chemical composition of citrate lyase from *Klebsiella pneumoniae*. *J. Biol. Chem.* 251:2911–15
73. Sparnins, V. L., Chapman, P. J., Dagley, S. 1974. Bacterial degradation of 4-hydroxyphenylacetic acid and homoprotocatechuic acid. *J. Bacteriol.* 120: 159–67
74. Srere, P. A. 1975. The enzymology of the formation and breakdown of citrate. *Adv. Enzymol. Relat. Areas Mol. Biol.* 43:57–101
75. Stanier, R. Y. 1947. Simultaneous adaptation: a new technique for the study of metabolic pathways. *J. Bacteriol.* 54:339–48
76. Stephenou, E., Giger, W. 1982. Persistent organic chemicals in sewage effluents. 2. Quantitative determinations of nonylphenols and nonylphenolethoxylates by capillary gas chromatography. *Environ. Sci. Technol.* 16:800–5
77. Suflita, J. M., Horowitz, A., Shelton, D. R., Tiedje, J. M. 1982. Dehalogenation: a novel pathway for the anaerobic biodegradation of haloaromatic compounds. *Science* 218:1115–17
78. Sykes, J., Metcalf, E., Pickering, J. D. 1977. The nature of the protein in "chloramphenicol particles" from *Escherichia coli* A19 (Hfr *rel met rns*). *J. Gen. Microbiol.* 98:1–27
79. Sze, I. S.-Y. 1985. *Novel oxygenases and metabolic pathways of microbial aromatic catabolism*. PhD thesis. Univ. Minn., St. Paul. 166 pp.
80. Tilley, F. W., Schaffer, J. M. 1926. Relation between the chemical constitution and germicidal activity of the monohydric alcohols and phenols. *J. Bacteriol.* 12:303–9

80a. Wackett, L. P., Shames, S. L., Venditti, C. P., Walsh, C. T. 1987. Bacterial carbon-phosphorus lyase: products, rates, and regulation of phosphonic and phosphinic acid metabolism. *J. Bacteriol.* 169:710–17

81. Walsh, C. 1979. *Enzymatic Reaction Mechanisms*. San Francisco: Freeman. 978 pp.
82. Weimberg, R., Doudoroff, M. 1955. The oxidation of L-arabinose by *Pseudomonas saccharophilia*. *J. Biol. Chem.* 217:606–24
83. Whittaker, J. W., Lipscomb, J. D. 1984. ^{17}O water and cyanide ligation by

the active site of protocatechuate 3,4-dioxygenase. *J. Biol. Chem.* 259:4487–95

84. Wierenga, R. K., Kalk, K. H., van der Laan, J. M., Drenth, J., Hofsteenge, J., et al. 1982. The structure of *p*-hydroxybenzoate hydroxylase. See Ref. 4, pp. 11–18.
85. Wood, H. G., Werkman, C. H. 1935. The utilization of carbon dioxide by the propionic acid bacteria in the dissimilation of glycerol. *J. Bacteriol.* 30:332
86. Wood, H. G., Werkman, C. H. 1940. The relationship of bacterial utilization of carbon dioxide to succinic acid formation. *Biochem. J.* 34:129–38
87. Young, L. Y. 1984. Anaerobic degradation of aromatic compounds. See Ref. 20, pp. 487–523

Ann. Rev. Microbiol. 1987. 41:25–49

THE TRANSIENT PHASE BETWEEN GROWTH AND NONGROWTH OF HETEROTROPHIC BACTERIA, WITH EMPHASIS ON THE MARINE ENVIRONMENT

S. Kjelleberg, M. Hermansson, and P. Mårdén

Department of Marine Microbiology, University of Göteborg, Carl Skottsbergs gata 22, S-413 19 Göteborg, Sweden

G. W. Jones

Department of Microbiology and Immunology, 6723-0620 Medical Science Building, University of Michigan, Ann Arbor, Michigan 48109

CONTENTS

0066-4227/87/1001-0025$02.00

INTRODUCTION

From the long-term perspective, all bacterial species must ultimately divide and grow in order to avoid extinction. Within the marine ecosystem, however, continuous growth of all bacterial cells between their rounds of division is unlikely to occur. This review focuses on some physiological and molecular processes of the bacterial downshift from growth to nongrowth induced by substrate limitations, with emphasis on the marine ecosystem. Bacteria are frequently exchanged to and from a broad range of discontinuities, such as interfaces; thus cells are transferred between widely different nutritional concentrations and substrate compositions and activities. Such fluctuations can induce an intermittent mode of growth, including periods of nongrowth during which the cell biomass does not increase. In fact, nutritionally distinct microenvironments exist at sediment floors, in river plumes, at the air-water interface, at inanimate and animate surfaces such as particles and the gastrointestinal tract of marine animals, and in areas of nutrient flux in near-shore habitats.

An understanding of the pathways of downshift and starvation survival, including the efficient substrate-scavenging capacity of nongrowing bacterial cells, is of fundamental importance not only in the area of microbial ecology but also in the area of public health related to survival of pathogens. The latter field encompasses study of both autochthonous and allochthonous bacteria as well as risk assessment of the release of engineered organisms. Starvation survival is also important in the area of biotechnology in industrial fermentation processes. This review strongly emphasizes that nongrowing bacteria are active and have the capacity to capture substrates at concentrations that do not allow for an increase in biomass. Substrates, exogenous as well as endogenous, are used for maintenance and reorganization of cellular processes during periods of nongrowth. Consequently the terms "dormant cells" and "dormancy" are misleading with regard to nongrowing cells in natural situations. Small, nongrowing, starved cells participate in the carbon flow in the environment.

While the net process in any environment is growth, measurements of average growth rates, which represent the activities of the active fraction of the population, are not likely to reflect intermittent periods of nongrowth of marine bacteria. This review presents research to date on the physiological responses of individual species to the environment; the goal of this research is to improve our understanding about bacterial survival, competition, and proliferation in some of the subecosystems of the highly discontinuous marine environment.

Starvation may have two important physiological characteristics (101): firstly, transient induction of responses to reduced availability of a range of substrates, including endogenous material; secondly, long-term synthesis of starvation-specific proteins at very low endogenous metabolic rates. Furthermore, we suggest that most if not all individual bacteria can use different survival strategies.

INTERMITTENT GROWTH OF MARINE BACTERIA

Bacterial Production and Predation

Heterotrophic bacteria in marine waters are producers as well as decomposers of particulate organic matter (145). Azam et al (10) stressed the importance of the "microbial loop," by which dissolved organic carbon, released primarily by phytoplankton but also by animals, is returned to the main food chain via bacteria, flagellates, and microzooplankton. In this model the bacterial biomass is limited by predation. Wright & Coffin (148) found a correlation between production of bacterial biomass and predation. Using *Escherichia coli* minicells as probes, Wikner et al (143) showed that between 27% and all of the bacteria produced were transferred to protozoa by predation.

It may be argued that the predation will remove any nongrowing bacteria and that, if so, starvation survival is not a useful strategy. It is unlikely, however, that all bacteria are grazed upon to an equal extent. At any given time the very small bacteria, some of which may be nongrowing, may not be under the same grazing pressure as larger cells. In fact, the decrease in size of bacteria upon exposure to nongrowing conditions may protect the cell from predation (e.g. 98, 112). The mean cell size of a mixed seawater culture decreased in response to predation (8).

Mean generation times of the total bacterial population have been calculated to range from a few hours to several days (139). Unlike bacterial growth in nutritious media, which is represented by a continuous increase in biomass, bacterial growth or nongrowth in the marine environment is basically a question of the length of the time period between two divisions and whether or not there is a constant increase in biomass during this time. During the generation times mentioned above cells can probably go through several periods of growth and nongrowth in response to changes in the nutritional conditions of the microenvironment.

Nutritional Diversity of Marine Bacteria

All studies of species or groups of species of marine bacterial populations have revealed high diversity of activity; bacteria range from nongrowing or starving to growing and highly active (e.g. 21, 82, 104, 136). Bacteria have frequently been divided into different groups based on their growth characteristics. Bacteria that are able to grow at the low ambient nutrient concentrations found in the sea have been defined as oligotrophs (e.g. 110), which resemble the K-strategists (64) and the planktobacteria (123). Copiotrophs (111) require much higher nutrient concentrations for growth, and may also be referred to as r-strategists (64) or epibacteria (123). Wright (147) has proposed three hypothetical bacterial plankton communities based on the effects of substrate limitation, grazing, and "dormancy": (*a*) active, grazer-controlled; (*b*) active, substrate-limited, perhaps in a transient state before the grazer community has become established; and (*c*) "dormant," substrate-limited. A natural sample may be a mixture of these three communities. "Dormancy" prevents the number of bacteria from declining excessively when sufficient substrate is not available for growth and guarantees a rapid response to a sudden input of nutrients. Furthermore, the grazing pressure on bacteria was less in offshore than in estuary samples, which indicates that the relative importance of the effects of grazing versus substrate concentration may shift at different locations and at different times (148). A situation of changing nutrient fluxes is apparent from measurements of short-term fluctuations of substrates, such as diurnal cycles, in the euphotic zone (e.g. 96). In summary, survival mechanisms during periods of nongrowth set the lower limit and grazing controls the upper limit of bacterial numbers. This explains the apparently small variations in bacterial concentrations in natural waters.

Influence of Surface Attachment

The importance of discontinuities for shifts in the activity of marine bacteria is mainly illustrated by the exchange of bacteria between interfaces and the water phase in relation to growth and nongrowth. The relative activity and biomass of attached versus free-living bacteria in the sea has been much debated (e.g. 11, 35). When bacteria were viewed mainly as decomposers of particulate matter, it was suggested that growth was only possible on particles and that free-living bacteria were "dormant" (e.g. 129). This view was challenged when it was shown that much of the heterotrophic activity, as measured by uptake of labeled substrates, was due to free-living cells (12, 144) and that few bacteria were associated with particles (12, 142). Ability to take up low concentrations of substrates, however, is not necessarily coupled with growth; substrate capture seems to be a common and significant feature of small starved nongrowing bacteria (see below).

With regard to the low percentage of attached cells on particles, Hermansson & Marshall (49) have suggested that reversibly attached cells may detach during handling. Attached cells may have different residence times at the surface (48) rather than being permanently, irreversibly bound. Reversibly attached bacteria are able to scavenge efficiently and benefit from the carbon source immobilized at surfaces. (48, 49, 66). Hermansson & Marshall (49) therefore argued that particles in the sea may support a loosely associated population, which is difficult to distinguish from the free-living bacteria because reversibly attached cells may be dislodged by conventional methods of sampling and size fractionation.

Furthermore, amounts of fragile aggregate particles or "marine snow" may be seriously underestimated using conventional water-sampling methods (33, 74, 125). These macroscopic aggregates have been found in oceanic (22, 33, 122) and coastal waters (74) and are inhabited by phytoplankton, cyanobacteria, protozoa, and bacteria at densities two to five orders of magnitude higher than the densities of free-living populations in the surrounding water (22, 28, 124, 125). The aggregates constitute zones of enriched organic matter (2) and nutrients (122). Bacteria attached to microscopic particles may also constitute a significant part of the total population and may be more active than the free-living cells (e.g. 53, 69, 80). Attached cells have also exhibited equal or lower activity than free-living bacteria in some instances (e.g. 3).

The air-water interface constitutes a large and important surface where bacteria are enriched (for reviews see 70, 103). Carlucci et al (20) concluded that the population enriched in the surface microlayers consists of bacteria of varying degrees of activity, from "dormant," inhibited, or damaged cells to a viable flora that flourishes in situ.

A Proposed Life Cycle

The concept of surface association as a dynamic process, with bacteria oscillating between the interface and the bulk phase (48, 49, 72), is the basis for a proposed life cycle of some marine bacteria: Small nongrowing cells are transported, by active or passive processes, from a fluctuating growth-deficient microenvironment in the bulk to a particle or its vicinity where growth or higher growth rates are possible. This model is supported by the observations that some of the small free-living cells in oceanic waters grow into larger cells upon cultivation (e.g. 82, 134) and that large cells are usually found on particles (22, 34, 53). After substrate utilization, the large cells may eventually detach by a change in surface character that affects adhesion (31, 71), or the progeny may detach after cell division (72). The latter process is in agreement with the suggestion that fecal pellets may act as "baby machines," producing free-living bacteria (63). Detachment involves a sharp and drastic decrease in concentration and a change in composition of substrates. The cellular responses to such a downshift are numerous and generally include a reduction in size.

The proposed surface-related life cycle is in clear agreement with Goldman's spinning wheel aggregate hypothesis. Goldman (42) concluded that aggregates may serve as sites of intense nutrient recycling and that association with the microzones of particles is a prerequisite for high growth rates and for the tight coupling between different trophic levels. Swimming and chemotaxis were suggested to be major survival mechanisms because they enable the cell to migrate between enriched aggregates. Azam & Ammerman (9) confirmed the importance of these mechanisms by demonstrating that motility allows bacteria to position themselves in spheres of diffusing dissolved organic matter surrounding phytoplankton cells. Goldman (42) suggested that "microbes may be very efficient nomads who know how to get to the oasis when they need to, but who spend a significant fraction of their time as wanderers." As we show in the following, they also seem to have the information they need to prepare themselves for the nomadic period upon departure from the oasis.

THE TRANSIENT PHASE BETWEEN GROWTH AND NONGROWTH

Many shifts in the cellular and population levels of nongrowing marine bacteria have been summarized in recent reviews by Morita (97, 98). Reports on miniaturization, i.e. formation of coccoid cells with a significant loss of the original volume, as well as on fragmentation, which results in an increase in the number of cells, can also be found in several original research publications (e.g. 56, 73, 106). This section briefly focuses on the viability of starving bacteria, the transient phase between growth and nongrowth, cellular alterations related to surface characteristics accompanying the transition, adhesion to substrata, and the utilization of immobilized surface substrates.

Formation and Presence of Small Cells

Starvation survival of marine bacteria is not unique to only a few selected species (5). Studies of starved marine bacteria report high degrees of viability. This indicates the evolutionary and ecological significance of periods of nongrowth in natural bacterial assemblages. Marine bacteria initially isolated on agar media containing low or no added carbon responded to added nutrients and formed larger cells with increased growth rates. These bacteria were also able to decrease in size upon starvation, becoming the ultramicrobacteria or "dwarfs" characteristic of the marine population. The starvation survival process was the same in nutrient-free artifical seawater and filtered natural seawater for a psychrophilic marine bacterium (107), an indigenous marine *Vibrio cholerae* (13), and five marine strains frequently used in starvation survival studies in our laboratory. Water samples from a Gulf Coast estuary were passed through a 0.2 μm filter and the filtered water was incubated with dilute nutrient broth for 21 days (82). This resuscitation procedure allowed for outgrowth of normally sized cells of *Vibrio, Pseudomonas, Aeromonas,* and *Alcaligenes* spp. on dilute nutrient agar. Several experiments have shown that bacteria of an average size of less than 0.03 μm^3 become large cells when grown on nutrient media (98). It is also clear that some of the dwarf cells in the marine environment do not become large cells when placed on media (82, 136, 141). It is also feasible, however, that these cells may demonstrate an increase in size when the correct nutrient is supplied (cf 89). The fact that normally less than 10% [varying from 0.01–90% (128)] of the total number of bacterial cells in natural waters show the capacity to form colonies on agar plates does not suggest that these bacteria are obligately oligotrophic, low-nutrient type bacteria or that they are unique prokaryotes. It is now well established that many bacteria that are initially unable to grow on agar plates can do so if a proper resuscitation method is used (24, 43).

Bacterial Surface Characteristics and Adhesion

The alteration in bacterial surface characteristics and adhesion during starvation survival is of ecological significance. Dawson et al (31) suggested that such nongrowth changes are a tactic in the survival strategy of marine bacteria. These workers noted the formation of fibrillar structures on the surface of a marine *Vibrio* during the initial phase of nongrowth and an apparent reduction in the amount of mucopeptide in the cell wall without a corresponding reduction in the outer membrane. Increased adhesion and surface hydrophobicity have since been observed as a consequence of the nutritional downshift of marine bacterial isolates (71; M. Hermansson, G. W. Jones & S. Kjelleberg, manuscript in preparation) and mixed natural bacterial assemblages (25).

Specific bacterial surface components may be altered; Wrangstadh et al (146) presented evidence of the production and release of a neutral exopolysaccharide by a marine *Pseudomonas* during starvation. The polymer renders the bacterial cell more hydrophilic and less liable for adhesion. After its release the cells become more adhesive. Using polymer-specific fluorescent antibodies it was demonstrated that the polymer is not formed on the bacterial surface prior to 4 hr of starvation. Synthesis of protein filaments would consume more energy than carbohydrate polymer production. It is significant, therefore, that of 16 bacterial strains isolated from the open ocean (5), most kept their flagella even after 8 mo of starvation, and that viable *E. coli* cells may have as many fimbriae after 9 days of starvation as cells at the onset of starvation (L. Rüden, personal communication).

Other apparent changes in the outer cell layer include folding and an increase in surface roughness (31, 71) and the formation of vesicles on the bacterial surface (56, 88). Based on specific degradation and the compositional shift of the carbohydrates of starved *V. cholerae,* it was suggested that the more hydrophilic O-side chains may be more readily utilized under starvation conditions (56). The number of three- and five-carbon sugars decreased, while the seven- and six-carbon sugars, probably making up the oligosaccharides of the R core, were relatively conserved. This would be in agreement with the fact that the outer membranes of certain marine bacteria become more hydrophobic during starvation (71; M. Hermansson, G. W. Jones & S. Kjelleberg, manuscript in preparation).

Direct evidence of the surface distribution of growing and nongrowing bacteria of a given species in the marine environment is basically available only for the marine autochthonous *V. cholerae* (e.g. 43, 83) and to some extent for luminescent marine bacteria (47). In addition to the nongrowing forms of *V. cholerae* there are cells that adhere and grow at a range of interfaces in marine waters (57), including interfaces within the gut of invertebrates, where adhesion may be specific (61). Free-living forms of marine luminescent bacteria do not grow but persist for long periods (47). Only with the use of cellular or genetic markers or specific antibodies will it be possible to obtain a precise picture of the distribution and growth of a marine bacterium in its natural habitat.

MECHANISMS OF STARVATION SURVIVAL

Bacteria have active control mechanisms that appear to facilitate the adjustment of the cell to changes in nutrient supply. The cost of maintenance for the organism will be detrimentally high if the cell remains in a high metabolic state during periods of low exogenous nutrient conditions. Control mechanisms affect the growth or nongrowth of bacteria at the low nutrient concentrations frequently encountered in the aquatic habitat. These mechanisms may modulate or be modulated by the systems used for uptake and transport, the coupling of these systems to sources of extracellular and intracellular metabolic energy reserves, the maintenance of osmotic integrity and membrane potential, and the turnover of essential cell components. The energy necessary to make appropriate shifts in synthetic pathways may be supplied by the endogenous metabolism of cellular constituents and storage polymers as well as by efficient scavenging of very low levels of exogenous substrate molecules. Recent studies of protein pattern shifts upon and during starvation should evoke considerable interest in the effects of the transient phase from growth to nongrowth and in the cellular makeup during prolonged starvation conditions.

Koch (76) performed an interesting series of experiments on slow-growing and starving *E. coli*. A significant aspect of these experiments, pertinent to microbial ecology, was that only two thirds of chemostat-grown cells with doubling times of 16 hr were actively engaged in protein synthesis at any instant, although all cells became capable of β-galactosidase synthesis within 3 hr. The experiment shows that a substantial part of the population was in a "dormant" state during slow chemostat culture. The overall conclusions that can be made from Koch's experiments are that virtually all bacteria in an assemblage, including those considered copiotrophs (111), may be able to use both K- and r-strategies during periods of their life cycle, and that the cells have sensing systems and switching mechanisms to select alternative responses. Whether or not the selected strategy depends on the stage of cell reproduction remains an open but intriguing question.

Nonmarine Bacteria

ENDOGENOUS METABOLISM AND CONSTITUENTS While several pathways of slow growth or starvation in aquatic and marine bacteria are presently being unraveled, the bulk of information must be extrapolated from the many studies on nonaquatic laboratory strains. Studies on endogenous metabolism and alterations of endogenous constituents have been performed using *Aerobacter, Arthrobacter, Bacillus, Brevibacterium, E. coli, Methanospirillum, Nocardia, Peptococcus, Pseudomonas, Salmonella typhimurium, Sarcia, Selenomonas, Staphylococcus,* and *Zymomonas,* (e.g. 15, 58, 94, 114, 133; reviewed in 29). Studies on specific protein alterations and stress responses are referred to below. Considerable variation in the behavior of strains is apparent. Therefore it is not possible to generalize, for example, about the association of the death of starved or nongrowing bacteria with the loss of any specific cellular constituent or reserve material. Nor is it possible to generalize about the influence of the growth history of the cells and their adaptability to starvation conditions (55, 93, 113). The exact role of storage polymers such as polyhydroxybutyrate and glycogen in the survival of the bacterial cell is unclear, as results are conflicting (29, 92, 114). Compounds such as RNA, ribosomes, protein, amino acids, and fatty acids, which are regarded as essential, are also used as substrates for endogenous metabolism.

A general conclusion is that viability is related to the rate at which constituents are metabolized rather than the absolute levels of a particular constituent at the time of starvation. A low rate of endogenous metabolism that matches the maintenance energy requirements improves the chances of prolonged viability. One of the better correlations between longevity and endogenous processes was observed for starved *Peptococcus prevotii* (95, 115), in which RNA was the only compound found to undergo substantial degradation. It is possible that the elevated levels of RNA found during slow growth of *E. coli* (75) allow for substantial RNA content to be lost without affecting viability during starvation (29).

The synthesis and degradation of DNA in starving laboratory strains are less clear. An initial increase in total DNA per cell has been observed for starved cells of *Arthrobacter* (121), *Aerobacter aerogenes* (132), *Brevibac-*

terium linens (15), *E. coli* (17, 59), *Methanospirillum hungatii* (18), and *Pseudomonas aeruginosa* (e.g. 84). This increase in DNA is often attributed to the completion of rounds of replication initiated before the onset of starvation. An increase in total cell numbers was generally reported for these experiments. The concept that DNA is synthesized at a rate proportional to the formation of the total biomass may not be supported by recent findings that increases in levels of DNA per milliliter of starvation regime were not related to increases in total numbers (see below). Inherent in the former concept is also the notion that DNA does not turn over. DNA is lost, however, during starvation of laboratory strains (84, 94); this is possibly due to the degradation of excess genome copies. Indeed, recent studies (119) have suggested that starving cells of *S. typhimurium* preferentially degrade supercoiled plasmid DNA while retaining copies of the plasmid in relaxed open circular forms.

PROTEINS The degradation and synthesis of protein (i.e. turnover) in nongrowing *E. coli* are roughly equal in extent and proceed at a rate of 4–5% for several hours; the rate is estimated as the amount of protein degraded per total amount of protein in the cell (86, 87). This value is higher than in growing cells. An even larger difference was found between nongrowing and growing *Bacillus cereus* (138). Horan et al (58) demonstrated that amino acids degraded from proteins are used as endogenous substrates. The role of protein degradation and synthesis in the survival of carbon-starved *E. coli* and *S. typhimurium* has recently been examined in more detail (116, 117). A series of peptidase-deficient mutants showed decreased stability during starvation. These cells were defective in protein synthesis not from an innate inability to synthesize proteins, but because of a lack of free amino acids for such production. Based on inhibition experiments using chloramphenicol and amino acid analogs, Reeve et al (116) also concluded that the proteins synthesized initially, during the first hours of starvation, were the most critical for survival; however, their continued activity was also necessary for survival after longer starvation times. A somewhat different picture was found during long-term (20 days) starvation of *S. typhimurium* (S. Kjelleberg, P. L. Conway & T.-A. Stenström, manuscript in preparation). Inhibitors of protein synthesis became more effective and resulted in greater loss of viability as the period of starvation increased.

Most starvation survival experiments suggest that the protein content is reduced with time of starvation, although a recent study showed that the protein content per biomass of living cell remained unchanged after 30 days of starvation (15). During growth under carbon limitations synthesis of catabolic enzymes is derepressed; this may improve the cell's capacity to use diverse substrates (91). These shifts may cause changes in chemotaxis and in uptake and transport of various substrates during starvation of marine bacteria, as discussed below. Various patterns of enzyme activity have been seen, however, for different laboratory strains during starvation. In *Selenomonas ruminantium,* cellular enzymes became stabilized (94), whereas in *Arthrobacter crystallopoietes* a loss in activity of inducible but not constitutive enzymes was observed (16).

Groat & Matin (44) recently analyzed more precisely the changes in protein synthesis of carbon-starved *E. coli*. Two-dimensional gel electrophoresis of in vivo pulse-labeled proteins resolved at least 30 polypeptides that were either new peptides or peptides synthesized at higher levels during the first 3–4 hr of starvation. Interestingly, several heat-shock proteins were among those synthesized, although the majority were considered to be starvation specific. The overlap in proteins induced by various shock treatments of *E. coli* is unclear. Conditions that evoke the stringent response (see below) induced synthesis of at least 5 of 17 heat-shock proteins (45), and some SOS-inducing agents evoked the heat-shock response (140). Various starvation conditions induced significant overlap in protein pattern of *S. typhimurium,* which indicates the synthesis of a core of general starvation-inducible proteins (127). Only a few proteins produced in this strain during heat shock or anaerobiosis were also indentified as starvation inducible, however.

THE STRINGENT RESPONSE Several of the observed shifts in endogenous constituents in slow-growing and nongrowing bacteria may be explained in terms of induction of the stringent response (37). This system for coordinating various mechanisms produces signal molecules that adjust the discrimination specificity of protein synthesis. In most bacteria the stringent response is initiated by amino acid starvation (26). The trigger is a lack of charged amino acyl tRNA molecules. This regulates the rate at which the *relA* gene expresses the product that catalyzes formation of the regulatory guanosine phosphates (50, 51). The compound ppGpp might also affect protein synthesis during carbon/energy starvation (44). The accumulation of ppGpp during this type of downshift may take place by a route independent of the *relA* gene product (37, 38). The induction of the stringent response or the formation and accumulation of ppGpp by other routes results in the following: The rate of RNA accumulation and synthesis is reduced, the rate of protein turnover is increased, and the membrane transport of various exogenous precursors is reduced, as is the endogenous synthesis of nucleotides, glycolytic intermediates, carbohydrates, lipids, fatty acids, polyamines, and peptidoglycans. In consequence, nongrowing bacteria rapidly become resistant to autolysis by cell wall synthesis inhibitors or chaotropic agents (137; S. Kjelleberg, P. L. Conway & T.-A. Stenström, manuscript in preparation).

THE PROTON-MOTIVE FORCE Enteric bacteria may be capable of utilizing acidic conditions as an electron chemical gradient to generate high-energy intermediates (126). Internal pH control is one of the functions of endogenous metabolism in starved cells (135). Uncoupling of the protein gradient immediately halted the survival process of two marine bacteria (73), and the gradual decrease in the proton-motive force found during starvation has been attributed to the gradual accumulation of nonviable cells of the culture (78). Indeed, the activation of the membrane appears to be a major regulatory mechanism of the starved cell (90).

ATP LEVELS The levels of ATP and adenylate energy charge have been related to viability during starvation in some instances (23, 30, 95, 115), whereas in others no such relationship exists (131). The selective excretion of AMP by starved *A. crystallopoietes,* for example (81), would maintain a high energy charge. The size of the adenylate pool is determined by a very complex balance between catabolic and incorporation pathways involving RNA and AMP catabolism, and RNA, DNA, and protein synthesis. While it is known that nucleotides turn over rapidly, high levels have been maintained in a number of experiments in which growth downshifts were induced (62). Ingraham et al (62) suggested that a specific regulatory circuit could control the levels of nucleoside phosphates such that reduced ATP levels will increase ADP, AMP, and pyrophosphate pools. The increase in pyrophosphates would then directly inhibit polymerization reactions and thereby reduce ATP consumption.

Marine Bacteria

It should be obvious from the above presentation of regulatory pathways that given that a microbe's environment is not constant during its life cycle and cannot provide for continuous growth, a series of active cellular responses will be established.

INITIAL PHASE OF STARVATION The formation of small cells is a significant feature of marine bacteria. It is not uncommon for intracellular integrity to decrease, for granules to disappear, and for the nuclear regions to become more compact during starvation (7, 105). The downshift response of marine bacteria has two phases, with a demarcation line dividing the initial period of pronounced cellular activities and the subsequent long-term survival when very low endogenous metabolism and small alterations in morphology, activity, and levels of endogenous constituents are apparent. During the initial transient phase of starvation, which may vary considerably in duration depending on the organisms and the environment, the following alterations have been observed: a temporary increase in endogenous respiration for both free-living (7, 88) and surface-bound cells (73); a temporary increase in amino acid uptake (P. Mårdén, T. Nyström & S. Kjelleberg, manuscript in preparation) and incorporation (108); temporary changes in activity upon supplementation with a complex medium (88); and nongradual alteration in cell surface composition and adhesion patterns (71, 146). Further support for the idea of an active initial period of metabolic and structural reorganization was derived using inhibitors of the proton flow, the electron transport chain, and membrane bound ATPase (73). Inhibition of the latter had a relatively small effect on starved cells after the initial phase (39). This was also observed for freshwater bacteria (54, 55).

ENERGY SUPPLY The psychrophilic marine *Vibrio* ANT 300 reduced biomass by 48% during 1 wk of starvation and a total of 56% over 3 wk (109). This reduction may have partly depended on an appreciable decrease in total lipid phosphate during the first week of starvation, since a 65% decrease in the amount of membrane phospholipids took place during the course of the experiment (109). Fifty percent of this loss occurred while the endogenous respiration was less than 1% of the initial rate. The phospholipids of *V. cholerae* declined 99.8% during the first week of starvation, and this decline was concomitant with visible loss of cell wall integrity (56). Total carbohydrates in this strain also exhibited a rapid and immediate decline. The rapid decline of total carbohydrates and lipids probably reflects the cell's initial utilization of easily available endogenous energy sources for the first rather active reorganization phase of starvation. Rapid initial consumption of poly-β-hydroxybutyrate (PHB), presumably for the energy needed for several of the temporary shifts in rates of uptake and synthesis (108), has been observed for the unidentified marine bacterium S14 (85, 88). Similar observations have been made for *V. cholerae* (56).

The effectiveness of a nongrowing cell should entail a compromise between biosynthetic and catabolic demands. It is possible that the cell's metabolism shifts from assimilation toward energy production during starvation. Several reports indicate that cells under starvation conditions preferentially use assimilated nutrients for energy production (27, 55, 79). Respiration of arginine by ANT 300 cells accounted for 35.9–59.4% of the total arginine transported after 35 days of starvation as compared to 7.2–19.5% at the onset of starvation (32). The relative respiration rate may depend on the substrate, however. ANT 300 cells starved for 4 mo showed 85% respiration with glutamic acid compared to 5% respiration with glucose (5).

ADENOSINE NUCLEOTIDES Between 15 and 20% of the ATP in growing cells may be used for transport (130). It is feasible that starved cells should maintain their ATP levels for substrate uptake. In most cases substrate uptake seems independent of ATP, however. ATP levels in respiring ANT 300 cells (7) and a marine *Pseudomonas* (79) steadily increased after the first 8 days of starvation. ATP concentrations in ANT 300 were significantly higher after 25 days of starvation than at the onset of starvation; a temporary rise in the ATP level was seen after 6 days of starvation (7). This was not found by Oliver & Stringer (109), however, who reported a 73% decrease in ATP per cell after 21 days of starvation. A downshift response may not provoke an immediate reduction of nucleotide triphosphates in accordance with the specific self-regulatory circuit control of nucleotide phosphates previously discussed for *E. coli*.

PROTEIN TURNOVER Generally, a net protein degradation is demonstrated for cells starved beyond the metabolic demarcation line (e.g. 7, 56). Endogenous respiration has been demonstrated to stabilize at a rate of less than 0.01% total carbon respired per hour after the initial period of metabolic activity (79, 106). In the initial starvation phase the total protein per biomass has been observed to increase, owing to degradation of other constituents, e.g. PHB (108). Protein profiles constitute a powerful tool for elucidating patterns of regulatory downshift responses. In four marine bacterial isolates examined by two-dimensional gel electrophoresis during starvation, the cellular proteins were not degraded uniformly (6, 65). *Vibrio* ANT 300 synthesized specific proteins after 7 days of starvation, and production increased in intensity within 30 days. In fact, dramatic changes in the protein fingerprint pattern were not detected until after 30 days of starvation. All three mesophilic isolates studied by Jouper-Jaan et al (65) exhibited different protein survival patterns during the first 24 hours of starvation. Two of these isolates synthesized new proteins or significantly increased amounts of a few proteins but exhibited significantly different degradation patterns. While the *Vibrio* strain DW1 lost only two proteins, 17–20 proteins were degraded in the unidentified gram-negative organism S14. Some of the new peptides synthesized in the latter strain are starvation-specific proteins because antibodies produced against three or four of the new proteins in cells starved for 24 hr agglutinated only starved cells (N. Albertson, G. W. Jones & S. Kjelleberg, manuscript in preparation). Immunoblotting demonstrated that starved S14 cells synthesized three or four new periplasmic space proteins and one new outer membrane protein. The increased rate of protein synthesis during the initial phase of starvation of DW1 and S14 may account for these alterations. Pulse labeling at various times of starvation clearly showed that new proteins were formed (N. Albertson, T. Nyström & S. Kjelleberg, manuscript in preparation). Protein synthesis during starvation was also confirmed (P. Mårdén, T. Nyström & S. Kjelleberg, manuscript in preparation) by the observation that starved cells remained viable after osmotic shock release of periplasmic proteins and subsequently resynthesized the shock-released proteins. Resynthesis also occurred after long-term starvation.

The formation of new proteins during starvation probably increases the chances for survival of the species in its natural habitat. In some isolates, however, these proteins do not appear to be directly related to the viability of the cell during the initiation of the survival process. Based on inhibition experiments with chloramphenicol during the initial 2 hr of starvation, Humphrey et al (60) suggested that cells of *Vibrio* strain DW1 have constitutive enzymes for the initiation of the dwarfing process. No loss in viability or change in the size reduction was observed during the first 2 hr of starvation. Hence, the possibility exists that the synthesis of new proteins is not a prerequisite for the initiation of the downshift process in some strains but is related to other, subsequent starvation-specific processes and to viability after the first few hours of starvation.

The protein response to starvation in marine bacteria does not appear to resemble the response to carbon starvation of *E. coli* and *Salmonella typhimurium* (44, 116, 117). The energy expenditure involved in the stress response exhibited by *E. coli* would not be an ecologically suitable survival tactic for marine bacteria because constant exposure to fluctuations in substrate levels and physical discontinuities would continually deplete limited intracellular reserves. We suggest that the depletion of exogenous energy and nutrients for marine bacteria does not resemble acknowledged bacterial stress factors. Marine bacteria may have a highly evolved mechanism by which they create a viable starvation state. It is of great interest, for example, that ribosome structures are conserved by starved marine vibrios (7, 56, 105); this survival strategy is perhaps similar to endospore formation.

THE STRINGENT RESPONSE AND cAMP Whether the marine isolates examined to date accumulate ppGpp during starvation is not known. The methionine-requiring marine *Vibrio* ANT 300, which had increased RNA content per respiring cell after 6 wk of starvation, did not exhibit the increased rate of RNA synthesis in response to amino acid deprivation (7) that would be expected if the stringent response was intact (137). Similarly, a decrease in cell wall synthesis would be expected in stringent, amino acid–deprived strains. The marine organism S14 increased its levels of D-alanine, typically found in the cell wall, during 48 hr of starvation. Degradation of proteins in this organism is not likely to supply sufficient amounts of amino acids that the stringent response would not be evoked during starvation; thus it appears that S14 also fails to exhibit the stringent response.

As with ppGpp, the levels of cAMP during nongrowth of marine bacteria have not been established. The immediate response of these organisms to suitable substrates indicates that there is an accumulation of cAMP. Ammerman & Azam (4) showed that large transient increases in cAMP occurred within a few minutes in natural populations of seawater bacteria. These populations took up cAMP dissolved in seawater by specific, energy-dependent, high-affinity transport systems with K_m values of 10^{-11} to 10^{-9} M. The authors speculated that cAMP released in conjunction with dissolved organic matter from algae may act as a metabolic cue. This would enable the cell to utilize a broader range of substrates.

DNA DNA synthesis and degradation are not identical in different nongrowing marine bacteria. While the starvation survival process in some strains appears unaffected by inhibitors of DNA synthesis (106), nalidixic acid and cadmium inhibited the incorporation of thymidine into DNA and the fragmentation process of S14 (P. Mårdén, M. Hermansson & S. Kjelleberg, manuscript in preparation; T. Nyström & S. Kjelleberg, manuscript in preparation). While a net synthesis of DNA has been observed for several strains (P. Mårdén, M. Hermansson & S. Kjelleberg, manuscript in preparation), including those without increased cell numbers during starvation (118), a continued decline in the amount of DNA per cell beyond the metabolic demarcation line of starvation has also been observed for several marine strains (56; P. Mårdén, M. Hermansson & S. Kjelleberg, manuscript in preparation). The division cycle of marine bacteria is an area that warrants further study. These bacteria have long mean generation times in their natural habitats, but may be grown at generation times of only a few hours or less. Presumably, DNA replication proceeds at a relatively constant rate in both the laboratory and the natural habitat. Could the long generation time reflect the time it takes for the cell to attain sufficient amounts of hitherto unknown DNA replication initiator molecules (cf 67)? One may also question how the size and biomass variation within the cell cycle influences DNA replication.

SUBSTRATE CAPTURE The ability of oligotrophic bacteria to function in low-nutrient habitats has primarily been ascribed to their high-affinity uptake systems (110). Oligotrophic bacteria have an affinity for organic substrates at concentrations less than 10 μg C liter^{-1} (21). Carlucci et al (21) suggested, however, that most low-nutrient bacteria are opportunistic and facultatively oligotrophic, and are capable of growing over a wide range of nutrient concentrations. This agrees with the observation of two uptake systems (0.007 μM and 0.11 μM) in a marine heterotrophic low-nutrient bacterium (52). High-affinity uptake has also been demonstrated for natural bacterial populations confronted with dissolved adenosine triphosphate (53) and different carbohydrates prepared from algae (14) in nutrient-dilute seawater. The evidence for multiphasic uptake kinetics in natural marine assemblages was recently reiterated (36). If a high-nutrient bacterium is capable of responding to low-nutrient concentrations, the transport systems are expected to operate at high affinity.

Nongrowing cells of the marine bacterium S14 increased their leucine uptake (calculated per milliliter of cell suspension) after 3 hr of starvation. The higher level of uptake remained during the experimental period of 96 hr of starvation (P. Mårdén, T. Nyström & S. Kjelleberg, manuscript in preparation). Cells of S14 maintained both high-affinity and low-affinity uptake systems. Cells grown in medium supplemented with leucine and other large nonpolar amino acids revealed an almost complete loss of uptake of leucine via the high-affinity system upon starvation; this indicates that the high-affinity system was repressed during growth in enriched medium. During starvation periods of up to 4 mo the marine *Vibrio* strain ANT 300 maintained its ability to take up a number of sugars and amino acids (5, 32, 39, 41). High-affinity uptake of mannitol was induced after several weeks of starvation and was maintained during prolonged starvation (27). The K_m for glutamic acid uptake by ANT 300 varied from 0.09 to 0.23 μM, with the lowest value after 20 days of starvation (41). Mannitol, glucose, and glutamate transport by a *Vibrio* species revealed both high- and low-affinity systems (27). The affinity for mannitol was 7.24 μM during the first 5 wk of starvation, and then a switch to an uptake system with higher affinity (K_m = 1.47 μM) occurred. A similar shift was noted for glutamate; the K_m value was initially 0.83 μM and decreased to 0.23 μM after 5 wk of starvation. These data demonstrate that bacteria that enter a nongrowing state in a nutrient-deficient environment maintain and regulate their nutrient uptake capacity and accordingly acquire an increased affinity for a given substrate. The substrate levels corresponding to the K_m values of the high-affinity uptake systems presented here are of the same order of magnitude as the concentration that was recently suggested as the upper limit for oligotrophic growth (10 μg C liter^{-1}) (21). This indicates that bacteria under nongrowing conditions are capable of obtaining and competing for solutes at concentration levels found in the marine environment. Button (19) reviewed additional aspects of microbial kinetics during nutrient-limited growth.

SPECIFICITY OF UPTAKE SYSTEMS It has been hypothesized that bacteria adapted to low-nutrient waters exhibit high affinity for substrates coupled with low substrate specificity within a single uptake system (1, 110). In this way a number of substrates could be transported through the same system. This would be advantageous for an organism utilizing a variety of nutrients at low concentrations. Bell (14) supported the hypothesis in a more general characterization of natural populations of marine bacteria in low-nutrient conditions. Akagi & Taga (1) examined the specificity of the high-affinity glucose and proline uptake systems of two marine bacterial isolates, one oligotroph and one copiotroph. The glucose uptake system was highly specific in both strains, whereas the proline uptake by the oligotrophic strain was inhibited by competition from nine other amino acids. Glick (41) stated that the uptake systems of ANT 300 did not become nonspecific as a response to starvation. Large nonpolar amino acids inhibited leucine uptake by S14 cells (P. Mårdén, T. Nyström & S. Kjelleberg, manuscript in preparation). This amino acid inhibition pattern was seen for all starvation times tested.

TRANSPORT MECHANISMS Under low-nutrient conditions, high-affinity transport seems to be mediated by binding (97) or carrier proteins (78). It is concievable that these systems are cooperative, although there is some conflicting evidence. Muir et al (99) demonstrated that in *E. coli* the energy cost was much higher for transport of maltose via a binding protein than for transport of lactose by proton symport (i.e. carrier proteins). They suggested that the higher energy cost for the binding-mediated transport was compensated by its higher affinity and higher capacity to concentrate substrate molecules. Carrier proteins usually also have high affinity for their solutes, however (78). One advantage provided by binding-protein transport seems to be the ability to transport several linked units of a substrate at the same time and cost, so substrates do not have to be hydrolyzed to monomers before transport (99). Kirchman & Hodson (68) suggested that natural aquatic bacteria take up polymers of an amino acid in preference to the corresponding monomer. Although few results have been reported, inhibition experiments indicate that the uptake capacity of starved cells may be dependent on the maintenance of a proton gradient (1, 32, 39, 41, 55).

Binding proteins have been suggested as the main substrate capture site for bacteria in low-nutrient waters (97). Binding protein–mediated uptake is normally demonstrated by an osmotic shock treatment for release of periplasmic proteins (reviewed in 120). Earlier in this review we described the recovery after an osmotic shock treatment of the leucine uptake system during starvation (P. Mårdén, T. Nyström & S. Kjelleberg, manuscript in preparation). Two proteins regained from the osmotic shock fluid had leucine-binding activity. Further evidence for binding protein–mediated transport in starved cells has been given by Glick (41), who demonstrated a 65% reduction of leucine uptake after shock treatment of ANT 300.

Binding proteins might also act as chemotactic receptors (41). Starved ANT 300 cells exhibited a chemotactic response toward arginine (39), and starved S14 cells exhibited a chemotactic response toward a mixture of low–molecular weight substances (K. Malmcrona-Friberg, A. Valeur, G. Odham & S. Kjelleberg, manuscript in preparation).

The uptake of solutes by a gram-negative bacterium could possibly be affected by changes in the composition of the cell membranes. Koch & Wang (77) discussed the diffusion resistance of the *E. coli* outer membrane. Large alterations in the fatty-acid profiles of the cell membranes have been observed in nongrowing bacteria (46, 56, 85, 109). An increased ratio of monounsaturated to saturated fatty acids (85) and an increased fraction of short-chain fatty acids (85, 109) may cause the increased fluidity of the membrane and the improved uptake capacity of starved cells. Guckert et al (46) found that the *trans/cis* ratio for the major monounsaturated fatty acids increased with starvation of *V. cholerae*. Since the *trans* isomer is less susceptible to enzymatic degradation than the corresponding *cis* isomer, the authors concluded that the observed alteration was related to increased membrane stability rather than maintenance of the fluidity of the membrane.

CONCLUDING REMARKS Nongrowing marine bacterial isolates direct the energy obtained from endogenous constituents toward a range of activities. These include the maintenance of intracellular high-energy charge levels (100, 102), high ATP levels (7, 79), and an energized membrane (39, 41, 73); increased endogenous metabolism during the initial phase of starvation (72, 73); synthesis of new proteins (6, 65; N. Albertson, G. W. Jones & S. Kjelleberg, manuscript in preparation; N. Albertson, T. Nyström & S. Kjelleberg, manuscript in preparation); an increase in the rate of protein synthesis during the initial phase of starvation (108); RNA production and accumulation (7); DNA degradation and synthesis (56; P. Mårdén, M. Hermansson & S. Kjelleberg, manuscript in preparation); increased uptake capacity (27, 32, 40, 41; P. Mårdén, T. Nyström & S. Kjelleberg, manuscript in preparation); cell wall synthesis (85); a shift in membrane lipid composition, including production of several lipid molecules (46, 56, 85, 109); the production and release of exopolysaccharides (146); increased adhesion to inanimate surfaces (25, 31, 71, 146); and increased chemotaxis (39, 41). Bacteria capable of responding to nutrient-deficient environments are clearly not inactive, although the overall metabolic activity is maintained at very low levels. The study of the physiological and molecular adjustments of nongrowing marine bacteria clearly warrants continued and increasing research efforts.

ACKNOWLEDGMENTS

We thank N. Albertson, P. Conway, K. Malmcrona-Friberg, G. Odham, T. Nyström, D. Robert, L. Rüden, and A. Valeur for contributing unpublished results included in this review. We are grateful to P. Conway and T. Nyström for helpful discussions and to A. Renås and M. Jehler for the assembly and typing of the text. Work from the authors' laboratories was supported in part by grants from the Swedish Natural Science Research Council and awards to GWJ from the National Institutes of Health (AI 19647) and the Environmental Protection Agency (CR 813568).

Literature Cited

1. Akagi, Y., Taga, N. 1980. Uptake of D-glucose and L-proline by oligotrophic and heterotrophic marine bacteria. *Can. J. Microbiol.* 26:454–59
2. Alldredge, A. L. 1979. The chemical composition of macroscopic aggregates in two neritic seas. *Limnol. Oceanogr.* 24:855–66
3. Alldredge, A. L., Cole, J. J., Caron, D. A. 1986. Production of heterotrophic bacteria inhabiting macroscopic organic aggregates (marine snow) from surface waters. *Limnol. Oceanogr.* 31:68–78
4. Ammerman, J. W., Azam, F. 1981. Dissolved cyclic adenosine monophosphate (cAMP) in the sea and uptake of cAMP by marine bacteria. *Mar. Ecol. Prog. Ser.* 5:85–89
5. Amy, P. S., Morita, R. Y. 1983. Starvation-survival patterns of sixteen freshly isolated open-ocean bacteria. *Appl. Environ. Microbiol.* 45:1109–15
6. Amy, P. S., Morita, R. Y. 1983. Protein patterns of growing and starved cells of a marine *Vibrio* sp. *Appl. Environ. Microbiol.* 45:1748–52
7. Amy, P. S., Pauling, C., Morita, R. Y. 1983. Starvation-survival processes of a marine vibrio. *Appl. Environ. Microbiol.* 45:1041–48
8. Andersson, A., Larsson, U., Hagström, Å. 1986. Size-selective grazing by a microflagellate on pelagic bacteria. *Mar. Ecol. Prog. Ser.* 33:51–57
9. Azam, F., Ammerman, J. W. 1984. Cycling of organic matter by bacterioplankton in pelagic marine ecosystems: microenvironmental considerations. In *Flows of Energy and Materials in Marine Ecosystems: Theory and Practice,* ed. M. J. R. Fasham, pp. 345–60. New York: Plenum
10. Azam, F., Fenchel, T., Field, J. G., Gray, J. S., Meyer-Reil, L. A., Thingstad, F. 1983. The ecological role of water-column microbes in the sea. *Mar. Ecol. Prog. Ser.* 10:257–63
11. Azam, F., Fuhrman, J. A. 1984. Measurement of bacterioplankton growth in the sea and its regulation by environmental conditions. In *Heterotrophic Activity in the Sea,* ed. J. E. Hobbie, P. J. leB. Williams, pp. 179–96. New York: Plenum
12. Azam, F., Hodson, R. E. 1977. Size distribution and activity of marine microheterotrophs. *Limnol. Oceanogr.* 22:492–501
13. Baker, R. M., Singleton, F. L., Hood, M. A. 1983. Effects of nutrient deprivation on *Vibrio cholerae*. *Appl. Environ. Microbiol.* 46:930–40
14. Bell, W. H. 1984. Bacterial adaptation to low-nutrient conditions as studied with algal extracellular products. *Microb. Ecol.* 10:217–30
15. Boyaval, P., Boyaval, E., Desmazeaud, M. J. 1985. Survival of *Brevibacterium linens* during nutrient starvation and intracellular changes. *Arch. Microbiol.* 141:128–32
16. Boylen, C. W., Ensign, J. C. 1970. Intracellular substrates for endogenous metabolism during long-term starvation of rod and spherical stage cells of *Arthrobacter crystallopoietes*. *J. Bacteriol.* 103:578–87
17. Brdar, B., Kos, E., Drakulic, M. 1965. Metabolism of nucleic acids and protein in starving bacteria. *Nature* 208:303–4
18. Breuil, C., Patel, G. B. 1980. Viability and depletion of cell constituents of *Methanospirillum hungatii* GP1 during starvation. *Can. J. Microbiol.* 26:887–92
19. Button, D. K. 1985. Kinetics of nutrient-limited transport and microbial growth. *Microbiol. Rev.* 49:270–97
20. Carlucci, A. F., Craven, D. B., Henrichs, S. M. 1985. Surface-film microheterotrophs: amino acid metabolism and solar radiation effects on their activities. *Mar. Biol.* 85:13–22
21. Carlucci, A. F., Shimp, S. L., Craven, D. B. 1986. Growth characteristics of low-nutrient bacteria from the northeast and central Pacific Ocean. *FEMS Microbiol. Ecol.* 38:1–10
22. Caron, D. A., Davies, P. G., Madin, L. P., Sieburth, J. M. 1982. Heterotrophic bacteria and bacterivorous protozoa in oceanic macroaggregates. *Science* 218:795–97
23. Chapman, A. G., Fall, L., Atkinson, D. E. 1971. Adenylate energy charge in *Escherichia coli* during growth and starvation. *J. Bacteriol.* 108:1072–86
24. Colwell, R. R., Brayton, P. R., Grimes, D. J., Roszak, D. B., Huq, S. A., Palmer, L. M. 1985. Viable but nonculturable *Vibrio cholerae* and related pathogens in the environment: Implications for release of genetically engineered microorganisms. *Biotechnology* 3:817–20
25. Conway, P. L., Maki, J., Mitchell, R., Kjelleberg, S. 1986. Starvation of marine flounder, squid and laboratory mice and its effect on the intestinal mi-

crobiota. *FEMS Microbiol. Ecol.* 38: 187–95

26. Cozzone, A. J. 1981. How do bacteria synthesize proteins during amino acid starvation? *Trends Biochem. Sci.* 6:108–10
27. Davis, C. L. 1985. *Physiological and ecological studies of mannitol utilizing marine bacteria*. PhD thesis. Univ. Cape Town, South Africa. 203 pp.
28. Davoll, P. J., Silver, M. W. 1986. Marine snow aggregates: life history sequence and microbial community of abandoned larvacean houses from Monterey Bay, California. *Mar. Ecol. Prog. Ser.* 33:111–20
29. Dawes, E. A. 1976. Endogenous metabolism and the survival of starved prokaryotes. In *The Survival of Vegetative Microbes,* ed. T. R. G. Gray, J. R. Postgate, pp. 19–53. Cambridge, UK: Cambridge Univ. Press
30. Dawes, E. A., Large, P. J. 1970. Effect of starvation on the viability and cellular constituents of *Zymomonas anaerobia* and *Zymomonas mobilis*. *J. Gen. Microbiol.* 60:31–42
31. Dawson, M. P., Humphrey, B. A., Marshall, K. C. 1981. Adhesion: a tactic in the survival strategy of a marine vibrio during starvation. *Curr. Microbiol.* 6:195–99
32. Faquin, W. C., Oliver, J. D. 1984. Arginine uptake by a psychrophilic marine *Vibrio* sp. during starvation-induced morphogenesis. *J. Gen. Microbiol.* 130:1331–35
33. Fellows, D. A., Karl, D. M., Knauer, G. A. 1981. Large particle fluxes and the vertical transport of living carbon in the upper 1500 m of the northeast Pacific Ocean. *Deep-Sea Res.* 28A:921–36
34. Ferguson, R. L., Rublee, P. 1976. Contribution of bacteria to standing crop of coastal plankton. *Limnol. Oceanogr.* 21:141–45
35. Fletcher, M., Marshall, K. C. 1982. Are solid surfaces of ecological significance to aquatic bacteria? *Adv. Microb. Ecol.* 6:199–230
36. Fuhrman, J. A., Ferguson, R. L. 1986. Nanomolar concentrations and rapid turnover of dissolved free amino acids in seawater: agreement between chemical and microbiological measurements. *Mar. Ecol. Prog. Ser.* 33:237–42
37. Gallant, J. A. 1979. Stringent control in *E. coli*. *Ann. Rev. Genet.* 13:393–415
38. Gallant, J., Lazzarine, R. 1976. The regulation of ribosomal RNA synthesis and degradation in bacteria. In *Protein Synthesis. A Series of Advances,* ed. E. McConkey, pp. 309–59. New York: Dekker
39. Geesey, G. G., Morita, R. Y. 1979. Capture of arginine at low concentrations by a marine psychrophilic bacterium. *Appl. Environ. Microbiol.* 38:1092–97
40. Geesey, G. G., Morita, R. Y. 1981. Relationship of cell envelope stability to substrate capture in a marine psychrophilic bacterium. *Appl. Environ. Microbiol.* 42:533–40
41. Glick, M. A. 1981. *Substrate capture, uptake, and utilization of some amino acids by starved cells of a psychrophilic marine vibrio*. MS thesis. Oregon State Univ., Corvallis. 61 pp.
42. Goldman, J. C. 1984. Conceptual role for microaggregates in pelagic waters. *Bull. Mar. Sci.* 35:462–76
43. Grimes, D. J., Attwell, R. W., Brayton, P. R., Palmer, L. M., Rollins, D. M., et al. 1986. Fate of enteric pathogenic bacteria in estuarine and marine environments. *Microbiol. Sci.* 3:324–29
44. Groat, R. G., Matin, A. 1986. Synthesis of unique proteins at the onset of carbon starvation in *Escherichia coli*. *J. Ind. Microbiol.* 1:69–73
45. Grossman, A. D., Taylor, W. E., Burton, Z. F., Burgess, R. R., Gross, C. A. 1985. Stringent response in *Escherichia coli* induces expression of heat shock proteins. *J. Mol. Biol.* 186:357–65
46. Guckert, J. B., Hood, M. A., White, D. C. 1986. Phospholipid ester-linked fatty acid profile changes during nutrient deprivation of *Vibrio cholerae:* Increases in the *trans*/*cis* ratio and proportions of cyclopropyl fatty acids. *Appl. Environ. Microbiol.* 52:794–801
47. Hastings, J. W. 1986. *Distribution, activities and functions of marine luminescent bacteria*. Presented at 4th Int. Symp. Microb. Ecol., Ljubljana, Yugoslavia
48. Hermansson, M., Dahlbäck, B. 1983. Bacterial activity at the air/water interface. *Microb. Ecol.* 9:317–28
49. Hermansson, M., Marshall, K. C. 1985. Utilization of surface localized substrate by non-adhesive marine bacteria. *Microb. Ecol.* 11:91–105
50. Hesseltine, W. A., Block, R. 1973. Synthesis of guanosine tetra- and pentaphosphate requires the presence of a codon specific uncharged transfer ribonucleoacid in the acceptor site of ribosomes. *Proc. Natl. Acad. Sci. USA* 70:1564–68
51. Hesseltine, W. A., Block, R., Gilbert, W., Weber, K. 1972. MSI and MSII are

made on ribosomes in an idling step of protein synthesis. *Nature* 238:381–84
52. Hodson, R. E., Carlucci, A. F., Azam, F. 1979. Glucose transport in a low nutrient marine bacterium. Abstract N-59. *Abstr. Annu. Meet. Am. Soc. Microbiol.* 1979:189
53. Hodson, R. E., Maccubbin, A. E., Pomeroy, L. R. 1981. Dissolved adenosine triphosphate utilization by free-living and attached bacterioplankton. *Mar. Biol.* 64:43–51
54. Hofle, M. G. 1983. Long-term changes in chemostat cultures of *Cytophaga johnsonae*. *Appl. Environ. Microbiol.* 46:1045–53
55. Hofle, M. G. 1984. Transient responses of glucose-limited cultures of *Cytophaga johnsonae* to nutrient excess and starvation. *Appl. Environ. Microbiol.* 47:356–62
56. Hood, M. A., Guckert, J. B., White, D. C., Deck, F. 1986. Effect of nutrient deprivation on lipid, carbohydrate, DNA, RNA, and protein levels in *Vibrio cholerae*. *Appl. Environ. Microbiol.* 52:788–93
57. Hood, M. A., Ness, G. E., Rodrick, G. E., Blake, N. J. 1984. The ecology of *Vibrio cholerae* in two Florida estuaries. In *Vibrios in the Environment,* ed. R. R. Colwell, pp. 309–409. New York: Wiley
58. Horan, N. J., Midgley, M., Dawes, E. A. 1981. Effect of starvation on transport, membrane potential and survival of *Staphylococcus epidermidis* under anaerobic conditions. *J. Gen. Microbiol.* 127:223–30
59. Horuichi, T. 1959. RNA degradation and DNA and protein synthesis of *E. coli* in phosphate deficient medium. *J. Biochem. Tokyo* 46:1467–80
60. Humphrey, B., Kjelleberg, S., Marshall, K. C. 1983. Responses of marine bacteria under starvation conditions at a solid-water interface. *Appl. Environ. Microbiol.* 45:43–47
61. Huq, A., Huq, S. A., Grimes, D. J., O'Brien, M., Chu, K. H., et al. 1986. Colonization of the gut of the blue crab *(Callinectes sapidus)* by *Vibrio cholerae*. *Appl. Environ. Microbiol.* 52:586–88
62. Ingraham, J. L., Maaløe, O., Neidhardt, F. C. 1983. *Growth of the Bacterial Cell*. Sunderland, Mass: Sinauer
63. Jacobsen, T. R., Azam, F. 1984. Role of bacteria in copepod fecal pellet decomposition: colonization, growth rates and mineralization. *Bull. Mar. Sci.* 35:495–503
64. Jannasch, H. W. 1974. Steady state and the chemostat in ecology. Comment. *Limnol. Oceanogr.* 19:716–20
65. Jouper-Jaan, Å., Dahlöf, B., Kjelleberg, S. 1986. Changes in the protein composition of three bacterial isolates from marine waters during short term energy and nutrient deprivation. *Appl. Environ. Microbiol.* 52:1419–21
66. Kefford, B., Kjelleberg, S., Marshall, K. C. 1982. Bacterial scavenging: Utilization of fatty acids localized at a solid-liquid interface. *Arch. Microbiol.* 133:257–60
67. Kepes, F. 1986. The cell cycle of *Escherichia coli* and some of its regulatory systems. *FEMS Microbiol. Rev.* 32: 225–46
68. Kirchman, D., Hodson, R. 1984. Inhibition by peptides of amino acid uptake by bacterial populations in natural waters: implications for the regulation of amino acid transport and incorporation. *Appl. Environ. Microbiol.* 47:624–31
69. Kirchman, D., Mitchell, R. 1982. Contribution of particle-bound bacteria to total microheterotrophic activity in five ponds and two marshes. *Appl. Environ. Microbiol.* 43:200–9
70. Kjelleberg, S. 1985. Mechanisms of bacterial adhesion at gas-liquid interfaces. In *Bacterial Adhesion,* ed D. C. Savage, M. M. Fletcher, pp. 163–94. New York: Plenum
71. Kjelleberg, S., Hermansson, M. 1984. Starvation-induced effects on bacterial surface characteristics. *Appl. Environ. Microbiol.* 48:497–503
72. Kjelleberg, S., Humphrey, B. A., Marshall, K. C. 1982. Effect of interfaces on small, starved marine bacteria. *Appl. Environ. Microbiol.* 43:1166–72
73. Kjelleberg, S., Humphrey, B. A., Marshall, K. C. 1983. Initial phases of starvation and activity of bacteria at surfaces. *Appl. Environ. Microbiol.* 46: 978–84
74. Knauer, G. A., Hebel, D., Cipriano, F. 1982. Marine snow: major site of primary production in coastal waters. *Nature* 300:630–31
75. Koch, A. L. 1971. The adaptive response of *Escherichia coli* to a feast and famine existence. *Adv. Microb. Physiol.* 6:147–217
76. Koch, A. L. 1979. Microbial growth in low concentrations of nutrients. In *Strategies of Microbial Life in Extreme Environments. Dahlem Konf., Berlin,* ed. M. Shilo, pp. 261–79. Weinheim, Germany: Verlag Chemie
77. Koch, A. L., Wang, C. H. 1982. How

close to the theoretical diffusion limit do bacterial uptake systems function? *Arch. Microbiol.* 131:36–42
78. Konings, W. N., Veldkamp, H. 1980. Phenotypic responses to environmental change. In *Contemporary Microbial Ecology*, ed. D. C. Ellwood, J. N. Hedger, M. J. Latham, J. M. Lynch, J. H. Slater, pp. 161–91. London: Academic
79. Kurath, G., Morita, R. Y. 1983. Starvation-survival physiological studies of a marine *Pseudomonas* sp. *Appl. Environ. Microbiol.* 45:1206–11
80. Laanbroek, H. J., Verplanke, J. C. 1986. Seasonal changes in percentages of attached bacteria enumerated in a tidal and a stagnant coastal basin: relation to bacterioplankton productivity. *FEMS Microbiol. Lett.* 38:87–98
81. Leps, W. T., Ensign, J. C. 1979. Adenylate nucleotide levels and energy charge in *Arthrobacter crystallopoietes* during growth and starvation. *Arch. Microbiol.* 122:69–76
82. MacDonell, M. T., Hood, M. A. 1982. Isolation and characterization of ultramicrobacteria from a Gulf Coast estuary. *Appl. Environ. Microbiol.* 43:566–71
83. MacDonell, M. T., Hood, M. A. 1984. Ultramicrovibrios in Gulf Coast estuarine waters: Isolation, characterization and incidence. See Ref. 57, pp. 551–62
84. MacKelvie, R. M., Campbell, J. J. R., Gronlund, A. F. 1968. Survival and intracellular changes of *Pseudomonas aeruginosa* during prolonged starvation. *Can. J. Microbiol.* 14:639–45
85. Malmcrona-Friberg, K., Tunlid, A., Mårdén, P., Kjelleberg, S., Odham, G. 1986. Chemical changes in cell envelope and poly-β-hydroxybutyrate during short term starvation of a marine bacterial isolate. *Arch. Microbiol.* 144:340–45
86. Mandelstam, J. 1958. Turnover of protein in growing and non-growing populations of *Escherichia coli*. *Biochem. J.* 69:110–19
87. Mandelstam, J. 1960. The intracellular turnover of protein and nucleic acids and its role in biochemical differentiation. *Bacteriol. Rev.* 24:289–308
88. Mårdén, P., Tunlid, A., Malmcrona-Friberg, K., Odham, G., Kjelleberg, S. 1985. Physiological and morphological changes during short term starvation of marine bacterial isolates. *Arch. Microbiol.* 142:326–32
89. Martin, P., MacLeod, R. A. 1984. Observations on the distinction between oligotrophic and eutrophic marine bacteria. *Appl. Environ. Microbiol.* 47:1017–22
90. Mason, C. A., Hamer, G., Bryers, J. D. 1986. The death and lysis of microorganisms in environmental processes. *FEMS Microbiol. Rev.* 39:373–401
91. Matin, A. 1979. Microbial regulatory mechanisms at low nutrient concentrations as studied in chemostat. See Ref. 76, pp. 323–39
92. Matin, A., Veldhuis, C., Stegeman, V., Veenhuis, M. 1979. Selective advantage of a *Spirillum* sp. in a carbon-limited environment. Accumulation of poly-β-hydroxy-butyric acid and its role in starvation. *J. Gen. Microbiol.* 112:349–55
93. Mink, R. W., Hespell, R. B. 1981. Long-term nutrient starvation of continuously cultured (glucose-limited) *Selenomonas ruminantium*. *J. Bacteriol.* 148:541–50
94. Mink, R. W., Patterson, J. A., Hespell, R. B. 1982. Changes in viability, cell composition, and enzyme levels during starvation of continuously cultured (ammonia-limited) *Selenomonas ruminantium*. *Appl. Environ. Microbiol.* 44:913–22
95. Montague, M. D., Dawes, E. A. 1974. The survival of *Peptococcus prévotii* in relation to the adenylate energy charge. *J. Gen. Microbiol.* 80:291–99
96. Mopper, K., Lindroth, P. 1982. Diel and depth variations in dissolved free amino acids and ammonium in the Baltic Sea determined by shipboard HPLC analysis. *Limnol. Oceanogr.* 27:336–47
97. Morita, R. Y. 1984. Substrate capture by marine heterotrophic bacteria in low nutrient waters. See Ref. 11, pp. 83–100
98. Morita, R. Y. 1985. Starvation and miniaturisation of heterotrophs, with special emphasis on maintenance of the starved viable state. In *Bacteria in Their Natural Environments*, ed. M. M. Fletcher, G. D. Floodgate, pp. 111–30. London: Academic
99. Muir, M., Williams, L., Ferenci, T. 1985. Influence of transport energization on the growth yield of *Escherichia coli*. *J. Bacteriol.* 163:1237–42
100. Nazly, N., Carter, I. S., Knowles, C. J. 1980. Adenine nucleotide pools during starvation of *Beneckea natriegens*. *J. Gen. Microbiol.* 116:295–303
101. Neidhardt, F. C., VanBogelen, R. A., Vaughn, V. 1984. The genetics and regulation of heat-shock proteins. *Ann. Rev. Genet.* 18:295–329
102. Niven, D. F., Collins, P. A., Knowles,

C. J. 1977. Adenylate energy charge during batch culture of *Beneckea natriegens*. *J. Gen. Microbiol*. 98:95–108

103. Norkrans, B. 1980. Surface microlayers in aquatic environments. *Adv. Microb. Ecol*. 4:51–85
104. Novitsky, J. A. 1986. Degradation of dead microbial biomass in a marine sediment. *Appl. Environ. Microbiol*. 52: 504–9
105. Novitsky, J. A., Morita, R. Y. 1976. Morphological characterization of small cells resulting from nutrient starvation of a psychrophilic marine vibrio. *Appl. Environ. Microbiol*. 32:617–22
106. Novitsky, J. A., Morita, R. Y. 1977. Survival of a psychrophilic marine vibrio under long-term nutrient starvation. *Appl. Environ. Microbiol*. 33:635–41
107. Novitsky, J. A., Morita, R. Y. 1978. Possible strategy for the survival of marine bacteria under starvation conditions. *Mar. Biol*. 48:289–95
108. Nyström, T., Mårdén, P., Kjelleberg, S. 1986. Relative changes in incorporation rates of leucine and methionine during starvation survival of two bacteria isolated from marine waters. *FEMS Microbiol. Ecol*. 38:285–92
109. Oliver, J. D., Stringer, W. F. 1984. Lipid composition of a psychrophilic marine *Vibrio* sp. during starvation-induced morphogenesis. *Appl. Environ. Microbiol*. 47:461–66
110. Poindexter, J. S. 1981. Oligotrophy: fast and famine existence. *Adv. Microb. Ecol*. 5:63–89
111. Poindexter, J. S. 1981. The caulobacters: ubiquitous unusual bacteria. *Microbiol. Rev*. 45:123–79
112. Pomeroy, L. R. 1984. Microbial processes in the sea: Diversity in nature and science. See Ref. 11, pp. 1–23
113. Postgate, J. R., Hunter, J. R. 1962. The survival of starved bacteria. *J. Gen. Microbiol*. 29:233–63
114. Preiss, J. 1984. Bacterial glycogen synthesis and its regulation. *Ann. Rev. Microbiol*. 38:419–58
115. Reece, P., Toth, D., Dawes, E. A. 1976. Fermentation of purines and their effect on the adenylate energy charge and viability of starved *Peptococcus prévotii*. *J. Gen. Microbiol*. 97:63–71
116. Reeve, C. A., Amy, P. S., Matin, A. 1984. Role of protein synthesis in the survival of carbon-starved *Escherichia coli* K-12. *J. Bacteriol*. 160:1041–46
117. Reeve, C. A., Bockman, A. T., Matin, A. 1984. Role of protein degradation in the survival of carbon-starved *Escherichia coli* and *Salmonella typhimurium*. *J. Bacteriol*. 157:758–63
118. Reichardt, W., Morita, R. Y. 1982. Survival stages of a psychrotrophic *Cytophaga johnsonae* strain. *Can. J. Microbiol*. 28:841–50
119. Robert, D. K. 1986. *Studies on the 63 megadalton cryptic plasmid of* Salmonella typhimurium. PhD thesis. Univ. Michigan, Ann Arbor. 174 pp.
120. Rosen, B. P., Heppel, L. A. 1973. Present status of binding proteins that are released from gram-negative bacteria by osmotic shock. In *Bacterial Membranes and Walls,* ed. L. Lieve, pp. 209–39. New York: Dekker
121. Scherer, C. G., Boylen, C. W. 1977. Macromolecular synthesis and degradation in *Arthrobacter* during periods of nutrient deprivation. *J. Bacteriol*. 132:584–89
122. Shanks, A. L., Trent, J. D. 1979. Marine snow: Microscale nutrient patches. *Limnol. Oceanogr*. 24:850–54
123. Sieburth, J. M. 1979. *Sea Microbes*. New York: Oxford Univ. Press
124. Silver, M. W., Alldredge, A. L. 1981. Bathy-pelagic marine snow: deep sea algal and detrital community. *J. Mar. Res*. 39:501–30
125. Silver, M. W., Shanks, A. L., Trent, J. D. 1978. Marine snow: microplankton habitat and source of small-scale patchiness in pelagic populations. *Science* 201:371–73
126. Sjogren, R. E., Gibson, M. J. 1981. Bacterial survival in a dilute environment. *Appl. Environ. Microbiol*. 41:1331–36
127. Spector, M. P., Aliabadi, Z., Gonzalez, T., Foster, J. W. 1986. Global control in *Salmonella typhimurium:* two-dimensional electrophoretic analysis of starvation-, anaerobiosis-, and heat shock–inducible proteins. *J. Bacteriol*. 168:420–24
128. Staley, J. T., Konopka, A. 1985. Measurement of *in situ* activities of nonphotosynthetic microorganisms in aquatic and terrestrial habitats. *Ann. Rev. Microbiol*. 39:321–46
129. Stevenson, L. H. 1978. A case for bacterial dormancy in aquatic systems. *Microb. Ecol*. 4:127–33
130. Stouthamer, H. A. 1973. A theoretical study of the amount of ATP required for synthesis of microbial cell material. *Antonie van Leeuwenhoek J. Microbiol. Serol*. 39:545–65
131. Strange, R. E., Wade, H. E., Dark, F. A. 1963. Effect of starvation on adenosine triphosphate concentration in *Aerobacter aerogenes*. *Nature* 199:55–57
132. Strange, R. E., Wade, H. E., Ness, A.

G. 1963. The catabolism of proteins and nucleic acids in starved *Aerobacter aerogenes*. *Biochem. J.* 86:197–203

133. Szewczyk, E., Mikucki, J. 1983. Protein A as a substrate of endogenous metabolism in staphylococci. *FEMS Microbiol. Lett.* 19:55–58
134. Tabor, P. S., Ohwada, K., Colwell, R. R. 1981. Filterable marine bacteria found in the deep sea: distribution, taxonomy, and response to starvation. *Microb. Ecol.* 7:67–83
135. Thomas, T. D., Batt, R. D. 1969. Degradation of cell constituents by starved *Streptococcus lactis* in relation to survival. *J. Gen. Microbiol.* 58:347–62
136. Torrella, F., Morita, R. Y. 1981. Microcultural study of bacterial size changes and microcolony and ultramicrocolony formation by heterotrophic bacteria in seawater. *Appl. Environ. Microbiol.* 41:518–27
137. Tuomanen, E., Tomasz, A. 1986. Induction of autolysis in nongrowing *Escherichia coli*. *J. Bacteriol.* 167:1077–80
138. Urba, R. C. 1959. Protein breakdown in *Bacillus cereus*. *Biochem. J.* 71:513–18
139. van Es, F. B., Meyer-Reil, L. A. 1982. Biomass and metabolic activity of heterotrophic marine bacteria. *Adv. Microb. Ecol.* 4:111–70
140. Walker, G. C. 1984. Mutagenesis and inducible responses to deoxyribonucleic acid damage in *Escherichia coli*. *Microbiol. Rev.* 48:60–93
141. Watson, S. W., Novitsky, T. J., Quinly, H. L., Valois, F. W. 1977. Determination of bacterial number and biomass in the marine environment. *Appl. Environ. Microbiol.* 33:940–46
142. Wiebe, W. J., Pomeroy, L. R. 1972. Microorganisms and their association with aggregates and detritus in the sea: a microscopic study. *Mem. Ist. Ital. Idrobiol. Dott Marco de Marchi Pallanza Italy* 29:325–52 (Suppl.)
143. Wikner, J., Andersson, A., Normark, S., Hagström, Å. 1986. Use of genetically marked minicells as a probe in measurement of predation on bacteria in aquatic environments. *Appl. Environ. Microbiol.* 52:4–8
144. Williams, P. J. leB. 1970. Heterotrophic utilization of dissolved organic compounds in the sea. 1. Size distribution of population and relationship between respiration and incorporation of growth substances. *J. Mar. Biol. Assoc. UK* 50:859
145. Williams, P. J. leB. 1981. Incorporation of microheterotrophic processes into the classical paradigm of the planktonic food web. *Kiel. Meeresforsch.* 5:1–28
146. Wrangstadh, M., Conway, P. L., Kjelleberg, S. 1986. The production and release of an extracellular polysaccharide during starvation of a marine *Pseudomonas* sp. and the effect thereof on adhesion. *Arch. Microbiol.* 145:220–27
147. Wright, R. T. 1984. Dynamics of pools of dissolved organic carbon. See Ref. 11, pp. 121–54
148. Wright, R. T., Coffin, R. B. 1984. Measuring microzooplankton grazing on planktonic marine bacteria by its impact on bacterial production. *Microb. Ecol.* 10:137–49

Ann. Rev. Microbiol. 1987. 41:51–75

THE BIOSYNTHESIS OF SULFUR-CONTAINING β-LACTAM ANTIBIOTICS

J. Nüesch, J. Heim, and H.-J. Treichler

Ciba-Geigy Limited, Biotechnology Department, CH-4002 Basel, Switzerland

CONTENTS

THE BIOSYNTHETIC ROUTE

During the past ten years it has become evident that sulfur-containing, so-called classical β-lactam antibiotics are all derived from the same amino acid precursors, which ultimately lead to the fermentation products penicillins, cephalosporins, and cephamycins.

Whereas penicillins with a hydrophobic side chain are only produced by filamentous fungi, the hydrophilic cephalosporins and cephamycins are made by different molds, by actinomycetes, and, as found very recently, by unicellular bacteria (reviewed in 47).

0066-4227/87/1001-0051$02.00

The current view of an overall biosynthetic scheme for classical β-lactam antibiotics is depicted in Figure 1 (for chemical structures compare Figure 2). In a first series of steps the three amino acid precursors L-α-aminoadipic acid, L-cysteine, and L-valine are condensed to form a tripeptide precursor; L-valine is converted to the D-isomer in the course of these reactions. Oxidative ring closure of the linear tripeptide leads to the bicyclic ring system of isopenicillin N. The pathway then branches. In the case of the hydrophobic penicillins, the α-aminoadipic acid side chain is exchanged in the final step for a hydrophobic acyl group, e.g. phenylacetyl in penicillin G. Biosynthesis in the hydrophilic branch of the pathway proceeds from isopenicillin N with a change in the configuration of the side chain followed by an oxidative ring expansion reaction. Further modifications lead to cephalosporin C and the cephamycins. The 7-methoxy derivatives of cephalosporin C, e.g. cephamycins, are produced only by actinomycetes; deacetylcephalosporin C and cephalosporin C serve as the biosynthetic precursors.

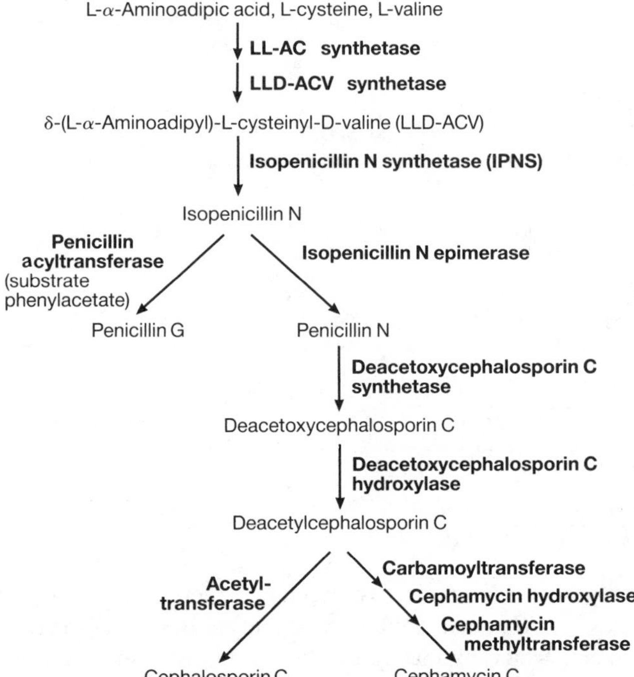

Figure 1 Biosynthetic pathways to penicillins and cephalosporins. The *left* branch at isopenicillin N leads to hydrophobic penicillins, the *right* branch to hydrophilic cephalosporins and cephamycins.

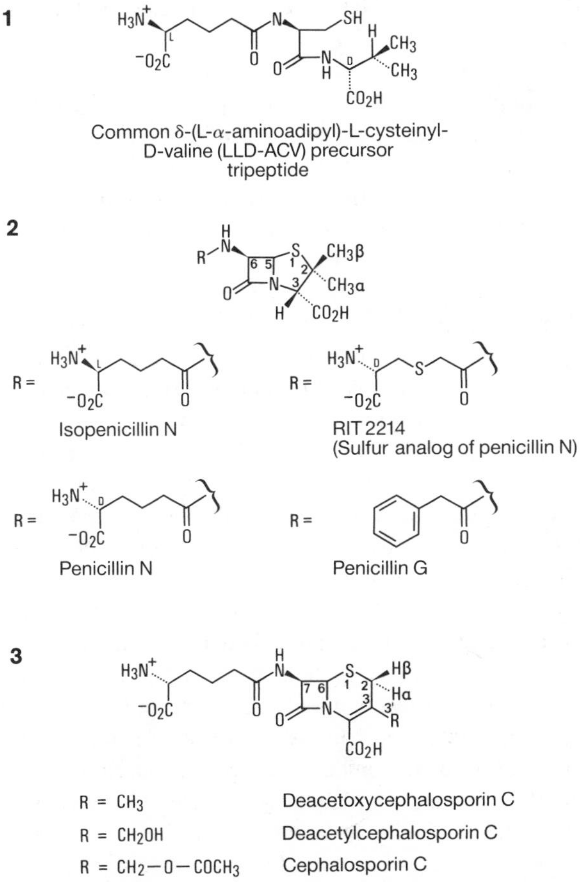

1

Common δ-(L-α-aminoadipyl)-L-cysteinyl-D-valine (LLD-ACV) precursor tripeptide

2

R = Isopenicillin N

R = RIT 2214 (Sulfur analog of penicillin N)

R = Penicillin N

R = Penicillin G

3

$R = CH_3$ Deacetoxycephalosporin C

$R = CH_2OH$ Deacetylcephalosporin C

$R = CH_2-O-COCH_3$ Cephalosporin C

Figure 2 Key compounds in the biosynthesis of penicillin-cephalosporin antibiotics.

Several excellent reviews have covered the elucidation of the common biosynthetic pathway in detail (34, 47, 103). In recent years intensive investigations have been concentrated on the central parts of the pathway, the formation of the bicyclic penam and the cephem nucleus. This review therefore updates the progress that has been made in purification of the biosynthetic enzymes involved and summarizes their properties in vitro. Owing to the availability of cell-free systems, regulatory mechanisms important for the initiation of β-lactam biosynthesis and for the precursor flow involved have now been studied in more detail. Selected regulatory elements are reviewed with emphasis on mechanisms found to be relevant for strain improvement of β-lactam–producing microorganisms.

CHARACTERIZATION OF β-LACTAM–SPECIFIC ENZYMES

The first two steps leading to the linear tripeptide are the least understood of the entire pathway. In the first step a peptide bond is formed between the δ-carboxyl group of L-α-aminoadipic acid and the α-amino group of L-cysteine. The reaction is carried out by the δ-(L-α-aminoadipyl)-L-cysteine synthetase (LL-AC-synthetase). Activity of just such an enzyme was recently shown in the soluble fraction of crude extracts of *Cephalosporium acremonium* (5, 25). The enzyme requires ATP and Mn^{2+} or Mg^{2+} (25). Adlington et al (5) demonstrated incorporation of L-valine into the tripeptide δ-(L-α-aminoadipyl)-L-cysteinyl-D-valine (LLD-ACV) using the same crude cell-free extracts of the *C. acremonium* β-lactam–negative mutant N-2. By analogy with the first step, this enzyme is called LLD-ACV-synthetase. Since valine is present in the tripeptide in its D-configuration, inversion of the L-form presumably takes place during valine activation prior to incorporation into the tripeptide. An enzyme-bound valine intermediate similar to that in gramicidin biosynthesis could be envisaged (41). In addition, Lopez-Nieto et al (64) observed that inhibitors of protein synthesis in vivo actually stimulate LLD-ACV biosynthesis in *Penicillium chrysogenum;* this observation further supports the idea of nonribosomal peptide formation as the initial step in penicillin biosynthesis. The cell-free systems described so far convert the precursor amino acids into the corresponding di- and tripeptide very inefficiently and require further optimization. What is very striking is the mechanistic similarity of the first two steps in penicillin and cephalosporin biosynthesis with the pathway of glutathionine (L-γ-glutamyl-L-cysteinyl-L-glycine) biosynthesis. Whether or not the two pathways are indeed carried out by the same enzymes, as reported by Lara et al (62) for LL-AC-synthetase in cell-free extracts of *P. chrysogenum,* should be confirmed with the better analytic tools available today.

Most of the recent work on β-lactam biosynthesis has been devoted to investigating the cyclization of the tripeptide to the penam nucleus, e.g. isopenicillin N. The reaction is catalyzed by the isopenicillin N synthetase (IPNS), an enzyme that has been purified from *C. acremonium, P. chrysogenum, Streptomyces clavuligerus,* and *Streptomyces lactamdurans* (31, 44, 47, 83, 86). The enzyme belongs to the group of dehydrogenases that, in the presence of Fe^{2+} and ascorbate, remove four hydrogen equivalents from the precursor tripeptide with consumption of one molecule of dioxygen (100). Molecular oxygen, Fe^{2+}, and ascorbate are required for all IPNS isolated from the different β-lactam producers. In general by using appropriate stabilizing agents the enzyme does not seem to be particularly unstable. Partially purified enzyme preparations usually lead to 80–100% ring cyclization of the precursor LLD-ACV within 10–30 min. The enzyme obeys Michaelis-Menten kinetics and shows an apparent K_m for the substrate LLD-ACV of 1.3×10^{-4} M in the case of *P. chrysogenum* and $1.6–2 \times 10^{-4}$ M in the case of *C. acremonium* (23, 86).

Purified IPNS consists of a single subunit with a molecular mass of ~39 kd in *C. acremonium* and *P. chrysogenum* (22, 44, 86) and 36.5 kd in *S. clavuligerus* (47). Whether the slight differences in molecular mass reflect differences in the purified eukaryotic and prokaryotic IPNS, in the strains used, or in the techniques used for estimation remains to be elucidated. Most of the data have been collected on IPNS of *C. acremonium* (22, 83). The enzyme exists in reduced and oxidized forms resulting from the formation of one easily accessible intramolecular disulfide bond. It has a pI of 5.05, and the N-terminal sequence of the first 50 amino acids has been determined. These sequence data enabled Samson et al (88) to prepare sets of oligonucleotides with which they recently isolated the gene encoding IPNS protein from a cosmid genome library of *C. acremonium*. The gene has an open reading frame that codes for a polypeptide with molecular mass of 38.4 kd. When this open reading frame was cloned into an *Escherichia coli* expression vector, recombinant *E. coli* cells produced a protein with authentic properties of IPNS, e.g. cyclization of the LLD-ACV precursor into isopenicillin N (88). The absence of any intron sequences and the unusually high G+C content of the *Cephalosporium* gene are striking. Isolation of the corresponding genes in *Streptomyces* and *Penicillium,* which now seems to be a feasible task, is highly desirable to shed some light on any evolutionary relationship between IPNS genes present in both prokaryotes and eukaryotes.

The availability of pure IPNS has also greatly facilitated studies on the peculiar biochemical mechanisms by which the bicyclic penam ring system, e.g. the β-lactam and thiazolidine rings, is elaborated in vivo. Loss of hydrogens during cyclization accompanies sulfur-carbon and nitrogen-carbon bond formation, with overall retention of configuration at each carbon atom (2). Attempts to find a free intermediate between LLD-ACV and isopenicillin N have all been unsuccessful. Neither a tripeptide containing a D-β-hydroxyvaline or *N*-hydroxy-D-valine in place of D-valine (25a) nor a seven-membered thiazepine serves as substrate for IPNS (2). However, feeding experiments are hampered by a serious drawback; the presumed intermediates often exist only in an enzyme-bound form and are not recognized when added in solution. From kinetic isotope effects, Baldwin et al (19) recently suggested that formation of the β-lactam ring, e.g. nitrogen-carbon bond formation, precedes that of the thiazolidine ring. The monocyclic β-lactam presumably exists as an enzyme-bound intermediate, whereby the sulfhydryl group of cysteine has a decisive role in binding the substrate to the active site of the enzyme IPNS (3, 20). Efforts to synthesize chemically the putative monocyclic β-lactam intermediate have been unsuccessful owing to the extremely short half-life of such a compound at or near neutral pH in solution (12).

Most of the data available on the terminal carbon-sulfur bond formation in biosynthesis of isopenicillin N appear to be consistent with a radical mechanism (24). Experiments with a modified tripeptide substrate, which are discussed in more detail in the section on mutasynthesis (13, 15, 18), as well as experiments with natural stereospecifically labeled LLD-ACV suggest intermediate carbon radical formation at position C-2 of valine. The active-site topology of the IPNS enzyme, however, puts strong restraints on rotational motion such as that created by formation of a radical; therefore the β-methyl isomer is preferentially formed (i.e. configuration is retained) when the substrate is natural LLD-ACV. With certain LLD-ACV analogs free rotation seems to be possible, and carbon-sulfur bonds are formed with both retention and inversion of configuration (18).

Compared with IPNS, much less is known about the acyltransferase that leads to hydrophobic penicillins or about isopenicillin N epimerase. The latter enzyme is only present in microorganisms that produce cephalosporins or cephamycins; thus it represents the point of divergence of penicillin and cephalosporin biosynthesis. It catalyzes the conversion of the L-α-aminoadipyl side chain in isopenicillin N to the D-α-aminoadipyl side chain in penicillin N. Isopenicillin N epimerase isolated from *C. acremonium* is remarkably labile (46, 58), whereas that from *Streptomyces* sp. appears to be stable (31, 49). Although the epimerase in *Streptomyces* sp. is not stimulated by pyridoxal-5-phosphate, a known cofactor of amino acid racemases, pyridoxal phosphate stabilizes its activity. Isopenicillin N epimerase seems to catalyze a 1 : 1 equilibrium mixture using either isopenicillin N or penicillin N as substrate (47). The molecular mass of the enzyme from *S. clavuligerus* is ~60 kd (49), while that of the *S. lactamdurans* enzyme is ~57 kd (31). Lübbe et al (65) recently demonstrated epimerase activity in a highly cephalosporin-producing strain of *C. acremonium*. In this organism pyridoxal phosphate slightly stimulates epimerase activity. With the development of strains that produce more plenteous or more stable biosynthetic enzyme, a detailed comparison of the eukaryotic epimerase with its bacterial counterpart can now be anticipated.

The precursor role of penicillin N for antibiotics containing the cephem nucleus was first shown by Kohsaka & Demain (56) and Yoshida et al (106). The reaction carried out by deacetoxycephalosporin C synthetase (expandase) involves an oxidative opening of the penam-thiazolidine ring to give upon reclosure the six-membered dihydrothiazine ring of deacetoxycephalosporin C. During ring expansion one of the methyl groups of penicillin N becomes incorporated into the cephem ring system. Expandase has been partially purified from *C. acremonium* (60), *S. clavuligerus* (50), and *S. lactamdurans* (31). The molecular masses were estimated to be 31 kd for the fungal enzyme (60), and 29 kd (50) and 27 kd (31) for the bacterial enzymes, respectively.

Expandase follows Michaelis-Menten kinetics with apparent K_m for penicillin N of 3×10^{-5} M *(C. acremonium)* (60) and 4.8×10^{-5} M *(S. lactamdurans)* (31). Interestingly, these constants are about an order of magnitude lower than the values obtained for IPNS, the preceding enzyme, which also carries out oxidative ring closure. Fungal and bacterial expandase shows an absolute requirement for Fe^{2+}, ascorbate, possibly ATP, molecular oxygen, and α-ketoglutarate. These unusual cofactor requirements assign expandase to the class of intermolecular dioxygenases. Enzymes of this class activate oxygen in the decomposition of equimolar amounts of α-ketoglutarate to form carbon dioxide and succinate; this reaction has been demonstrated with the enzyme from *C. acremonium* (60). However, whereas one atom of molecular oxygen usually appears in the oxidized substrate, no oxygen is incorporated in the final product, deacetoxycephalosporin C. On the basis of current knowledge, one can tentatively postulate the formation of an oxidized enzyme-bound intermediate, which is then in the course of the reaction broken down to the cephem and water.

Townsend (95) recently followed the stereochemical fate of chiralmethyl valine in whole-cell incorporation studies. He concluded that ring expansion takes place with complete epimerization. One possible explanation is again the formation of an intermediate radical during ring expansion. This intermediate could by analogous scavenging then lead to the shunt metabolite 3-β-hydroxycepham, which is described as a minor component in commercial cephalosporin fermentations (76). (For alternative interpretations see the section on mutasynthesis, below.) Cell-free ring expansion experiments (76) have shown that the cepham is not a precursor of deacetoxycephalosporin C.

Very similar to the expandase is the next enzyme in the pathway, deacetoxycephalosporin C hydroxylase, which catalyzes hydroxylation at the C-3' methyl group to give deacetoxycephalosporin C. This enzyme also belongs to the group of α-ketoglutarate–linked dioxygenases. Scheidegger et al (90) have provided evidence that ring expansion and hydroxylation reactions both reside on the same 33-kd protein in *C. acremonium*. In *S. clavuligerus,* however, both enzymes can be separated by anion-exchange chromatography (52); this may be the first major difference between fungal and prokaryotic cephalosporin biosynthesis. Adlington et al (4) have shown that 3-exomethylene cephalosporin C is also converted to deacetylcephalosporin C by cell-free extracts of *C. acremonium*. It would be interesting to use this compound as substrate for further discrimination of the hydroxylases presumably involved.

No major recent work has been published on the terminal acetylation leading to cephalosporin C or on the terminal steps in cephamycin biosynthesis. The reader is referred to previous reviews (33, 47) for complete information.

MECHANISMS CONTROLLING ONSET OF β-LACTAM BIOSYNTHESIS

Supply of α-Aminoadipic Acid

The biosynthesis of penicillins and cephalosporins depends on the metabolic supply of the precursor amino acids valine, cysteine, and α-aminoadipic acid (α-AAA). Although only the former two amino acids are finally incorporated in the β-lactam nucleus, the requirement for α-AAA has long been established. A recent review by Martin & Aharonowitz (70) summarizes the regulatory aspects of lysine metabolism, from which α-AAA is derived. *Penicillium chrysogenum* requires only low amounts of α-AAA, because the amino acid is reused after it is released from isopenicillin N in the final step of benzylpenicillin synthesis. In contrast, the β-lactam antibiotics of *C. acremonium* and actinomycetes are characterized by the obligatory attachment of α-AAA throughout the entire biosynthetic pathway. In fungi, the carbon flow over the lysine pathway is controlled by inhibition and/or repression of homocitrate synthase (72). Lysine both inhibits homocitrate synthase activity and represses its synthesis in *P. chrysogenum* (69). This restricts the availability of α-AAA, an intermediate in the eukaryotic pathway. As in *P. chrysogenum,* high lysine concentrations also reduce the synthesis of cephalosporin C in *C. acremonium,* and α-AAA relieves lysine inhibition (74).

In actinomycetes the pathway leading to lysine and α-AAA differs from that in fungi. Lysine is derived from 2,6-diaminopimelic acid (DAP). There may be at least two targets for regulation of the flow of carbon to α-AAA: DAP decarboxylase (55) and lysine-aminotransferase (53). Any interference in their activities will also affect antibiotic formation. Even more important in terms of regulation is the concerted feedback inhibition by lysine and threonine of aspartokinase, the first enzyme in the lysine biosynthetic pathway, which phosphorylates aspartate (75). Because of the different pathways to α-AAA, the isolation of analog-resistant mutants in the fungal and actinomycete systems led to opposite results. *P. chrysogenum* mutants resistant to lysine antimetabolites overproduced lysine and were underproducers of penicillin (73). On the other hand, lysine analog-resistant mutants of *S. clavuligerus* accumulated DAP and overproduced cephamycin (75).

The internal pool of α-AAA in strains of *P. chrysogenum* was recently analyzed (45). Since α-AAA is a substrate for two different branch-point enzymes (α-AAA reductase in the lysine pathway and LL-AC-synthetase, the first step in β-lactam biosynthesis), its intracellular concentration is an important parameter in the regulation of flux. The onset of α-AAA-cysteinyl dipeptide formation essentially depends on the availability of α-AAA and the affinity of the two enzymes for α-AAA. In fact, a correlation of penicillin production rate with intracellular α-AAA concentration was almost linear for four strains of increasing antibiotic potential. The relative capacity of *P. chrysogenum* to produce penicillin may mainly be due to the buildup of an internal pool of α-AAA sufficient for a high-velocity LL-AC synthetase reaction.

Role of Sulfur Metabolism

Conversion of methionine to cysteine and of sulfate to cysteine is one of the crucial factors in the economics of β-lactam formation. Queener et al (85) have summarized the recent knowledge of the regulatory mechanisms governing the flow of sulfur metabolites through these interconnecting pathways. Figure 3 depicts according to present knowledge this complex set of enzymatic steps as it exists in fungi and yeasts and to a certain degree also in actinomycetes. Owing to the different sulfur supply sources used in the selection for cephalosporin C product improvement, various lineages of altered sulfur metabolism have been obtained in *C. acremonium* (85). Wild-type strains of *C. acremonium* prefer methionine to sulfate for optimal cephalosporin C synthesis; this natural preference was exploited and successfully extended for strain improvement. In this lineage the sulfur for the antibiotic biosynthesis is almost exclusively derived from methionine by means of reverse transsulfuration (28, 82). Inorganic and organic sulfur compounds such as homocysteine and cystathionine did not stimulate cephalosporin C production. The conversion of L-methionine to L-homoscysteine involves Steps 13, 14, and 15 of Figure 3. The enzymes leading from L-homocysteine to L-cysteine (Steps 16 and 17) were designated cystathionine-β-synthase and cystathionine-γ-lyase. Cephalosporin C formation depends on a functional cystathionine-γ-lyase (97). Mutants that have impaired cystathionine-γ-lyase are also impaired in β-lactam formation and are unresponsive to methionine additions that were stimulatory to the parents. Mutants of uniform evolutionary lineage that differed in amount of antibiotic production showed a proportional increase in cephalosporin C titer and cystathionine-γ-lyase activity (J. Nüesch & A. Scheidegger, unpublished results).

Mutants defective in cystathionine-γ-lyase also exhibited a reduced ability to produce cephalosporin C on sulfate-supplemented, methionine-free media

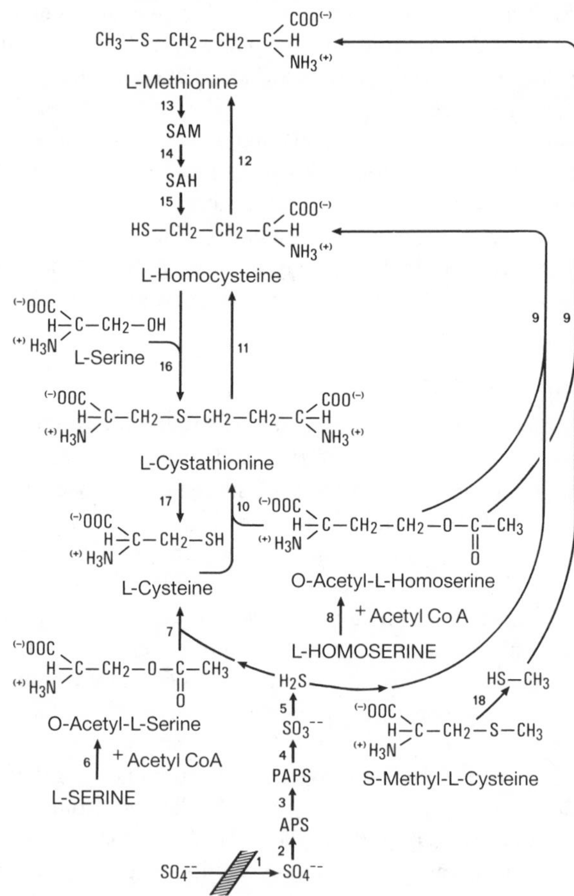

Figure 3 Sulfur metabolism pathways in yeast and fungi, including *C. acremonium* and *P. chrysogenum*.

(98). It appears that the formation of cystathionine is a prerequisite for efficient antibiotic synthesis. The cleavage of cystathionine may constitute an essential step in primary metabolism, and this reaction can be considered an inducer of the transfer of the cysteine moiety into the pathway of secondary metabolism.

In the 1970s, *C. acremonium* mutants were isolated that could efficiently use inorganic sulfate for antibiotic production (38, 57, 80). Some of these mutants could use either methionine or sulfate with equal efficiency. They resembled *P. chrysogenum* in their ability to use inorganic sulfate for antibiotic synthesis.

The formation of cystathionine from inorganic sulfide as the end product of the sulfur assimilation pathway can take place via two alternative routes. Both routes require two enzymatic steps, either Reactions 7 and 10 or Reactions 9 and 16 in Figure 3. Treichler et al (97) obtained a monogenic mutant in which both routes of cystathionine formation were blocked by a single mutational event. The mutant was characterized as defective in the acetylation of homoserine (Step 8 in Figure 3). It had a unique requirement for *O*-acetylhomoserine (OAH). Because of its inability to synthesize OAH as the common acceptor substrate for both metabolic routes leading to cystathionine, this mutant could not convert cysteine or any form of inorganic sulfur to cystathionine. The fact that the parent efficiently synthesized β-lactam on cysteine or sulfate while the mutant did not supports the contention that cystathionine is metabolically closer to and therefore more readily converted to β-lactam than cysteine.

The stimulation of antibiotic synthesis by methionine and its nonsulfur analog norleucine is an important feature in *C. acremonium*. The results of Drew & Demain (39) with sulfur amino acid auxotrophs support the role of methionine as a regulatory effector and emphasize the importance of the endogenous methionine supply. In addition, the increase in both IPNS and deacetoxycephalosporin C synthetase activities in *C. acremonium* grown with methionine or norleucine (89) points to a regulatory role for methionine. This regulatory effect of methionine is accompanied by a pronounced increase in the degree of fragmentation into unicells, with concomitant induction of cephalosporin biosynthesis. It is not clear, however, whether the antibiotic is an effector required for morphological differentiation or whether cephalosporin production is merely a consequence of morphological differentiation. In this respect the enhanced antibiotic potential from excess sulfate observed in mutant OAS^-, in which Step 7 is genetically blocked, might be due to a diversion of sulfur to methionine via Steps 9 and 12 (98). The simultaneous diversions to methionine and cystathionine in this mutant may contribute synergistically to enhanced antibiotic synthesis. The methionine effect in *C. acremonium* seems to be more complex than previously thought.

The efficiency with which these alternative sulfur pathways operate in different organisms, or even in different lineages of the same organism, is genetically determined. It is conceivable that the exceptionally pronounced regulatory effects of methionine in *C. acremonium* can outweigh even the unique functioning of cystathionine-γ-lyase as an effector of dipeptide initiation (37). In each system, control mechanisms will determine the preferred substrate and pathway. Therefore, expansion of the study of biosynthetic cysteine routes to other β-lactam–forming organisms will be of great interest. Expression of cystathionine-γ-lyase was investigated in *P. chrysogenum* (42). When this organism was grown on excess sulfate, enzyme activity was as high as in *C. acremonium,* but enzymatic activity and antibiotic production did not respond to methionine. In addition, mutants of *P. chrysogenum* defective in OAH-sulfhydrylase were unable to grow on inorganic sulfur sources, which indicates that only the second pathway is operative (36).

Cystathionine-γ-Lyase in Actinomycetes That Produce Sulfur-Containing Antibiotics

Although numerous studies have been conducted with *Penicillium* and *Cephalosporium,* the metabolism of cysteine and methionine by β-lactam–producing actinomycetes is a topic that has been virtually ignored. Kern & Inamine (54) have examined for the first time the possible importance of reverse transsulfuration in the cephamycin C producer *S. lactamdurans*. Extracts of this prokaryotic organism were found to have cystathionine-γ-lyase activity. The specific activity profile indicates that the enzyme is present during the period of rapid antibiotic synthesis. The enzyme is inhibited by cysteine, which suggests that feedback inhibition may be an important regulatory mechanism in vivo. Abeles & Walsh (1) showed that cystathionine-γ-lyase from rat liver is inactivated by incubation with propargyl-glycine (PGly). The potent inhibition of cephamycin C production by low levels of PGly indicates that this compound inhibits a biosynthetic enzyme that is directly involved in cephamycin C synthesis. The dual activities of PGly as an inhibitor of both cystathionine-γ-lyase and cephamycin C synthesis also indicate that the cysteinyl moiety of the antibiotic may be derived in large part via reverse transsulfuration (54). On the other hand, the persistent production of low but significant levels of cephalosporin C at relatively high levels of PGly may indicate that *S. lactamdurans* is also capable of deriving some cysteine for cephamycin C synthesis from the sulfate reduction pathway.

Cystathionine-γ-lyase activity was also demonstrated in *Streptomyces cattleya,* a producer of thienamycin (102). Thienamycin is a novel compound that contains carbon instead of sulfur in the five-membered ring. The sulfur atom is adjacent to the bicyclic ring system in a cysteaminyl side chain (Figure 4). Of the two known producers of thienamycin, only *S. cattleya* also synthesizes penicillin N and cephamycin C. Recent biosynthetic studies with *S. cattleya* suggest that cephamycin C and thienamycin are formed by entirely different secondary pathways (102). Results from these studies have established that cysteine is the precursor of the cysteaminyl side chain of thienamycin. The efficient incorporation of cystine into the cysteaminyl side chain supports a direct role for cysteine itself. The fact that cysteamine is not a precursor suggests that decarboxylation occurs after its incorporation into the thienamycin nucleus. Cystathionine and methionine, as well as cysteine, served as efficient sulfur donors for thienamycin and cephamycin C biosynthesis (101). In contrast, sulfate was found to be a poor source of sulfur. On the other hand, PGly, the suicide substrate for cystathionine-γ-lyase, inhibits to a similar extent the synthesis of both thienamycin and cephamycin C by *S. cattleya*. This inhibition can be substantially reversed with cystathionine or cysteine. We thus conclude that the reverse transsulfuration pathway is the major operating route for both biosyntheses. Cystathionine-γ-lyase thereby functions in vivo to provide cysteine for antibiotic synthesis.

Nagasawa et al (78) reported the wide distribution of cystathionine-γ-lyase in seven genera of actinomycetes. The enzyme was purified from *Streptomyces phaeochromogenes*. Cystathionine-β-synthase and cystathionine-γ-synthase activies were also detected in crude extracts of *S. phaeochromogenes*. Accordingly, the reverse transsulfuration pathway in actinomycetes may be similar to that in yeast and molds.

Recently, a *Streptomyces* species, *Streptomyces antibioticus* strain DSM 1865, was found to produce the novel siderophore desferrithiocin (Figure 4) (77). Investigations on its biosynthesis revealed the incorporation of cysteine (J. Nüesch & U. Suter, unpublished results). Cystathionine may be a trigger, since methionine, homocysteine, cystathionine, and cysteine were optimal sulfur sources for high productivity. ^{35}S from methionine was incorporated as well as ^{35}S from cysteine. Virtually no ^{14}C from methionine was incorporated. Cystathionine-γ-lyase activity was also established. Direct proof of γ-cystathionine–dependent product formation is not available, because cystathionine-γ-lyase–defective mutants have not yet been obtained. However, cystathionine-γ-lyase might be an important effector for cysteine-derived compounds other than β-lactams.

Role of S*-Sulfocysteine and Cystathionine*

The previous sections have shown that prokaryotic and eukaryotic organisms are capable of deriving some cysteine for β-lactam synthesis directly from the anabolic sulfate reduction pathway. In the course of strain improvement of *C. acremonium,* Suzuki et al (94) showed that in contrast to the addition of cysteine or sulfate, the addition of thiosulfate or *S*-sulfocysteine was effective

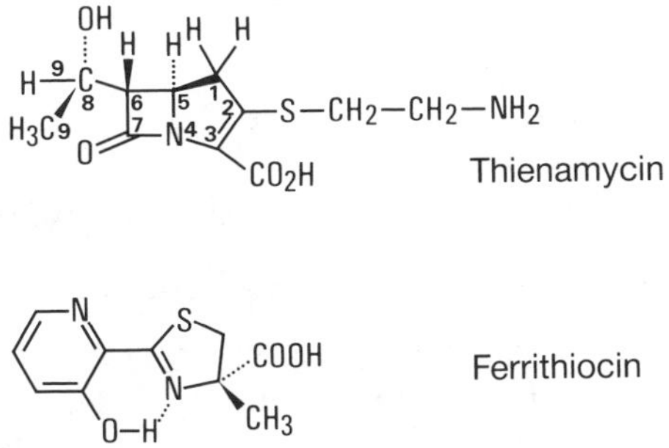

Figure 4 Sulfur-containing antibiotics from actinomycetes.

for the production of cephalosporin C. They concluded from this finding that the sulfur of cephalosporin C may be derived directly from the sulfate reduction pathway:

$$\text{sulfate} \rightarrow \text{thiosulfate} \xrightarrow{\text{L-serine}} S\text{-sulfocysteine} \xrightarrow{\text{NADPH glutathione oxidoreductase NAD}} \text{cysteine.}$$

To test for the existence of this pathway, Suzuki et al (94) further characterized mutant OAH^- of *C. acremonium,* which was previously described by its inability to utilize inorganic sulfur sources, cysteine, and *S*-methylcysteine. In a chemically defined medium containing *S*-sulfocysteine supplemented with a low level of methionine sufficient to support optimal growth but not antibiotic production, the mutant produced cephalosporin C in proportion to the amount of *S*-sulfocysteine added (94). Woodin & Segel (105) reported that cell-free extracts of *P. chrysogenum* converted *S*-sulfocysteine to cysteine. The driving force of the reaction was the NADPH-dependent reduction of oxidized glutathione by glutathione reductase. Activity of this enzyme was also detected in cell-free extracts of *C. acremonium* mutant 8650^+/OAH. Woodin & Segel (105) also confirmed the presence of *S*-sulfocysteine in cells of *C. acremonium*. Furthermore, enzyme preparations that catalyze the condensation of thiosulfate and serine to form *S*-sulfocysteine are well known in higher fungi (105).

These findings indicate that *S*-sulfocysteine is equivalent to cystathionine as a compound that is enzymatically convertable to cysteine. Since the sulfhydryl group of cysteine is known to be reactive and unstable in any cellular compartment, it is reasonable to assume that it must be generated de novo from a more stable compound, even if peptide initiation takes place in a rather redox-protected cellular compartment. Cystathionine is such a stable compound, and it can generate free cysteine by the action of cystathionine-γ-lyase. On the other hand, the conversion of *S*-sulfocysteine to cysteine by a specific oxidoreductase is analogous to the cleavage of cystathionine.

MECHANISMS THAT AFFECT FLUX IN THE PATHWAY

Carbon and Nitrogen Catabolite Regulation

Carbon and nitrogen sources that support high growth rates, such as glucose and ammonium, respectively, greatly reduce β-lactam production in prokaryotes and eukaryotes. The molecular basis for this regulation is unknown, although most of the data point to a regulation at the transcriptional and/or translational level resulting in repressed formation of crucial enzymes (reviewed in 70). Cortés et al (32) suggested that the intracellular effector that mediates glucose control over cephamycin biosynthesis in *S. lactamdurans* is a phosphorylated intermediate of sugar metabolism. Inorganic phosphate in the medium, which also negatively influences β-lactam formation in *C. acremonium,* exerts its effect indirectly by regulating the rate of glucose uptake (59; A. Scheidegger, unpublished results). Martin et al (71) described the complex pattern of glucose regulation of penicillin biosynthesis in *P. chrysogenum*. On the one hand, glucose stimulated homocitrate synthetase, the first enzyme in lysine biosynthesis; on the other hand, it repressed the formation of β-lactam biosynthetic enzymes. An increased flux of intermediates through the lysine pathway could be the reason for the decreased availability of the intermediate α-AAA for β-lactam biosynthesis. In *C. acremonium,* pronounced repression at the ring expansion step shows that glucose affects the biosynthetic enzymes directly (43). The coordinate severalfold increase of IPNS, epimerase, and expandase in a highly producing strain and the simultaneous loss of these enzymes in the LLD-ACV–accumulating mutant N-2 (87, 93) point to the existence of a gene that coordinately regulates these central steps in cephalosporin biosynthesis. Such a gene could then be the target for carbon and/or nitrogen repression. Interestingly, the highly cephalosporin-producing strain *C. acremonium* C-10 is much less sensitive to glucose repression at the ring expansion step (93). Castro et al (29) have demonstrated coordinate regulation of the above-mentioned central steps by free ammonium in *S. lactamdurans*. In *C. acremonium,* repression of expandase by nitrogen can be relieved by ammonium-trapping agents such as tribasic magnesium phosphate (91). Whether glutamine synthetase, the key target for nitrogen regulation of primary metabolism, has any role in regulating β-lactam biosynthesis remains to be established.

Compartmental Localization of End Products and Intermediates in the Pathway

Flux in the β-lactam biosynthetic pathway also seems to be influenced by the localization of the respective intermediates or end products. Addition of hydrophobic *N*-acyl side chain precursors to fermentations of *P. chrysogenum* leads to preferential formation of the corresponding hydrophobic penicillins, which are easily secreted into the medium (30). When they are not supplemented with precursors, fermentations of even industrial *P. chrysogenum* strains do not accumulate more than 100 mg liter^{-1} of the precursor isopenicillin N. Moreover, most of the isopenicillin N is found inside the mycelium (H.-J. Treichler, unpublished observation). Isopenicillin N has also never been reported in the fermentation broth of *C. acremonium,* although small amounts could have been overlooked owing to the presence of penicillin N, from which it is not readily distinguished chemically (S. Queener, personal communication). Penicillin N, the D-epimer of isopenicillin N, is always found in *Cephalosporium* fermentations. Presumably owing to oxygen limitation and the strong regulation of the ring expansion step, the precursor penicillin N is accumulated, secreted, and thereby withdrawn from any further biosynthetic conversion. Felix et al (40) recently described a mutant blocked at the ring expansion step whose secretion of penicillin N is almost as great as the parental strain's secretion of cephalosporin C. Mutants with blocks in the subsequent steps of cephalosporin biosynthesis were also obtained. The mutant with a block in the terminal acetyl transferase reaction secreted deacetylcephalosporin C in an amount equal to half of the cephalosporin C produced by the parental strain (81). In contrast, when a block was introduced in the preceding hydroxylation the precursor deacetoxycephalosporin C remained intracellular (possibly not liberated from the enzyme) and was present in very low amounts. The nonsecretion of deacetoxycephalosporin C in this mutant leads directly to cytoplasmic accumulation and subsequent *N*-acetylation (96). In summary, end products and certain intermediates in the biosynthetic pathways of both penicillin and cephalosporin exhibit certain properties that facilitate their secretion. End-product elimination creates a gradient that greatly enhances biosynthetic capacity.

MUTASYNTHESIS AND IN VITRO SUBSTRATE SPECIFICITIES OF PURIFIED ENZYMES

One of the well established characteristics that distinguishes secondary metabolism from primary metabolism is the diversity of products from the same set of biosynthetic enzymes. Usually a whole family of structurally related antibiotics, such as actinomycins, is found in a producing organism, rather than a single component (107). In the biosynthesis of classical β-lactam antibiotics only the formation of hydrophobic penicillins leads to significant variability. Owing to the broad substrate specificity of the terminal penicillin acyltransferase, more than 100 precursor acids have been incorporated, via their coenzyme A esters, into penicillins, which have different side-chain substitutions (30). As discussed in the previous section, addition of favored precursors such as phenylacetic acid directs the biosynthesis toward one particular product, thereby greatly enhancing overall productivity. In contrast, the cephalosporin and cephamycin biosyntheses seem to be extremely stringent. No major cephalosporin C analog has been reported to be produced in vivo by wild-type strains. Only a lysine auxotroph of *C. acremonium* produced a new penicillin following the addition of an aminoadipic acid analog (99). This strain, which is blocked in an early step of lysine biosynthesis, produces penicillin N and cephalosporin C only when fed with α-aminoadipic acid. The new antibiotic, called RIT 2214, was obtained after supplementation of the culture with L-*S*-carboxylmethylcysteine. Chemical analysis revealed that it is analogous to penicillin N, with carboxymethylcysteine incorporated in the side chain in its D-configuration (Figure 2). No corresponding cephalosporin was detected. In a similar experiment an isoleucine-valine mutant of *S. lipmanii* was exposed to the valine analog 3-bromovaline (J. Heim, unpublished observation). There was no indication of a new penicillin- or cephalosporin-type antibiotic. These results lead to the conclusion that in nature the biosynthetic pathways of hydrophilic cephalosporin and cephamycin have been optimized for the three amino acid precursors and that little room is left for modified substrates in vivo.

With the availability of cell-free systems, partially purified enzyme preparations, and homogeneously purified IPNS, a considerable effort has recently been focused on the behavior of modified substrates in vitro. Results have been somewhat unexpected and are summarized in the following. Again, most of the experiments have been carried out with partially or homogeneously purified IPNS from different β-lactam producers and by feeding modifications of the precursor tripeptide. New penicillins have now indeed been produced in vitro with IPNS isolated from *C. acremonium* and *S. clavuligerus* (8, 48, 61). However, changes are tolerated in only two of the three moieties varied, the α-aminoadipyl (AAA) and the valinyl (V) parts, to yield on ring closure the corresponding penicillin. Analogs with serine or alanine in place of cysteine (C) are not cyclized and do not act as inhibitors of the cyclization of the natural LLD-ACV. We therefore postulate that the serine- and alanine-containing tripeptides that have been isolated from the fermentation broth of *P. chrysogenum* (79) are shunt metabolites that presumably arise from an overflow of the first peptide-synthesizing steps in an industrial strain. The findings also illustrate the unequivocal role of the cysteinyl-sulfur atom in the course of ring cyclization.

Based on the successful cyclization of adipyl-CV, DLD-ACV, and L-*S*-carboxymethylcysteinyl-CV, the minimum requirements for incorporation of an α-aminoadipic acid analog into a penicillin have been summarized as follows (10): (*a*) The analog must have a six-carbon or equivalent chain, and (*b*) it must terminate in a carboxy group. Any shorter or longer chain fails to be incorporated, as does the glycyl-aminoadipyl-CV tetrapeptide, which is also a natural fermentation product (63). These fundamental requirements might result from a separation between a basic binding site on the IPNS enzyme and the catalytically active site so that the carboxy side chain terminus is located at a distance equivalent to the length of the adipic acid chain (9). The only analog known so far that is smoothly converted to a penicillin comparable to the natural precursor has a sulfur replacing one methylene group in the chain, e.g. carboxymethylcysteine. However, whereas in vivo the D-epimer is produced (99), cell-free extracts of *C. acremonium* produce only the L-epimer, owing to the inherent lability of the isopenicillin N epimerase (66, 104).

A rather surprising finding with other α-aminoadipic acid substitutes was published recently by two groups (14, 67, 68). On adding the linear heteropeptide phenylacetyl-CV to cell-free extracts of *C. acremonium, P. chrysogenum,* or *S. clavuligerus,* Luengo et al (67, 68) obtained an antibiotic (although in very low yield) corresponding to penicillin G. Baldwin et al (14) examined, in addition to phenylacetyl-CV, a phenoxyacetyl-CV dipeptide, and they obtained penicillin V with purified IPNS from *C. acremonium*. The conversion rate was less than 0.2%, compared to 100% with the natural substrate LLD-ACV. Analysis of Michaelis-Menten constants and maximum velocity of the unexpected reaction revealed a difference of three orders of magnitude in V_{max} (14). It is therefore unlikely that the described cell-free cyclization of phenylacetyl/phenoxyacetyl-CV has any significance for the natural biosynthesis of penicillin G in *P. chrysogenum* or other organisms. On the contrary, these results emphasize once more the importance of the terminal carboxy group in the side chain for efficient cyclization.

Concerning modifications of the valine moiety, the first results (8) indicated that tripeptide analogs with isoleucine, alloisoleucine, or α-aminobutyric acid as carboxy terminal amino acid are cyclized to new penicillins in vitro. Continuations of those studies with IPNS purified to homogeneity produced unexpected results (11, 21). In a 3:1 penam:cepham ratio (or higher) not only the corresponding penam-type compound but also several cepham-type compounds are formed simultaneously by the in vitro activity of IPNS. One example with isoleucine as analog is shown in Figure 5 (21). In addition to the predominant single penam-type compound, two cepham compounds that are epimeric at C-2 have been found. Very interesting in this respect is the recent detection of a penam-type antibiotic and a cepham-type antibiotic in cultures of a *Streptomyces* strain whose precursor could presumably be the α-aminobutyric acid analog of LLD-ACV (6). With unsaturated analogs such as allylglycine or vinylglycine the situation becomes even more complex. Not only penams and cephams but also seven-membered homoceph-3-em and hydroxy-homocephem ring systems are generated on cyclization with IPNS (16, 17). Hydroxylation of the product indicates that in this case IPNS has even switched its function from a dehydrogenase to a mixed dehydrogenase-monooxygenase incorporating molecular oxygen into the substrate (9). Mechanistic considerations of valine analogs again support the idea that production of an enzyme-bound monocyclic β-lactam intermediate is followed by removal of a hydrogen atom, which leads to formation of an intermediate carbon radical. Depending on the position where the radical is created, several cyclization types can be envisaged (9).

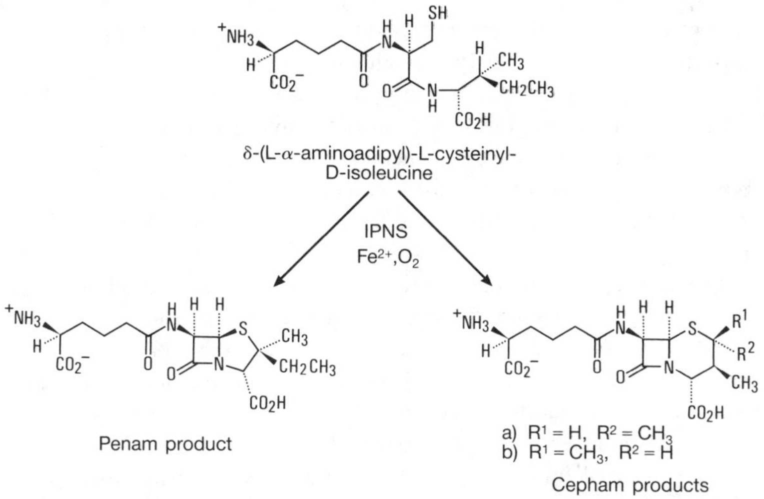

δ-(L-α-aminoadipyl)-L-cysteinyl-D-isoleucine

IPNS
Fe^{2+},O_2

Penam product

a) $R^1 = H$, $R^2 = CH_3$
b) $R^1 = CH_3$, $R^2 = H$

Cepham products

Figure 5 Enzymatic conversion of the isoleucine-tripeptide analog to penam and cepham products.

Very little is known about substrate specificities of all steps following IPNS in cephalosporin-cephamycin biosynthesis. Addition of DLD-ACV to cell-free extracts of *C. acremonium* circumvents the absence of the labile epimerase and leads to immediate formation of deacetoxycephalosporin C with penicillin N as intermediate (35, 92). Cell-free extracts of *S. clavuligerus* that contain all relevant enzymes in an active form are in this respect very interesting. In vitro stepwise formation of the corresponding penicillins and cephalosporins has been achieved, starting from carboxymethylcysteinyl-CV as well as from certain valinyl analogs (27, 47). Whether these results indicate a stringent requirement of the ring expansion step for a D-α-aminoadipyl or equivalent side chain remains to be proved with more extensive studies using purified enzyme preparations.

FUTURE OUTLOOK

During the past few years progress in the field of penicillin-cephalosporin biosynthesis has been extremely rapid. Most exciting are the results on the broad substrate ranges for the formation of the bicyclic ring systems. We can foresee the development of in vitro enzyme reactors to carry out penicillin and cephalosporin synthesis of compounds that are not easily accessible by chemical derivatization. Jensen et al (51) demonstrated the first successful operation of such a reactor with immobilized enzymes from *S. clavuligerus*. Now that the gene encoding IPNS from *C. acremonium* has been isolated, the isolation of genes coding for deacetoxycephalosporin C synthetase and further enzymes from this mold as well as from *Streptomyces* can be expected. Cloning of the genes will certainly facilitate production of large quantities of the biosynthetic enzymes to further elucidate their unique mode of action. With the development of efficient transformation systems for actinomycetes (26) and filamentous fungi (84), we can also envisage the reintroduction and possible amplification of certain genes to overcome presumed bottlenecks in the biosynthetic pathways. In vitro mutagenesis could lead to enzymes with altered substrate ranges and consequently to the production of novel β-lactams even in vivo.

Biosynthesis of β-lactam antibiotics is tightly controlled, both at the level of precursor availability and at the level of the biosynthetic enzymes themselves. Advances in the understanding of the negative and positive control mechanisms will depend largely on the isolation of regulatory elements or genes affecting the pathway. The results will show whether there are any specific genes controlling β-lactam biosynthesis or whether control is exerted via overall metabolism of the cell. If there are specific regulatory genes, it may be possible to amplify or eliminate them to stimulate precursor flow in the pathway. In this context one enzyme that seems particularly important is cystathionine-γ-lyase. Its preeminent role as a trigger of classical β-lactam biosynthesis has been shown. It is, however, also involved in the biosynthesis of thienamycin and possibly other metabolites that originate from cysteine. Cloning of the corresponding gene is highly desirable to throw some light on its ubiquitous activity.

Literature Cited

1. Abeles, R. H., Walsh, C. T. 1973. Inactivation of cystathionine-γ-lyase, in vitro and in vivo by propargylglycine. *J. Am. Chem. Soc.* 95:6124–25
2. Abraham, E. P. 1984. Aspects of the biosynthesis of β-lactam antibiotics. *Biochem. Soc. Trans.* 12:580–82
3. Adlington, R. M., Aplin, R. T., Baldwin, J. E., Chakravarti, B., Field, L. D., et al. 1983. Conversion of $^{17}O/^{18}O$ labelled δ-(L-α-aminoadipyl)-L-cysteinyl-D-valine to $^{17}O/^{18}O$ labelled isopenicillin N in a cell-free extract of *Cephalosporium acremonium*. A study by ^{17}O-NMR spectroscopy and mass spectrometry. *Tetrahedron* 39:1061–68
4. Adlington, R. M., Baldwin, J. E., Chakravarti, B., Jung, M., Moroney, S. E., et al. 1983. Conversion of 3-exomethylene cephalosporin C into deacetyl cephalosporin C in a cell-free extract from *Cephalosporium acremonium* (CW-19). *J. Chem. Soc. Chem. Commun.* 1983:153–54
5. Adlington, R. M., Baldwin, J. E., Lopez-Nieto, M., Murphy, J. A., Patel, N. 1983. A study of the biosynthesis of the tripeptide δ-(L-α-aminoadipyl)-L-cysteinyl-D-valine in a β-lactam-negative mutant of *Cephalosporium acremonium*. *Biochem. J.* 213:573–76
6. Aldridge, D. C., Carr, D. M., Davies, D. H., Hudson, A. J., Nolan, R. D., et al. 1985. Antibiotics 13285 A 1 and A 2: Novel cepham and penam metabolites from a *Streptomyces* species. *J. Chem. Soc. Chem. Commun.* 1985:1513–14
7. Deleted in proof
8. Bahadur, G. A., Baldwin, J. E., Usher, J. J., Abraham, E. P., Jayatilake, G. S., et al. 1981. Cell-free biosynthesis of penicillins. Conversion of peptides into new β-lactam antibiotics. *J. Am. Chem. Soc.* 103:7650–51
9. Baldwin, J. E. 1985. Recent advances in penicillin biosynthesis. In *Recent Advances in the Chemistry of β-Lactam Antibiotics. R. Soc. Chem. Spec. Publ.*

No. 52, ed. A. G. Brown, S. M. Roberts, pp. 62–85. London: R. Soc. Chem.

10. Baldwin, J. E., Abraham, E. P., Adlington, R. M., Bahadur, G. A., Chakravarti, B., et al. 1984. Penicillin biosynthesis: Active site mapping with aminoadipoylcysteinylvaline variants. *J. Chem. Soc. Chem. Commun.* 1984: 1225–27
11. Baldwin, J. E., Abraham, E. P., Adlington, R. M., Chakravarti, B., Jerome, A. E., et al. 1983. Penicillin biosynthesis. Dual pathways from a modified substrate. *J. Chem. Soc. Chem. Commun.* 1983:1317–19
12. Baldwin, J. E., Abraham, E. P., Adlington, R. M., Crimmin, M. J., Field, L. D., et al. 1984. The synthesis and reactions of a monocyclic β-lactam tripeptide, 1[(1*R*)-carboxy-2-methylpropyl]-(3*R*)-[(5*S*)-5-amino-5-carboxypentanamido]-(4*R*)- mercaptoazetidin-2-one, a putative intermediate in penicillin biosynthesis. *Tetrahedron* 40:1907–18
13. Baldwin, J. E., Abraham, E. P., Adlington, R. M., Murphy, J. A., Green, N. B., et al. 1983. Penicillin biosynthesis. On the stereochemistry of carbon-sulphur bond formation with modified substrates. *J. Chem. Soc. Chem. Commun.* 1983:1319–20
14. Baldwin, J. E., Abraham, E. P., Burge, G. L., Ting, H.-H. 1985. Penicillin biosynthesis. Direct biosynthetic formation of penicillin V and penicillin G. *J. Chem. Soc. Chem. Commun.* 1985: 1808–9
15. Baldwin, J. E., Abraham, E. P., Lovel, C. G., Ting, H.-H. 1984. Inhibition of penicillin biosynthesis by δ-(L-α-amino-δ-adipyl)-L-cysteinylglycine. Evidence for initial β-lactam ring formation. *J. Chem. Soc. Chem. Commun.* 1984:902–3
16. Baldwin, J. E., Adlington, R. M., Basak, A., Flitsch, S. L., Forrest, A. K., et al. 1986. Penicillin biosynthesis: Structure-reactivity profile of unsaturated substrates for isopenicillin N synthetase. *J. Chem. Soc. Chem. Commun.* 1986:273–75
17. Baldwin, J. E., Adlington, R. M., Derome, A. E., Ting, H.-H., Turner, N. J. 1984. Penicillin biosynthesis: Multiple pathways from a modified substrate. *J. Chem. Soc. Chem. Commun.* 1984: 1211–14
18. Baldwin, J. E., Adlington, R. M., Domayne-Hayman, B. P., Ting, H.-H., Turner, N. J. 1986. Stereospecificity of carbon-sulphur bond formation in penicillin biosynthesis. *J. Chem. Soc. Chem. Commun.* 1986:110–13
19. Baldwin, J. E., Adlington, R. M., Moroney, S. E., Field, L. D., Ting, H.-H. 1984. Stepwise ring closure in penicillin biosynthesis. Initial β-lactam formation. *J. Chem. Soc. Chem. Commun.* 1984:984–86
20. Baldwin, J. E., Adlington, R. M., Ting, H.-H., Arigoni, D., Graf, P., et al. 1985. Penicillin biosynthesis: The immediate origin of the sulphur atom. *Tetrahedron* 41:3339–43
21. Baldwin, J. E., Adlington, R. M., Turner, N. J., Domayne-Hayman, B. P., Ting, H.-H., et al. 1984. Penicillin biosynthesis: Enzymatic synthesis of new cephams. *J. Chem. Soc. Chem. Commun.* 1984:1167–70
22. Baldwin, J. E., Gagnon, J., Ting, H.-H. 1985. N-terminal amino-acid sequence and some properties of isopenicillin N synthetase from *Cephalosporium acremonium. FEBS Lett.* 188:253–56
23. Baldwin, J. E., Moroney, S. E., Ting, H.-H. 1985. A coupled enzyme assay for isopenicillin N synthetase. *Anal. Biochem.* 145:183–87
24. Baldwin, J. E., Wan, T. S. 1981. Penicillin biosynthesis. Retention of configuration at C-3 of valine during its incorporation into the Arnstein tripeptide. *Tetrahedron* 37:1589–95
25. Banko, G., Wolfe, S., Demain, A. L. 1986. Cell-free synthesis of δ-(L-α-aminoadipyl)-L-cysteine, the first intermediate of penicillin and cephalosporin biosynthesis. *Biochem. Biophys. Res. Commun.* 137:528–35

25a. Baxter, R. L., Thomson, G. A., Scott, A. I. 1984. Synthesis and biological activity of δ-(L-α-aminoadipyl)-L-cysteinyl-*N*-hydroxy-D-valine: a proposed intermediate in the biosynthesis of the penicillins. *J. Chem. Soc. Chem. Commun.* 1984:32–34

26. Bibb, M., Chater, K. F., Hopwood, D. A. 1983. Developments in *Streptomyces* cloning. In *Experimental Manipulation of Gene Expression,* ed. M. Inouye, pp. 53–82. New York: Academic
27. Bowers, R. J., Jensen, S. E., Lyubechansky, L., Westlake, D. W. S., Wolfe, S. 1984. Enzymatic synthesis of the penicillin and cephalosporin nuclei from an acyclic peptide containing carboxymethylcysteine. *Biochem. Biophys. Res. Commun.* 120:607–13
28. Caltrider, P. G., Niss, H. F. 1966. Role of methionine in cephalosporin synthesis. *Appl. Microbiol.* 14:746–53
29. Castro, J. M., Liras, P., Cortés, J.,

Martin, J. F. 1985. Regulation of α-aminoadipyl-cysteinyl-valine, isopenicillin N synthetase, isopenicillin N isomerase, and deacetoxycephalosporin C synthetase in *Streptomyces lactamdurans*. *Appl. Microbiol. Biotechnol.* 22: 32–40
30. Cole, M. 1966. Microbial synthesis of penicillins. *Process Biochem.* 1:334–38
31. Cortés, J., Castro, J. M., Laiz, L., Liras, P., Martin, J. F. 1986. Purification, characterization and regulation of isopenicillin N synthetase, isopenicillin N epimerase, and deacetoxycephalosporin C synthetase of *Streptomyces lactamdurans*. *5th Int. Symp. Genet. Ind. Microorg., Split, Yugoslavia,* p. 188. Int. Comm Genet. Ind. Microorg.
32. Cortés, J., Liras, P., Castro, J. M., Romero, J., Martin, J. F. 1984. Regulation of the biosynthesis of cephamycin C by *Streptomyces lactamdurans*. *Biochem. Soc. Trans.* 12:863–64
33. Demain, A. L. 1981. Biosynthetic manipulations in the development of β-lactam antibiotics. In *β-Lactam Antibiotics: Mode of Action, New Developments and Future Prospects,* ed. M. R. J. Salton, G. D. Shockman, pp. 567–83. New York: Academic
34. Demain, A. L. 1983. Biosynthesis of β-lactam antibiotics. In *Antibiotics Containing the Beta-Lactam Structure, Part I,* ed. A. L. Demain, N. A. Solomon, pp. 189–228. Berlin: Springer. 358 pp.
35. Demain, A. L., Shen, Y.-Q., Jensen, S. E., Westlake, D. W. S., Wolfe, S. 1986. Further studies on the cyclization of the unnatural tripeptide δ-(D-α-aminoadipyl)-L-cysteinyl-D-valine to penicillin N. *J. Antibiot.* 39:1007–10
36. Doebeli, H., Nüesch, J. 1980. Regulatory properties of *O*-acetyl-L-serine sulfhydrylase of *C. acremonium*. *Antimicrob. Agents Chemother.* 18:111–17
37. Drew, S. W., Demain, A. L. 1975. Production of cephalosporin C by single and double sulfur auxotrophic mutants of *Cephalosporium acremonium*. *Antimicrob. Agents Chemother.* 8:5–10
38. Drew, S. W., Demain, A. L. 1975. The obligatory role of methionine in the conversion of sulfate to cephalosporin C. *Eur. J. Appl. Microbiol.* 2:121–28
39. Drew, S. W., Demain, A. L. 1975. Stimulation of cephalosporin production by methionine peptides in a mutant blocked in reverse transsulfuration. *J. Antibiot.* 28:889–95
40. Felix, H. R., Peter, H. H., Treichler, H. J. 1981. Microbiological ring expansion of penicillin N. *J. Antibiot.* 34:567–75
41. Gevers, W., Kleinkauf, H., Lipmann, F. 1969. Peptidyl transfers in gramicidin S biosynthesis from enzyme-bound thioester intermediates. *Proc. Natl. Acad. Sci. USA* 63:1335–42
42. Gygax, D., Doebeli, N., Nüesch, J. 1980. Correlation between β-lactam antibiotics production and γ-cystathionine activity. *Experientia* 36:487
43. Heim, J., Shen, Y.-Q., Wolfe, S., Demain, A. L. 1984. Regulation of isopenicillin N synthetase and deacetoxycephalosporin C synthetase by carbon source during the fermentation of *Cephalosporium acremonium*. *Appl. Microbiol. Biotechnol.* 19:232–36
44. Hollander, I. J., Shen, Y.-Q., Heim, J., Demain, A. L., Wolfe, S. 1984. A pure enzyme catalyzing penicillin biosynthesis. *Science* 224:610–12
45. Jaklitsch, W. M., Hampel, W., Röhr, M., Kubicek, C. P. 1986. α-Aminoadipate pool concentration and penicillin biosynthesis in strains of *Penicillium chrysogenum*. *Can. J. Microbiol.* 32: 473–80
46. Jayatilake, G. S., Huddleston, J. A., Abraham, E. P. 1981. Conversion of isopenicillin N into penicillin N in cell-free extracts of *Cephalosporium acremonium*. *Biochem. J.* 194:645–47
47. Jensen, S. E. 1985. Biosynthesis of cephalosporins. *CRC Crit. Rev. Biotechnol.* 3:277–301
48. Jensen, S. E., Westlake, D. W. S., Bowers, R. J., Ingold, C. F., Jouany, M. 1984. Penicillin formation by cell-free extracts of *Streptomyces clavuligerus*. Behaviour of aminoadipyl-modified analogs of the natural peptide precursor δ-(L-α-aminoadipyl)-L-cysteinyl-D-valine (ACV). *Can. J. Chem.* 62:2712–20
49. Jensen, S. E., Westlake, D. W. S., Wolfe, S. 1983. Partial purification and characterization of isopenicillin N epimerase activity from *Streptomyces clavuligerus*. *Can. J. Microbiol.* 29:1526–31
50. Jensen, S. E., Westlake, D. W. S., Wolfe, S. 1983. Analysis of penicillin N ring expansion activity from *Streptomyces clavuligerus* by ion-pair high-pressure liquid chromatography. *Antimicrob. Agents Chemother.* 24:307–12
51. Jensen, S. E., Westlake, D. W. S., Wolfe, S. 1984. Production of penicillins and cephalosporins in an immobilized enzyme reactor. *Appl. Microbiol. Biotechnol.* 20:155–60
52. Jensen, S. E., Westlake, D. W. S., Wolfe, S. 1985. Deacetoxycephalosporin C synthetase and deacetoxycephalosporin C hydroxylase are two separate

enzymes in *Streptomyces clavuligerus*. *J. Antibiot*. 38:263–65

53. Kern, B. A., Hendlin, D., Inamine, E. 1980. L-lysine ε-aminotransferase involved in cephamycin C synthesis in *Streptomyces lactamdurans*. *Antimicrob. Agents Chemother*. 17:679–85
54. Kern, B. A., Inamine, E. 1981. Cystathionine-γ-lyase activity in the cephamycin C producer *Streptomyces lactamdurans*. *J. Antibiot*. 36:583–89
55. Kirpatrick, J. R., Doolin, J. L., Godfrey, O. W. 1973. Lysine biosynthesis in *Streptomyces lipmanii*. Implication in antibiotic biosynthesis. *Antimicrob. Agents Chemother*. 4:542–50
56. Kohsaka, M., Demain, A. L. 1976. Conversion of penicillin N to cephalosporin(s) by cell-free extracts of *Cephalosporium acremonium*. *Biochem. Biophys. Res. Commun*. 70:465–73
57. Komatsu, K. J., Mizuno, M., Kodaira, R. 1975. Effect of methionine on cephalosporin C and penicillin N production by a mutant of *Cephalosporium acremonium*. *J. Antibiot*. 28:881–88
58. Konomi, T., Herchen, S., Baldwin, J. E., Yoshida, M., Hunt, N. A., Demain, A. L. 1979. Cell-free conversion of δ-(L-aminoadipyl)-L-cysteinyl-D-valine to an antibiotic with properties of isopenicillin N in *Cephalosporium acremonium*. *Biochem. J*. 184:427–30
59. Küenzi, M. 1980. Regulation of cephalosporin C synthesis in *Cephalosporium acremonium* by phosphate and glucose. *Arch. Microbiol*. 128:78–83
60. Kupka, J., Shen, Y.-Q., Wolfe, S., Demain, A. L. 1983. Partial purification and properties of the α-ketoglutarate–linked ring expansion enzyme of β-lactam biosynthesis. *FEMS Microbiol. Lett*. 16:1–6
61. Kupka, J., Shen, Y.-Q., Wolfe, S., Demain, A. L. 1983. Partial purification and comparison of ring-cyclization and ring expansion enzymes of β-lactam biosynthesis in *Cephalosporium acremonium*. *Can. J. Microbiol*. 29:488–96
62. Lara, F., Mateos, R. del C., Vázquez, G., Sánchez, S. 1982. Induction of penicillin biosynthesis by L-glutamate in *Penicillium chrysogenum*. *Biochem. Biophys. Res. Commun*. 105:172–78
63. Loder, P. B., Abraham, E. P. 1971. Isolation and nature of intracellular peptides from a cephalosporin C producing *Cephalosporium* sp. *Biochem. J*. 123: 471–76
64. López-Nieto, M. J., Ramos, F. R., Luengo, J. M., Martin, J. F. 1985. Characterization of the biosynthesis in vivo of α-aminoadipyl-cysteinyl-valine in *Penicillium chrysogenum*. *Appl. Microbiol. Biotechnol*. 22:343–51
65. Lübbe, C., Wolfe, S., Demain, A. L. 1986. Isopenicillin N epimerase activity in a high cephalosporin producing strain of *Cephalosporium acremonium*. *Appl. Microbiol. Biotechnol*. 23:367–68
66. Lübbe, C., Wolfe, S., Shields, J. E., Queener, S. W., Neuss, N., Demain, A. L. 1985. Side-chain configuration of the sulfur-analog of penicillin produced by *Cephalosporium acremonium*. *J. Antibiot*. 38:1792–94
67. Luengo, J. M., Alemany, M. T., Salto, F., Ramos, F., López-Nieto, M. J., et al. 1986. Direct enzymatic synthesis of penicillin G using cyclases of *Penicillium chrysogenum* and *Acremonium chrysogenum*. *Bio-Technology* 4:44–47
68. Luengo, J. M., López-Nieto, M. J., Salto, F. 1986. Cyclization of phenylacetyl-L-cysteinyl-D-valine to benzylpenicillin using cell-free extracts of *Streptomyces clavuligerus*. *J. Antibiot*. 40:1144–47
69. Luengo, J. M., Revilla, G., Villanueva, J. R., Martin, J. F. 1979. Lysine regulation of penicillin biosynthesis in low-producing and industrial strains of *Penicillium chrysogenum*. *J. Gen. Microbiol*. 115:207–11
70. Martin, J. F., Aharonowitz, Y. 1983. Regulation of biosynthesis of β-lactam antibiotics. See Ref. 34, pp. 229–54
71. Martin, J. F., Revilla, G., López-Nieto, M. J., Ramos, F. R., Cantoral, J. M. 1984. Carbon catabolite regulation of penicillin biosynthesis at the α-aminoadipic acid and α-aminoadipyl-cysteinyl-valine levels. *Biochem. Soc. Trans*. 12:866–67
72. Masurekar, P. S., Demain, A. L. 1972. Lysine control of penicillin biosynthesis. *Can. J. Microbiol*. 18:1045–48
73. Masurekar, P. S., Demain, A. L. 1974. Impaired penicillin production in lysine regulatory mutants of *Penicillium chrysogenum*. *Antimicrob. Agents Chemother*. 6:366–68
74. Mehta, R. J., Speth, J. L., Nash, C. H. 1979. Lysine stimulation of cephalosporin C synthesis in *Cephalosporium acremonium*. *Eur. J. Appl. Microbiol. Biotechnol*. 8:177–82
75. Mendelovitz, S., Aharonowitz, Y. 1980. Cephalosporin production by *Streptomyces clavuligerus* mutant possessing a deregulated aspartokinase activity. *6th Int. Ferment. Symp., London, Canada*. Ottowa: Natl. Res. Counc. Publ. (Abstr.)
76. Miller, R. D., Huckstep, L. L., McDermott, J. P., Queener, S. W., Kukolja, S., et al. 1981. High performance liquid

chromatography (HPLC) of natural products. IV. The use of HPLC in biosynthetic studies of cephalosporin C in the cell-free system. *J. Antibiot.* 34:984–93

77. Naegeli, H. U., Zaehner, H. 1980. The structure of desferri-ferrithiocin. *Helv. Chim. Acta* 63:1400–8
78. Nagasawa, T., Kanzaki, H., Yamada, H. 1984. Cystathionine-γ-lyase of *Streptomyces phaeochromogenes. J. Biol. Chem.* 259:10393–403
79. Neuss, N., Miller, R. D., Affolder, C. A., Nakatsukosa, W., Mabe, J., et al. 1980. High performance liquid chromatography (HPLC) of natural products. 3. Isolation of new tripeptides from the fermentation broth of *P. chrysogenum. Helv. Chim. Acta* 63:1119–29
80. Niss, H. F., Nash, C. H. 1973. Synthesis of cephalosporin C from sulfate by mutants of *Cephalosporium acremonium. Antimicrob. Agents Chemother.* 4:474–78
81. Nüesch, J., Hinnen, A., Liersch, M., Treichler, H. J. 1976. A biochemical and genetical approach to the biosynthesis of cephalosporin C. *2nd Int. Symp. Genet. Ind. Microorg., Sheffield, England, 1974,* pp. 451–72. London: Academic
82. Nüesch, J., Treichler, H. J., Liersch, M. 1973. The biosynthesis of cephalosporin C. In *Genetics of Industrial Microorganisms,* Vol. 2, ed. Z. Vanek, Z. Hostalek, J. Cudlin, pp. 309–34. Prague: Academia. 340 pp.
83. Pang, C.-P., Chakravarti, B., Adlington, R. M., Ting, H.-H., White, R. L., et al. 1984. Purification of isopenicillin N synthetase. *Biochem. J.* 222:789–95
84. Queener, S. W., Ingolia, T. D., Skatrud, P. L., Chapman, J. L., Kaster, K. R. 1985. A system for genetic transformation of *Cephalosporium acremonium.* In *Microbiology 1985,* ed. L. Leive, pp. 468–72. Washington, DC: Am. Soc. Microbiol. 486 pp.
85. Queener, S. W., Wilkerson, S., Tunin, D. R., McDermott, J. P., Chapman, J. L., et al. 1984. Cephalosporin C fermentation: Biochemical and regulatory aspects of sulfur metabolism. In *Drugs and the Pharmaceutical Sciences,* Vol. 22, *Biotechnology of Industrial Antibiotics,* ed. E. J. Vandamme, pp. 141–70. New York: Dekker
86. Ramos, F. R., López-Nieto, M. J., Martin, J. F. 1985. Isopenicillin N synthetase of *Penicillium chrysogenum,* an enzyme that converts δ-(L-aminoadipyl)-L-aminoadipyl)-L-cysteinyl-D-valine to isopenicillin N. *Antimicrob. Agents Chemother.* 27:380–87
87. Ramos, F. R., López-Nieto, M. J., Martin, J. F. 1986. Coordinate increase of isopenicillin N synthetase, isopenicillin N epimerase and deacetoxycephalosporin C synthetase in a high cephalosporin-producing mutant of *Acremonium chrysogenum* and simultaneous loss of the three enzymes in a non-producing mutant. *FEMS Microbiol. Lett.* 35:123–27
88. Samson, S. M., Belagaje, R., Blankenship, D. T., Chapman, J. L., Perry, D., et al. 1985. Isolation, sequence determination and expression in *E. coli* of the isopenicillin N synthetase gene from *Cephalosporium acremonium. Nature* 318:191–94
89. Sawada, Y., Konomi, T., Solomon, N. A., Demain, A. L. 1980. Increase in activity of β-lactam synthetases after growth of *Cephalosporium acremonium* with methionine or norleucine. *FEMS Microbiol. Lett.* 9:281–84
90. Scheidegger, A., Küenzi, M. T., Nüesch, J. 1984. Partial purification and catalytic properties of a bifunctional enzyme in the biosynthetic pathway of β-lactams in *Cephalosporium acremonium. J. Antibiot.* 37:522–31
91. Shen, Y.-Q., Heim, J., Solomon, N. A., Wolfe, S., Demain, A. L. 1984. Repression of β-lactam production in *Cephalosporium acremonium* by nitrogen sources. *J. Antibiot.* 37:503–11
92. Shen, Y.-Q., Wolfe, S., Demain, A. L. 1984. Enzymatic conversion of the unnatural tripeptide δ-(D-α-aminoadipyl)-L-cysteinyl-D-valine to β-lactam antibiotics. *J. Antibiot.* 37:1044–48
93. Shen, Y.-Q., Wolfe, S., Demain, A. L. 1986. Levels of isopenicillin N synthetase and deacetoxycephalosporin C synthetase in *Cephalosporium acremonium* producing high and low levels of cephalosporin C. *Bio-Technology* 4:61–64
94. Suzuki, M., Fujisawa, Y., Uchida, M. 1980. *S*-Sulfocysteine as a source of the sulfur atom of cephalosporin C. *Agric. Biol. Chem.* 44:1995–97
95. Townsend, C. A. 1985. The stereochemical fate of chiral-methyl valines in cephalosporin C biosynthesis. *J. Nat. Prod.—Lloydia* 48:708–24
96. Traxler, P., Treichler, H. J., Nüesch, J. 1975. Synthesis of *N*-acetyldeacetoxycephalosporin C by a mutant of *Cephalosporium acremonium. J. Antibiot.* 28:605–6
97. Treichler, H. J., Liersch, M., Nüesch,

J. 1978. Genetics and biochemistry of cephalosporin biosynthesis. In *Antibiotics and Other Secondary Metabolites. 5th FEMS Symp.*, ed. R. Hütter, T. Leisinger, J. Nüesch, W. Wehrli, pp. 177–99. London: Academic. 255 pp.

98. Treichler, H. J., Liersch, M., Nüesch, J., Döbeli, H. 1979. Role of sulfur metabolism in cephalosporin C and penicillin biosynthesis. In *Genetics of Industrial Microorganisms*, ed. O. K. Sebek, A. I. Laskin, pp. 97–104. Washington, DC: Am. Soc. Microbiol. 283 pp.
99. Troonen, H., Roelants, P., Boon, B. 1976. RIT 2214, a new biosynthetic penicillin produced by a mutant of *Cephalosporium acremonium*. *J. Antibiot.* 29:1258–67
100. White, R. L., John, E. M., Baldwin, J. E., Abraham, E. P. 1982. Stoichiometry of oxygen consumption in the biosynthesis of isopenicillin N from a tripeptide. *Biochem. J.* 203:791–93
101. Whitney, J. G., Brannon, D. R., Mabe, J. A., Wicker, K. J. 1972. Incorporation of labelled precursors into A 16886B, a novel β-lactam antibiotic produced by *Streptomyces clavuligerus*. *Antimicrob. Agents Chemother.* 1:247–51
102. Williamson, J. M., Meyer, R., Inamine, E. 1985. Reverse transsulfuration and its relationship to thienamycin biosynthesis in *Streptomyces cattleya*. *Antimicrob. Agents Chemother.* 28:478–84
103. Wolfe, S., Demain, A. L., Jensen, S. E., Westlake, D. W. S. 1984. Enzymatic approach to syntheses of unnatural beta-lactams. *Science* 226:1386–92
104. Wolfe, S., Hollander, I. J., Demain, A. L. 1984. Enzymatic synthesis of a sulfur-analog of penicillin using the "cyclase" of *Cephalosporium acremonium*. *Bio-Technology* 2:635–36
105. Woodin, T., Segel, J. 1968. Glutathionine-reductase–dependent metabolism of cysteine-*S*-sulfate by *Penicillium chrysogenum*. *Biochim. Biophys. Acta* 167:78–88
106. Yoshida, M., Konomi, T., Kohsaka, M., Baldwin, J. E., Herchen, S., et al. 1978. Cell-free ring expansion of penicillin N to deacetoxycephalosporin C by *Cephalosporium acremonium* CW-19 and its mutants. *Proc. Natl. Acad. Sci. USA* 75:6253–57
107. Zähner, H. 1978. The search for new secondary metabolites. See Ref. 97, pp. 1–17

Ann. Rev. Microbiol. 1987. 41:77–101

INCOMPATIBILITY GROUP P PLASMIDS: Genetics, Evolution, and Use in Genetic Manipulation

Christopher M. Thomas and Christopher A. Smith

Department of Genetics, University of Birmingham, P.O. Box 363, Birmingham B15 2TT, England

CONTENTS

INTRODUCTION

Bacterial plasmids that share either replication control or partitioning functions compete for stable inheritance, are termed incompatible, and are placed in the same incompatibility group. The subject of this review is one such group, classified in *Escherichia coli* as IncP and in *Pseudomonas* as IncP-1 (for previous reviews see 61, 123, 146). This group is of particular interest because its members are promiscuous, i.e. are capable of conjugal transfer

0066-4227/87/1001-0077$02.00

among and stable maintenance in almost all gram-negative bacterial species. This property is exhibited by only a few other groups such as IncC, IncN, IncQ, and IncW. Such promiscuity is of interest owing to its role in the spread of antibiotic resistance and its application to genetic manipulation in diverse bacterial species. The IncP host range includes *Acetobacter xylinum* (160), *Achromobacter parvulus* (144), *Acinetobacter* spp. (104), *Aeromonas* spp. (144), *Agrobacterium* spp. (27, 28), *Alcaligenes* spp. (104), *Anabaena* spp. (171a), *Azospirillum brazilense* (112), *Azotobacter* spp. (104), *Bordetella* spp. (67), *Caulobacter* spp. (2), *Chromobacterium violaceum* (28), Enterobacteriaceae (22, 23, 28, 29, 36, 65, 71), *Hypomicrobium* X (33), *Legionella pneumophila* (39), *Methylobacterium organophilum* (3, 80), *Methylococcus* spp. (164), *Methylophillus methylotrophus* (170), *Methylosinus trichosporium* (164), *Myxococcus xanthus* (16, 108), *Neisseria* spp. (104), *Paracoccus denitrificans* (107), *Pseudomonas* spp. (27, 28, 104), *Rhizobium* spp. (11, 27, 28), *Rhodopseudomonas* spp. (104), *Rhodospirillum* spp. (104), *Thiobacillus* spp. (30, 89), *Vibrio cholerae* (104), *Xanthomonas* spp. (90, 158), and *Zymomonas mobilis* (17). Like plasmids of other groups (13, 56), IncP plasmids can transfer to species in which they cannot replicate (58, 131). *Bacteroides* is the only documented gram-negative genus that is nonpermissive for IncP plasmid maintenance (58, 131). Some strains of other genera are also incapable of supporting IncP plasmid replication (16, 91). A general explanation for strain-specific effects is provided by the observation that in *Pseudomonas aeruginosa* prophage B3 inhibits IncP replication (142).

A list of naturally occurring plasmids of the *E. coli* group IncP is shown in Table 1. The plasmids have been found in diverse geographical locations, usually as agents conferring antibiotic resistance in clinical isolates. However, in a study of plasmids in enteric species collected before the clinical use of antibiotics no IncP plasmids were found (29). Table 1 divides IncP plasmids into three subgroups: The α and β subgroups are evolutionary branches, while the third subgroup contains plasmids assigned to neither α nor β. Plasmids of this last subgroup have a narrow host range (51, 74, 103); for pHH502 this is probably due to its lack of a complete IncP replicon (C. A. Smith, unpublished). These last plasmids, which all show asymmetric incompatiblity [displacing but not being displaced by other IncP plasmids, probably owing to the possession of additional replicon(s) with different incompatibility (51, 103)], are not discussed futher.

MOLECULAR ANALYSIS OF RK2/RP4/RP1

IncPα plasmids RK2, RP1, RP4, R18, and R68, which were all isolated in the same Birmingham (UK) hospital in 1969 (76, 77), are the best studied IncP plasmids. They are indistinguishable by gross restriction map (26, 143) and

Table 1 Naturally occurring IncP plasmids

Plasmid	Resistance markers[a] Km	Tc	Ap	Sm	Su	Hg	Cm	Gm	Tp	Ox	Size (kb)	Country of origin	Ref.
α													
RK2	+	+	+								60	UK	77
RP4	+	+	+								60	UK	28
RP1	+	+	+								60	UK	77
R18	+	+	+								60	UK	76
R68	+	+	+								60	UK	76
R26	+	+	+	+	+	+	+	+			72	Spain	142
R527	+	+	+	+	+	+	+	+			72	Spain	142
R702	+	+		+	+	+					77	USA	66
R839	+	+	+	+	+	+					87	UK	70
R938	+	+	+	+	+	+	+				84	France	69
R995	+	+									57	Hong Kong	162
R934	+	+	+			+					72	France	70
R1033	+	+	+	+	+	+	+	+			75	Spain	139
pMG22[b]	+	+			+	+		+			—	USA	78
pUZ8	+	+				+					58	Spain	68
β													
R751									+		53	UK	81
R772	+										61	USA	23
R906				+	+	+				+	58	Japan	67
pJP4[c]						+					80	Australia	37
Other													
pAV1					+						—	UK	74
pHH502		+				+	+		+		71	UK	103
pMU700–pMU707				+	+						—	Australia	51

[a] Abbreviations: Km, Kanamycin; Tc, Tetracycline; Ap, Ampicillin; Sm, Streptomycin; Su, Sulphonamide; Hg, Merbromin; Cm, Chloramphenicol; Gm, Gentamycin; Tp, Trimethoprim; Ox, Oxacillin.

[b] Provisionally assigned to *α* on basis of Km^r and Tc^r.

[c] Plasmid pJP4 also confers (in *Alcaligenes*) the ability to use 2,4-dichlorophenoxyacetic acid and 3-chlorobenzoate.

heteroduplex analysis (19). Until evidence should be found to the contrary we assume that information obtained with any of these also applies to the others. Figure 1 shows the organization of these plasmids.

The restriction map reveals clustering of sites in regions where phenotypic markers and/or transposable elements disrupt the backbone of maintenance and transfer functions. This backbone is remarkably deficient in restriction sites for many enzymes. This deficiency may reflect selection against susceptibility to restriction in a plasmid that has undergone multiple interspecies

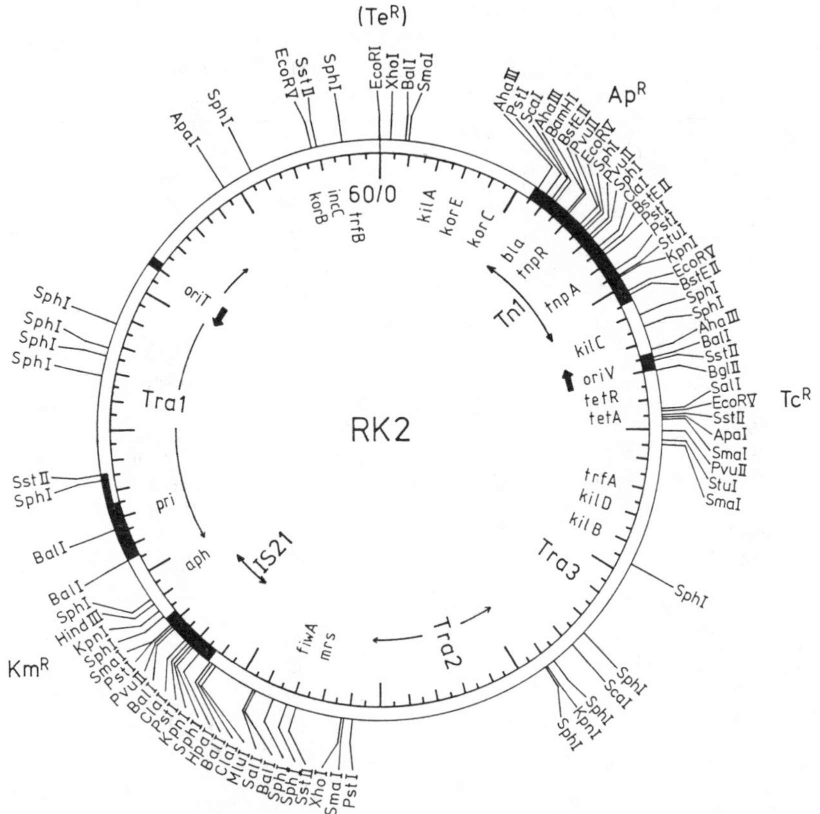

Figure 1 Physical and genetic map of RK2/RP4/RP1 (60 kb). The restriction map is based on coordinates supplied by E. Lanka (personal communication). Genetic loci: *aph,* aminoglycoside phosphotransferase; *bla,* β-lactamase; *fiw,* fertility inhibition toward IncW plasmids; *inc,* incompatbility toward IncP plasmids; *kil,* host-lethal or plasmid-inhibitory function; *kor,* suppression of *kil* effects; *mrs,* multimer resolution system; *oriT,* origin of conjugal DNA transfer; *oriV,* origin of vegetative replication; *pri,* primase; *tet,* tetracycline resistance; *tnp,* transposition function; *trf, trans*-acting replication functions (*trfB* is the same as *korA* and *korD*). Ap^r, Km^r, Tc^r, and (Te^r) show (cryptic) genes conferring resistance to ampicillin, kanamycin, tetracycline, and tellurite. The DNA sequence has been determined for kb coordinates 9.3–19.3 (135, 141, 165; M. Pinkney, C. A. Smith, J. P. Ibbotson & C. M. Thomas, unpublished); ~38.0–55.5 (E. Lanka, personal communication); 55.5–2.5 (154; B. D. M. Theophilus, C. A. Smith & C. M. Thomas, unpublished); IS*21* (R. Moore & N. Willetts, personal communication).

transfers, since these plasmids are subject to restriction during conjugation if they possess susceptible sites (57). The high G+C content of the plasmid backbone (18) may also be responsible for some of the deficiency. The abundance of restriction sites in segments such as Tn*1* may be due both to their recent acquisition by IncP plasmids and to their lower G+C content.

The *bla* gene is part of Tn*1* (also designated Tn*801*). Tn*1* is closely related to Tn*3* (72) but confers resistance to a higher level of ampicillin. This distinction and some restriction-site differences reflect minor sequence variations between the two transposons. The inducible tetracycline resistance genes (*tetA*, resistance protein; *tetR*, regulator) are not part of a transposon, but their DNA sequence is so closely related to that of the tetracycline resistance genes of Tn*1721* (165) that an ancestor of one must have acquired them from a relative of the other. The *aph* gene is related to that of Tn*903* but has a different distribution of restriction sites. The *Hin*dIII site is very close to the start of the gene, which is transcribed clockwise (E. Lanka, personal communication). The *aph* gene is not part of a transposable element, but the defective transposable element IS*21* to one side of *aph* [also designated IS*8* (32, 169)] could originally have formed part of a transposon responsible for its acquisition. A region that spans the *Eco*RI site encodes a cryptic tellurite resistance locus, *tel*, of which active variants can easily be selected (15, 144a). A segment that contains *tel* and the flanking loci *kilA* and *korA*, designated Tn*521*, has been reported to transpose (D. E. Bradley & D. E. Taylor, personal communication), although its ends have not been precisely defined. Deletions in RP4 with ends at about coordinate 2.0 can arise spontaneously (120), which is consistent with the presence of such an element in this region.

Other phenotypic properties of these plasmids include reduction in host virulence (172), susceptibility to pilus-specific phages PRR1, PR3, PRD1, and Pf3 (142), resistance to female-specific phage G101 (142), resistance to aeruginocin AP41 (142), fertility inhibition toward IncW plasmids (177), inhibition of natural chromosome mobilization in certain species (113), and inhibition of ColE1 amplification when ColE1 is *cis*-linked (42).

Replication Functions

Only two loci are essential for replication: *oriV*, from which vegetative replication proceeds unidirectionally (98; Figure 1), and *trfA* (*trans*-acting replication function), which can be *cis* or *trans* and whose polypeptide product (TrfA) is essential for initiation at *oriV* (41, 129, 147, 153). However, the *trfA* operon is part of a coordinately regulated set of genes (Figure 2) that are either host lethal (*kilA, -B,* and *-C;* 41, 43) or inhibitory to plasmid maintenance (*kilD*/*kilB*$_1$; 111, 134) when not controlled by *kor* (*kil*-override) genes. Since the KilD$^+$ phenotype is associated with the *trfA* operon, only KilD$^-$ mutant plasmids (110, 147) can be maintained without a third plasmid region, which suppresses *kilD*. This region was originally designated *trfB* because of its apparent involvement in replication (153). KilD$^-$ miniplasmids with *oriV* and *trfA* replicate in species representative of the parental host range, which indicates that a single replicon is sufficient for promiscuity (124,

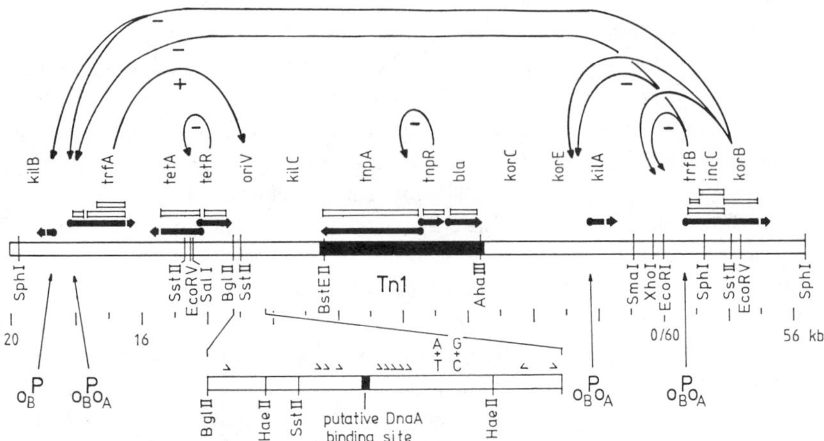

Figure 2 The region of RK2 encoding replication functions and the coordinately regulated *kil* and *kor* loci. Only those interactions whose molecular basis is clear are shown *(curved arrows)*. The main elements of *oriV* are shown on the expanded segment. *Filled arrows:* identified promoters and transcripts; *open bar segments:* known open reading frames; ⊿: origin repeats. Key restriction sites are shown as reference points.

125). However, the stability of such plasmids is influenced by their exact structure as well by the presence of the additional loci *trfB*, *korB*, and *oriT* (125).

THE *oriV* REGION DNA sequence analysis (141) of the 393-bp minimal *oriV* (25, 156) shows it to consist of neighboring G+C-rich and A+T-rich regions of 70 and 50 bp, respectively, and a series of five 17-bp tandem repeats of TGACA^{C}/A^{T}/GTGAGGGGC^{A}/G^{G}/C with a repeating distance of 22–23 bp (Figure 2). By analogy with other replication origins it is likely that TrfA binds to these repeats, either causing a structural change in the DNA or facilitating the binding of additional proteins. Upstream of these repeats is an inverted repeat of a sequence homologous to the DnaA consensus binding site (137). While RK2 can integratively suppress certain *dnaA*(Ts) strains (96), this is not a reliable indicator of DnaA independence since plasmid R1 is both capable of integrative suppression and dependent on DnaA for its replication (105). RK2 may therefore normally require DnaA. Both small and large insertions close to the upstream end of the group of repeats inactivate *oriV* in all species tested except *P. aeruginosa* (24, 25, 87). This may indicate differences either in the DnaA protein of different species or in the dependence on DnaA in different species. A number of putative weak promoters were also identified by DNA sequence analysis but attempts to detect transcription from them have been unsuccessful (M. A. Cross, unpublished).

Outside the minimal *oriV* are additional copies of the direct repeats (Figure 2) (137, 165; C. M. Thomas, unpublished). Those upstream are arranged as a single copy and an unevenly spaced set of three; the region containing them contributes to plasmid incompatibility *(incA, incB)* (151; M. A. Cross, unpublished) and copy number control *(copA, copB)* (151). A single Tn*7*-generated deletion upstream of the single repeat inhibits replication in *Pseudomonas stutzeri* (87), but the reason for this effect is unclear. The downstream repeats are degenerate and occur in both direct and inverted orientation; their role is uncertain. If TrfA bound to the different sets of repeats can associate, then the A+T- and G+C-rich regions may be enclosed in a looped structure, which could modulate their activity.

IncP PLASMID REPLICATION IN VITRO The first in vitro system developed for IncP plasmid DNA replication involved plasmid DNA-membrane complexes isolated from *E. coli* minicells (45). The incorporation of [^{3}H]TTP into the DNA of such complexes has many of the properties expected of plasmid replication. However, the active TrfA appears to be insufficient for much replication, since de novo protein synthesis is required (45, 85). Unfortunately, electron microscopic analysis of replicative intermediates revealed D-loops not only at *oriV* but also in the two antibiotic resistance genes of the mini-RK2 plasmid template (44). This raises doubt about the occurrence of normal replication in this system.

We have recently observed replication of a mini-RK2 plasmid in soluble extracts of *E. coli*. This activity differs in some important respects from that in the membrane system (M. Pinkney, unpublished). When plasmid DNA is added to extracts of plasmid-free bacteria, incorporation of [^{3}H]TTP depends on the addition of extracts containing TrfA. In contrast to the activity in the membrane system, the activity in soluble extracts is not inhibited by chloramphenicol or rifampicin. This suggests that transcription in the membrane system is required for de novo protein synthesis but that RNA polymerase is not needed for the initiation of RK2 replication.

THE *trfA* REGION The *trfA* gene is the second open reading frame (ORF) in the *trfA* operon (Figure 2) (129, 135). The first ORF encodes a polypeptide of 116 amino acids that may be responsible for the KilD$^+$ phenotype, although it has not been proven. The *trfA* gene encodes two polypeptides of 382 and 285 amino acids (P_{382} and P_{285}) and uses two translational starts in the same ORF (86, 129, 135). This might provide flexibility in the interaction of *oriV* and TrfA with host products. While P_{285} is sufficient for initiation of replication at *oriV* in *E. coli* (129) and *Pseudomonas putida,* establishment and stable maintenance is seriously impaired in *P. aeruginosa* when production of P_{382} is abolished (V. Shingler & C. M. Thomas, unpublished; R. Durland & D. R. Helinski, personal communication). This confirms the importance of the *trfA* gene structure in the host range of RK2. Both TrfA proteins bind DNA (M. Pinkney, unpublished). Their predicted amino acid sequences indicate that both are basic (135) and contain a region that could adopt the α-helix–turn–α-helix structure characteristic of the DNA-binding domain of a number of proteins (106).

The *trfA* operon promoter has been defined by footprinting of *E. coli* RNA polymerase (133). In addition, reverse transcriptase mapping in *E. coli, P. putida,* and *P. aeruginosa* has shown that *trfA* mRNA has the same 5' end in all three species (155). Its DNA sequence fits the *E. coli* consensus promoter well at both the -10 and -35 regions, and a T-to-C transition that reduces this homology reduces promoter strength 10- to 20-fold in all three species; these characteristics indicate that in each case *trfA* is transcribed by an RNA polymerase with similar recognition properties (M. Pinkney, B. D. M. Theophilus & C. M. Thomas, unpublished). Thus the *trfA* operon appears to be transcribed from an efficient broad-host-range promoter. The importance of this is emphasized by a Tn*7* insertion separating the *trfA* ORF from the *trfA* promoter (24). This allows replication in *P. aeruginosa* but not in *E. coli,* because the Tn*7* promoter from which *trfA* is transcribed is too weak in *E. coli* to give sufficient TrfA for replication (J. Nash & V. Krishnapillai, personal communication). It has been argued that the $kilB_1/kilD$ locus is positively required for replication (126) since certain KilD$^-$ mutants cannot replicate when *trfB* is present, as it is in RK2 (126, 145). However, since the only sequence reported for a KilD$^-$ mutation (133, 145) is a *trfA*-promoter down mutation, we propose that it remains to be demonstrated that a *kilD* locus distinct from *trfA* or its promoter is required for replication.

A second promoter with the regulatory properties expected of the *kilB* promoter has also been identified in the *trfA* region (Figure 2) (129, 130, 133).

THE *trfB-incC-korB* REGION Normally the *trfA* promoter is repressed by TrfB and KorB, the protein products of *trfB* and *korB,* which are part of the *trfB* operon (Figure 2) (130). The first gene is designated *trfB, korA,* or *korD* because it is responsible for various phenotypes: TrfB$^+$ complements pRK2501ts3, a mini-RK2 plasmid that is temperature sensitive for maintenance (147); KorA$^+$ suppresses KilA$^+$ (43); KorD$^+$ suppresses KilD$^+$ (134). The *trfB* ORF encodes a polypeptide of 101 amino acids (9, 136, 154). Its predicted amino acid sequence reveals a basic protein with a region that could adopt the α-helix–turn–α-helix characteristic of the DNA-binding domain of a number of proteins (106). Sequence analysis of a TrfB$^-$ mutant revealed a mutation in this region (154). KorB, a 49-kd polypeptide, is encoded by the third ORF in the *trfB* operon (10, 136). While it was first recognized by its regulation of *kilB,* KorB also regulates transcription of a number of other operons (10, 130, 145). The effects of TrfB and KorB on the *trfA*, *trfB*, *kilA*, and putative *kilB* promoters have been analyzed using promoter probe plasmids (10, 130, 145, 176). TrfB alone represses the *trfA* and *trfB* promoters, which are both very strong, approximately tenfold; KorB alone represses them two- to fourfold, and TrfB and KorB together repress these promoters 100- to 200-fold (130, 145). Comparison of these promoters with the *kilA* promoter, which is repressed by both TrfB and KorB, and the putative *kilB* promoter, which is repressed by KorB (15-fold) but not TrfB, has identified two inverted repeats. These have been tentatively designated O_A (TrfB operator), found in TrfB-repressed promoters, and O_B (KorB operator), found in KorB-repressed promoters (133, 176) (Figure 2). A T-to-C transition at the center of O_A results in loss of sensitivity to TrfB (145); this confirms the importance of O_A in regulation by TrfB.

The ORF for *incC*, a third gene in the *trfB* operon, overlaps the *trfB* ORF in a different phase and encodes polypeptides of 364 and 259 amino acids (154). Its first translational start precedes that of *trfB* by ten nucleotides, while the second start overlaps the *trfB* stop codon. $IncC^+$ is at least partly responsible for the IncP-1(II) incompatibility phenotype (99), which is only expressed strongly against plasmids that carry a functional *korB* gene as well as the *trfA*, *trfB*, and *oriV* regions (152). While IncC does not appear to repress transcription of the *trfA* operon, it does reduce the level of its polypeptide products (150), possibly by affecting mRNA stability, translational efficiency, or protein stability. A derivative of RP4, pRP761, which has a Tn*76* insertion in *incC* that reduces *korB* expression owing to polarity, is very unstable in *E. coli* and cannot be maintained at all in *P. aeruginosa* (6). This effect may reflect host differences in the expression of *kil*, *kor*, and *incC* genes, as well as host sensitivity to Kil functions; the effect emphasizes the importance of the *trfB* operon in controlling plasmid maintenance in diverse species.

PLASMID COPY NUMBER CONTROL RK2 has a copy number of 4–7 per chromosome equivalent in *E. coli* (42). However, mini-RK2 plasmids that lack *korB* have a copy number of 10–11 (151, 152). In such miniplasmids copy number is primarily controlled by elements *(copA, copB)* adjacent to *oriV*. It is probably the direct repeat sequences that are involved, since deletion of either the single repeat or two of the three repeats increases copy number to 18–20, while a combination of both deletions gives a copy number of 35–40 (151). These elements may act by titrating TrfA or by promoting a nonproductive complex between TrfA and the *oriV* region. Alternatively, they might act by producing an inhibitory RNA or protein product, although we have been unable to demonstrate the production of either (M. A. Cross, unpublished). When the entire *trfB-incC-korB* operon is present in the mini-plasmid, copy number is similar to that of RK2 and the deletions adjacent to *oriV* then have very little effect on copy number (152). IncC as well as KorB appears to be involved in limiting the level of TrfA, since inactivation of *incC* by a small deletion also allows significant increase in copy number in the deletions adjacent to *oriV* (C. M. Thomas, unpublished). Therefore RK2 possesses at least two replication control circuits, one provided by the action of TrfB, IncC, and KorB on TrfA level, and the other provided by elements in the *oriV* region.

Additional Stable Inheritance Functions

The stability of naturally occurring IncP plasmids suggests that they carry auxilliary inheritance functions. A function encoded by *Pst*I fragment C (coordinates 30.8 to 37.1) acts in *cis* to resolve multimers that form owing to homologous recombination between plasmid monomers (N. J. Grinter, G. Brewster & P. T. Barth, personal communication). RP4 *pri* also appears to reduce multimerization (N. J. Grinter, G. Brewster & P. T. Barth, personal communication). Tn*1* can also resolve multimers, but part of *Pst*I fragment C is still necessary for plasmid stability even when Tn*1* is present; this suggests that this region encodes a partitioning apparatus as well as a resolution system (121). This function increases the stability not only of pBR322 and pACYC177 but also of unstable broad-host-range plasmids in a wide range of species (121).

It has been suggested that the *kil* genes may code for proteins required for partitioning (99, 155), since overproduction of a plasmid-specified membrane protein essential for partitioning to daughter cells might inhibit host-cell growth. In addition, the similarities of *kil* and *kor* to *ccdB* and *ccdA* of plasmid F (79) and to *sok* and *hok* of plasmid R1 (49) suggest a role in lethality to plasmid-free segregant bacteria, although activation of the Kil$^+$ phenotype in bacteria that have lost the plasmid could not be explained by the transcriptional repression of *kil* genes by Kor proteins described above. However, IncC acts at a posttranscriptional level and the means by which KorC and KorA suppress the KilC$^+$ phenotype (43, 175) and KorE suppresses the KilA$^+$ phenotype (D. Figurski, personal communication) remain unknown. These or other as yet undiscovered interactions may therefore represent regulatory circuits that could allow such an effect. Phenotypically, IncP plasmids behave as though they have *ccd*-like functions. IncP plasmids such as pTH10 [RP4*trfA*(Ts)] (63; C. M. Thomas, unpublished) inhibit the increase of viable cells once the plasmid has been diluted to one copy per cell after shift to 42°C. This lack of plasmid replication results in a mixed population of bacteria apparently similar to that produced in analogous circumstances by the *ccd* functions of mini-F (79).

Conjugal Transfer Functions

IncP plasmids are self-transmissible at high frequency. Mating occurs most efficiently on a solid surface such as agar or nitrocellulose filters (31). Poor mating in liquid is probably associated with the rigid sex pili produced by the plasmids (14). Conjugal transfer requires both a *cis*-acting site, *oriT*, where transfer replication is initiated, and a number of *trans*-acting functions that are necessary either for cell-cell contact and mating-bridge formation, for initiation and continuation of transfer replication in donor and recipient, or for control of these two processes (Figure 1). Transfer replication initiated at *oriT* proceeds clockwise with respect to the plasmid map as standardly drawn (1), so the antibiotic resistance markers are transferred in the order Pn^r, Tc^r, Km^r.

THE *oriT* REGION RK2 *oriT*, located at coordinate 51.0, was cloned as a fragment capable of allowing the RK2 transfer system to mobilize pBR322 (59, 60). The *cis*-acting site spans two adjacent *Hpa*II fragments of 112 and 138 bp; the 112-bp fragment alone allows mobilization at a much lower efficiency, while the 138-bp fragment alone allows none (60; D. G. Guiney, personal communication). The most prominent feature of the DNA sequence of this region is a large inverted repeat (60). RK2 can be isolated as a DNA-protein "relaxation complex" (59), which on treatment with SDS or pronase causes the CCC DNA to be nicked to OC DNA. The nick, originally mapped roughly to *oriT* (59), lies at one end of the inverted repeat (E. Lanka, personal communication). The complex of one or more proteins with plasmid DNA in the vicinity of *oriT*, termed the relaxosome, represents a distinct component of the conjugation machinery. It has a narrower specificity than the IncP cell-cell contact and DNA transfer machinery, which can mobilize a wide range of plasmids. Thus, while the segment from coordinate 50 to 52.5, which encodes both *oriT* and a number of proteins, can be mobilized by R751 (IncPβ), the minimal *oriT* on its own cannot (173). Three polypeptides, of 11, 15, and 26 kd, have been identified in this region; the first of these binds to DNA (48; E. Lanka & D. Guiney, personal communication). In addition, a *trans*-acting function encoded between coordinates 52.85 and 53.65 is necessary in the bacterial recipient for transfer from *P. aeruginosa* or *P. stutzeri* to *E. coli, Salmonella typhimurium,* and *Pseudomonas maltophilia* (122).

OTHER *tra* FUNCTIONS The other *tra* functions are encoded in two major segments (Figure 1; 4, 7). Complementation analyses between different Tra^- mutants indicate a minimum of eight cistrons, with at least four in the Tra1 region and two in the Tra2/3 region (8, 143, 167). The Tra1 region has been completely sequenced and the proteins it encodes have been purified (E. Lanka, personal communication). It appears to be transcribed divergently from two promoters near *oriT* (48), but there may be secondary internal promoters. The *pri* gene is not the last in this unit; two small proteins are encoded between *pri* and *aph*. Tra3 and one of the Tra2 cistrons are involved in surface exclusion (4), while cistrons in both Tra1 and Tra2 are involved in the structure and/or assembly of pili (7, 8). Tra2 and Tra3 may form a single block, although certain transposon insertions between them do not abolish transfer (4, 7, 8). Accurate mapping of the pRP101 Tn*7* insertion that defines Tra3 clearly shows that it is not in the *trfA-kilD-kilB* region (94; C. A. Smith, unpublished). Golub & Low (50) reported that RP4 shows homolgy to the F *ssb* gene, but more rigorous testing has indicated that this is not the case (E. Golub, personal communication).

THE *pri* GENE RP4 *pri* produces two polypeptides, of 118 and 80 kd, by using alternative translational starts within a single ORF (95). Both polypeptides catalyze the synthesis of short oligoribonucleotides with CpA or pCpA at their 5' termini using single-stranded DNA (ssDNA) template (92), and both can weakly suppress *E. coli dnaG* mutations. Insertions in *pri* significantly lower conjugal transfer frequencies, which suggests that Pri is normally involved in conjugation (92, 97). This effect is dependent on the particular donor and recipient and is particularly marked for transfer from *P. aeruginosa* to *P. stutzeri* (88, 92, 97). The defect can be complemented by the RP4 *pri* cloned in either the donor or the recipient; this suggests that Pri is normally carried across with plasmid ssDNA during conjugation (88, 97). It is unclear, however, why *P. stutzeri* primase can function to prime complementary strand synthesis after intraspecies but not interspecies transfer. RP4 primase must normally avoid these problems. Recent experiments in which a mutant M13 phage was used to test DNA fragments for DNA priming signals have shown that Pri acts on sequences in the *oriT* region to prime DNA synthesis. The *oriV* region does not have such signals; this confirms the association between Pri and transfer rather than vegetative replication (D. Guiney, personal communication).

CONTROL OF THE Tra GENES Mutations that increase the expression of the Tra1 genes can be isolated by screening for more efficient complementation of chromosomal *dnaG*(Ts) mutants, which results from increased levels of Pri (95). It is unclear whether these mutations inactivate a repressor gene; recent identification of the Tra1 and *oriT* region promoters should allow this question to be answered.

IncP plasmids are subject to fertility inhibition by a range of plasmids (101, 171). In the case of plasmid F the product of the *pifC* gene probably inhibits fertility by binding specifically to RP4 DNA (101). Examination of the Tra1 DNA sequence has identified a partial PifC operator site that might be responsible for the observed inhibition (M. Malamy & E. Lanka, personal communication).

OTHER IncP PLASMIDS

Division of the IncP Plasmids into Two Subgroups

The division of the IncP plasmids into the IncPα and IncPβ subgroups (Table 1) was proposed by Yakobson & Guiney (173). They used Southern blotting to compare the *Hae*II fragments carrying homology to the *oriT* region of RK2 and found only two patterns of hybridization among the ten IncP plasmids investigated. Most of the plasmids showed a single band of the same size as that from RK2 and were designated IncPα, while R751, R906, and R772 showed two bands of smaller size and lower homology and were designated IncPβ. Studies using probes derived from the *trfA* and *oriV* regions of RK2 (21) and the *pri* locus of RP4 (93) confirmed the validity of this subgrouping, which is also supported by heteroduplex analysis (119, 162) and complementation studies (173). We have assigned a fourth plasmid, pJP4, to the IncPβ subgroup (138).

The IncPα Subgroup

Heteroduplex analysis (19, 162) has shown that the plasmids designated IncPα have extensive regions of homology covering the segments encoding the replication and conjugal transfer functions, Tc^r and Km^r. This subgroup consists of the indistinguishable isolates RP1, RP4, RK2, R18, and R68 (described in detail above) and a series of plasmids related to them by the insertion or deletion of a small number of segments (162). Many of these segments carry additional resistance determinants and most are known or putative transposable elements.

Besides conferring Tc^r and Km^r, the IncPα plasmids share other phenotypic markers that are absent from the IncPβ subgroup. R26, R527, and RP1 (IncPα), but not R751 and R906 (IncPβ), confer on *P. aeruginosa* insensitivity to aeruginosin AP41 (142). RP1 and six other IncPα plasmids inhibit the fertility of a coresident IncW plasmid, but R751 and R906 lack this property. They also lack homology to the *Pst*I fragment of RP1 that carries *fiwA* (Figure 1), which is partly responsible for this phenotype (177). RP4, pUZ8, and R26 (IncPα), but not R772 and R751 (IncPβ), possess a cryptic tellurite resistance locus, *tel* (15). It is therefore probable that the IncPα plasmids are derived from a common ancestor that conferred Km^r, Tc^r, insensitivity to aeruginosin AP41, and resistance to tellurite and that inhibited the fertility of IncW plasmids.

The IncPβ Subgroup

Figure 3 shows a map of R751, the most thoroughly studied member of the IncPβ subgroup. The restriction sites on the genome of R751 are clustered in two regions. One cluster interrupts the segment of the plasmid involved in conjugal transfer between the Tra1 and Tra2 regions; this cluster contains

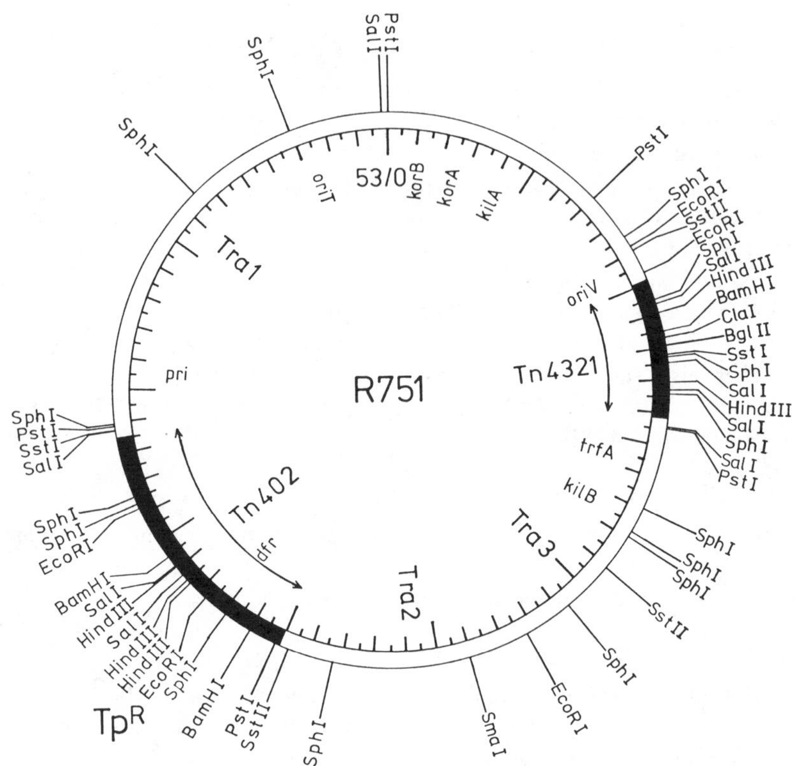

Figure 3 Physical and genetic map of R751 (53 kb). There is some restriction site polymorphism among laboratory strains of R751 (75, 100, 131, 163); the restriction map shown is based on one supplied by E. Lanka (personal communication) with minor modifications (C. A. Smith, unpublished). The genetic map is derived from the results of several studies (100, 118, 119, 138, 162; C. A. Smith, unpublished). The genetic loci are as defined for RK2 (Figure 1), except for *dfr,* trimethoprim-resistant dihydrofolate reductase. The end points indicated for Tn*402* (100) and Tn*4321* (118) are only approximate.

Tn*402* (100), carrying the Tp^r determinant, which shows high homology to the Tp^r determinants of plasmids R338 (IncW) and R67 (IncM) (46). The other cluster of restriction sites separates the vegetative-replication loci *oriV* and *trfA* and contains the putative transposable element Tn*4321* (118).

Like the map of R751, those of R906 and R772 (73, 118, 138, 163) show two major clusters of restriction sites containing the resistance loci, but there is no similarity in the restriction maps of these clusters. Heteroduplexes between R751 and R906 revealed extensive regions of homology covering the segments that carry the genes involved in replication and conjugal transfer (118, 119). The restriction maps of R751, R906, and R772 all correspond closely in these regions, although in R772 the relative orientation of these segments differs from that in the other two plasmids, which implies that an inversion has occurred in the lineage of this plasmid (138). The restriction map of pJP4 (38) shows a different but equally uneven distribution, with the sites concentrated in the segment of the plasmid that encodes Hg^r and degradation of 2,4-dichlorophenoxyacetic acid and 3-chlorobenzoate. Unlike the members of the IncPα subgroup, the IncPβ plasmids do not confer a common resistance phenotype on their hosts. However, both R906 and pJP4 carry a common Hg^r determinant, which is also present in an inactive form in R772 but is absent from R751 (138). This determinant (which is closely related to the Hg^r region of Tn*501*) is located at the same position adjacent to the *trfA* region in R906, R772, and pJP4; moreover, the DNA sequences of all three plasmids are identical for at least 150 bp spanning one end of the Hg^r region (138). It is probable that these three plasmids share a common ancestor that carried this Hg^r region. It is not clear whether R751 is also derived from this ancestor and has lost the entire Hg^r region by deletion, or whether it represents a separate line of descent from a more distant common ancestor of the IncPβ plasmids which had not yet acquired the Hg^r region. Comparison of the DNA sequences of the *oriV* regions of the IncPβ plasmids provides some support for the second hypothesis; those of R906 and R772 (C. A. Smith, unpublished) are almost identical, but show only 76% overall homology to that of R751 (137).

Relationship Between the Subgroups

Although the overall genetic organization of RK2 and R751 is very similar, there is no similarity in the restriction maps of the two plasmids (Figures 1 and 3). Heteroduplex analysis (162) showed that R751 and R906 (IncPβ) have extensive regions of homology to RP4 (IncPα); thus these plasmids are probably derived from a common ancestor. However, the formation of small symmetrical single-stranded loops in some heteroduplexes in regions that were double stranded in others implies that there has been substantial sequence divergence between the two subgroups (162). There is no evidence that the putative common ancestor of both subgroups conferred any phenotype on its host other than sensitivity to phages PRR1, PR3, PRD1, and Pf3 and resistance to phage G101 (142).

SEQUENCE COMPARISONS The only locus for which nucleotide sequences have been published for plasmids of both subgroups is *oriV*. The *oriV* regions of RK2 (141) and R751 (137) show only 65% overall homology; this suggests that sequence divergence between the plasmids through the accumulation of base changes and short deletions or insertions has proceeded for a considerable period. However, all the features in the *oriV* region of RK2 indicated in

Figure 2 are also present in R751 and form regions of high sequence conservation between the two plasmids. The single copy of the direct repeat found upstream of the *oriV* region in RK2 is not adjacent to the *oriV* region of R751, but is separated from the group of three repeats by a region homologous to one end of Tn*501* and by the putative Tn*4321* (118, 138). However, the single repeat is adjacent to the *oriV* regions in R906 and R772, which show 64% overall homology to the *oriV* region of RK2, with the same features conserved as in R751 (C. A. Smith, unpublished).

INTERCHANGEABILITY OF HOMOLOGOUS FUNCTIONS Despite the substantial sequence divergence between the IncPα and IncPβ plasmids many regions involved in replication or conjugal transfer have remained functionally interchangeable between the two subgroups. The *trfA* gene of RK2 provided in *trans* allows replication from the *oriV* regions of R751 (137) and R906 (21), and the *trfA* gene of R751 allows replication from the *oriV* region of RK2 (C. A. Smith, unpublished). The necessity of replication may account for the conservation of the specificity of this interaction. Similarly, the IncPβ plasmids R751, R906, and R772 are each able to supress the host-lethal effects of the loci *kilA* and *kilB* of RK2 (43), and the *korA* and *korB* loci of RK2 will supress the effects of a locus on R751 analogous to *kilB* (C. A. Smith, unpublished). The Tra1/*oriT* region of RK2 will form a functional conjugal transfer system with the Tra2/3 region of R751, as will the Tra1/*oriT* region of R751 with the Tra2/3 region of RK2 (100). Conservation between the conjugal transfer systems of the two subgroups is also revealed by the sensitivity of both subgroups to inhibition of fertility by IncP-2 plasmid R38 (142).

There are, however, limits to this functional interchangeability. While the IncPβ plasmids R751 and R772 can both mobilize a plasmid carrying a segment of RK2 (IncPα) that includes both *oriT* and the genes necessary for the formation of the relaxosome, only other IncPα plasmids can mobilize a plasmid carrying *oriT* alone (173).

USE OF IncP PLASMIDS IN GENETIC MANIPULATION

Chromosome Mobilization

IncP plasmids are capable of mobilizing chromosomal genes (chromosome mobilization ability: CMA) and of forming R' plasmids and have been used to construct linkage maps in many species (61). Derivatives with enhanced CMA (ECMA) arise spontaneously. The best characterized is R68.45 (62), which although isolated in *P. aeruginosa* gives ECMA in many species (61). In R68.45, IS*21* (IS*8*) is duplicated (32, 169). One copy of IS*21* is active owing to transcription into it from a promoter either situated in the other copy or created at the junction between the two copies (127; R. Moore & N. Willetts, personal communication). IS*21*-mediated mobilization results in cointegrated molecules flanked by direct repeats of IS*21;* resolution leaves one copy in the mobilized genome (116). ECMA may also arise by acquisition of host IS elements (102). Other strategies to improve IncP CMA include integration of cloned fragments with homology to the host chromosome (5), insertion of bacteriophage λ DNA to allow integration at chromosomal attachment site(s) (109, 168), or insertion of a mini-Mu derivative that transposes at high frequency (161). Integration into the host chromosome may be forced by selection for the retention of markers carried by a temperature-sensitive plasmid. Most of such IncP plasmids [pMR5 (117), RP1(Ts)12 (159), pME301 (166), pTH10 (63)] are *trfA*(Ts) mutants (147; C. M. Thomas, unpublished). However, the temperature sensitivity varies from host to host (76a, 157), even when the bacteria are able to grow at the nonpermissive temperature. Hfr strains thus derived may be unstable on return to the permissive temperature, but stable derivatives can be isolated and often result from inactivation of the IncP replicon (54, 114). Finally, a Tn*5*-$oriT_{\mathrm{RK2}}$ hybrid can be introduced randomly into the chromosome and allows mobilization when other *tra* functions are supplied in *trans* (174).

Transposon Mutagenesis

The IncP conjugation system is widely used for transfer of transposons between diverse gram-negative species. Most commonly the transfer system is used to deliver transposon-carrying plasmid DNA to a host where the plasmid either cannot replicate (40, 132, 158) or is deleterious to the host (12, 20); survival of the transposon markers indicates transposition into the recipient genome. Temperature-sensitive IncP plasmids (described above under *Chromosome Mobilization*) can also be used, since they can be eliminated after transposition by growth at the nonpermissive temperature (115, 157). Such methodology has been extended to allow stable insertion of cloned DNA into the chromosomes of diverse species (53).

Cloning Vectors

RK2 is too large for easy handling as a vector, and its self-transmissibility is undesirable for biological containment of hybrid plasmids. However, a variety of IncP plasmids useful as cloning vectors have been created; many of these have recently been well summarized (123). Two plasmids in particular, pRK2501 [11 kb (82)] and pRK290 [20 kb (35)], have formed the basis of a range of derivatives. The complete nucleotide sequence of pRK2501 is now known (135, 141, 154, 165; M. Pinkney, C. A. Smith & C. M. Thomas, unpublished). It is significantly unstable (150) despite having a copy number of about ten (151), and it is nonmobilizable. Plasmid pRK290 is a mobilizable

Tra$^-$ deletion derivative of RK2 that was chosen for its stability of inheritance, particularly in *Rhizobium* spp. (35). These plasmids have been modified to provide an increased number of unique restriction sites, such that insertion into some of them can be detected chromogenically (34); to form cosmid vectors packageable into phage λ (3, 47, 84, 128) or P22 (140); and to form promoter-probe vectors (34, 83).

An understanding of IncP maintenance functions can explain the behavior of IncP vectors and aid in the design of new vectors. For example, the instability of pRK2501 (150) and the greater but not complete stability of pRK290 (125) may result from their retention of only some of the loci proposed to be involved in stable inheritance; pRK2501 has only *kilD*, while pRK290 has not only *kilD*, *kilB*, and *korB* but also *oriT*, whose presence has been shown to influence plasmid stability (125). High-copy-number derivatives of pRK2501 are very unstable with respect to both segregation and structure, possibly owing to the increased gene dosage of *kilD* (148). However, joining a *copA,copB* mutant *oriV* region to a KilD$^-$ *trfA* gene under the control of the *trpE* promoter produces a more stable plasmid whose copy number is regulated by the concentration of tryptophan in the medium (149). Insertion of the same *oriV* into a pSC101 vector allows its copy number, which is normally about seven, to rise to about 40 when *trfA* is supplied in *trans* (64).

Instability or a reduction in copy number can result from insertion of promoter-containing fragments either downstream or upstream of *oriV* such that transcription is directed across *oriV* (25). Instability might also result when a fragment cloned at the *Eco*RI site directs high level transcription toward *trfB*, since although there is a putative terminator prior to the *trfB* operon promoter (154) any extra transcription of this operon would reduce *trfA* transcription and inhibit replication.

Plasmids consisting of the IncP transfer system joined to a narrow-host-range vector (most commonly pRK2013) (41) are very useful for introducing mobilizable but non–self-transmissible IncP, IncQ, and IncW vectors into bacterial species for which transformation procedures have not yet been developed. However, IncP plasmids can only be transiently established with pRK2013 (35) because TrfB, IncC, and KorB produced by pRK2013 inhibit *trfA* expression and thus inhibit replication. Use of a strain with chromosomally integrated IncP *tra* genes (132) may avoid this incompatibility.

Acknowledgments

We would like to thank all those who responded to our request for information about IncP plasmids, especially those who generously provided unpublished information. Work on IncP plasmids in this laboratory is currently supported in part by the UK Medical Research Council.

Literature Cited

1. Al-Doori, Z., Watson, M., Scaife, J. 1982. The orientation of transfer of the plasmid RP4. *Genet. Res.* 39:99–103
2. Alexander, J. L., Jollick, J. D. 1977. Transfer and expression of pseudomonas plasmid RP1 in caulobacter. *J. Gen. Microbiol.* 99:325–31
3. Allen, L. N., Hanson, R. S. 1985. Construction of broad-host-range cosmid cloning vectors: identification of genes necessary for growth of *Methylobacterium organophilum* on methanol. *J. Bacteriol.* 161:955–63
4. Barth, P. T. 1979. RP4 and R300B as wide host-range cloning vehicles. In *Plasmids of Medical, Environmental and Commercial Importance,* ed. K. N. Timmis, A. Pühler, pp. 399–410. Amsterdam: Elsevier/North Holland
5. Barth, P. T. 1979. Plasmid RP4, with *Escherichia coli* DNA inserted in vitro, mediates chromosomal transfer. *Plasmid* 2:130–36
6. Barth, P. T., Ellis, K., Bechhofer, D. H., Figurski, D. H. 1984. Involvement of *kil* and *kor* genes in the phenotype of a host-range mutant of RP4. *Mol. Gen. Genet.* 197:236–43
7. Barth, P. T., Grinter, N. J. 1977. Map of plasmid RP4 derived by insertion of transposon C. *J. Mol. Biol.* 113:455–74
8. Barth, P. T., Grinter, N. J., Bradley, D. E. 1978. Conjugal transfer system of plasmid RP4: analysis by transposon 7 insertion. *J. Bacteriol.* 133:43–52
9. Bechhofer, D. H., Figurski, D. H. 1983. Map location and nucleotide sequence of *korA,* a key regulatory gene of promiscuous plasmid RK2. *Nucleic Acids Res.* 11:7453–69
10. Bechhofer, D. H., Kornacki, J. A., Firshein, W. B., Figurski, D. H. 1986. Gene control in broad host range plasmid RK2: expression, polypeptide product and multiple regulatory functions of *korB*. *Proc. Natl. Acad. Sci. USA* 83:394–98
11. Beringer, J. E. 1974. R factor transfer in *Rhizobium leguminosarum*. *J. Gen. Microbiol.* 84:188–98
12. Beringer, J. E., Beynon, J. L., Buchanon-Wollaston, A. V., Johnston, A. W. B. 1978. Transfer of the drug-resistance transposon Tn*5* to *Rhizobium*. *Nature* 276:633–34
13. Boulnois, G. J., Varley, J. M., Sharpe, G. S., Franklin, F. C. H. 1985. Transposon donor plasmids, based on ColIb-P9, for use in *Pseudomonas putida* and a variety of other gram negative bacteria. *Mol. Gen. Genet.* 200:65–67
14. Bradley, D. E. 1980. Morphological and serological relationships of conjugative pili. *Plasmid* 4:155–69
15. Bradley, D. E. 1985. Detection of tellurite-resistance determinants in IncP plasmids. *J. Gen. Microbiol.* 131: 3135–37
16. Breton, A. M., Jaoua, S., Guespin-Michel, J. 1985. Transfer of plasmid RP4 to *Myxococcus xanthus* and evidence for its integration into the chromosome. *J. Bacteriol.* 161:523–28
17. Browne, G. M., Skotnicki, M. L., Goodman, A. E., Rogers, P. L. 1984. Transformation of *Zymomonas mobilis* by a hybrid plasmid. *Plasmid* 12:211–14
18. Burkardt, H. J., Pühler, A., Wohlleben, W. 1980. Adenine + thymine content of different genes located on the broad host range plasmid RK2. *J. Gen. Microbiol.* 177:135–40
19. Burkardt, H. J., Riess, G., Pühler, A. 1979. Relationship of group P1 plasmids revealed by heteroduplex experiments: RP1, RP4, R68 and RK2 are identical. *J. Gen. Microbiol.* 114:341–48
20. Casey, C., Bolton, E., O'Gara, F. 1983. Behaviour of bacteriophage Mu-based IncP suicide vector plasmids in *Rhizobium* spp. *FEMS Microbiol. Lett.* 20:217–23
21. Chikami, G. K., Guiney, D. G., Schmidhauser, T. J., Helinski, D. R. 1985. Comparison of ten IncP plasmids: homology in the regions involved in plasmid replication. *J. Bacteriol.* 162:656–60
22. Cho, J. J., Panopoulos, N. J., Schroth, M. N. 1975. Genetic transfer of *Pseudomonas aeruginosa* R factors to plant pathogenic *Erwinia* species. *J. Bacteriol.* 122:1141–56
23. Coetzee, J. N. 1978. Mobilization of the *Proteus mirabilis* chromosome by R plasmid R772. *J. Gen. Microbiol.* 108:103–9
24. Cowan, P., Krishnapillai, V. 1982. Tn*7* insertion mutations affecting the host range of the promiscuous IncP-1 plasmid R18. *Plasmid* 8:164–74
25. Cross, M. A., Warne, S. R., Thomas, C. M. 1986. Analysis of the vegetative replication origin of broad host range plasmid RK2 by transposon mutagenesis. *Plasmid* 15:132–46
26. Currier, T. C., Morgan, M. K. 1981. Restriction endonuclease analyses of the incompatibility group P-1 plasmids RK2, RP1, RP4, R68 and R68.45. *Curr. Microbiol.* 5:323–27

27. Datta, N., Hedges, R. W. 1972. Host range of R factors. *J. Gen. Microbiol.* 70:453–60
28. Datta, N., Hedges, R. W., Shaw, E. J., Sykes, R. B., Richmond, M. H. 1971. Properties of an R-factor from *Pseudomonas aeruginosa*. *J. Bacteriol.* 108:1244–49
29. Datta, N., Hughes, V. 1983. Plasmids of the same Inc groups in enterobacteria before and after the medical use of antibiotics. *Nature* 306:616–17
30. Davidson, M. S., Summers, A. O. 1983. Wide-host-range plasmids function in the genus *Thiobacillus*. *Appl. Environ. Microbiol.* 46:565–72
31. Dennison, S., Baumberg, S. 1975. Conjugational behaviour of N plasmids in *Escherichia coli* K12. *Mol. Gen. Genet.* 138:323–31
32. Depicker, A., De Block, M., Inze, D., Van Montagu, M., Schell, J. 1980. IS-like element IS*8* in RP4 plasmid and its involvement in cointegration. *Gene* 10:329–38
33. Dijkhuizen, L., Harder, W., De Boer, L., Van Boven, A., Clement, W., et al. 1984. Genetic manipulation of the restricted facultative methylotroph *Hypomicrobium* X by the R-plasmid mediated introduction of the *Escherichia coli pdh* genes. *Arch. Microbiol.* 139:311–18
34. Ditta, G., Schmidhauser, T., Yakobson, E., Lu, P., Liang, X.-W., et al. 1985. Plasmids related to the broad host range vector, pRK290, useful for gene cloning and for monitoring gene expression. *Plasmid* 13:149–53
35. Ditta, G., Stanfield, S., Corbin, D., Helinski, D. R. 1980. Broad host range DNA cloning system for gram negative bacteria: Construction of a gene bank of *Rhizobium meliloti*. *Proc. Natl. Acad. Sci. USA* 77:7347–51
36. Dixon, R. A., Cannon, F. C., Kondorosi, A. 1976. Construction of a P plasmid carrying nitrogen fixation genes in *Escherichia coli*. *Nature* 260:268–71
37. Don, R. H., Pemberton, J. M. 1981. Properties of six pesticide degradation plasmids isolated from *Alcaligenes paradoxus* and *Alcaligenes eutrophus*. *J. Bacteriol.* 145:681–86
38. Don, R. H., Pemberton, J. M. 1985. Genetic and physical map of the 2,4-dichlorophenoxyacetic acid–degradative plasmid pJP4. *J. Bacteriol.* 161:466–68
39. Dreyfus, I. A., Iglewski, B. H. 1985. Conjugation-mediated genetic exchange in *Legionella pneumophila*. *J. Bacteriol.* 161:80–84
40. Ely, B. 1985. Vectors for transposon mutagenesis of non-enteric bacteria. *Mol. Gen. Genet.* 200:302–4
41. Figurski, D., Helinski, D. R. 1979. Replication of an origin-containing derivative of plasmid RK2 dependent on a plasmid function provided in trans. *Proc. Natl. Acad. Sci. USA* 76:1648–52
42. Figurski, D., Meyer, R., Helinski, D. R. 1979. Suppression of ColE1 replication properties by the IncP-1 plasmid RK2 in hybrid plasmids constructed *in vitro*. *J. Mol. Biol.* 133:295–318
43. Figurski, D. H., Pohlman, R. F., Bechhofer, D. H., Prince, A. S., Kelton, C. A. 1982. The broad host range plasmid RK2 encodes multiple *kil* genes potentially lethal to *Escherichia coli* host cells. *Proc. Natl. Acad. Sci. USA* 79:1935–39
44. Firshein, W., Caro, L. 1984. Detection of displacement ("D") loops with the properties of a replicating intermediate synthesized by a DNA/membrane complex derived from the low-copy-number plasmid RK2. *Plasmid* 12:227–32
45. Firshein, W., Strumph, P., Benjamin, P., Burnstein, K., Kornacki, J. 1982. Replication of a low-copy-number plasmid by a plasmid DNA–membrane complex extracted from minicells of *Escherichia coli*. *J. Bacteriol.* 150:1234–43
46. Flensburg, J., Steen, R. 1986. Nucleotide sequence analysis of the trimethoprim resistant dihydrofolate reductase encoded by R plasmid R751. *Nucleic Acids Res.* 14:5933
47. Friedman, A. M., Long, S. R., Brown, S. E., Buikema, W. J., Ausubel, F. M. 1982. Construction of a broad host range cloning vector and its use in the genetic analysis of *Rhizobium* mutants. *Gene* 18:289–96
48. Fürste, J. P., Ziegelin, G., Pansegrau, W., Lanka, E. 1986. Conjugative transfer of promiscuous plasmid RP4: plasmid specified functions essential for formation of relaxosomes. *UCLA Symp. Mol. Cell. Biol.* (NS) 47:In press
49. Gerdes, K., Rasmussen, P. B., Molin, S. 1986. Unique type of plasmid maintenance function: postsegregational killing of plasmid-free cells. *Proc. Natl. Acad. Sci. USA* 83:3116–20
50. Golub, E. I., Low, K. B. 1985. Conjugative plasmids of enteric bacteria from many different incompatibility groups have similar genes for single-stranded DNA-binding proteins. *J. Bacteriol.* 162:235–41
51. Grant, A. J., Bird, P. I., Pittard, J. 1980. Naturally occurring plasmids exhibiting incompatibility with members of incompatibility groups I and P. *J. Bacteriol.* 144:758–65

52. Deleted in proof
53. Grinter, N. J. 1983. A broad-host-range cloning vector transposable to various replicons. *Gene* 21:133–43
54. Grinter, N. J. 1984. Replication defective RP4 plasmids recovered after chromosomal integration. *Plasmid* 11:65–73
55. Deleted in proof
56. Guiney, D. G. 1982. Host range of conjugation and replication functions of the *Escherichia coli* sex plasmid F*lac*. *J. Mol. Biol.* 162:699–703
57. Guiney, D. G. 1984. Promiscuous transfer of drug resistance in gram-negative bacteria. *J. Infect. Dis.* 149:320–29
58. Guiney, D. G., Hasegawa, P., Davis, C. E. 1984. Plasmid transfer from *Escherichia coli* to *Bacteroides fragilis:* differential expression of antibiotic resistance phenotypes. *Proc. Natl. Acad. Sci. USA* 81:7203–6
59. Guiney, D. G., Helinski, D. R. 1979. The DNA-protein relaxation complex of the plasmid RK2: location of the site-specific nick in the region of the proposed origin of transfer. *Mol. Gen. Genet.* 176:183–89
60. Guiney, D. G., Yakobson, E. 1983. Location and nucleotide sequence of the transfer origin of the broad host range plasmid RK2. *Proc. Natl. Acad. Sci. USA* 80:3595–98
61. Haas, D. 1983. Genetic aspects of biodegradation by pseudomonads. *Experientia* 39:1199–213
62. Haas, D., Holloway, B. W. 1978. Chromosome mobilization by the R plasmid R68.45: a tool in *Pseudomonas* genetics. *Mol. Gen. Genet.* 158:229–37
63. Harayama, S., Tsuda, M., Iino, T. 1980. High frequency mobilization of the chromosome of *Escherichia coli* by a mutant of plasmid RP4 temperature-sensitive for maintenance. *Mol. Gen. Genet.* 180:47–56
64. Hasnain, S., Thomas, C. M. 1986. Construction of a novel gene bank of *Bacillus subtilis* using a low copy number vector in *Escherichia coli*. *J. Gen. Microbiol.* 132:1863–74
65. Hedges, R. W. 1975. R factors from *Proteus mirabilis* and *P. vulgaris*. *J. Gen. Microbiol.* 87:301–11
66. Hedges, R. W., Jacob, A. 1974. Transposition of ampicillin resistance from RP4 to other replicons. *Mol. Gen. Genet.* 132:31–40
67. Hedges, R. W., Jacob, A., Smith, J. T. 1974. Properties of an R factor from *Bordetella bronchiseptica*. *J. Gen. Microbiol.* 84:199–204
68. Hedges, R. W., Matthew, M. 1979. Acquisition by *Escherichia coli* of plasmid-borne β-lactamases normally confined to *Pseudomonas* spp. *Plasmid* 2:269–78
69. Hedges, R. W., Matthew, M., Smith, D. I., Cresswell, J. M., Jacob, A. E. 1977. Properties of a transposon conferring resistance to penicillins and streptomycin. *Gene* 1:241–53
70. Hedges, R. W., Rodriguez-Lemoine, L., Datta, N. 1975. R factors from *Serratia marsescens*. *J. Gen. Microbiol.* 86:88–92
71. Heeseman, J., Laufs, R. 1983. Construction of a mobilizable *Yersinia enterocolitica* virulence plasmid. *J. Bacteriol.* 155:761–67
72. Heffron, F., McCarthy, B. J., Ohtsubo, H., Ohtsubo, E. 1979. DNA sequence analysis of the transposon Tn*3*: three genes and three sites involved in the transposition of Tn*3*. *Cell* 18:1153–63
73. Hille, J., van Kan, J., Klasen, I., Schilperoort, R. 1983. Site-directed mutagenesis in *Escherichia coli* of a stable R772::Ti cointegrate plasmid from *Agrobacterium tumefaciens*. *J. Bacteriol.* 154:693–701
74. Hinchliffe, E., Vivian, A. 1980. Naturally occurring plasmids in *Acinetobacter calcoaceticus:* a P class R factor of restricted host range. *J. Gen. Microbiol.* 116:75–80
75. Hirsch, P. R., Beringer, J. E. 1984. A physical map of pPH1JI and pPH4JI. *Plasmid* 12:139–41
76. Holloway, B. W., Richmond, M. H. 1973. R-factors used for genetic studies in strains of *Pseudomonas aeruginosa* and their origin. *Genet. Res.* 21:103–5
76a. Hooykaas, J. J., Dulk-Ras, H. D., Schilperoort, R. A. 1982. Phenotypic expression of mutations in a wide-host-range R plasmid in *Escherichia coli* and *Rhizobium meliloti*. *J. Bacteriol.* 150:395–97
77. Ingram, L. C., Richmond, M. H., Sykes, R. B. 1973. Molecular characterization of the R factors implicated in the carbenicillin resistance of a sequence of *Pseudomonas aeruginosa* strains isolated from burns. *Antimicrob. Agents Chemother.* 3:279–88
78. Jacoby, G. A. 1980. Plasmid determined resistance to carbenicillin and gentamicin in *Pseudomonas aeruginosa*. In *Plasmids and Transposons, Environmental Effects and Maintenance Mechanisms,* ed. C. Stuttard, K. R. Rozee, pp. 83–93. New York: Academic
79. Jaffe, A., Ogura, T., Hiraga, S. 1985. Effects of the *ccd* function of the F plasmid on bacterial growth. *J. Bacteriol.* 163:841–49

80. Jeyaseelan, K., Guest, J. R. 1979. Transfer of antibiotic resistance to facultative methylotrophs with plasmid R68.45. *FEMS Microbiol. Lett.* 6:87–89
81. Jobanputra, R. S., Datta, N. 1974. Trimethoprim R factors in enterobacteria from clinical specimens. *J. Med. Microbiol.* 7:169–77
82. Kahn, M., Kolter, R., Thomas, C., Figurski, D., Meyer, R., et al. 1979. Plasmid cloning vehicles derived from plasmids ColE1, F, R6K and RK2. *Methods Enzymol.* 68:268–80
83. Kahn, M. L., Timblin, C. R. 1984. Gene fusion vehicles for the analysis of gene expression in *Rhizobium meliloti*. *J. Bacteriol.* 158:1070–77
84. Knauf, V. C., Nester, E. W. 1982. Wide host range cloning vectors: a cosmid clone bank of an *Agrobacterium* Ti plasmid. *Plasmid* 8:45–54
85. Kornacki, J. A., Firschein, W. 1986. Replication of plasmid RK2 in vitro by a DNA-membrane complex: evidence for initiation of replication and its coupling to transcription and translation. *J. Bacteriol.* 167:319–26
86. Kornacki, J. A., West, A. H., Firshein, W. 1984. Proteins encoded by the *trans*-acting replication and maintenance regions of broad host range plasmid RK2. *Plasmid* 11:48–57
87. Krishnapillai, V. 1986. Genetic analysis of bacterial plasmid promiscuity. *J. Genet.* 65:103–20
88. Krishnapillai, V., Nash, J., Lanka, E. 1984. Insertion mutations in the promiscuous IncP-1 plasmid R18 which affect its host range between *Pseudomonas* species. *Plasmid* 12:170–80
89. Kulpa, C. F., Roskey, M. T., Travis, M. T. 1983. Transfer of plasmid RP1 into chemolithotrophic *Thiobacillus neapolitanus*. *J. Bacteriol.* 156:434–36
90. Lai, M., Panopoulos, N. J., Shaffer, S. 1977. Transmission of R plasmids among *Xanthomonas* spp. and other plant pathogenic bacteria. *Phytopathology* 67:1044–50
91. Lam, S. T., Lam, B. S., Stroebel, G. 1985. A vehicle for the introduction of transposons into plant-associated pseudomonads. *Plasmid* 13:200–4
92. Lanka, E., Barth, P. T. 1981. Plasmid RP4 specifies a deoxyribonucleic acid primase involved in its conjugal transfer and maintenance. *J. Bacteriol.* 148:769–81
93. Lanka, E., Fürste, J. P., Yakobson, E., Guiney, D. G. 1985. Conserved regions at the DNA primase locus of the *Inc*Pα and *Inc*Pβ plasmids. *Plasmid* 14:217–23
94. Lanka, E., Lurz, R., Fürste, J. P. 1983. Molecular cloning and mapping of *Sph*I restriction fragments of plasmid RP4. *Plasmid* 10:303–7
95. Lanka, E., Lurz, R., Kröger, M., Fürste, J. P. 1984. Plasmid RP4 encodes two forms of a DNA primase. *Mol. Gen. Genet.* 194:65–72
96. Martin, R. R., Thornton, C. L., Ungler, L. 1981. Formation of *Escherichia coli* Hfr strains by integrative suppression with P group plasmid RP1. *J. Bacteriol.* 145:713–21
97. Merryweather, A., Barth, P. T., Wilkins, B. M. 1986. Role and specificity of plasmid RP4–encoded DNA primase in bacterial conjugation. *J. Bacteriol.* 167:12–17
98. Meyer, R., Helinski, D. R. 1977. Unidirectional replication of the P-group plasmid RK2. *Biochim. Biophys. Acta* 487:109–13
99. Meyer, R., Hinds, M. 1982. Multiple mechanisms for expression of incompatibility by broad host range plasmid RK2. *J. Bacteriol.* 152:1078–90
100. Meyer, R. J., Shapiro, J. A. 1980. Genetic organization of the broad-host-range IncP-1 plasmid R751. *J. Bacteriol.* 143:1362–73
101. Miller, J. F., Lanka, E., Malamy, M. H. 1985. F factor inhibition of broad-host-range plasmid RP4: requirement for the protein product of Pif operon regulatory gene *pifC*. *J. Bacteriol.* 163:1067–73
102. Nayudu, M., Holloway, B. W. 1981. Isolation and characterization of R-plasmids with enhanced chromosomal mobilization ability in *Escherichia coli* K12. *Plasmid* 6:53–66
103. Nugent, M. E., Ellis, K., Datta, D. E. 1982. pHH502, a plasmid with IncP and Inclα characters, loses the latter by a specific *recA*-independent deletion event. *J. Gen. Microbiol.* 128:2781–90
104. Olsen, R. H., Shipley, P. 1973. Host range of the *Pseudomonas aeruginosa* R factor R1822. *J. Bacteriol.* 113:772–80
105. Ortega, S., Lanka, E., Diaz, R. 1986. The involvement of host replication proteins and of specific origin sequences in the in vitro replication of mini plasmid R1 DNA. *Nucleic Acids Res.* 14:4865–80
106. Pabo, C. O., Sauer, R. T. 1984. Protein-DNA recognition. *Ann. Rev. Biochem.* 17:293–321
107. Paraskeva, C. 1979. Transfer of kanamycin resistance mediated by plasmid R68.45 in *Paracoccus denitrificans*. *J. Bacteriol.* 139:1062–64

108. Parish, J. H. 1975. Transfer of drug resistance to *Myxococcus* from bacteria carrying drug resistance factors. *J. Gen. Microbiol.* 87:198–210
109. Pastrana, R., Brammar, W. J. 1979. In vitro insertion of the lambda attachment site into the plasmid RP4. *Mol. Gen. Genet.* 177:163–68
110. Pohlman, R. F., Figurski, D. H. 1983. Conditional lethal mutants of the *kilB* determinant of broad host range plasmid RK2. *Plasmid* 10:82–95
111. Pohlman, R. F., Figurski, D. H. 1983. Essential genes of plasmid RK2 in *Escherichia coli: trfB* region controls a *kil* gene near *trfA*. *J. Bacteriol.* 156: 584–91
112. Polsinelli, M., Baldanzi, E., Bazzicalupo, M., Gallori, E. 1980. Transfer of pRD1 from *Escherichia coli* to *Azospirillum brazilense*. *Mol. Gen. Genet.* 178:709–11
113. Pühler, A., Burkardt, H.-J. 1978. Fertility inhibition in *Rhizobium lupini* by the resistance plasmid RP4. *Mol. Gen. Genet.* 162:163–71
114. Reimmann, C., Haas, D. 1986. IS*21* insertion in the *trfA* replication control gene of chromosomally integrated plasmid RP1: a property of stable *Pseudomonas aeruginosa* Hfr strains. *Mol. Gen. Genet.* 203:511–19
115. Rella, M., Mercenier, A., Haas, D. 1985. Transposon insertion mutagenesis of *Pseudomonas aeruginosa* with a Tn*5* derivative: application to the physical mapping of the *arc* cluster. *Gene* 33:293–303
116. Riess, G., Maespohl, B., Puehler, A. 1983. Analysis of IS*21*-mediated mobilization of plasmid pACYC184 by R68.45 in *Escherichia coli*. *Plasmid* 10:111–18
117. Robinson, M. K., Bennett, P. M., Falkow, S., Dodd, H. M. 1980. Isolation of a temperature-sensitive derivative of RP4. *Plasmid* 3:343–49
118. Sakanyan, V. A., Azaryan, N. G., Krupenko, M. A. 1985. Molecular organisation of the plasmid R906. *Mol. Biol. Moscow* 4:964–73
119. Sakanyan, V. A., Krupenko, M. A., Alikhanyan, S. I. 1983. Homology of broad host range plasmids. *Genetika* 19:1409–18
120. Sakanyan, V. A., Yakubor, L. L., Alikhanian, S. I., Stepanov, A. I. 1978. Mapping of RP4 plasmid deletion mutants of pAS8 hybrid (RP4-ColE1). *Mol. Gen. Genet.* 165:331–41
121. Saurugger, P. N., Hrabak, O., Schwab, H., Lafferty, R. M. 1986. Mapping and cloning of the *par*-region of broad-host-range plasmid RP4. *J. Biotechnol.* 4:333–43
122. Schilf, W., Krishnapillai, V. 1986. Genetic analysis of insertion mutants of the promiscuous IncP-1 plasmid R18 mapping near *oriT* which affect its host range. *Plasmid* 15:48–56
123. Schmidhauser, T. J., Ditta, G., Helinski, D. R. 1987. Broad host range plasmid cloning vehicles in gram negative bacteria. In *Vectors: A Survey of Molecular Cloning Vectors and Their Uses*, ed. R. Rodriguez. Woburn, Mass: Butterworth. In press
124. Schmidhauser, T. J., Filutowicz, M., Helinski, D. R. 1983. Replication of derivatives of the broad host-range plasmid RK2 in two distantly related bacteria. *Plasmid* 9:325–30
125. Schmidhauser, T. J., Helinski, D. R. 1985. Regions of broad-host-range plasmid RK2 involved in replication and stable maintenance in nine species of gram-negative bacteria. *J. Bacteriol.* 164:446–55
126. Schreiner, H. C., Bechhofer, D. H., Pohlman, R. F., Young, C., Borden, P. A., Figurski, D. H. 1985. Replication control in promiscuous plasmid RK2: *kil* and *kor* functions affect expression of the essential replication gene *trfA*. *J. Bacteriol.* 163:228–37
127. Schurter, W., Holloway, B. W. 1986. Genetic analysis of promoters in the insertion sequence IS*21* of plasmid R68.45. *Plasmid* 15:8–18
128. Selveraj, G., Iyer, V. N. 1985. A small mobilizable IncP group plasmid vector packageable into bacteriophage capsids in vitro. *Plasmid* 13:70–74
129. Shingler, V., Thomas, C. M. 1984. Analysis of the *trfA* region of broad host range plasmid RK2 by transposon mutagenesis and identification of polypeptide products. *J. Mol. Biol.* 175:229–50
130. Shingler, V., Thomas, C. M. 1984. Transcription in the *trfA* region of broad host range plasmid RK2 is regulated by *trfB* and *korB*. *Mol. Gen. Genet.* 195:523–29
131. Shoemaker, N. B., Getty, C., Gardener, J. F., Salyers, A. A. 1986. Tn*4351* transposes in *Bacteroides* spp. and mediates the integration of plasmid R751 into the *Bacteroides* chromosome. *J. Bacteriol.* 165:929–36
132. Simon, R., Priefer, U., Pühler, A. 1983. A broad host range mobilization system for in vivo genetic engineering: transposon mutagenesis in gram negative bacteria. *Bio-Technol.* 2: 784–91

133. Smith, C. A., Shingler, V., Thomas, C. M. 1984. The *trfA* and *trfB* promoter regions of broad host range plasmid RK2 share common potential regulatory sequences. *Nucleic Acids Res*. 12:3619–30
134. Smith, C. A., Thomas, C. M. 1983. Deletion mapping of *kil* and *kor* functions in the *trfA* and *trfB* regions of broad host range plasmid RK2. *Mol. Gen. Genet*. 190:245–54
135. Smith, C. A., Thomas, C. M. 1984. Nucleotide sequence of the *trfA* gene of broad host range plasmid RK2. *J. Mol. Biol*. 175:251–62
136. Smith, C. A., Thomas, C. M. 1984. Molecular genetic analysis of the *trfB* and *korB* region of broad host range plasmid RK2. *J. Gen. Microbiol*. 130:1651–63
137. Smith, C. A., Thomas, C. M. 1985. Comparison of the nucleotide sequences of the vegetative replication origins of broad host range IncP plasmids R751 and RK2 reveals conserved features of probable functional importance. *Nucleic Acids Res*. 13:557–72
138. Smith, C. A., Thomas, C. M. 1987. Comparison of the organisation of the genomes of phenotypically diverse plasmids of incompatibility group P: members of the IncPβ sub-group are closely related. *Mol. Gen. Genet*. 206:419–27
139. Smith, D. I., Gomez, L. R., Rubio, C. M., Datta, N., Jacob, A. E., Hedges, R. W. 1975. Third type of plasmid conferring gentamycin resistance in *Pseudomonas aeruginosa*. *Antimicrob. Agents Chemother*. 8:227–30
140. Somerville, J. E., Kahn, M. L. 1983. Cloning of the glutamine synthetase I gene from *Rhizobium meliloti*. *J. Bacteriol*. 156:168–76
141. Stalker, D. M., Thomas, C. M., Helinski, D. R. 1981. Nucleotide sequence of the region of the origin of replication of the broad host range plasmid RK2. *Mol. Gen. Genet*. 181:8–12
142. Stanisich, V. A., Ortiz, J. M. 1976. Similarities between plasmids of the P-incompatbility group derived from different bacterial genera. *J. Gen. Microbiol*. 94:281–89
143. Stokes, H. W., Moore, R. J., Krishnapillai, V. 1981. Complementation analysis in *Pseudomonas aeruginosa* of the transfer genes of the wide host range R plasmid R18. *Plasmid* 5:202–12
144. Tardiff, G., Grant, R. B. 1980. Characterization of the host range of the N plasmids. See Ref. 78, pp. 351–59
144a. Taylor, D. E., Bradley, D. E. 1987. Location on RP4 of a tellurite-resistance determinant not normally expressed in IncPα plasmids. *Antimicrob. Agents Chemother*. In press
145. Theophilus, B. D. M., Cross, M. A., Smith, C. A., Thomas, C. M. 1985. Regulation of the *trfA* and *trfB* promoters of broad host range plasmid RK2. Identification of sequences essential for regulation by *trfB/korA/korD*. *Nucleic Acids Res*. 13:8129–42
146. Thomas, C. M. 1981. Molecular genetics of broad host range plasmid RK2. *Plasmid* 5:10–19
147. Thomas, C. M. 1981. Complementation analysis of replication and maintenance functions of broad host range plasmids RK2 and RP1. *Plasmid* 5:277–91
148. Thomas, C. M. 1983. Instability of a high copy number mutant of a miniplasmid derived from broad host range IncP plasmid RK2. *Plasmid* 10:184–95
149. Thomas, C. M. 1984. Genetic evidence for the direction of transcription of the *trfA* gene of broad host range plasmid RK2. *J. Gen. Microbiol*. 130:1641–50
150. Thomas, C. M. 1986. Evidence for the involvement of the *incC* locus of broad host range plasmid RK2 in plasmid maintenance. *Plasmid* 16:15–29
151. Thomas, C. M., Cross, M. A., Hussain, A. A. K., Smith, C. A. 1984. Analysis of copy number control elements in the region of the vegetative replication origin of broad host range plasmid RK2. *EMBO J*. 3:57–63
152. Thomas, C. M., Hussain, A. A. K. 1984. The *korB* gene of broad host range plasmid RK2 is a major copy number control element which may act together with *trfB* by limiting *trfA* expression. *EMBO J*. 3:1513–19
153. Thomas, C. M., Meyer, R., Helinski, D. R. 1980. Regions of the broad host-range plasmid RK2 which are essential for replication and maintenance. *J. Bacteriol*. 141:213–22
154. Thomas, C. M., Smith, C. A. 1986. The *trfB* region of broad host range plasmid RK2: the nucleotide sequence reveals *incC* and key regulatory gene *trfB/korA/korD* as overlapping genes. *Nucleic Acids Res*. 14:4453–69
155. Thomas, C. M., Smith, C. A., Shingler, V., Cross, M. A., Hussain, A. A. K., Pinkney, M. 1985. Regulation of replication and maintenance functions of broad host range plasmid RK2. In *Plasmids in Bacteria*, ed. D. R. Helinski, S. N. Cohen, D. B. Clewell, D. A. Jackson, A. Hollaender, pp. 261–76. New York: Plenum

156. Thomas, C. M., Stalker, D. M., Helinski, D. R. 1981. Replication and incompatibility properties of the origin region of replication of broad host range plasmid RK2. *Mol. Gen. Genet.* 181:1–7
157. Tsuda, M., Harayama, S., Iino, T. 1984. Tn*501* insertion mutagenesis in *Pseudomonas aeruginosa* PAO. *Mol. Gen. Genet.* 196:494–500
158. Turner, P., Barber, C., Daniels, M. 1984. Behaviour of the transposons Tn*5* and Tn*7* in *Xanthomonas campestris* pv. *campestris*. *Mol. Gen. Genet.* 195:101–7
159. Urlapova, S. V., Myakinin, V. B., Stepanov, A. I. 1979. Temperature-sensitive mutant of the plasmid RP1. *Genetika* 15(3):433–43
160. Valla, S., Coucheron, D. H., Kjosbakken, J. 1986. Conjugative transfer of the naturally occurring plasmids of *Acetobacter xylinum* by IncP-plasmid–mediated mobilization. *J. Bacteriol.* 165:336–39
161. Van Gijsegem, F., Toussaint, A. 1982. Chromosome transfer and R-prime formation by the RP4::mini-Mu derivative in *Escherichia coli, Salmonella typhimurium, Klebsiella pneumoniae* and *Proteus mirabilis*. *Plasmid* 7:30–44
162. Villarroel, R., Hedges, R. W., Maenhaut, R., Leemans, J., Engler, G., et al. 1983. Heteroduplex analysis of P-plasmid evolution: the role of insertion and deletion of transposable elements. *Mol. Gen. Genet.* 189:390–99
163. Ward, J. M., Grinsted, J. 1982. Analysis of the IncP-1 group plasmids R906 and R751 and their relationship to RP1. *Plasmid* 8:244–52
164. Warner, P. J., Higgins, I. J., Drozd, J. W. 1980. Conjugative transfer of antibiotic resistance to methylotrophic bacteria. *FEMS Microbiol. Lett.* 7:181–85
165. Waters, S. H., Grinsted, J., Rogowsky, P., Altenbuchner, J., Schmitt, R. 1983. The tetracycline resistance determinants of RP1 and Tn*1721:* nucleotide sequence analysis. *Nucleic Acids Res.* 11:6089–105
166. Watson, J. 1980. Replication mutants of the IncP-1 plasmid RP1. *Experientia* 36:1451
167. Watson, J., Schmidt, L., Willetts, N. 1980. Cloning of the Tra1 region of RP1. *Plasmid* 4:175–83
168. Watson, M. D., Scaife, J. G. 1978. Chromosomal transfer promoted by the promiscuous plasmid RP4. *Plasmid* 1:226–37
169. Willetts, N. S., Crowther, C., Holloway, B. W. 1981. The insertion sequence IS*21* of R68.45 and the molecular basis for mobilization of the bacterial chromosome. *Plasmid* 6:30–52
170. Windass, J. D., Worsey, M. J., Pioli, E. M., Pioli, D., Barth, P. T., et al. 1980. Improved conversion of methanol to single-cell protein by *Methylophilus methylotrophus*. *Nature* 287:396–401
171. Winnans, S. C., Walker, G. C. 1985. Fertility inhibition of RP1 by IncN plasmid pKM101. *J. Bacteriol.* 161:425–27
171a. Wolk, C. P., Vonshak, A., Kehoe, P., Elhai, J. 1984. Construction of shuttle vectors capable of conjugative transfer from *Escherichia coli* to nitrogen-fixing filamentous cyanobacteria. *Proc. Natl. Acad. Sci. USA* 81:1561–65
172. Wretlind, B., Becker, B., Haas, D. 1985. IncP-1 R plasmids decrease the serum resistance and the virulence of *P. aeruginosa*. *J. Gen. Microbiol.* 131:2701–4
173. Yakobson, E., Guiney, D. 1983. Homology in the transfer origins of broad host range *Inc*P plasmids: definition of two subgroups of P plasmids. *Mol. Gen. Genet.* 192:436–38
174. Yakobson, E. A., Guiney, D. G. 1984. Conjugal transfer of bacterial chromosomes mediated by the RK2 plasmid transfer origin cloned into transposon Tn*5*. *J. Bacteriol.* 160:451–53
175. Young, C., Bechhofer, D. H., Figurski, D. H. 1984. Gene regulation in plasmid RK2: Positive control by *korA* in the expression of *korC*. *J. Bacteriol.* 157:247–52
176. Young, C., Prince, A. S., Figurski, D. H. 1985. *korA* function of promiscuous plasmid RK2: an autorepressor that inhibits expression of host-lethal gene *kilA* and replication gene *trfA*. *Proc. Natl. Acad. Sci. USA* 82:7374–78
177. Yusoff, K., Stanisich, V. A. 1984. Location of a function on RP1 that fertility inhibits IncW plasmids. *Plasmid* 11:178–81

Ann. Rev. Microbiol. 1987. 41:103–26

CELL-FREE IMMUNITY IN INSECTS

Hans G. Boman and Dan Hultmark

Department of Microbiology, University of Stockholm, S-10691 Stockholm, Sweden

CONTENTS

INTRODUCTION

Disease in insects is a classical subject much older than the science of microbiology. In fact the first records go back to Aristotle's description of a pest affecting honey bees. In the last century, first Bassi (reviewed in 74) and

0066-4227/87/1001-0103$02.00

then Pasteur (reviewed in 24) settled the nature of two diseases in silkworms; these were pioneering studies, since they were the first cases in which diseases were shown to be caused by microbes. These early investigations were motivated by the economic importance of the silk industry and the losses caused by diseases of silkworms.

An important aspect of insect pathogens is their use as agents for biological control of insect pests. Many different microorganisms have been used for this purpose, but *Bacillus thuringiensis* is without comparison the organism that has found the most applications. There is vast literature on biological control of insects (14), insect pathogenic bacteria (64), and *B. thuringiensis* (6).

The number of insects is very impressive. Nearly one million species are recorded, and estimates indicate that the number of individual insects at a given time may be as high as 10^{18} (112). These high numbers imply that insects must have very effective means to defend themselves against infections. Moreover, since insects have never been found to possess lymphocytes or immunoglobulins, and since their immunity lacks the high degree of specificity that is found in vertebrates, one must expect their immune system to be designed in an entirely different way.

Research on immunization of insects can be divided into three periods. The first paralleled the early development of immunology in general, and included a very active phase in the 1920s. For a review of early work up to 1945 see Reference 99. The second period began in 1958–1959 with Briggs' and Chadwick-Stephens' in vitro demonstration of an inducible cell-free immunity caused by antibacterial substances (reviewed in 17). Finally, the third period started in 1980 with the isolation of the first pure antibacterial factors, later named cecropins.

In this review we focus on recent work on cell-free immunity and related phenomena, only briefly describing cellular immunity. Reviews covering both cellular and humoral reactions have been published earlier (13, 38, 82). Other reviews have dealt with cell-free immunity only (12) and with biochemical aspects of insect immunity (25). Since much is known about the molecular basis for the reaction against bacteria, it is the main topic of this review. In contrast, the question of antiviral mechanisms in insects has not yet been convincingly treated. In addition, relatively little is known about immune reactions against higher protists such as fungi and protozoa. However, their potential susceptibility to phenoloxidase and lectins was a main reason to include these factors in the present review.

Some scientists tend to reserve the word "immunity" for phenomena involving either lymphocytes or immunoglobulins. We think this is an unfortunate limitation because (*a*) a broad interpretation is in agreement with the historical usage of the term as well as with the definition given in Webster's dictionary: "resistance to or protection against a specified disease; power to resist infection." (*b*) The term immunity has long been used in connection with invertebrates. Therefore "insect immunity" is already widely listed as a key word in the computerized documentation of scientific papers. (*c*) The alternative expression, "defense reactions," would have a double meaning since it is also used for reactions by which insects defend themselves against other insects and against predators. The stinging of bees and wasps is a defense reaction, and melittin, the main component of bee venom, is a defense substance.

CELLULAR IMMUNITY

Different types of blood cells have important roles in the protection of insects against invading microorganisms. Small agents such as bacteria or unicellular fungi are phagocytized, while larger aggregates of foreign cells or parasites such as nematodes are encapsulated by the cooperation of many hemocytes. Finally, nodule formation occurs as a response to a large number of foreign cells. The latter process resembles coagulation and starts around a clot in which the foreign cells are entrapped.

Price & Ratcliffe (76) designed a classification scheme for hemocytes, based on an investigation of 15 insect orders. They recognized six cell types: prohemocytes, plasmatocytes, granulocytes (granular cells), coagulocytes (cystocytes), spherulocytes (spherule cells), and oenocytoids. This terminology is now widely accepted, although an alternative classification into nine groups has been devised (13). A method for the physical separation of granular cells, plasmatocytes, and spherule cells was recently published (65).

The main types of cells involved in phagocytosis are plasmatocytes and granular cells. Plasmatocytes also participate in capsule and nodule formation. However, the core of a nodule is formed from material released by granular cells. The lack of uniformity in nomenclature for hemocytes accounts for reports of phagocytosis involving at least six other types of cells, which presumably were either plasmatocytes or granular cells.

A detailed treatment of cellular immune reactions is beyond the scope of this review. The reader searching for an introduction to this area is referred to a review by Ratcliffe (81). The specialist is referred to a book on insect hemocytes (39) and to a review by Ratcliffe et al (82).

ANTIBACTERIAL FACTORS INDUCIBLE BY AN INFECTION

The Overall Response to an Infection

Bacteria injected into the hemolymph of an insect are rapidly removed from circulation by phagocytosis and nodule formation (25). In some cases investigated, phagocytized bacteria were not killed, but continued to live in a steady state for up to 1 wk. Granulocytes with bacteria were found in aggregates with other hemocytes; these complexes have been associated with fat body and pericardial cells (1, 30).

The next step in the development of immunity is the synthesis of RNA and specific proteins, which give rise to increasing antibacterial activity in the hemolymph. This activity continues to rise for a period that depends on the species. In the two giant silkworm moths, *Hyalophora cecropia* and *Antheraea pernyi,* the peak time was 7–8 days, while in the somewhat smaller *Samia cynthia* it was about half that time. After the peak the activity gradually declines, and it disappears after a time about equal to that required for reaching the maximum. The overall protein pattern and the profile of newly synthesized proteins was demonstrated as nine bands (P1–P9) in SDS polyacrylamide gel electrophoresis. Two of these bands, P5 and P9, were found to contain families of proteins later named attacins and cecropins, respectively.

So far 15 different proteins have been purified from cecropia immune hemolymph. The largest part of the antibacterial activity against a variety of bacteria can be accounted for by the cecropins. Protein P7 was identified as a lysozyme. In *Manduca sexta,* the tobacco hornworm, injection of bacteria induced more than 25 different proteins, and there were differences in the responses in larvae and pupae (47). The antibacterial activity in *M. sexta* seems to be due to the same three groups of proteins found in cecropia, namely cecropins, attacins, and lysozyme (94). Similar immune responses have also been found in several flies (34, 51, 55, 70, 86) and in a beetle (93). Some heterometabolous insects appear to lack the immune response of the lepidopteran type, with a massive synthesis of bacteriolytic proteins. However, antibacterial factors of low molecular weight have been reported in a locust (60) and in the assassin bug (22).

Lysozymes

The first antibacterial factor purified from insect hemolymph was lysozyme (75). This enzyme has also been found in the gut of several insects (68, 84), in hemocytes of *Spodoptera eridania* (2) and *Locusta* (116), and in a hemocyte-like cell line (61). The cricket *Gryllus bimaculatus* was found to have two lysozymes with different pH optima (88). Evolutionary aspects of insect lysozymes were discussed by Jollès et al (50) and Engström et al (28). An enzyme referred to as a lysozyme was purified from eggs of *Ceratitis capitata* and was found to have a molecular weight of 23×10^3, which is about 30% larger than other lysozymes (33). However, the substrate specificity indicated that the enzyme may in fact be a chitinase rather than a true lysozyme.

Hemolymph from nonimmunized cecropia larvae contains some lysozyme, while pupae might be deficient in the enzyme. However, injection of nonpathogenic bacteria causes lysozyme induction both in larvae and in pupae. The cecropia lysozyme was isolated in connection with the purification of cecropins A and B (46). The complete amino acid sequence of the enzyme was recently worked out, and a cDNA clone containing the lysozyme information was isolated and sequenced (28). Cecropia lysozyme is composed of 120 amino acids, has a molecular weight of 13.8×10^3, and shows great similarity to vertebrate lysozymes of the chicken type. The amino acid residues responsible for the catalytic activity, the substrate binding groove, and the four S-S bridges are essentially conserved. Purification of cecropia lysozyme often gives two peaks, and structural studies show that the cecropia population contains three variants (28).

The ceropia lysozyme is bactericidal only to a few gram-positive bacteria such as *Bacillus megaterium* and *Micrococcus luteus*. It also acts on *Escherichia coli* D22, an *envA* mutant with defective cell division. However, cecropins A and B are also active against these bacteria, and no bacterium has yet been found to be lysozyme sensitive and cecropin resistant. Thus, a main function of the lysozyme may be not to kill sensitive bacteria but to remove the murein sacculus which is left after the action of cecropins and attacins. In addition, it is likely that lysozyme works in synergy with cecropins and attacins (29).

Cecropins

The cecropins constitute a family of inducible, strongly basic proteins with molecular weights around 4×10^3 and potent antibacterial activity. These properties make it easy to demonstrate cecropins in immune hemolymph using polyacrylamide gel electrophoresis at pH 4. When the gel is overlaid with a suitable indicator bacterium and incubated overnight, the different cecropins and the attacin family are visualized as spots without bacterial growth (46). This method was used to demonstrate cecropinlike molecules in eight different lepidopteran species (42), in *Drosophila* (34, 86), in tsetse flies (51), and in darkling beetle (93).

H. cecropia pupae contain three major cecropins, A, B, and D, which are easy to separate. *H. cecropia* also has four minor antibacterial components (45), two of which may be cecropin precursors with an additional C-terminal glycine. In the Chinese oak silkworm moth, *Antheraea pernyi,* cecropin D is the main antibacterial factor, while the A and B forms are present in lower concentrations (78). Our group has purified and sequenced cecropins A, B, and D from *H. cecropia* (45, 98) and cecropins B and D from *A. pernyi* (78). Several cecropins similar to the A and B forms were also isolated and sequenced from *Bombyx mori* (106; X.-m. Qu, personal communication). Some of these forms contain one residue of the modified amino acid hydroxylysine (106). Okada & Natori (70) isolated and later sequenced (71) a cecropin from larvae of a meat fly, *Sarcophaga peregrina*.

Cecropins isolated from insects other than cecropia have in several instances been given new names such as lepidopteran (106), bactericidin (26), and sarcotoxin (71). These multiple names are unfortunate because they confuse the literature, and we propose that all antibacterial compounds with a convincing homology should be named cecropins. Before a sequence is available a compound may be referred to as "cecropinlike" or given a letter-number designation.

The known sequences for the major cecropins (Figure 1) show that the N-terminal parts are strongly basic while the C-terminal regions are neutral and contain long hydrophobic stretches. In all cases the cecropins have an amidated C-terminal residue. In cecropia the B and D forms show 65 and 62% homology, respectively, with cecropin A. However, about half the amino acid substitutions are strictly conservative, and the different cecropins are therefore very similar molecules. The high degree of homology among cecropins A, B, and D suggests that they have arisen through gene duplications.

Model building, theoretical predictions, and circular dichroism spectra indicate that cecropins can form nearly perfect amphipathic α-helices, i.e.

```
                     1       5         10        15        20        25        30        35
Hyalophora  A        K W K L F K K I E K V G Q N I R D G I I K A G P A V A V V G Q A T Q I A K*
            B        K W K V F K K I E K M G R N I R N G I V K A G P A I A V L G E A K A L*
            D          W N P F K E L E K V G Q R V R D A V I S A G P A V A T V A Q A T A L A K*
Antheraea   B        K W K I F K K I E K V G R N I R N G I I K A G P A V A V L G E A K A L*
            D          W N P F K E L E R A G Q R V R D A I I S A G P A V A T V A Q A T A L A K*
Bombyx      CM_IV    R W K I F K K I E K V G Q N I R D G I V K A G P A V A V V G Q A A T I*
            A        R W K I F K K I E K M G R N I R D G I V K A G P A I E V L G S A K A I*
Sarcophaga  IA   G W L K K I G K K I E R V G Q H T R D A T I Q G L G I A Q Q A A N V A A T A R*
```

Hydrophilic residues

- Basic: **K** Lysine, **R** Arginine, **H** Histidine
- Acidic: **D** Aspartic acid, **E** Glutamic acid
- Neutral: **N** Asparagine, **Q** Glutamine, **S** Serine, **T** Threonine

Hydrophobic residues

- A Alanine, F Phenylalanine, G Glycine, I Isoleucine, L Leucine
- M Methionine, P Proline, V Valine, W Tryptophan

Figure 1 Sequences of the major forms of cecropins, with N-terminals to the *left*. Hydrophilic residues are in *boldface*. An *asterisk* indicates that the C-terminal is amidated. References: in *Hyalophora,* cecropin A (98), B (111), D (45); in *Antheraea,* cecropin B (19), D (78); in *Bombyx,* "antibacterial peptide CM_{IV}" (X.-m. Qu, personal communication), "lepidopteran A" (106); in *Sarcophaga,* "sarcotoxin IA" (71). The latter protein was first reported to have a free C-terminal, but is now known to be amidated (S. Natori, unpublished). Most amino acid replacements in *Hyalophora, Antheraea,* and *Bombyx* can be explained by single-base shifts. In addition to the major forms, several minor cecropins have been described: C, E, and F from *Hyalophora* (45) and IB and IC from *Sarcophaga* (71). They differ by single amino acid substitutions or by absence of the C-terminal amide. Variants B and C from *Bombyx* contain the modified amino acid hydroxylysine (106).

cylindrical molecules with charged groups on one longitudinal side and hydrophobic side chains on the opposite side (5, 66, 96). Proteins with amphipathic helices are often associated with membranes (52), and this secondary structure may be of importance for the membrane-disrupting activity of the cecropins.

The antibacterial activity of the cecropins has been recorded in different ways. Three techniques depend on the use of bacteria suspended in buffer; the action of the cecropins is monitored either as a decrease in viable count (31), a decrease in light absorption (46), or an increase in ultraviolet-absorbing substances released from the cells (98). Finally, it is possible to measure the activity of cecropins on growing bacteria either by measuring inhibition zones on thin agar plates (42) or by measuring the minimum inhibitory concentration (MIC) with twofold dilution steps of the cecropins. For routine measurements the inhibition-zone assay may be the method of choice; formulae have been derived for the calculation of a minimum lethal concentration (44).

Cecropins A and B are highly active against several gram-positive and gram-negative bacteria, while the D form shows high activity only against *E. coli* and *Acinetobacter calcoaceticus* (Table 1). Several of the bacteria tested, e.g. *Serratia marcescens, Pseudomonas aeruginosa,* and *Enterococcus faecalis,* have been encountered as occasional insect pathogens. Two of the others, *Xenorhabdus nematophilus* and *B. thuringiensis,* are obligate insect pathogens. The former, which is cecropin sensitive, lives symbiotically with a nematode that destroys the cecropins (37). The latter is the organism most

Table 1 Antibacterial spectra of the three major cecropins, A, B, and D[a]

Bacterial species	Strain	Lethal concentration for cecropin (μM) A	B	D
Escherichia coli	D21	0.3	0.3	0.4
	D31	0.2	0.3	0.4
Acinetobacter calcoaceticus	Ac11	0.3	0.7	1.4
Xenorhabdus nematophilus	Xn21	1.4	1.6	19
Pseudomonas aeruginosa	OT97	2.6	1.9	100
Serratia marcescens	Db11	4.2	4.5	14
Micrococcus luteus	Ml11	1.4	1.3	21
Bacillus megaterium	Bm11	0.6	0.4	41
Bacillus subtilis	Bs11	61	18	>95
Bacillus thuringiensis	Bt11	>80	>133	>95
Enterococcus faecalis	AD-4	15	7.3	>95

[a] Thin agar plates were seeded with the appropriate test bacterium. Small wells were punched in the plates and loaded with a dilution series of each sample. After overnight incubation at 30°C, inhibition zones were recorded and lethal concentrations were calculated as described by Hultmark et al (45).

commonly used for biological control of insects (14) and is totally resistant to all cecropins. However, protoplasts of *B. thuringiensis* are sensitive to cecropins; this means that the cell wall acts as a barrier that excludes the cecropins (A. Heierson & I. Sidén, unpublished results). It is interesting that all three cecropins act on *A. calcoaceticus,* an organism that is known to be resistant to many antibiotics.

Melittin, the main component of bee venom, has activity against *E. coli, Bacillus megaterium,* and *Bacillus subtilis* (10). The general design of melittin is similar to that of the cecropins, but the polarity is reversed because in melittin the C-terminus is basic and the N-terminus is hydrophobic. Furthermore, both proteins have one proline in the middle, a single tryptophan in front of a basic sequence stretch, and amidated C-termini that are formed from glycine residues (102, 111). Cecropin A was shown to lyse *E. coli* but not Chang liver cells, while melittin lysed both (98). Cecropins have no effect on other mammalian cell lines tested or on yeast. Therefore cecropins seem to be specific for prokaryotic membranes, while melittin can destroy both prokaryotic and eukaryotic cells. Cecropin from *Sarcophaga* inhibits uptake of proline in bacteria and causes leakage of potassium ions and ATP (72). Cecropin A lyses a variety of artificial liposomes (H. Steiner, unpublished observation).

SOLID-PHASE SYNTHESIS OF CECROPINS AND ANALOGS The structure of cecropins A and B, including the amide groups, was confirmed by solid-phase synthesis (4, 111). A number of analogs were also prepared (5, 66) in which residues 2, 6, and 8 were altered in such a way that the side chain was changed from hydrophobic to hydrophilic or vice versa. In addition, proline was substituted for residue 4 or 8 to break the α-helix.

Table 2 shows that two of the truncated analogs, [2–37] and [1–33]COOH, are fully active against *E. coli,* while the complete molecule is needed for good activity against all of the test bacteria. The Phe-2 analog showed only a fivefold reduction in activity against *Micrococcus luteus,* while the Glu-2 analog showed a tenfold reduction of activity against *E. coli* and almost no activity against the three other test bacteria. The truncated analog [3–37] also showed drastically reduced activity. Therefore Trp-2 seems to be an essential residue; this is supported by the fact that it is also conserved in all cecropins (Figure 1). The Pro-4 and Pro-8 analogs are also fully active against *E. coli.* Thus, the antibacterial activity does not require an extended α-helix. However, the two proline analogs did not show any activity against *M. luteus,* so the broad spectrum was lost when the α-helix was disrupted. The overall conclusion from these experiments is that the cecropin structure was selected for broad-spectrum activity and that this is the first property lost by amino acid

Table 2 Lethal concentrations (μM) of cecropin A and synthetic analogs[a]

Cecropin A analog	*E. coli* D21	*P. aeruginosa* OT97	*B. megaterium* Bm11	*M. luteus* Ml11
[1–37]Natural	0.3	2.6	0.6	1.4
[1–37]Synthetic	0.3	1.5	0.6	1.5
[1–37]COOH	0.5	8.2	10.8	17.8
[2–37]	0.4	8.6	1.5	7.3
[3–37]	2.6	90	13	>110
[1–33]COOH	0.4	13	24	27
[Phe–2]	0.3	3.5	0.78	7.4
[Glu–2]	3.2	170	39	>170
[Pro–4]	0.4	8.1	11	87
[Leu–6]	0.6	120	0.8	7.3
[Glu–6]	0.6	34	2.2	4.7
[Pro–8]	0.5	15	31	80

[a] All compounds have amidated C-terminals except where a carboxyl is indicated. Cecropin A contains 37 residues (see Figure 1), and analogs are identical except for the amino acid residue inserted at the indicated position. Lethal concentrations were determined as described in footnote to Table 1.

substitutions. The broad-spectrum activity also requires an amidated C-terminus.

CECROPIN PRECURSORS A cDNA clone corresponding to cecropin B contains information for 26 amino acid residues in the N-terminus that are not present in the mature cecropin B (111). The signal peptide probably makes up the first 22 of these 26 amino acids, because the Ala-Ala bond after position 22 is a likely cleavage point for the signal peptidase (110). The tetrapeptide left before the start of the mature cecropin B, Ala-Pro-Glu-Pro, is identical to the first four amino acid residues of the prosequence of melittin (58). Since cecropia hemolymph contains an Ala-Pro-R-recognizing peptidase (H. G. Boman, unpublished observation), it is also likely that cecropins are made from a proform by sequential cleavage of proline-containing dipeptides. Recently, Matsumoto et al (64a) isolated a cecropin cDNA clone from *Sarcophaga*. Its sequence did not contain the Ala-Pro-Glu-Pro motif. However, the N-terminus of the proposed signal peptide showed a striking homology (11 of 15 residues) to cecropin B. This is surprising, since signal peptides are usually regarded as rapidly evolving structures. The same region is also strongly conserved in prepro-cecropin A (D. Lidholm & G. Gudmundsson, unpublished observation).

Secondly, the amide group in the C-terminus of the mature cecropin B must be derived from the glycine residue that terminates the coding part of the DNA sequence. The mechanism of amidation is thus analogous to that found for melittin (102) and several brain hormones (85). For production of mature cecropin B the precursor molecule must therefore be processed in two or three steps at both ends.

Attacins

Attacins were first isolated as the inducible protein P5 (12), but at that time neither heterogeneity nor antibacterial activity was detected. The attacins were then rediscovered as an antibacterial fraction with molecular weight considerably larger than that of the cecropins (44). Subsequent studies revealed as many as six different components (A–F) which all cross-reacted with antisera prepared against protein P5. Five different attacinlike peptides were detected in *Manduca sexta* by two-dimensional polyacrylamide gel electrophoresis (47).

The N-terminal sequences for five of the attacins indicated that three of the basic forms were identical, while the two more acidic forms had a slightly different sequence (44). These data strongly suggested the existence of only two different genes. This was supported by the isolation and sequencing of two attacin cDNA clones that showed 76% homology in the coding region, in contrast to only 36% in the region beyond the stop signal (56). As in the case of the cecropins, it seems probable that the two attacin genes have arisen through gene duplication.

How can two genes give rise to six different proteins? The most likely mechanism is terminal modification. Engström et al (27) worked out the complete sequence of 184 amino acid residues for attacin F. They showed that this protein is likely to be derived from attacin E by proteolytic cleavage of a C-terminal extension. In agreement with these protein data, the cDNA clone corresponding to the acidic attacin codes for an extended protein of 188 amino acid residues (attacin E). The difference corresponds to a C-terminal tetrapeptide, Ser-Lys-Tyr-Phe, which is also encoded in the clone for the basic attacin. Since this peptide contains one positive charge, an incomplete proteolytic removal could account for four attacins with different charges (56). It is not yet known if such processing is the result of an artificial proteolytic cleavage or if it has a natural function.

Several lines of evidence indicate that the attacins are made in a preproform (56, 62). In addition there may be a further modification at the N-terminus because evidence for a pyroglutamate group was found at the N-terminus of a basic attacin (27). A partial cyclization of the N-terminal glutamine residue would involve a partial loss of one positive charge. Thus, together with the removal of the tetrapeptide at the C-terminus, the different steps of processing could account for all six forms of attacins observed.

The antibacterial spectra of the attacins seems rather narrow, with good activity only against *E. coli* and two other bacteria originating from the gut of an *Antheraea* larva (44). A study of the mechanism of action on *E. coli* demonstrated that both main attacins act on the outer membrane (29). In particular, attacin facilitates the action of cecropin and lysozyme, thereby enabling these three immune proteins to work in consonance.

Attacins have been found in two other moths besides cecropia: *A. pernyi* (Å. Engström, personal communication) and *M. sexta* (94). Recently, a group of attacinlike proteins was also purified from *S. peregrina* (3a).

Diptericins

A series of inducible antibacterial proteins was detected in larvae of the fly *Phormia terranovae*. One protein was purified and found to be about 9 kd and to have a pI of 7.8 (55). Recently, further components were isolated, and a total of seven closely related proteins have now been purified to homogeneity (D. Hoffmann, personal communication). Amino acid composition and sequence data show that these so-called diptericins constitute a novel family of antibacterial proteins, clearly different from cecropins and attacins.

OTHER INDUCIBLE FACTORS

Immune protein P4 is one of the major components induced by injection of bacteria in cecropia pupae. The purified protein is a single polypeptide of 48kd with a basic isoelectric point (80). Newly synthesized P4 was also found to be the main labeled component in the fat body. No antibacterial activity or other function has yet been found for P4. Andersson & Steiner (3) recently demonstrated two forms of P4 and showed that the protein is present in the basal membrane of the fat body as well as on certain hemocytes. A possibly related protein was also identified in *M. sexta* as four different isoproteins, believed to be glycosylated (47). There was no evidence of glycosylation of *H. cecropia* P4.

Karp and coworkers have reported that the American cockroach, *Periplaneta americana,* can induce a specific protection against bee or snake venom (54, 83). Cockroaches immunized by bee venom toxoid were protected against bee venom but not against a snake venom, and vice versa. The authors claimed that the cockroaches had a secondary response resembling that in mammals (54). However, since no quantitative assay was used and since no evidence was presented for a clonal selection mechanism, the phenomenon observed could well have been a repeated primary response. Rheins & Karp (referenced in 53) reported that the major toxic component in bee venom was phospholipase A2. Karp (53) later observed an altered protein pattern in hemolymph after toxoid injection. The results of Karp and coworkers constitute the only observation to date of a specific immune response in an insect. However, these results will remain difficult to evaluate until they have been substantiated by further experiments in other laboratories.

IMMUNITY-RELATED FACTORS NORMALLY PRESENT IN INSECT HEMOLYMPH

Some factors relevant to immunity are normally present either in the hemolymph or in hemocytes. In the latter case they can be released without de novo synthesis of proteins. Two main factors normally present in insect hemolymph are prophenoloxidase and lectins. Inducibility has been reported for the synthesis of lectins but not for that of prophenoloxidase.

Croft et al (20) have reported an antitrypanosomal factor in tsetse fly hemolymph, which abolished the mobility of *Trypanosoma brucei brucei*. Nothing is known about the biochemistry of this factor.

Prophenoloxidase and its Activation

Phenoloxidases are widely distributed both in plants and in invertebrates, and the literature is extensive (for reviews see 13, 92). The enzyme is very reactive and is probably always stored in the form of an inactive prophenoloxidase. The end product of the phenoloxidase reaction is melanin, a dark insoluble material deposited around microorganisms or parasites in nodules and capsules. A similar reaction is also seen after wounding, and it may have a protective function. Together with nodule-forming components, the melanin might form a mechanical barrier, preventing growth of entrapped parasites. It is also possible that intermediates in melanin formation could be toxic to invading organisms. Stephens & Marshal (100) observed 25 years ago that immunized insects make an inhibitor of phenoloxidase activation. Such an inhibitor could possibly have a protective function against toxic products formed during melanization. Namihira et al (69) reported partial purification of an inhibitor from house flies, and Saul & Sugumaran (87a) purified an inhibitor from *M. sexta* by affinity chromatography.

Ashida and coworkers (7, 115) have studied the prophenoloxidase from the silkworm *Bombyx mori* in detail. The purified proenzyme was found to have a molecular weight of 80×10^3, and it is probably a dimer (7). It is activated by a specific serine protease, the prophenoloxidase-activating enzyme (PPAE). The activation of PPAE in the hemolymph is preceded by the appearance of another serine enzyme (115). The activation reaction is started by bacterial peptidoglycan but not by lipopolysaccharide. The activating enzyme removes a 5-kd peptide from each subunit of the prophenoloxidase. The activated phenoloxidase binds strongly to various substrates, and during the reaction it is found as aggregates of progressively higher molecular weight. Purified phenoloxidase from the house fly, *Musca domestica,* was found to have a molecular weight of about 340 × 10^3, while that of latent phenoloxidase was only 178 × 10^3 (109). This enzyme was stabilized by EDTA during isolation. Prophenoloxidase in crayfish can be activated by laminarin, a β-1,3-glucan, and this reaction leads to increased phagocytosis (91). Insect prophenoloxidase can also be activated by laminarin, with a simultaneous increase in phagocytosis (63).

Lectins

Bernheimer (9) had already demonstrated 35 years ago that the hemolymph from 10 of 46 species of lepidopteran larvae had the capacity to agglutinate mammalian erythrocytes. The agglutination of red blood cells has in many cases been attributed to lectins, proteins with a highly specific multivalent capacity to bind to certain sugar residues on erythrocyte membranes. Such proteins have long been known in many plants (36) and invertebrates (13, 113). However, the natural function of most of the lectins has so far remained a puzzle.

Natori's group has purified a lectin from the hemolymph of larvae of the meat fly *Sarcophaga peregrina* (57). They found that the protein is inducible by injury in larvae but is made constitutively in pupae. The lectin can also be induced by injecting larvae with sheep red blood cells (104). *Sarcophaga* lectin is composed of six subunits (four α-subunits and two β-subunits), and it has shown a specificity for galactose. Recent cloning data indicate that both subunits of the lectin are encoded by the same gene, and that the β-subunit is obtained from the α-subunit by proteolytic cleavage (104). The lectin in *Sarcophaga* is synthesized during embryogenesis and pupation (105). *Sarcophaga* lectin has also induced a mouse-macrophage cell line to synthesize an antitumor factor, which has been purified (48).

Lectins have also been purified from the cricket *Teleogryllus commodus* (40) and from the grasshopper *Melanoplus sanguinipes* (95). The latter lectin had a molecular weight of 590 × 10^3, with a 70-kd subunit composed of two polypeptide chains. Like the *Sarcophaga* lectin, the grasshopper lectin is specific for galactose.

Castro et al (16) recently purified a series of isolectins from *H. cecropia* by affinity chromatography. These lectins were tetramers with two types of subunits, the A and B chains, in different proportions. The 41-kd A chain showed specificity for galactose, while the B chain of 38 kd was associated with ability to agglutinate rabbit erythrocytes. Attempts to induce synthesis of the cecropia lectins have been unsuccessful (16, 113).

A possible function of some lectins could be to agglutinate invading microorganisms that carry surface components containing the corresponding sugar residues. The agglutinated foreign cells would then be more easily phagocytosed or encapsulated, and melanization would follow. However, in only one case has an insect lectin been shown to agglutinate a microorganism carried by that insect. Pereira et al (73) demonstrated that the assassin bug, a vector for *Trypanosoma cruzi,* contains two lectins that agglutinate trypanosomes.

Another function linked to coagulation was recently suggested for M13, a lectinlike protein induced by bacteria in larvae of the tobacco hornworm (47). Purified M13 was needed in hemocyte-mediated coagulation, and this function was inhibited by glucose (67).

A very large (260 kd) lectin in the common silkworm *B. mori* was found in peak amounts in fifth instar larvae just before pupation (103). The lectin showed specificity for glucuronic acid, but it could not be purified. The *Bombyx* and *Sarcophaga* lectins could perhaps be involved in the recognition of larval tissue in connection with metamorphosis.

Humoral Encapsulation

Capsule formation is an immune reaction in which many hemocytes cooperate in the formation of layers that completely cover parasites and other foreign bodies. However, certain dipterans with few hemocytes can form capsule layers without the participation of any cells; this phenomenon is called humoral encapsulation (reviewed in 13). The reaction is very fast and can conveniently be followed under a microscope, but very little is known about the biochemistry. Humoral encapsulation is effective against bacteria and fungi as well as against nematodes. Götz (36a) has made a scientific film of the humoral encapsulation of a nematode.

THE INJURY REACTION

Anyone who has collected insects in the field knows that they are often damaged. The injured parts are likely entry points for infections; it is therefore not surprising that injuries have been found to activate the humoral immune system. It is also clear that hemocytes have an important role in wound healing. Thus, the response to injury shows up as a complex mixture of cellular and humoral reactions. Wound responses in insects were reviewed by Yeaton (114).

Insect surgery and organ transplantations have been important tools in physiological research. Sham operations showed early that an injury triggers the onset of many biosynthetic reactions in insects (8, 18). Even an injury as small as that inflicted by a fine injection needle is enough to cause significant protein synthesis (31). Thus, the immune response to injected bacteria must always be compared to the response to a sham injection with saline (11, 31).

The relationship between the humoral responses to injuries and to infections in insects is not yet understood. The current state of knowledge has recently been summarized (38).

INDUCTION OF IMMUNE PROTEINS

How can an unspecific signal like the injection of bacteria be translated to a specific signal that selectively activates 15–20 genes in the main organ of synthesis? If the first step is phagocytosis of the bacteria, then there must be a chain of signals from the bacteria to the DNA of the genes for the cecropins and the attacins. The problem may be simpler in the case of hemocytes that are capable of both phagocytosis and lysozyme synthesis (26). Experiments with isolated fat body from *H. cecropia* pupae as well as with intact pupae show clearly that the fat body must be the major site of synthesis for the immune proteins (1, 26, 32). Thus, a signal must pass from hemocytes to fat body cells. The same must be true in the response to injury.

Early Events During an Infection

In *H. cecropia,* immune-specific RNA appears 2–4 hr after infection (D. Lidholm, unpublished observation). After 5 hr, actinomycin D can no longer block the expression of immunity (11, 31). A somewhat shorter lag period of 3–4 hr was found in *Galleria* (42). Since both actinomycin D and cycloheximide block the appearance of cecropins and attacins, and since these proteins can be labeled with radioactive amino acids, they must be formed by de novo synthesis and they are not released from a reservoir. Antibacterial activity in *Phormia* appeared in only 4 hr (55), and in *Sarcophaga* there was an almost immediate synthesis of the inducible lectin (105). In both cases there was evidence of a requirement for RNA synthesis.

The immune response in insects can be induced by the injection of a variety of substances into the hemocoel (25). The strongest response has consistently been observed after injection of live nonpathogenic bacteria or heat-killed pathogens. In cecropia the response was shown to be unspecific because identical polypeptide patterns were induced by the injection of gram-positive and gram-negative bacteria (31). Poly(IC) can also induce antibacterial activity (79). A mechanical injury can elicit a similar but much weaker response. Treatment with ultrasound (34, 77) or pricking with a needle (70) have also been used for convenient induction of immune proteins in large numbers of individual insects.

Induction of Immune Proteins In Vitro

Early attempts to study the induction mechanism in vitro using fat body in tissue culture were hampered by the fact that dissection alone was enough to trigger a dramatic wound response in the explanted tissue (23, 32). However, Dunn et al (26) recently developed a system with uninduced fat body from *Manduca* larvae, which could be induced in vitro to synthesize lysozyme and other antibacterial factors. A soluble peptidoglycan fragment produced from bacterial cell walls by lysozyme digestion was found to work as an efficient inducer in this system. The researchers suggested that in vivo the immune response is elicited by peptidoglycan fragments that are released from phagocytized bacteria. A system has also been developed for in vitro induction of cecropia fat body (M. Trenczek & H. Bennich, unpublished observation).

PASSIVE AND ACTIVE RESISTANCE TO IMMUNITY

All forms of life need immune systems to protect themselves against invading microorganisms or parasites. Since this is a universal phenomenon, one can expect that there has been strong evolutionary pressure on pathogens to develop mechanisms to escape immune reactions. Different groups of protozoa have evolved especially sophisticated means to avoid mammalian immune recognition.

Two main types of avoidance mechanisms have been found to counteract insect immunity; they have been referred to as "passive" and "active" resistance. Passive resistance is said to occur if the envelope of a bacterium can in some way prevent the insect's antibacterial activity. If this is so, bacterial mutants may exist that are deficient in this protection. This was first shown for a strain of *Serratia marcescens* that is pathogenic for *Drosophila* and highly resistant to *H. cecropia* immunity. Certain spontaneous phage-resistant mutants were isolated and found to have lost their virulence in *Drosophila* and to be sensitive to immune hemolymph from *H. cecropia* (35). Some phage-resistant mutants of *E. coli* were shown to acquire increased sensitivity to cecropin D (89). In addition, phage-resistant *B. thuringiensis* mutants were found which had lost their virulence and had become sensitive to *H. cecropia* immunity (41).

Active resistance is due to destruction of immunity by an active process. So far the only mechanism found has been proteolytic digestion of the cecropins and attacins by so-called "immune inhibitors." The first such inhibitor, InA, was isolated from *B. thuringiensis* (90), and its specificity has been determined (21). However, in *H. cecropia* InA is produced only late after infection when the insect is already dead or dying (97). Furthermore, a barrier in the cell wall makes *B. thuringiensis* totally insensitive to both cecropins and attacins (A. Heierson & I. Sidén, unpublished information). Thus, the survival value of InA has not yet been explained. A very complex situation was found in a nematode with a symbiotic bacterium, *Xenorhabdus nematophilus* (see Table 1). This bacterium is highly sensitive to cecropins A and

B, but the nematode helps it by producing an immune inhibitor that quickly destroys the cecropins and attacins (37). Passive and active resistance may also be involved in the complex interplay between parasitic wasps, their symbiotic viruses, and their hosts (101).

THE FUNCTION OF INSECT IMMUNITY

The Natural Flora

There is extensive literature on insect microbiology. In his classic book, Steinhaus (99) lists about 250 identified bacteria that in one way or another have been associated with insects. Of these, about 40–50% are small gram-negative rods and 15–25% are gram-positive spore-forming bacteria. Insects, like other animals, are infected by organisms of three different kinds: (*a*) harmless nonpathogenic organisms, (*b*) occasional pathogens, and (*c*) obligate pathogens. Representative bacteria belonging to these three groups have been assayed for their susceptibility to different cecropins (Table 1). These and other data show that the immune system always works against nonpathogenic bacteria, while it works against only some of the pathogens. We therefore assume that the main function of the immune system is to control the natural flora that insects meet. This is a very important role in insects such as houseflies and cockroaches, which live in decomposing organic matter. However, as a consequence of the ecological equilibrium between insects and microbes a number of pathogens have evolved on which the immune system does not work. An additional fact to consider is that many insects are dependent on intracellular symbionts (108) which in some cases seem to be gram-negative rods. For such insects it is important that the immune system be compatible with the symbionts.

How Insects are Infected

Free-living insects are probably most often infected by the intake of contaminated food, but infections through the trachea may also be common. In addition, insects are infected through wounds caused by accidents or by the attack of other animals. Some insects may also be infected by release of the intestinal flora during metamorphosis. This is so far only a hypothetical possibility for which there are no data. Few recent attempts have been made to study the gut flora of insects (for older references see 99). Jarosz (49) found that the gut of the greater wax moth, *Galleria mellonella,* only contained a single bacterial species, *Streptococcus faecium*. Only two gram-negative bacteria, *A. calcoaceticus* and *P. maltophilia,* could be isolated from the gut of the Chinese oak silkworm moth, *A. pernyi* (44). A few studies have also been performed on antibacterial mechanisms in the gut (68, 87).

COMPARATIVE ASPECTS

Comparing Different Insects

As a group of animals, insects are extremely diverse, showing adaptation to very different ecological niches. It is therefore quite likely that various insects will be found to have immune systems that are very different in their design. An immune system composed of antibacterial proteins like the cecropins has so far only been convincingly demonstrated in flies (Diptera) and moths (Lepidoptera); it probably also occurs in beetles (Coleoptera) and bugs (Hemiptera). It is possible that lysozyme, phenoloxidase, and lectins are more universal than cecropins and attacins, but much more data are needed before a safe judgment can be passed on how the immune systems of the entire group of insects are designed.

Comparing Insects and Vertebrates

Comparison of DNA and protein sequences and fossil records indicate that insects and vertebrates diverged about 600 million years ago. During this evolutionary time the differences in the basic structure of the two types of animals, as well as other factors, must have selected very different immune mechanisms.

First of all, it is important how well the animals are protected from invading microbes. Insects have very resistant exoskeletons, and no epithelia are directly exposed to invaders since the cuticle completely covers the outside as well as the foregut, hindgut, and tracheal tubes. Even the intestinal epithelium of the midgut is protected in most insects by the peritrophic membrane, a structure that encloses the gut contents. Vertebrates, in contrast, offer greater access for invading organisms, which can easily penetrate the mucous epithelia of the intestinal system, the respiratory system, the excretory system, and the genital organs.

Secondly, the duration of an inducible immune response in any animal must be regarded in terms of both the life span and the reproductive rate of the animal. Morover, since the rates for DNA, RNA, and protein synthesis are probably similar in all animals, a fast response will have to be built on RNA and protein synthesis, while a slower response can rely on cell proliferation. Therefore, in long-lived vertebrates with relatively few offspring per generation there has been strong selection pressure for the development of a lasting secondary response based on cell proliferation. In contrast, short-lived insects with thousands of progeny have developed fast and unspecific responses based on RNA and protein synthesis.

As a result of these differences and because of the way it has been adapted to avoid self-destruction, the mammalian immune system has become very specific, in fact so specific that in certain situations it may create biological disadvantages. Insect immunity, on the other hand, does not show the same degree of specificity, and this is an advantage as long as self-destruction is avoided. Insects exhibit no immune reactions against homologous hemocytes and allografts; they have only limited reactions against xenografts (15, 59, 107).

CONCLUDING REMARKS

About five years have passed since our previous reviews on insect immunity were written (12, 38). During this time the field has grown considerably, and the perspective has changed from physiology and emerging biochemistry to molecular biology and genetics.

The structure of the cecropia genes for antibacterial proteins are currently under intensive study in our laboratory. We hope that within the next five years it will become clear how duplications have arranged the immune genes, whether these genes share some unique promotor sequences, and whether specific transcription factors can recognize these structures. These data will then be linked to new information that will no doubt come from further study of in vitro induction.

ACKNOWLEDGMENTS

This work was supported by grants from the Swedish Natural Science Research Council. We thank Hans Bennich, Ingrid Faye, and Håkan Steiner for critical readings of our manuscript. We also thank many colleagues in other laboratories for sending re- and preprints and for giving us access to unpublished data. In order to keep a reasonable relation between the text and the literature cited, however, we have had to sacrifice 30–40 references submitted to us.

Literature Cited

1. Abu-Hakima, R., Faye, I. 1981. An ultrastructural and autoradiographic study of the immune response in *Hyalophora cecropia* pupae. *Cell Tissue Res.* 217:311–20
2. Anderson, R. S., Cook, M. L. 1979. Induction of lysozymelike activity in the hemolymph and hemocytes of an insect, *Spodoptera eridania*. *J. Invertebr. Pathol.* 33:197–203
3. Andersson, K., Steiner, H. 1987. Structure and properties of protein P4, the major bacteria-inducible protein in pupae of *Hyalophora cecropia*. *Insect Biochem.* 17:133–40

3a. Ando, K., Okada, M., Natori, S. 1987. Purification of sarcotoxin II, antibacterial proteins from *Sarcophaga peregrina* (flesh fly) larvae. *Biochemistry* 26:226–30

4. Andreu, D., Merrifield, R. B., Steiner, H., Boman, H. G. 1983. Solid-phase synthesis of cecropin A and related peptides. *Proc. Natl. Acad. Sci. USA* 80:6475–79
5. Andreu, D., Merrifield, R. B., Steiner, H., Boman, H. G. 1985. N-Terminal analogues of cecropin A: synthesis, antibacterial activity, and conformational properties. *Biochemistry* 24:1683–88
6. Aronson, A. I., Beckman, W., Dunn, P. 1986. *Bacillus thuringiensis* and related insect pathogens. *Microbiol. Rev.* 50:1–24

7. Ashida, M., Dohke, K. 1980. Activation of pro-phenoloxidase by the activating enzyme of the silkworm, *Bombyx mori*. *Insect Biochem*. 10:37–47
8. Barth, R. H. Jr., Bunyard, P. P., Hamilton, T. H. 1964. RNA metabolism in pupae of the oak silkworm, *Antheraea pernyi:* The effects of diapause, development and injury. *Proc. Natl. Acad. Sci. USA* 52:1572–80
9. Bernheimer, A. W. 1952. Hemagglutinins in caterpillar bloods. *Science* 115:150–51
10. Boman, H. G. 1982. Humoral immunity in insects and the counter defence of some pathogens. *Fortschr. Zool*. 27:211–22
11. Boman, H. G., Boman, A., Pigon, A. 1981. Immune and injury responses in *cecropia* pupae—RNA isolation and comparison of protein synthesis in vivo and in vitro. *Insect Biochem*. 11:33–42
12. Boman, H. G., Hultmark, D. 1981. Cell-free immunity in insects. *Trends Biochem. Sci*. 6:306–9
13. Brehélin, M., ed. 1986. *Immunity in Invertebrates. Cells, Molecules, and Defense Reactions*. Heidelberg: Springer-Verlag. 233 pp.
14. Burges, H. D., ed. 1981. *Microbial Control of Pests and Plant Diseases 1970–1980*. London/New York: Academic. 949 pp.
15. Carton, Y. 1976. Isogenic, allogenic and xenogenic transplants in an insect species. *Transplantation* 21:17–22
16. Castro, V. M., Boman, H. G., Hammarström, S. 1987. Isolation and characterization of a group of isolectins with galactose/*N*-acetylgalactosamine specificity from hemolymph of the giant silk moth *Hyalophora cecropia*. *Insect Biochem*. 17:513–23
17. Chadwick, J. M., Aston, W. P. 1979. An overview of insect immunity. In *Animal Models of Comparative and Developmental Aspects of Immunity and Disease*, ed. M. E. Gershwin, E. L. Cooper, pp. 1–14. New York: Pergamon. 396 pp.
18. Cherbas, L. 1973. The induction of an injury reaction in cultured haemocytes from saturniid pupae. *J. Insect Physiol*. 19:2011–23
19. Craig, A. G., Engström, Å., Bennich, H., Kamensky, I. 1987. Plasma desorption mass spectrometry coupled with conventional peptide sequencing techniques. *Biochem. Mass Spectrosc*. In press
20. Croft, S. L., East, J. S., Molyneux, D. H. 1982. Anti-trypanosomal factor in the haemolymph of *Glossina*. *Acta Trop*. 39:293–302
21. Dalhammar, G., Steiner, H. 1984. Characterization of inhibitor A, a protease from *Bacillus thuringiensis* which degrades attacins and cecropins, two classes of antibacterial proteins in insects. *Eur. J. Biochem*. 139:247–52
22. de Azambuja, P., Freitas, C. C., Garcia, E. S. 1986. Evidence and partial characterization of an inducible antibacterial factor in the haemolymph of *Rhodnius prolixus*. *J. Insect Physiol*. 32:807–12
23. De Verno, P. J., Chadwick, J. S., Aston, W. P., Dunphy, G. B. 1984. The in vitro generation of an antibacterial activity from the fat body and hemolymph of non-immunized larvae of *Galleria mellonella*. *Dev. Comp. Immunol*. 8:537–46
24. Dubos, R. 1986. *Louis Pasteur. Free Lance of Science*. New York: Da Capo. 420 pp.
25. Dunn, P. E. 1986. Biochemical aspects of insect immunology. *Ann. Rev. Entomol*. 31:321–39
26. Dunn, P. E., Dai, W., Kanost, M. R., Geng, C. 1985. Soluble peptidoglycan fragments stimulate antibacterial protein synthesis by fat body from larvae of *Manduca sexta*. *Dev. Comp. Immunol*. 9:559–68
27. Engström, Å., Engström, P., Tao, Z.-j., Carlsson, A., Bennich, H. 1984. Insect immunity. The primary structure of the antibacterial protein attacin F and its relation to two native attacins from *Hyalophora cecropia*. *EMBO J*. 3: 2065–70
28. Engström, Å., Xanthopoulos, K. G., Boman, H. G., Bennich, H. 1985. Amino acid and cDNA sequences of lysozyme from *Hyalophora cecropia*. *EMBO J*. 4:2119–22
29. Engström, P., Carlsson, A., Engström, Å., Tao, Z.-j., Bennich, H. 1984. The antibacterial effect of attacins from the silk moth *Hyalophora cecropia* is directed against the outer membrane of *Escherichia coli*. *EMBO J*. 3:3347–51
30. Faye, I. 1978. Insect immunity: Early fate of bacteria injected in saturniid pupae. *J. Invertebr. Pathol*. 31:19–26
31. Faye, I., Pye, A., Rasmuson, T., Boman, H. G., Boman, I. A. 1975. Insect immunity II. Simultaneous induction of antibacterial activity and selective synthesis of some hemolymph proteins in diapausing pupae of *Hyalophora cecropia* and *Samia cynthia*. *Infect. Immun*. 12:1426–38

32. Faye, I., Wyatt, G. R. 1980. The synthesis of antibacterial proteins in isolated fat body from cecropia silkmoth pupae. *Experientia* 36:1325–26
33. Fernandez-Sousa, J. M., Gavilanes, J. G., Municio, A. M., Perez-Aranda, A., Rodriguez, R. 1977. Lysozyme from the insect *Ceratitis capitata* eggs. *Eur. J. Biochem.* 72:25–33
34. Flyg, C., Dalhammar, G., Rasmuson, B., Boman, H. G. 1987. Insect immunity. Inducible antibacterial activity in *Drosophila. Insect Biochem.* 17:153–60
35. Flyg, C., Kenne, K., Boman, H. G. 1980. Insect pathogenic properties of *Serratia marcescens:* Phage-resistant mutants with a decreased resistance to *cecropia* immunity and a decreased virulence to *Drosophila. J. Gen. Microbiol.* 120:173–81
36. Goldstein, I. J., Etzler, M. E., eds. 1983. *Chemical Taxonomy, Molecular Biology, and Function of Plant Lectins.* New York: Liss. 298 pp.
36a. Götz, P. 1976. Parasit-Wirt-Beziehungen zwischen dem Nematoden *Hydromermis contorta* und der Zuckmücke *Chironomus thummi. Publ. Wiss. Filmen* 9:67–87
37. Götz, P., Boman, A., Boman, H. G. 1981. Interactions between insect immunity and an insect-pathogenic nematode with symbiotic bacteria. *Proc. R. Soc. London Ser. B* 212:333–50
38. Götz, P., Boman, H. G. 1985. Insect immunity. In *Comprehensive Insect Physiology Biochemistry and Pharmacology,* Vol. 3, ed. G. A. Kerkut, L. I. Gilbert, pp. 453–85. Oxford/New York: Pergamon. 625 pp.
39. Gupta, A. P. 1979. *Insect Hemocytes. Development, Forms, Functions and Techniques.* London/New York: Cambridge Univ. Press. 614 pp.
40. Hapner, K. D., Jermyn, M. A. 1981. Haemagglutinin activity in the haemolymph of *Teleogryllus commodus* (Walker). *Insect Biochem.* 11:287–95
41. Heierson, A., Sidén, I., Kivaisi, A., Boman, H. G. 1986. Bacteriophage-resistant mutants of *Bacillus thuringiensis* with decreased virulence in pupae of *Hyalophora cecropia. J. Bacteriol.* 167:18–24
42. Hoffmann, D., Hultmark, D., Boman, H. G. 1981. Insect immunity: *Galleria mellonella* and other lepidoptera have cecropia-P9-like factors active against gram negative bacteria. *Insect Biochem.* 11:537–48
43. Horohov, D. W., Dunn, P. E. 1983. Phagocytosis and nodule formation by hemocytes of *Manduca sexta* larvae following injection of *Pseudomonas aeruginosa. J. Invertebr. Pathol.* 41: 203–13
44. Hultmark, D., Engström, Å., Andersson, K., Steiner, H., Bennich, H., Boman, H. G. 1983. Insect immunity. Attacins, a family of antibacterial proteins from *Hyalophora cecropia. EMBO J.* 2:571–76
45. Hultmark, D., Engström, Å., Bennich, H., Kapur, R., Boman, H. G. 1982. Insect immunity: Isolation and structure of cecropin D and four minor antibacterial components from cecropia pupae. *Eur. J. Biochem.* 127:207–17
46. Hultmark, D., Steiner, H., Rasmuson, T., Boman, H. G. 1980. Insect immunity. Purification and properties of three inducible bactericidal proteins from hemolymph of immunized pupae of *Hyalophora cecropia. Eur. J. Biochem.* 106:7–16
47. Hurlbert, R. E., Karlinsey, J. E., Spence, K. D. 1985. Differential synthesis of bacteria-induced proteins of *Manduca sexta* larvae and pupae. *J. Insect Physiol.* 31:205–15
48. Itoh, A., Ohsawa, F., Natori, S. 1986. Purification of a cytotoxic protein produced by the murine macrophage-like cell line J774.1 in response to *Sacrophaga* lectin. *J. Biochem.* 99:9–15
49. Jarosz, J. 1983. *Streptococcus faecium* in the intestine of the greater wax moth, *Galleria mellonella. Microbios Lett.* 23:125–28
50. Jollès, J., Schoentgen, F., Croizier, G., Croizier, L., Jollès, P. 1979. Insect lysozymes from three species of lepidoptera: their structural relatedness to the C (chicken) type lysozyme. *J. Mol. Evol.* 14:267–71
51. Kaaya, G. P., Flyg, C., Boman, H. G. 1987. Insect immunity. Induction of cecropin and attacin-like antibacterial factors in the haemolymph of *Glossina morsitans morsitans. Insect Biochem.* 17:309–15
52. Kaiser, E. T., Kézdy, F. J. 1984. Amphiphilic secondary structure: Design of peptide hormones. *Science* 223:249–55
53. Karp, R. D. 1985. Preliminary characterization of the inducible humoral factor in the American cockroach *(Periplaneta americana). Dev. Comp. Immunol.* 9:569–75
54. Karp, R. D., Rheins, L. A. 1980. Induction of specific humoral immunity to soluble proteins in the American cockroach *(Periplaneta americana).* II. Nature of the secondary response. *Dev. Comp. Immunol.* 4:629–39

55. Keppi, E., Zachary, D., Robertson, M., Hoffmann, D., Hoffmann, J. A. 1986. Induced antibacterial proteins in the haemolymph of *Phormia terranovae* (Diptera). *Insect Biochem.* 16:395–402
56. Kockum, K., Faye, I., von Hofsten, P., Lee, J.-Y., Xanthopoulos, K. G., Boman, H. G. 1984. Insect immunity. Isolation and sequence of two cDNA clones corresponding to acidic and basic attacins from *Hyalophora cecropia*. *EMBO J.* 3:2071–75
57. Komano, H., Mizuno, D., Natori, S. 1980. Purification of lectin induced in the hemolymph of *Sarcophaga peregrina* larvae on injury. *J. Biol. Chem.* 255:2919–24
58. Kreil, G., Haiml, L., Suchanek, G. 1980. Stepwise cleavage of the pro part of promelittin by dipeptidylpeptidase IV. *Eur. J. Biochem.* 111:49–58
59. Lackie, A. M. 1986. Hemolymph transfer as an assay for immunorecognition in the cockroach *Periplaneta americana* and the locust *Schistocerca gregaria*. *Transplantation* 41:360–63
60. Lambert, J., Hoffmann, D. 1985. Mise en évidence d'un facteur actif contre des bactéries gram négatives dans le sang de *Locusta migratoria*. *C. R. Seances Acad. Sci. Ser. III* 300:425–30 (In French)
61. Landureau, J. C. 1976. Insect cell and tissue culture as a tool for developmental biology. In *Invertebrate Tissue Culture*, ed. E. Kurstak, K. Maramorosch, pp. 101–30. New York: Academic. 398 pp.
62. Lee, J.-Y., Edlund, T., Ny, T., Faye, I., Boman, H. G. 1983. Insect immunity. Isolation of cDNA clones corresponding to attacins and immune protein P4 from *Hyalophora cecropia*. *EMBO J.* 2:577–81
63. Leonard, C., Ratcliffe, N. A., Rowley, A. F. 1985. The role of prophenoloxidase activation in non-self recognition and phagocytosis by insect blood cells. *J. Insect. Physiol.* 31:789–99
64. Lysenko, O. 1985. Non-sporeforming bacteria pathogenic to insects: Incidence and mechanisms. *Ann. Rev. Microbiol.* 39:673–95
64a. Matsumoto, N., Okada, M., Takahashi, H., Ming, Q. X., Nakajima, Y., et al. 1986. Molecular cloning of a cDNA and assignment of the C-terminal of sarcotoxin IA, a potent antibacterial protein of *Sarcophaga peregrina*. *Biochem. J.* 239:717–22
65. Mead, G. P., Ratcliffe, N. A., Renwrantz, L. R. 1986. The separation of insect haemocyte types on percoll gradients; methodology and problems. *J. Insect. Physiol.* 32:167–77
66. Merrifield, R. B., Vizioli, L. D., Boman, H. G. 1982. Synthesis of the antibacterial peptide cecropin A(1–33). *Biochemistry* 21:5020–30
67. Minnick, M. F., Rupp, R. A., Spence, K. D. 1986. A bacterial-induced lectin which triggers hemocyte coagulation in *Manduca sexta*. *Biochem. Biophys. Res. Commun.* 137:729–35
68. Mohrig, W., Messner, B. 1968. Immunreaktionen bei Insekten. II. Lysozym als antimikrobielles Agens im Darmtrakt von Insekten. *Biol. Zentralbl.* 87:705–18 (In German)
69. Namihira, G., Ejima, T., Inaba, T., Funatsu, M. 1979. A protein factor inhibiting activation of prophenoloxidase with natural activator and its purification. *Agric. Biol. Chem.* 43:471–76
70. Okada, M., Natori, S. 1983. Purification and characterization of an antibacterial protein from haemolymph of *Sarcophaga peregrina* (flesh-fly) larvae. *Biochem. J.* 211:727–34
71. Okada, M., Natori, S., 1985. Primary structure of sarcotoxin I, an antibacterial protein induced in hemolymph of *Sarcophaga peregrina* (flesh fly) larvae. *J. Biol. Chem.* 260:7174–77
72. Okada, M., Natori, S. 1985. Ionophore activity of sarcotoxin I, a bactericidal protein of *Sarcophaga peregrina*. *Biochem. J.* 229:453–58
73. Pereira, M. E. A., Andrade, A. F. B., Ribeiro, J. M. C. 1981. Lectins of distinct specificity in *Rhodnius prolixus* interact selectively with *Trypanosoma cruzi*. *Science* 211:597–600
74. Porter, J. R. 1973. Agostino Bassi bicentennial (1773–1973). *Bacteriol. Rev.* 37:284–88
75. Powning, R. F., Davidson, W. J. 1976. Studies on insect bacteriolytic enzymes—II. Some physical and enzymatic properties of lysozyme from haemolymph of *Galleria mellonella*. *Comp. Biochem. Physiol. B* 55:221–28
76. Price, C. D., Ratcliffe, N. A. 1974. A reappraisal of insect haemocyte classification by the examination of blood from fifteen insect orders. *Z. Zellforsch. Mikrosk. Anat.* 147:537–49
77. Qi, G., Zhou, Q., Qu, X.-m., Huang, Z. 1984. Some inducible antibacterial substances developed from the heamolymph [sic] of oak silkworm, *Antheraea pernyi* pupae by ultrasonic treatment. *Kexue Tongbao* 29:670–74
78. Qu, X.-m., Steiner, H., Engström, Å., Bennich, H., Boman, H. G. 1982. Insect immunity: Isolation and structure of

cecropins B and D from pupae of the Chinese oak silk moth, *Antheraea pernyi*. *Eur. J. Biochem.* 127:219–24

79. Qu, X.-m., Wu, K.-z., Qiu, X.-z., Zhang, S.-q., Lee, S.-y. 1986. Isolation and characterization of six antibacterial peptides from the hemolymph of immunized *Bombyx mori* pupae by injection of poly I:C. *Acta Biochim. Biophys. Sin.* 18:284–91 (In Chinese with English summary)
80. Rasmuson, T., Boman, H. G. 1979. Insect immunity—V. Purification and some properties of immune protein P4 from haemolymph of *Hyalophora cecropia* pupae. *Insect Biochem.* 9:259–64
81. Ratcliffe, N. A. 1985. Invertebrate immunity—A primer for the non-specialist. *Immunol. Lett.* 10:253–70
82. Ratcliffe, N. A., Rowley, A. F., Fitzgerald, S. W., Rhodes, C. P. 1985. Invertebrate immunity: Basic concepts and recent advances. *Int. Rev. Cytol.* 97:183–350
83. Rheins, L. A., Karp, R. D., Butz, A. 1980. Induction of specific humoral immunity to soluble proteins in the American cockroach *(Periplaneta americana)*. I. Nature of the primary response. *Dev. Comp. Immunol.* 4:447–58
84. Ribeiro, J. M. C., Pereira, M. E. A. 1984. Midgut glycosidases of *Rhodnius prolixus*. *Insect Biochem.* 14:103–8
85. Richter, D. 1983. Vasopressin and oxytocin are expressed as polyproteins. *Trends Biochem. Sci.* 8:278–81
86. Robertson, M., Postlethwait, J. H. 1986. The humoral antibacterial response of *Drosophila* adults. *Dev. Comp. Immunol.* 10:167–79
87. Rupp, R. A., Spence, K. D. 1985. Protein alterations in *Manduca sexta* midgut and haemolymph following treatment with a sublethal dose of *Bacillus thuringiensis* crystal endotoxin. *Insect Biochem.* 15:147–54
87a. Saul, S. J., Sugumaran, M. 1986. Protease inhibitor controls prophenoloxidase activation in *Manduca sexta*. *FEBS Lett.* 208:113–16
88. Schneider, P. M. 1985. Purification and properties of three lysozymes from hemolymph of the cricket, *Gryllus bimaculatus* (de Geer). *Insect Biochem.* 15:463–70
89. Sidén, I., Boman, H. G. 1983. *Escherichia coli* mutants with an altered sensitivity to cecropin D. *J. Bacteriol.* 154:170–76
90. Sidén, I., Dalhammar, G., Telander, B., Boman, H. G., Somerville, H. 1979. Virulence factors in *Bacillus thuringiensis:* Purification and properties of a protein inhibitor of immunity in insects. *J. Gen. Microbiol.* 114:45–52
91. Smith, V. J., Söderhäll, K. 1983. β-1,3 glucan activation of crustacean hemocytes in vitro and in vivo. *Biol. Bull. Woods Hole Mass.* 164:299–314
92. Söderhäll, K. 1982. Prophenoloxidase activating system and melanization—A recognition mechanism of arthropods? A review. *Dev. Comp. Immunol.* 6:601–11
93. Spies, A. G., Karlinsey, J. E., Spence, K. 1986. The immune proteins of the darkling beetle, *Eleodes* (Coleoptera: Tenebrionidae). *J. Invertebr. Pathol.* 47:234–35
94. Spies, A. G., Karlinsey, J. E., Spence, K. D. 1986. Antibacterial hemolymph proteins of *Manduca sexta*. *Comp. Biochem. Physiol. B* 83:125–33
95. Stebbins, M. R., Hapner, K. D. 1985. Preparation and properties of haemagglutinin from haemolymph of Acrididae (grasshoppers). *Insect Biochem.* 15:451–62
96. Steiner, H. 1982. Secondary structure of the cecropins: Antibacterial peptides from the moth *Hyalophora cecropia*. *FEBS Lett.* 137:283–87
97. Steiner, H. 1985. Role of the exoprotease InA in the pathogenicity of *Bacillus thuringiensis* in pupae of *Hyalophora cecropia*. *J. Invertebr. Pathol.* 46:346–47
98. Steiner, H., Hultmark, D., Engström, Å., Bennich, H., Boman, H. G. 1981. Sequence and specificity of two antibacterial proteins involved in insect immunity. *Nature* 292:246–48
99. Steinhaus, E. A. 1967. *Insect Microbiology*. New York/London: Hafner. 763 pp. (Facsimile of the edition of 1947)
100. Stephens, J. M., Marshall, J. H. 1962. Some properties of an immune factor isolated from the blood of actively immunized wax moth larvae. *Can. J. Microbiol.* 8:719–25
101. Stoltz, D. B., Vinson, S. B. 1979. Viruses and parasitism in insects. *Adv. Virus Res.* 24:125–71
102. Suchanek, G., Kreil, G. 1977. Translation of melittin messenger RNA in vitro yields a product terminating with glutaminylglycine rather than with glutaminamide. *Proc. Natl. Acad. Sci. USA* 74:975–78
103. Suzuki, T., Natori, S. 1983. Identification of a protein having hemagglutinating activity in the hemolymph of the silkworm, *Bombyx mori*. *J. Biochem.* 93:583–90
104. Takahashi, H., Komano, H., Kawaguchi, N., Kitamura, N., Nakanishi, S.

Natori, S. 1985. Cloning and sequencing of cDNA of *Sarcophaga peregrina* humoral lectin induced on injury of the body wall. *J. Biol. Chem.* 260:12228–33

105. Takahashi, H., Komano, H., Natori, S. 1986. Expression of the lectin gene in *Sarcophaga peregrina* during normal development and under conditions where the defence mechanism is activated. *J. Insect Physiol.* 32:771–79
106. Teshima, T., Ueki, Y., Nakai, T., Shiba, T., Kikuchi, M. 1986. Structure determination of lepidopteran, self-defense substance produced by silkworm. *Tetrahedron* 42:829–34
107. Thomas, I. G., Ratcliffe, N. A. 1982. Integumental grafting and immunorecognition in insects. *Dev. Comp. Immunol.* 6:643–54
108. Trager, W. 1974. Some aspects of intracellular parasitism. *Science* 183:269–73
109. Tsukamoto, T., Ishiguro, M., Funatsu, M. 1986. Isolation of latent phenoloxidase from prepupae of the housefly, *Musca domestica. Insect Biochem.* 16:573–81
110. von Heijne, G. 1986. A new method for predicting signal sequence cleavage sites. *Nucleic Acids Res.* 14:4683–90
111. von Hofsten, P., Faye, I., Kockum, K., Lee, J.-Y., Xanthopoulos, K. G., et al. 1985. Molecular cloning, cDNA sequencing, and chemical synthesis of cecropin B from *Hyalophora cecropia. Proc. Natl. Acad. Sci. USA* 82:2240–43
112. Wigglesworth, V. B. 1964. *The Life of Insects*. New York/London: New Am. Libr./New Engl. Libr. 383 pp.
113. Yeaton, R. W. 1981. Invertebrate lectins: II. Diversity of specificity, biological synthesis and function in recognition. *Dev. Comp. Immunol.* 5:535–45
114. Yeaton, R. W. 1983. Wound responses in insects. *Am. Zool.* 23:195–203
115. Yoshida, H., Ashida, M. 1986. Microbial activation of two serine enzymes and prophenoloxidase in the plasma fraction of hemolymph of the silkworm, *Bombyx mori. Insect Biochem.* 16:539–45
116. Zachary, D., Hoffmann, D. 1984. Lysozyme is stored in the granules of certain haemocyte types in *Locusta. J. Insect Physiol.* 30:405–11

Ann. Rev. Microbiol. 1987. 41:127–51

COMPARTMENTATION OF CARBOHYDRATE METABOLISM IN TRYPANOSOMES[1]

Fred R. Opperdoes

Research Unit for Tropical Diseases, International Institute of Cellular and Molecular Pathology, Avenue Hippocrate 74, B-1200 Brussels, Belgium

---

CONTENTS

[1]Abbreviations used: HK, hexokinase; PGI, phosphoglucose isomerase; PFK, phosphofructokinase; ALD, aldolase; TIM, triosephosphate isomerase; GAPDH, glyceraldehyde-phosphate dehydrogenase; PGK, phosphoglycerate kinase; GK, glycerol kinase; GPDH, glycerol 3-phosphate dehydrogenase; MDH, malate dehydrogenase; PEPCK, phosphoenolpyruvate carboxykinase; DHAP, dihydroxyacetone phosphate; G-3-P, glycerol 3-phosphate; PGA, 3-phosphoglycerate; G-1,3-P_2, 1,3-diphosphoglycerate; PEP, phosphoenolpyruvate; G-6-P, glucose 6-phosphate; G-1,6-P_2, glucose 1,6-bisphosphate; F-1,6-P_2, fructose 1,6-bisphosphate; F-2,6-P_2, fructose 2,6-bisphosphate; P_i, inorganic phosphate; SHAM, salicylhydroxamic acid.

0066-4227/87/1001-0127$02.00

INTRODUCTION

The protozoan hemoflagellates belonging to the order Kinetoplastida, which comprise the genera *Trypanosoma* and *Leishmania,* are the causative agents of a number of important diseases of humans, such as sleeping sickness in tropical Africa, Chagas' disease in South America, and the different clinical manifestations of leishmaniasis in almost all tropical and subtropical parts of the world (53).

Over the last decade these parasitic protozoans have drawn much attention because they combine a number of unusual characteristics such as antigenic variation, kinetoplast DNA, glycosomes, and discontinuous transcription; the latter three are common to all representatives of the family Trypanosomatidae (63). Several of these characteristics have already been extensively reviewed in the literature (3, 7, 20, 21, 81), but only little attention has been given to the unique compartmentation of carbohydrate metabolism in these organisms.

This chapter summarizes the recent advances in our understanding of carbohydrate metabolism in trypanosomes and closely related species, with special reference to the glycosome. This organelle, which is unique to the Kinetoplastida, contains several enzymes of glycolysis and glycerol metabolism as well as enzymes involved in such diverse pathways as carbon dioxide fixation, pyrimidine biosynthesis, ether-lipid biosynthesis, and purine salvage. These latter pathways and their relation to the glycosome are also discussed. For earlier reviews on the glycosome and its role in carbohydrate metabolism in Trypanosomatidae the reader should consult References 24, 60, 61, and 71.

BIOLOGY AND LIFE CYCLE OF THE TRYPANOSOMATIDAE

The family Trypanosomatidae consists of a great variety of genera that infect birds, mammals *(Trypanosoma, Leishmania),* fish, amphibia, insects (*Crithidia* spp.), and even plants. Many have complex life cycles that may comprise both extracellular and intracellular forms. Transmission to the vertebrate host always takes place via bloodsucking vectors (flies, bugs, fleas, or leeches). The parasites of medical importance all belong to the two genera *Trypanosoma* and *Leishmania*. The remainder of this chapter is concerned mainly with the African trypanosome *Trypanosoma brucei;* where appropriate, other species of *Trypanosoma* as well as *Leishmania* and *Crithidia* are dealt with. For more information on the biology of these protozoans the reader is referred to References 44, 45, and 91.

The life cycle of *T. brucei* is one of the most complex cycles described for these hemoflagellates. In the bloodstream of the mammalian host this species shows a wide and continuous variation in form (pleomorphism) ranging from dividing long, slender trypomastigotes with a long free flagellum, via intermediate forms, to nondividing short, stumpy forms with no free flagellum (89–91). In the long, slender form the single mitochondrion is reduced to a peripheral canal with almost no cristae. Cytochromes are absent and the Krebs cycle is nonfunctional (9).

Transformation from the long, slender form to the stumpy form is marked by a swelling of the mitochondrial canal and the development of tubular cristae. At the same time the mitochondrion acquires some participation in cellular metabolism (9, 26).

When the trypanosome is ingested by the bloodfeeding tsetse fly the short, stumpy form develops into a midgut stage or procyclic trypomastigote. This form has a well-developed mitochondrion with an extensively branched network of prominent platelike cristae (89). Cytochromes are present, and respiration is sensitive to cyanide. This form mainly grows on amino acids rather than on glucose and excretes acetate and succinate rather than pyruvate (8). The midgut trypanosome then migrates to the salivary glands of the fly, where it transforms into a metacyclic form, which in turn is infectious to the vertebrate host. To date nothing is known about the metabolism of this latter stage of the life cycle.

GLYCOLYSIS

Owing to the absence of a functional mitochondrion, the bloodstream form of *T. brucei* is entirely dependent on glycolysis for its production of energy. Because this form lacks significant polysaccharide reserves or high-energy phosphate stores such as creatine phosphate or polyphosphates, it relies entirely on an exogenous source of carbohydrate (67). Glucose is the preferred energy source but fructose, mannose, and glycerol can also support motility and respiration (80).

Glucose metabolism in the bloodstream-form trypanosome differs from glycolysis in other eukaryotes in a number of respects: (*a*) Lactate dehydrogenase is absent, and therefore the reducing equivalents generated in glycolysis are indirectly reoxidized by molecular oxygen via a dihydroxyacetone phosphate (DHAP):glycerol 3-phosphate (G-3-P) shuttle plus a cyanide-insensitive mitochondrial oxidase:G-3-P oxidase. (*b*) Under aerobiosis, pyruvate is the sole end product of glycolysis and is excreted into the host's bloodstream. (*c*) Under anaerobic conditions glucose is quantitatively converted into equimolar amounts of pyruvate and glycerol, which are excreted (25). This glucose dismutation proceeds with net ATP synthesis (67), which can only be explained by the compartmentation of glycolysis in these cells (see below).

The glycolytic flux of carbon through the *T. brucei* cell is very high; 0.08 μmol glucose are consumed per minute per milligram protein, and for each molecule of pyruvate produced one molecule of ATP is synthesized (22, 67). After the trypanosome's transformation from bloodstream form to insect stage its consumption of glucose occurs at a much lower rate. This substrate is now converted to succinate, acetate, and CO_2, which are the main end products excreted. The oxidative metabolism of amino acids by the functional Krebs cycle and the respiratory chain serves as the major source of energy in this stage of the life cycle.

THE GLYCOSOME

Several early reports alluded to a particle-bound nature for glycolytic enzymes in the Trypanosomatidae (77–80). However, this particulate nature was not fully understood until 1977, when Opperdoes & Borst (65) described for the first time the association of nine enzymes with a microbodylike organelle in *T. brucei*. Seven of these enzymes are involved in glycolysis [i.e. hexokinase (HK), phosphoglucose isomerase (PGI), phosphofructokinase (PFK), aldolase (ALD), triosephosphate isomerase (TIM), glyceraldehyde-phosphate dehydrogenase (GAPDH), and phosphoglycerate kinase (PGK)], while two are involved in glycerol metabolism [i.e. glycerol kinase (GK) and glycerol 3-phosphate dehydrogenase (GPDH)]. Because of the high degree of specialization and the lack of typical microbody-marker enzymes associated with this organelle, the name "glycosome" was coined.

Glycosomes have since been found in other members of the Trypanosomatidae (12, 41, 85), but they have been studied best in *T. brucei* (Figure 1). Morphometric analysis (64) has shown that the organelles are homogeneous in size, with an average diameter of 0.27 μm. The particle has an electron-dense granular matrix and sometimes contains a crystalloid core; it is surrounded by a normal bilayer membrane 6–7 nm thick. On average a bloodstream-form trypanosome contains 240 glycosomes, which represent between 4.3% (64) and 8% (4) of its total volume and a similar percentage of its protein content (49, 64). In other genera glycosomes may be less abundant. Promastigote stages of *Leishmania tropica* contain only 50–100 glycosomes (D. T. Hart, personal communication), while in the amastigote stage of *Leishmania mexicana* not more than 10 glycosomes were found (16).

Highly purified organelles obtained by a combination of differential centrifugation and sequential gradient centrifugation in Percoll- and sucrose-containing media have been analyzed for their chemical composition (40, 64). Seventy percent of the glycosomal protein has been found in the matrix, whereas the remainder seems to be associated with the membrane (64). A limited number of polypeptides are associated with the bloodstream-form

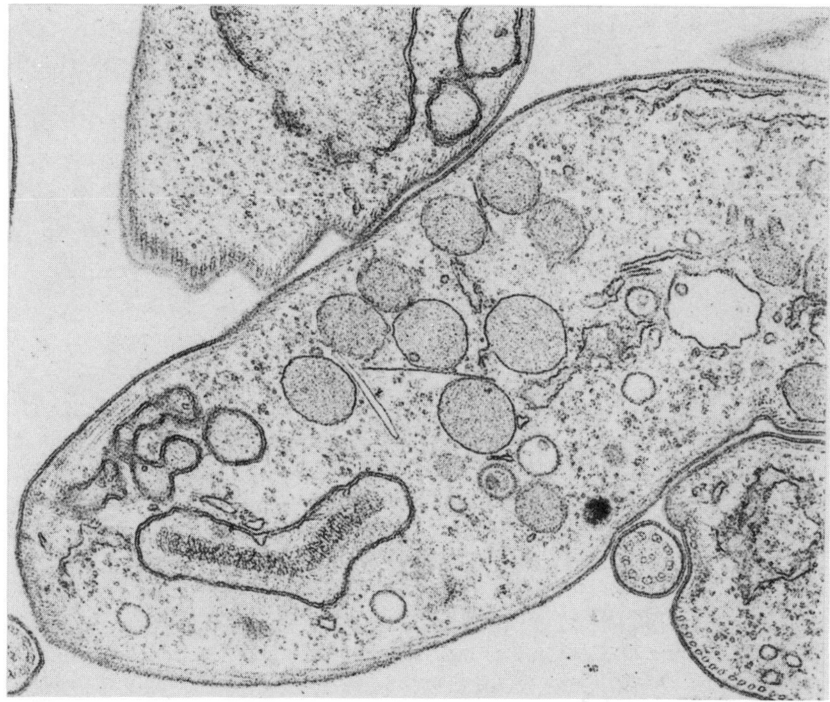

Figure 1 Electron micrograph of a cross section of *Trypanosoma brucei* bloodstream forms showing a cluster of glycosomes in the vicinity of the kinetoplast region of the mitochondrion containing DNA fibers. Photograph kindly provided by Ms. I. Coppens.

glycosome; the major polypeptides have all been identified as the subunits of the glycolytic enzymes, which together account for more than 87% of the glycosome's total protein (1, 49) (Figure 2). In the procyclic stage the relative abundance of the glycosomal polypeptides differs significantly, and a few additional polypeptide bands have been found (40). However, the total amount of polypeptides is still very limited.

The glycosomal membrane has a phospholipid content of 600 nmol (mg protein)$^{-1}$, which is essentially all the phospholipid present in the organelle. The phospholipid consists mainly of phosphatidyl choline and phosphatidyl ethanolamine; other phospholipids were not detected in highly purified organelle fractions from the bloodstream form (64), although similar preparations from procyclic insect stages contained in addition some phosphatidyl serine and phosphatidyl inositol (40).

Only trace amounts of DNA were detected in glycosomal fractions, and there has been no indication of a glycosome-specific class of DNA (64).

The organelles have an isotonic density of 1.09 g cm^{-3}. In sucrose

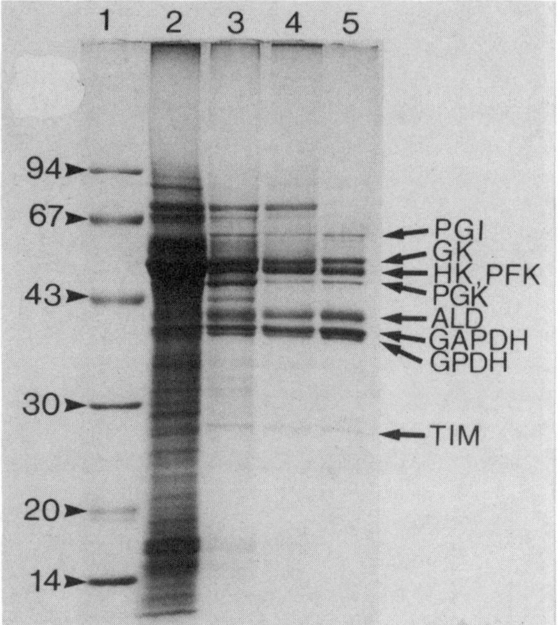

Figure 2 Polypeptide composition of the crude cell extract *(Lane 2)* and various fractions obtained during the purification of glycosomes. *Lane 5* shows 12-fold purified glycosomes. Molecular weights ($\times$ 10^3) of markers are indicated. Abbreviations: *PGI*, phosphoglucose isomerase; *GK*, glycerol kinase; *HK*, hexokinase; *PFK*, phosphofructokinase; *PGK*, phosphoglycerate kinase; *ALD*, aldolase; *GAPDH*, glyceraldehyde-P dehydrogenase; *GPDH*, glycerol 3-phosphate dehydrogenase; *TIM*, triosephosphate isomerase. Reprinted from Reference 49, with permission.

gradients they band at a density of 1.23 g cm^{-3}, which is similar to the banding density of the microbodies from other eukaryotic cells (64).

Unlike other microbodies, glycosomes in general contain neither hydrogen peroxide–producing oxidases nor catalase. Some catalase, however, was found by both cytochemical (56) and biochemical techniques (66) in the microbodies of *Crithidia* spp. An as-yet-unidentified peroxidase has been described in the organelles of *Trypanosoma cruzi* (19).

In addition to the early enzymes of the glycolytic sequence, phosphomannose isomerase (34; 0. Misset & F. R. Opperdoes, unpublished), the last two enzymes of the pyrimidine biosynthetic pathway (orotate phosphoribosyl transferase and orotidine carboxylase), the purine salvage enzyme hypoxanthine guanine phosphoribosyl transferase, and the enzymes adenylate kinase, malate dehydrogenase (MDH), and phosphoenolpyruvate carboxykinase (PEPCK) have also been demonstrated in glycosomes. The organelles are also involved in the synthesis of ether lipids; they contain the enzymes DHAP acyltransferase, acyl CoA reductase, and G-3-P:NADP$^+$ oxidoreductase (see below).

COMPARTMENTATION OF GLYCOLYSIS

When the glycolytic enzymes in glycosomes of *T. brucei* are isolated in isotonic sucrose buffers all exhibit latencies, which can be as high as 92%. From this observation Opperdoes & Borst (65) concluded that the glycosomal membrane acts as a permeability barrier which prevents the dilution of the glycolytic intermediates into the cytosol. What actually happens, however, is more complex. After the glycosomal membrane is disrupted by repeated cycles of freezing and thawing, some of the enzymes (especially HK and PFK) continue to exhibit a considerable degree of latency, which is only abolished by further addition of detergents and salt (92; F. R. Opperdoes, unpublished). Several investigators have shown that under conditions of low ionic strength all enzymes except PGI stick together and behave as members of a large multienzyme complex, even after complete removal of the surrounding membrane (58, 59, 72). Such complexes can be dissociated into their constituent components by increasing the ionic strength, while subsequent reassociation is obtained by slow removal of the salt through dialysis (59). These observations suggest that the latency observed for some of the glycolytic enzymes after the disruption of the glycosome is due to a limited diffusion of substrates and products through a multienzyme complex of glycolytic proteins.

The permeability properties of the glycosomal membrane are still not well understood. Arguments against a true permeability barrier are as follows. Firstly, no evidence has been presented for the existence of specific transport proteins in the glycosomal membrane. Secondly, isolated glycosomes are capable of catalyzing part of the glycolytic pathway using as substrate phosphorylated intermediates, which are not supposed to permeate the intact glycosomal membrane. Thirdly, such glycolytic activity is stimulated by the cofactors NAD and ATP, which also do not freely permeate through membranes. Fourthly, Hammond et al (34) detected some permeability to intermediates but also compartmentation of ATP. Finally, Patthey & Deshusses (76) have reported that the membranes of isolated glycosomes from insect-stage *T. brucei* are permeable to small solutes but not to larger molecules. Most of these observations can be explained, as already pointed out above, by the fact that latency of the glycosomal enzymes in organelle preparations, by itself, may not guarantee the intactness of the organellar membrane.

Strong support for true impermeability of the membrane comes from in vivo pulse-labeling experiments with [^{14}C]glucose. Visser et al (93) have shown that 20–30% of the glycolytic intermediates are in a separate compartment. These intermediates were completely labeled within 15 s; labeled pyruvate appeared in the external medium after a similar interval. This pool, which probably represents the glycosomal compartment, only slowly equilibrated with the remainder of the cell.

In view of the latter observations it is most likely that in the intact cell the glycosomal membrane constitutes a true permeability barrier to most of the glycolytic intermediates and cofactors. Any evidence to the contrary, when obtained from isolated organelles, should therefore be considered with caution, since artifacts resulting from the purification procedure can not at present be excluded.

Such extreme compartmentation gives the trypanosome several important advantages over other eukaryotic cells. In *T. brucei* the intracellular concentration of glycolytic intermediates is not dissimilar from that of other cells (92). Since glycosomes represent only 4% of the cellular volume and contain 20–30% of the glycolytic intermediates, the concentration of these intermediates in the glycosomal compartment must be at least five times higher than that in the cytosol (49, 50).

Misset and coworkers (49, 50) have estimated the concentration of both active sites and metabolites for three of the glycolytic steps in the glycosome. They concluded that these concentrations were similar and in the millimolar range. Since each of these enzymes has a K_m for its respective substrate of the same order of magnitude, a high proportion of the glycosomal metabolites must be in a protein-bound form. This has important implications for our understanding of glycosomal glycolysis. Kinetic data obtained on the purified enzymes may not simply be extrapolated to the situation inside the glycosome, since they apply to free rather than bound metabolite concentrations.

High intraglycosomal metabolite and active-site concentrations (49) allow the trypanosome to maintain an extreme glycolytic flux with a relatively small amount of protein. In this respect a tunneling of metabolites through the multienzyme complex has been suggested, but evidence for such tunneling has only been obtained for intact cells (93), not for isolated glycosomes (2).

The extreme glycolytic efficiency of the trypanosome is probably best illustrated by the fact that in muscle tissues, which have a high demand for glucose, GAPDH constitutes over 10% of the soluble proteins, while in trypanosomes this enzyme represents only 0.5% of the total (49). Nevertheless the glycolytic rate of the trypanosome is by far superior to that of other eukaryotic cell types.

The evidence available to date does not favor the existence of a multienzyme complex in the sense of a group of spatially well-ordered enzymes. Firstly, the cross-linking experiments carried out by Aman et al (1) do not allow us to conclude that such a multienzyme complex exists in situ. The observed cross-linking of all enzymes into one sedimentable fraction can simply be explained by the relatively high enzyme concentration and the close proximity of the enzymes in the glycosome (1, 49). Secondly, there exists no evidence for a tunneling of the metabolites through the multienzyme complex itself (1, 2). Thirdly, the crystalloid cores occasionally seen inside glycosomes have a lamellar structure with a regular spacing of 10 nm (64). This is in better agreement with the estimated Stokes' diameter of the two largest enzymes, HK and PFK (9 and 7.8 nm, respectively), than with that of a functional multienzyme complex, which has been estimated to be 27 nm (49). The fact that the latter two enzymes tend to form heavy aggregates especially under conditions of low ionic strength (72; O. Misset and F. R Opperdoes, unpublished) and the low frequency at which crystalloid cores are observed support the idea that the crystalloid core of the glycosomes could be an artifact.

Aerobic Glycolysis

The glycosome catalyzes the following two overall reactions (65):

$$\text{Glucose} + 2\ P_i + 2\ \text{DHAP} \rightarrow 2\ \text{PGA} + 2\ \text{G-3-P}$$
$$\text{Glycerol} + P_i + 2\ \text{DHAP} \rightarrow \text{PGA} + 2\ \text{G-3-P}.$$

Since the production and consumption of ATP and NAD are balanced, the glycosome itself is not involved in net ATP synthesis. This occurs in the cytosol, where phosphoglyceric acid (PGA) is converted to pyruvate. The G-3-P produced by the glycosome is reoxidized to DHAP by a mitochondrial terminal oxidase, *sn*-glycerol 3-phosphate oxidase; this process is not linked to oxidative phosphorylation (29, 66). In the above reactions substrates and products must cross the membrane at high rates. Since pulse-labeling experiments and latency studies on the glycosomal enzymes (see above) suggested that the glycosomal membrane is impermeable to the bulk of the phosphorylated intermediates, Opperdoes & Borst (65) proposed that specific translocaters for DHAP, G-3-P, PGA, and P_i facilitate their diffusion over the membrane. Direct evidence that the membrane is selectively permeable to these compounds but impermeable to other glycolytic intermediates is still lacking.

Anaerobic Glycolysis

Another important advantage of the compartmentation of glycolysis is the possibility of anaerobic glycolysis in the absence of a functional lactate dehydrogenase. Under anaerobic conditions, or when the glycerol 3-phosphate oxidase is inhibited by salicyl hydroxamic acid (SHAM), glucose is metabolized at the same rate as under aerobic conditions, forming equimolar amounts of pyruvate and glycerol (25, 67). Originally it was thought that glycerol was formed from G-3-P by the action of a phosphatase (9), but Borst and coworkers (5, 67) have pointed out that no net production of ATP would occur under such conditions; consequently the trypanosome would not be able to survive anaerobically, as it lacks energy reserves. In vitro experiments showed that ATP levels were maintained in the trypanosome even though respiration was blocked by SHAM (67). A thermodynamic calculation shows that the standard free-energy change of the overall dismutation of glucose into pyruvate and glycerol is sufficiently negative to allow the synthesis of one molecule of ATP:

$$\begin{array}{lr} \text{glucose} \rightarrow \text{pyruvate} + \text{glycerol} & -77\ \text{kJ mol}^{-1} \\ \text{ADP} + \text{P}_i \rightarrow \text{ATP} + \text{H}_2\text{O} & +31\ \text{kJ mol}^{-1} \\ \hline \text{glucose} + \text{ADP} + \text{P}_i \rightarrow \text{pyruvate} + \text{glycerol} + \text{ATP} + \text{H}_2\text{O} & -46\ \text{kJ mol}^{-1}. \end{array}$$

Several hypothetical schemes have been proposed to account for the observed overall reaction (14). The proposal that net ATP synthesis is achieved as a result of the reversal of the glycerol kinase–catalyzed reaction under anaerobic conditions (67) is still the most likely one. Such a reversal is facilitated by the drastic increase in intracellular G-3-P (35, 36, 92) and by the ability of the ATP/ADP ratio inside the glycosome to change independently from that of the cytosol. These two events, together with relatively high specific activity of glycerol kinase (36, 67), would then allow the synthesis of glycerol and ATP from G-3-P and ADP. Such a reaction would be facilitated even more by the presence of powerful naturally occurring ATP traps inside the glycosome, in the form of glucose plus HK and fructose 6-phosphate (F-6-P) plus PFK (34).

Several lines of evidence support the above hypothesis. Firstly, no novel phosphorylated intermediates or alternative enzymes required for the other schemes have been found. Secondly, a glycerol kinase of high specific activity is only found in trypanosomes capable of synthesizing glycerol under anaerobic conditions (36). Thirdly, a G-3-P:ADP transphosphorylase activity catalyzed by glycerol kinase has been found in *T. brucei* (34–36; F. R. Opperdoes, unpublished results). Fourthly, pulse-labeling experiments with [^{14}C]glycerol have shown that under anaerobic conditions glycerol kinase indeed functions under true equilibrium conditions, rather than catalyzing the phosphorylation of glycerol to G-3-P (35). Finally, Gruenberg et al (30) have shown that in *T. brucei* glycerol is removed by an asymmetric carrier-mediated process with a K_m for efflux of glycerol (20 μM) about one order of magnitude lower than for entry. This may help keep the intracellular glycerol concentration low and thus prevent the inhibition of the reversed glycerol kinase reaction by mass action.

OTHER ENZYMES AND PATHWAYS OF GLYCOSOMES

Glycolysis in the insect stage of *T. brucei* runs at a reduced rate, since mitochondrial activity has been derepressed in these cells. Consequently, other substrates such as amino acids and fatty acids are used for energy generation in this stage, while most of the glycolytic enzymes are largely reduced in activity (40, 69) (Figure 3).

Glucose consumption in the insect stage is accompanied by the fixation of CO_2. This results in the excretion of significant amounts of succinate under aerobic as well as anaerobic conditions (8). Important differences between the bloodstream and the procyclic forms are that the latter has (*a*) highly elevated PEPCK and MDH activities associated with the glycosome (68, 69), (*b*) most of the PGK in the cytosol and variable but reduced amounts in the glycosome (11, 69, 75), and (*c*) no active pyruvate kinase activity (10, 11, 69). These properties, together with the regulatory properties of the two CO_2-fixing enzymes PEPCK and malic enzyme, suggest that glycosomal CO_2 fixation by PEPCK serves the reoxidation of glycolytic NADH by the glycosomal malate dehydrogenase (MDH) rather than by some anaplerotic pathway (10, 68).

PEPCK has also been found in the glycosomes of *T. cruzi* epimastigotes (12), *Leishmania* promastigotes (41), and *Crithidia* spp. (12), but only in the first has it been found unequivocally together with the enzyme MDH (12).

Examination of the pathways in Figure 3 indicates that if no net change in ATP occurs in the glycosomal compartment during glucose metabolism, all the diphosphoglycerate (G-1,3-P_2) that leaves the glycosome must reenter it as phosphoenol pyruvate (PEP). To fulfill this requirement pyruvate kinase should be inactive (68) (see below). Similarly, if no net change in NAD^+ occurs, then all NADH produced in the GAPDH reaction should be available for the reduction of oxaloacetate to malate; otherwise PEP would accumulate. Another consequence of maintaining zero net internal balance of ATP and NAD^+ in the glycosome is that the glycerol 3-phosphate shuttle must be inoperative. Indeed, studies on isolated glycosomes from procyclic stages have shown that these organelles are capable of catalyzing the conversion of glucose to 1,3-diphosphoglycerate without production of G-3-P (10), provided that PEP is present. However, when PEP is replaced by ATP, G-3-P is produced. The in vivo functioning of this pathway in *Crithidia fasciculata* has recently been confirmed by [^{13}C]NMR experiments (13).

Changes in metabolism that occur from one stage of the life cycle to another do not affect the relative volume contribution of the glycosome (4,

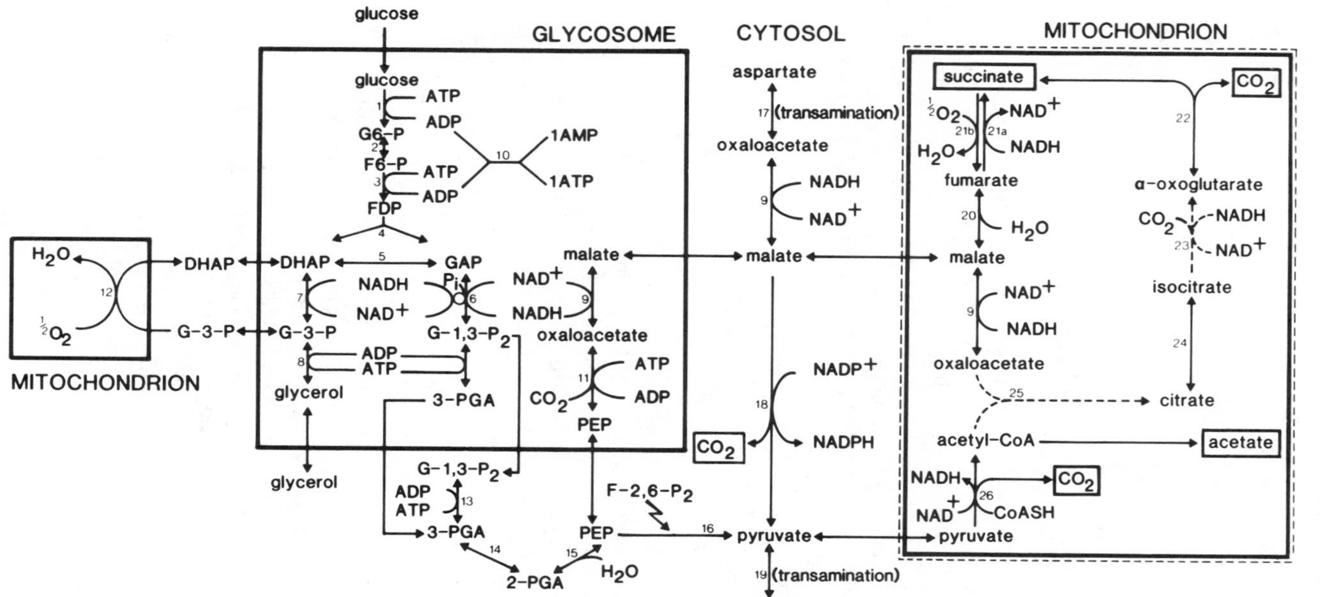

Figure 3 Pathways of glucose metabolism in procyclic trypomastigotes. The enzyme locations in the glycosome and cytosol have been established, but the location of the mitochondrial enzymes (except for glycerophosphate oxidase and malate dehydrogenase) have not yet been clearly demonstrated. End products of aerobic or anaerobic metabolism are enclosed in *boxes*. The *dashed lines* indicate enzymes whose presence remains uncertain. *1*, hexokinase; *2*, phosphoglucose isomerase; *3*, phosphofructokinase; *4*, aldolase; *5*, triosephosphate isomerase; *6*, glyceraldehyde-3-phosphate dehydrogenase; *7*, glycerol-3-phosphate dehydrogenase; *8*, glycerol kinase; *9*, malate dehydrogenase; *10*, adenylate kinase; *11*, phosphoenolpyruvate carboxykinase; *12*, glycerol-3-phosphate oxidase; *13*, phosphoglycerate kinase; *14*, phosphoglycerate mutase; *15*, enolase; *16*, pyruvate kinase; *17*, aspartate aminotransferase; *18*, malic enzyme; *19*, alanine aminotransferase; *20*, fumarate hydratase; *21a*, fumarate reductase; *21b*, succinate dehydrogenase; *22*, α-oxoglutarate decarboxylase; *23*, isocitrate dehydrogenase; *24*, aconitase; *25*, citrate synthetase; *26*, pyruvate dehydrogenase. (Modified from Reference 24.)

43) but do affect its appearance. In the bloodstream form glycosomes are round or ellipsoid and homogeneous in size (64), and they have a well-defined equilibrium density in sucrose gradients (65). In the insect stages glycosomes are more heterogeneous in both size and density and have a bacilliform appearance (87; M. Veenhuis & F. R. Opperdoes, unpublished). The relative protein content of glycosomes is similar in both stages of the life cycle (40).

Adenylate kinase has been found in glycosomes of *Trypanosoma rhodesiense* (46), *T. brucei* bloodstream and procyclic trypomastigotes (69), and *Leishmania* spp. promastigotes (41). Since glycosomes contain ATP-consuming as well as ATP-forming enzymes without themselves being involved in net ATP synthesis, the function of this glycosomal adenylate kinase is probably the regulation of the internal ATP level to prevent inhibition of glycolysis by a total lack of ATP. McLaughlin (47) has suggested that the glycosomal adenylate kinase is an integral membrane protein. The enzyme does not show latency, and it has an absolute dependence on phospholipids for activity (47).

Pyrimidine and Purine Metabolism

In the Trypanosomatidae, as in higher eukaryotes, the first three enzymes in the pyrimidine biosynthetic pathway are cytoplasmic enzymes. In the mammalian host the fourth enzyme is dihydro-orotate dehydrogenase, a membrane-bound enzyme associated with the mitochondrial respiratory chain. In all members of the family Trypanosomatidae studied, this enzyme is a cytosolic dihydro-orotate oxidase that utilizes molecular oxygen (37, 74, 84). In all other cells studied to date the last two enzymes of this pathway, orotate phosphoribosyl transferase and orotidine-5'-phosphate decarboxylase, are cell sap enzymes. In contrast, in *T. cruzi, Leishmania* spp., *Crithidia* spp., and *T. brucei* both enzymes are tightly associated with glycosomes (37–39). Unlike the glycolytic enzymes in the glycosome, orotate phosphoribosyl transferase does not show latency (38), either because it is situated on the outer surface of the glycosome or because the membrane does not constitute a permeability barrier to phosphoribosyl pyrophosphate owing to the presence of a translocater specific for this substrate (38).

Only one enzyme involved in the purine salvage pathway, the hypoxanthine-guanine phosphoribosyltransferase, has been found in the glycosomes of *T. cruzi* (32), *L. mexicana* (42), and *T. brucei* (34). In other eukaryotic cells this enzyme is soluble. The other phosphoribosyl transferases were all found in the soluble portion of the cell.

Ether-Lipid Biosynthesis

Considerable amounts of glycerol phospholipids containing ether linkages have been found in several representatives of the Trypanosomatidae (cf 62), and evidence is available that these organisms are capable of de novo synthesis of ether lipids.

DHAP acyltransferase and alkyl/acyl DHAP reductase, two key enzymes of ether-lipid biosynthesis (33), were recently found in the microbodies of *T. brucei* procyclics (62) and *Leishmania* promastigotes (41). Furthermore, the glycosomes of *T. brucei* also contain an acyl CoA synthetase (D. T. Hart & F. R. Opperdoes, unpublished) and acyl CoA reductase (62), two more enzymes of the ether-lipid pathway; these enzymes convert long-chain fatty acids into their corresponding alcohols, which may serve as substrates for ether-lipid synthesis. Purified glycosomes are capable of converting DHAP and acyl CoA to acyl DHAP and lysophosphatidic acid, but not to phosphatidic acid; G-3-P, on the other hand, could not be used as a substrate (62). This suggests that glycosomes are not involved in the synthesis of normal phospholipids. These observations, together with the presence of glycerol kinase, GPDH, TIM, and acyl CoA synthetase in glycosomes, suggest that the entire machinery for synthesis of ether lipids from glycerol and fatty acids is associated with the microbodies of trypanosomatids.

REGULATION OF GLYCOLYSIS

The individual activities of the glycolytic enzymes are all in excess of the overall glycolytic rate. An inspection of the relative maximal activities (49, 58) suggests that the aldolase and phosphoglycerate mutase reactions might be among the rate-limiting steps in glycolysis. Hexokinase and PFK, two enzymes that fulfill an important regulatory role in most eukaryotic cell types, do not have such function in *T. brucei*. Hexokinase is insensitive to glucose 6-phosphate (G-6-P) and glucose 1,6-bisphosphate (G-1,6-P_2) (57), while PFK, although an allosteric enzyme, is not regulated by compounds such as fructose 1,6-bisphosphate (F-1,6-P_2), fructose 2,6-bisphosphate (F-2,6-P_2), and citrate (17, 57, 86). Opperdoes et al (67) inferred that glycolysis is unregulated from the fact that a transition from aerobic to anaerobic conditions did not change the glycolytic rate in *T. brucei* bloodstream forms. In addition, cross-over analysis of the levels of glycolytic intermediates at such a transition revealed the absence of a cross-over point between G-6-P and PEP. Instead the level of all metabolites decreased upon anaerobiosis (92), which suggests that the rate-limiting step is located before G-6-P. This is supported by Gruenberg et al's suggestion (31) that the transport of glucose over the plasma membrane might be the overall rate-limiting step in glycolysis.

Nevertheless, some recent evidence suggests that the glycolytic pathway may be regulated at the level of pyruvate kinase. Fructose 2,6-bisphosphate, which is known as the most potent regulator of PFK in eukaryotes but which is totally inactive on the trypanosomal enzyme, was detected in extracts of *T. brucei;* it turned out to be a potent activator of trypanosomatid pyruvate kinase activity (88). To date this F-2,6-P_2 regulation of pyruvate kinase seems to be unique to the Trypanosomatidae, since it has also been found in *Leishmania* spp. and *Crithidia* spp. but not in *Euglena,* yeast, or mammalian liver (88). F-2,6-P_2 is present in the bloodstream form but could not be found in the insect stages (E. Van Schaftingen & F. R. Opperdoes, unpublished). This probably explains why in these stages pyruvate kinase is inactive even though the enzyme could be detected (51, 68). Whether the compound has an important role in the regulation of the flux of carbon from PEP, either through the pyruvate kinase step toward pyruvate or back into the glycosome toward the formation of malate, remains to be seen (see Figure 3).

PROPERTIES OF THE GLYCOSOMAL ENZYMES

Several reports on the purification of glycosomal enzymes have appeared (49, 50, and references therein), but the tendency of the glycolytic enzymes to stick together has severely hampered the purification of the individual enzymes. Misset and coworkers (49, 50) succeeded in the simultaneous purification of nine glycosomal enzymes using detergent-treated glycosomes and hydrophobic interaction chromatography under conditions of high ionic strength. Cronin & Tipton (17) described the purification of PFK from *T. brucei* by a different method. The purification of MDH and PEPCK from *Leishmania* has also been reported (54).

The properties of the glycolytic enzymes are summarized in Table 1. Hexokinase has a subunit size of 50 kd, similar to that of yeast but half the size of the mammalian enzymes (49, 50). Trypanosomal HK is hexameric. This is unique, since only monomeric hexokinases have been described for animals, insects, and plants and since yeast hexokinase is dimeric. Trypanosomal phosphofructokinase, although a tetramer like all other PFKs studied to date, has a subunit molecular weight of 50 kd (17, 49); this is quite different from that of the enzyme from other eukaryotes (75–120 kd) and from bacteria (35 kd). With respect to its regulatory properties PFK resembles the bacterial enzymes rather than the mammalian ones in that it is activated by neither fructose 1,6-bisphosphate nor fructose 2,6-bisphosphate, nor is it inhibited by either citrate or phosphate.

The other seven enzymes shown in Table 1 resemble their homologs from other organisms in that they have similar or slightly larger subunit molecular weights and similar subunit compositions (49).

BIOGENESIS OF GLYCOSOMES

Gene Organization

No evidence has been found for the presence of DNA in glycosomes (64). This renders it highly unlikely that glycosomes contain a separate protein-synthesizing machinery, as is the case for mitochondria or chloroplasts.

Table 1 A comparison of subunit and native molecular weights of the glycolytic enzymes from *Trypanosoma brucei* glycosomes (49) and mammalian cytosol

Enzyme	Native M_r ($\times 10^3$)	Subunit M_r ($\times 10^3$)		Subunit composition	
	T. brucei	*T. brucei*	Mammalian	*T. brucei*	Mammalian
Hexokinase	295 ± 12 (4)[a]	50.3 ± 0.8 (10)	97–110	6	1
Glucose-P isomerase	105 ± 6 (5)	62.4 ± 1.6 (8)	59–61	2	2
Phosphofructokinase	196 ± 15 (3)	50.3 ± 0.8 (10)	79–84	4	4
Aldolase	157 ± 3 (3)	40.5 ± 0.5 (9)	39–40	4	4
Triose-P isomerase	55 ± 0 (3)	27.0 ± 0.5 (8)	26–28	2	2
Glyceraldehyde-P dehydrogenase	139 (1)	38.5 ± 0.5 (10)	36–37	4	4
Phosphoglycerate kinase	48 ± 4 (2)	47.0 ± 0.7 (11)	43–46	1	1
Glycerol-3-P dehydrogenase	66 ± 6 (3)	37.2 ± 0.7 (9)	33–35	2	2
Glycerol kinase	82 (1)	52.5 ± 0.4 (8)	?	2	?

[a] Number of determinations indicated in parentheses.

Glycolytic enzymes have been highly conserved throughout evolution. This has allowed the identification of the genes coding for several of these enzymes in *T. brucei* using heterologous probes originating from yeast and mammals. The genes for aldolase (P. A. M. Michels, unpublished), GAPDH (48), TIM (83), and PGK (73) have all been identified, and each of these genes has been located on a nuclear chromosome (28; P. A. M. Michels, personal communication).

Two tandemly linked, completely identical genes have been found for both the glycosomal GAPDH (gGAPDH) (48) and aldolase (P. A. M. Michels, unpublished). For TIM only one gene is present (83). In the case of PGK three tandemly linked but not identical genes have been described; two of these code for the cytosolic (cPGK) and glycosomal (gPGK) isoenzymes, respectively (73).

Site of Synthesis

The sequence of events involved in the synthesis of the glycolytic enzymes and their subsequent integration into the glycosome has been approached using pulse-chase experiments with [^{35}S]methionine and live procyclic trypomastigotes of *T. brucei* (39a, 70). The glycosomal enzymes gGAPDH, aldolase, and GPDH appeared first in the soluble fraction of the trypanosome (the cytosol) and only a few minutes later in the glycosome-enriched fraction. The half-life of these polypeptides in the cytosol varied from 1.0 to 3.4 min, which indicates that their transfer to the glycosome is a very efficient process. Similarly, the half-life of the glycosomal polypeptides once inside the glycosome was very short; values varying from 30 min to 1 hr have been found, depending on the method used (39a, 70). The fact that three different polypeptides have similar glycosomal half-lives may indicate that the organelle itself turns over at a high rate. A turnover time of less than 2 hr is in complete agreement with the calculated rate of synthesis of the glycosomal polypeptides, which is sufficient for the replacement of 1% of the total enzyme pool per minute.

The mRNA for glycolytic enzymes was predominantly found on free polysomes rather than on membrane-bound polysomes (39a). Each of the products when synthesized in vitro had the same apparent molecular weight as that of the corresponding native enzyme isolated from the glycosome. Similarly, the polypeptides in the cytosol did not change in size upon their transfer to the glycosome (39a).

The size of the native polypeptides is in complete agreement with that predicted from the nucleotide sequence of the respective genes. Protein sequencing of the N-terminal part of GAPDH (48) and three-dimensional structure determination of TIM (94) have revealed that the N-terminus of each of these proteins is not subjected to any form of processing. No evidence has been found for covalent modification of native proteins in the form of glycosylation or phosphorylation (39a, 94).

These observations together indicate that the glycosomal polypeptides are synthesized in the cytosol and posttranslationally transferred to the glycosomes by a process that involves neither the removal of a cleavable leader sequence nor secondary modification.

Properties of the Glycolytic Polypeptides

The recent purification of nine glycosomal enzymes (49, 50) and two of their cytosolic counterparts (cGAPDH and cPGK) (51, 52) has revealed a number of remarkable differences between the glycosomal enzymes and their soluble homologs.

Measurements on SDS polyacrylamide gels have shown that glycosomal GAPDH is 5 kd larger than its cytosolic counterpart (cGAPDH) and at least 2 kd larger than any other GAPDH studied thus far (52). This larger size has been confirmed by sequence data (48). The gGAPDH contains several unique insertions and a short C-terminal extension not found in ten other GAPDHs whose sequences are known. It has an isoelectric point of 9.3, while that of the cytosolic isoenzyme is 7.9 (52). One of the unique insertions, located in the N-terminal part of gGAPDH, contains three positively charged residues; this insertion is absent from the N-terminal part of its cytosolic counterpart, for which the sequence of the first 85 residues was established by protein sequencing (52). The absence of this positively charged insertion probably explains the difference in pI between these two isoenzymes.

Glycosomal PGK has a subunit molecular mass of 47 kd, while that of its cytosolic counterpart is 45 kd. Moreover, the two isoenzymes differ considerably in pI, with values of 9.3 and 6.3 for the glycosomal and cytosolic isoenzymes, respectively (51). However, apart from a moderately hydrophobic 20–amino acid C-terminal extension in the glycosomal isoenzyme, the two sequences are 93% homologous at the amino acid level (73). The difference in size on SDS gels can be accounted for by the presence of the C-terminal extension, while the striking difference in charge is entirely due to only 12 of 30 amino acid substitutions. The 7% difference in sequence and/or the C-terminal extension on the glycosomal isoenzyme should, therefore, be responsible for the different topogenesis of these two enzymes (6, 73, 94).

Several other glycolytic enzymes from *T. brucei* also have subunit molecular weights significantly larger than those of their homologs from other organisms (49). The *T. brucei* PGI is 5 kd larger than that from yeast and rabbit muscle; the *T. brucei* GPDH is 2 kd larger than that isolated from rabbit muscle, and *T. brucei* aldolase is 2 kd larger than that of rabbit muscle (15). However, TIM does not differ in size from its counterparts from other organisms (49, 83).

When these glycosomal enzymes are subjected to isoelectric focusing, all except PGI have a pI in the range of 8.8–10.2. These values are 1–4 units higher than those of mammalian glycolytic enzymes and 3–6 units higher than those of the glycolytic enzymes of other unicellular organisms (49).

The complete amino acid sequences of aldolase (15; P. A. M. Michels, unpublished) and TIM (83) show that neither of the two enzymes contains a C-terminal extension. This suggests that such an extension is not the common denominator responsible for import into glycosomes. Both enzymes, however, like the others discussed above, have one unique insertion carrying one or more positively charged residues that are not found in any other counterpart whose sequence is available (94).

The cumulative data on biosynthesis, sequence, and properties of the glycosomal enzymes suggest that glycosomal enzymes, unlike most mitochondrial, chloroplast, and secretory proteins, do not require a cleavable leader sequence for their import into the organelle. However, they must contain an internal topogenic signal instead. Such signals may have arisen from individual mutations, or may have integrated as blocks into the primary sequence and may therefore be located in the insertions unique to these polypeptides (see 6).

Wierenga et al (94) have recently modeled the amino acid sequences of three glycosomal enzymes, TIM, PGK, and GAPDH, into the respective three-dimensional structures of homologous enzymes from other organisms using the known structural coordinates. They have presented evidence that all of these enzymes have in common two "hot spots" about 40 Å apart, of which one is always associated with a unique insertion. These hot spots include a pair of positively charged side chains pointing into the solution, separated by 7 Å. The sequence of the glycosomal aldolase (15), for which no homologous three-dimensional information is available, is compatible with the presence of the same conformation on its surface. Based on these striking similarities, Wierenga et al (94) have proposed that this feature has an important role in the import of enzymes into glycosomes.

The hot spot hypothesis (94) does not, however, take into account all the unique properties of the glycosomal enzymes and therefore may be an oversimplification. One should also keep in mind that some of the unique properties of the enzymes may contribute to the internal organization of the particles. Some of the elements that are only present in glycosomal enzymes may serve as areas of interaction between the constituent enzymes of a multienzyme complex and thus may significantly increase its stability.

ARE GLYCOSOMES RELATED TO OTHER MICROBODIES?

The microbody class of eukaryotic organelles comprises the peroxisomes found in most plant and animal cells as well as the glyoxysomes, which are typical for plant cells. Both types of organelles are generally supposed to be involved in peroxide metabolism; i.e. they contain hydrogen peroxide–producing oxidases together with catalase. For some trypanosomatid species, typical peroxisomal enzymes such as D-amino acid oxidase, α-hydroxyacid oxidase (*Crithidia* spp.) (56), and glyoxylate-cycle enzymes (*Leishmania* spp.) (55) have been reported. However, none of these enzymes have ever been found in association with the microbody fraction of these organisms.

Because glycosomes are extremely specialized in carrying out glycolysis and also contain enzymes involved in pyrimidine biosynthesis, de Duve (18) raised the question of whether peroxisomes and glycosomes are related. There are now strong indications that glycosomes share several important enzymes and pathways with the peroxisomes of higher eukaryotes. Catalase has been demonstrated in two representatives of the Trypanosomatidae, i.e. *Crithidia* spp. (56, 66) and *Leptomonas samueli* (82). In addition, several enzymes involved in fatty acid metabolism and the biosynthesis of ether lipids and regarded to be typical of the peroxisomes of various organisms have been found in glycosomes. Glycosomal proteins, like peroxisomes and glyoxysomes, seem to be synthesized on free ribosomes and then transported post-translationally into the organelle without removal of a cleavable leader sequence. These characteristics, together with the morphological resemblance of the two types of organelles, suggest that glycosomes and peroxisomes are indeed related.

CONCLUSIONS AND OUTLOOK

Trypanosomes possess many unusual features in carbohydrate metabolism which go together with an extreme degree of compartmentation of this pathway. Seven of the glycolytic enzymes and two enzymes of glycerol metabolism are localized inside glycosomes. Since the discovery of these organelles in *Trypanosoma brucei,* they have been found in all major representatives of the trypanosomatid family. Glycosomes also contain adenylate kinase, two enzymes of de novo pyrimidine biosynthesis, one enzyme involved in purine salvage, and, depending on the metabolic function of the organelle, malate dehydrogenase and phosphoenolpyruvate carboxykinase. In addition, they have an important role in the biosynthesis of ether lipids.

The bloodstream forms of many African trypanosomes in particular rely entirely on glycolysis for energy production (60, 61). The inhibition of this pathway by a combination of drugs (25) leads to rapid death of the trypanosome and its complete elimination from the bloodstream of the infected organism (14). Two of the major drugs presently used in the treatment of African sleeping sickness, melarsoprol (27) and suramin (22), are thought to kill this organism by inhibiting glycolysis. This illustrates that the glycolytic pathway of the trypanosome is an excellent target for the development of new and effective trypanocidal drugs (61). Individual glycolytic enzymes differ from their homologs from other organisms in that at least some of them contain unique insertions and additional positive charges on their surface, separated by about 40 Å, which may function as the topogenic signal for import into the glycosome.

Irrespective of their topogenic functions, common features specific to glycosomal enzymes may be of major interest for the design of drugs, not only against the agent of sleeping sickness, but also against other pathogenic Trypanosomatidae, such as *T. cruzi* and *Leishmania* spp., which all possess glycosomes. The trypanocide suramin is a symmetrical molecule with two clusters of negative charges separated by a distance of about 40 Å. Several of the glycosomal enzymes from *T. brucei* have a much higher affinity for this drug than their homologs from other organisms (23, 51). Therefore, it is possible that this drug, by its high affinity for glycosomal enzymes, interferes with the import of the polypeptides into the glycosome and thus with the biogenesis of an entire organelle.

ACKNOWLEDGMENTS

I would like to thank Ms. Françoise Mylle for excellent secretarial assistance. The work by the author presented in this chapter received financial support from the UNDP/World Bank/WHO Special Programme for Research and Training in Tropical Diseases and from the Science and Technology for Development Programme of the Commission of the European Communities.

Literature Cited

1. Aman, R. A., Kenyon, G. L., Wang, C. C. 1985. Cross-linking of the enzymes in the glycosomes of *Trypanosoma brucei*. *J. Biol. Chem.* 260:6966–73
2. Aman, R. A., Wang, C. C. 1986. Absence of substrate channeling in the glycosome of *Trypanosoma brucei*. *Mol. Biochem. Parasitol.* 19:1–10
3. Benne, R. 1985. Mitochondrial genes in trypanosomes. *Trends Genet.* 1:117–21
4. Böhringer, S., Hecker, H. 1975. Quantitative ultrastructural investigations of the life cycle of *Trypanosoma brucei:* a morphometric analysis. *J. Protozool.* 22:463–67
5. Borst, P. 1977. Metabolism and chemotherapy of African trypanosomes. *Trans. R. Soc. Trop. Med. Hyg.* 71:3–4
6. Borst, P. 1986. How proteins get into microbodies (peroxisomes, glyoxysomes, glycosomes). *Biochim. Biophys. Acta* 866:179–203
7. Borst, P. 1986. Discontinuous transcription and antigenic variation in trypanosomes. *Ann. Rev. Biochem.* 55:701–32
8. Bowman, I. B. R. 1974. Intermediary metabolism of pathogenic flagellates. *Ciba Found. Symp.* 20:255–71
9. Bowman, I. B. R., Flynn, I. W. 1976. Oxidative metabolism of trypanosomes. In *Biology of the Kinetoplastida*, ed. W. H. R. Lumsden, D. A. Evans, 1:435–76. New York: Academic
10. Broman, K., Knupfer, A.-L., Ropars, M., Deshusses, J. 1983. Occurrence and role of phosphoenolpyruvate carboxykinase in procyclic *Trypanosoma brucei* glycosomes. *Mol. Biochem. Parasitol.* 8:79–87
11. Broman, K., Ropars, M., Deshusses, J. 1982. Subcellular location of glycolytic enzymes in *Trypanosoma brucei* culture form. *Experientia* 38:533–34
12. Cannata, J. J. B., Valle, E., Docampo, R., Cazullo, J. J. 1982. Subcellular localization of phosphoenol pyruvate carboxykinase in the trypanosomatids *Trypanosoma cruzi* and *Crithidia fasciculata*. *Mol. Biochem. Parasitol.* 6:151–60
13. Cazzulo, J. J., De Cazzulo, B. M. F., Engel, J. C., Cannata, J. J. B. 1985.

End products and enzyme levels of aerobic glucose fermentation in Trypanosomatids. *Mol. Biochem. Parasitol.* 16:329–43
14. Clarkson, A. B., Brohn, F. H. 1976. Trypanosomiasis: an approach to chemotherapy by the inhibition of carbohydrate catabolism. *Science* 194:204–6
15. Clayton, C. E. 1985. Structure and regulated expression of genes encoding fructose bisphosphate aldolase in *Trypanosoma brucei. EMBO J.* 4:2997–3003
16. Coombs, G. H., Tetley, L., Moss, V. A., Vickerman, K. 1986. Three-dimensional structure of the *Leishmania* amastigote as revealed by computer-aided reconstruction from serial sections. *Parasitology* 92:13–23
17. Cronin, C. N., Tipton, K. F. 1985. Purification and regulatory properties of phosphofructokinase from *Trypanosoma (Trypanozoon) brucei brucei. Biochem. J.* 227:113–24
18. de Duve, C. 1982. Peroxisomes and related particles in historical perspective. *Ann. NY Acad. Sci.* 386:1–4
19. DoCampo, R., Deboiso, J. F., Boveris, A., Stoppani, A. O. M. 1976. Localization of peroxidase activity in *Trypanosoma cruzi* microbodies. *Experientia* 32:972–75
20. Englund, P. T. 1981. Kinetoplast DNA. In *Biochemistry and Physiology of Protozoa,* ed. M. Levandowsky, S. H. Hutner, pp. 333–83. New York: Academic
21. Englund, P. T., Hajduk, S. L., Marini, J. C. 1982. The molecular biology of trypanosomes. *Ann. Rev. Biochem.* 51:695–726
22. Fairlamb, A. H., Bowman, I. B. R. 1980. Uptake of the trypanocidal drug suramin by bloodstream forms of *Trypanosoma brucei* and its effects on respiration and growth rate in vivo. *Mol. Biochem. Parasitol.* 1:315–33
23. Fairlamb, A. H., Oduro, K. K., Bowman, I. B. R. 1979. Action of the trypanocidal drug suramin on the enzymes of aerobic glycolysis of *Trypanosoma brucei in vivo. FEBS Spec. Meet. Enzymes, Dubrovnik,* Abstr. S4–10
24. Fairlamb, A. H., Opperdoes, F. R. 1986. Carbohydrate metabolism in African trypanosomes with special reference to the glycosome. In *Carbohydrate Metabolism in Cultured Cells,* ed. M. J. Morgan, pp. 183–224. New York/London: Plenum
25. Fairlamb, A. H., Opperdoes, F. R., Borst, P. 1977. New approach to screening drugs for activity against African trypanosomes. *Nature* 265:270–71
26. Flynn, I. W., Bowman, I. B. R. 1973. The metabolism of carbohydrate by pleomorphic African trypanosomes. *Comp. Biochem. Physiol.* 45B:25–42
27. Flynn, I. W., Bowman, I. B. R. 1974. The action of trypanocidal arsenical drugs on *Trypanosoma brucei* and *Trypanosoma rhodesiense. Comp. Biochem. Physiol.* 48B:261–73
28. Gibson, W. C., Osinga, K. A., Michels, P. A. M., Borst, P. 1985. Trypanosomes of subgenus *Trypanozoon* are diploid for housekeeping genes. *Mol. Biochem. Parasitol.* 16:231–42
29. Grant, P. T., Sargent, J. R. 1960. Properties of L-α-glycerophosphate oxidase and its role in the respiration of *Trypanosoma rhodesiense. Biochem. J.* 76:229–37
30. Gruenberg, J., Schwendimann, B., Sharma, P. R., Deshusses, J. 1980. Role of glycerol permeation in the bloodstream form of *Trypanosoma brucei. J. Protozool.* 27:484–91
31. Gruenberg, J., Sharma, P. J., Deshusses, J. 1978. D-Glucose transport in *Trypanosoma brucei.* D-Glucose transport is the rate limiting step of its metabolism. *Eur. J. Biochem.* 89:461–69
32. Gutteridge, W. E., Davies, M. J. 1982. Properties of the purine ribosyltransferases of *Trypanosoma cruzi. FEMS Microbiol. Lett.* 13:207–12
33. Hajra, A. K., Bishop, J. E. 1982. Glycerolipid biosynthesis in peroxisomes via the acyl dihydroxyacetone phosphate pathway. *Ann. NY Acad. Sci.* 386:170–82
34. Hammond, D. J., Aman, R. A., Wang, C. C. 1985. The role of compartmentation and glycerol kinase in the synthesis of ATP within the glycosome of *Trypanosoma brucei. J. Biol. Chem.* 260:5646–54
35. Hammond, D. J., Bowman, I. B. R. 1980. *Trypanosoma brucei:* the effect of glycerol on the anaerobic metabolism of glucose. *Mol. Biochem. Parasitol.* 2:63–75
36. Hammond, D. J., Bowman, I. B. R. 1980. Studies on glycerol kinase and its role in ATP synthesis in *Trypanosoma brucei. Mol. Biochem. Parasitol.* 2:77–91
37. Hammond, D. J., Gutteridge, W. E. 1980. Enzymes of pyrimidine biosynthesis in *Trypanosoma cruzi. FEBS Lett.* 118:259–62
38. Hammond, D. J., Gutteridge, W. E. 1982. UMP synthesis in the Kinetoplastida. *Biochim. Biophys. Acta* 718: 1–10

39. Hammond, D. J., Gutteridge, W. E., Opperdoes, F. R. 1981. A novel location for two enzymes of de novo pyrimidine biosynthesis in trypanosomes and *Leishmania*. *FEBS Lett*. 128:27–29
39a. Hart, D. T., Baudhuin, P., Opperdoes, F. R., de Duve, C. 1987. Biogenesis of the glycosome in *Trypanosoma brucei:* The synthesis, translocation and turnover of glycosomal polypeptides. *EMBO J*. In press
40. Hart, D. T., Misset, O., Edwards, S., Opperdoes, F. R. 1984. A comparison of the glycosomes (microbodies) isolated from *Trypanosoma brucei* bloodstream form and cultured procyclic trypomastigotes. *Mol. Biochem. Parasitol.* 12:25–35
41. Hart, D. T., Opperdoes, F. R. 1984. The occurrence of glycosomes (microbodies) in the promastigote stage of four *Leishmania* species. *Mol. Biochem. Parasitol.* 13:159–72
42. Hassan, H. F., Mottram, J. C., Coombs, G. H. 1985. Subcellular localization of purine metabolizing enzymes in *Leishmania mexicana mexicana*. *Comp. Biochem. Physiol.* 81B:1037–40
43. Hecker, H. 1980. Application of morphometry to pathogenic trypanosomes (protozoa, mastigophora). *Pathol. Res. Pract.* 166:203–17
44. Hoare, C. A., ed. 1972. *The Trypanosomes of Mammals. Zoological Monograph*. Oxford: Blackwell
45. McGhee, M. B., Cosgrove, W. B. 1980. Biology and physiology of the lower Trypanosomatidae. *Microbiol. Rev.* 44:140–73
46. McLaughlin, J. 1981. Association of adenylate kinase with the glycosome of *Trypanosoma rhodesiense*. *Biochem. Int.* 2:345–53
47. McLaughlin, J. 1985. The presence of alpha-glycerophosphate dehydrogenase (NAD^+-linked) and adenylate kinase as core and integral membrane enzymes respectively in the glycosomes of *Trypanosoma rhodesiense*. *Mol. Biochem. Parasitol.* 14:219–30
48. Michels, P. A. M., Poliszczak, A., Osinga, K. A., Misset, O., Van Beeumen, J., et al. 1986. Two tandemly linked identical genes code for the glycosomal glyceraldehyde-phosphate dehydrogenase in *Trypanosoma brucei*. *EMBO J*. 5:1049–56
49. Misset, O., Bos, O. J. M., Opperdoes, F. R. 1986. Glycolytic enzymes of *Trypanosoma brucei*. Simultaneous purification, intraglycosomal concentrations and physical properties. *Eur. J. Biochem*. 157:441–53
50. Misset, O., Opperdoes, F. R. 1984. Simultaneous purification of hexokinase, class-I fructose-bisphosphate aldolase, triosephosphate isomerase and phosphoglycerate kinase from *Trypanosoma brucei*. *Eur. J. Biochem*. 144:475–83
51. Misset, O., Opperdoes, F. R. 1987. The phosphoglycerate kinases from *Trypanosoma brucei:* a comparison of the glycosomal and the cytosolic isoenzyme and their sensitivity towards suramin. *Eur. J. Biochem*. 162:493–500
52. Misset, O., Van Beeumen, J., Lambeir, A.-M., Van der Meer, R., Opperdoes, F. R. 1987. Glyceraldehyde-phosphate dehydrogenase from *Trypanosoma brucei:* comparison of the glycosomal and cytosolic isoenzymes. *Eur. J. Biochem*. 162:501–7
53. Molyneux, D. H., Ashford, R. W. 1983. *The Biology of* Trypanosoma *and* Leishmania, *Parasites of Man and Domestic Animals*. London: Taylor & Francis
54. Mottram, J. C., Coombs, G. H. 1985. Purification of particulate malate-dehydrogenase and phosphoenol pyruvate carboxykinase from *Leishmania mexicana mexicana*. *Biochim. Biophys. Acta* 827:310–19
55. Mukkada, A. J. 1977. Tricarboxylic acid and glyoxylate cycle enzymes in the *Leishmania*. *Acta Trop*. 34:167–75
56. Muse, K. E., Roberts, J. F. 1973. Microbodies in *Crithidia fasciculata*. *Protoplasma* 78:343–48
57. Nwagwu, M., Opperdoes, F. R. 1981. Regulation of glycolysis in *Trypanosoma brucei:* hexokinase and phosphofructokinase activity. *Acta Trop*. 39:61–72
58. Oduro, K. K., Bowman, I. B. R., Flynn, I. W. 1980. *Trypanosoma brucei:* activities and subcellular distribution of glycolytic enzymes from differently disrupted cells. *Exp. Parasitol.* 50:123–35
59. Oduro, K. K., Bowman, I. B. R., Flynn, I. W. 1980. *Trypanosoma brucei:* preparation and some properties of a multienzyme complex catalysing part of the glycolytic pathway. *Exp. Parasitol*. 50:240–50
60. Opperdoes, F. R. 1983. In *Parasitology, A Global Perspective,* ed. K. S. Warren, J. Z. Bowers, pp. 191–202. New York: Springer-Verlag
61. Opperdoes, F. R. 1983. Glycolysis as target for the development of new trypanocidal drugs. In *Mechanism of Drug*

Action, ed. T. P. Singer, T. E. Mansour, R. N. Undarza, P. 121–32. New York: Academic

62. Opperdoes, F. R. 1984. Localization of the initial steps in alkoxyphospholipid biosynthesis in glycosomes (microbodies) of *Trypanosoma brucei. FEBS Lett.* 169:35–39
63. Opperdoes, F. R. 1985. Biochemical peculiarities of trypanosomes, African and South-American. *Br. Med. Bull.* 41:130–36
64. Opperdoes, F. R., Baudhuin, P., Coppens, I., De Roe, C., Edwards, S. W., et al. 1984. Purification, morphometric analysis and characterization of the glycosomes (microbodies) of the protozoan hemoflagellate. *J. Cell Biol.* 98:1178–84
65. Opperdoes, F. R., Borst, P. 1977. Localization of nine glycolytic enzymes in a microbody-like organelle in *Trypanosoma brucei:* the glycosome. *FEBS Lett.* 80:360–64
66. Opperdoes, F. R., Borst, P., Bakker, S., Leene, W. 1977. Localization of glycerol-3-phosphate oxidase in the mitochondrion and NAD^+-linked glycerol-3-phosphate dehydrogenase in the microbodies of the bloodstream form of *Trypanosoma brucei. Eur. J. Biochem.* 76:29–39
67. Opperdoes, F. R., Borst, P., Fonk, K. 1976. The potential use of inhibitors of glycerol-3-phosphate oxidase for chemotherapy of African trypanosomiasis. *FEBS Lett.* 62:169–72
68. Opperdoes, F. R., Cottem, D. 1982. Involvement of the glycosome of *Trypanosoma brucei* in carbon dioxide fixation. *FEBS Lett.* 143:60–64
69. Opperdoes, F. R., Markos, A., Steiger, R. F. 1981. Localisation of malate dehydrogenase, adenylate kinase and glycolytic enzymes in glycosomes and the threonine pathway in the mitochondrion of cultured trypomastigotes of *Trypanosoma brucei. Mol. Biochem. Parasitol.* 4:291–309
70. Opperdoes, F. R., Michels, P. A. M., Misset, O., Hart, D. T., Van Beeumen, J. 1987. Biogenesis, properties and sequence of glyceraldehyde-phosphate dehydrogenase from *Trypanosoma brucei. UCLA Symp. Mol. Cell. Biol.* (NS) 42:543–52
71. Opperdoes, F. R., Misset, O., Hart, D. T. 1983. Metabolic pathways associated with the glycosomes (microbodies) of the Trypanosomatidae. *John Jacob Abel Symp. Mol. Parasitol.,* pp. 63–75. Baltimore, Md: Academic
72. Opperdoes, F. R., Nwagwu, M. 1980. Suborganellar localization of glycolytic enzymes in the glycosome of *Trypanosoma brucei.* In *The Host-Invader Interplay,* ed. H. Van den Bossche, pp. 683–86. Amsterdam: Elsevier/North Holland Biomed.
73. Osinga, K. A., Swinkels, B. W., Gibson, W. C., Borst, P., Veeneman, G. H., et al. 1985. Topogenesis of microbody enzymes: a sequence comparison of the genes for the glycosomal (microbody) and cytosolic phosphoglycerate kinases of *Trypanosoma brucei. EMBO J.* 4:3811–17
74. Pascal, R. A., Letrang, N., Cerami, A., Walsh, C. 1983. Purification and properties of dihydroorotate oxidase from *Crithidia fasciculata* and *Trypanosoma brucei. Biochemistry* 22:171–78
75. Patthey, J. P., Deshusses, J. 1985. Multiple forms of 3-phosphoglycerate kinase in *Trypanosoma brucei brucei. Experientia* 41:547
76. Patthey, J. P., Deshusses, J. 1986. Accessibility of *Trypanosoma brucei* glycosomal enzymes to labeling agents of various sizes. *Experientia* 42:646
77. Reynolds, C. H. 1975. The NAD-linked α-glycerophosphate dehydrogenase of trypanosomes. *Biochem. Biophys. Res. Commun.* 67:538–43
78. Risby, E. L., Seed, J. R. 1969. Purification and properties of purified hexokinase from the African trypanosomes and *Trypanosoma equiperdum. J. Protozool.* 16:193–97
79. Risby, E. L., Seed, T. M., Seed, J. R. 1969. *Trypanosoma gambiense, T. rhodesiense, T. brucei, T. equiperdum* and *T. lewisi:* purification and properties of phosphohexose isomerase. *Exp. Parasitol.* 25:101–6
80. Ryley, J. F. 1962. Studies on the metabolism of the protozoa. 9. Comparative metabolism of the bloodstream and culture forms of *T. rhodesiense. Biochem. J.* 85:211–23
81. Simpson, L. 1986. Kinetoplast DNA in trypanosomatid flagellates. *Int. Rev. Cytol.* 99:119–79
82. Souto-Padron, T., de Souza, W. 1982. Fine structure and cytochemistry of peroxisomes (microbodies) in *Leptomonas samueli. Cell Tissue Res.* 222:153–58
83. Swinkels, B. W., Gibson, W. C., Osinga, K. A., Kramer, R., Veeneman, G. H., et al. 1986. Characterization of the gene for the microbody (glycosomal) triosephosphate isomerase of *Trypanosoma brucei. EMBO J.* 5:1291–98
84. Tampitag, S., O'Sullivan, W. J. 1986.

Enzymes of pyrimidine biosynthesis in *Crithidia luciliae*. *Mol. Biochem. Parasitol.* 19:125–34

85. Taylor, M. B., Berghausen H., Heyworth, P., Messenger, N., Rees, L. J., Gutteridge, W. E. 1979. Subcellular localisation of some glycolytic enzymes in parasitic and flagellated protozoa. *J. Biochem.* 11:117–20
86. Taylor, M., Gutteridge, W. E. 1986. The regulation of phosphofructokinase in epimastigote *Trypanosoma cruzi*. *FEBS Lett.* 201:262–66
87. Tetley, L., Vickerman, K. 1985. Differentiation in *Trypanosoma brucei:* host-parasite cell junctions and their persistence during acquisition of the variable antigen coat. *J. Cell Sci.* 74:1–19
88. Van Schaftingen, E., Opperdoes, F. R., Hers, H. G. 1985. Stimulation of *Trypanosoma brucei* pyruvate kinase by fructose 2,6-bisphosphate. *Eur. J. Biochem.* 153:403–6
89. Vickerman, K. 1965. Polymorphism and mitochondrial activity in sleeping sickness trypanosomes. *Nature* 208: 762–66
90. Vickerman, K. 1971. Morphological and physiological considerations of extracellular blood protozoa. In *Ecology and Physiology of Parasites*, ed. A. M. Fallis, pp. 58–91. Toronto: Univ. Toronto Press
91. Vickerman, K., Preston, T. M. 1976. Comparative cell biology of the kinetoplastid flagellates. See Ref. 9, pp. 35–130
92. Visser, N., Opperdoes, F. R. 1980. Glycolysis in *Trypanosoma brucei*. *Eur. J. Biochem.* 103:623–32
93. Visser, N., Opperdoes, F. R., Borst, P. 1981. Subcellular compartmentation of glycolytic intermediates in *Trypanosoma brucei*. *Eur. J. Biochem.* 118:521–26
94. Wierenga, R. K., Swinkels, B., Michels, P. A. M., Osinga, K., Misset, O., et al. 1987. Common elements on the surface of the glycolytic enzymes from *Trypanosoma brucei* may serve as topogenic signals for import into glycosomes. *EMBO J.* 6:215–21

Ann. Rev. Microbiol. 1987. 41:153–80

THE ATTENUATION OF POLIOVIRUS NEUROVIRULENCE

J. W. Almond

Department of Microbiology, University of Reading, London Road, Reading RG1 5AQ, England

---

CONTENTS

INTRODUCTION

The poliovirus is the causative agent of poliomyelitis, an acute disease of the central nervous system of humans that may result in paralysis. References to atrophied limbs and palsies in early records supported by archaeological excavation suggest that poliomyelitis has plagued mankind since antiquity (71). It was only toward the end of the nineteenth century, however, that the clinical and epidemiological features of the disease were studied in detail. Around this time the epidemiological pattern of poliomyelitis changed from a primarily endemic form to the dramatically epidemic form that, in the first half of the twentieth century, made this one of the most feared of infectious diseases (51). Nowadays in industrialized countries the general public may be forgiven for regarding poliomyelitis as a disease of the past. Widespread vaccination with either of the two very good vaccines available (80, 82) has so

0066-4227/87/1001-0153$02.00

effectively controlled the disease for almost a quarter of a century that infectious acute spinal paralysis is now extremely rare. Poliomyelitis has not been totally eradicated, however, and in countries using the live attenuated vaccines a low level of disease persists (9). Paradoxically, much of this seems to be caused by vaccination (8). Moreover, regular importation of wild poliovirus strains from countries where the virus is still endemic makes continued vaccination essential (10). In the endemic areas themselves, poliomyelitis remains a significant public health problem. Lameness surveys have indicated a high incidence in many Third World countries, with attack rates comparable to those in the United States and Western Europe during the peak years before vaccines were available (11). The World Health Organization (WHO) estimates that worldwide there are at least 400,000 cases of the disease per annum, and the actual figure may be substantially higher (51).

Against this background of poliomyelitis, work on the causative agent, the poliovirus, has proceeded apace over the past five or six years. During this period, mainly through the application of recombinant DNA and modern immunological and biophysical techniques, the poliovirus has become probably the best studied of viruses that affect humans. This review is concerned with recent work aimed at determining the genetic and molecular basis of poliovirus neurovirulence. Much of this work has centered on detailed molecular analyses of Sabin's live attenuated vaccines and comparison with their neurovirulent progenitor strains.

Poliomyelitis: The Disease

Much of what is presently accepted knowledge concerning the pathogenesis of poliovirus has come from work carried out during the two decades up to the mid 1950s by workers such as Bodian (12), Paul (71), and Sabin (77). Before that time, mainly because of the predominance of ideas propagated by workers such as Flexner, it was believed that poliovirus was strictly neurotropic and that it entered the body through the nasal route and, via nerve endings, proceeded directly to the central nervous system. This dogma held until the late 1930s, even though as early as 1912 Kling had reported the isolation of polioviruses not only from the throat but also from the feces of both paralytic and nonparalytic patients (reviewed in 51).

It is currently believed that poliovirus is transmitted via the fecal oral route. Like other enteroviruses it is acid stable and can survive transit through the stomach. The virus is capable of replication in the oropharynx, however, and it is therefore possible that it may occasionally be spread by droplets. Whether replication in the throat normally precedes replication in the alimentary tract after natural contact remains unclear. After establishing itself in the gastrointestinal tract, possibly in gut epithelioid cells, the virus multiplies in mesenteric lymph nodes and Peyer's patches (Figure 1; reviewed in 14). This results in dissemination of virus to systemic lymph nodes and other sites of virus replication, which may include brown fat (13). Release of the virus from these sites probably accounts for the observed viremia, during which the major targets involved in clinical illness, namely cells of the nervous system, become infected. There is evidence that the virus gains access to the nervous system both through penetration of the reticular endothelial system and by spreading through peripheral nerve fibers (14). Cellular damage from virus infection ranges from mild chromatolysis to neuronophagia and complete destruction. Cells that are not killed but have edema and lose their function may recover completely. There is also an inflammatory response in the spinal cord, and it has been suggested that the virus multiplies productively in invading lymphocytes (11a). The cells of the anterior horn of the spinal cord are those most affected with gross destruction of motor neurons. In severe cases intermediate grey ganglia and even posterior horn lesions can be found. That the virus spreads to the brain is shown by the presence of lesions in the hypothalamus, thalamus, reticular formation, vermis cerebelli, and deep cerebellar nuclei. The cortex is virtually spared, with the exception of the motor cortex along the precentral gyrus (12, 13).

The period from infection to the onset of clinical paralytic poliomyelitis is usually around 11 days, although it is important to note that this is the outcome of only about 1% of infections. In the vast majority of individuals the infection is associated with only a minor illness, which may take the form of

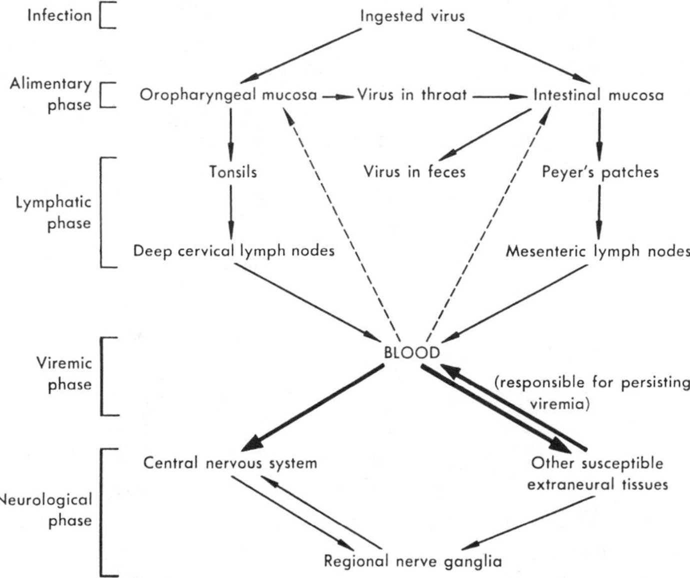

Figure 1 Pathogenesis of poliomyelitis. The model is based on a synthesis of data obtained from humans and chimpanzees. Adapted in Reference 21a from References 13 and 80.

fever, malaise, drowsiness, headache, nausea, vomiting, and sore throat. Like many other enteroviruses, the poliovirus may occasionally be associated with an aseptic meningitis from which there is usually complete recovery unless there is progression to paralysis. Paralytic poliomyelitis, when it occurs, follows some days after the minor illness and is usually of the form of a flaccid paralysis that most frequently involves the lower limbs. Occasionally the bulbar form of paralysis is observed, which involves involuntary muscles of the pharynx and larynx (52).

Poliomyelitis: The Virus

STRUCTURE On the basis of serum neutralization tests, polioviruses can be divided into three serotypes. The recently adopted nomenclature identifies the type of virus (P1, P2, or P3), the isolate number, its country (or city) of origin, and its year of isolation, e.g. P3/2672/France/1980 (97). Polioviruses are members of the Enterovirus genus of the family Picornaviridae, a large group of human pathogens associated with a fascinating array of disease syndromes. The poliovirus particle is composed of 60 copies of each of four virus proteins, VP1–VP4, which form an icosahedral shell of ~30-nm diameter surrounding a single-stranded positive-sense RNA genome of approximately 7450 nucleotides (29). The RNA has a 22–amino acid protein (VPg) covalently linked to its 5' terminus and a poly A tract at its 3' terminus (reviewed in 41). The protein structure of the virus has recently been determined to near-atomic resolution by X-ray crystallography (29). Three of the proteins, VP1, VP2, and VP3 (33.5, 30, and 26.5 kd, respectively) have a similar three-dimensional structure: a core made up of eight antiparallel strands of β-pleated sheet forming a "β barrel"–type structure, plus two short α-helical regions lying at right angles to each other. The major differences among these three proteins are in the size and conformations of their termini and of the loops joining the domains that contribute to the common core structures. The similarity of VP1, VP2, and VP3 is such that they can be considered quasi-equivalent in their contribution to the $T = 3$ virion architecture. VP4, a somewhat smaller protein of 7.4 kd, occupies a position on the interior of the shell that is analogous to the N-terminal domains of VP1 and VP3. Nucleotide sequence analysis of numerous antigenic variants of all three serotypes of polio, selected on the basis of their resistance to neutralizing monoclonal antibodies (56, 60), together with studies on synthetic peptides (95) and proteins expressed in *Escherichia coli* (93) have revealed the major antigenic domains on the virus particles. As might be expected, these are located on the high points of a virion's surface and are primarily composed of random-coil structures. It has been suggested that a cleft or canyon 24 Å deep and 12–30 Å wide, which separates a peak on the fivefold axes of the virion formed by VP1 molecules from a plateau formed by VP2 and VP3, may be the region of the virus that binds to the cellular receptor (74). Since the ability to recognize a specific cellular receptor is believed to be crucial to the poliovirus's tissue tropism and thereby to its pathogenesis (19), it is likely that the shape and properties of this canyon constitute the essential feature that distinguishes a poliovirus from other enteroviruses.

REPLICATION The replication cycle of the poliovirus begins with the binding of the virus particle to a specific cellular receptor. The virus then enters the cell, probably through receptor-mediated endocytosis. Following uncoating, a step that is still poorly understood, the RNA genome is liberated into the cytoplasm. The genome acts directly as a messenger RNA, coding for a polyprotein of ~250 kd. The polyprotein is cleaved in a precise manner by at least two viral proteases (27, 92) to give rise to structural and nonstructural proteins. Comparison of amino- and carboxy-terminal sequences of individual proteins with the polyprotein sequence as deduced from the complete nucleotide sequence of the genome has led to a comprehensive genetic map for poliovirus type 1 as illustrated in Figure 2 (70 and references therein; reviewed in 41). Although the corresponding data have not been obtained for poliovirus types 2 and 3, there is sufficient homology in the polyprotein to allow a fairly precise alignment and the identification of likely cleavage signals for these two serotypes also (83, 91). Regions of the RNA genome that are believed not to code for protein include a relatively long stretch of 742 bases at the 5' terminus and 72 bases and a poly A tract at the 3' terminus (39). The functions of these noncoding regions are unknown but are likely to include, for example, the control of translation, replication, and packaging of the RNA. Replication of the RNA genome proceeds via the synthesis of a negative-stranded template that directs the synthesis of new positive strands. The complexities of these events are still controversial (18, 88), but it is known that these syntheses require host as well as viral proteins (21). New positive strands may serve either as messenger RNAs or as new virion RNAs. New virus particles are assembled in the cytoplasm of infected cells, from which they are eventually liberated following lysis and destruction of the cell (reviewed in 75).

Few of the features of replication as determined in vitro can be correlated directly with pathogenic processes in vivo. However, the initial event of receptor recognition seems to be of paramount importance. This is the major determinant of tissue tropism in these viruses and therefore determines the sites of virus-induced tissue damage (20), although other factors influence the fine tuning of tissue tropism (see below). In fact receptor specificity groups the viruses of the whole Picornaviridae family according to their original classification based on disease and histopathology in humans and animals (19). In the late 1950s and early 1960s Holland et al (31, 49) established that in picornaviruses receptors on the surface of cells are a prerequisite for

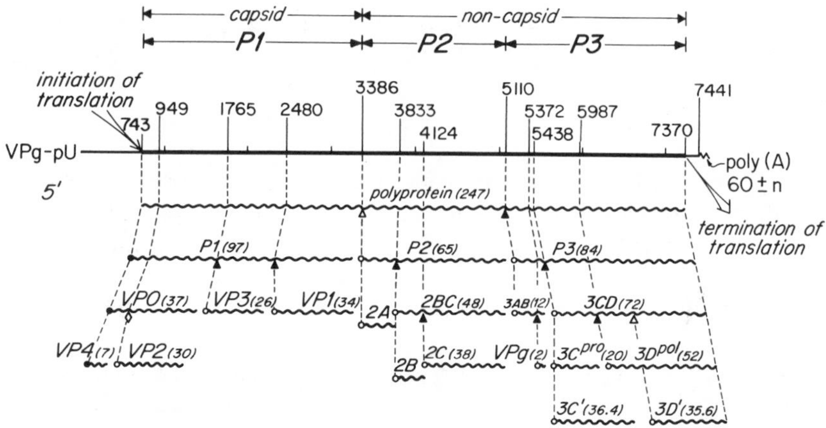

Figure 2 Gene organization of poliovirus RNA. Virion RNA, terminated at the 5' end with the genome-linked protein VPg and at the 3' end with poly(A), is shown as a *solid line*, with the translated region more pronounced than the noncoding regions. The numbers above the virion RNA refer to the first nucleotide of the codon specifying the N-terminal amino acid for the virus-specific proteins. The coding region has been divided into three regions *(P1, P2, P3)*, corresponding to rapid cleavages of the polyprotein. The newly adopted nomenclature of polypeptides is used according to Reference 75a. *Numbers in parentheses* are calculated molecular mass in kd. *Open circles* indicate that the terminal amino acid has been experimentally determined. The N-terminus is glycine in all regions except for VP2, where it is serine. The C-terminal amino acid of protein 3D is phenylalanine. *Filled circles* indicate that the N-termini are known to be blocked. *Filled triangles* correspond to Gln-Gly pairs that are cleaved during proteolytic processing of a polypeptide by the virus-coded proteinase 3C (27). *Open triangles* correspond to Tyr-Gly pairs cleaved by viral proteinase 2A (92). The *open diamond* corresponds to an Asn-Ser pair cleaved only during morphogenesis. Polypeptides 3C' and 3D' are products of an alternative cleavage whose biological significance is unknown. [Reproduced with permission from Nomoto & Wimmer (66a)].

infection by intact virus but that naked RNA has a broader host range (31, 49). Holland also observed a good correlation between the tissues that possessed receptors and the known sites of virus replication (e.g. brain, spinal cord, intestine). Receptors could not be detected on cells of the heart, spleen, liver, and lung (30). There is currently great interest in the poliovirus receptor. Several groups have isolated monoclonal antibodies that are able to block the entry of all three serotypes of poliovirus into susceptible cells; these are thought to be directed against the receptor (58, 63). The antibodies are presently being used in attempts to clone the gene encoding the receptor (53). Precise definition of the tissue locations of the receptor and its full biological characterization are eagerly awaited and will be crucial to our understanding of poliovirus pathogenesis.

ATTENUATION OF POLIOVIRUSES

Historical

Attenuation is defined as the process of weakening the characteristic force of an entity (4). For polioviruses we are concerned with the weakening of neurovirulence. The idea that infectious agents could be purposefully attenuated for vaccines was established toward the end of the last century by Pasteur during his pioneering work on diseases such as rabies and anthrax. The first attenuation of poliovirus was probably that described by Theiler (89) in 1941. He found that extensive passage of the virus in mouse brain resulted in a variant that, unlike the original isolate, was unable to cause paralysis in rhesus monkeys after intracerebral inoculation. This work and similar work on other viruses during this period raised the prospect of developing a live attenuated vaccine against poliomyelitis. Another major landmark in poliovirology came in 1949 when Enders and colleagues demonstrated that polioviruses could be cultivated in cultures of human cells of nonneural origin (24). Multiple passage of the P1/Brunhilde strain of poliovirus in cells in culture gave rise to a virus of markedly lower neurovirulence for monkeys (25). This observation illustrated the great potential of using multiple passage in vitro and paved the way for the isolation of attenuated strains of all three serotypes of poliovirus.

At about the same time, studies on the effects of poliovirus in whole animals were also proceeding rapidly. In addition to establishing the route of infection and means of spread within the body, comparison of the effects of poliovirus in different animal species established that neurovirulence depends not only on the strain of virus but also on the type of host and the route of inoculation. It became clear, for example, that neurovirulence in mice and monkeys may vary independently and that primates show a range of motor neuron susceptibility that seems to be inversely related to their susceptibility to infection via the alimentary tract (77). Quantitative assays of neurovirulence in monkeys were developed following the realization that neurovirulence was not an all or none phenomenon. Because virus titers could now be measured in culture (24), the neurovirulence of a given strain of virus could for the first time be related to the administered dose measured in terms of $TCID_{50}$.

The study of poliovirus biology, pathogenesis, and immunity reached its zenith in this period with the development of two vaccines: first the formalin-inactivated vaccine of Salk (82) and subsequently the live attenuated oral vaccine developed by Sabin (80). The live attenuated strains have been extensively compared with their wild-type progenitors in neurovirulence studies.

The WHO Neurovirulence Test

Although the ideal quantitation of the neurovirulence of a given strain of poliovirus is the determination of its LD_{50} in a primate host, the large number of monkeys required and the expense involved have caused such tests to be generally discontinued. An alternative quantitation of neurovirulence is afforded by inoculation of a small group of standardized animals with a uniformly sized inoculum, followed by clinical observation and histological examination of the extent of damage and pathological spread within the CNS (96). In poliovirus research the route of inoculation, administered dose, and species of monkey used for testing may vary from laboratory to laboratory, so comparison of results is difficult. There is greater uniformity, however, in the tests used by manufacturers for assessing vaccine safety. For this the WHO recommends the sensitive method of intraspinal inoculation of cynomolgus monkeys with an inoculum of $TCID_{50}$ between $10^{5.5}$ and $10^{6.5}$ in 0.1 ml (98). The results are compared with those of an accepted reference strain. A standard scoring system is used to evaluate virus activity in hemisections of the spinal cord and brain. The values assigned are: 1 = cellular infiltration only, 2 = cellular infiltration with minimal neuronal damage, 3 = cellular infiltration with extensive neuronal damage, and 4 = massive neuronal damage with or without cellular infiltration. A mean lesion score is calculated from hemisection readings of the lumbar cord, cervical cord, and brain. Although acceptance of the vaccine depends on favorable comparison with the concurrently tested reference, the level of neurovirulence of any virus preparation can be gauged from its mean lesion score together with observations on clinical paralysis. Broadly speaking, a score of <1.0 is typical of an attenuated virus, 1.0–2.0 is intermediate, and >2.0 indicates neurovirulence (98).

The Sabin Vaccine Strains

DERIVATION OF THE SABIN VACCINE STRAINS The live attenuated strains developed by Sabin gained widespread acceptance as vaccines following large-scale trials in the early 1960s, and they are now in regular use throughout much of the world. All three strains were derived from wild-type isolates by protracted passage in monkey tissue in vitro and in vivo (81). The conditions of passage were important. Low multiplicity of infection followed by incubation for long enough to allow multiple rounds of replication had no effect on neurovirulence. High multiplicity of infection and rapid passage, on the other hand, gave rise to variant viruses with reduced neurovirulence (76). These conditions presumably allowed the preferential expansion of subpopulations of rare variants that were better adapted than the original stock for growth in tissue culture. For reasons that are still not clear in molecular terms, these variants were found to have reduced neurovirulence for primates. Table 1 summarizes the passages involved in the isolation of each of the three vaccine strains.

The type 1 vaccine strain P1/LS-c,2ab was derived from a neurovirulent strain, P1/Mahoney/41, isolated from the feces of a healthy child. After the

Table 1 The derivation of Sabin's attenuated vaccine strains (81)

Isolate	Derivation
Type 1	
P1/Mahoney/41 (fecal isolate, healthy children)	14 passages in vivo (monkeys) 2 passages in vitro (monkey testicle)
P1/Mahoney/Monk 14 T2	24 passages in vitro (monkey testicle) 18 passages in vitro (monkey kidney) 10 alternate passages: 5 in vivo (intradermally in monkeys) 5 in vitro (monkey kidney)
P1/Mahoney/LS-c	5 passages in vitro (monkey kidney) 3 plaque purifications (monkey kidney) 2 passages in vitro (preparative) (monkey kidney)
	P1/LS-c,2ab/KP$_2$/56 Sabin vaccine strain
Type 2	
P2/P712/56 (fecal isolate, healthy children)	4 passages in vitro (monkey kidney) 3 plaque purifications (monkey kidney) 1 passage in vivo (orally in chimpanzees) 3 plaque purifications (monkey kidney)
	P2/P712,Ch,2ab/KP$_2$/56 Sabin vaccine strain
Type 3	
P3/Leon/37 (isolate from fatal paralytic case)	21 passages in vivo (intracerebrally in monkeys) 8 passages in vitro (monkey testicle)
Sabin (1954)	39 passages in vitro (monkey kidney) 3 plaque purifications (monkey kidney) 3 passages in vitro (preparative) (monkey kidney)
	P3/Leon 12a$_1$b KP$_3$/56 Sabin vaccine strain

extensive passages in monkey tissue shown, P1/LS-c,2ab was plaque-purified from the passaged pool and selected for vaccine use on the basis of its low neurovirulence in cynomolgus monkeys.

The type 2 vaccine strain P2/P712,Ch,2ab was derived from the wild-type P2/P712/56, which was also a fecal isolate from a healthy child. In this case, however, the wild type was a naturally occurring strain of low neurovirulence. Relatively few passages were required to produce a strain with a sufficiently attenuated phenotype.

The type 3 vaccine strain P3/Leon 12a$_1$b was derived from the strain P3/Leon/37, which, unlike the other two progenitor wild-type strains, was isolated from the brain and spinal cord of a victim of a fatal case of paralytic poliomyelitis. P3/Leon 12a$_1$b was selected after extensive passage in culture on the basis of its having the lowest neurovirulence of several candidates isolated by plaque purification.

The final stages of the derivation of all three vaccine strains (81) incorporated three plaque purifications, which were presumably intended to improve the genetic homogeneity of the virus population. The preparations of virus derived from these gave rise to the "Sabin original" (SO) stocks now stored at WHO Headquarters in Geneva, from which vaccine manufacturers' SO+1 to SO+3 seed stocks are derived. For the type 3 virus many manufacturers now use derivative stocks prepared by plaque purification of virus recovered from infectious RNA. This seed is at the SO+5 passage level and is designated SOR to differentiate it from the SO type 3 seed (98). Although the SOR viruses are very similar in neurovirulence to the Sabin original stocks, there is evidence of minor biological differences; for example, they are more temperature sensitive than the original SO viruses (P. D. Minor, personal communication).

The biological and physical properties of the Sabin vaccine strains have been studied in great detail in attempts to identify in vitro correlates of neurovirulence. All three vaccine strains are temperature sensitive for replication in culture; replication capacity at 40°C is much lower than that at 35°C (79). This character, measured by the rct (replicative capacity at supraoptimal temperature) marker test, correlates well with attenuation and has been suggested as a determinant of attenuation per se (47). Other markers that correlate to some extent with attenuation are the d marker (for delayed growth under agar containing low concentrations of bicarbonate) and plaque size (35).

REVERSION TO NEUROVIRULENCE Over the past 20 years or so a wealth of information has been accumulated, mainly under the auspices of the WHO, on the transmissibility, efficacy, and safety of the Sabin vaccines (8). Although they are generally regarded as very safe and highly effective, the vaccines are not perfect. For example, there are occasional problems with take rates in tropical countries for reasons that are not clear (50). Because of the genetic instability of stocks each new batch of vaccine must be rigorously tested using the expensive monkey neurovirulence test (15, 61). Perhaps the most serious problem, however, is the fact that the vaccines are associated with paralytic disease, albeit at a low level, in vaccinees (17, 54, 55, 67). From the results of a 10-yr enquiry (1970–1979) carried out in 13 well-vaccinated countries, it was estimated that for children under 3 yr the risk of taking the vaccine is 0.84 cases of paralysis per million recipients (8). Where viral diagnosis was made in this study, most vaccine-associated cases of paralysis were caused by type 3 virus, and these cases were found among both recipients and contacts. Of type 2 cases, more were in contacts than in recipients. Although a few cases of paralysis were associated with the type 1 vaccine, the viruses isolated were subsequently found not to be "vaccinelike" by accepted criteria.

To be vaccine associated, cases of spinal paralysis with acute onset (typical of poliomyelitis) and persistence of more than 6 wk must begin 7–30 days after vaccination or 7–60 days after contact with a known vaccinee (23). The isolation of virus from these cases and its characterization as vaccinelike is crucial to establishing a causal link between the vaccine and the disease. This evidence is particularly convincing when the virus is isolated from the CNS as opposed to feces and is then shown to be highly neurovirulent in monkeys. Viruses may be designated "vaccinelike" on the basis of testing with strain-specific adsorbed sera or according to T_1 oligonucleotide fingerprinting (59). In one instance a poliovirus type 3 isolate from a fatal case of poliomyelitis that fulfilled the above criteria was shown by complete nucleotide sequence analysis to be a bona fide revertant of the vaccine (16; see below). Revertant strains of this type are proving useful in studies on the genetic basis of attenuation.

THE GENETIC BASIS OF ATTENUATION

After a decade of relative quiescence, the past few years have seen a renewed interest in the development of vaccines, mainly thanks to the application of recombinant DNA and hybridoma technologies to the study of pathogenic microorganisms. There is promise for a new generation of vaccines that, rather than being derived empirically, will be constructed with precisely defined genetic properties (99). This prospect has rekindled an interest in the attenuation of pathogenesis and a desire to understand its molecular basis. Attenuation studies also offer an opportunity to define those properties of a microorganism that are essential for its pathogenicity. Care must be taken, however, to avoid the conclusion that experiments on attenuation will necessarily provide direct information on mechanisms of virulence or pathogenicity. For example, it is likely that a majority of the mutations that affect the efficiency of replication of a pathogen will attenuate it to some degree, and mutations of this sort could occur in all essential and even some nonessential genes (4). For polioviruses, then, attenuating mutations may not necessarily locate and define regions of the genome that confer neurovirulence if they do not affect functions that are specific for cells of the CNS.

Since attenuated variants were first isolated it has been a goal of poliovirologists to understand precisely by what mechanism(s) their neurovirulence is reduced. This section deals with recent work aimed at defining the genetic basis of attenuation, mainly using Sabin's vaccines. For the most part the effects of the mutations in molecular terms are still far from clear.

Studies on Poliovirus Type 1

An early indication of the extent of the nucleotide sequence difference between the Sabin type 1 strain P1/LS-c,2ab and its neurovirulent progenitor

P1/Mahoney/41 was provided by T_1 oligonucleotide fingerprinting (64). A comparison of these two strains revealed that 7 of the approximately 69 major oligonucleotides differed in their migration rates. Since the characteristic spots in an oligonucleotide fingerprint represent about 18% of the genome, this finding suggested that the viruses differed by about 35 mutations in total. Although this interpretation correlated well with an estimate based on tryptic peptide mapping of the coat proteins (38), it turned out to be a slight underestimate. Later the complete nucleotide sequences of these two viruses showed that there are 57 nucleotide sequence differences in the 7441-base genome, all of which are point mutations and 21 of which give rise to amino acid changes (66; Table 2). Although the coding changes are generally scattered throughout the genome, a cluster of 7 in VP1, 5 of which occur within a stretch of 18 amino acids, suggested that modification of this protein may be involved in attenuation. Nomoto and his colleagues later turned their attention to defining which mutations may be important for attenuation by constructing defined recombinants between the type 1 vaccine and P1/Mahoney/41 via infectious cDNA (40, 68, 69; see below).

In the meantime, Agol and his colleagues in the USSR had attempted to map the location of mutations important to the attenuation phenotype not only of type 1 but also of the type 3 vaccine. Their approach was to allow neurovirulent and attenuated viruses of different serotypes to undergo recombination during mixed infection of cell cultures and to select for recombinants on the basis of antigenicity and guanadine hydrochloride resistance (3, 90). Oligonucleotide fingerprinting confirmed that the strains selected were indeed recombinants and showed that recombination had occurred almost exclusively in a central region of the genome so that recombinants possessed the capsid half from one strain and the noncapsid half from another. Neurovirulence tests were carried out (by intracerebral inoculation of cercopithecus monkeys) on a series of recombinants derived from a mixed infection with a guanidine-sensitive type 3 poliovirus [a vaccine derivative (Leon 2) or a wild type 3 strain (452/62 31)] and a guanidine-resistant type 1 poliovirus, after selection for isolates with type 3 antigenicity. The results indicated that the phenotype of any particular recombinant was determined by the origin of its 5' or capsid half (2, 3). This indicated that attenuating mutations of the type 3 strain (and by implication the vaccine) reside in the 5' half of the genome, which agrees with the conclusion presented in the next section.

To shed more light on the location of attenuating mutations in the type 1 strain, Agol and colleagues performed backcrosses between these type 3/type 1 recombinants and either the Sabin type 1 vaccine or a neurovirulent type 1 strain (2). Of the recombinants obtained, two were studied further: v1/a1-6, which derived the 5' half of its genome from the virulent type 1 parent and its 3' half from the avirulent type 1 parent, and a1/v1-7, which had the opposite configuration. Neurovirulence tests indicated that v1/a1-6 was virulent,

Table 2 Comparison of genome sequences between P1/Mahoney/41 and P1/LS-c,2ab[a]

Nucleotide position	Substitutions: Base M	Base Sab	Amino acid M	Amino acid Sab	Nucleotide position	Substitutions: Base M	Base Sab	Amino acid M	Amino acid Sab
21	U	C			3445	C	U (3)		
26	A	G			3460	U	A (3)	Asp	Glu
189	C	U			3492	G	A (2)	Ser	Asn
355	C	U			3766	C	A (3)		
480	A	G			3785	U	A (1)	Ser	Thr
649	C	U			3896	A	G (1)		
674	C	U			3898	C	A (3)	Ser	Gly
935	G	U (1)	Ala	Ser	3919	C	U (3)		
1208	A	C (1)			4003	C	U (3)		
1228	G	C (3)			4116	U	C (2)	Ile	Thr
1442	A	G (1)	Asn	Asp	4444	U	C (3)		
1465	C	U (3)			4789	A	G (3)		
1490	C	U (1)	Leu	Phe	5107	U	C (3)		
1507	G	A (3)			5137	A	G (3)		
1747	C	U (3)			5420	C	A (1)		
1885	A	U (3)			5440	A	G (3)		
1942	C	A (3)			6143	G	A (1)	Asp	Asn
1944	C	A (2)	Thr	Lys	6203	U	C (1)	Tyr	His
2353	U	C (3)			6373	C	U (3)		
2438	U	A (1)	Leu	Met	6616	G	A (3)		
2545	A	G (3)			6679	U	C (3)		
2585	A	G (1)	Thr	Ala	6734	A	G (1)	Lys	Glu
2741	A	G (1)	Thr	Ala	6853	C	U (3)		
2749	G	A (3)	Met	Ile	7071	C	U (2)	Thr	Ile
2762	C	U (1)	Pro	Ser	7198	U	A (3)		
2775	C	A (2)	Thr	Lys	7243	U	A (3)		
2795	G	A (1)	Ala	Thr	7410	U	C		
2879	C	U (1)	Leu	Phe	7441	A	G		
3163	U	C (3)							

[a] M and Sab represent P1/Mahoney/41 and P1/LS-c,2ab, respectively. Numbers in parentheses represent positions of in-phase codon. Predicted amino acid changes resulting from base substitutions are also shown. Reproduced with permission from Reference 66.

whereas a1/v1-7 was attenuated, which confirms that attenuating mutations also map to the 5' half of the genome in type 1. The fact that v1/a1-6 was not as neurovirulent as the wild type suggested that mutations in the 3' half of the genome may also have a weakly attenuating effect. These results are in good agreement with those obtained by the more sophisticated procedures of Nomoto and colleagues described below. Agol and colleagues discussed the importance of recovering virus from diseased animals to ascertain its identity with the virus inoculum. Of five viruses isolated from spinal cords, three retained the temperature-sensitive RNA characteristic of the parental inoculum strain and showed minimal changes according to T_1 oligonucleotide fingerprinting (1). Unfortunately however, the characterization of the recovered virus was incomplete, as neurovirulence tests were not carried out.

Nomoto and colleagues (40, 68, 69) have used a more refined method of mapping attenuating mutations of poliovirus type 1. Their approach was made possible by a seminal piece of work by Racaniello & Baltimore, which must rank as a major landmark in picornavirus genetics: the recovery of infectious virus from a cloned cDNA copy of the genome (73). Because of this important discovery, Nomoto and colleagues were able to construct recombinants from the neurovirulent P1/Mahoney/41 and the vaccine strain P1/LS-c,2ab via convenient common restriction sites in the cDNA and to recover virus by transfection of cells in culture (40). These studies have produced results that are somewhat complicated, but a number of firm conclusions can be drawn (68). Firstly, mutations scattered over at least the first 5601 bases of the genome seem to contribute to the attenuation of P1/LS-c,2ab. This can be seen simply from the extent of attenuation of the recombinants mapped in Figure 3. Recombinant 4a, which contained Sabin sequences only at positions 1–1122 of its genome, was significantly attenuated compared with P1/Mahoney/41 (Table 3); the recombinant had a lower lesion score and lower spread value and caused paralysis in just one of eight animals. Recombinant 3b, which contained Sabin sequences at positions 70–3664, was more attenuated than 4a. And recombinant 3a, which contained Sabin sequences at positions 1–70 and from position 3665 to the 3' end, was also significantly attenuated. The question of whether mutations in the region from position 5602 to the 3' end also contribute to attenuation is less clear. Recombinant 2b, which was Mahoney-like apart from this region, appeared less neurovirulent than P1/Mahoney/41, as judged by the fact that it caused no paralysis in four animals. This is a surprising result, however, in that its lesion score and spread value were, within standard errors, the same as those of P1/Mahoney/41. Moreover, the reciprocal recombinant, 2a, was certainly no less attenuated than P1/Ls-c,2ab. A result of zero paralysis for recombinant 2b seems additionally questionable in light of the observation that recombinant 3a, which derived significantly less of its genome from P1/Mahoney/41 and which had lower lesion and spread scores, paralyzed one of the monkeys. Further tests, preferably with larger numbers of animals, should be performed on recombinant 2b to clarify its neurovirulence.

Secondly, comparison of the results for the pair of reciprocal recombinants 4a and 4b suggests that attenuating mutations contained in the first 1122 bases (which include only one mutation that affects coding) are at least as strong as those in the remainder of the genome. This is concluded from the fact that these two recombinants had similar lesion scores and spread values. Further analysis of the 5' 1112 nucleotides suggests that a strongly attenuating mutation is located at position 480 (65; A. Nomoto, personal communica-

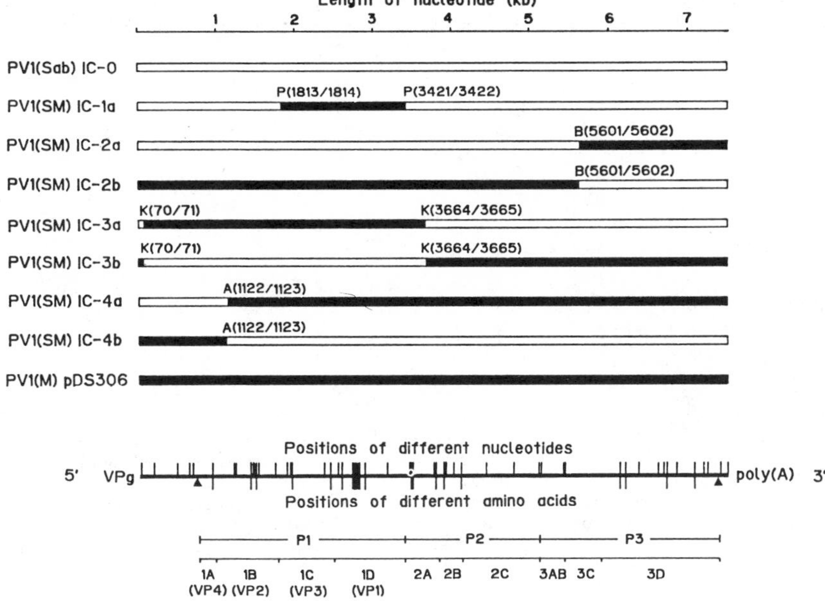

Figure 3 Genome structure of recombinant type 1 polioviruses and locations of nucleotide and amino acid differences between P1/Mahoney/41 and P1/LS-c,2ab strains. The expected genome structures of the recombinant viruses are shown by the combination of sequences of P1/Mahoney/41 *(filled bars)* and P1/LS-c,2ab *(open bars)*. *K, A, P,* and *B* represent cleavage sites of the restriction enzymes *Kpn*I, *Aat*II, *Pst*I, and *Bgl*II, respectively. *Numbers in parentheses* following the restriction sites indicate nucleotide positions, numbered from the 5' end of the viral genome. Length of the entire genome of poliovirus type 1 is indicated at the top of the figure in kilobases (kb) from the 5' terminus. Genomic RNA and its gene organization are shown at the bottom of the figure using the recently adopted nomenclature (75a). The positions of initiation and termination of virus polyprotein synthesis are indicated by *filled triangles*. The locations of nucleotide and amino acid differences between the progenitor, P1/Mahoney/41, and the vaccine strain, P1/LS-c,2ab, are indicated by *lines over* and *under* the genome RNA, respectively. Reproduced from Reference 68 with permission.

tion). In this respect polio type 1 is very similar to poliovirus type 3, which has an attenuating mutation at position 472 (see below).

A further inconsistency in the type 1 data concerns the recombinant 1a (68; Figure 3) and its reciprocal, 1b (65). Recombinant 1a was attenuated as expected. However, according to the results discussed above, 1b should have been only weakly attenuated since it lacked the strong attenuating mutations in the 5' 1813 bases and in the region of nucleotides 3664–5602. In fact, recombinant 1b was as attenuated as the vaccine itself. This unexpected result may have been due to the recombinant's acquisition of additional attenuating mutations, unrelated to those in the vaccine, during its propagation in culture. This is always a possibility with viruses with RNA genomes because of their high genetic variability (32).

Table 3 Monkey neurovirulence tests and in vitro markers of poliovirus type 1 recombinant viruses

Virus[a]	Lesion score (mean ± SE)	Spread value (mean ± SE)	Incidence of paralysis (no. paralyzed/no. injected)	rct[b]	d[c]	Plaque siz (mm in diameter)[d]
PV1(Sab)IC-0	0.07 ± 0.01	2.0 ± 0.0	0/4	>6.64	3.96	7.0
PV1(SM)IC-1a	0.15 ± 0.05	5.0 ± 1.9	0/4	5.70	1.88	8.5
PV1(SM)IC-2a	0.05 ± 0.01	1.5 ± 0.3	0/4	4.09	4.26	8.5
PV1(SM)IC-2b	2.03 ± 0.38	37.3 ± 0.6	0/4	2.77	0.66	14.5
PV1(SM)IC-3a	0.42 ± 0.17	11.1 ± 4.7	1/8	2.36	0.27	13.5
PV1(SM)IC-3b	0.11 ± 0.04	4.3 ± 1.3	0/4	4.73	4.63	6.5
PV1(SM)IC-4a	0.72 ± 0.27	18.4 ± 5.7	1/8	3.55	1.06	14.0
PV1(SM)IC-4b	0.80 ± 0.27	20.1 ± 6.1	0/8	5.09	3.18	4.5
PV1(M)pDS306	2.48 ± 0.34	37.0 ± 1.0	3/4	0.56	0.07	16.0
PV1(SM)IC-1b[e]	0.07		0/3	>7.64		

[a] Nomenclature of strains: PV1(Sab)IC-0 = P1/LS-c,2ab recovered from infectious cDNA; PV1(SM)IC = recombinan between P1/LS-c,2ab and P1/Mahoney/41, constructed via cDNA; PV1(M)pDS306 = recovered P1/Mahoney/41.

[b] rct marker values shown here are the logarithmic differences of virus titers obtained at 36 and 39.5°C (except f PV1(M)pDS306, for which the rct marker values of titers obtained at 36 and 40°C are given).

[c] d marker values shown here are the logarithmic differences of virus titers obtained at two different sodium bicarbona concentrations.

[d] Plaque size on day 5 of growth.

[e] Experiments with this recombinant, the reciprocal construct of IC-1a, were carried out in a different test series (65).

A further piece of information relevant to this analysis of attenuation is the result obtained with a monoclonal antibody–selected mutant of P1/Mahoney/41 (22). This mutant had a change from threonine to lysine in VP3, precisely the same as the change caused by the mutation at position 1944 in the vaccine; the mutant was therefore vaccinelike at this position (66). It turned out to be fully neurovirulent, which indicates that this change, at least on its own, does not attenuate the virus to any degree.

The series of Mahoney/Sabin type 1 recombinants constructed by Nomoto and colleagues has also been used to map the mutations responsible for the rct, d, and plaque-size markers. The results indicate that temperature-sensitive (rct) mutations are scattered throughout the genome, even in the region from position 5602 to the 3' terminus. Plaque-size and d markers both mapped to capsid proteins, which indicates that these characteristics are not a secondary manifestation of all of the attenuating mutations (68; see Table 3).

This approach to mapping attenuating mutations is limited by the availability of convenient restriction endonuclease cleavage sites within the cDNA through which recombinants can be constructed. A more precise identification of all of the mutations that contribute to the attenuation of P1/LS-c,2ab will probably require site-directed mutagenesis on P1/Mahoney/41 cDNA. The overall picture gained from the above studies, however, namely that several mutations contribute to attenuation, is in accord with the observed stability of this vaccine strain (8). The type 1 vaccine is rarely if ever associated with paralysis in vaccinees and is generally regarded as one of the safest vaccines used on humans.

Studies on Poliovirus Type 3

As mentioned above, the Sabin strain most associated with paralysis in vaccinees is the serotype 3 strain, P3/Leon $12a_1b$ (8). Results of early trials with this virus suggested that it was the least stable of the three, and this is supported by results in tissue culture (15, 61). The first biochemical evidence of its involvement in vaccine-associated paralysis came from serological and oligonucleotide fingerprinting studies. Several isolates were observed to have fingerprints that were clearly vaccinelike (17, 54, 55, 67). Studies on the genetic and molecular basis of attenuation in the Sabin type 3 vaccine have been based on several approaches.

Comparative nucleotide sequence analyses revealed that the vaccine strain P3/Leon $12a_1b$ and its neurovirulent progenitor P3/Leon/37 differ by just 10 point mutations in their 7432 nucleotides (83, 86; Figure 4). These mutations must therefore account for the attenuated phenotype of the vaccine. Just three of these mutations give rise to amino acid substitutions: a change from serine to phenylalanine in virus protein VP3, a change from lysine to arginine in VP1, and a change from threonine to alanine in the nonstructural protein P2-A, identified recently as a tyrosine-glycine–specific protease. On the basis of its likely effects on protein conformation, the change from serine to phenylalanine is the most drastic and therefore seems likely to be involved in attenuation.

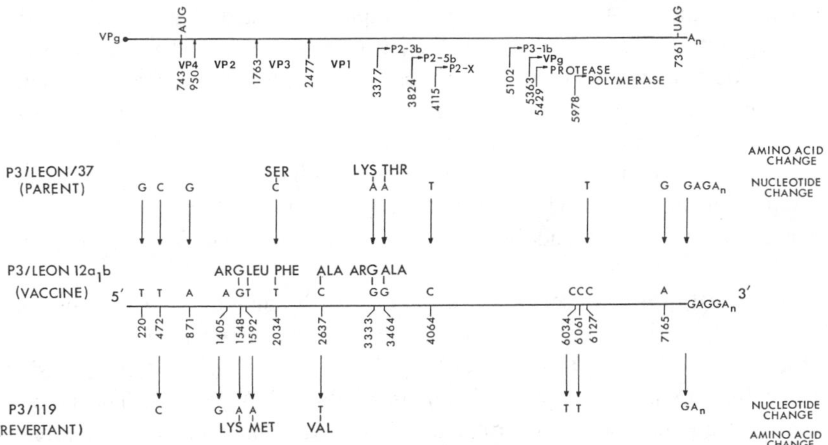

Figure 4 Nucleotide and predicted amino acid sequence differences between poliovirus type 3 strains P3/Leon/37 (parent), P3/Leon $12a_1b$ (vaccine), and P3/119 (revertant).

Additional information has been provided by the sequencing of P3/119, a vaccinelike strain isolated from the brain and spinal cord of a victim of fatal paralytic poliomyelitis temporally associated with vaccination (16, 84). This strain showed just seven changes from the vaccine but, significantly, was vaccinelike at eight of the ten positions where the vaccine differed from its progenitor. The sequence therefore provided conclusive evidence that P3/119 is on a direct genealogical lineage from the progenitor and the vaccine. This is the strongest possible evidence that P3/119 is a bona fide revertant of the vaccine. The mutations in P3/119 are also indicated in Figure 4. Again, three of the mutations give rise to amino acid changes: from arginine to lysine and from leucine to methionine in VP2, and from alanine to valine in VP1. The availability of the P3/119 sequence allowed reassessment of the likely contribution to attenuation of the mutations observed in the vaccine. Thus mutations that were silent with respect to protein coding and that showed no evidence of back mutation were considered unlikely to have a role. The coding change affecting P-2A was also tentatively ruled out, as there are no amino acid changes at all in the nonstructural proteins of P3/119 that could possibly compensate for this change (16). This conclusion was reached before the role of P-2A as a protease was established (92).

Cann et al (16) concluded that attentuation is likely to be effected by one of three genetic changes: the base change at position 472, which directly back-mutated in P3/119; the structural protein changes; or the mutation(s) just prior to the poly A tract. Further evidence for the involvement of the cytidine-uridine mutation at position 472 was provided by sequencing studies on excreted viruses (26). The virus was isolated from fecal samples collected from children after their routine vaccination (57; P. D. Minor, personal communication). Analysis of these samples indicated that growth in the human gut provided a strong selection for viruses that had reverted to the wild-type sequence with cytidine at position 472. Moreover, this back mutation was associated with an increase in neurovirulence. One of the virus isolates, DM4, showed intermediate neurovirulence in terms of both its lesion score and its induction of paralysis in cynomolgus monkeys. A partial de-attenuation of vaccine virus after passage through the human gut had been noted previously (48), and the back mutation at position 472 seems the likely explanation for this.

The most direct information bearing on which mutation(s) are responsible for the attenuation phenotype of the poliovirus type 3 vaccine has come from studies analogous to those described above for poliovirus type 1, using recombinant viruses prepared from infectious cDNA (5, 94). The recombinants were designed for examination of the individual and combined effects of the mutations that, based on sequence comparisons, were deemed likely to be important for attenuation. Viruses recovered by transfection were examined by the standard WHO safety test of intraspinal inoculation in cynomolgus monkeys. The results (G. D. Westrop et al, unpublished), which are summarized in Figure 5, suggest that the attenuated phenotype of the vaccine is conferred by just two mutations: the change from cytidine to uridine at position 472, as discussed above, and the change from cytidine to uridine at position 2034, which causes a serine-to-phenylalanine substitution in VP3. The mutation at position 472 is the stronger of the two; on its own it causes a 70–80% reduction in the number of animals paralyzed and a reduction in

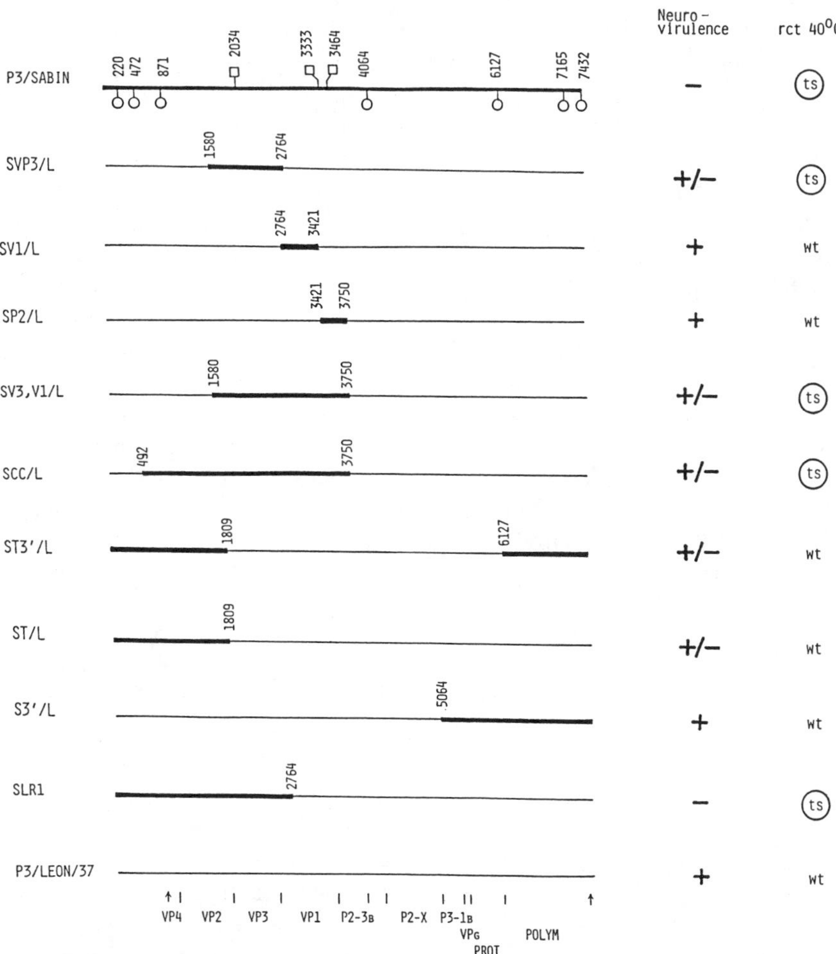

Figure 5 Progenitor/vaccine poliovirus type 3 recombinants constructed via cDNA. Sites of mutations on the genome of the Sabin vaccine strain P3/Leon $12a_1b$ are indicated by *open squares* (mutations that affect coding) and *open circles* (silent mutations). Positions of crossover points are indicated and correspond to restriction endonuclease cleavage sites. Structures of all viruses recovered by transfection of recombinant cDNAs have been verified by primer extension RNA sequencing at each site of crossover points (94; G. D. Westrop et al, unpublished).

histological lesion score from ~2.5 to ~1.3 (G. D. Westrop et al, unpublished). The mutation at position 2034 is less attenuating, but its combination with that at position 472 probably accounts for the fully attenuated phenotype of the vaccine. The serine-to-phenylalanine substitution that results from this mutation confers a temperature-sensitive phenotype on the virus. One of the attenuating mutations is therefore linked to the rct marker. Further experiments to verify these conclusions, including the testing of a virus that resembles P3/Leon/37 at positions 472 and 2034 but is otherwise entirely vaccinelike, are in progress. These results agree well with those of Agol and coworkers discussed above, namely that attenuating mutations in poliovirus type 3 map to the 5' half of the genome (3).

It seems, then, that reversion of the type 3 vaccine to neurovirulence requires at least two point mutations: the back mutation to cytidine at position 472, which probably occurs in all vaccinees within a few days of vaccination, plus at least one further mutation, which counteracts the attenuating effects of the change from serine to phenylalanine in VP3. This may be a direct back mutation at nucleotide 2034 or, as seems more likely from partial sequence analysis of other revertants, a suppressor mutation at a second site (5; D. M. A. Evans et al, unpublished). Analysis of the likely effects of the serine-to-phenylalanine mutation, based on Hogle et al's three-dimensional model of poliovirus (29), suggests that it may act by destabilizing protomer interactions along the fivefold axes of the particle. It is interesting that a number of neurovirulent revertants that retain phenylalanine at position 2034 but are temperature resistant carry mutations that have a propinquity to this region and may therefore restabilize the protomer interaction (D. M. A. Evans et al, unpublished).

Studies on Poliovirus Type 2

Although the poliovirus type 2 vaccine, P2/P712,Ch,2ab, has been sequenced (91), its progenitor P2/P712 has not. Thus identification of potential attenuating mutations in this vaccine strain has not been possible to date. In any event, even if the P2/P712 sequence were available, a comparison might not be informative in this case, as this wild strain is naturally attenuated (81). Genetic analysis of the Sabin type 2 vaccine has therefore lagged behind that of types 1 and 3. This is unfortunate, since type 2 is associated with paralysis in vaccinees (8) and there is a need to understand the molecular basis of its reversion to, or in this case acquisition of, a neurovirulent phenotype. Evidence that the type 2 vaccine causes paralysis is similar to the evidence that implicated type 3 prior to sequencing studies; i.e. serological and T_1 oligonucleotide fingerprinting tests have identified isolates from paralytic cases temporally associated with vaccination as "vaccinelike" (37). As mentioned above, most cases of type 2 vaccine–related paralysis are in contacts rather than in the vaccinees themselves (8). This may imply that the type 2 vaccine requires more mutations for the virus to assume a neurovirulent phenotype than the type 3 vaccine. Sequencing studies on revertants of type 2 will be informative in this regard and are in progress in several laboratories.

Careful T_1 oligonucleotide fingerprint analyses of type 2 revertants from trivalent vaccine recipients suggest that many of the revertants may be intertypic recombinants (36; O. M. Kew, personal communication). This raises the question of whether recombination has a direct role in this strain's acquisition of a neurovirulent phenotype.

Racaniello and colleagues (43–45) recently initiated a series of interesting and potentially very informative experiments on poliovirus neurovirulence using type 2 as a model. The P2/Lansing strain of poliovirus, when inoculated intracerebrally into mice, causes a fatal paralytic disease that resembles human poliomyelitis both clinically and histopathologically (33, 34). The mouse is therefore an experimentally versatile animal model in which at least some of the fundamental aspects of poliovirus neurovirulence can be studied. P2/Lansing was originally isolated from a fatal case of human poliomyelitis and was adapted initially to cotton rat brain. The virus was found to induce paralysis in white mice (6, 7; reviewed in 72). To identify the sequences of P2/Lansing that confer the mouse neurovirulence phenotype, Racaniello and colleagues constructed recombinants between P2/Lansing and the mouse-avirulent P1/Mahoney/41 using cloned infectious cDNAs of the two strains (45). Recombinants recovered by cDNA transfection of cell cultures were inoculated into 18–21-day-old Swiss-Webster mice and an LD_{50} was calculated for each virus. These experiments indicated that mouse neurovirulence maps to the capsid proteins of P2/Lansing, since a recombinant that was P1/Mahoney-like apart from bases 631–3413 caused disease identical to that caused by P2/Lansing.

Further information on the determinants of neurovirulence in mice was provided by the isolation of antigenic variants of P2/Lansing using neutralizing monoclonal antibodies (43). Several of these variants had considerably reduced neurovirulence for mice and showed reduced replication (100-fold lower titers than the wild type) in the mouse brain, although their replicative capacity in HeLa cells was mainly unaffected. Sequencing of the variants revealed mutations in the region of bases 93–105 of virus protein VP1. The level of attenuation of a given variant (if any) depended on the location and nature of the amino acid change. Racaniello et al (45) postulated that the original adaptation of P2/Lansing to mice involved the selection of variants that were able to recognize receptors on cells of the mouse brain. One might expect, therefore, that the attenuated antigenic variants would have a reduced ability to bind to or enter cells of the mouse CNS, although this has not yet been reported. The monoclonal antibodies used to select these variants were known to be directed against epitopes contained in this region of VP1 (56). It will be of interest to see whether monoclonals directed at other antigenic sites will similarly select variants of reduced neurovirulence.

Another potentially very important observation from the mouse model concerned the attenuating mutation in the 5' noncoding region of poliovirus type 3 at position 472 (discussed above). This mutation can also attenuate a mouse-virulent P2/Lansing derivative, as shown by the construction via cDNA of intertypic recombinants between P2/Lansing and either P3/Leon 37 or the type 3 vaccine, P3/Leon $12a_1b$ (44). The recombinants derived their 5' terminal 786 bases from one of the type 3 strains but otherwise entirely resembled P2/Lansing. The recombinants therefore differed at only two positions, 220 and 472 (refer to Figure 4). When tested in mice, the recombinant that derived its 5' terminus from P3/Leon/37 was found to be as neurovirulent as P2/Lansing, with an LD_{50} of $<6 \times 10^2$ pfu, whereas the counterpart derived from the vaccine was attenuated, with an LD_{50} of $>2 \times 10^7$ pfu. The two viruses also differed significantly in their ability to replicate in mouse brain, whereas their replication in HeLa cell cultures was very similar. The importance of position 472 was further illustrated by the isolation of virus from the brain and spinal cord of one animal that had developed paralysis after having been given a high dose (2×10^7 pfu) of the attenuated recombinant. Sequencing of these isolates revealed that position 472 had back-mutated to the wild type (cytidine), whereas base 220 remained uridine. Moreover, these isolates showed a level of neurovirulence comparable to that of P2/Lansing when injected into other mice. It seems, then, that the single nucleotide change of cytidine to uridine at position 472 renders the virus incapable of replicating and incapable of causing pathological damage in the mouse brain. This is in direct accord with the observations for poliovirus type 3 in primates described above (26, 94). Thus the mouse model accurately reflects the results in primates that identified an attenuating mutation in the 5' noncoding region of the genome. This indicates that at least some of the factors that affect the neurovirulence of poliovirus are similar in the CNS of primates and mice. Thus the mouse may provide a useful animal model in which to study poliovirus pathogenesis and in which to develop alternative, safer vaccines.

SUMMARY AND DISCUSSION

Despite the fact that polioviruses are probably the best studied of all animal viruses, our understanding of the molecular mechanisms by which they cause neurological disease is still rudimentary. However, the studies carried out over the past few years, which have mainly been concerned with the attenuation phenotype of Sabin's vaccines, have added significantly to the body of knowledge established in the years up to the mid 1950s. Detailed comparisons of attenuated and virulent strains have been made to identify those features of the virus that are responsible for its neurovirulence. The results have confirmed that neurovirulence is a complex phenomenon that is influenced by many facets of the virus-host interaction. However, the experiments have identified several individual point mutations that can significantly reduce the neurovirulence of the virus. In at least one case the mutations may identify a function of the genome of the virus that is dependent on cell type.

Perhaps the most intriguing finding is that mutations in the 5' noncoding region of the genome (position 472 in type 3 and position 480 in type 1) can dramatically affect the level of neurovirulence. Evidence to date suggests that this mutation affects virus replication in vivo, in both the gut and the spinal cord, but does not affect replication in cultured cells (26, 44). This observation also holds for P2/Lansing-derived virus in mice, which suggests that the factors involved in the limitation of virus growth are the same in mice and primates. An understanding of the molecular mechanisms involved requires better knowledge of the function of the 5' noncoding region. This is the most highly conserved part of the genome, both among enteroviruses and between enteroviruses and rhinoviruses (85). There is evidence that it has extensive secondary structure in both polioviruses and other members of the Picornaviridae family (46, 62). Whether an alteration to secondary rather than primary structure is the basis of attenuation is not yet clear; however, it is noteworthy that computer predictions of folding suggest that the mutation at position 472 should produce a significant change in the secondary structure of this region of the molecule (26). Intriguingly, the corresponding region of type 1 can be folded into a structure similar to that proposed for type 3, and in this configuration the attenuating mutation at position 480 forms part of the adjacent base pair (M. A. Skinner et al, unpublished observation). The only available data on the function of the region suggest that these mutations may affect the translatability of the RNA (87), although why this should be cell type–specific is unknown. This observation should be followed up using defined recombinants or site-directed mutants. A thorough understanding of the function of the 5' noncoding region and an explanation of the effects of the mutation at position 472 might suggest how stable derivatives could be prepared via cDNA (e.g. by deletion). This knowledge might form the basis of new attenuated vaccines incapable of reversion to neurovirulence.

Little is known about the molecular effects of the remaining attenuating mutations. In type 3 the only other mutation that affects neurovirulence also confers a temperature-sensitive phenotype (94; G. D. Westrop et al, unpublished). Temperature-sensitive mutations also map to the same region as weakly attenuating mutations in type 1 (68). Since all the vaccine strains share the Rct$^-$ phenotype, it is likely that the temperature-sensitive mutations are in themselves attenuating, as has been noticed for numerous other viruses (reviewed in 4). It is unclear, however, why the temperature-sensitive mutations should selectively block replication in the CNS while apparently allowing replication in the gut to proceed normally when the temperatures of these two sites are very similar. Future studies would be aided greatly by the development of cell culture systems that accurately reflect the conditions of nerve and gut cells in vivo.

One attractive and much-discussed hypothesis for attenuation is that the vaccine viruses may be less neurotropic than wild strains because they have a reduced affinity for the nerve-cell receptor (30, 42, 78). This could indeed be one of the mechanisms by which polioviruses are attenuated, and it involves a function that must be essential for neurovirulence. It is possible that such a mechanism could be operative in Sabin type 1 vaccine, where weakly attenuating mutations map to the structural protein region and where there are several mutations clustered in VP1 (67, 68). However, the available evidence suggests that there is probably no difference in affinity for nerve-cell receptor between the Sabin type 1 vaccine strain and P1/Mahoney/41 (28, 30, 42, 78), although this conclusion may be worthy of reexamination using modern methods. For Sabin type 3, the single attenuating mutation that maps to the capsid region lies at position 92 in VP3. This is well away from the canyon, i.e. the putative receptor-binding domain of the virus particle (74). The essential difference between the P2/Lansing virus and the naturally mouse-avirulent P1/Mahoney/41 could well be in receptor affinity (45). If this is the case, P1/Mahoney/41 could be thought of as having attenuating mutations for mice in the region of the capsid involved with this property. Whether reduced receptor affinity is also the basis of attenuation in the antigenic variants of P2/Lansing is not clear (43), although there is no direct evidence as yet that the region to which these map (VP1 amino acids 89–101) is involved in receptor interaction.

Poliovirus neurovirulence will remain a topic of interest in the years ahead. Many questions remain to be answered. What are the extraneural sites of virus replication and are these limited by the availability of the receptor and/or by factors that interact with the 5' noncoding region? What is the mechanism of motor neuron destruction and by what route are these cells infected? What is the role of cellular immunity, if any, in pathogenesis? Why do only 1% of infected people develop paralysis? Why is there a limited number of serotypes of poliovirus in spite of the fact that these RNA genomes are capable of rapid evolution? Pursuit of the answers to these questions will be interesting and challenging. The practical outcome of success could be not merely worldwide control of poliomyelitis but its eventual elimination.

ACKNOWLEDGMENTS

I would like to thank my colleagues Philip Minor, Kate Wareham, Gareth Westrop, and Michael Sullivan for their helpful suggestions and criticisms. I am grateful to Akio Nomoto and Vincent Racaniello for providing material prior to publication and to Eckard Wimmer for providing Figure 2.

Literature Cited

1. Agol, V. I., Drozdov, S. G., Frolova, M. P., Grachev, V. P., Kolesnikova, M. S., et al. 1985. Neurovirulence of the intertypic poliovirus recombinant v3/a1-25: characterization of strains isolated from the spinal cord of diseased monkeys and evaluation of the contribution of the 3' half of the genome. *J. Gen. Virol.* 65:309–16
2. Agol, V. I., Drozdov, S. G., Grachev, V. P., Kolesnikova, M. S., Kozlov, V. G., et al. 1985. Recombinants between attenuated and virulent strains of poliovirus type 1: derivation and characterization of recombinants with centrally located crossover points. *Virology* 143:467–77
3. Agol, V. I., Grachev, V. P., Drozdov, S. G., Kolesnikova, M. S., Kozlov, V. G., et al. 1984. Construction and properties of intertypic poliovirus recombinants: first approximation mapping of the major determinants of neurovirulence. *Virology* 136:41–55
4. Almond, J. W., Cann, A. J. 1984. Attenuation. In *Immune Intervention,* ed. I. M. Roitt, 1:13–56. New York: Academic
5. Almond, J. W., Westrop, G. D., Cann, A. J., Stanway, G., Evans, D. M. A., et al. 1985. Attenuation and reversion to neurovirulence of the Sabin poliovirus type 3 vaccine. In *Vaccines '85,* ed. R. A. Lerner, R. M. Chanock, F. Brown, pp. 271–83. New York: Cold Spring Harbor Lab.
6. Armstrong, C. 1939. The experimental transmission of poliomyelitis to the eastern cotton rat, *Sigmodon hispidus hispidus*. *Public Health Rep.* 54:1719–21
7. Armstrong, C. 1939. Successful transfer of the Lansing strain of poliomyelitis virus from the cotton rat to the white mouse. *Public Health Rep.* 54:2302–5
8. Assaad, F., Cockburn, W. C. 1982. The relation between acute persisting spinal paralysis and poliomyelitis vaccine—results of a ten-year enquiry. *Bull. WHO* 60:231–42
9. Assaad, F. A., Cockburn, W. C., Perkins, F. T. 1972. The relation between acute persisting spinal paralysis and poliomyelitis vaccine (oral): results of a WHO enquiry. *WHO Tech. Rep. Ser.* 486:20–58
10. Assaad, F. A., Ljungars-Esteves, K. 1984. World overview of poliomyelitis: regional patterns and trends. *Rev. Infect. Dis.* 6:S302–7
11. Bernier, R. H. 1984. Some observations on lameness surveys. *Rev. Infect. Dis.* 6:S371–75

11a. Blinzinger, I., Simon, J., Magrath, D., Boulger, L. 1969. Poliovirus crystals within the endoplasmic reticulum of endothelial and mononuclear cells in the monkey spinal cord. *Science* 163:1336–37

12. Bodian, D. 1949. Histopathologic basis of clinical findings in poliomyelitis. *Am. J. Med.* 6:563–78
13. Bodian, D. 1955. Emerging concept of poliomyelitis infection. *Science* 122: 105–8
14. Bodian, D., Horstmann, M. D. 1965. Polioviruses. In *Viral and Ricksettsial Infections of Man,* ed. F. L. Hosfall, I. Tamm, pp. 430–73. Philadelphia/Toronto: Lippincott
15. Boulger, L. R., Magrath, D. I. 1973. Differing neurovirulence of three Sabin attenuated type 3 vaccine seed pools and their progeny. *J. Biol. Stand.* 1:139–47
16. Cann, A. J., Stanway, G., Hughes, P. J., Minor, P. D., Evans, D. M. A., et al. 1984. Reversion to neurovirulence of the live-attenuated Sabin type 3 oral poliovirus vaccine. *Nucleic Acids Res.* 12:7787–92
17. Cossart, Y. 1977. Evolution of poliovirus since introduction of attenuated vaccine. *Br. Med. J.* 1:1621–23
18. Crawford, N. M., Baltimore, D. 1983. Genome-linked protein VPg of poliovirus is present as free VPg and VPg-pUpU in poliovirus-infected cells. *Proc. Natl. Acad. Sci. USA* 80:7452–55
19. Crowell, R. L., Landau, B. J. 1983. Receptors in the initiation of picornavirus infections. *Compr. Virol.* 18:1–42
20. Crowell, R. L., Landau, B. J., Siak, J. 1981. Picornavirus receptors in pathogenesis. In *Virus Receptors, Part 2,* ed. K. Lonberg-Holm, L. Philipson, pp. 171–84. London: Chapman & Hall
21. Dasgupta, A., Hollingshead, P., Baltimore, D. 1982. Antibody to a host protein prevents initiation by the poliovirus replicase. *J. Virol.* 42:1114–17

21a. Davis, B. D., Dulbecco, R., Eisen, H. N., Ginsberg, H. S. 1980. *Microbiology*. New York: Harper & Row. 3rd ed.

22. Diamond, D. C., Jameson, B. A., Bonin, T., Kohara, M., Abe, S., et al. 1985. Antigenic variation and resistance to neutralization in poliovirus type 1. *Science* 229:1090–93
23. Domok, I. 1981. Markers of poliovirus strains isolated from cases temporally associated with the use of live poliovirus vaccine: report on a WHO collaborative study. *J. Biol. Stand.* 9:163–84
24. Enders, J. F., Weller, T. H., Robbins, F. C. 1949. Cultivation of the Lansing

strain of poliomyelitis virus in cultures of various human embryonic tissues. *Science* 109:85–87

25. Enders, J. F., Weller, T. H., Robbins, F. C. 1952. Alterations in pathogenicity for monkeys of Brunhilde strain of poliomyelitis virus following cultivation in human tissues. *Fed. Proc.* 11:467–75
26. Evans, D. M. A., Dunn, G., Minor, P. D., Schild, G. C., Cann, A. J., et al. 1985. Increased neurovirulence associated with a single nucleotide change in a non-coding region of the Sabin type 3 poliovaccine genome. *Nature* 314:548–50
27. Hanecak, R., Semler, B. L., Anderson, C. W., Wimmer, E. 1982. Proteolytic processing of poliovirus polypeptides: antibodies to polypeptide P3-7c inhibit cleavage at glutamine-glycine pairs. *Proc. Natl. Acad. Sci. USA* 79:3973–97
28. Harter, D. H., Choppin, P. W. 1965. Adsorption of attenuated and neurovirulent poliovirus strains to central nervous system tissues of primates. *J. Immunol.* 95:730–36
29. Hogle, J. M., Chow, M., Filman, D. J. 1985. The three-dimensional structure of poliovirus at 2.9 Å resolution. *Science* 229:1358–65
30. Holland, J. J. 1961. Receptor affinities as major determinants of enterovirus tissue tropisms in humans. *Virology* 15:312–26
31. Holland, J. J., McLaren, L. C., Syverton, J. T. 1959. The mammalian cell–virus relationship: IV. Infection of naturally insusceptible cells with enterovirus ribonucleic acid. *J. Exp. Med.* 110:65–80
32. Holland, J., Spindler, K., Horodyski, F., Grabau, E., Nichol, S., et al. 1982. Rapid evolution of RNA genomes. *Science* 215:1577–85
33. Jubelt, B., Gallez-Hawkins, B., Narayan, O., Johnson, R. T. 1980. Pathogenesis of human poliovirus infection in mice. 1. Clinical and pathological studies. *J. Neuropathol. Exp. Neurol.* 39:138–48
34. Jubelt, B., Narayan, O., Johnson, R. T. 1980. Pathogenesis of human poliovirus infection in mice. 2. Age dependency of paralysis. *J. Neuropathol. Exp. Neurol.* 39:149–58
35. Kantoch, M. 1978. Markers and vaccines. *Adv. Virus Res.* 22:259–327
36. Kew, O. M., Nottay, B. K. 1986. Evolution of the oral polio vaccine strains in humans occurs by both mutation and intramolecular recombination. In *Vaccines '86,* ed. R. A. Lerner, R. M. Chanock, F. Brown, pp. 357–62. New York: Cold Spring Harbor Lab.
37. Kew, O. M., Nottay, B. K., Hatch, M. H., Nakano, J. H., Obijeski, J. F. 1981. Multiple genetic changes can occur in the oral poliovaccines upon replication in humans. *J. Gen. Virol.* 56:337–47
38. Kew, O. M., Pallansch, D. R., Omilianowski, D. R., Rueckert, R. R. 1980. Changes in three of the four coat proteins of an oral polio vaccine strain derived from type 1 polioviruses. *J. Virol.* 33:256–63
39. Kitamura, N., Semler, B, Rothberg, P. G., Larsen, G. R., Adler, C. J., et al. 1981. Primary structure, gene organization and polypeptide expression of poliovirus RNA. *Nature* 291:547–53
40. Kohara, M., Omata, T., Kameda, A., Semler, B. L., Itoh, H., et al. 1985. In vitro phenotypic markers of a poliovirus recombinant constructed from infectious cDNA clones of the neurovirulent Mahoney strain and the attenuated Sabin 1 strain. *J. Virol.* 53:786–92
41. Kuhn, R. J., Wimmer, E. 1987. The replication of picornaviruses. In *The Molecular Biology of Positive Strand RNA Viruses,* ed. D. J. Rowlands, B. W. J. Mahy, M. Mayo. London: Academic. In press
42. Kunin, C. M. 1961. Virus-tissue union and the pathogenesis of enterovirus infections. *J. Immunol.* 88:556–69
43. La Monica, N., Almond, J. W., Racaniello, V. R. 1987. Genetic analysis of poliovirus neurovirulence in mice. In *Vaccines '87,* ed. R. A. Lerner, R. M. Chanock, F. Brown. New York: Cold Spring Harbor Lab. In press
44. La Monica, N., Almond, J. W., Racaniello, V. R. 1987. A mouse model for poliovirus neurovirulence identifies mutations that attenuate the virus for man. Submitted for publication
45. La Monica, N., Meriam, C., Racaniello, V. R. 1986. Mapping of sequences required for mouse neurovirulence of poliovirus type 2 Lansing. *J. Virol.* 57:515–25
46. Larsen, G. R., Semler, B. L., Wimmer, E. 1981. Stable hairpin structure within the 5' terminal 85 nucleotides of poliovirus RNA. *J. Virol.* 37:328–35
47. Lwoff, A. 1959. Factors influencing the evolution of viral diseases at the cellular level and in the organism. *Bacteriol. Rev.* 23:109–24
48. Magrath, D. I., Boulger, L. R., Hartley, E. G. 1964. Strains of poliovirus from individuals fed Sabin vaccine studied by in vitro markers and monkey neurovirulence tests. *10th Eur. Symp. Eur. Assoc. Against Poliomyelitis Allied Dis., Warsaw,* pp. 334–45. Paris: Masson & Cie
49. McLaren, L. C., Holland, J. J., Syver-

ton, J. T. 1959. The mammalian cell–virus relationship. I. Attachment of poliovirus to cultivated cells of primate and non-primate origin. *J. Exp. Med.* 109:475–85

50. Melnick, J. L. 1980. Poliomyelitis vaccines: an appraisal after 25 years. *Compr. Ther.* 5:6–14
51. Melnick, J. L. 1984. Paralytic polio in the 20th century. *Karger Gaz.* 46–47: 1–5
52. Melnick, J. L. 1985. Enteroviruses: polioviruses, coxsackieviruses, echoviruses, and newer enteroviruses. In *Virology,* ed. B. N. Fields, D. M. Knipe, R. M. Chanock, J. L. Melnick, B. Roizman, R. E. Shope, pp. 739–94. New York: Raven
53. Mendelsohn, C., Johnson, B., Lionetti, K. A. L., Novis, P., Wimmer, E., et al. 1986. Transformation of a human poliovirus receptor gene into mouse cells. *Proc. Natl. Acad. Sci. USA* 83:1–5
54. Minor, P. D. 1980. Comparative biochemical studies of type 3 polioviruses. *J. Virol.* 34:73–84
55. Minor, P. D. 1982. Characterization of strains of type 3 poliovirus by oligonucleotide mapping. *J. Gen. Virol.* 59:307–17
56. Minor, P. D., Ferguson, M., Evans, D. M. A., Almond, J. W., Icenogle, J. P. 1986. Antigenic structure of polioviruses of serotypes 1, 2 and 3. *J. Gen. Virol.* 67:1283–91
57. Minor, P. D., John, A., Ferguson, M., Icenogle, J. P. 1986. Antigenic and molecular evolution of the vaccine strain of type 3 poliovirus during the period of excretion by a primary vaccinee. *J. Gen. Virol.* 67:693–706
58. Minor, P. D., Pipkin, P. A., Hockley, D., Schild, G. C., Almond, J. W. 1984. Monoclonal antibodies which block cellular receptors of poliovirus. *Virus Res.* 1:203–12
59. Minor, P. D., Schild, G. C. 1981. Identification of the origin of poliovirus isolates. *Lancet* 2:968–69
60. Minor, P. D., Schild, G. C., Bootman, J., Evans, D. M. A., Ferguson, M., et al. 1983. Location and primary structure of a major antigenic site for poliovirus neutralization. *Nature* 301:674–79
61. Nakano, M., Matsukura, T., Yoshii, K., Komatsu, T., Mukoyama, A. 1979. Genetical analysis on the stability of poliovirus type 3 Leon $12a_1b$. *J. Biol. Stand.* 7:157–68
62. Newton, S. E., Carroll, A. R., Campbell, O., Clarke, B. E., Rowlands, D. J. 1985. The sequence of foot-and-mouth disease virus RNA to the 5' side of the poly (C) tract. *Gene* 40:331–36
63. Nobis, P., Zibirre, R., Meyer, G., Kuhne, J., Warnecke, G., et al. 1985. Production of a monoclonal antibody against an epitope on HeLa cells that is the functional poliovirus binding site. *J. Gen. Virol.* 66:2563–69
64. Nomoto, A., Kitamura, N., Lee, J. J., Rothberg, P. G., Nobumasa, I., et al. 1981. Identification of point mutations in the genome of the poliovirus Sabin vaccine LS-c,2ab, and catalogue of RNAse T_1- and RNAse A–resistant oligonucleotides of poliovirus type 1 (Mahoney) RNA. *Virology* 112:217–27
65. Nomoto, A., Kohara, M., Kuge, S., Kawamura, N., Arita, M., et al. 1987. Study on virulence of poliovirus type 1 using in vitro modified viruses. *UCLA Symp. Mol. Cell. Biol.* In press
66. Nomoto, A., Omata, T., Toyoda, H., Kuge, S., Horie, H. et al. 1982. Complete nucleotide sequence of the attenuated poliovirus Sabin 1 strain genome. *Proc. Natl. Acad. Sci. USA* 79:5793–97

66a. Nomoto, A., Wimmer, E. 1987. Genetic studies of the antigenicity and the attenuation phenotype of poliovirus. *Symp. Soc. Gen. Microbiol.* 40:In press

67. Nottay, B. K., Kew, O., Hatch, M., Heyward, J., Obijesi, T. 1981. Molecular variation of type 1 vaccine related and wild polioviruses during replication in humans. *Virology* 108:405
68. Omata, T., Kohara, M., Kuge, S., Komatsu, T., Abe, S., et al. 1986. Genetic analysis of the attenuation phenotype of poliovirus type 1. *J. Virol.* 58:348–58
69. Omata, T., Kohara, M., Sakai, Y., Kameda, A., Imura, N., et al. 1984. Cloned infectious complementary DNA of the poliovirus Sabin 1 genome: biochemical and biological properties of the recovered virus. *Gene* 32:1–10
70. Pallansch, M. A., Kew, O. M., Semler, B. L., Omilianowski, D. R., Anderson, C. W. et al. 1984. Protein processing map of poliovirus. *J. Virol.* 49:873–80
71. Paul, J. R. 1971. *A History of Poliomyelitis.* New Haven/London: Yale Univ. Press
72. Racaniello, V. R. 1986. Poliovirus neurovirulence. *Adv. Virus Res.* In press
73. Racaniello, V. R., Baltimore, D. 1981. Cloned poliovirus complementary DNA is infectious in mammalian cells. *Science* 214:916–19
74. Rossman, M. G., Arnold, E., Erickson, J. W., Frankenberger, E. A., Griffith, J. P., et al. 1985. Structure of a human common cold virus and functional relationship to other picornaviruses. *Nature* 317:145–53

75. Rueckert, R. R. 1985. Picornaviruses and their replication. See Ref. 52, pp. 705–38

75a. Rueckert, R. R., Wimmer, E. 1984. Systematic nomenclature of picornavirus proteins. *J. Virol.* 50:957–59

76. Sabin, A. B. 1955. Characteristics and genetic potentialities of experimentally produced and naturally occurring variants of poliomyelitis virus. *Ann. NY Acad. Sci.* 61:924–38

77. Sabin, A. B. 1956. Pathogenesis of poliomyelitis: reappraisal in the light of new data. *Science* 123:1151–57

78. Sabin, A. B. 1957. Properties of attenuated poliovirus and their behavior in human beings. *Spec. Publ. NY Acad. Sci.* 5:113–37

79. Sabin, A. B. 1961. Reproductive capacity of polioviruses of diverse origins at various temperatures. *Perspect. Virol.* 2:90–100

80. Sabin, A. B. 1965. Oral poliovirus vaccine: history of its development and prospects for eradication of poliomyelitis. *J. Am. Med. Assoc.* 194:130–34

81. Sabin, A. B., Boulger, L. R. 1973. History of Sabin attenuated poliovirus oral live vaccine strains. *J. Biol. Stand.* 1:115–18

82. Salk, J. E. 1960. Persistence of immunity after administration of formalin-treated poliovirus vaccine. *Lancet* 2: 715–23

83. Stanway, G., Cann, A. J., Hauptmann, R., Hughes, P., Clarke, L. D., et al. 1983. The nucleotide sequence of poliovirus type 3 Leon $12a_1b$: comparison with poliovirus type 1. *Nucleic Acids Res.* 11:5629–43

84. Stanway, W., Cann, A. J., Hauptmann, R., Mountford, R. C., Clarke, L. D., et al. 1983. Nucleic acid sequence of the region of the genome encoding capsid protein VP1 of neurovirulent and attenuated type 3 polioviruses. *Eur. J. Biochem.* 135:529–33

85. Stanway, G., Hughes, P. J., Mountford, R. C., Minor, P. D., Almond, J. W. 1984. The complete nucleotide sequence of a common cold virus: human rhinovirus 14. *Nucleic Acids Res.* 12:7859–75

86. Stanway, G., Hughes, P. J., Mountford, R. C., Reeve, P., Minor, P. D. 1984. Comparison of the complete nucleotide sequences of the genomes of the neurovirulent poliovirus P3/Leon/37 and its attenuated Sabin vaccine derivative P3/Leon $12a_1b$. *Proc. Natl. Acad. Sci. USA* 81:1539–43.

87. Svitikin, Y. V., Maslova, S. V., Agol, V. I. 1985. The genomes of attenuated and virulent poliovirus strains differ in their in vitro translation efficiencies. *Virology* 147:243–52

88. Takeda, N., Kuhn, R. J., Young, C. F., Takegami, T., Wimmer, E. 1986. Initiation of polivirus plus-strand RNA synthesis in a membrane complex of infected HeLa cells. *J. Virol.* 60:43–53

89. Theiler, M. 1941. Studies on poliomyelitis. *Medicine Baltimore* 20:443–62

90. Tolskaya, E. A., Romanova, L. A., Kolesnikova, M. S., Agol, V. I. 1983. Intertypic recombination in poliovirus: genetic and biochemical studies. *Virology* 124:121–32

91. Toyoda, H., Kohara, M., Kataoka, Y., Suganuma, T., Omata, T., et al. 1984. Complete nucleotide sequences of all three poliovirus serotype genomes. Implications for genetic relationship, gene function and antigenic determinants. *J. Mol. Biol.* 174:561–85

92. Toyoda, H., Nicklin, M. J. H., Murray, M. G., Anderson, C. W., Dunn, J. F., et al. 1986. A second virus-encoded proteinase involved in proteolytic processing of poliovirus polyprotein. *Cell* 45:761–70

93. Van der Werf, S., Wynchowski, C., Bruneau, P., Blondel, B., Crainic, R., et al. 1983. Localization of a poliovirus type 1 neutralization epitope in viral capsid polypeptide VP1. *Proc. Natl. Acad. Sci. USA* 80:5080–84

94. Westrop, G. D., Evans, D. M. A., Minor, P. D., Magrath, D., Schild, G. C., et al. 1986. Investigation of the molecular basis of attenuation in the Sabin type 3 vaccine using novel recombinant polioviruses constructed from infectious cDNA. See Ref. 41. In press

95. Wimmer, E., Emini, E., Diamond, D. C. 1986. Mapping neutralization domains of viruses. In *Concepts in Clinical Pathogenesis II,* ed. A. L. Notkins, M. B. A. Oldstone, pp. 159–73. New York: Springer-Verlag

96. Winter, M. M., Boulger, L. R. 1963. The sensitivity of neurovirulence tests. *Extr. 9e Symp. Assoc. Eur. Contre Polio. Mal. Assoc., Stockholm,* pp. 341–51. Paris: Masson & Cie

97. World Health Organization. 1981. A system of nomenclature for poliovirus strains: a WHO memorandum. *Bull. WHO* 59:853–54

98. World Health Organization. 1983. Requirements for poliomyelitis vaccine (oral). *WHO Tech. Rep. Ser.* 687:107–75

99. World Health Organization. 1983. *The WHO Programme for Vaccine Development,* pp. 1–24. Geneva: WHO

Ann. Rev. Microbiol. 1987. 41:181–208

REPETITIVE PROTEINS AND GENES OF MALARIA

D. J. Kemp, R. L. Coppel, and R. F. Anders

The Walter and Eliza Hall Institute of Medical Research, Victoria 3050, Australia

---

CONTENTS

MALARIA STAGES EXPOSED TO THE HOST IMMUNE SYSTEM

Malaria is a mosquito-borne infection caused by protozoa of the genus *Plasmodium*. Presently about one sixth of the world's population lives in areas where malaria is endemic; it is estimated that there are 200 million cases annually, and malaria is assumed to contribute to at least two million deaths per year. Four species of malaria infect humans: *P. falciparum, P. vivax, P. ovale,* and *P. malariae*. The malaria life cycle is complex and has two distinct phases, sexual and asexual. The asexual phase in humans commences when

0066-4227/87/1001-0181$02.00

sporozoites are released from the salivary glands of biting mosquitoes. Within minutes of being injected the sporozoites enter liver cells. Here they develop over several days into thousands of merozoites. On release, the merozoites invade red blood cells, where they develop into immature or ring-stage trophozoites (rings), mature trophozoites, and schizonts. The daughter merozoites produced by the schizont-infected cell are released from the erythrocyte and can invade other red blood cells in a number of successive cycles. After invasion some parasites differentiate into the sexual forms. When these gametocytes are ingested by another mosquito they initiate the sexual phase and develop through a number of stages into sporozoites, thereby completing the cycle.

Control measures directed at the parasite itself (drug therapy) or the mosquito (insecticides or measures to prevent breeding) have eradicated or controlled malaria in some countries, but have failed in much of the tropical world. The advent of drug-resistant parasite strains and insecticide-resistant mosquito strains has contributed to a serious resurgence of malaria in some parts of the world. Thus there is considerable interest in developing a vaccine. This has been given considerable impetus by recent technical developments. First, a system for the continuous culture of *P. falciparum* in vitro (139) has facilitated study of the parasite in the laboratory. Second, recombinant DNA (5, 8, 27, 28, 40, 76, 84, 99) and monoclonal antibody (22, 62, 155) techniques have been applied to ascertain which antigens are potential components of a molecular vaccine. Most of our knowledge of malaria genes comes from these recent studies, predominantly on *P. falciparum*.

Most forms of the parasite are intracellular and therefore not directly accessible to the human immune system. To understand the immunobiology of the host-parasite relationship it is helpful to categorize antigens of malaria parasites according to the stage(s) of the life cycle in which they are expressed and according to their cellular location. The first such category includes a single major protein, the *circumsporozoite protein,* which covers the sporozoite surface. Nothing is currently known about antigens expressed by liver stages that develop after sporozoite invasion, but antigens on the surface of these infected cells may be targets of protective immunity.

Merozoites are transitorily free in the bloodstream, and therefore *merozoite surface proteins* are potential targets of protective immune responses. The invasion of red cells by merozoites is a complicated process that involves discharge of a major secretory organelle, the rhoptry. *Rhoptry antigens* are currently being characterized because there is evidence that they can induce protective immune responses. After invading the red cell, the parasite, enclosed by its plasma membrane, resides within a vacuole (the parasitophorous vacuole) bounded by the parasitophorous vacuole membrane derived from the red cell. *Secreted antigens* can be secreted into this vacuole and released into the plasma on rupture of the schizont or can be released from the red cell surface after being transported across the red cell cytoplasm. Although their functions are unknown, secreted antigens may have important immunological consequences. A number of *erythrocyte surface antigens* are associated with modifications of the red cell surface such as the "knobs" in the membranes of red cells infected with mature parasites. The remaining categories are antigens that appear to be *soluble cytoplasmic components,* which are of little interest in relation to vaccine development, and antigens that cannot yet be assigned to the above categories.

Gametocyte antigens, targets of transmission-blocking antibodies, have been identified by reaction with antibodies (22), but so far no structural data have been reported for any of these antigens.

Sequencing studies have shown that many antigens contain extensive arrays of tandemly repeating short amino acid sequences (37). Although similar repetitive polypeptides are common in nature, the interest in this observation lies in the fact that much of the antibody response induced by malaria infections is directed against epitopes encoded by these repeats. Although the exact functions of the repeats of any one molecule are in the realm of speculation, in toto the immune response against them must have important consequences in the host-parasite relationship. We present some views on these consequences. Noncoding repetitive sequences in the parasite genome are not considered in this review. Several recent reviews on related aspects of malaria immunology have appeared (3, 17, 26, 77, 104, 138).

Nomenclature

The earliest malaria antigens identified were named according to their structure or function and were referred to by the corresponding acronyms, e.g. the schizont-infected cell agglutination (SICA) antigen (20) and the histidine-rich protein (HRP) (78). More recently, antigens have been designated by their apparent molecular weights. However, the apparent molecular weights are usually in error by 10–30% because of anomalous migration on SDS-polyacrylamide gel electrophoresis. Furthermore, many of the polypeptides vary in size from isolate to isolate, and many exist as series of processed fragments, often overlapping in size with other antigens. For these reasons we believe a system of acronyms is less confusing.

SPOROZOITE ANTIGENS

The Circumsporozoite Protein

Little is known about sporozoite proteins with the notable exception of the circumsporozoite (CS) protein. The CS protein found on the surface of mature sporozoites ranges from 40 to 60 kd in different species and varies in immunological reactivity (103, 122). The structure of the CS protein precursor has been determined for several species of *Plasmodium* (9, 40, 43, 45, 46, 108, 127). All of the precursors possess a typical signal sequence at the N-terminus and a transmembrane anchor sequence at the C-terminus. The CS proteins are synthesized as a precursor from which 50–100 amino acids are removed prior to insertion in the membrane (103, 156). Upon reaction with antibodies the CS protein is shed as a tail-like precipitate; this phenomenon is termed the circumsporozoite reaction.

The central region of the CS protein consists of tandem repeats that are highly conserved within the molecule. The protein differs among species in length, sequence, and number of repeats (Figure 1), although the overall length of the molecule is relatively conserved. For example, *P. knowlesi* (H strain) has 12 copies of the 12–amino acid sequence GQPQAQGDGANA (108), whereas the *P. falciparum* CS protein (clone 7G8) contains 37 copies of the sequence NANP and four copies of a variant repeat, NVDP (40). Conservation of the repetitive structure is less pronounced at the nucleotide

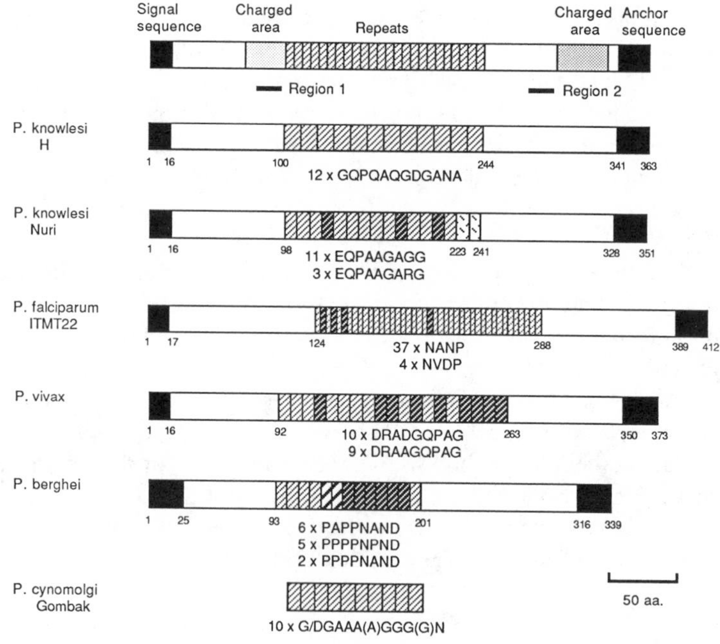

Figure 1 Structures of CS proteins from different species. At the top the general structure is shown; different shadings represent the signal sequence, charged areas, repeats, and anchor sequence. Regions 1 and 2 are indicated. The next five molecules are from the species indicated at the left, taken from (from the top down) references 108, 127, 40, 9, and 43. The repeats shown on the bottom line are from a partial sequence (45). The number and sequence of repeats are shown below each molecule. The *numbers* indicate amino acid positions.

level; for example, the 37 copies of the dodecamer encoding NANP are coded by 11 different nucleotide sequences. In the CS proteins of *P. cynomolgi* and *P. yoelii* the repetitive region is composed of two continuous blocks of tandem repeats of differing sequence (43, 52). In monoclonal antibody studies of the CS protein Zavala et al (157) detected a single immunodominant epitope repeated many times. Repeats are the targets of neutralizing monoclonal antibodies (158); In *P. falciparum* a synthetic peptide that contained three copies of NANP bound monoclonal antibodies as effectively as did intact sporozoites (158).

Variation in the repeats of the CS protein within a species does occur, but the extent differs from species to species. In the Nuri strain of *P. knowlesi* the CS protein contains 11 repeats of EQPAAGAGG and three repeats of the closely related sequence EQPAAGARG (127); this repeat differs greatly from that in the H strain of *P. knowlesi* (see above). The *P. cynomolgi* species complex consists of six isolates, five of which have CS proteins with markedly different repetitive regions (52). In contrast, the repetitive region of *P. falciparum* is much more conserved; as yet no strain has been detected that lacks the NANP repeat sequence (148), although the number of repeats changes. In some *P. falciparum* strains the 5' flanking region contains two to four copies of a five–amino acid sequence rich in asparagine (41, 89). At the 5' boundary of the repeats there is a region with 27 of 48 amino acids charged in *P. knowlesi,* and 27 of 53 are charged in this region in *P. falciparum* (Figure 1). A second less extensive charged area occurs near the anchor sequence.

There is little sequence homology between *P. falciparum* and *P. knowlesi* CS proteins other than two short sequences: Region I, where 9 of 15 amino acids are identical, and Region II, where 12 of 13 amino acids are identical (40). Region I lies within the 5' charged sequence. Region II, adjacent to the 3' charged sequence, contains two cysteine residues which may be important in tertiary structure. The nucleotide sequence of Region II is also highly conserved, with 25 of 27 positions identical in *P. falciparum* and *P. knowlesi* (40). It is therefore useful as a hybridization probe to identify the gene for other CS proteins, including that of *P. vivax* (97). The *P. vivax* CS protein is closely related to the CS protein of the simian malarias *P. knowlesi* and *P. cynomolgi,* with extensive homology in the regions outside the repeat (9). This finding is consistent with reports that *P. vivax* is more closely related to the simian malarias than *P. falciparum* (96).

The repeat region may represent a ligand that binds to a molecule on the hepatocyte and thereby mediates sporozoite attachment; the many copies may increase the avidity of the interaction (55). Evidence for this is provided by the observation that antibodies blocked hepatocyte binding in vitro (10) and diminished sporozoite infectivity in vivo (103). Since direct binding of the repeats to hepatocytes has not been detected, other regions of the CS protein may be important for hepatocyte recognition. Indeed, a nonapeptide from the N-terminal, nonrepetitive region of the *P. falciparum* CS protein and a related decapeptide from the *P. knowlesi* CS protein bound to human hepatoma cells under conditions suitable for invasion (1). Alternatively, the CS protein may serve as an immune defense mechanism (54), as it is readily shed from the sporozoite surface and presents multiple copies of the repetitive epitope per molecule to the immune system.

ASEXUAL BLOOD-STAGE ANTIGENS

Secreted Antigens

S-ANTIGEN The S-antigens are soluble, heat-stable antigens found in plasma that exhibit much more extensive size and serological diversity than any other malaria antigen found so far (4, 151–153). The sequences of clones that encode four antigenically different S-antigens of *P. falciparum* have been determined (30, 36; R. B. Saint, R. L. Coppel, A. F. Cowman, G. B. Brown, P. T. Shi, et al, manuscript submitted; H. Brown, N. Barzaga, P. E. Crewther, G. V. Brown, D. J. Kemp, et al, in preparation); each consists of a single exon. There is a central block of tandem repeats (Figure 2) flanked, as in the CS proteins, by nonrepeat sequences containing regions rich in charged amino acids that show considerable homology. The S-antigen of *P. falciparum* isolate FC27 contains about 100 repeats of the undecamer PAKASQGGLED, whereas that of NF7 contains about 40 repeats of an octamer with two variants, ARKSDEAE and ALKSDEAE. There are also two repeats of a pentadecamer, SDAGTEGPKGTGGPG, near the C-terminus in NF7. The FC27 and NF7 repeats are so different that they do not cross-hybridize (30, 33, 36). Studies with synthetic peptides showed that natural antibody responses were directed against epitopes encoded by the repeats (6, 7), and the repeats of FC27 and NF7 did not cross-react immunologically.

The major repeats from isolate V1 were undecamers, again of distinctly different sequence (H. Brown, in preparation). However, the S-antigens of isolates V1 and K1 contained copies of the 15–amino acid sequence located near the C-terminus as in NF7, and the N- and C-terminal portions were almost identical to those of NF7. The pentadecamer contains two internally homologous regions, and recombination within these would explain the origin of the major V1 undecamer. The repeat of K1 was deleted extensively in *Escherichia coli,* but studies with antibodies against the predicted synthetic peptide demonstrated that the major repeats of K1 were also related in sequence to the pentadecamer, although they differed in reading frame (R. B. Saint, R. L. Coppel, A. F. Cowman, G. V. Brown, P. T. Shi, et al, manuscript submitted). Thus the serological diversity of S-antigens reflects

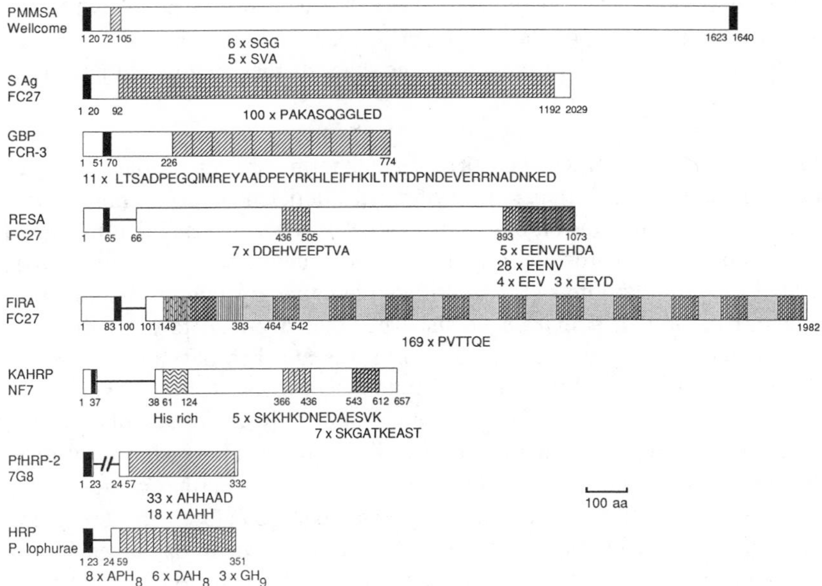

Figure 2 Structures of genes for blood-stage antigens. The genes indicated at the left are taken from (from the top down) references 65, 36, 83, 48, 134 and H. D. Stahl, unpublished, 141, 150, and 123. The *boxed regions* represent exons and the *lines* represent introns. *Filled segments* represent hydrophobic segments of signal sequences and, for the PMMSA, a C-terminal anchor segment. *Shaded areas* represent repetitive regions, and the number and the consensus sequence of repeats is shown below each molecule. The *numbers* represent amino acid positions. Minor variants are not shown; note that in some molecules, e.g. FIRA, most repeats are variants.

the occurrence of different repeats that may vary in sequence, size, number, or reading frame.

The S-antigens behave as an allelic series of variant genes at a single locus (36; R. B. Saint, R. L. Coppel, A. F. Cowman, G. V. Brown, P. T. Shi, et al, manuscript submitted). There is no evidence for unique or repeat S-antigen sequences other than those at the single S-antigen gene of the haploid *P. falciparum* genome. Thus the generation of S-antigen diversity presumably does not involve any sequence rearrangements.

At present the function of the S-antigen is unknown. The extreme serological diversity presumably results from strong immunological selection, so the S-antigen may somehow be involved in immune evasion. Although there is no evidence that the S-antigen is attached to merozoites, merozoites that escape during schizont rupture must be bathed in S-antigen, because it is located in the parasitophorous vacuole (30). One possibility is that S-antigen becomes transiently associated with the merozoite as a peripheral membrane protein. Although an anti–S-antigen monoclonal antibody inhibited merozoite invasion in vitro (125), there is no other evidence that host-protective strain-specific immune responses are directed against S-antigens. The extreme diversity of the repeats in these antigens suggests that they would have little value as vaccine components.

GLYCOPHORIN-BINDING PROTEIN Glycophorins are major sialoglycoproteins of the red cell surface. As erythrocytes deficient in glycophorins were not susceptible to invasion by the merozoite, it was suggested that the initial interaction of the merozoite with the red cell could be via glycophorin (101, 109, 110, 112). Perkins (112) identified a putative glycophorin-binding protein (GBP) of *P. falciparum* on the basis that it bound to an aminoethyl acrylamide column to which glycophorin was attached but not to a control column containing a different glycoprotein. However, van Schravendijk & Newbold (144) found that the GBP bound in the absence of glycophorin and attributed this to ion exchange effects. Analogous studies using a different support for glycophorin yielded different results (75). Further confusing the issue, a ring-infected erythrocyte surface antigen (see below) was reported to bind to glycophorin columns (115), which is most easily explained by ion exchange effects. There is no evidence that the GBP binds to intact red cells. Evidence that the GBP is located on the merozoite surface has been obtained by electron microscopy (113, 120), but other studies have found anti-GBP antibodies reacting with antigen within the red cell cytoplasm (13). It is difficult to establish if the molecule is loosely bound to the merozoite surface and can readily be washed off, or if it is a contaminant of the surface because it is secreted into the space surrounding the parasite. It is important to recognize that isolated *P. falciparum* merozoite preparations contain a large proportion of noninvasive merozoites. As it is agreed that the GBP molecule is initially secreted, we prefer to consider it in the category of secreted antigens and to remain open-minded about its possible function.

cDNA clones that encode the putative GBP have been isolated (13, 27, 28, 76, 77, 124). The protein is a soluble polypeptide of 105–130 kd. It is secreted from schizonts and released into culture supernatant, where it is found as heat-stable molecules. These molecules were predominantly 110 and 84 kd in isolate FC27 (13). All tested isolates were immunologically indistinguishable.

The GBP from isolate FCR3 contained a 50–amino acid sequence that was repeated 11 times with a low degree of variation. In the GBP from isolate FC27 the single repeat studied lay within the consensus sequence found for isolate FCR3 (13). The GBP gene (83) had a single exon, containing a 13–amino acid hydrophobic core of a signal sequence commencing at amino acid 51. Following this was a highly charged region of 155 amino acids, of which 35% were basic, followed by 11 repeats of the 50–amino acid sequence, continuing to the C-terminus. The absence of a hydrophobic membrane anchor sequence at the C-terminus accords with the evidence that this is a secreted soluble molecule. Segments of the GBP expressed in *E. coli* bound to a glycophorin-acrylamide column (83), but this evidence of glycophorin binding is subject to the criticisms above. Like the S-antigen, however, the GBP could possibly associate with the merozoites as a peripheral membrane protein, and there is some evidence that the S-antigen and GBP exist as a complex (M. E. Perkins, personal communication). Such a model would resolve some of the apparently contradictory data.

Merozoite Surface Antigens

PRECURSOR TO THE MAJOR MEROZOITE SURFACE ANTIGENS A molecule with promise as a vaccine candidate is the precursor to the major merozoite surface antigens (PMMSA), otherwise known as the 195K protein (63, 64), P190 (59), the 200-kd protein (118), or the polymorphic schizont antigen (94). Protective immune responses directed against the PMMSA have been demonstrated in several systems (47, 62). In studies with the *P. falciparum* PMMSA a small number of *Saimiri* monkeys were immunized and some protection was obtained (23, 58, 118, 129). The PMMSA is synthesized at the schizont stage and has molecular mass of 180–220 kd (61, 63, 69, 111). During schizont maturation it is processed to three fragments of 83, 42, and 19 kd (64). The 83-kd and 19-kd components are apparently shed during invasion, but the 42-kd antigen remains associated with the ring-stage parasites (64).

A panel of monoclonal antibodies has provided evidence that the protein has variable and constant epitopes (60, 94, 95). The complete sequences of the PMMSA gene from four isolates of *P. falciparum* (65, 92, 119a, 137) and partial sequences from two other isolates (24, 149) have now been reported. These include representatives of the two major serotypes defined by monoclonal antibodies. The single exon (Figure 2) commences with an N-terminal signal peptide followed near the N-terminus by a relatively short region of variable tripeptide repeats. Near the C-terminus there are a hydrophobic anchor sequence and a cluster of cysteine residues. Detailed comparison of the complete sequences and a partial sequence from the CAMP isolate revealed that diversity in these allelic genes is confined to specific parts of the polypeptide chain. The gene can be divided into 17 discrete blocks ranging in homology from 10–87% at the amino acid level (137). Blocks 1 and 17, the two most highly conserved regions, encode the N- and C-terminal sequences, respectively.

The outstanding feature of the variable blocks is that each has only two versions among the four sequences compared. There are almost no differences between homologous versions within these blocks, but there are many differences between the two versions; thus the antigen is dimorphic. If isolates MAD20 and K1 are considered to be the two dimorphic prototypes, then the CAMP and Wellcome types could be readily generated by intragenic recombination (137).

The only substantial exception to the dimorphism rule is in the tripeptide repeat region. It contains varying numbers of repeats of general type S-X-X, and up to four different amino acids occur in some positions of the six sequences that can be compared. Hence this region is reminiscent of the S-antigen. A substantial proportion of the repeats appeared to be deleted during culture in vitro (77).

The 83-kd fragment from the Wellcome strain has been localized to the N-terminal region by amino acid sequencing commencing at residue 25 (65). The 42-kd fragment is most likely the C-terminus (126) and bears an acyl group (57) that may function as a membrane anchor. The locations of several epitopes recognized by monoclonal antibodies have been defined within a few hundred base pairs (65, 91), and a tentative processing pathway has been described (91). The relationships of the observed repetitive and nonrepetitive variable regions to the immunologically variable regions should soon be known.

Erythrocyte Surface Antigens

RING-INFECTED ERYTHROCYTE SURFACE ANTIGEN Antibodies against clone Ag13 (76) reacted with an antigen associated with the membrane of erythrocytes containing ring-stage parasites (29). Perlmann and coworkers (114) detected the same antigen, which they designated Pf155, using sera from individuals living in areas where malaria is endemic. Antibodies to the ring-infected erythrocyte surface antigen (RESA) inhibit merozoite invasion in vitro (12, 147). Immunoelectronmicroscopy with rabbit affinity–purified human antibodies (19) localized the RESA to micronemes in merozoites. As the merozoite surface was free of label, it appeared that the RESA was transferred to the erythrocyte surface from within merozoites at or shortly after invasion and was not involved in the initial interaction between the merozoite surface and the erythrocyte membrane. The RESA is currently of particular interest because immunization with fused polypeptides containing RESA fragments protected *Aotus* monkeys against overwhelming infection (25).

The RESA in merozoites and in ring-stages has a molecular mass of 155 kd (19, 29). The polypeptide in merozoites is soluble in the nonionic detergent Triton X-100, but after transfer to the membrane of the ring-infected erythrocyte it is largely Triton insoluble (19). Thus it seems likely that the RESA interacts with the erythrocyte membrane skeleton. Whether the RESA penetrates the membrane lipid bilayer is not yet clear.

The RESA contains two exons separated by a short intervening sequence (48) (Figure 2). Exon 1 commences with a hydrophilic sequence followed by a hydrophobic stretch of 13 residues commencing at amino acid 52, and terminates with an intervening sequence that follows this signal core. The major open reading frame of 1008 amino acids is encoded by exon 2, which commences 203 bases downstream. The polypeptide has a predicted molecular mass of 125 kd, which is considerably lower than the 155 kd determined using SDS polyacrylamide gel electrophoresis. The RESA contains two blocks of tandem repeats within exon 2 (35, 48). A 3' block of repeats contains closely related acidic sequences of eight amino acids (EENVEHDA), four amino acids (EENV), and three amino acids (EEV). Approximately 600 bases 5' to this block there is a second block of repeats encoding degenerate sequences related to the acidic sequence DDEHVEEPTVA. The two blocks of repeats encode cross-reacting antigenic epitopes (7, 35). Naturally immunogenic epitopes in nonrepeat regions of the RESA polypeptide have not been identified.

Although the function of the RESA has not been established, it seems likely that its secretion from merozoites and subsequent association with the erythrocyte membrane and/or cytoskeleton is an integral part of the invasion process. However, the persistence of the RESA long after merozoite invasion may indicate alternative functions. Further, it is not known whether the repeats serve such a primary function or a secondary immunological function. However, the pattern of sequence variation in the 5' repeats is strikingly different from that of the S-antigens; the RESA 5' repeats are degenerate along the molecule but highly conserved between two different isolates. Restriction fragment–length polymorphisms defining two alleles of the RESA were observed at both its 5' and 3' ends (48), and all five isolates studied belonged to one or the other of these two alleles. The sequences of representatives of these two alleles (isolates FC27 and NF7) were almost identical over the region that could be compared (35, 48).

KNOB-ASSOCIATED HISTIDINE-RICH PROTEIN Erythrocytes infected with mature stages of *P. falciparum* attach to the endothelial cells lining venules of the deep tissues. This "cytoadherence" leads to the sequestration of mature asexual stages out of the peripheral circulation (90, 100). The presence of "knobs" (140), electron-dense protruberances of the plasma membrane of red cells infected with mature parasites, is implicated in cytoadherence of infected cells (11, 85, 142). Knobs may be composed of several proteins, of which one or more is responsible for the electron-dense material whereas others may be involved in adherence (143).

A protein of 85–105 kd, designated the knob-associated histidine-rich protein (KAHRP) (79), has been found in knobby but not in knobless lines (56, 79, 80, 87). Isolation of cDNA clone SD17, which encodes KAHRP, has allowed the unambiguous demonstration by immunoelectronmicroscopy that KAHRP is localized in the knobs (39). In several knobless isolates of *P. falciparum* the gene that encodes KAHRP, normally located on chromosome 2, was partially (121) or completely (34, 39) deleted from the genome. When part of the gene remained following the deletion of ~20% of chromosome 2, the KAHRP gene sequences were located near the telomere (121). Loss of expression of KAHRP following such deletions may result in the knobless phenotype.

The complete structure (Figure 2) of KAHRP from isolate NF7 has recently been determined (141), as has a partial sequence for isolate FCR-3 (82). The single intervening sequence is located immediately after a signal sequence core of 11 hydrophobic residues commencing at amino acid 21 (141). The predicted sequence includes only ~8% histidine, and this is mainly contributed by strings near the 5' end of exon 2; the strings contain 11, 6, and 10 histidine residues in NF7 and 7, 6, and 9 histidine residues in FCR-3. The sequences of NF7 and FCR-3 are remarkably similar over the region that can be compared. The structure of the molecule is reminiscent of the RESA in that exon 2 contains two blocks of repeats, designated the 5' and 3' repeats (141). The 5' repeats consist of five relatively degenerate copies of a 13–16–amino acid sequence related to the tetradecamer SKKHKDNEDAESVK. The 3' repeats contain seven copies of the 10–amino acid sequence SKGATKEAST; substitutions of glycine for glutamic acid at position 3 and vice versa at position 7 are most common. Studies with precisely defined expression clones and with synthetic peptides have demonstrated that the histidine-rich region and the 5' and 3' repeats all bear non–cross-reacting immunological epitopes recognized during natural infection in humans (141).

MATURE PARASITE–INFECTED ERYTHROCYTE SURFACE ANTIGEN A high–molecular weight mature-stage antigen that contains the hexamer repeat GESKET has been identified by screening a *λamp*3 expression library (31). The gene shows dramatic restriction fragment–length polymorphism, but in contrast to S-antigens all isolates tested so far cross-hybridize strongly. The polypeptide varies in mass between about 250 and 280 kd and differs antigenically among different isolates, although it has fewer distinct forms than S-antigens. The antigen appears to be associated with the cytoskeleton of erythrocytes infected with trophozoites and schizonts, although the degree of cytoskeletal association appears to vary among isolates. As it is localized at the surface of erythrocytes infected with mature stages, it has been termed the mature parasite–infected erythrocyte surface antigen (MESA). Although the properties of the MESA are similar to those of the strain-specific molecule believed to be responsible for sequestration (88), there is at present no evidence that the MESA is involved in cytoadherence. However, it certainly contributes to antigenic diversity among *P. falciparum* strains. Furthermore, another high–molecular weight membrane-associated antigen that may correspond to the MESA has recently been described (73).

PF11-1 A genomic DNA clone expressing a repetitive portion of a *P. falciparum* antigen was identified by its reaction with a pool of human sera from African adults and a rabbit serum against Triton X-100–soluble proteins of the asexual stages (84). The clone encoded 22 complete repeats of a 27-bp sequence. The consensus repeat was EELVEEVIP, often with the leucine replaced by valine or isoleucine and the isoleucine replaced by valine. The corresponding *P. falciparum* antigen is believed to be a schizont antigen associated with the membrane of the infected erythrocyte (84). The mass of the protein is unknown but is estimated to be 250 kd from the size of the mRNA. The repeats of Pf11-1 are quite similar in sequence to the repeat regions at the C-terminus of the RESA, EENVEENV and EENVEHDA (29). Not surprisingly, in a number of assays polyclonal antibodies to Pf11-1 reacted with the RESA and vice versa (145). This is but one of many cross-reactivities among *P. falciparum* antigens (2).

Rhoptry Proteins

Evidence from several species of *Plasmodium* indicates that the rhoptries contain antigens capable of inducing protective immune responses. A 235-kd rhoptry protein from *P. yoelii* protected mice against this species (62, 106). The analogous protein in *P. falciparum* has not been identified, but a number of smaller rhoptry antigens have been described (21, 70). A 41-kd rhoptry antigen, frequently detected as a doublet, has been identified by use of monoclonal antibodies (116). Some monoclonal antibodies to this polypeptide also react with an 82-kd polypeptide, which is processed during schizogony into two smaller products (16). Evidence that the 41-kd antigen may have a role in protective immunity came from studies on inhibition of growth in vitro (119) and immunization studies in *Saimiri* monkeys (117). To date, no structural studies have been reported for this potentially important molecule.

Clone Ag44 (77) encodes a distinct rhoptry antigen. Antibodies purified on Ag44 recognize three polypeptides (107, 105, and 103 kd), which are variably represented in the different stages of the asexual cycle (38). The relationship of these antigens to those identified by monoclonal antibodies is not known. The Ag44 cDNA fragment contained no repeats (27a).

Other Repetitive Antigens

THE HISTIDINE-RICH PROTEIN OF *PLASMODIUM LOPHURAE* Kilejian (78) isolated a protein of 35–40 kd from cytoplasmic granules of the avian malaria parasite *P. lophurae* that was remarkably rich in histidine residues (~73%).

Although its function remains unknown, this histidine-rich protein (HRP) was the first malaria protein known to have an unusual structure. Studies in which ducklings immunized with HRP were protected against fatal infections with *P. lophurae* have proved controversial (81, 98, 128). The HRP gene (74, 123) is interrupted by a single intervening sequence located after the signal sequence (Figure 2). After a short stretch of hydrophilic amino acids the remainder of the polypeptide is composed primarily of repeats, of which the predominant varieties are the decamers APH_8 and DAH_8 (74, 123).

Because of the early interest in this protein, there were attempts to find the corresponding gene product by labeling *P. falciparum* with [^{3}H]histidine (80). Some polypeptides relatively rich in histidine were found (see below), but sequence data has shown no other relationship of these genes to the HRP of *P. lophurae*.

HISTIDINE- AND ALANINE-RICH PROTEINS A cDNA encoding a small histidine- and alanine-rich protein (SHARP) was isolated as an antigen-positive clone (135). This protein was polymorphic among different strains (~28–33 kd) and contained two blocks of repeats, one composed of the hexamer AHHAAN and the other composed of the pentamer HHDGA. Antibodies purified on the cloned antigen cross-reacted with a second, larger polypeptide as well (~60 kd), although these two polypeptides were separate gene products. The gene for the larger polypeptide cross-hybridized with SHARP cDNA but was deleted from a cloned *P. falciparum* line (D10) that does not express the larger polypeptide. As the gene for the SHARP was not present in clone HB3, neither gene is necessary for growth of blood stages in vitro. However, it remains to be seen whether both genes can be deleted simultaneously. The genes for both the SHARP (renamed PfHRP-III) and the larger protein (designated PfHRP-II) were recently cloned (150). The two genes (Figure 2) are both interrupted by intervening sequences located after the signal core region and they show high levels of homology, which suggests that they have originated from a common ancestral sequence. The repeat of the larger polypeptide is most commonly AHHAAD, which differs only in the sixth position from that in the SHARP (AHHAAN). The flanking regions are less divergent than the repeat.

In studies with a monoclonal antibody approximately 50% of the PfHRP-II molecules were secreted from intact parasitized erythrocytes between 2 and 24 hr of culture (72). Immunoelectronmicroscopy localized PfHRP-II to several cell compartments including the parasite cytoplasm, concentrated "packets" in the host erythrocyte cytoplasm, and the infected red blood cell membrane. It is not known whether these polypeptides are targets of protective immunity.

FALCIPARUM INTERSPERSED REPEAT ANTIGEN The >300-kd falciparum interspersed repeat antigen (FIRA) was first defined by cDNA clone Ag231 (134). All isolates of *P. falciparum* tested so far express this antigen. As antibodies to FIRA are abundant in the majority of individuals living in endemic areas, the antigen is unlikely to be a target of protective antibody responses. The FIRA contains blocks of 13 tandemly repeated hexapeptide sequences related to the consensus sequence VTTEEP (Figure 2). These blocks of hexapeptide repeats alternate with an 81–amino acid sequence. The entire unit forms a higher-order repeat, which is present about 10 times in the FIRA molecule. The gene shows marked restriction fragment–length polymorphisms in different isolates, which suggests that the higher-order repeat number varies. Studies with synthetic polypeptides demonstrated that the hexapeptide repeats encode naturally immunogenic epitopes. It is not yet known, however, whether the 81–amino acid repeat, the longest so far found in a malaria antigen, encodes epitopes recognized by antibodies from individuals exposed to malaria.

The nucleotide sequence of a chromosomal FIRA clone (H. D. Stahl, P. E. Crewther, R. F. Anders & D. J. Kemp, in preparation) is organized in a manner analogous to those of the RESA, KAHRP, and HRPs: It contains a short exon 1, a much longer exon 2 (which contains the repeats), and a hydrophobic signal core segment at the boundary of the two exons. So far 31 different varieties of the hexapeptide repeat have been found in the FC27 sequence, but this is almost certainly an underestimate because of deletions. Immunofluorescence studies have revealed that the FIRA is external to the parasite but within the parasitized erythrocyte, even in the ring stage. This is consistent with its presence in the supernatant following saponin lysis of infected erythrocytes. The high degree of variability along the molecule makes the FIRA a prime candidate as an immunological "smokescreen" molecule (see below).

ASPARAGINE-RICH PROTEINS A cDNA clone that encodes part of an antigen containing 40% asparagine (132) was designated the asparagine-rich protein (ARP). It contains repeats of the tetramer NNNM and the octamer NNNMNHNM but is also rich in asparagine in nonrepetitive regions of the polypeptide. Immunoblotting studies revealed a complex pattern, and it was not clear whether the multiple bands represented different gene products or fragments generated by proteolysis. A clone encoding another asparagine-rich protein, designated the clustered asparagine-rich protein (CARP) has been described (146). No repeats were found in this clone and it appears to be a different gene product from the ARP. A number of cDNA expression clones that react with anti-ARP antibodies but do not cross-hybridize with ARP (77), as well as a cDNA clone containing a different tetramer-based asparagine-rich repeat that does not hybridize to the above classes (G. Woodrow, personal communication), have been isolated. Studies with fluorescent antibodies suggest that one or more ARPs are located within the blood-stage parasite. The antibodies also react with sporozoites (132), but probably do not with the NANP repeat, even though the repeat region of the CS protein of *P. falciparum* is rich in asparagine.

Cytoplasmic Antigens

PfHSP70 One antigen that appears to be an internal component of the parasite is an abundant 75-kd polypeptide represented by λ*amp*3 clone Ag63 (5, 18). It is highly conserved in size and sequence in all isolates examined so far (14). A search of the DNA sequence databases (14) revealed that Ag63 is closely homologous to the 70-kd major heat-shock protein (Hsp70) first identified in *Drosophila melanogaster* and subsequently found to have counterparts in many other organisms. It contains repeats of the tetrapeptide GGMP inserted into a poorly conserved region of Hsp70. Ag63 is closely related to the "cognate Hsp70," a noninducible member of the same sequence family (107). An antigen that was thought to be a merozoite surface protein is the same gene product (50).

Fragmentary data on a number of other repetitive antigens have been reported (86, 131).

GENERATION OF REPEAT DIVERSITY AND ANTIGENIC VARIATION

The molecular basis of serological diversity in the S-antigen, CS protein, and PMMSA systems of different *P. falciparum* isolates or species has now been studied in some detail. Sequencing studies have shown that in S-antigens and CS proteins antigenic diversity results entirely from differences in the repeats. In the PMMSA a variant repeat system similar to but more limited than that of the S-antigen is found; however, in addition, recombination generates diversity outside the repeats (137).

The different repeats in these genes could have arisen by a variety of mechanisms. The two extreme possibilities are that (*a*) the different repetitive genes are a series of diverse alleles at a single locus or (*b*) there is a multigene family in the genome that can be activated by DNA rearrangement or transcriptional mechanisms. However, all the available data support the belief that the S-antigen genes and the PMMSA are both series of alleles at single loci.

The most remarkable property of the S-antigen repeats is that even though they can differ so drastically from isolate to isolate, they can be precisely conserved along the molecule. Unequal crossing over or gene conversion can readily explain how a mutation, deletion, or insertion can be spread along a repeat array. This subject has been extensively studied in the case of transcriptionally inert repetitive satellite DNA in eukaryotes (130). The data suggest that as repeats mutate, the mutations spread through the repeat array by unequal crossing over. As a result, parasite populations emerge with S-antigens that differ in sequence and in antigenic properties. However, many of these populations coexist at any one time and within a restricted geographical area. The same principle presumably applies to all the repetitive antigens, except that the outcome depends on the balance of selective forces for each gene.

A phenomenon similar to antigenic variation in trypanosomes may also occur in *Plasmodium*. Brown & Brown (20) found that the schizont-infected cell agglutination (SICA) antigen on the surface of cells infected with *P. knowlesi* schizonts was immunologically variable. Changes in SICA antigen specificity could occur in a cloned *P. knowlesi* line (71). The SICA antigen appeared as a set of high–molecular weight polypeptides on the surface of the schizont-infected cell; SICAs differed in size and antigenicity in the cloned progeny of a single cell (71). A similar phenomenon may occur in *P. falciparum* (66). Although this phenomenon is equivalent to antigenic variation in trypanosomes, the underlying molecular mechanisms are not necessarily the same. In trypanosomes individual sequences encoding these variants can be switched on transcriptionally, usually by DNA rearrangements that move the gene to be expressed to transcriptionally active telomeric sites (15). The molecular mechanisms for antigenic variation of the SICA antigen are unknown. One possibility is that antigenic specificity is determined by a repetitive sequence of a single gene and can be rapidly varied by the spread of mutations through the repeat array. However, a number of other recombination mechanisms involving single or multiple genes can be envisaged (51).

REPETITIVE ANTIGENS AND IMMUNE EVASION

Polypeptides containing repetitive sequences are widely distributed in nature and include well-studied fibrous proteins such as silks, collagen, and keratins. More obscure examples are the antifreeze protein in the bloodstream of some Antarctic fish (42) and the glue proteins secreted by the salivary gland of *D. melanogaster* (53, 102). There are many other similar examples, including proteins of viruses, bacteria, and plants.

As many polypeptides in nature contain extensive sequence repeats, why are the *P. falciparum* repetitive polypeptides of special interest? First, the extent of the repeats is sometimes extraordinary; for example, over 90% of the S-antigen of *P. falciparum* isolate FC27 is composed of the 11–amino acid repeat. Second, these antigens are dominant natural immunogens; much of the antibody response induced by malaria infections is directed against repeat epitopes. FIRA in particular is a dominant antibody specificity in the sera of many individuals exposed to malaria (133), and the repeats are the immunodominant parts of the molecule. Third, many of these repetitive antigens undergo specific processing or fragmentation. Collectively the primary translation products and proteolytic fragments account for a substantial proportion of the naturally immunogenic polypeptides detectable by two-dimensional electrophoresis. Fourth, the remarkable level of antigenic diversity among different strains of *P. falciparum* (49, 136) reflects, at least in part, the repeat structure of these antigens.

Ycas (154) and Ohno (105) have concluded that repeats may have a major role in protein evolution. Rapid evolution is presumably important for polypeptides that must adapt to rapidly evolving host-parasite relationships. The genes for some antigens may have evolved from the grafting of a common ancestral 5' exon to the repetitive regions. The modular structure and repetitive nature of the *P. falciparum* antigens suggest that they have evolved to serve a number of purposes using widely employed evolutionary strategies. However, because some of the repeats are highly immunogenic, the antigens compete against host-protective immune responses. In such a situation of immunological selection, gene duplication and variation from the spread of mutations in repetitive sequences would increase this competition and allow elaboration of the resulting gene products into a "smokescreen" as described below. Hence the repeats in some of these gene products may now have no function other than in immune evasion, whereas other repetitive genes may retain critical functions. The RESA and the CS protein of *P. falciparum* appear to be highly conserved, which suggests that their repeats are under considerable selection, presumably because they have functions apart from interacting with the host immune system.

THE SMOKESCREEN MODEL FOR IMMUNE EVASION

The immune response to malaria differs from the immune response to viral or bacterial infections in several ways. One is that protective immunity develops only after the individual is exposed to many infections over several years. This may reflect the need for involvement of multiple strains of parasites because of antigenic variation and/or strain differences, or it may reflect the parasites' use of other strategies to evade specific immune responses. Despite the inefficient development of protective immune responses there are dramatic immunological consequences of repeated infection with malaria parasites. For example, many nonimmune individuals have very high levels of antibodies to many repetitive antigens such as the FIRA and exhibit an associated marked hypergammaglobulinemia. The magnitude of this ineffectual B-cell response suggests that hyperstimulation of irrelevant B-cell responses leads to less effective responses against critical "protective" epitopes. In this sense these antigens would act as a "smokescreen," helping to camouflage the parasite from the host immune system.

Immune competition resulting from the presentation to the host immune system of large amounts of many immunogenic repeat antigens may be important in this evasion strategy. However, the finding of multiple cross-reactivities among many of these antigens has suggested an additional, more subtle mechanism, which is mediated by the effects of these repeats on B-cell proliferation and maturation of antibody affinity.

The cross-reactions among malaria antigens may be intramolecular or intermolecular. Intramolecular cross-reactions occur between different epitopes within a single block of repeats or between epitopes in different repetitive regions of the one polypeptide. Intermolecular cross-reactions may involve two different asexual-stage antigens, antigens of different life-cycle stages, or antigens of different strains of *P. falciparum* or species of *Plasmodium*.

Intramolecular cross-reactions have been best documented in the RESA polypeptide. Antisera raised against a fused polypeptide produced in *E. coli* and encoding 3' repeat sequences reacted with a fused polypeptide encoding 5' repeat sequences (35). Subsequently, the cross-reactivity between these two RESA repeat structures, which have some sequence homology, was confirmed with affinity-purified human antibodies and monoclonal antibodies (7). The same monoclonal antibodies have been used to document cross-reactivities within the block of 3' repeats. Some of these monoclonals were reactive with both the four– and eight–amino acid repeat sequences that make up the 3' repeats, whereas others reacted with only the four–amino acid repeat (6). Such intramolecular cross-reactions presumably occur in numerous other antigens where there is a degree of degeneracy in the sequence repeats, e.g. the FIRA, some S-antigens, and Pf11.1 (84).

The first intermolecular cross-reactions of *P. falciparum* to be reported involved the CS protein and an asexual blood-stage antigen that both reacted with a single monoclonal antibody (67). The asexual blood-stage antigen contained a short degenerate tetrameric repeat related to the repeat in the CS protein (32, 68). Antibodies to this CS protein–related antigen (CRA), affinity-purified from human sera, reacted with the surface of sporozoites. Thus, antisporozoite immunity may be induced or boosted by this asexual blood-stage antigen.

A major network of intra- and intermolecular cross-reactivities involves the RESA, the FIRA, Pf11.1, and probably several other molecules (145). Antibodies purified on clones encoding any one of these antigens are frequently reactive with two or more of them. The sequences have similar repeats; many contain pairs of acidic residues that may be involved in the cross-reacting epitopes. Other documented intermolecular cross-reactivities involve histidine-rich and asparagine-rich proteins (38, 132).

These unusual characteristics of malaria antigens may be critical to any role of the repeats in modifying the host immune response. Furthermore, they may be responsible for distinctive features of the immune response in malaria such as marked hypergammaglobulinemia, slow development of unstable immunity, and autoantibody production.

The role of somatic mutation in the antibody response has suggested how such a network of cross-reacting epitopes may cause some of these effects. High-affinity antibody responses probably depend on the selection of B cells with somatic mutations accumulated in their surface immunoglobulin that cause the B-cell receptors to have higher affinities (reviewed in 93). These B cells proliferate preferentially to produce antibody-secreting plasma cells. In normal immune responses somatic mutants are selected by the epitope that triggered the parental naive B cell to proliferate. However, malaria antigens have a large number of structurally related cross-reacting epitopes in addition to the triggering epitope. As a result, a mutated cell that fails to react with the triggering epitope with sufficient affinity may react with one of the cross-reacting epitopes with higher affinity and, as a consequence, undergo further proliferation.

The above considerations have led to the hypothesis that in antimalarial immune responses an abnormally high proportion of the B cells that accumulate somatic mutations in their surface immunoglobulin are preserved during clonal expansion (2). This may be responsible, in part, for the hypergammaglobulinemia associated with malaria and may also increase the incidence of autoantibodies. More importantly, the development of high-affinity antibody responses, which could be important for immunity, may be delayed because of limits to B-cell proliferation imposed by the size of the B-cell pool.

OUTLOOK FOR A MALARIA VACCINE

There is evidence that repetitive epitopes within the RESA and the CS protein can induce host-protective immune responses. Current trials in humans and monkeys are assessing the efficacy of these antigens as vaccines. However, the repeats in many molecules may be capable of rapid variation, particularly under the selection pressure of a strong immune response induced by immunization. This variation and the patterns of diversity common to repetitive antigens such as S-antigens, PMMSAs, and some CS proteins suggest that even if such vaccines are successful in early trials they may fail in the field. Although few naturally occurring immune responses to antigens of *P. falci-*

parum have been mapped to nonrepetitive regions of the polypeptides, the identification of such sequences and the assessment of their efficacy as vaccine components are of great importance. However, the evidence that recombination generates diversity in the nonrepetitive regions of the PMMSA suggests further caution.

If the many related repetitive sequences do favor parasite survival because of the delayed maturation of antibody affinity as suggested above, there are reasons to be optimistic about a vaccine based on critical repetitive sequences. Immunization with these epitopes in isolation from their many structural analogs should enable more rapid development of protective immunity. One important conclusion from the recent studies of malaria antigens and a further reason for optimism is that antigenic diversity in plasmodia differs fundamentally from that in African trypanosomes. The variant-specific glycoprotein (VSG) of the trypanosomes is the target of host-protective immune responses but also mediates evasion by varying antigenically (15). Effective immunity against one variant antigenic type is therefore useless to the host, and the number of different VSGs makes improbable the development of an effective vaccine. Although some antigenic variation does occur in plasmodia, there clearly are other antigens that are targets of protective immune responses. If we can understand the role of the repeat structures in these antigens it may be possible to develop a vaccine that will be effective against most strains of *P. falciparum*.

ACKNOWLEDGMENTS

We thank Graham Mitchell, Graham Brown, and Irwin Sherman for criticizing the manuscript. We also thank them and many members of The Walter and Eliza Hall Institute for stimulating discussions. This work was supported by the Australian National Health and Medical Research Council, the John D. and Catherine T. MacArthur Foundation, the Rockefeller Foundation Great Neglected Diseases Network, the Australian National Research Grants Scheme, and the Australian Malaria Vaccine Joint Venture. We thank the members of the Papua New Guinea Institute of Medical Research for parasites and serum samples. Thanks also to Heather Saunders and Diana Viti for typing the manuscript.

Literature Cited

1. Aley, S. B., Bates, M. D., Tam, J., Nussenzweig, V., Nussenzweig, R. S., Hollingdale, M. R. 1986. Synthetic peptide from *Plasmodium falciparum* circumsporozoite protein has affinity for human hepatoma cell line HEPG2-A16. In *Parasitology—Quo Vadit? ICOPA VI Handbook,* ed. M. J. Howell, p. 81. Canberra, Australia: Aust. Acad. Sci.
2. Anders, R. F. 1986. Multiple cross-reactivities amongst antigens of *Plasmodium Falciparum* impair the development of protective immunity against malaria. *Parasite Immunol.* 8:529–39
3. Anders, R. F. 1985. Candidate antigens for an asexual blood-stage vaccine. *Parasitol. Today* 1:153–56
4. Anders, R. F., Brown, G. V., Edwards,

A. E. 1983. Characterization of an S-antigen synthesized by several isolates of *Plasmodium falciparum*. *Proc. Natl. Acad. Sci. USA* 80:6652–56

5. Anders, R. F., Coppel, R. L., Brown, G. V., Saint, R. B., Cowman, A. F., et al. 1984. *Plasmodium falciparum* complementary DNA clones expressed in *Escherichia coli* encode many distinct antigens. *Mol. Biol. Med.* 2:177–91
6. Anders, R. F., Shi, P. T., Brown, G. V., Stahl, H. D., Favaloro, J., et al. 1985. Immune recognition of repeat structures in malaria antigens. In *Immune Recognition of Protein Antigens. Current Communications in Molecular Biology, Cold Spring Harbor Laboratory,* ed. W. G. Laver, G. M. Air, pp. 193–97. Cold Spring Harbor, NY: Cold Spring Harbor Lab.
7. Anders, R. F., Shi, P. T., Scanlon, D. B., Leach, S. J., Coppel, R. L., et al. 1986. Antigenic repeat structures in proteins of *Plasmodium falciparum*. *Ciba Found. Symp.* 119:164–75
8. Ardeshir, F., Flint, J. E., Reese, R. T. 1985. Expression of *Plasmodium falciparum* surface antigens in *Escherichia coli*. *Proc. Natl. Acad. Sci. USA* 82: 2518–22
9. Arnot, D. E., Barnwell, J. W., Tam, J. P., Nussenzweig, V., Nussenzweig, R. S., Enea, V. 1985. Circumsporozoite protein of *Plasmodium vivax:* Gene cloning and characterization of the immunodominant epitope. *Science* 230: 815–18
10. Ballou, W. R., Rothbard, J., Wirtz, R. A., Gordon, D. M., Williams, J. S., et al. 1985. Immunogenicity of synthetic peptides from circumsporozoite protein of *Plasmodium falciparum*. *Science* 228:996–99
11. Barnwell, J. W., Howard, R. J., Coon, H. G., Miller, L. H. 1983. Splenic requirement for antigenic variation and expression of the variant antigen on the erythrocyte membrane in cloned *Plasmodium knowlesi* malaria. *Infect. Immun.* 40:985–94
12. Berzins, K., Perlmann, H., Udomsangpetch, R., Wahlin, B., Wahlgren, M., et al. 1985. Pf 155, a candidate for a blood stage vaccine in *Plasmodium falciparum* malaria. *Dev. Biol. Stand.* 62:99–106
13. Bianco, A. E., Culvenor, J. G., Coppel, R. L., Crewther, P. E., McIntyre, P., et al. 1987. A putative glycophorin-binding protein is secreted from schizonts of *Plasmodium falciparum*. *Mol. Biochem. Parasitol.* In press
14. Bianco, A. E., Favaloro, J. M., Burkot, T. R., Culvenor, J. G., Crewther, P. E., et al. 1986. A repetitive antigen of *Plasmodium falciparum* that is homologous to heat shock protein 70 of *Drosophila melanogaster*. *Proc. Natl. Acad. Sci. USA* 83:8713–17
15. Boothroyd, J. C. 1985. Antigenic variation in African trypanosomes. *Ann. Rev. Microbiol.* 39:475–502
16. Braun-Breton, C., Jendoubi, M., Brunet, E., Perrin, L., Scaife, J., Pereira da Silva, L. 1986. In vivo time course of synthesis and processing of major schizont membrane polypeptides in *Plasmodium falciparum*. *Mol. Biochem. Parasitol.* 20:33–43
17. Brown, G. V. 1986. Prospects for a vaccine against malaria. *Med. J. Aust.* 144:703–4
18. Brown, G. V., Anders, R. F., Coppel, R. L., Saint, R. B., Cowman, A. F., et al. 1984. The expression of *Plasmodium falciparum* blood-stage antigens in *Escherichia coli*. *Philos. Trans. R. Soc. London Ser. B* 307:179–87
19. Brown, G. V., Culvenor, J. G., Crewther, P. E., Bianco, A. E., Coppel, R. L., et al. 1985. Localization of the Ring-infected Erythrocyte Surface Antigen (RESA) of *Plasmodium falciparum* in merozoites and ring-infected erythrocytes. *J. Exp. Med.* 162:774–79
20. Brown, K. N., Brown, I. N. 1965. Immunity to malaria: antigenic variation in chronic infections of *Plasmodium knowlesi*. *Nature* 208:1286–88
21. Campbell, G. H., Miller, L. H., Hudson, D., Franco, E. L., Andrysiak, P. M. 1984. Monoclonal antibody characterization of *Plasmodium falciparum* antigens. *Am. J. Trop. Med. Hyg.* 33:1051–54
22. Carter, R., Miller, L. H., Rener, J., Kaushall, D. C., Kumar, N., et al. 1984. Target antigens in malaria transmission blocking immunity. *Philos. Trans. R. Soc. London Ser. B* 307:201–13
23. Cheung, A., Leban, J., Shaw, A. R., Merkli, B., Stocker, J., et al. 1986. Immunization with synthetic peptides of a *Plasmodium falciparum* surface antigen induces antimerozoite antibodies. *Proc. Natl. Acad. Sci. USA* 83:8328–32
24. Cheung, A., Shaw, A. R., Leban, J., Perrin, L. H. 1985. Cloning and expression in *Escherichia coli* of a surface antigen of *Plasmodium falciparum* merozoites. *EMBO J.* 4:1007–11
25. Collins, W. E., Anders, R. F., Pappaioanou, M., Campbell, G. H., Brown, G. V., et al. 1986. Immunization of *Aotus* monkeys with recombinant proteins of an erythrocyte surface anti-

gen of *Plasmodium falciparum*. *Nature* 323:259–62

26. Coppel, R. L. 1986. Prospects for a malaria vaccine. *Microbiol. Sci.* 3:292–95
27. Coppel, R. L., Anders, R. F., Brown, G. V., Cowman, A. F., Lingelbach, K. R., et al. 1984. Detection and analysis of *Plasmodium falciparum* antigens expressed in *Escherichia coli*. In *Molecular Parasitology,* ed. J. T. August, pp. 103–15. New York: Academic

27a. Coppel, R. L., Bianco, A. E., Culvenor, J. G., Crewther, P. E., Brown, G. V., et al. 1987. cDNA clone expressing a rhoptry protein of *Plasmodium falciparum*. *Mol. Biochem. Parasitol.* In press

28. Coppel, R. L., Brown, G. V., Mitchell, G. F., Anders, R. F., Kemp, D. J. 1984. Identification of a cDNA clone encoding a mature blood stage antigen of *Plasmodium falciparum* by immunization of mice with bacterial lysates. *EMBO J.* 3:403–7
29. Coppel, R. L., Cowman, A. F., Anders, R. F., Bianco, A. E., Saint, R. B., et al. 1984. Immune sera recognize on erythrocytes a *Plasmodium falciparum* antigen composed of repeated amino acid sequences. *Nature* 310:789–91
30. Coppel, R. L., Cowman, A. F., Lingelbach, K. R., Brown, G. V., Saint, R. B., et al. 1983. Isolate-specific S-antigen of *Plasmodium falciparum* contains a repeated sequence of eleven amino acids. *Nature* 306:751–56
31. Coppel, R. L., Culvenor, J. G., Bianco, A. E., Crewther, P. E., Stahl, H. D., et al. 1986. Variable antigen associated with the surface of erythrocytes infected with mature stages of *Plasmodium falciparum*. *Mol. Biochem. Parasitol.* 20:265–77
32. Coppel, R. L., Favaloro, J. M., Crewther, P. E., Burkot, T. R., Bianco, A. E., et al. 1985. A blood stage antigen of *Plasmodium falciparum* shares determinants with the sporozoite coat protein. *Proc. Natl. Acad. Sci. USA* 82:5121–25
33. Coppel, R. L., Saint, R. B., Stahl, H. D., Langford, C. J., Brown, G. V., et al. 1985. *Plasmodium falciparum:* differentiation of isolates with DNA hybridization using antigen gene probes. *Exp. Parasitol.* 60:82–89
34. Corcoran, L. M., Forsyth, K. P., Bianco, A. E., Brown, G. V., Kemp, D. J. 1986. Chromosome size polymorphisms in *Plasmodium falciparum* can involve deletions and are frequent in natural parasite populations. *Cell* 44:87–95
35. Cowman, A. F., Coppel, R. L., Saint, R. B., Favaloro, J., Crewther, P. E., et al. 1984. The ring-infected erythrocyte surface antigen (RESA) polypeptide of *Plasmodium falciparum* contains two separate blocks of tandem repeats encoding antigenic epitopes that are naturally immunogenic in man. *Mol. Biol. Med.* 2:207–21
36. Cowman, A. F., Saint, R. B., Coppel, R. L., Brown, G. V., Anders, R. F., Kemp, D. J. 1985. Conserved sequences flank variable tandem repeats in two S-antigen genes of *Plasmodium falciparum*. *Cell* 40:775–83
37. Cowman, A. F., Saint, R. B., Coppel, R. L., Brown, G. V., Favaloro, J., et al. 1985. Repeat structures in protein antigens of asexual erythrocyte stages of *Plasmodium falciparum*. In *Vaccines 85,* ed. R. Lerner, R. Chanock, pp. 13–18. Cold Spring Harbor, NY: Cold Spring Harbor Lab.
38. Crewther, P. E., Bianco, A. E., Brown, G. V., Coppel, R. L., Stahl, H. D., et al. 1986. Affinity purification of human antibodies directed against cloned antigens of *Plasmodium falciparum*. *J. Immunol. Methods* 86:257–64
39. Culvenor, J. G., Langford, C. J., Crewther, P. E., Saint, R. B., Coppel, R. L., et al. 1987. *Plasmodium falciparum:* Identification and localization of a knob protein antigen expressed by a cDNA clone. *Exp. Parasitol.* 63:58–67
40. Dame, J. B., Williams, J. L., McCutchan, T. F., Weber, J. L., Wirtz, R. A., et al. 1984. Structure of the gene encoding the immunodominant surface antigen on the sporozoite of the human malaria parasite *Plasmodium falciparum*. *Science* 225:593–99
41. de la Cruz, V. F., McCutchan, T. F. 1986. Heterogeneity at the 5' end of the circumsporozoite protein gene of *Plasmodium falciparum* is due to a previously undescribed repeat sequence. *Nucleic Acids Res.* 14:4695
42. DeVries, A. L., Vandenheede, J., Feeney, R. E. 1971. Primary structure of freezing point–depressing glycoproteins. *J. Biol. Chem.* 246:305–8
43. Eichinger, D. J., Arnot, D. E., Tam, J. P., Nussenzweig, V., Enea, V. 1986. The circumsporozoite protein of *Plasmodium falciparum:* gene cloning and identification of the immunodominant epitope. *Mol. Cell. Biol.* 6:3965–72
44. Ellis, J., Ozaki, L. S., Gwadz, R. W., Cochrane, A. H., Nussenzweig, V., et al. 1983. Cloning and expression in *E. coli* of the malarial sporozoite surface antigen gene from *Plasmodium knowlesi*. *Nature* 302:536–38
45. Enea, V., Arnot, D., Schmidt, E. C., Cochrane, A., Gwadz, R., Nussen-

zweig, R. S. 1984. Circumsporozoite gene of *Plasmodium cynomolgi* (Gombak): cDNA cloning and expression of the repetitive circumsporozoite epitope. *Proc. Natl. Acad. Sci. USA* 81:7520–24

46. Enea, V., Ellis, J., Zavala, F., Arnot, D. E., Asavanich, A., et al. 1984. DNA cloning of *Plasmodium falciparum* circumsporozoite gene: amino acid sequence of repetitive epitope. *Science* 225:628–30
47. Epstein, N., Miller, L. H., Kaushel, D. C., Udeinya, I. J., Rener, J., et al. 1981. Monoclonal antibodies against a specific surface determinant on malarial *(Plasmodium knowlesi)* merozoites block erythrocyte invasion. *J. Immunol.* 127:212–17
48. Favaloro, J. M., Coppel, R. L., Corcoran, L. M., Foote, S. J., Brown, G. V., et al. 1986. Structure of the RESA gene of *Plasmodium falciparum. Nucleic Acids Res.* 14:8265–77
49. Fenton, B., Walker, A., Walliker, D. 1985. Protein variation in clones of *Plasmodium falciparum* detected by two dimensional electrophoresis. *Mol. Biochem. Parasitol.* 16:173–83
50. Flint, J. E., Ardeshir, F., Reese, R. T. 1986. A merozoite surface antigen of *Plasmodium falciparum* cloned and expressed in *Escherichia coli.* In *Vaccines 86. New Approaches To Immunization,* ed. F. Brown, R. M. Chanock, R. A. Lerner, pp. 175–79. Cold Spring Harbor, NY: Cold Spring Harbor Lab.
51. Forney, J. D., Epstein, L. M., Preer, L. B., Rudman, B. M., Widmayer, D. J., et al. 1983. Structure and expression of genes for surface proteins of *Paramecium. Mol. Cell. Biol.* 3:466–74
52. Galinski, M. R., Arnot, D. A., Cochrane, A. H., Barnwell, J. W., Nussenzweig, R. S., Enea, V. 1986. The circumsporozoite genes of *Plasmodium cynomolgi:* an evolutionary analysis. See Ref. 1, p. 10
53. Garfinkel, M. D., Pruitt, R. E., Meyerowitz, E. M. 1983. DNA sequences, gene regulation and modular protein evolution in the *Drosophila* 68C glue gene cluster. *J. Mol. Biol.* 168: 765–89
54. Godson, G. N., Ellis, J., Ozaki, L. S., Svec, P., Lupski, J. R. 1984. Structure and organization of genes for sporozoite surface antigens. *Philos. Trans. R. Soc. London Ser. B* 307:129–39
55. Godson, G. N., Ellis, J., Svec, P., Schlesinger, D. H., Nussenzweig, V. 1983. Identification and chemical synthesis of a tandemly repeated immunogenic region of *Plasmodium knowlesi* circumsporozoite protein. *Nature* 305:29–33
56. Hadley, T. J., Leech, J. H., Green, T. J., Daniel, W. A., Wahlgren, M., et al. 1983. A comparison of knobby (K+) and knobless (K−) parasites from two strains of *Plasmodium falciparum. Mol. Biochem. Parasitol.* 9:271–78
57. Haldar, K., Ferguson, M. A. J., Cross, G. A. M. 1985. Acylation of a *Plasmodium falciparum* merozoite surface antigen via *sn*-1,2-diacyl glycerol. *J. Biol. Chem.* 260:4969–74
58. Hall, R., Hyde, J. E., Goman, M., Simmons, D. L., Hope, I. A., et al. 1984. Major surface antigen gene of a human malaria parasite cloned and expressed in bacteria. *Nature* 311:379–82
59. Hall, R., McBride, J., Morgan, G., Tait, A., Zolg, J. W., et al. 1983. Antigens of the erythrocytic stages of the human malaria parasite *Plasmodium falciparum* detected by monoclonal antibodies. *Mol. Biochem. Parasitol.* 7: 247–65
60. Hall, R., Osland, A., Hyde, J. E., Simmons, D. L. C., Hope, I. A., Scaife, J. G. 1984. Processing, polymorphism, and biological significance of P190, a major surface antigen of the erythrocytic forms of *Plasmodium falciparum. Mol. Biochem. Parasitol.* 11:61–80
61. Heidrich, H. G., Strych, W., Mrema, J. E. K. 1983. Identification of surface and internal antigens from spontaneously released *Plasmodium falciparum* merozoites by radioiodination and metabolic labelling. *Z. Parasitenkd.* 69:715–25
62. Holder, A. A., Freeman, R. R. 1981. Immunization against blood-stage rodent malaria using purified parasite antigens. *Nature* 294:361–64
63. Holder, A. A., Freeman, R. R. 1982. Biosynthesis and processing of a *Plasmodium falciparum* schizont antigen recognized by immune serum and a monoclonal antibody. *J. Exp. Med.* 156:1528–38
64. Holder, A. A., Freeman, R. R. 1984. The three major antigens on the surface of *Plasmodium falciparum* merozoites are derived from a single high molecular weight precursor. *J. Exp. Med.* 160: 624–29
65. Holder, A. A., Lockyer, M. J., Odink, K. G., Sandhu, J. S., Riveros-Moreno, V., et al. 1985. Primary structure of the precursor to the three major surface antigens of *Plasmodium falciparum* merozoites. *Nature* 317:270–73
66. Hommel, M., David, P. H., Oligino, L. D. 1983. Surface alterations of erythrocytes in *Plasmodium falciparum* malaria. *J. Exp. Med.* 157:1137–48

67. Hope, I. A., Hall, R., Simmons, D. L., Hyde, J. E., Scaife, J. G. 1984. Evidence for immunological cross-reaction between sporozoites and blood stages of a human malaria parasite. *Nature* 308: 191–94
68. Hope, I. A., Mackay, M., Hyde, J. E., Goman, M., Scaife, J. 1985. The gene for an exported antigen of the malaria parasite *Plasmodium falciparum* cloned and expressed in *Escherichia coli*. *Nucleic Acids Res.* 13:369–79
69. Howard, R. F., Reese, R. T. 1984. Synthesis of merozoite proteins and glycoproteins during the schizogony of *Plasmodium falciparum*. *Mol. Biochem. Parasitol.* 10:319–34
70. Howard, R. F., Stanley, H. A., Campbell, G. H., Reese, R. T. 1984. Proteins responsible for a punctate fluorescence pattern in *Plasmodium falciparum* merozoites. *Am. J. Trop. Med. Hyg.* 33:1055–59
71. Howard, R. J., Barnwell, J. W. 1984. The detergent solubility properties of a malarial *(Plasmodium knowlesi)* variant antigen expressed on the surface of infected erythrocytes. *J. Cell. Biochem.* 24:297–306
72. Howard, R. J., Uni, S., Aikawa, M., Aley, S. B., Leech, J. H., et al. 1986. Secretion of a malarial histidine-rich protein (Pf HRP II) from *Plasmodium falciparum*-infected erythrocytes. *J. Cell Biol.* 103:1269–77
73. Howard, R. J., Uni, S., Lyon, J. A., Taylor, D. W., Daniel, W., Aikawa, M. 1987. Export of *Plasmodium falciparum* proteins to the host erythrocyte membrane: special problems of protein trafficking and topogenesis. In *Host-Parasite Cellular and Molecular Interactions in Protozoan Diseases,* ed. K.-P. Chang. New York: Plenum. In press
74. Irving, D. O., Cross, G. A. M., Feder, R., Wallach, M. 1986. Structure and organization of the histidine-rich protein gene of *Plasmodium lophurae*. *Mol. Biochem. Parasitol.* 18:223–34
75. Jungery, M., Boyle, D., Patel, T., Pasvol, G., Weatherall, D. J. 1983. Lectin-like polypeptides of *P. falciparum* bind to red cell sialoglycoproteins. *Nature* 301:704–5
76. Kemp, D. J., Coppel, R. L., Cowman, A. F., Saint, R. B., Brown, G. V., Anders, R. F. 1983. Expression of *Plasmodium falciparum* blood-stage antigens in *Escherichia coli:* Detection with antibodies from immune humans. *Proc. Natl. Acad. Sci. USA* 80:3787–91
77. Kemp, D. J., Coppel, R. L., Stahl, H. D., Bianco, A. E., Corcoran, L. M., et al. 1986. Genes for antigens of *Plasmodium falciparum*. *Parasitology* 91:S83–S108
78. Kilejian, A. 1974. A unique histidine-rich polypeptide from the malaria parasite, *Plasmodium lophurae*. *J. Biol. Chem.* 249:4650–55
79. Kilejian, A. 1979. Characterization of a protein correlated with the production of knob-like protrusions on membranes of erythrocytes infected with *Plasmodium falciparum*. *Proc. Natl. Acad. Sci. USA* 76:4650–53
80. Kilejian, A. 1980. Homology between a histidine-rich protein from *Plasmodium lophurae* and a protein associated with the knob-like protrusions on membranes of erythrocytes infected with *Plasmodium falciparum*. *J. Exp. Med.* 151: 1534–38
81. Kilejian, A. 1981. *Plasmodium lophurae:* immunogenicity of a histidine-rich protein. *Exp. Parasitol.* 52:291
82. Kilejian, A., Sharma, Y. D., Karoui, H., Naslund, L. 1986. Histidine-rich domain of the knob protein of the human malaria parasite *Plasmodium falciparum*. *Proc. Natl. Acad. Sci. USA* 83: 7938–41
83. Kochan, J., Perkins, M., Ravetch, J. V. 1986. A tandemly repeated sequence determines the binding domain for an erythrocyte receptor binding protein of *P. falciparum*. *Cell* 44:689–96
84. Koenen, M., Scherf, A., Mercereau, O., Langsley, G., Sibilli, L., et al. 1984. Human antisera detect a *Plasmodium falciparum* genomic clone encoding a nonapeptide repeat. *Nature* 311: 382–85
85. Langreth, S. G., Reese, R. T., Moytl, M. R., Trager, W. 1979. *Plasmodium falciparum:* Loss of knobs on the infected erythrocyte surface after long-term cultivation. *Exp. Parasitol.* 48: 213–19
86. Langsley, G., Scherf, A., Mercereau-Puijalon, O., Koenen, M., Kahane, B., et al. 1985. Characterization of *P. falciparum* antigenic determinants isolated from a genomic expression library by differential antibody screening. *Nucleic Acids Res.* 13:4191–202
87. Leech, J. H., Barnwell, J. W., Aikawa, M., Miller, L. H., Howard, R. J. 1984. *Plasmodium falciparum* malaria: Association of knobs on the surface of infected erythrocytes with a histidine-rich protein and the erythrocyte skeleton. *J. Cell Biol.* 98:1256–64
88. Leech, J. H., Barnwell, J. W., Miller, L. H., Howard, R. J. 1984. Identification of a strain-specific malarial antigen exposed on the surface of *Plasmodium*

falciparum-infected erythrocytes. *J. Exp. Med.* 159:1567–75

89. Lockyer, M. J., Schwarz, R. T. 1987. Strain variation in the circumsporozoite protein of *Plasmodium falciparum. Mol. Biochem. Parasitol.* 22:101–8
90. Luse, S. A., Miller, L. H. 1971. *Plasmodium falciparum* malaria ultrastructure of parasitized erythrocytes in cardiac vessels. *Am. J. Trop. Med. Hyg.* 20:655–60
91. Lyon, J. A., Geller, R. H., Haynes, J. D., Chulay, J. D., Weber, J. L. 1986. Epitope map and processing scheme for the 195,000-dalton surface glycoprotein of *Plasmodium falciparum* merozoites deduced from cloned overlapping segments of the gene. *Proc. Natl. Acad. Sci. USA* 83:2989–93
92. Mackay, M., Goman, M., Bone, N., Hyde, J. E., Scaife, J., et al. 1985. Polymorphism of the precursor for the major surface antigens of *Plasmodium falciparum* merozoites: studies at the genetic level. *EMBO J.* 4:3823–29
93. Manser, T., Wysocki, L. J., Gridley, T., Near, R. I., Gefter, M. L. 1985. The molecular evolution of the immune response. *Immunol. Today* 6:94–101
94. McBride, J. S., Newbold, C. I., Anand, R. 1985. Polymorphism of a high molecular weight schizont antigen of the human malaria parasite *Plasmodium falciparum. J. Exp. Med.* 161:160–80
95. McBride, J. S., Walliker, D., Morgan, G. 1982. Antigenic diversity in the human malaria parasite *Plasmodium falciparum. Science* 217:254–57
96. McCutchan, T. F., Dame, J. B., Miller, L. H., Barnwell, J. 1984. Evolutionary relatedness of *Plasmodium* species as determined by the structure of DNA. *Science* 225:808–11
97. McCutchan, T. F., Lal, A. A., de la Cruz, V. F., Miller, L. H., Maloy, W. L., et al. 1985. Sequence of the immunodominant epitope for the surface protein on sporozoites of *Plasmodium vivax. Science* 230:1381–83
98. McDonald, V., Hannon, M., Tanigoshi, L., Sherman, I. W. 1981. *Plasmodium lophurae:* immunization of Pekin ducklings with different antigen preparations. *Exp. Parasitol.* 51:195–203
99. McGarvey, M. J., Sheybani, E., Loche, M. P., Perrin, L., Mach, B. 1984. Identification and expression in *Escherichia coli* of merozoite stage-specific genes of the human malarial parasite *Plasmodium falciparum. Proc. Natl. Acad. Sci. USA* 81:3690–94
100. Miller, L. 1969. Distribution of mature trophozoites and schizonts of *Plasmodium falciparum* in the organs of *Aotus trivirgatus,* the night monkey. *Am. J. Trop. Med. Hyg.* 18:860–65
101. Miller, L. H., McAuliffe, F. M., Mason, S. J. 1977. Erythrocyte receptors for malaria merozoites. *Am. J. Trop. Med. Hyg.* 26:204–8
102. Muskavitch, M. A. T., Hogness, D. S. 1982. An expandable gene that encodes a Drosophila glue protein is not expressed in variants lacking remote upstream sequences. *Cell* 29:1041–51
103. Nardin, E. H., Nussenzweig, V., Nussenzweig, R. S., Collins, W. E., Harinasuta, K. T., et al. 1982. Circumsporozoite proteins of human malaria parasites *Plasmodium falciparum* and *Plasmodium vivax. J. Exp. Med.* 156: 20–30
104. Nussenzweig, V., Nussenzweig, R. S. 1985. Circumsporozoite proteins of malaria parasites. *Cell* 42:401–3
105. Ohno, S. 1984. Birth of a unique enzyme from an alternative reading frame of the preexisted, internally repetitious coding sequence. *Proc. Natl. Acad. Sci. USA* 81:2421–25
106. Oka, M., Aikawa, M., Freeman, R. R., Holder, A. A., Fine, E. 1984. Ultrastructural localization of protective antigens of *Plasmodium yoelii* merozoites by the use of monoclonal antibodies and ultrathin cryomicrotomy. *Am. J. Trop. Med. Hyg.* 33:342–46
107. O'Malley, K., Mauron, A., Barchas, J. D., Kedes, L. 1985. Constitutively expressed rat mRNA encoding a 70-kilodalton heat-shock-like protein. *Mol. Cell. Biol.* 5:3476–83
108. Ozaki, L. S., Svec, P., Nussenzweig, R. S., Nussenzweig, V., Godson, G. N. 1983. Structure of the *Plasmodium knowlesi* gene coding for the circumsporozoite protein. *Cell* 34:815–22
109. Pasvol, G. J., Wainscoat, J. S., Weatherall, D. J. 1982. Erythrocytes deficient in glycophorin resist invasion by the malarial parasite *Plasmodium falciparum. Nature* 297:64–66
110. Perkins, M. 1981. Inhibitory effects of erythrocyte membrane proteins on the in vitro invasion of the human malarial parasite *(Plasmodium falciparum)* into its host cell. *J. Cell Biol.* 90:563–67
111. Perkins, M. 1982. Surface proteins of schizont-infected erythrocytes and merozoites of *Plasmodium falciparum. Mol. Biochem. Parasitol.* 5:55–64
112. Perkins, M. E. 1984. Surface proteins of *Plasmodium falciparum* merozoites binding to the erythrocyte receptor, glycophorin. *J. Exp. Med.* 160:788–98
113. Perkins, M. E., Ravetch, J. V. 1986.

Interaction of *Plasmodium falciparum* merozoite proteins with the erythrocyte surface. See Ref. 50, pp. 157–60

114. Perlmann, H., Berzins, K., Wahlgren, M., Carlsson, J., Bjorkman, A., et al. 1984. Antibodies in malarial sera to parasite antigens in the membrane of erythrocytes infected with early asexual stages of *Plasmodium falciparum*. *J. Exp. Med.* 159:1686–704
115. Perlmann, P., Berzins, K., Carlsson, J., Perlmann, H., Sjoberg, K., et al. 1986. Specificity and inhibitory activity of antibodies to a *Plasmodium falciparum* antigen (Pf 155) and its major amino acid repeat sequence. See Ref. 50, pp. 149–55
116. Perrin, L. H., Dayal, R. 1982. Immunity to asexual erythrocytic stages of *Plasmodium falciparum:* role of defined antigens in the humoral response. *Immunol. Rev.* 61:245–69
117. Perrin, L. H., Merkli, B., Gabra, M. S., Stocker, J. W., Chizzolini, C., Richle, R. 1985. Immunization with a *Plasmodium falciparum* merozoite surface antigen induces a partial immunity in monkeys. *J. Clin. Invest.* 75:1718–21
118. Perrin, L. H., Merkli, B., Locke, M., Chizzolini, C., Smart, J., Richle, R. 1984. Antimalarial immunity in Saimiri monkeys. Immunization with surface components of asexual blood stage. *J. Exp. Med.* 160:441–51
119. Perrin, L. H., Ramirez, E., Lambert, P. H., Miescher, P. A. 1981. Inhibition of *P. falciparum* growth in human erythrocytes by monoclonal antibodies. *Nature* 289:301–3

119a. Peterson, G., Coppel, R. L., McIntyre, P., Woodrow, G., Brown, G. V., et al. 1987. Two forms of variation in the precursor to the major merozoite surface antigen of *P. falciparum*. Submitted

120. Pirson, P. J., Perkins, M. E. 1985. Characterization with monoclonal antibodies of a surface antigen of *Plasmodium falciparum* merozoites. *J. Immunol.* 134:1946–51
121. Pologe, L. G., Ravetch, J. V. 1986. A chromosomal rearrangement in a *P. falciparum* histidine-rich protein gene is associated with the knobless phenotype. *Nature* 322:474–77
122. Potocnjak, P., Yoshida, N., Nussenzweig, R. S., Nussenzweig, V. 1980. Monovalent fragments (Fab) of monoclonal antibodies to a sporozoite surface antigen (Pb44) protect mice against malarial infection. *J. Exp. Med.* 151: 1504–13
123. Ravetch, J. V., Feder, R., Pavlovec, A., Blobel, G. 1984. Primary structure and genomic organization of the histidine-rich protein of the malaria parasite *Plasmodium lophurae*. *Nature* 312:616–20
124. Ravetch, J. V., Kochan, J., Perkins, M. 1985. Isolation of the gene for a glycophorin-binding protein implicated in erythrocyte invasion by a malaria parasite. *Science* 227:1593–97
125. Saul, A., Cooper, J., Ingram, L., Anders, R. F., Brown, G. V. 1985. Invasion of erythrocytes in vitro by *Plasmodium falciparum* can be inhibited by a monoclonal antibody directed against an S antigen. *Parasite Immunol.* 7:587–93
126. Schwarz, R. T., Riveros-Moreno, V., Lockyer, M. J. Nicholls, S. G., Davey, L. S., et al. 1986. Structural diversity of the major surface antigen of *Plasmodium falciparum* merozoites. *Mol. Cell. Biol.* 6:964–68
127. Sharma, S., Svec, P., Mitchell, G. H., Godson, G. N. 1985. Diversity of circumsporozoite antigen genes from two strains of the malarial parasite *Plasmodium knowlesi*. *Science* 229:779–82
128. Sherman, I. W. 1981. *Plasmodium lophurae:* protective immunogenicity of the histidine-rich protein? *Exp. Parasitol.* 52:292–95
129. Siddiqui, W. A., Tam, L. Q., Kan, S.-C., Kramer, K. J., Case, S. E., et al. 1986. Induction of protective immunity to monoclonal-antibody-defined *Plasmodium falciparum* antigens requires strong adjuvant in Aotus monkeys. *Infect. Immun.* 52:314–18
130. Smith, G. P. 1976. Evolution of repeated DNA sequences by unequal crossover. *Science* 191:528–35
131. Stahl, H. D., Bianco, A. E., Crewther, P. E., Anders, R. F., Kyne, A. P., et al. 1986. Sorting large numbers of clones expressing *Plasmodium falciparum* antigens in *Escherichia coli* by differential antibody screening. *Mol. Biol. Med.* 3:351–68
132. Stahl, H. D., Bianco, A. E., Crewther, P. E., Burkot, T., Coppel, R. L., et al. 1986. An asparagine-rich protein from blood stages of *Plasmodium falciparum* shares determinants with sporozoites. *Nucleic Acids Res.* 14:3089–102
133. Stahl, H. D., Coppel, R. L., Brown, G. V., Saint, R. B., Lingelbach, K., et al. 1984. Differential antibody screening of cloned *Plasmodium falciparum* sequences expressed in *Escherichia coli:* Procedure for isolation of defined antigens and analysis of human antisera. *Proc. Natl. Acad. Sci. USA* 81:2456–60
134. Stahl, H. D., Crewther, P. E., Anders,

R. F., Brown, G. V., Coppel, R. L., et al. 1985. Interspersed blocks of repetitive and charged amino acids in a dominant immunogen of *Plasmodium falciparum*. *Proc. Natl. Acad. Sci. USA* 82:543–47

135. Stahl, H. D., Kemp, D. J., Crewther, P. E., Scanlon, D., Woodrow, G., et al. 1985. Sequence of a cDNA encoding a small polymorphic histidine- and alanine-rich protein from *Plasmodium falciparum*. *Nucleic Acids Res.* 13:7837–46
136. Tait, A. 1981. Analysis of protein variation in *Plasmodium falciparum* by two-dimensional gel electrophoresis. *Mol. Biochem. Parasitol.* 2:205–18
137. Tanabe, K., Mackay, M., Goman, M., Scaife, J. G. 1987. Allelic dimorphism in a surface antigen gene of the malarial parasite *Plasmodium falciparum*. *J. Mol. Biol.* In press
138. Targett, G. A. T., Sinden, R. E. 1985. Transmission blocking vaccines. *Parasitol. Today* 1:156–59
139. Trager, W., Jensen, J. B. 1976. Human malaria parasites in continuous culture. *Science* 193:673–75
140. Trager, W., Rudzinska, M. A., Bradbury, P. C. 1966. The fine structure of *Plasmodium falciparum* and its host erythrocytes in natural malarial infections in man. *Bull. WHO* 35:883–85
141. Triglia, T., Stahl, H. D., Crewther, P. E., Scanlon, D., Brown, G. V., et al. 1987. The complete structure of the knob-associated histidine-rich protein from *Plasmodium falciparum*. *EMBO J.* 6:1413–19
142. Udeinya, I. J., Miller, L. H., McGregor, I. A., Jensen, J. B. 1983. *Plasmodium falciparum* strain-specific antibodies block binding of infected erythrocytes to amelanotic melanoma cells. *Nature* 303:429–31
143. Udeinya, I. J., Schmidt, J. A., Aikawa, M., Miller, L. H., Green, I. 1981. Falciparum malaria-infected erythrocytes specifically bind to cultured human endothelial cells. *Science* 213:555–57
144. van Schravendijk, M., Newbold, C. J. 1987. Possible pitfalls in the identification of glycophorin-binding of *Plasmodium falciparum*. *J. Exp. Med.* In press
145. Wahlgren, M., Aslund, L., Franzen, L., Sundvall, M., Berzins, K., et al. 1986. Serological cross-reactions between genetically distinct *Plasmodium falciparum* antigens. See Ref. 50, pp. 169–73
146. Wahlgren, M., Aslund, L., Franzen, L., Sundvall, M., Wahlin, B., et al. 1986. A *Plasmodium falciparum* antigen containing clusters of asparagine residues. *Proc. Natl. Acad. Sci. USA* 83:2677–81
147. Wahlin, B., Wahlgren, M., Perlmann, H., Berzins, K., Bjorkman, A., et al. 1984. Human antibodies to a M_r 155,000 *Plasmodium falciparum* antigen efficiently inhibit merozoite invasion. *Proc. Natl. Acad. Sci. USA* 81:7912–16
148. Weber, J. L., Hockmeyer, W. T. 1985. Structure of the circumsporozoite protein gene in 18 strains of *Plasmodium falciparum*. *Mol. Biochem. Parasitol.* 15:305–16
149. Weber, J. L., Leininger, W. M., Lyon, J. A. 1986. Variation in the gene encoding a major merozoite surface antigen of the human malaria parasite *Plasmodium falciparum*. *Nucleic Acids Res.* 14:3311–23
150. Wellems, T. E., Howard, R. J. 1986. Homologous genes encode two distinct histidine-rich proteins in a cloned isolate of *Plasmodium falciparum*. *Proc. Natl. Acad. Sci. USA* 83:6065–69
151. Wilson, R. J. M. 1980. Serotyping *Plasmodium falciparum* malaria with S-antigens. *Nature* 284:451–52
152. Wilson, R. J. M., McGregor, I. A., Hall, P., Williams, K., Bartholomew, R. 1969. Antigens associated with *Plasmodium falciparum* infections in man. *Lancet* 2:201–5
153. Wilson, R. J. M., McGregor, I. A., Williams, K. 1975. Occurrence of S-antigens in serum in *Plasmodium falciparum* infections in man. *Trans. R. Soc. Trop. Med. Hyg.* 69:453–59
154. Ycas, M. 1972. De novo origin of periodic proteins. *J. Mol. Evol.* 2:17–27
155. Yoshida, N., Nussenzweig, R. S., Potocnjak, P., Nussenzweig, V., Aikawa, M. 1980. Hybridoma produces protective antibodies directed against the sporozoite stage of malaria parasite. *Science* 207:71–73
156. Yoshida, N., Potocnjak, P., Nussenzweig, V., Nussenzweig, R. S. 1981. Biosynthesis of Pb44, the protective antigen of sporozoites of *Plasmodium berghei*. *J. Exp. Med.* 154:1225–36
157. Zavala, F., Cochrane, A. H., Nardin, E. H., Nussenzweig, R. S., Nussenzweig, V. 1983. Circumsporozoite proteins of malaria parasites contain a single immunodominant region with two or more identical epitopes. *J. Exp. Med.* 157:1947–57
158. Zavala, F., Tam, J. P., Hollingdale, M. R., Cochrane, A. H., Quakyi, I., et al. 1985. Rationale for development of a synthetic vaccine against *Plasmodium falciparum* malaria. *Science* 228:1436–40

Ann. Rev. Microbiol. 1987. 41:209–25

SPECIFIC AND RAPID IDENTIFICATION OF MEDICALLY IMPORTANT FUNGI BY EXOANTIGEN DETECTION[1]

Leo Kaufman and Paul G. Standard

Division of Mycotic Diseases, Center for Infectious Diseases, Centers for Disease Control, Atlanta, Georgia 30333

CONTENTS

INTRODUCTION AND HISTORICAL ASPECTS

Exoantigen Tests

Exoantigens have proven valuable for immunoidentification of fungal pathogens and for resolving taxonomic problems. Most fungi produce or carry unique antigens, and the recognition of these results in the specific identification of the elaborating fungus. Exoantigens are defined as antigens or soluble immunogenic macromolecules produced by fungi early in their development. These antigens are readily detected in culture broths or aqueous extracts of slant cultures. The exoantigen test depends on interaction between concentrated or unconcentrated antigens produced by fungi in culture and homologous antibodies that are specifically generated to precipitate them. The complex(es) or precipitate(s) formed are readily checked for fusion with preselected reference precipitates in counterimmunoelectrophoretic or immunodiffusion tests to establish the identity of the fungus producing the antigen(s). Specific identification of fungi may thus be accomplished within 2–5 days of receipt of mature cultures. Conventional identification of a dimorphic fungus has been known to take weeks or months.

Our experience indicates that fungi, which produce specific antigens, can be immunologically identified in their typical, atypical, or nonsporulating state. Mold-form antigens are excellent markers, and the exoantigen technique obviates the need for time-consuming, temperature-dependent conversion or in vivo inoculation studies to isolate a mold suspected of being a dimorphic pathogen. Exoantigen(s) can be detected even with contaminated or nonviable fungi (3).

Exoantigen analyses are extremely useful for determining antigenic relationships among various isolates of a particular species as well as between different species of fungi. The data derived are useful for determining taxonomic relatedness among morphologically similar and dissimilar fungi.

The technique reduces the potential for exposure to biohazardous fungi by eliminating the need for the in vitro conversion or cultural manipulations used in animal inoculations.

Rapid exoantigen tests eliminate costs arising from time-consuming and laborious morphologic, physiologic, and cultural studies. In addition, rapid exoantigen identification of a pathogen benefits the patient by permitting administration of appropriate therapy soon after receipt of a mature isolate.

Evolution of a Practical Technique

The earliest attempt to identify fungi by their soluble antigens was that of Manych & Sourek (19). They used a single diffusion tube precipitin test and inoculated cultures of *Blastomyces dermatitidis, Coccidioides immitis, Histoplasma capsulatum,* and *Paracoccidioides brasiliensis* on the surface of serum agar. They were able to detect precipitin bands as the antigens produced by the proliferating fungi were precipitated. However, the test was limited by the fact that it took up to 21 days for bands to develop. In addition, these bands could not be specifically identified because of the absence of control antigens. Without specific antibodies, control antigens are necessary for a successful test, since many fungi share extra generic antigens. The inability to recognize these common antigens could lead to false identifications.

The problems encountered by Manych & Sourek (19) were overcome by Standard & Kaufman (35) through the use of a reverse agar gel double diffusion technique incorporating control antigens. Using this approach they successfully distinguished the specific antigens of *H. capsulatum* from those shared with other fungi and thus developed the exoantigen technique for specifically identifying this species and its varieties.

Since fungal pathogens may produce nonspecific as well as specific antigens, the currently used exoantigen tests necessitate the use of control antigens and monospecific antibody or polyclonal antibodies rich in a specific antibody.

The specific antibodies needed for the exoantigen tests have been produced by a number of techniques. Most commonly, antisera are produced by the immunization of rabbits with crude soluble antigens or precipitin arcs (39). In some cases, reagent antisera may be produced by immunization of rabbits with purified antigens obtained by adsorption of dried soluble antigens with cross-reacting antibodies. Antibodies produced with crude soluble antigens rarely contain only specific antibodies. More often, sera with both specific and cross-reacting antibodies or with only cross-reacting antibodies are obtained. The precipitin arc technique has been most useful for producing monospecific antibodies (7, 8). This technique consists of immunizing animals with antigen-antibody complexes harvested from immunodiffusion or counterimmunoelectrophoresis reactions that have been emulsified in Freund's complete adjuvant. Antisera produced by the latter method may require little or no adsorption.

Exoantigens may be obtained from fungal cultures in two ways. They may be extracted from slant cultures with an aqueous merthiolate solution or produced in shaken broth cultures.

Materials and Methods

A properly performed exoantigen test requires that the quality of reference antigen and antibody be controlled. Certain antigen and antibody systems have been thoroughly evaluated for sensitivity and specificity. Excellent examples are the H and M reagents for *H. capsulatum,* the A reagent for *B. dermatitidis,* and the F, HL, and HS reagents for *C. immitis*. Although laboratories strive to produce consistently high-quality reagents, this goal is not always achieved. Consequently, we recommend that each new lot of a reagent be titrated and evaluated as to sensitivity and specificity with known homologous and heterologous antigens and/or antisera. The precipitates formed by the interaction of reference reagents should be distinct. New lots of reagents should yield reactions similar to or better than the established reagents. Evaluated working reagents should be continuously reevaluated each time the test is performed by scrutiny of the quality of the reference bands.

REFERENCE ANTIGENS Protocols for the preparation of control or reference antigens of *B. dermatitidis, C. immitis, H. capsulatum,* and *P. brasiliensis* have been published (37, 39). Unfortunately, the antigens of *Cladosporium carrionii, Cladosporium bantianum, Phialophora verrucosa, Exophiala jeanselmei,* and *Wangiella dermatitidis* have not yet been standardized, nor have the means of their preparation been extensively evaluated (5, 6, 9, 16, 33).

TEST ANTIGENS Antigens from suspected pathogenic fungi are obtained in slant culture extracts or filtrates from shaken broth cultures. The source of antigen depends on the etiologic agent suspected or the quality of the antigen preparation obtained. Antigens derived from shaken broth cultures are recommended for the *Exophiala* spp. and *W. dermatitidis* because the slant extracts of many of the so-called black yeasts are frequently viscous and are difficult to concentrate. This method is also used with some *B. dermatitidis* isolates that produce inadequate quantities of the A antigen on slants.

The slant extraction procedure is rather simple. A mature slant culture is covered with an aqueous solution of 0.02% merthiolate. After 24 hr the extract is removed, filter sterilized, and appropriately concentrated (13, 39). This method is most applicable for preparing test antigens from isolates of *B. dermatitidis, C. immitis, H. capsulatum, P. brasiliensis,* and *Aspergillus* spp. (14, 32, 39).

Exoantigens are obtained from shaken broth cultures by placing a heavy mycelial inoculum (at least 2 cm^2) into a 50-ml flask containing 15 ml of brain heart infusion broth (BHIB). The culture is incubated at 25°C while rotating at 150 rpm. After 3 days, 5 ml is removed and treated with 1% merthiolate solution to give a final concentration of 0.02%. After 24 hr the broth is filter sterilized and appropriately concentrated (31, 39). Shaken broth cultures are recommended as a source of exoantigens whenever a false-negative is suspected with the slant extraction procedure.

REFERENCE ANTISERA Quality reference antisera can be produced by immunization of rabbits with purified or unpurified culture filtrates or electrophoretic or immunodiffusion precipitin arcs (7–9, 33, 34, 39). Precipitin arcs are highly recommended as immunogens, since they elicit less cross-reactive antisera. To date, monoclonal antibodies that react specifically against the established exoantigens have not been produced. Consequently the value of monoclonal antibodies for specifically detecting exoantigens has yet to be determined.

IMMUNIZATION PROCEDURES Reference antisera against antigens consisting of immunoelectrophoretic or immunodiffusion precipitin arcs or against culture-filtrate antigens are readily prepared in rabbits following absorption with air-dried heterologous antibodies. Antisera to arc antigens are prepared by injecting mixtures of equal quantities of emulsified arcs and Freund's complete adjuvant subcutaneously in the dorsal area of a rabbit. (Freund's complete adjuvant can cause intermittent fever, localized swelling, and sometimes necrosis at the injection site. Other adjuvants are becoming available which may facilitate antibody production with fewer side effects.) Multiple injections are given about a month apart, and serum samples are checked for specific precipitins approximately 3 wk after the second injection. If precipitin responses are not optimal, injections are continued every week with continuous serum checks until antibody yields and quality are optimal (7, 8, 39).

Acceptable antisera may also be obtained within 3 wk by injecting rabbits intramuscularly with culture-filtrate antigens mixed with equal volumes of Freund's incomplete adjuvant and then administering intravenous injections of the antigen alone. If antibody yields are not satisfactory, additional intravenous injections are administered to achieve the desired quality (35).

COMMERCIAL REAGENTS Commercial exoantigen reagents are available for identifying *B. dermatitidis, C. immitis,* and *H. capsulatum*. These may be obtained from Immuno-Mycologics, Inc. (Norman, Oklahoma) and the Nolan-Scott Biological Labs, Inc. (Tucker, Georgia). If properly titered, the commercial fungal reagents used in ID tests for detecting antibody in sera can be used in exoantigen tests. Our experience indicates that the serodiagnostic reagents must be checked for appropriate antibodies and must be rebalanced for the exoantigen test for *B. dermatitidis* and *H. capsulatum*. Coccidioidomycosis serodiagnostic reagents must also be carefully checked, since they are frequently devoid of the HS antibody homologous to the earliest specific exoantigen elaborated by *C. immitis*.

CRITERIA FOR IMMUNOIDENTIFICATION The only antigen considered positive is antigen that forms a distinct line of identity with the reference antigen-antibody precipitate. Lines of nonidentity or lines of partial identity are considered negative (3, 14, 17, 35–37, 39).

Although false-positive reactions have not been encountered, false-negative reactions may occur. They may be attributable to technical error, absence of specific homologous antibody in the reference antiserum, or inadequate antigen concentration due to insufficient growth or to improper growth conditions. Tests should be repeated when mycological or clinical judgment suggests that a negative culture should be positive. We also recommend that cultures that are negative by the exoantigen test be further examined by conventional morphologic tests to make certain that they are not known pathogens.

We prefer to test unknown antigens in duplicate in adjacent lateral wells rather than to test different antigens in adjacent wells. This is necessary because some isolates, such as *C. immitis,* produce antigen so abundant that precipitates can extend in front of an adjacent well. Such an extended line in front of a well with a negative extract could be misinterpreted as a positive result (17, 39).

The exoantigen test procedure has been evaluated extensively with Centers for Disease Control reference reagents and commercial reagents (2, 4, 10, 29). It has been found to be extremely sensitive and specific for identifying systemic fungal pathogens. Positive results are recognized as the equivalent of cultural isolation and identification by conventional tests. For a detailed review of the test protocol, concentration procedures, and reagent preparation the reader is referred elsewhere (39).

SAFETY CONSIDERATIONS To assure that laboratory workers do not become infected, we recommend that all fungal cultures be handled in a biological safety hood and that broth cultures or extracts be treated with a preservative. The preservative chosen should be one that renders the culture nonviable without destroying its exoantigens. Although 0.2% formaldehyde can effectively kill the mycelial forms of *B. dermatitidis, C. immitis,* and *H. capsulatum,* it unfortunately also inactivates the *B. dermatitidis* A and *C. immitis* HL antigens. Solutions of 0.02% thimerosol have consistently killed these fungi in broth culture but not in slant extracts. We routinely use 0.02% merthiolate and treat broth cultures and slant extracts for 24 hr. We have noted that on rare occasions viable elements can survive in the thimerosol-treated extract. To our knowledge, however, no accidental infections among laboratory workers have occurred when this procedure has been used. Thimerosol does not deleteriously affect the exoantigens. For those workers who prefer to use 0.02% thimerosol and yet be assured that they are working with sterile extracts, we recommend sterilization of the extracts by passage through a 0.45-μm membrane filter (38).

APPLICABILITY TO IDENTIFICATION OF FUNGAL PATHOGENS

Definitive identification of pathogenic fungi has long depended on extensive morphologic and physiologic studies. The early detection of relevant exoantigens in culture extracts or growth media has rendered unnecessary such prolonged cultural and biochemical studies for the identification of many fungal pathogens. With the properly performed exoantigen technique, only a brief, direct microscopic examination is required to determine whether an unidentified fungal culture produces conidia or is devoid of conidia or conidiophores. This preliminary examination allows selection of appropriate reference exoantigen test reagents, avoids the use of irrelevant reagents, and economizes on time.

Identification of Dimorphic Fungi

The proper identification of a dimorphic fungal pathogen depends on isolation of its mold form, demonstration of its diagnostic conidia, and in vitro conversion to its tissue form. As indicated earlier, these steps are time-consuming and tedious. In certain instances, in vitro conversion fails and in vivo studies must be performed. Exoantigen tests have proven to be extremely valuable for identifying dimorphic pathogens, particularly those that are present as atypical cultures or are difficult to convert.

BLASTOMYCES DERMATITIDIS *Blastomyces dermatitidis* is the causative agent of blastomycosis, a systemic disease of North America (except Mexico), Africa, the Middle East, and India. At 25°C the fungus grows as a white to tan mold on Sabouraud dextrose agar (SDA). Microscopic examination may reveal round to oval conidia ranging in size from 3 to 12 μm. These conidia, however, are not distinct since many pathogenic and saprophytic fungi such as *Chrysosporium* spp. and atypical isolates of *H. capsulatum* may bear similar conidia. Proper identification of *B. dermatitidis* requires demonstration of its characteristic tissue form by conversion on culture media or in vivo. The exoantigen test is a rapid and reliable alternative for these procedures. In 1973 (12), a specific precipitin, designated A, was found in patients with blastomycosis. Later, using an anti-A serum, we showed the production of a specific exoantigen homologous to the A antibody in slant extracts of 56 isolates of *B. dermatitidis* (14). Although extracts of *Chrysosporium* spp., *C. immitis,* and *H. capsulatum* may cross-react with non-A antibodies in anti-A and anti–*B. dermatitidis* sera, in no instance have they produced an A precipitate. More recently, Kaufman et al (18) evaluated the value of antibody to *B. dermatitidis* A precipitin arcs for identifying *B. dermatitidis* isolates from the United States, Canada, Africa, India, and Israel. The specific A antigen was found among isolates of *B. dermatitidis* from all parts of the world, with the exception of some African isolates of *B. dermatitidis*. Our studies also revealed that all isolates of *B. dermatitidis* shared an antigen designated K. Thus, it is apparent that at least two serotypes of the fungus exist, one that possesses the A antigen and one that does not. The A-deficient serotype appears to be prevalent only in Africa. The predominant K antigen, produced in the early growth stages of *B. dermatitidis,* may be useful for screening suspected *B. dermatitidis* cultures, although it is also found in *Chrysosporium parvum* and *C. parvum* var. *crescens* (P. G. Standard & L. Kaufman, unpublished data).

B. dermatitidis yeast cells generally range from 6 to 15 μm in diameter, but in some cases of blastomycosis yeast cells have measured 2–4 μm in diameter. These are generally confused with the tissue-form cells of *H. capsulatum* var. *capsulatum*. The exoantigen technique has proven reliable for identifying the mycelial anamorph of the small tissue form of *B. dermatitidis*.

Recently, we and others (2, 17) have noted that some isolates of *B. dermatitidis* fail to produce sufficient A antigen in the slant extraction method but invariably produce this diagnostic antigen in shaken broth cultures. Consequently, we recommend that cultures that are suspected of being *B. dermatitidis* but that fail to produce A exoantigen by the slant extraction procedure be retested using supernatant antigens from shaken broth cultures. In fact, antigens from cultures whose characteristics are compatible with *B. dermatitidis* might best be derived only from shaken broth cultures. Antigen yields from broth cultures grown for approximately one week at 37°C are higher than yields from cultures grown at 25°C (D. W. McLaughlin & P. G. Standard, unpublished data).

COCCIDIOIDES IMMITIS *C. immitis* is the causative agent of coccidioidomycosis, a disease of the New World. Definitive diagnosis of this disease depends on (*a*) isolation of the mold form of the fungus, which typically produces alternate, barrel-shaped arthroconidia, and (*b*) its successful conversion to the parasitic or endosporulating spherule form. This pathogen is not always identified without problems, since several members of the family Gymnoascaceae, notably species of *Arachniotus, Auxarthron, Uncinocarpus,* and the form genus *Malbranchea* produce superficially similar arthroconidia. Furthermore, atypical, nonsporulating, and pigmented forms of *C. immitis* can also be encountered. The exoantigen test has proven to be completely sensitive and specific for identifying *C. immitis* in its typical or atypical mold forms (10, 36). Unconcentrated slant extracts concentrated 25 times are excellent sources of diagnostic exoantigens. These exoantigens may be either heat stable (HS) (remaining stable after boiling for 10 min) or heat labile (HL or F) (losing activity after heating at 60°C). Detection of either of these antigens identifies the pathogen.

The HS antigen is usually the earliest of the three diagnostic antigens to appear. It can usually be detected in unconcentrated slant extracts. The HL antigen is generally detected in later growth stages, and the F antigen is detected last. Reference reagents for the immunologic identification of *C. immitis* must contain the HS antigen and preferably contain a combination of the three antigens. Commercial reagents should meet this standard. Our experience indicates that false-negative results ensue when reagents used do not meet this standard. It is important to note that the serodiagnostically valuable *C. immitis* heat-stable tube precipitinogen (TP) is of limited value for identifying *C. immitis* isolates because this antigen is also common to certain gymnoascaceous saprophytes such as *Arachniotus, Auxarthron,* and *Malbranchea* spp. (15).

HISTOPLASMA CAPSULATUM The monotypic genus *Histoplasma* consists of *H. capsulatum* var. *capsulatum* (the causative agent of histoplasmosis capsulati), *H. capsulatum* var. *duboisii* (the causative agent of histoplasmosis duboisii), and *H. capsulatum* var. *farciminosum* (the agent of histoplasmosis farciminosum) (40). The mycelial forms of these three varieties are indistinguishable from each other. Their typical mycelial forms are characterized by the production of tuberculate macroconidia. Detection of these macroconidia, however, provides data that are only suggestive of an *H. capsulatum* variety, since a number of saprophytic fungi in the genera *Arthroderma, Chrysosporium, Corynascus, Renispora,* and *Sepedonium* have mycelial forms that grossly and microscopically resemble those of the *Histoplasma* varieties. To further complicate identification, atypical nonsporulating (14) and red-pigmented isolates (22) of *H. capsulatum* have been encountered. Our studies have shown that the soluble H and M antigens are elaborated only by the three varieties of *H. capsulatum*. The antigenic and morphologic similarities exhibited by *H. capsulatum* var. *capsulatum, H. capsulatum* var. *duboisii,* and *H. capsulatum* var. *farciminosum* are the basis for their classification as varieties of *H. capsulatum* (14).

The H and M antigens are unique to the *Histoplasma* varieties and are reliable for their identification. The antigens are readily produced by the typical and atypical varieties of the genus. *Renispora flavissima,* a bearer of tuberculate macroconidia, was originally thought to be *H. capsulatum,* but negative H and M exoantigen tests correctly suggested that the fungus was not *H. capsulatum*. This fungus was subsequently described and properly classified as a new species (34).

Although the exoantigen technique does not allow differentiation of the three varieties of *H. capsulatum,* this does not pose a diagnostic problem in the United States, since the only H and M producer in the Americas is *H. capsulatum* var. *capsulatum*. In Africa, however, an isolate derived from an animal could be any one of the three varieties, and specific identification would depend on the morphology of the parasitic and mold forms.

PARACOCCIDIOIDES BRASILIENSIS Protein antigen E (or its equivalent 1) and antigens 2 and 3 are all specific for *P. brasiliensis*. Their detection permits specific identification of the fungus in its mycelial form (37). Our studies do not support the report of Restrepo & Moncada (28) that one of these diagnostic antigens is shared with *H. capsulatum*. None of the extracts derived from numerous isolates of *H. capsulatum, B. dermatitidis, C. im-*

mitis, Chrysosporum spp., and *Sepedonium* spp. contained antigens identical to the three antigens of *P. brasiliensis*.

SPOROTHRIX SCHENCKII Polonelli & Morace (25) found that antisera prepared in rabbits against yeast-form cells of *S. schenckii* produced three precipitin bands upon reaction with antigens from the mycelial form of this fungus. Interestingly, antigens of *Ceratocystis minor*, the purported perfect state of *S. schenckii*, and antigens of *Graphium penicillioides* produced bands similar to those of *S. schenckii*, whereas numerous other fungal antigens were either negative or produced bands of nonidentity with the reference system. The latter reaction was noted with exoantigen from a *Sporotrichum* species.

Identification of Nondimorphic Fungi

Nondimorphic pathogenic fungi such as hyaline aspergilli and dermatophytes or dematiaceous or "black" fungi are not always easily identified. The fungi frequently fail to sporulate or to produce the typical morphological or biochemical characteristics that establish their identity. The dematiaceous fungi responsible for chromoblastomycosis and phaeohyphomycosis are often difficult to identify because they are polymorphic. Exoantigens, however, have proven useful for the identification of many such fungi.

HYALINE FUNGI

Aspergillus *groups* The identification of aspergilli, including the medically important species, is dependent on morphologic studies. These species can usually be accurately identified without difficulty by experienced laboratory workers. However, identification involves the use of standardized media and the time-consuming preparation of slide cultures. In some instances nonsporulating isolates, such as the albino type of *Aspergillus fumigatus*, are encountered; these require even more extensive studies. Antisera prepared against extracts of *A. fumigatus, A. flavus, A. nidulans, A. niger*, and *A. terreus* demonstrated variable intrageneric cross-reactions, which were eliminated by adsorptions with antigens of selected heterologous *Aspergillus* species. Accordingly, group-specific antisera were developed for specific identification of *Aspergillus* species of medical importance (32). Exoantigens have proven reliable for identifying the sterile albino-type isolates of *A. fumigatus*. To date, exoantigen analyses have supported the conventional classification of the *Aspergillus* species that are pathogenic for animals and humans into various series (27).

Basidiobolus *and* Conidiobolus *spp.* Members of the order Entomophthorales are ubiquitous fungi that form flat, waxy colonies. Within the order are two genera, *Basidiobolus* and *Conidiobolus*, with pathogenic species that cause subcutaneous zygomycosis. To date, exoantigen studies have not been carried out on *Conidiobolus* spp. Some investigators have resorted to exoantigen analyses in the hope of resolving the taxonomic problems of *Basidiobolus* spp. that have not been solved by morphologic and physiologic studies. Polonelli & Morace (25), using polysaccharide and concentrated soluble filtrate antigens, found that *B. haptosporus, B. meristosporus, B. microsporus,* and *B. ranarum* were closely related. However, distinct antigens were recognized with the use of adsorbed antisera. These antisera permitted the antigenic separation of the four species. More recently, Yangco et al (41) described two heat-stable exoantigens common to *B. haptosporus* and *B. ranarum,* designated N and Y. *B. meristosporus, B. microsporus, C. coronatus,* and *C. incongruus* share only the N antigen. Other heterologous zygomycetes of the order Mucorales and dimorphic fungi studied produce neither antigen. The data suggest that *B. haptosporus* and *B. ranarum* are antigenically similar to each other but distinct from *B. meristosporus* and *B. microsporus*. The sharing of the N antigen by the *Basidiobolus* and *Conidiobolus* species implies a taxonomic relationship between the two genera and supports their taxonomic classification in the order Entomophthorales.

Dermatophytes Pleomorphism and inconsistent or absent conidial production present major problems to mycologists attempting to identify certain of the dermatophytes. Identification is currently based on morphologic, thermotolerance, and biochemical tests. Theoretically, exoantigen tests should provide an ideal means for rapidly and specifically identifying dermatophytes. Unfortunately, this has not been the case. Philpot (23) and later Kaufman & Lopez (11) detected species-specific antigens for several of the *Microsporum* and *Trichophyton* species; however, they did not successfully develop exoantigen tests for these dermatophytes, perhaps because they could not produce specific factor sera or consistently demonstrate specific antigens. Extensive antigenic variation was noted among isolates of a given species and between species. Kaufman & Lopez (11) noted more antigenic similarities than differences among the dermatophyte species. Similarly, Polonelli & Morace (26), working with monoclonal antibodies against *Microsporum canis,* reported extensive reactions with homologous and heterologous dermatophyte antigens. One type of monoclonal antibody, although nonspecific, failed to react with three different isolates of *M. canis;* this suggests that serotypes may exist within the species. In an extension of this study, Polonelli et al (24) produced monoclonal antibodies that proved useful for identifying *M. canis* isolates and for differentiating strains within the species.

Penicillium marneffei Most species of *Penicillium* are saprophytes, but a few have been reported to cause disease in humans. *Penicillium marneffei* is such a species. It is a dimorphic pathogen currently only known to occur in Southeast Asia. It produces two distinct exoantigens, which permit its differentiation from other species of *Penicillium* as well as from other fungi (30).

Pseudallescheria boydii *P. boydii* is the most common cause of eumycotic mycetoma in the United States. It can also cause other forms of disease. Polonelli & Morace (25) noted, as expected, that cultures of *P. boydii* produced antigens in common with those of its anamorph *Scedosporium apiospermum,* but not with those of other morphologically or taxonomically related fungi. These antigens would be useful for the rapid identification of *P. boydii*.

DEMATIACEOUS FUNGI

Cladosporium bantianum *and* Cladosporium carrionii Over 50 species have been described in the genus *Cladosporium*. The majority are ubiquitous and occur in nature as saprophytes. The various species generally produce long chains of conidia acropetally from the tip of branched or unbranched erect conidiophores. Differentiation of species mostly depends on the shape, size, and number of conidia. *C. bantianum,* a causative agent of cerebral phaeohyphomycosis, and *C. carrionii,* an agent of chromoblastomycosis, are among the few human pathogens in the genus. Separation of these pathogens from the saprophytic species and identification have required time-consuming morphologic and biochemical tests. Honbo et al (9) demonstrated that exoantigens could be used to accurately identify such pathogens. They found specific antigens among isolates of *C. bantianum* and *C. carrionii,* which permit their rapid, specific identification and separation from the common saprophytic species, i.e. *C. cladosporoides* and *C. herbarum*.

In 1981, McGinnis & Borelli (20) concluded that *C. bantianum* and *C. trichoides* were conspecific. The exoantigen studies of Honbo et al (9) revealed that these fungi are antigenically identical and thus support the synonymy of *C. bantianum* and *C. trichoides*.

Additional studies have been carried out on other dematiaceous fungi. Espinel-Ingroff et al (5) recently described an exoantigen test that could be used for rapid identification or differentiation of *C. bantianum, Fonsecaea pedrosoi,* and *Phialophora verrucosa* from one another and from other saprophytic and pathogenic dematiaceous genera and species.

Phialophora verrucosa Until recently, *P. verrucosa* and *P. americana* were both considered etiologic agents of chromoblastomycosis. Some mycologists considered them two distinct species because *P. americana* was thought to produce phialides with deeper collarettes. However, Shoji et al (33) studied

32 isolates of *P. americana* and *P. verrucosa* and found the depth of phialidic collarettes to be variable. This character is thus unreliable for differentiation of these fungi. They also found that these species produce exoantigens that are antigenically identical. These findings support their contention that the fungi are conspecific. The antigenic relationship of *P. verrucosa* to other medically important species of *Phialophora* such as *P. parasitica, P. repens,* and *P. richardsiae* and its relationship to the common saprophytes found in this genus have not been investigated.

Exophiala jeanselmei, Exophiala *spp., and* Wangiella dermatitidis *Exophiala jeanselmei* and *Wangiella dermatitidis* are common agents of phaeohyphomycosis. Because of their polymorphic nature and the superficial similarities of their conidiogenous cells, they may be difficult to identify and to differentiate from one another as well as from other *Exophiala* species such as *E. moniliae, E. spinifera,* and *E. werneckii.* All of these species are frequently recovered in the clinical laboratory as black yeasts and may be confused with other fungi that manifest this property. Specific identification is time-consuming and requires accurate knowledge of their conidial ontogeny and demonstration of their mold forms with typical conidia and conidia-bearing structures.

McGinnis & Padhye (21) demonstrated that *Phialophora jeanselmei* and *P. gougerotii* are morphologically similar and should be considered a single species in the genus *Exophiala,* i.e. *E. jeanselmei.* Exoantigen studies have supported this conclusion. Kaufman et al (16) showed that *E. jeanselmei* has three serotypes. Isolates that cause mycetomas have been identified as serotype 1, whereas those that cause phaeohyphomycosis and that were originally identified as *P. gougerotii* belong to serotype 1, 2, or 3. As a result of these studies, exoantigen reagents have been developed for specifically identifying 6-day-old cultures of *E. jeanselmei* and *W. dermatitidis.* Espinel-Ingroff et al (6) have confirmed the antigenic distinction noted between these fungi and the diagnostic value of the exoantigens.

FUTURE CONSIDERATIONS AND CONCLUDING REMARKS

Use of Monoclonal Antibodies

Currently, most exoantigen studies are performed with adsorbed polyclonal or purified antibodies. These heterogenous reagents are reliable and may recognize an antigen via a number of different epitopes. In contrast, monoclonal antibodies represent a homogenous population of immunoglobulins directed against a single antigenic determinant. Monoclonal reagents could be extremely valuable not only for diagnostic purposes but for characterizing the fine structure of species exoantigens and for elucidating intraspecific antigen relationships. Successful development of refined reagents could eliminate cross-reactions and obviate the need for multiple adsorptions, which frequently weaken a reagent. These reagents would also eliminate the need for highly purified immunizing antigens and would minimize the use of control antigens currently required in every exoantigen test pattern. Polonelli et al (24) successfully produced precipitating monoclonal antibodies capable of serotyping *M. canis*. The full impact of monoclonal antibodies on exoantigen tests has not yet been realized.

Use of Purified Antigens

The interpretation of exoantigen tests is complicated by the fact that the fungal antigens used are impure and frequently demonstrate cross-reactivity. Use of polyacrylamide gel electrophoresis, immunoadsorption, and other techniques could permit separation and purification of species-specific antigens. Availability of purified antigens will increase opportunities for successful production of more specific polyclonal antibodies or specific monoclonal antibodies. More importantly, the availability of pure antigens should aid in the characterization and standardization of reagents, which should make the tests more specific, allow better comparison of results from different laboratories, and establish standards for the production of quality commercial products.

Application to Study of Antigenic and Taxonomic Relationships Among Fungi

Exoantigen studies can be helpful in establishing taxonomic relationships. Antigenic characterization is worthy of taxonomic consideration because the antigens appear to be a constant manifestation of fungal growth, whereas morphologic qualities frequently vary depending on the growth medium and the age of the fungus. The occurrence of specific antigens can help separate fungal pathogens from morphologically similar nonpathogenic fungi and can justify their taxonomic position. For example, all varieties of *H. capsulatum* produce H and M antigens, whereas the morphologically similar nonpathogenic *Chrysosporium* spp., *R. flavissima,* and *Sepedonium* spp. are devoid of these antigens (14).

Exoantigen analysis established the existence of serotypes among *B. dermatitidis* isolates (18). The serotypic differences among isolates of this species appear to be associated with their geographic distribution and morphologic differences. Carmichael (1) considers the genus *Blastomyces* a synonym of *Chrysosporium*. Studies by Sekhon et al (31) indicate that *Chrysosporium* spp. and *Geomyces (Chrysosporium) pannorus* produce antigens that are shared by the morphologically similar mold forms of *B. der-*

matitidis and *H. capsulatum*. Much stronger antigenic relationships, however, were noted with *B. dermatitidis* than with *H. capsulatum*. These antigenic analyses support Carmichael's treatment of *B. dermatitidis* as a *Chrysosporium* species.

Undoubtedly, antigenic structure can contribute to the proper classification of dimorphic fungi. Recent antigenic studies with the pathogenic aspergilli (32) have substantiated the traditional classification of aspergilli into group species based on color of the colonies and morphologic characteristics. The value of exoantigens in classifying numerous other fungi awaits investigation.

Summary

Within a decade of its introduction, the exoantigen technique has won general acceptance for accurate and rapid identification of fungal pathogens. This acceptance is emphasized by the fact that positive exoantigen results obtained with the dimorphic pathogenic fungi *B. dermatitidis, C. immitis, H. capsulatum* varieties *capsulatum, duboisii,* and *farciminosum,* and *P. brasiliensis* are no longer considered presumptive evidence but are considered definitive data for species identification. Technical problems associated with poor sensitivity and false-positives in some of the early tests have been resolved. The test expands the diagnostic capabilities of the laboratory. We encourage the establishment of libraries of antisera for species identification and, where appropriate, for serotyping.

Since the test is simple and reagents for many of the pathogens are commercially available, the test can be performed in most laboratories. As highly defined antigens are produced, more standardized and specific tests will be developed. Hybridoma technology may provide the means for producing specific antibodies without the need for highly purified antigens, which have been difficult to produce.

The identification of numerous fungi could be facilitated by application of exoantigen techniques. Specific antisera should be developed to achieve this goal and to obtain antigenic data useful for elucidating taxonomic relationships. Some fungi cannot be classified on the basis of morphologic and biochemical qualities alone. Supplementary data obtained with exoantigen analyses could undoubtedly aid in resolving such problems.

Acknowledgments

The authors are grateful to the many medical mycologists who, through independent and collaborative studies, evaluated the exoantigen technique and helped transform it from a presumptive procedure to a recognized practical and definitive one. We are also indebted to Drs. L. Ajello and A. A. Padhye for their critical review of the manuscript and to Mrs. S. Stephenson for her secretarial assistance.

Literature Cited

1. Carmichael, J. W. 1952. *Chrysosporium* and some other aleuriosporic hyphomycetes. *Can. J. Bot.* 40:1137–73
2. Denys, G. A., Newman, M. M., Standard, P. G. 1983. Evaluation of a commercial exoantigen test system for the rapid identification of systemic fungal pathogens. *Am. J. Clin. Pathol.* 79:379–81
3. DiSalvo, A. F. 1985. Identification of mycotic pathogens using antisera to exoantigens. *Clin. Immunol. Newsl.* 6:24–29
4. DiSalvo, A. F., Sekhon, A. S., Land, G. A., Fleming, W. H. 1980. Evaluation of the exoantigen test for identification of *Histoplasma* species and *Coccidioides immitis* cultures. *J. Clin. Microbiol.* 11:238–41
5. Espinel-Ingroff, A., Shadomy, S., Dixon, D., Goldson, P. 1986. Exoantigen test for *Cladosporium bantianum, Fonsecaea pedrosoi* and *Phialophora verrucosa. J. Clin. Microbiol.* 23:305–10
6. Espinel-Ingroff, A., Shadomy, S., Kerkering, T. M., Shadomy, H. J. 1984. Exoantigen test for differentiation of *Exophiala jeanselmei* and *Wangiella dermatitidis* isolates from other dematiaceous fungi. *J. Clin. Microbiol.* 20:23–27
7. Green, J. H., Harrell, W. K., Gray, S. B., Johnson, J. E., Bolin, R. C., et al. 1976. H and M antigens of *Histoplasma capsulatum:* Preparation of antisera and location of these antigens in yeast-phase cells. *Infect. Immun.* 14:826–31
8. Green, J. H., Harrell, W. K., Johnson, J. E., Benson, R. 1979. Preparation of reference antisera for laboratory diagnosis of blastomycosis. *J. Clin. Microbiol.* 10:1–7
9. Honbo, S., Standard, P. G., Padhye, A. A., Ajello, L., Kaufman, L. 1984. Antigenic relationships among *Cladosporium* species of medical importance. *Sabouraudia* 22:301–10
10. Huppert, M., Sun, S. H., Rice, E. H. 1978. Specificity of exoantigens for identifying cultures of *Coccidioides immitis. J. Clin. Microbiol.* 8:346–48
11. Kaufman, L., Lopez, R. B. 1980. Immunodiffusion studies of morphologically similar dermatophyte species, *Proc. 5th Int. Conf. Mycoses, Caracas, Venezuela,* pp. 159–73. Washington, DC: Pan Am. Health Org.
12. Kaufman, L., McLaughlin, D. W., Clark, M. J., Blumer, S. 1973. Specific immunodiffusion test for blastomycosis. *Appl. Microbiol.* 26:244–47
13. Kaufman, L., Standard, P. 1978. Improved version of the exoantigen test for identification of *Coccidioides immitis* and *Histoplasma capsulatum* cultures. *J. Clin. Microbiol.* 8:42–45
14. Kaufman, L., Standard, P. G. 1978. Immuno-identification of cultures of fungi pathogenic to man. *Curr. Microbiol.* 1:135–40
15. Kaufman, L., Standard, P. G., Huppert, M., Pappagianis, D. 1985. Comparison and diagnostic value of the coccidioidin heat-stable (HS and tube precipitin) antigens in immunodiffusion. *J. Clin. Microbiol.* 22:515–18
16. Kaufman, L., Standard, P., Padhye, A. A. 1980. Serologic relationship among isolates of *Exophiala jeanselmei* (*Phialophora jeanselmei, P. gougerotii*) and *Wangiella dermatitidis*. See Ref. 11, pp. 252–58
17. Kaufman, L., Standard, P. G., Padhye, A. A. 1983. Exoantigen tests for the immunoidentification of fungal cultures. *Mycopathologia* 82:3–12
18. Kaufman, L., Standard, P. G., Weeks, R. J., Padhye, A. A. 1983. Detection of two *Blastomyces dermatitidis* serotypes by exoantigen analysis. *J. Clin. Microbiol.* 18:110–14
19. Manych, J., Sourek, J. 1966. Diagnostic possibilities of utilizing precipitation in agar for the identification of *Histoplasma capsulatum, Coccidioides immitis, Blastomyces dermatitidis,* and *Paracoccidioides brasiliensis. J. Hyg. Epidemiol. Microbiol. Immunol.* 10:74–84
20. McGinnis, M. R., Borelli, D. 1981. *Cladosporium bantianum* and its synonym *Cladosporium trichoides. Mycotaxon* 13:127–36
21. McGinnis, M. G., Padhye, A. A. 1977. *Exophiala jeanselmei,* a new combination for *Phialophora jeanselmei. Mycotaxon* 5:341–52
22. Morris, P. R., Terreni, A. A., DiSalvo, A. F. 1986. Red pigmented *Histoplasma capsulatum*—an unusual variant. *J. Med. Vet. Mycol.* 24:231–33
23. Philpot, C. M. 1978. Serological differences among dermatophytes. *Sabouraudia* 16:247–56
24. Polonelli, L., Castagnola, M., Morace, G. 1986. Identification and serotyping of *Microsporum canis* isolates by monoclonal antibodies. *J. Clin. Microbiol.* 23:609–15
25. Polonelli, L., Morace, G. 1984. Rapid immunoidentification of pathogenic fungi. *Proc. Eur. Symp. New Horiz. Microbiol., Rome,* ed. A. Sanna, G. Morace,

pp. 203–19. New York: Elsevier

26. Polonelli, L., Morace, G. 1985. Serological analysis of dermatophyte isolates with monoclonal antibodies produced against *Microsporum canis*. *J. Clin. Microbiol.* 21:138–39
27. Raper, K. B., Fennell, D. I. 1965. *The Genus* Aspergillus. Baltimore, Md: Williams & Wilkins. 686 pp.
28. Restrepo, A., Moncada, L. H. 1974. Characterization of the precipitin bands detected in the immunodiffusion test for paracoccidioidomycosis. *Appl. Microbiol.* 28:138–44
29. Sekhon, A. S., DiSalvo, A. F., Standard, P. G., Kaufman, L, Terreni, A. A., Garg, A. K. 1984. Evaluation of commercial reagents to identify the exoantigens of *Blastomyces dermatitidis, Coccidioides immitis* and *Histoplasma* species cultures. *Am. J. Clin. Pathol.* 82:206–9
30. Sekhon, A. S., Li, J. S. K., Garg, A. K. 1982. *Penicillosis marneffei:* serological and exoantigen studies. *Mycopathologia* 77:51–57
31. Sekhon, A. S., Standard, P. G., Kaufman, L., Garg, A. K. 1986. Reliability of exoantigens for differentiating *Blastomyces dermatitidis* and *Histoplasma capsulatum* from *Chrysosporium* and *Geomyces* species. *Diagn. Microbiol. Infect. Dis.* 4:215–21
32. Sekhon, A. S., Standard, P. G., Kaufman, L., Garg, A. K., Cifuentes, P. 1986. Grouping of *Aspergillus* species with exoantigens. *Diagn. Immunol.* 4:112–16
33. Shoji, A., Padhye, A. A., Standard, P. G., Kaufman, L., Ajello, L. 1986. The relationship of *Phialophora verrucosa* to *Phialophora americana*. *J. Med. Vet. Mycol.* 24:23–34
34. Sigler, L., Gaur, P. K., Lichtwart, R. W., Carmichael, J. W. 1979. *Renispora flavissima,* a new gymnoascaceous fungus with tuberculate chrysosporium conidia. *Mycotaxon* 10:133–41
35. Standard, P. G., Kaufman, L. 1976. Specific immunological test for rapid identification of members of the genus *Histoplasma*. *J. Clin. Microbiol.* 3:191–99
36. Standard, P. G., Kaufman, L. 1977. Immunological procedures for the rapid and specific identification of *Coccidioides immitis* cultures. *J. Clin. Microbiol.* 5:149–53
37. Standard, P. G., Kaufman, L. 1980. A rapid and specific method for the immunological identification of mycelial form cultures of *Paracoccidioides brasiliensis*. *Curr. Microbiol.* 4:297–300
38. Standard, P. G., Kaufman, L. 1982. Safety considerations in handling exoantigen extracts from pathogenic fungi. *J. Clin. Microbiol.* 15:663–67
39. Standard, P. G., Kaufman, L., Whaley, S. D. 1985. *Rapid Identification of Pathogenic Isolates by Immunodiffusion. CDC Lab Manual.* Atlanta, Ga: US Dep. Health Hum. Serv. Publ. Health Serv., Cent. Dis. Control
40. Weeks, R. J., Padhye, A. A., Ajello, L. 1985. *Histoplasma capsulatum* variety *farciminosum:* a new combination for *Histoplasma farciminosum*. *Mycologia* 77:964–70
41. Yangco, B. G., Nettlow, A., Okafor, J. I., Park, J., Strake, D. T. 1986. Comparative antigenic studies of species of *Basidiobolus* and other medically important fungi. *J. Clin. Microbiol.* 23:679–82

Ann. Rev. Microbiol. 1987. 41:227–58

GENETICS OF AZOTOBACTERS: Applications to Nitrogen Fixation and Related Aspects of Metabolism

Christina Kennedy and Aresa Toukdarian

AFRC Unit of Nitrogen Fixation, University of Sussex, Brighton BN1 9RQ, England

CONTENTS

INTRODUCTION

This review is about genetic developments in the genus *Azotobacter*, which belongs to the family Azotobacteraceae. As the name implies, all members of the family can fix atmospheric nitrogen. *Bergey's Manual* limits the genera in

0066-4227/87/1001-0227$02.00

Azotobacteraceae to two, *Azotobacter* and *Azomonas,* and excludes the previous members *Beijerinckia* and *Derxia* mainly on the basis of dissimilarity of rRNA cistrons (149).

Azotobacters are aerobic, mainly soil-dwelling organisms with a unique array of metabolic capabilities in addition to nitrogen fixation. Different species and strains have various abilities to synthesize alginates, poly-β-hydroxybutyrate, pigments, and plant hormones (see 40, 47, 62, 117, 152 for recent work and/or references). Some strains associate with roots of certain plants; a notable case is the association of *Azotobacter paspali* with the tropical grass *Paspalum notatum* (6, 19). Azotobacters form small resting-phase cysts which are less differentiated than spores and which contain novel lipids (118, 147 and references therein). Small filterable cells, less metabolically active than cysts, have been found in aging cultures (5).

Azotobacters are the first nitrogen-fixing organisms found to contain more than one type of nitrogenase enzyme (14). The well-known molybdenum nitrogenase is synthesized if the metal is present in the environment. If molybdenum concentrations are less than about 10 nM, a vanadium-based enzyme is available for nitrogen fixation in both *Azotobacter chroococcum* and *Azotobacter vinelandii* (57, 127). The latter species may synthesize a third, as yet uncharacterized, nitrogenase under these conditions (64). The other important and unique feature of azotobacters is their extreme tolerance to oxygen while fixing nitrogen; mechanisms for protecting nitrogenase against O_2 damage exist alongside a dependence on aerobic metabolism for energy and growth. Additionally, the H_2 evolved during nitrogen fixation can be catabolized by an uptake hydrogenase.

Developments in molecular biology and methods available for genetic manipulation in azotobacters have made possible the genetic analysis of nitrogen fixation and related aspects of nitrogen, oxygen, and hydrogen metabolism. These are the subjects of this review. The biochemistry and physiology of these processes are not discussed at length here. With the exception of reference to a few papers on *Azotobacter beijerinckii,* the review concerns two species, *A. vinelandii* and *A. chroococcum.* The strains used in most of the cited work on *A. vinelandii* are strain MS (ATCC12837), strain O (ATCC12518), and the nongummy strain O derivative OP or UW (ATCC13705). For *A. chroococcum,* strain MCD1, a nongummy derivative of NCIB8003 (ATCC44122) cured of three of its five plasmids, has been widely used (126). While the two species exhibit minor differences in habitat and metabolism (9, 101, 155), an important difference relevant to genetic studies is that *A. vinelandii* is generally a better recipient of DNA introduced by transformation and of wide–host range plasmids introduced by conjugation. This difference may be related to the presence of indigenous plasmids in *A. chroococcum* (126) but not in *A. vinelandii.*

GENETIC TECHNIQUES APPLIED TO AZOTOBACTERS

Recombinant DNA methodology has been used to construct *A. chroococcum* and *A. vinelandii* gene banks in plasmid and phage vectors (20, 67, 70, 92) and in numerous cloning and hybridization experiments. Strategic application of recombinant and classical genetic techniques in azotobacters depends on aspects of genome organization, gene expression, mutability, transformability, receptiveness in conjugation, and ability to maintain plasmids.

Analysis of Gene Organization and Expression

Although azotobacters contain more DNA than most other bacteria, the size of their genome is typical of prokaryotes. Denatured DNA from *A. vinelandii* strain ATCC 12837 (strain MS) renatures at a rate similar to that of denatured DNA from *Escherichia coli* (132). DNAs isolated gently from both organisms as folded chromosomes also sediment at similar rates (132). Restriction-enzyme digests of DNA from *A. chroococcum* strain MCD1 or CW8 separated by electrophoresis in two dimensions yielded fragments that summed to about 2000 kb, approximately half the size of the *E. coli* genome (126). The total amount of DNA per azotobacter cell is, however, much higher than that in *E. coli;* the number of chromosome equivalents per cell during exponential growth has been estimated as at least 40 in *A. vinelandii* (132) and 20–25 in *A. chroococcum* (126). The reason for such a high DNA content in azotobacters is not known, but it may be related to the fact that the azotobacter cell is approximately 10-fold larger than other bacteria (132). The G+C content of azotobacter DNA is 65–68% (9).

Core RNA polymerase isolated from *A. vinelandii* has the same $\alpha_2\beta\beta'$ component structure as the *E. coli* enzyme (80). Also similar in both organisms are the conventional sigma factor and another recognition factor, the *ntrA* gene product (also called *glnF* or *rpoN*), which is required for transcription of certain genes involved in nitrogen assimilation (33, 159; M. J. Merrick, J. R. Gibbins & A. E. Toukdarian, manuscript submitted). Complementation studies, described later in this review, have shown that some azotobacter genes are expressed in *E. coli* and *Klebsiella pneumoniae* (67, 72, 92, 159) and conversely that genes from other bacteria are expressed in azotobacters (69, 70, 116, 156, 166). However, except for *nif,* too few azotobacter promoters have been sequenced to allow general features to be compared to promoters of other organisms.

The modified tRNA species 2-methyl-*trans*-zeatin is found in *A. vinelandii*. Although the proportion varies with culture conditions, including fixed nitrogen supply, its significance to translation or gene expression is unknown (2).

The laboratory strains of *A. vinelandii* do not contain native plasmids. On the other hand, of eight *A. chroococcum* strains studied, including six new soil isolates, all have two to six plasmids ranging in size from 5.5–200 Md (126). No specific properties encoded by these plasmids have been found.

Gene Transfer Systems

Genetic information can be transferred into azotobacters by conjugation or by transformation. Temperate phages with the potential for transduction have also been characterized.

A number of plasmids with a broad host range can be maintained in azotobacters. Plasmids of incompatibility groups P and Q replicate in *A. vinelandii* (26, 45, 69–71, 158, 159) and *A. chroococcum* (69, 71, 116, 156, 166). Only IncP and IncW plasmids will replicate in *A. beijerinckii* (97). These plasmids can be conjugally transferred from *E. coli* to azotobacters (26, 69, 97, 156). Transfer between azotobacters or from azotobacters to other bacteria has also been reported (26, 69, 158).

The self-transmissible IncP plasmids RP4 or R68.45 transfer from *E. coli* to *A. vinelandii* at a frequency of 10^{-3}–10^{-4} per recipient, while they transfer between *A. vinelandii* strains at a frequency greater than 10^{-1} per recipient (26, 158). RP4 transfers from *E. coli* to *A. beijerinckii* at a frequency of 10^{-3} per recipient (97). The transfer of certain IncP plasmids and all IncQ plasmids requires that *tra* gene functions be provided by a helper plasmid. Using RP4 as the helper, IncQ plasmids transfer from *E. coli* to *A. vinelandii* at a frequency of 10^{-4} per recipient (26). With pRK2013 as helper, the frequency of IncQ transfer to *A. vinelandii* is 10^{-1}–10^{-2} and to *A. chroococcum* is 10^{-3}–10^{-4} (69, 71, 116). IncQ plasmids can be transferred from *A. vinelandii* to *E. coli* in three parent crosses using pRK2013 carried by a second *E. coli* strain (69). Although pRK2013 is not maintained in azotobacters, its *tra* genes, derived from RP4, are apparently transiently expressed in *A. vinelandii*. IncP plasmids that require helper functions can be transferred from *E. coli* to azotobacters using pRK2013, although they cannot be similarly transferred from azotobacters to *E. coli* (69, 156, 166). The low frequencies (10^{-5}–10^{-6}) originally reported for the transfer of IncP plasmids into *A. chroococcum* using pRK2013 (116) have been improved to about 10^{-2} per recipient by a new method for matings (156).

The copy number of IncP or IncW plasmids in *E. coli* is 1–10 per cell, while IncQ plasmids are maintained at 20–50 copies per cell (157). Since replication functions are intrinsic to such plasmids, it is likely that the copy number per cell will be similar in multichromosomal azotobacters. Therefore the copy number of plasmid genes per azotobacter chromosome may be much lower than in *E. coli*. These considerations are relevant to regulation studies in which gene dosage effects can be important (67, 69, 71, 156).

Chromosome mobilization mediated by conjugative plasmids occurs in *A. vinelandii* and *A. beijerinckii*. The frequency of *nif* gene transfer from wild-type *A. vinelandii* to *A. vinelandii nif* mutants is 10^{-4}–10^{-5} per recipient using R68.45 and 10^{-8} using RP4 (158). Transfer of chromosomally encoded rifampicin resistance by RP4 occurs at a frequency of 10^{-8} per recipient in *A. beijerinckii,* while the transfer of *nif* genes could not be demonstrated (97). Chromosomal mobilization experiments showed that ten independently isolated mutants of *A. vinelandii* that are unable to grow on N_2 or nitrate have mutations in the same linkage group (133).

Transformation, although originally reported for both *A. vinelandii* and *A. chroococcum* (137), has been studied and used only in *A. vinelandii*. Page & Sadoff (104) described the first reliable procedure for inducing competence and for transforming *A. vinelandii* cultures; they combined both steps in a plate method. After a procedure for inducing competence in liquid using iron limitation was discovered (105), it was possible to investigate the factors that affect the development of competence (34, 99, 103, 105) separately from those that affect DNA uptake, recombination, and expression (34, 106). Page and coworkers' studies have culminated in a method in which molybdenum limitation and iron limitation are used to induce competence, which significantly increases the transformation frequency of *A. vinelandii* UW (100). This procedure, with slight modification, has been used to transform plasmids into *A. vinelandii* (45) at a higher frequency than that obtained with an earlier method based on Ca^{2+} shock to induce competence (26).

A. vinelandii can be transformed with heterologous as well as homologous DNA. Various *A. vinelandii* Nif^- mutants are transformed to Nif^+ by DNA from other *Azotobacter* species (34, 100) and from various *Rhizobium* species (12, 34, 89, 98, 100).

Transformation has been used to determine the linkage of mutations (11, 98, 104); to construct strains (16, 17, 122, 159); to show that cloned DNA carries the genetic information corresponding to a particular mutation (17); and, after transposon mutagenesis, to show linkage of the mutant phenotype with the drug resistance marker of the transposon (66).

Plasmids with a narrow host range limited to *E. coli* and related bacteria can be transferred either by conjugation or by transformation into azotobacters, but they do not replicate. This property has been exploited genetically for obtaining suicide vectors for transposon mutagenesis and for introducing mutated genes into azotobacters (see following section).

Introduction of certain broad–host range plasmids can result in observable changes in the azotobacter cell (46, 97). Colonies of *A. vinelandii* MS carrying the IncP plasmid pRK2501 or the IncQ plasmid RSF1010 or pGSS15 differ in size and morphology from the plasmid-free parent; cells are smaller, produce more capsular exopolysaccharide, and are less able to fix nitrogen or synthesize siderophores (46). *A. beijerinckii* carrying RP4, but not other IncP or any IncW plasmids, also has a colony morphology significantly different from that of the parent (97). The reasons for these effects are not known.

Phages that specifically infect many strains of *A. vinelandii* and/or *A. chroococcum* have been described (18, 37). There has been one brief report that certain of these phages transduced DNA from the former to the latter organism (164). However, they have mainly been used to confirm that isolates from mutagenesis or gene-transfer experiments were azotobacters and not contaminants (12, 26, 42, 61, 91, 93, 145).

An interesting property of one group of phages is an ability to induce pseudo-lysogenic conversion in *A. vinelandii* strain O (109, 153, 154). Properties of a pseudo-lysogen include loss of the polysaccharide coat, acquisition of flagellae and motility, and yellow pigmentation (153). Repeated subculturing of a pseudo-lysogen gives rise to permanently converted cells that have the phenotype of a pseudo-lysogen but from which free phage or parental-type cells can no longer be isolated (153). The commonly used laboratory strain UW (OP) of *A. vinelandii,* which is an unencapsulated variant of strain O, has many of the properties of a permanently converted isolate (154). Page (99) has suggested that lysogeny may have a role in the ability of *A. vinelandii* UW to be transformed at high frequency.

Mutagenesis

Mutants of azotobacters have been isolated with varying degrees of success. It is possible to isolate spontaneous mutants of *A. vinelandii* or *A. chroococcum* (50, 121, 144). Mutants can also be isolated after mutagenesis with *N*-methyl-*N'*-nitro-*N*-nitrosoguanidine (NTG) (10, 42, 61, 87, 115, 141, 143, 166), ethylmethanesulfonate (EMS) (83, 93), ICR 191 (86, 133), hydroxylamine (50), UV light (133, 150, 151, 165), or transposons (66, 70, 97, 115). The conditions used for mutagenesis are typical for prokaryotes. Treatment of mutagenized cultures with penicillin, with or without cycloserine, has been reported to enrich certain phenotypes (61, 83, 86, 91, 115). Since mutagenesis is possible in azotobacters, the reported difficulties in isolating certain types of mutants probably reflect problems of chromosome segregation and selection.

Azotobacter mutants resistant to antibiotics or toxic compounds can be isolated directly by plating a mutagenized culture on medium containing the inhibitory substance. Antibiotic-resistant mutants that have been isolated include rifampicin-resistant *A. vinelandii* (11, 26, 104) and *A. beijerinckia* (97) and a streptomycin- and nalidixic acid–resistant *A. chroococcum* (126); all of these have proved useful for genetic studies. However, mutants of *A. beijerinkii* resistant to streptomycin, sulphonamides, or six other drugs could not be isolated (97). Azotobacters are extremely sensitive to many antibiotics, and the level of resistance conferred by a chromosomal mutation or by a resistance gene encoded by a transposon or plasmid is up to 100-fold less than in *E. coli* (45, 69, 97, 116, 126).

The isolation of azotobacters resistant to toxic compounds has resulted in a number of interesting mutants. Such *A. vinelandii* mutants include chlorate-resistant strains with little or no nitrate reductase activity (133, 144), methyl-alanine-resistant strains that fix nitrogen in the presence of NH_4^+ (50, 143), methylammonium-resistant strains that overproduce nitrogenase (50), and L-methionine-DL-sulfoximine (MSX)–resistant strains with altered glutamine synthetase (87). Selection for citrate resistance in *A. chroococcum* has resulted in mutants with increased sensitivity to oxygen while fixing nitrogen (Fos^-) (115), while selection for azide resistance has resulted in mutants that are NH_4^+ constitutive for nitrogen fixation (141).

The isolation of mutants that cannot be directly selected requires several generations of outgrowth of the mutagenized culture on nonselective medium prior to screening (42, 86, 93, 165). This is presumably because the recessive mutation must be present on most or all of the 20–40 chromosomes in order to be expressed. Such mutants that have been successfully isolated include Nif^- mutants of *A. vinelandii* (10, 42, 93, 98, 139, 145, 165), respiratory mutants of *A. vinelandii* (60, 61, 91), Fos^- mutants of *A. chroococcum* (115), and hydrogen-uptake (Hup^-) mutants of *A. chroococcum* (113, 166).

Auxotrophs, however, have been more difficult to isolate (83, 93, 97, 132). Some mutations may be lethal because the required metabolite cannot be transported (93). This idea is supported by the fact that many presumptive auxotrophs of *A. vinelandii* that are able to grow on rich but not minimal medium cannot be fed by any one compound (70, 86, 93). Kennedy et al (70) have postulated that azotobacters can transport oligopeptides but not all amino acids. Auxotrophs that have been isolated include methionine-, uracil-, and hypoxanthine- or adenine-requiring strains of *A. vinelandii* (70, 86, 93, 104), and adenine and leucine auxotrophs of *A. beijerinckii* (97).

The instability of mutant phenotypes is frequently a problem with azotobacters. Mutants often revert to wild type even after repeated subculturing (70, 91, 93, 150, 165). The fact that some mutations are stable and others are not is probably related to the presence of multiple chromosomes in the cell, which can mask potentially lethal consequences of certain recessive mutations.

Transposon mutagenesis has facilitated mutant isolation in *A. vinelandii* and *A. chroococcum* (66, 70, 115) and has been the only successful method reported for inducing mutations in *A. beijerinckii* (97). The advantage of transposon mutagenesis is that the antibiotic resistance encoded by the transposon gives a selectable marker for nonselectable mutations. Narrow–host range, mobilizable plasmids carrying transposons have generated mutants in *A. vinelandii* (with Tn*5*) (66, 70) and in *A. chroococcum* (with Tn*1*) (115). Streptomycin resistance, as well as kanamycin resistance, is conferred by Tn*5* in *A. vinelandii* (70). The broad–host range plasmid RP4 carrying

Tn*76* can behave as a suicide vector, generating mutants in *A. beijerinckii* (97).

Several methods for constructing specific mutants using cloned genes have been developed. *A. vinelandii* strains with part or all of the *nifHDK* genes deleted were constructed by cotransformation of the wild type with a plasmid that carried *A. vinelandii nif* DNA with the desired deletion along with chromosomal DNA isolated from a rifampicin-resistant mutant (16, 17, 122). Rifampicin-resistant transformants were positively selected and then screened for the Nif phenotype. Those that were Nif$^-$ should also have been transformed with the plasmid DNA and must have incorporated the mutated *nif* region by recombination.

Robson (125) constructed a *nifHDK*-deleted strain of *A. chroococcum* using a slightly different approach. A gene encoding resistance to kanamycin was cloned into the deletion site of the *nif* fragment so that the deletion could be positively selected. This altered fragment was then cloned into the narrow–host range IncN Tra$^+$ plasmid pCU101. After transfer of the plasmid into *A. chroococcum,* kanamycin-resistant transconjugants were isolated. Recombination between the homologous *nif* regions of the plasmid and chromosome resulted in introduction of the deletion mutation, marked by Km^r, into the chromosome.

A third technique was used to construct nitrogen-regulatory mutants *(ntr)* of *A. vinelandii* (159). Plasmids carrying the cloned *A. vinelandii ntr* genes were mutagenized with Tn*5* in *E. coli*. The sites of Tn*5* insertion were mapped, and the plasmids of interest were used to transform kanamycin resistance into *A. vinelandii*.

In all three procedures, isolates in which the plasmid was integrated by a double crossover event are distinguished from single crossover recombinants by screening for antibiotic markers encoded by the plasmid replicon. These markers appear only at low frequency, possibly because they are unstable. Confirmation that the desired mutation has been made is obtained by hybridization analysis of DNA isolated from the presumptive mutant (16, 17, 125, 159).

NITROGEN FIXATION IN AZOTOBACTERS

The development of azotobacter *nif* genetics falls into three phases. Initial work involved only *A. vinelandii;* from 1950 to 1971 a few Nif$^-$ mutants were isolated (10, 42, 53, 93, 165), and from 1973 to 1982 Brill and coworkers isolated the UW series of mutants, identified the biochemical lesions, and constructed a *nif* linkage map by transformation analysis (11, 96, 139, 140). Phenotypic characterization of *nif* point mutants led to the proposal in 1980 of the second ("alternative") enzyme system for nitrogen fixation in

A. vinelandii (14). The third and molecular era began in 1983–1984 with the first reports of recombinant plasmids carrying *nif* genes from *A. vinelandii* (92) and from *A. choococcum* (67). An important feature of both papers was the recognition of reiterations of *nifH*-like sequences in azotobacters; in *A. chroococcum* these repeats were later correlated with the alternative, vanadium-based nitrogenase. This section focuses on recent advances in understanding the genetic basis for both the conventional and alternative nitrogenase systems in *A. chroococcum* and *A. vinelandii*.

In *K. pneumoniae* a cluster of 17 contiguous *nif* genes encodes the three nitrogenase polypeptides and other proteins necessary for FeMo cofactor synthesis, Fe protein activation, electron transfer to nitrogenase, and regulation (see References 32 and 96a and Table 1). There is no apparent reiteration of *nif* genes in *K. pneumoniae,* nor is there evidence of a nitrogenase other than the conventional Mo-containing enzyme. Other genes in diverse locations of the chromosome are involved in *nif* gene regulation (32, 55a) or in Mo assimilation (63).

The *K. pneumoniae* system served as a starting point for analyzing *nif* and related genes in azotobacters, but differences would be expected because of the unique oxygen tolerance of azotobacters, their alternative nitrogenase systems, and also the distinct phylogenetic differences between the organisms. The similarities and differences that have emerged are discussed in the following paragraphs.

Table 1 *nif* gene structure and function in *K. pneumoniae*

QB ← AL ← F → MVSU ⟵ XNE ⟵ YKDH ← J[a] →

Gene	Nature of product	Gene	Nature of product	Gene	Nature of product
A	Transcription activator	J	Pyruvate flavodoxin oxidoreductase	S	Unknown
B	FeMo cofactor biosynthesis	K	β Subunit Kp1	U	Unknown
D	α Subunit Kp1	L	Blocks *nifA* product activity in NH_4^+, O_2	V	FeMo cofactor biosynthesis
E	FeMo cofactor biosynthesis	M	Activates nascent Kp2	X	Unknown
F	Flavodoxin biosynthesis	N	FeMo cofactor biosynthesis	Y	Unknown
H	Kp2 subunit	Q	FeMo cofactor biosynthesis		

[a]Arrows refer to direction of transcription. Kp1 and Kp2 refer to the MoFe protein and Fe protein components of *K. pneumoniae* nitrogenase.

Major nif *Gene Region*

A group of genes spanning 25–30 kb of DNA was characterized in *A. chroococcum* (67) after the isolation from a gene bank of cosmids with 45-kb inserts that hybridized to the *nifHD* genes of *A. vinelandii* cloned in pBR325 (17). Hybridization to *nif* gene probes from *K. pneumoniae* coupled with DNA sequencing and complementation analysis has revealed that the *nif* genes *FMVSUNEKDH* are present in this region (41, 67). Thus it bears some resemblance to the *nif* cluster of *K. pneumoniae*, but the homologous *nif* genes are spread over a longer stretch of DNA and appear to lack the five flanking genes *nifQBAL* and *nifJ*. Although no corresponding fragment has been isolated from *A. vinelandii*, analysis of other cloned fragments suggests that this organism has a similar arrangement of *nif* genes (20, 28, 59, 70). The organization of *nif* genes in this region may therefore be common to all azotobacters.

*nif*HDK The genes best characterized for structure and function in *A. vinelandii* and *A. chroococcum* are *nifHDK*. These genes encode the single protein subunit for the nitrogenase Fe protein (Av2 or Ac2) and the α and β subunits of the MoFe protein (Av1 or Ac1), respectively. Two other regions in *A. chroococcum* of unknown distance from the *nifHDK* cluster hybridize to *nifH* and *nifK* gene probes, respectively (67, 125). In *A. vinelandii*, two other regions distinct from *nifHDK* hybridize to *nifH* (16, 64). These are discussed in the following section on genes for the alternative nitrogenase.

The DNA coding sequences of *A. vinelandii* and *A. chroococcum nifHDK* closely resemble those of the *nifHDK* genes in *K. pneumoniae* (20, 128). Plasmids carrying *nifHDK* genes of *A. chroococcum* restore significant acetylene-reducing activity to *nifH*, *nifD*, and *nifK* mutants of both *A. vinelandii* and *K. pneumoniae* (67, 70). Since one of the complemented *K. pneumoniae* mutants was a *nifH*::Tn*5* insertion, as was one of the *A. vinelandii* mutants, and since the *nifD* and *nifK* products are absent owing to polarity of Tn*5*, the activity in these complemented organisms must be exclusively from *A. chroococcum* components. In the case of the *K. pneumoniae nifD*::Tn*5* and *nifK*::Tn*7* mutants (67), it is not known whether there is productive association of *A. chroococcum* MoFe protein with *K. pneumoniae* Fe protein or whether the activity is solely from *A. chroococcum* nitrogenase components. In any event, it can be concluded from these experiments that the *K. pneumoniae nifM* product activates the *A. chroococcum* Fe protein and also that the *nifQBVNE* products can form active *K. pneumoniae* FeMo cofactor, which associates with the *A. chroococcum* MoFe protein. This is the first unequivocal demonstration that nitrogenase polypeptides from one organism can be substrates for modification by *nif* gene products in another organism.

DNA sequences with potential secondary structure have been found in the intergenic regions between *nifH* and *nifD*, between *nifD* and *nifK*, and between *nifK* and the next open reading frame (ORF) (as yet not identified as a gene) in *A. vinelandii* and *A. chroococcum* (20, 128). These sequences of 20–40 bp can form single stem and loop structures that may influence gene expression or mRNA stability. Analysis of transcripts from this region in both organisms shows that following removal of NH_4^+ from cultures, three RNA species are synthesized (67, 81). Their lengths of 1.1, 2.6, and 4.3 kb in *A. chroococcum* (67) correspond to the sizes expected for transcripts ending at the intergenic regions. The consequence of these three different mRNAs to nitrogen fixation is not known, but they could serve to adjust the balance of nitrogenase components in different growth conditions. The ratio of the amount of Ac1 to the amount of Ac2 synthesized ranged from 0.3 to 1.0 under different growth-limiting conditions in chemostat cultures of *A. chroococcum* (161); these ratios influence the electron flow through nitrogenase. Krol et al (81) detected three transcripts from the *nifHDK* region of *A. vinelandii* that were of similar size to transcripts characterized from *K. pneumoniae*. Since the intergenic regions of *K. pneumoniae* are smaller and show no potential secondary structure (61a, 63a, 135), the relationship of these regions in azotobacters to regulation of transcription or to stabilization of mRNA is uncertain.

A fourth transcript from the *nifHDK* region of *A. chroococcum* is 6.4 kb and extends from *nifHDK* downward but not into *nifEN* (41). It is a minor species, representing only 1–2% of the total mRNA from the *nifHDK* region. Sequencing data suggest that more than one and probably two ORFs are encoded in the region downstream from *nifK* (D. Evans, personal communication), but it is not known if one of them is the equivalent of the *nifY* gene, which lies between *nifK* and *nifE* in *K. pneumoniae*. Three open reading frames downstream of *nifHDK* and preceding *nifEN* were reported for *A. vinelandii* (29).

nifEN The azotobacter *nifEN* genes are separated by about 2.5 kb from *nifK* (28, 41). Sequencing of these genes from *A. vinelandii* led to the discovery that they are significantly homologous to the *nifDK* genes; *nifE* is like *nifD* and *nifN* is similar to *nifK* (28, 29). As in *K. pneumoniae*, the *A. vinelandii* *nifEN* gene products are necessary for FeMo cofactor biosynthesis. This fact was demonstrated with an *A. vinelandii* *nifN*::Tn*5* mutant, which was identified genetically using Nif$^-$ derivatives of the IncP plasmid pRD1 (which carries the entire *K. pneumoniae* *nif* gene cluster) (70). The insertion site correlates physically with the sequenced gene. Nitrogenase activity was restored to extracts of the mutant by the addition of purified *K. pneumoniae* FeMo cofactor. Thus the homologous sectors of *nifE* and *nifD* and of *nifN* and *nifK* may encode portions of the proteins that interact with the cofactor and/or its biosynthetic precursors. In *A. chroococcum* a second region of the genome with *nifEN* homology has been identified by hybridization (R. Robson, personal communication).

nifUSVM Although the *nifUSVM* operon is contiguous with *nifENX* in *K. pneumoniae,* a gap of 6–8 kb separates the two regions in *A. chroococcum.* The *nifUSV* genes have been located by hybridization to the corresponding *K. pneumoniae* genes (41). A *K. pneumoniae nifM* probe did not hybridize to the *A. chroococcum nif* cluster, but the presence of this gene has been indicated by complementation experiments. A cosmid that carries 22 kb of the *nif* cluster complements *nifM* mutants of *K. pneumoniae* (72). This cosmid, isolated by its ability to complement the Tn*5*-induced $NifN^-$ and $NifM^-$ mutants of *A. vinelandii* (70), also complements *nifV* and *nifS* mutants of *K. pneumoniae* for acetylene reduction and growth on N_2 (72).

The $NifM^-$ mutant MV21 of *A. vinelandii* has been characterized genetically by complementation experiments using Nif^- derivatives of pRD1 (70). Only *nifM* mutants of pRD1 fail to restore a Nif^+ phenotype to MV21. The biochemical phenotype of MV21 is the same as that of *K. pneumoniae nifM* mutants: The nitrogenase Fe protein is synthesized in an inactive form. Thus in both organisms the *nifM* product activates nascent Fe protein.

nifF Another gene identified in *K. pneumoniae* may be present in the major *nif* region of *A. chroococcum.* A *K. pneumoniae nifF* probe hybridized to a fragment from the *A. chroococcum* cosmid just adjacent to the *nifUSVM* region (41). The *K. pneumoniae nifF* gene product is a flavodoxin that couples electron transfer between the *nifJ*-encoded pyruvate oxidoreductase and oxidized nitrogenase Fe protein. NH_4^+-repressible flavodoxins have been purified from *A. vinelandii* and *A. chroococcum* (30, 78, 79). Several physical and biochemical parameters of these proteins are similar to those of the *K. pneumoniae nifF* gene product (30, 36). It is possible that the *nifF* hybridizing region adjacent to *nifUSVM* in *A. chroococcum* encodes the flavodoxin used for nitrogenase reduction.

Genes for Alternative Nitrogenase Systems

Bishop et al (14) presented genetic evidence for a second pathway for nitrogen fixation in azotobacters in 1980. They found that *nifD* and *nifK* mutants of *A. vinelandii* yielded Nif^+ derivatives at low frequency that were pseudo-revertants: The derivatives retained the original mutations and had secondary mutations that gave the Nif^+ phenotype. These strains had low acetylene reduction activity, synthesized four NH_4^+-repressible proteins not previously observed, and were resistant to tungstate, an inhibitor of nitrogenase. A number of other Nif$^-$ mutants also grew on N_2 in medium depleted of molybdenum and, under these conditions, synthesized the four new proteins. Bishop and coworkers thus proposed that *A. vinelandii* produced a second nitrogenase in response to molybdenum deprivation.

In complementary studies, Riddle et al (121) reported that a spontaneous tungsten-resistant (W^r) mutant of *A. vinelandii* lacked MoFe protein subunits but synthesized some NH_4^+-repressible proteins that are not present in the wild type. Page & Collinson (102) found protein patterns in the *nif* mutant UW3 that varied with regimes of molybdenum (and vanadium) addition and removal. Furthermore, extracts of a *nifH* mutant (UW91) had Fe protein activity if the organism was grown without molybdenum, but not if molybdenum was present in the medium (114). Since all of these experiments involved organisms that retained intact, although mutated, *nifHDK* genes, it was possible that a second mutation or low levels of molybdenum led to the suppression of the Nif$^-$ phenotype by enabling residual or leaky activity of the mutant proteins to become expressed. In work with chemostat cultures, Eady & Robson (38) observed that molybdenum stimulated growth of *A. vinelandii* in medium purified to remove metals. While this seemed to indicate a dependence on molybdenum at low concentrations for nitrogen fixation, Bishop et al (13) subsequently showed that the effect of molybdenum was to enhance sulfate uptake, thereby relieving sulfate limitation in these cultures.

Proof of the existence of an alternative nitrogenase was provided by the construction in Bishop's laboratory of mutants deleted for the *nifHDK* genes. Mutations made in vitro by deleting specific restriction fragments were introduced into the *A. vinelandii* genome (16). The mutants grew on N_2 in the absence of molybdenum but not in its presence, incorporated $^{15}N_2$ into cell components, and failed to grow on Ar/O_2 mixtures, demonstrating a dependence on N_2 for growth (13). Nitrogenase has been purified from similar *nifHDK*-deletion mutants of *A. chroococcum* (125, 127) and *A. vinelandii* (57, 58). The structure of these enzymes is similar to that of the conventional nitrogenases, but in both cases vanadium, not molybdenum, is present. Each has a dimeric Fe protein (Ac2* and Av2') with subunit molecular weight of about 31×10^3 and a tetrameric VaFe protein (Ac1* and Av1') with molecular weights of about 52×10^3 and 55×10^3 for the two subunits. The presence of vanadium in the enzymes accords with the *A. chroococcum* mutant's dependence on vanadium for growth and with the stimulation of enzyme activity in the *A. vinelandii* deletion strain when $NaVaO_3$ is added to the growth medium. The Va enzymes reduce substrates, particularly acetylene, less efficiently than the Mo nitrogenases, and their ESR spectra are different.

A second *nifH* hybridizing region of *A. chroococcum* has two contiguous genes that are transcribed only in Mo-deprived organisms (130). The first of these, *nifH**, encodes a protein whose amino acid sequence is 90% identical to that of Fe protein from conventional Mo nitrogenase. The amino acid composition of Ac2* (39) agrees well with that predicted from the sequence of *nifH**. Thus it is fairly certain that *nifH** encodes Ac2*. The ORF adjacent to *nifH** encodes a 63–amino acid polypeptide whose sequence is most like that of a small low-potential ferredoxin (130). However, its role in nitrogen fixation has not been assigned.

The *A. chroococcum nifK* and *nifEN* genes hybridize to other regions of the genome (125; R. Robson, personal communication). Since probes carrying *nifK* or the ferredoxinlike gene each hybridize to an 8-kb *Eco*RI fragment, it is possible that an Ac1* subunit gene is near the Ac2* gene. The extra *nifEN* region may encode proteins involved in formation of the FeV cofactor associated with Ac1* (3).

A. vinelandii has a second *nifH* region that also hybridizes to a DNA probe carrying the small ferredoxinlike gene of *A. chroococcum* (64). It too is transcribed only under conditions of molybdenum depletion. Unlike *A. chroococcum,* however, this species has a third region in its genome that hybridizes to a *nifH* probe (16). During the course of derepression in Mo-free medium, *A. vinelandii* apparently synthesizes two different patterns of proteins, one during the early stages (pattern B1) and another in the late growth stages (pattern B2) (15, 102). Chisnell & Bishop (25) presented biochemical evidence that two different nitrogenase Fe proteins are produced under molybdenum deficiency. It is therefore possible that *A. vinelandii* produces two nitrogenase enzymes in response to molybdenum depletion, of which one is the V enzyme and the other has not yet been characterized.

At least one of the genes encoded in the major *nif* cluster is required for growth of *A. vinelandii* on N_2 in the absence of molybdenum. The NifM$^-$ mutant MV21 is unable to grow on N_2 whether molybdenum is present or not (70). Although the Tn*5* insert in MV21 may affect the structure or expression of other genes besides *nifM,* the only discernible biochemical phenotype of mutants grown in Mo-sufficient medium is that of an inactive nitrogenase Fe protein. It therefore seems likely that the *nifM* gene product activates nascent Fe proteins of all systems. The other Tn*5* mutant, MV22, can grow on agar medium without molybdenum. Thus the *nifN* gene product is apparently not necessary for production of the cofactor used by the alternative nitrogenase system(s) of *A. vinelandii*.

A series of Nif$^-$ Tn*5* mutants of *A. vinelandii* confirms that some genes are necessary for both systems while others are needed solely for fixation in the absence of molybdenum (66). One Nif$^-$ mutant that lacks FeMo cofactor under Mo sufficiency is also unable to grow on Mo-deficient medium; if both phenotypes are due to the absence of a single gene product, then the mutated gene is probably involved in biosynthesis of both FeMo and FeV cofactors. The phenotype of another mutant is similar to that of MV21, the NifM$^-$

strain; however, Tn*5* is inserted in a different fragment than in MV21. Other mutants have significant nitrogenase activity in the presence but not in the absence of molybdenum. These mutations map in three different locations; one is linked to the *nifHDK* region, another is near the site of the genes that are mutated in UW1 and UW45 (regulatory and FeMo cofactor mutations, respectively), and the third is linked to neither. One other mutant expressed the Mo enzyme under Mo sufficiency or deficiency, but did not express the alternative system when Mo was absent. In the latter condition, Mo enzyme was detected in the mutant as polypeptides on two-dimensional gel electrophoresis. Thus, at least six different genes in addition to *nifH** are involved in the alternative nitrogenase system(s) of *A. vinelandii*.

Homologies to Rhizobium Genes

Genes in rhizobia that are necessary for nitrogen fixation in nodules are called *fix* or *nif*. Some rhizobial *nif* genes are sufficiently similar to those in *A. vinelandii* to repair Nif$^-$ mutants after transformation with wild-type rhizobium DNA (12, 34, 89, 98, 100). Those repaired include *nifD, nifK,* and *nifA*-like mutants. It is not known whether restoration to Nif$^+$ involves recombination to replace the defective azotobacter sequences, or whether establishment of the rhizobium DNA occurs without removal of the original mutation, e.g. by maintenance of a rhizobium plasmid. Among Nif$^+$ UW10 *(nifD)* derivatives transformed with *Rhizobium trifolii* or *Bradyrhizobium japonicum* DNA, about 10% of the transformants agglutinated after treatment with trifoliin or antiserum against whole *B. japonicum* cells, respectively (12, 89). Thus, *A. vinelandii* is also able to incorporate and express rhizobium genes involved in cell-surface components. Interestingly, DNA from *Rhizobium* species that are capable of fixing nitrogen under free-living conditions transforms UW1, a *nifA*-like mutant, to Nif$^+$ at much higher frequencies than DNA from fast-growing strains that depend on the nodule environment for nitrogen fixation (98).

Some of the *fix/nif* genes of rhizobia hybridize to azotobacter genomic fragments. For two of these, *nifA (fixY)* and *nifB (fixZ),* the corresponding genes from *K. pneumoniae* do not hybridize to restriction fragments of *A. chroococcum* DNA (67). However, *nifA* probes from both *Rhizobium leguminosarum* and *Rhizobium meliloti* hybridize to several restriction fragments in *A. chroococcum* and *A. vinelandii* (72). The number and size of these fragments suggest that there are probably two regions with *nifA* homology. Genetic experiments described below indicate that at least one of these regions is required for *nif* gene expression. A *nifB* probe from *R. leguminosarum* apparently hybridizes to only one region of *A. chroococcum* (R. Robson, personal communication). The *nifB* and *nifA* genes are contiguous in *R. leguminosarum;* in the azotobacter hybridization experiments, one or two similarly sized fragments hybridized to both the *nifA* and *nifB* probes, so linkage is probable.

The *fixABC* genes common to a number of rhizobia have no apparent counterparts in *K. pneumoniae,* but do hybridize to one region in both *A. vinelandii* (54) and *A. chroococcum* (R. Robson, personal communication). Since both rhizobia and azotobacters are obligate aerobes, the *fixABC* genes may encode proteins involved in aerobic nitrogen fixation, e.g. specific electron-transfer proteins.

REGULATION OF NITROGEN FIXATION

Environmental factors that influence nitrogenase synthesis in azotobacters include, as for other diazotrophs, the supply of NH_4^+, O_2, and molybdenum. In azotobacters the molybdenum supply apparently determines which nitrogenase enzyme will be available for nitrogen fixation. As for *nif* genes, regulation has been best studied in *K. pneumoniae*. The genetic elements involved in responses to the environment, including promoters and regulatory genes, show similarities to and differences from those in *K. pneumoniae*.

nif *Promoters and Regulatory Genes*

The consensus DNA sequence found near the start of *nif* transcriptional units in *K. pneumoniae* and also in various *Rhizobium* species is CTGG-N_8-TGCA, with the invariant GC located at −12,−13 and the invariant GG at −23,−24 nucleotides before the site where transcription starts (33). Similar −12 and −24 sequences are located at appropriate distances before the translation initiation codons of the *nifH* and *nifE* genes in both *A. vinelandii* and *A. chroococcum* (20, 28, 128; D. Evans, personal communication). An upstream motif of TGT-N_{10}-ACA is thought to be the binding site for the *nifA* activator in *K. pneumoniae;* this too is found at appropriate distances in the azotobacter *nifH* and *nifE* promoter regions (20, 28, 128). These promoter structures evidently require a *nifA*-like protein for their expression. Other evidence for a *nifA* gene in azotobacters is as follows.

(*a*) The *K. pneumoniae nifA* gene cloned on wide–host range plasmids under a constitutive promoter activated expression of nitrogen fixation in regulatory mutants of both *A. vinelandii* and *A. chroococcum* (71). (*b*) The *A. chroococcum nifH* promoter fused to *lacZ* on a wide–host range vector required for its expression in *E. coli nifA* genes from *K. pneumoniae, R. meliloti,* or *B. japonicum* cloned on multicopy plasmids and expressed from constitutive promoters (68, 72). (*c*) The multicopy effect, whereby the *nifH* promoter cloned on a high–copy number plasmid inhibits nitrogen fixation by sequestering *nifA* product and preventing activation of other *nif* promoters in *K. pneumoniae,* occurs in both *A. vinelandii* and *K. pneumoniae* that carry many copies of the *A. chroococcum nifH* promoter (67). (*d*) Inhibition is partially relieved in *A. vinelandii* and *K. pneumoniae* by provision of additional copies of *K. pneumoniae nifA* (67).

Shah et al (139) isolated *nifA*-like regulatory mutants after chemical mutagenesis. These mutants of *A. vinelandii* include UW1, which has been widely used in gene mapping, transformation, and complementation experiments (11, 71, 98). Its mutation is closely linked to that in UW45, a Nif^- mutant that lacks FeMo cofactor, but is only distantly linked to *nifH, nifD*, or *nifK* mutations (11). More recently, a Tn*5*-induced Nif^- mutant of *A. vinelandii* complemented by *K. pneumoniae nifA* plasmids was shown to have Tn*5* inserted in a region that hybridizes to the two rhizobium *nifA* probes mentioned previously (68). Mapping suggests that it is linked to the UW1 mutation (E. Santero, personal communication). In addition, 12 kb of DNA from the Tn*5* insert region hybridized to the *nifA* genes of *R. meliloti* and *R. leguminosarum*. This region hybridizes to *A. chroococcum* DNA distinct from the *nif* genes described previously, which indicates that it is not closely linked to *nifFUSVM* as it is in *K. pneumoniae*. This *nifA* region of *A. chroococcum* corresponds to one of the regions that hybridizes to the rhizobium *nifA* probes (C. Kennedy & R. Humphrey, unpublished results).

In *K. pneumoniae, nif* expression also requires the *ntrA* sigma-like recognition factor (33). *A. vinelandii* contains a similar protein, and mutations in its *ntrA* gene result in a Nif^- phenotype (133, 159). In addition to *ntrA, K. pneumoniae* requires the *ntrC* gene product for expression of *nifA* from the *nifL* promoter. Although *A. vinelandii* has a *ntrC* gene, which complements *K. pneumoniae ntrC* mutants, this gene is apparently not involved in nitrogen fixation, since *ntrC*::Tn*5* mutants are Nif^+, at least under conditions of molybdenum sufficiency. However, such mutants are unable to express the *K. pneumoniae nifL* promoter (159). The *ntr* genes and their involvement in other aspects of nitrogen assimilation are discussed below.

Little is known about activators of gene expression that might be involved in the alternative systems. The region just upstream of nifH* in *A. chroococcum* contains -12 and -24 consensus sequences, but no upstream activator-binding motif is evident (130). The *nifA* gene of *K. pneumoniae* does not activate expression of *nifH*-lacZ* fusion in an *E. coli* background (130). Therefore an activator other than *nifA* is indicated for expression of *A. chroococcum* V nitrogenase. The matter is more complicated in *A. vinelandii*, which may, as described previously, have two alternative nitrogenases expressed in response to molybdenum depletion. While the *ntrA* mutants fail to grow on N_2 in the presence or absence of molybdenum (159), the *nifA* mutants fail to fix nitrogen only in the former condition (68). Nevertheless, *nifA* could be necessary for one but not both of the proposed alternative systems.

A report (21) that the *K. pneumoniae nif* plasmid pRD1 complemented structural *nif* mutants of *A. vinelandii* was withdrawn (22); the reconsidered view is consistent with more detailed studies on expression of *K. pneumoniae nif* promoters in *A. vinelandii,* which reveal that the *nifL, nifF, nifU,* and *nifE* promoters can be expressed but that the *nifH* promoter cannot (69; A. Toukdarian, unpublished results). This is also consistent with successful complementation of *A. vinelandii nifN* and *nifM* but not *nifH* mutants by the *K. pneumoniae* genes (70). The failure of the *nifH* promoter to be expressed is intriguing since the promoter features discussed above are similar in *K. pneumoniae* and *A. vinelandii nifH* genes. Therefore, either the azotobacter *nifA* product activates promoters with features that have not yet been determined, or there is a negative control mechanism in *A. vinelandii* that overrides *nifA* activation of the *K. pneumoniae nifH* promoter.

Repression and Inhibition by NH_4^+

In azotobacters, excessive concentrations of NH_4^+ repress synthesis of *nif* gene products (84, 138) and also prevent nitrogenase activity. Of the other azotobacter nitrogen sources such as nitrate, urea, and amino acids, only nitrate apparently represses nitrogenase synthesis (35). Both NH_4^+ and nitrate can inhibit nitrogenase activity in vivo (24, 27, 35, 43, 73, 82). In *A. chroococcum,* inhibition is prevented by substances that block activity of glutamine synthetase, such as L-methionine DL sulfoximine (MSX); this indicates that NH_4^+ or NO_3^- must be assimilated first (23, 24). The degree to which NH_4^+ inhibits activity varies considerably with growth conditions such as O_2 input, respiratory rate, pH, and growth stage (52, 77). In cells of *A. vinelandii* respiring at low rates, NH_4^+ can even stimulate nitrogenase activity; addition of NH_4^+ leads to increased respiration rates, which overcome O_2 inhibition of nitrogenase activity (77) (see below).

In *K. pneumoniae,* NH_4^+ represses nitrogenase synthesis via the *nifL* gene product, which under these conditions prevents the *nifA* product from activating expression of other *nif* genes (33). In azotobacters, NH_4^+ prevents transcription of *nif* genes (67, 81), but a gene homologous to *nifL* has not been detected in hybridization experiments. The *nifL* product of *K. pneumoniae,* when introduced on a multicopy plasmid, had no effect on nitrogenase activity in *A. vinelandii* but prevented nitrogen fixation in *K. pneumoniae* (69). Multicopy *nifL* did, however, prevent the *K. pneumoniae nifA* product from restoring nitrogenase activity to an *A. vinelandii nifA* mutant, which indicates that the *K. pneumoniae nifL* product is synthesized and can function in *A. vinelandii* (69). Whatever the mechanism of NH_4^+ repression in azotobacters, it is prevented by *K. pneumoniae nifA* plasmids (71) or by treatment with MSX (49).

A number of azotobacter mutants that fix nitrogen in the presence of NH_4^+

have been reported, and some of these excrete excess NH_4^+ into the growth medium. Mutants have been isolated by screening of Nif^+ revertants of a Nif^- regulatory mutant (48), of oddly colored colonies from media with redox indicator dyes (151), and of mutants resistant to methylammonium (50) or azide (141). The nature of the mutations in these strains is not known.

NITROGEN ASSIMILATION

Azotobacters grow on fewer nitrogen compounds than most other diazotrophs. Thompson & Skerman (155) reported some strain variations in the nitrogen sources utilized, but in general azotobacters grow on N_2, ammonium salts, urea, nitrate, and a few amino acids. Studies in our laboratory have shown that *A. chroococcum* MCD1 grows on the amino acids glutamine and asparagine (F. Cejudo, personal communication) and that *A. vinelandii* UW grows on glutamine, asparagine, glutamate, and aspartate (A. Toukdarian, unpublished results). Asparaginase from *A. vinelandii* has been characterized (44). In comparison to current knowledge about genes for nitrogen fixation, very little is known about the genes required for the assimilation of other nitrogen sources or about the regulation of those genes.

ntr *Genes*

The genes *ntrA* and *ntrC* were isolated on separate cosmids from an *A. vinelandii* gene library by complementation of *E. coli* mutants (159). As in enteric bacteria, the *ntrC* gene in *A. vinelandii* was linked to *glnA*, the gene encoding glutamine synthetase, and was not linked to *ntrA*. The presence of *ntrB* in *A. vinelandii*, between *glnA* and *ntrC*, was shown by hybridization with a *K. pneumoniae ntrB* gene probe (A. Toukdarian, unpublished results).

The *ntrA* gene of *A. vinelandii* has been sequenced and the predicted amino acid sequence has been compared to that of the *K. pneumoniae ntrA* gene (M. J. Merrick, J. R. Gibbins & A. E. Toukdarian, manuscript submitted). The two proteins are 72% homologous if conservative replacements of residues are considered. In the highly conserved C-terminus a potential DNA-binding site has been identified; this may be important in promoter recognition.

We constructed *ntrA* and *ntrC* mutants by transforming *A. vinelandii* with cloned genes mutagenized with Tn*5* (159). In addition, Santero et al (133) isolated *ntrA* mutants after chemical mutagenesis of *A. vinelandii* by screening chlorate-resistant mutants for the ability to utilize N_2, nitrate, or nitrite as sole nitrogen source (133). The *ntrA* mutants of *A. vinelandii* could not grow on N_2 or nitrate, while the *ntrC* mutants could not grow on nitrate. Growth on urea or on amino acids was not affected.

NH_4^+ Assimilation: Glutamine Synthetase

Ammonia is assimilated in many bacteria by glutamate dehydrogenase (GDH) when the extracellular concentration of ammonia is high and by glutamine synthetase (GS) and glutamate synthase (GOGAT) at low ammonia concentrations.

An early report of GDH activity in *A. vinelandii* strain O grown on N_2 (88) has been contradicted by a failure to detect this activity in cultures of strain UW grown on N_2 or NH_4^+ (74). GDH, GS, and GOGAT activities have all been reported in *A. chroococcum* (35). The level of GDH or GOGAT activity does not vary with the nitrogen source (N_2, nitrate, or NH_4^+). GOGAT was found in both soluble and membrane-bound forms in *A. vinelandii*. The proportion of enzyme activity in each form can be varied by changing the rate of aeration of the culture, which also affects the intracytoplasmic membrane content of the cell (131). Synthesis of glutamate synthase is not repressed by NH_4^+ (85).

The GS of *A. vinelandii* is biochemically very similar to the GS of other gram-negative bacteria. The enzyme is dodecameric, with a single subunit size of about 53 kd (75, 142). Each subunit of the enzyme can be adenylylated, and this results in decreased biosynthetic activity (75, 142). Adenylylation in vivo occurs in response to ammonia (65, 75, 76, 142). The gene encoding GS, *glnA*, has been cloned from *A. vinelandii* and is expressed in both *E. coli* and *K. pneumoniae* (159).

A. vinelandii GS differs in some respects from the enteric enzyme. Most importantly, the absolute amount of GS protein does not seem to vary with nitrogen source in *A. vinelandii* (76, 85). The variation in amount of GS protein in enterics is due to control of *glnA* expression by the *ntr* genes. Expression of *A. vinelandii glnA* requires neither *ntrA* nor *ntrC* function (133, 159).

A. vinelandii mutants with altered GS activity have been isolated (87, 95). Attempts to construct mutants that wholly lack GS activity have not been successful. Hybridization analysis of DNA purified from kanamycin-resistant isolates of *A. vinelandii* transformed with *glnA*::Tn*5* DNA indicates that both wild-type and mutated genes are maintained (A. Toukdarian, unpublished results). This would suggest that a *glnA* mutation carried on all the chromosomes of a cell is lethal. Indeed, experiments indicate that glutamine is one of several amino acids that are not taken up by *A. vinelandii* (D. Dean, personal communication). Utilization of glutamine as a nitrogen source in azotobacters may therefore involve a periplasmic or extracellular glutaminase.

A. vinelandii has a specific system for the active transport of NH_4^+ (7, 8, 51, 74). The transport system appears to be regulated via direct feedback inhibition of the transporter by intracellular metabolites (such as glutamine) rather than by repression of gene expression (65). Full GS activity is required for maximal rates of uptake (95), although GS is not solely responsible for transport (8, 65).

Utilization of Nitrate and Nitrite

Assimilatory nitrate and nitrite reductases have been characterized from *A. vinelandii* and *A. chroococcum* (55, 146, 148, 160). Enzyme activities are induced by nitrate or nitrite in the growth medium. The induction of nitrate reductase requires the *ntrA* and *ntrC* gene products in *A. vinelandii* (133, 159). Nitrite reductase also requires *ntrA* for induction (133); the requirement for *ntrC* has not been examined. NH_4^+ prevents induction but has no effect on the activity of either enzyme (119, 120, 148, 160). A mutant lacking nitrate reductase activity, isolated by chlorate resistance, has been described (144). The uptake of nitrate by *A. chroococcum* has been studied; the uptake system is inducible by nitrate or nitrite (120) and is transiently inhibited by ammonium (119).

OXYGEN TOLERANCE

Azotobacters are very tolerant of oxygen while fixing nitrogen and exhibit several physiological responses to O_2 that are relevant to nitrogen fixation. These include two mechanisms for protecting nitrogenase from irreversible damage. A third response, O_2 repression (124), spares the energy that might be lost by futile biosynthesis of nitrogenase. The biochemistry and physiology of these processes have been reviewed by Robson & Postgate (129) and are only summarized here.

Azotobacters respire at a rate that results in faster consumption than solution of oxygen. They have uniquely high rates of respiration coupled with a multitude of cytochromes and redox proteins (56). Thus nitrogenase is apparently maintained in an essentially anoxic environment inside cells that nevertheless derive energy from aerobic metabolism. This is referred to as respiratory protection. Although the significance of respiratory protection has been argued (31, 110, 111), the conflict seems more related to semantics than to concepts.

If O_2 enters the culture more rapidly than it is removed by respiration, O_2 stress is imposed, and the second means of protection becomes important. During O_2 stress nitrogenase binds to a 2Fe-2S protective protein, also called Fe/S II, to give an O_2-stable complex that is protected from O_2 damage but that is inert to nitrogenase substrates. Protective proteins from both *A. vinelandii* (134) and *A. chroococcum* (123) have been characterized. Oxygen stress is generally transitory because rates of respiration increase to meet the new demand, after which the protected complex dissociates to give active nitrogenase. Another response to increased O_2 levels is the formation of intracytoplasmic membranes, which although indistinguishable from the cytoplasmic membrane in terms of dehydrogenase profile and specific respiratory activity, may be related to increased respiratory capacity (31, 112). Very high levels of superoxide dismutase in azotobacters have been reported (94), but their significance to nitrogenase is not known.

The study of mutants has shed light on the mechanisms underlying some aspects of oxygen tolerance in azotobacters, especially respiratory protection. After chemical or Tn*1* mutagenesis of *A. chroococcum,* Ramos & Robson (115) isolated nearly 300 mutants that were more sensitive than the wild type to oxygen when fixing nitrogen. These Fos^- (fixation oxygen sensitive) mutants were unable to grow in air on solid medium with N_2 as N source and sucrose as C source. Growth was restored by decreasing the ambient O_2 concentration to 0.2–1% or, in most mutants, by adding acetate, 1 mM Ca^{2+}, or TCA cycle intermediates. Ten mutants, which fell into three classes, were studied in more detail.

The two group RI mutants failed to synthesize nitrogenase proteins when growing in air. Introduction of *K. pneumoniae nifA* expressed constitutively on a high–copy number plasmid restored the Fos^+ phenotype (115). In these mutants an O_2-sensing molecule either directly or indirectly affecting *nif* gene expression was mutated. It is relevant that in wild-type *A. chroococcum,* constitutive *nifA* plasmids prevent O_2 repression of nitrogenase synthesis (71).

Group RII mutants (six) had respiration rates lower than those of the wild type and differed in their response to added TCA cycle intermediates (115). In general, both these and group RIII mutants seemed unable to metabolize sugars through to carboxylic acids as efficiently as the wild type. Their O_2 sensitivity was similar to that of organisms that are C-limited in chemostat cultures. One RII mutant, Fos242, had low citrate synthase activity; it metabolized both glucose and acetate slowly (116). All associated phenotypes of Fos242 were complemented by the *E. coli* citrate synthase gene cloned on a wide–host range plasmid.

The two RIII group mutants had normal respiration rates but lower apparent affinity for O_2 than the wild type. One mutant, Fos189, was selected for failure to grow on pyruvate after characterization of earlier Fos^- mutants had shown the link with carboxylic acids. This mutant's inability to grow on pyruvate has been correlated with low PEP carboxylase activity (116a). Respiration efficiency is thus limited during growth on sugars. Ramos & Robson (116a) isolated the wild-type PEP carboxylase gene of *A. chroococcum* from a gene bank by exploiting its ability to complement PEP carboxylase-deficient (Pcp^-) mutants of *E. coli.*

While Fos^- mutants indicate that an active TCA cycle is necessary for efficient respiratory protection of nitrogenase in *A. chroococcum,* little is known about which components of the complex azotobacter respiratory chains, consisting of major and minor pathways for electron flow (56), contribute to adaptation to increases in oxygen concentration. Mutants of *A. vinelandii* that lack or cannot reoxidize cytochromes c_4 or c_5 are Nif^+ and have normal respiration rates, which suggests that the minor respiratory chain is not involved (60, 61, 90). Mutants with elevated levels of cytochrome *d*, a component of the major pathway, are also unaffected in nitrogen fixation (91).

Mutants lacking protective protein have not been reported. Since its synthesis is not repressed by NH_4^+ (123), the protein is apparently not *nif* specific and thus may have a physiological role in addition to that of protecting nitrogenase from damage by O_2.

HYDROGEN UPTAKE

The H_2 that evolves during nitrogen fixation as a consequence of nitrogenase activity can be oxidized by hydrogenase to recoup electrons and gain energy. Hydrogenases in rhizobia are reviewed by Evans et al in this volume. In *A. vinelandii,* this Ni-containing enzyme is a dimer with two subunits of about 60 and 30 kd (4, 136); it has low H_2-evolving activity in addition to the major uptake function. In *A. chroococcum,* enzyme synthesis depends on nickel (108); activity is higher in N_2-grown organisms than in NH_4^+-grown organisms and is highest under N_2-fixing carbon-limited conditions (107). H_2 metabolism can support nitrogen fixation, although not as well as (for example) mannitol metabolism (162). *A. vinelandii* exhibits mixotrophic growth involving H_2 and mannose (163).

Mutants of *A. chroococcum* that were defective in hydrogen uptake (Hup^-) were isolated by NTG mutagenesis (166). These mutants can be divided into three classes, a, b, and c, on the basis of their residual H_2-uptake and H_2-evolving activities. Representative mutants did not grow as well as the wild type during N_2 fixation under C-limited chemostat conditions; this indicates that hydrogenase can be beneficial (1). Type a and c mutants were complemented to full activity by the plasmid pHU1, which carries *hup* genes from *B. japonicum* (156a). Type b mutants were complemented poorly by pHU2, which carries a fragment that partially overlaps pHU1. Interestingly, the hydrogenase in these complemented b mutants had a pH optimum characteristic of a soluble hydrogenase. It is possible that the *B. japonicum* hydrogenase is unable to associate with the membrane of *A. chroococcum*. A fourth mutant, type d, had high hydrogenase activity, but the enzyme, unlike that in the wild type, was not associated with the cell membrane (166).

A molecular analysis of *hup* genes was possible after cosmids were isolated from an *A. chroococcum* gene library by their ability to hybridize to *hup* DNA from *B. japonicum* (156). Tibelius et al characterized 15 kb of DNA and defined the regions according to their ability, when subcloned on IncP plasmids, to complement Hup^- mutants. Type a and c mutants are complemented by DNA regions separable from but contiguous with the region that complements type b mutants. As expected, all three mutants are complemented by a plasmid that encompasses both regions (156a). The hydrogenase structural genes must be located in the b region because a 3.6-kb *SalI* fragment from this region hybridizes to oligonucleotide probes with sequences encoding the two different N-termini of the *A. vinelandii* hydrogenase subunits (D. Arp, personal communication; K. H. Tibelius, M. Buck & C. M. Ford, personal communication). Consistent with this is recent evidence that the *B. japonicum* hydrogenase subunits are encoded by DNA that mainly hybridizes to the b *hup* region of *A. chroococcum* (167).

So far the benefits of hydrogenases in azotobacters have been examined only under conditions of molybdenum sufficiency in which the conventional nitrogenase has provided fixed nitrogen for growth. It is now of interest to determine whether hydrogenase provides a growth advantage to organisms that use the Va nitrogenases, because more H_2 may be evolved by these enzymes than by Mo nitrogenase (39, 127).

CONCLUDING REMARKS AND FUTURE PROSPECTS

A significant feature of nitrogen fixation in azotobacters is the wide dispersal of genes that are directly involved in, or that influence the physiology of, this process. Dispersal of *nif* genes also occurs in rhizobia and in the photosynthetic nitrogen fixer *Rhodopseudomonas capsulata* (41a). Thus, dispersal of *nif* genes may be the norm for nitrogen-fixing organisms, rather than tight clustering of *nif* genes as is found in *K. pneumoniae*.

The discovery of vanadium nitrogenase in azotobacters, underpinned by the construction of mutants in which the conventional *nifHDK* genes are deleted, is important and heralds a new period in the study of the ecology, physiology, and genetics of diazotrophs. Nitrogen-fixing organisms that have only a Va-based enzyme will probably be discovered; these may include *Rhizobium* species that nodulate legumes in Mo-deficient soils. The biochemistry of the vanadium nitrogenases and the chemistry of vanadium complexes will undoubtedly lead to a greater understanding of the mechanisms by which nitrogen can be made reactive.

The feature that initially distinguished nitrogen fixation in azotobacters was oxygen tolerance. Further analysis of genes involved in the responses of azotobacters to O_2 may provide a means for making other nitrogen-fixing organisms less vulnerable to its effects.

ACKNOWLEDGMENTS

We are grateful to many people for their generous response to our requests for reprints and preprints, and in particular to Rob Robson, David Evans, Karl Tibelius, Geoff Yates, and Dennis Dean for communicating unpublished results. We thank John Postgate, Richard Pau, and Paul Russell for criticizing the manuscript, Beryl Scutt for typing it, and Richard Humphrey for maintaining experimental work in the laboratory during the final weeks of its preparation.

Literature Cited

1. Aguilar, O. M., Yates, M. G., Postgate, J. R. 1985. The beneficial effect of hydrogenase in *Azotobacter chroococcum* under nitrogen-fixing, carbon-limiting conditions in continuous and batch cultures. *J. Gen. Microbiol.* 131:3141–45
2. Ajitkumar, P., Cherayil, J. D. 1985. Presence of 2-methylthioribsoyl-*trans*-zeatin in *Azotobacter vinelandii* transfer RNA. *J. Bacteriol.* 162:752–55
3. Arber, J. M., Dobson, B. R., Eady, R. R., Stevens, P., Hasnain, S. S., et al. 1987. Vanadium K-edge X-ray absorption spectra of the VFe protein of the vanadium nitrogenase of *Azotobacter chroococcum*. *Nature* 325:372–74
4. Arp, D. J., McCollum, L. C., Seefeldt, L. C. 1985. Molecular and immunological comparison of membrane-bound H_2-oxidizing hydrogenases of *Bradyrhizobium japonicum, Alcaligenes eutrophus, Alcaligenes latus,* and *Azotobacter vinelandii*. *J. Bacteriol.* 163:15–20
5. Ballesteros, F., Gonzalez-Lopez, J., de la Rubia, T., Martinez-Toledo, M. V., Ramos-Cormensana, A. 1986. Some biological characteristics of *Azotobacter vinelandii* (filtrable cells) grown in dialysed soil media. *Microbios* 46:159–64
6. Barea, J. M., Brown, M. G. 1974. Effect of plant growth produced by *Azotobacter paspali* related to synthesis of plant growth regulating substances. *J. Appl. Bacteriol.* 37:583–93
7. Barnes, E. M., Zimniak, P. 1981. Transport of ammonium and methylammonium ions by *Azotobacter vinelandii*. *J. Bacteriol.* 146:512–16
8. Barnes, E. M., Zimniak, P., Jayakumar, A. 1983. Role of glutamine synthetase in the uptake and metabolism of methylammonium by *Azotobacter vinelandii*. *J. Bacteriol.* 156:752–57
9. Becking, J. H. 1981. The family Azotobacteraceae. In *The Prokaryotes*, ed. M. P. Starr, H. Stolp, H. G. Trüper, A. Balows, H. G. Schlegel, pp. 794–817. Berlin/Heidelberg/New York: Springer-Verlag
10. Benemann, J. R., Sheu, C. W., Valentine, R. C. 1971. Temperature sensitive nitrogen fixation mutants of *Azotobacter vinelandii*. *Arch. Mikrobiol.* 79:49–58
11. Bishop, P. E., Brill, W. J. 1977. Genetic analysis of *Azotobacter vinelandii* mutant strains unable to fix nitrogen. *J. Bacteriol.* 130:954–56
12. Bishop, P. E., Dazzo, F. B., Appelbaum, E. R., Maier, R. J., Brill, W. J. 1977. Intergeneric transfer of genes involved in the *Rhizobium*-legume symbiosis. *Science* 198:938–40
13. Bishop, P. E., Hawkins, M. E., Eady, R. R. 1986. N_2-fixation in Mo-deficient continuous culture of a strain of *Azotobacter vinelandii* carrying a deletion of the structural genes for nitrogenase (*nifHDK*). *Biochem. J.* 238:437–42
14. Bishop, P. E., Jarlenski, D. M. L., Hetherington, D. R. 1980. Evidence for an alternative nitrogen fixation system in *Azotobacter vinelandii*. *Proc. Natl. Acad. Sci. USA* 77:7342–46
15. Bishop, P. E., Jarlenski, D. M. L., Hetherington, D. R. 1982. Expression of an alternative nitrogen fixation system in *Azotobacter vinelandii*. *J. Bacteriol.* 150:1244–51
16. Bishop, P. E., Premakumar, R., Dean, D. R., Jacobson, M. R., Chisnell, J. R., et al. 1986. Nitrogen fixation by *Azotobacter vinelandii* strains having deletions in structural genes for nitrogenase. *Science* 232:92–94
17. Bishop, P. E., Rizzo, T. M., Bott, K. F.

1985. Molecular cloning of *nif* DNA from *Azotobacter vinelandii*. *J. Bacteriol.* 162:21–28

18. Bishop, P. E., Supiano, M. A., Brill, W. J. 1977. Technique for isolating phage for *Azotobacter vinelandii*. *Appl. Environ. Microbiol.* 33:1007–8
19. Boddey, R. M., Chalk, P. M., Victoria, R. L., Matsui, E., Dobereiner, J. 1983. The use of the ^{15}N isotope dilution technique to estimate the contribution of associated biological nitrogen fixation to the nitrogen nutrition of *Paspalum notatum* cv. batatais. *Can. J. Microbiol.* 29:1036–45
20. Brigle, K. E., Newton, W. E., Dean, D. R. 1985. Complete nucleotide sequence of the *Azotobacter vinelandii* nitrogenase structural gene cluster. *Gene* 37:37–44
21. Cannon, F. C., Postgate, J. R. 1976. Expression of *Klebsiella* nitrogen fixation genes *(nif)* in *Azotobacter*. *Nature* 260:271–72
22. Cannon, F., Postgate, J. 1983. Expression of *Klebsiella* nitrogen fixation genes in *Azotobacter*—A caution. *Nature* 306:290
23. Cejudo, F. J., Delatorre, A., Paneque, A. 1984. Short-term ammonium inhibition of nitrogen fixation in *Azotobacter*. *Biochem. Biophys. Res. Commun.* 123:431–37
24. Cejudo, F. J., Paneque, A. 1986. Short-term nitrate (nitrite) inhibition of nitrogen fixation in *Azotobacter chroococcum*. *J. Bacteriol.* 165:240–43
25. Chisnell, J. R., Bishop, P. E. 1985. Partial purification of an alternative nitrogenase from a *nifHDK* deletion strain of *Azotobacter vinelandii*. See Ref. 41a, p. 623 (Abstr.)
26. David, M., Tronchet, M., Denarie, J. 1981. Transformation of *Azotobacter vinelandii* with plasmids RP4 (IncP-1 group) and RSF1010 (IncQ group). *J. Bacteriol.* 146:1154–57
27. Davis, L. C., Shah, V. K., Brill, W. J., Orme-Johnson, W. H. 1972. Nitrogenase II. Changes in the epr signal of component I (iron-molybdenum protein) of *Azotobacter vinelandii* nitrogenase during repression and derepression. *Biochim. Biophys. Acta* 256:512–23
28. Dean, D. R., Brigle, K. E. 1985. *Azotobacter vinelandii nifD*-encoded and *nifE*-encoded polypeptides share structural homology. *Proc. Natl. Acad. Sci. USA* 82:5720–23
29. Dean, D. R., Brigle, K. E. 1985. The *Azotobacter vinelandii nifD* and *nifK* encoded polypeptides share striking homology with the *nifE* and *nifN* encoded polypeptides. See Ref. 41a, p. 513 (Abstr.)
30. Deistung, J., Thorneley, R. N. F. 1986. Electron transfer to nitrogenase. Characterization of flavodoxin from *Azotobacter chroococcum* and comparison of its redox potentials with those of flavodoxins from *Azotobacter vinelandii* and *Klebsiella pneumoniae* (*nifF*-gene product). *Biochem. J.* 239:69–75
31. Dingler, C., Oelze, J. 1985. Reversible and irreversible inactivation of cellular nitrogenase upon oxygen stress in *Azotobacter vinelandii* growing in oxygen controlled continuous culture. *Arch. Microbiol.* 141:80–84
32. Dixon, R. A. 1984. The genetic complexity of nitrogen fixation. *J. Gen. Microbiol.* 130:2745–55
33. Dixon, R. A., Austin, S., Buck, M., Drummond, M., Hill, S., et al. 1987. Genetics and regulation of *nif* and related genes in *Klebsiella pneumoniae*. *Philos. Trans. R. Soc. London,* In press
34. Doran, J. L., Page, W. J. 1983. Heat sensitivity of *Azotobacter vinelandii* genetic transformation. *J. Bacteriol.* 155:159–68
35. Drozd, J. W., Tubb, R. S., Postgate, J. R. 1972. A chemostat study of the effect of fixed nitrogen sources on nitrogen fixation, membranes and free amino acids in *Azotobacter chroococcum*. *J. Gen. Microbiol.* 73:221–32
36. Drummond, M. 1986. Structure predictions and surface charge of nitrogenase flavodoxins from *Klebsiella pneumoniae* and *Azotobacter vinelandii*. *Eur. J. Biochem.* 159:549–53
37. Duff, J. T., Wyss, O. 1961. Isolation and classification of a new series of *Azotobacter* bacteriophages. *J. Gen. Microbiol.* 24:273–89
38. Eady, R. R., Robson, R. L. 1984. Characteristics of N_2 fixation in Mo-limited batch and continuous cultures of *Azotobacter vinelandii*. *Biochem. J.* 224:853–62
39. Eady, R. R., Robson, R. L., Richardson, T. H., Miller, R. W., Hawkins, M. 1987. The vanadium nitrogenase of *Azotobacter chroococcum*. Purification and properties of the VFe protein. *Biochem. J.* 244:197–207
40. Elessawy, A. A., Elsayed, M. A., Mohamed, Y. A. H., Elshanshoury, A. 1984. Effect of combined nitrogen in the production of plant growth regulators by *Azotobacter chroococcum*. *Zentralbl. Mikrobiol.* 139:327–33
41. Evans, D., Jones, R., Woodley, P., Kennedy, C., Robson, R. 1985. *nif* gene

organization in *Azotobacter chroococcum*. See Ref. 41a, p. 506 (Abstr.)

41a. Evans, H. J., Bottomley, P. J., Newton, W. E., eds. 1985. *Nitrogen Fixation Research Progress*. Dordrecht/ Boston/Lancaster: Nijhoff. 731 pp.

42. Fisher, R. J., Brill, W. J. 1969. Mutants of *Azotobacter vinelandii* unable to fix nitrogen. *Biochim. Biophys. Acta* 184: 99–105

43. Gadkari, D., Stolp, H. 1974. Influence of nitrogen source on growth and nitrogenase activity in *Azotobacter vinelandii*. *Arch. Microbiol.* 96:135–44

44. Gaffar, S. A., Shethna, Y. I. 1977. Purification and some biological properties of asparaginase from *Azotobacter vinelandii*. *Appl. Environ. Microbiol.* 33:508–14

45. Glick, B. R., Brooks, H. E., Pasternak, J. J. 1985. Transformation of *Azotobacter vinelandii* with plasmid DNA. *J. Bacteriol.* 162:276–79

46. Glick, B. R., Brooks, H. E., Pasternak, J. J. 1986. Physiological effects of plasmid DNA transformation on *Azotobacter vinelandii*. *Can. J. Microbiol.* 32:145–48

47. Gonzalez-Lopez, J., Salmeron, V., Martinez-Toledo, M. V., Ballesteros, F., Ramos-Cormenzana, A. 1986. Production of auxins, gibberellins and cytokinins by *Azotobacter vinelandii* ATCC12837 in chemically defined media and dialyzed soil media. *Soil Biol. Biochem.* 18:119–20

48. Gordon, J. K., Brill, W. J. 1972. Mutants that produce nitrogenase in the presence of ammonia. *Proc. Natl. Acad. Sci. USA* 69:3501–3

49. Gordon, J. K., Brill, W. J. 1974. Derepression of nitrogenase synthesis in the presence of excess NH_4^+. *Biochem. Biophys. Res. Commun.* 59:967–71

50. Gordon, J. K., Jacobson, M. R. 1983. Isolation and characterization of *Azotobacter vinelandii* mutant strains with potential as bacterial fertilizer. *Can. J. Microbiol.* 29:973–78

51. Gordon, J. K., Moore, R. A. 1981. Ammonium and methylammonium transport by the nitrogen-fixing bacterium *Azotobacter vinelandii*. *J. Bacteriol.* 148:435–42

52. Gordon, J. K., Shah, V. K., Brill, W. J. 1981. Feedback inhibition of nitrogenase. *J. Bacteriol.* 148:884–88

53. Green, M., Alexander, M., Wilson, P. W. 1953. Mutants of Azotobacter unable to use N_2. *J. Bacteriol.* 66:623–24

54. Guber, M., Hennecke, H. 1986. *fixA*, *B* and *C* are essential for symbiotic and free-living, microaerobic nitrogen fixation. *FEBS Lett.* 200:186–92

55. Guerrero, M. G., Vega, J. M., Leadbetter, E., Losada, M. 1973. Preparation and characterization of a soluble nitrate reductase from *Azotobacter chroococcum*. *Arch. Mikrobiol.* 91:287–304

55a. Gussin, G. N., Ronson, C. W., Ausubel, F. M. 1986. Regulation of nitrogen fixation genes. *Ann. Rev. Genet.* 20:567–91

56. Haddock, B. A., Jones, C. W. 1977. Bacterial respiration. *Bacteriol. Rev.* 41:47–99

57. Hales, B. J., Case, E. E., Morningstar, J. E., Dzeda, M. F., Mauterer, L. A. 1986. Isolation of a new V-containing nitrogenase from *Azotobacter vinelandii*. *Biochemistry* 25:7251–55

58. Hales, B. J., Langosch, D. J., Case, E. E. 1986. Isolation and characterization of a second nitrogenase Fe-protein from *Azotobacter vinelandii*. *J. Biol. Chem.* 261:15301–6

59. Helfrich, R. J., Ligon, J. M., Upchurch, R. G. 1985. Identification and organization of *nif* genes of *Azotobacter vinelandii*. See Ref. 41a, p. 507 (Abstr.)

60. Hoffman, P. S., Morgan, T. V., Dervartanian, D. V. 1979. Respiratory chain characteristics of mutants of *Azotobacter vinelandii* negative to tetramethyl-*para*-phenylene-diamine oxidase. *Eur. J. Biochem.* 110:19–27

61. Hoffman, P. S., Morgan, T. V., Dervartanian, D. V. 1980. Respiratory properties of cytochrome-*c*-deficient mutants of *Azotobacter vinelandii*. *Eur. J. Biochem.* 110:349–54

61a. Holland, D., Zilberstein, A., Zamir, A., Sussman, J. L. 1987. A quantitative approach to sequence comparisons of nitrogenase MoFe protein and subunits including the newly sequenced *nifK* from *K. pneumoniae*. *Biochem. J.* In press

62. Horan, N. J., Jarman, T. R., Dawes, E. A. 1983. Studies on some enzymes of aliginic acid biosynthesis in *Azotobacter vinelandii* grown in continuous culture. *J. Gen. Microbiol.* 129:2985–90

63. Imperial, J., Ugalde, R. A,. Shah, V. K., Brill, W. J. 1985. Mol^- mutants of *Klebsiella pneumoniae* requiring high levels of molybdate for nitrogenase activity. *J. Bacteriol.* 163:1285–87

63a. Ioannidis, I., Buck, M. 1987. Nucleotide sequence of the *Klebsiella pneumoniae nifD* gene and predicted amino acid sequence of the subunit of nitrogenase MoFe protein. *Biochem J.* In press

64. Jacobson, M. R., Premakumar, R., Bishop, P. E. 1986. Transcriptional

regulation of nitrogen fixation by molybdenum in *Azotobacter vinelandii*. *J. Bacteriol*. 167:480–86
65. Jayakumar, A., Barnes, E. M. 1984. The role of glutamine in regulation of ammonium transport in *Azotobacter vinelandii*. *Arch. Biochem. Biophys*. 231:95–101
66. Joerger, R. D., Premakumar, R., Bishop, P. E. 1986. Tn5-induced mutants of *Azotobacter vinelandii* affected in nitrogen fixation under Mo-deficient and Mo-sufficient conditions. *J. Bacteriol*. 168:673–82
67. Jones, R., Woodley, P., Robson, R. 1984. Cloning and organization of some genes for nitrogen fixation from *Azotobacter chroococcum* and their expression in *Klebsiella pneumoniae*. *Mol. Gen. Genet*. 197:318–27
68. Kennedy, C., Buck, M., Evans, D., Humphrey, R., Jones, R., et al. 1987. The genetic analysis of nitrogen fixation, oxygen tolerance, and hydrogen uptake in azotobacters. *Philos. Trans. R. Soc. London Ser. B* In press
69. Kennedy, C., Drummond, M. H. 1985. The use of cloned *nif* regulatory elements from *Klebsiella pneumoniae* to examine *nif* regulation in *Azotobacter vinelandii*. *J. Gen. Microbiol*. 131: 1787–95
70. Kennedy, C., Gamal, R., Humphrey, R., Ramos, J., Brigle, K., Dean, D. 1986. *nifH, nifM* and *nifN* genes of *Azotobacter vinelandii*. Characterisation by Tn*5* mutagenesis and isolation from pLAFR1 gene banks. *Mol. Gen. Genet*. 205:318–25
71. Kennedy, C., Robson, R. L. 1983. Activation of *nif* expression in azotobacter by the *nifA* gene product of *Klebsiella pneumoniae*. *Nature* 301:626–28
72. Kennedy, C., Robson, R., Jones, R., Woodley, P., Evans, D., et al. 1985. Genetic and physical characterisation of *nif* and *ntr* genes in *Azotobacter chroococcum* and *A. vinelandii*. See Ref. 41a, pp. 469–76
73. Kleiner, D. 1974. Quantitative relations for the repression of nitrogenase synthesis in *Azotobacter vinelandii* by ammonia. *Arch. Microbiol*. 101:153–59
74. Kleiner, D. 1975. Ammonium uptake by nitrogen-fixing bacteria. 1. *Azotobacter vinelandii*. *Arch. Microbiol*. 104:163–69
75. Kleinschmidt, J. A., Kleiner, D. 1978. The glutamine-synthetase from *Azotobacter vinelandii:* purification, characterization, regulation and localization. *Eur. J. Biochem*. 89:51–60
76. Kleinschmidt, J. A., Kleiner, D. 1981. Relationships between nitrogenase glutamine-synthetase, glutamine, and energy-charge in *Azotobacter vinelandii*. *Arch. Microbiol*. 128:412–15
77. Klugkist, J., Haaker, H. 1984. Inhibition of nitrogenase activity by ammonium chloride in *Azotobacter vinelandii*. *J. Bacteriol*. 157:148–51
78. Klugkist, J., Haaker, H., Veeger, C. 1986. Studies on the mechanism of electron transport to nitrogenase in *Azotobacter vinelandii*. *Eur. J. Biochem*. 155:41–46
79. Klugkist, J., Voorberg, J., Haaker, H., Veeger, C. 1986. Characterization of three different flavodoxins from *Azotobacter vinelandii*. *Eur. J. Biochem*. 155:33–40
80. Krakow, J. S. 1975. On the role of sulfhydryl groups in the structure and function of *Azotobacter vinelandii* RNA polymerase. *Biochemistry* 14:4522–27
81. Krol, A. J. M., Hontelez, J. G. J., Roozendaal, B., vanKammen, A. 1982. On the operon structure of the nitrogenase genes of *Rhizobium leguminosarum* and *Azotobacter vinelandii*. *Nucleic Acids Res*. 10:4147–57
82. Kurz, W. G. W., LaRue, T. A., Chatson, K. B. 1975. Nitrogenase in synchronized *Azotobacter vinelandii* OP. *Can. J. Microbiol*. 21:984–88
83. Leach, C. K., Battikhi, M. 1978. Isolation of auxotrophs from duplex-forming strains of *Azotobacter vinelandii*. *Proc. Soc. Gen. Microbiol*. 5:110 (Abstr.)
84. Lepo, J. E., Wyss, O. 1974. Derepression of nitrogenase in *Azotobacter*. *Biochem. Biophys. Res. Commun*. 60: 76–80
85. Lepo, J. E., Wyss, O., Tabita, F. R. 1982. Regulation and biochemical characterization of the glutamine-synthetase of *Azotobacter vinelandii*. *Biochim. Biophys. Acta* 704:414–21
86. Lugue, F., Santero, E., Medina, J. R., Tortolero, M. 1987. An effective mutagenic method in *Azotobacter vinelandii*. *Microbiologia SEM* 3:45–48
87. Lugue, F., Santero, E., Tortolero, M., Medina, J. R. 1986. Genetic alteration of glutamine synthetase in *Azotobacter vinelandii*. *Microbios Lett*. 33:29–32
88. Magee, W. E., Burns, R. H. 1956. Oxidative activity and nitrogen fixation in cell-free preparations from *Azotobacter vinelandii*. *J. Bacteriol*. 71:635–43
89. Maier, R. J., Bishop, P. E., Brill, W. J. 1978. Transfer from *Rhizobium japonicum* to *Azotobacter vinelandii* of genes required for nodulation. *J. Bacteriol*. 134:1199–201

90. McInerney, M. J., Holmes, K. S., Dervartanian, D. V. 1982. Effect of O_2 limitation on growth and respiration of the wild-type and an ascorbate tetramethyl-P-phenylenediamine oxidase negative mutant strain of *Azotobacter vinelandii*. *J. Bioenerg. Biomembr.* 14:451–56
91. McInerney, M. J., Holmes, K. S., Hoffman, P., Dervartanian, D. V. 1984. Respiratory mutants of *Azotobacter vinelandii* with elevated levels of cytochrome *d*. *Eur. J. Biochem.* 141:447–52
92. Medhora, M., Phadnis, S. H., Das, H. K. 1983. Construction of a gene library from the nitrogen-fixing aerobe *Azotobacter vinelandii*. *Gene* 25:355–60
93. Mishra, A. K., Wyss, O. 1968. Induced mutations in *Azotobacter* and isolation of an adenine requiring mutant. *Nucleus Calcutta* 11:96–105
94. Moore, E. R. B., Norrod, E. P., Jurtshuk, P. 1984. Superoxide dismutases of *Azotobacter vinelandii* and other aerobic, free-living nitrogen-fixing bacteria. *FEMS Microbiol. Lett.* 24:261–65
95. Moore, R. A., Gordon, J. K. 1984. Metabolism of methylammoniuim by *Azotobacter vinelandii*. *Arch. Microbiol.* 140:193–97
96. Nagatani, H. H., Shah, V. K., Brill, W. J. 1974. Activation of inactive nitrogenase by acid-treated component I. *J. Bacteriol.* 120:697–701
96a. Orme-Johnson, W. H. 1985. Molecular basis of biological nitrogen fixation. *Ann. Rev. Biophys. Biophys. Chem.* 14:419–59
97. Owen, D. J., Ward, A. C. 1985. Transfer of transposable drug-resistance elements Tn*5*, Tn*7*, and Tn*76* to *Azotobacter beijerinckii:* use of plasmid RP4, Tn*76* as a suicide vector. *Plasmid* 14:162–6
98. Page, W. J. 1978. Transformation of *Azotobacter vinelandii* strains unable to fix nitrogen with *Rhizobium* spp. DNA. *Can. J. Microbiol.* 24:209–14
99. Page, W. J. 1982. Optimal conditions for induction of competence in nitrogen-fixing *Azotobacter vinelandii*. *Can. J. Microbiol.* 28:389–97
100. Page, W. J. 1985. Genetic transformation of molybdenum starved *Azotobacter vinelandii*—increased transformation frequency and recipient range. *Can. J. Microbiol.* 31:659–62
101. Page, W. J. 1986. Sodium dependent growth of *Azotobacter chroococcum*. *Appl. Environ. Microbiol.* 51:510–14
102. Page, W. J., Collinson, S. K. 1982. Molybdenum enhancement of nitrogen fixation in a Mo-starved *Azotobacter vinelandii nif* mutant. *Can. J. Microbiol.* 28:1173–80
103. Page, W. J., Doran, J. L. 1981. Recovery of competence in calcium-limited *Azotobacter vinelandii*. *J. Bacteriol.* 146:33–40
104. Page, W. J., Sadoff, H. L. 1976. Physiological factors affecting transformation of *Azotobacter vinelandii*. *J. Bacteriol.* 125:1080–87
105. Page, W. J., vonTigerstrom, M. 1978. Induction of transformation competence in *Azotobacter vinelandii* iron-limited cultures. *Can. J. Microbiol.* 24:1590–94
106. Page, W. J., vonTigerstrom, M. 1979. Optimal conditions for transformation of *Azotobacter vinelandii*. *J. Bacteriol.* 139:1058–61
107. Partridge, C. D. P., Walker, C. C., Yates, M. G., Postgate, J. R. 1980. The relationship between hydrogenase and nitrogenase in *Azotobacter chroococcum:* effect of nitrogen sources on hydrogenase activity. *J. Gen. Microbiol.* 119:313–19
108. Partridge, C. D. P., Yates, M. G. 1982. Effect of chelating agents on hydrogenase in *Azotobacter chroococcum:* evidence that nickel is required for hydrogenase synthesis. *Biochem. J.* 204:339–44
109. Pike, L., Wyss, O. 1975. Isolation and characterization of phage-resistant strains of *Azotobacter vinelandii*. *J. Gen. Microbiol.* 89:182–86
110. Post, E., Golecki, J. R., Oelze, J. 1982. Morphological and ultrastructural variations in *Azotobacter vinelandii* growing in oxygen-controlled continuous culture. *Arch. Microbiol.* 133:75–82
111. Post, E., Kleiner, D., Oelze, J. 1983. Whole cell respiration and nitrogenase activities in *Azotobacter vinelandii* growing in oxygen controlled continuous culture. *Arch. Microbiol.* 134:68–72
112. Post, E., Vakalopoulou, E., Oelze, J. 1983. On the relationship of intracytoplasmic to cytoplasmic membranes in nitrogen-fixing *Azotobacter vinelandii*. *Arch. Microbiol.* 134:265–69
113. Postgate, J. R., Partridge, C. D. P., Robson, R. L., Simpson, F. B., Yates, M. G. 1982. A method for screening for hydrogenase negative mutants of *Azotobacter chroococcum*. *J. Gen. Microbiol.* 128:905–8
114. Premakumar, R., Lemos, E. M., Bishop, P. E. 1984. Evidence for two

dinitrogenase reductases under regulatory control by molybdenum in *Azotobacter vinelandii*. *Biochim. Biophys. Acta* 797:64–70
115. Ramos, J. L., Robson, R. L. 1985. Isolation and properties of mutants of *Azotobacter chroococcum* defective in aerobic nitrogen fixation. *J. Gen. Microbiol.* 131:1449–58
116. Ramos, J. L., Robson, R. L. 1985. Lesions in citrate synthase that affect aerobic nitrogen fixation by *Azotobacter chroococcum*. *J. Bacteriol.* 162:746–51
116a. Ramos, J. L., Robson, R. L. 1987. Cloning of the gene for phosphoenol pyruvate carboxylase from *Azotobacter chroococcum,* an enzyme important in aerobic N_2 fixation. *Mol. Gen. Genet.* In press
117. Reusch, R. N., Sadoff, H. L. 1983. D-(−)-poly-β-hydroxybutyrate in membranes of genetically competent bacteria. *J. Bacteriol.* 156:778–88
118. Reusch, R. N., Sadoff, H. L. 1983. Novel lipid components of the *Azotobacter vinelandii* cyst membrane. *Nature* 302:268–70
119. Revilla, E., Cejudo, F. J., Llobell, A., Paneque, A. 1986. Short-term ammonium inhibition of nitrate uptake by *Azotobacter chroococcum*. *Arch. Microbiol.* 144:187–90
120. Revilla, E., Llobell, A., Paneque, A. 1985. The assimilatory nitrate uptake in *Azotobacter chroococcum:* induction by nitrate and by cyanate. *J. Plant Physiol.* 118:165–76
121. Riddle, G. D., Simonson, J. G., Hales, B. J., Braymer, H. D. 1982. Nitrogen fixation system of tungsten-resistant mutants of *Azotobacter vinelandii*. *J. Bacteriol.* 152:72–80
122. Robinson, A., Burgess, B. K., Dean, D. R. 1986. Activity, reconstitution and accumulation of nitrogenase components in *Azotobacter vinelandii* mutant strains containing defined deletions within the nitrogenase structural gene cluster. *J. Bacteriol.* 166:180–86
123. Robson, R. L. 1979. Characterization of an oxygen-stable nitrogenase complex isolated from *Azotobacter chroococcum*. *Biochem. J.* 181:569–75
124. Robson, R. L. 1979. O_2-repression of nitrogenase synthesis in *Azotobacter chroococcum*. *FEMS Microbiol. Lett.* 5:259–62
125. Robson, R. L. 1986. Nitrogen fixation in strains of *Azotobacter chroococcum* bearing deletions of a cluster of genes coding for nitrogenase. *Arch. Microbiol.* 146:74–79
126. Robson, R. L., Chesshyre, J. A., Wheeler, C., Jones, R., Woodley, P. R., Postgate, J. R. 1984. Genome size and complexity in *Azotobacter chroococcum*. *J. Gen. Microbiol.* 130:1603–12
127. Robson, R. L., Eady, R. R., Richardson, T. H., Miller, R. W., Hawkins, M., Postgate, J. R. 1986. The alternative nitrogenase of *Azotobacter chroococcum* is a vanadium enzyme. *Nature* 322:388–90
128. Robson, R., Jones, R., Woodley, P., Evans, D. 1985. The DNA sequence of nitrogenase genes from *Azotobacter chroococcum*. See Ref. 41a, p. 514 (Abstr.)
129. Robson, R. L., Postgate, J. R. 1980. Oxygen and hydrogen in biological nitrogen fixation. *Ann. Rev. Microbiol.* 34:183–207
130. Robson, R., Woodley, P., Jones, R. 1986. Second gene *(nifH*)* coding for a nitrogenase iron protein in *Azotobacter chroococcum* is adjacent to a gene coding for a ferredoxin-like protein. *EMBO J.* 5:1159–63
131. Rockel, D., Hernando, J. J., Vakalopoulou, E., Post, E., Oelze, J. 1983. Localization and activities of nitrogenase, glutamine-synthetase and glutamate synthase in *Azotobacter vinelandii* grown in oxygen-controlled continuous culture. *Arch. Microbiol.* 136:74–78
132. Sadoff, H. L., Shimei, B., Ellis, S. 1979. Characterization of *Azotobacter vinelandii* deoxyribonucleic acid and folded chromosomes. *J. Bacteriol.* 138:871–77
133. Santero, E., Lugue, F., Medina, J. R., Tortolero, M. 1986. Isolation of *ntrA*-like mutants of *Azotobacter vinelandii*. *J. Bacteriol.* 166:541–44
134. Scherings, G., Haaker, H., Wassink, H., Veeger, C. 1983. On the formation of an oxygen-tolerant three component nitrogenase complex from *Azotobacter vinelandii*. *Eur. J. Biochem.* 135:591–99
135. Scott, K. F., Rolfe, B. G., Shine, J. 1981. Biological nitrogen fixation: primary structure of the *K. pneumoniae nifH* and *nifD* genes. *J. Mol. Appl. Genet.* 1:71–81
136. Seefeldt, L. C., Arp, D. J. 1986. Purification to homogeneity of *Azotobacter vinelandii* hydrogenase: a nickel and iron containing αβ dimer. *Biochimie* 68:25–34
137. Sen, H., Pal, T. K., Sen, S. P. 1969. Intergenic transformation between *Rhizobium* and *Azotobacter*. *Antonie van*

Leeuwenhoek J. Microbiol. Serol. 34:533–40

138. Shah, V. K., Davis, L. C., Brill, W. J. 1972. Nitrogenase I. Repression and derepression of the iron-molybdenum and iron proteins of nitrogenase in *Azotobacter vinelandii*. *Biochim. Biophys. Acta* 256:498–511
139. Shah, V. K., Davis, L. C., Gordon, J. K., Orme-Johnson, W. H., Brill, W. J. 1973. Nitrogenase III. Nitrogenaseless mutants of *Azotobacter vinelandii:* activities, cross-reactions and epr spectra. *Biochim. Biophys. Acta* 292:246–55
140. Shah, V. K., Davis, L. C., Stieghorst, M., Brill, W. J. 1974. Mutant of *Azotobacter vinelandii* that hyperproduces nitrogenase component II. *J. Bacteriol.* 117:971–79
141. Sharma, P. K., Chahal, V. P. S. 1985. NH_4^+ derepression of azide resistant mutant of *Azotobacter chroococcum*. *Zentralbl. Mikrobiol.* 140:575–78
142. Siedel, J., Shelton, E. 1979. Purification and properties of *Azotobacter vinelandii* glutamine-synthetase. *Arch. Biochem. Biophys.* 192:214–22
143. Sorger, G. J. 1968. Regulation of nitrogen fixation in *Azotobacter vinelandii* OP and an apparently partially constitutive mutant. *J. Bacteriol.* 95:1721–26
144. Sorger, G. J. 1969. Regulation of nitrogen fixation in *Azotobacter vinelandii* OP: the role of nitrate reductase. *J. Bacteriol.* 98:56–61
145. Sorger, G. J., Trofimenkoff, D. 1970. Nitrogenaseless mutants of *Azotobacter vinelandii*. *Proc. Natl. Acad. Sci. USA* 65:74–80
146. Spencer, D., Takahashi, H., Nason, A. 1957. Relationship of nitrite and hydroxylamine reductases to nitrate assimilation and nitrogen fixation in *Azotobacter agile*. *J. Bacteriol.* 73:553–62
147. Stockall, A. M., Edwards, C. 1985. Changes in respiratory activity during encystment of *Azotobacter vinelandii*. *J. Gen. Microbiol.* 131:1403–10
148. Taniguchi, S., Ohmachi, K. 1960. Particulate nitrate reductase of *Azotobacter vinelandii*. *J. Biochem.* 48:50–62
149. Tchan, Y.-T. 1984. Azotobacteraceae. In *Bergey's Manual of Systematic Bacteriology,* Vol. 1, ed. N. Krieg, J. G. Holt, pp. 219–25. Baltimore/London: Williams & Wilkins
150. Terzaghi, B. E. 1980. Ultraviolet sensitivity and mutagenesis of *Azotobacter*. *J. Gen. Microbiol.* 118:271–73
151. Terzaghi, B. E. 1980. A method for the isolation of *Azotobacter* mutants derepressed for *nif*. *J. Gen. Microbiol.* 118:275–78
152. Terzaghi, B. E., Terzaghi, E. 1986. *Azotobacter* biology, biochemistry, and molecular biology. In *Nitrogen Fixation,* Vol. 4, ed. W. J. Broughton, A. Pühler, pp. 127–63. Oxford: Clarendon
153. Thompson, B. J., Domingo, E., Warner, R. C. 1980. Pseudolysogeny of azotobacter phages. *Virology* 102:267–77
154. Thompson, B. J., Wagner, M. S., Domingo, E., Warner, R. C. 1980. Pseudo-lysogenic conversion of *Azotobacter vinelandii* by phage A21 and the formation of a stably converted form. *Virology* 102:278–85
155. Thompson, J. P., Skerman, V. B. D. 1979. *Azotobacteraceae: The Taxonomy and Ecology of the Aerobic Nitrogen Fixing Bacteria*. London: Academic
156. Tibelius, K. H., Robson, R. L., Yates, M. G. 1987. Cloning and characterization of hydrogenase genes from *Azotobacter chroococcum*. *Mol. Gen. Genet.* 206:285–90

156a. Tibelius, K. H., Yates, M. G. 1987. Complementation of *Azotobacter chroococcum* hydrogenase mutants by cloned *hup* determinants from *Rhizobium japonicum* and *Azotobacter chroococcum*. *J. Gen. Microbiol.* In press

157. Timmis, K. N., Pühler, A. 1979. *Plasmids of Medical, Environmental and Commercial Importance*. Amsterdam: Elsevier/North-Holland
158. Tortolero, M., Santero, E., Casadesus, J. 1983. Plasmid transfer and mobilization of *nif* markers in *Azotobacter vinelandii*. *Microbios Lett.* 22:31–35
159. Toukdarian, A., Kennedy, C. 1986. Regulation of nitrogen metabolism in *Azotobacter vinelandii:* isolation of *ntr* and *glnA* genes and construction of *ntr* mutants. *EMBO J.* 5:399–407
160. Vega, J. M., Guerrero, M. G., Leadbetter, E., Losada, M. 1973. Reduced nicotinamide-adenine dinucleotide-nitrate reductase from *Azotobacter chroococcum*. *Biochem. J.* 133:701–8
161. Walker, C. C., Partridge, C. D. P., Yates, M. G. 1981. The effect of nutrient limitation in hydrogen production by nitrogenase in continuous cultures of *Azotobacter chroococcum*. *J. Gen. Microbiol.* 124:317–27
162. Walker, C. C., Yates, M. G. 1978. Hydrogen cycle in nitrogen-fixing *Azotobacter chroococcum*. *Biochimie* 60:225–31

163. Wong, T. Y., Maier, R. J. 1985. H_2-dependent mixotrophic growth of N_2-fixing *Azotobacter vinelandii*. *J. Bacteriol.* 163:528–33
164. Wyss, O., Nimeck, M. W. 1962. Interspecific transduction in *Azotobacter*. *Fed. Proc.* 21:456 (Abstr.)
165. Wyss, O., Wyss, M. B. 1950. Mutants of azotobacter that do not fix nitrogen. *J. Bacteriol.* 59:287–91
166. Yates, M. G., Robson, R. L. 1985. Mutants of *Azotobacter chroococcum* defective in hydrogenase activity. *J. Gen. Microbiol.* 131:1459–66
167. Zuber, M., Harker, A. R., Sultana, M. A., Evans, H. J. 1986. Cloning and expression of *Bradyrhizobium japonicum* uptake hydrogenase structural genes in *Escherichia coli*. *Proc. Natl. Acad. Sci. USA* 83:7668–72

Ann. Rev. Microbiol. 1987. 41:259–89

BIOSYNTHESIS OF PEPTIDE ANTIBIOTICS

H. Kleinkauf and H. von Döhren

Institut für Biochemie und Molekulare Biologie, Technische Universität Berlin, D-1000 Berlin 10, Federal Republic of Germany

---

CONTENTS

INTRODUCTION

Although the term "antibiotic" is just aimed at one set of effects exerted by many peptides, this review directs the reader to an exciting general mech-

0066-4227/87/1001-0259$02.00

anism of peptide biosynthesis directed by polyenzymes. This polymerization mechanism has been evaluated in connection with the classical peptide antibiotics, and we treat it together with recent results on ribosomal and enzymic routes.

Structural variation among peptide antibiotics appears to be vast, and the diversity of constituent amino acids seems almost unlimited. Still, there are few structural types; these include linear, cyclic, and branched cyclic peptides, lactones, and depsipeptides. Quite a few of these peptide backbones contain other than α,α-peptide linkages. The close link to polyketide biosynthesis in the addition of activated carboxyl compounds is reflected in frequent acylation reactions or acetate introductions. The enzymology of peptide formation is far more advanced than that of polyacetates. However, the biosynthetic origins of many of the fascinating structures unraveled in recent years are unsolved. We discuss structural features of these compounds in some detail and connect these, when possible, to enzymes or multienzymes. Biosynthetic pathways and reactions have been reviewed and discussed frequently (30, 73, 86, 88–91, 93, 106, 107, 190, 199). Here emphasis is given to recent results and developments.

PEPTIDE CHAIN DERIVATION

Recent compilations of peptide antibiotics or bioactive peptides (12, 92, 147, 187) list several hundred compounds, including modified amino acids, linear and cyclic peptides, lactones, depsipeptides, and polypeptides. Biosynthetic pathways can be divided into peptide chain derivation and modification reactions. Modification reactions include the formation of constituents as well as chain modifications such as cyclizations.

Ribosomal and Nonribosomal Pathways

Compounds containing nonprotein constituents in the polypeptide chain could either derive from modification of the chain or originate from nonribosomal systems. In such systems the chain is assembled by enzymatic steps; assembly is not directed by an RNA template, although aminoacyl-tRNAs may participate. A template or templates are provided by protein surfaces that originate from mRNA-coded structures. Thus the information transfer indirectly conforms to Crick's so-called central dogma and can be traced to chromosomal or nonchromosomal elements.

Ribosomally Formed Antibiotics

Certain polypeptides produced by strains of streptomycetes, such as neocarzinostatin (43), macromomycin (154), actinoxanthin (80), and the AN-peptides, are close in size to proteins with antibiotic or other bioactive properties. (125–128, 184), The region above 50 amino acids is neglected in screens for reasons of antigenicity; thus polypeptides in this size range contain only protein amino acids (128). Important for various actions, including antitumor and mutagenic properties, is a chromophore constituent (25). This heterogeneous group of polypeptides, also referred to as protein antibiotics (12), is certainly formed via ribosomal mechanisms, but no studies on possible pre- or propeptides have yet been carried out.

The modification of ribosomally derived polypeptides, which in a sense extends the use of pre- or propeptides connected to peptide bond cleavage, has been found in several bacterial thioether-containing peptides: subtilin *(Bacillus subtilis)* (137, 160), nisin *(Streptococcus lactis)* (48, 193), epidermin and pep-5 *(Staphylococcus epidermidis)* (3, 4, 152), and ancovenin (*Streptomyces* sp.) (194). These peptides contain dehydroamino acids, D-amino acids, and thioether bridges of various sizes, and are formed by proteolytic cleavage of modified precursors, as is implied by recent studies on the formation of subtilin. A subtilin precursor was detected using subtilin-directed antibodies (137), and subsequently a prepropeptide of 129 residues (compared to 32 in the antibiotic) was isolated. In addition to the action of specific proteinases, the postulated biosynthetic events are the formation of dehydropeptides from serine or threonine and thioether formation. Thiothers are formed by the addition of Cys-thiol to the C–C double bonds in the dehydropeptides, with configurational inversion leading to rings of various sizes (for dehydroalanine(aa)$_n$Cys, $n = 3$ or 4; for dehydroaminobutyrate(aa)$_n$Cys, $n = 2, 3,$ or 5). Even more complicated cyclization has been found in ancovenin, an inhibitor of angiotensin-converting enzyme (Figure 1). Although no enzymes involved have been characterized so far, preliminary observations of cell-free conversions have been carried out in the nisin system (149). Translational inhibitors affect nisin formation in vivo, and serine, threonine, and cysteine are precursors of the thioether amino acids lanthionine and β-methyllanthionine. A nonproducing mutant of *S. lactis* formed low-molecular-weight pro-nisin, which could be transformed to a nisinlike substance by an extract of the producing strain. An enzyme fraction has been obtained by a short microbead treatment of cells, indicating a possible cell-surface location. This transformation is apparently the final proteolytic processing of pro-nisin. The idea of repeated sequences in some prepropeptides appears reasonable from the composition of prepro-subtilin (K. Kurahashi, personal communication) as well as from the detection of several structurally similar peptides in pep-5 preparations (152; H.-G. Sahl, personal communication).

Evidence for an extrachromosomal location of biosynthetic genes of nisin or epidermin has come from the detection of plasmids in producer strains and from the correlation of plasmid curing and nonproductivity. However, many

1

Leu
Dha Ala Met
Ile Leu Pro → Gly Gly Gly
Ile → Dhb → DAla Ala → DAbu Ala → Lys → DAbu Ala → Asn
S S S
Met
S
His ← Ala DAbu ← Lys
Lys — Dha ← Val ← His ← Ile ← Ser ← Ala DAbu ← Ala
S

2

Leu
Dha Ala Gln
Glu Leu Pro → Gly Gly Dhb
Trp → Lys → DAla Ala → DAbu Ala → Val → DAbu Ala → Phe
S S S
Leu
S
Asn ← Ala DAbu ← Gln
Lys ← Dha ← Ile ← Lys ← Ala DAbu ← Leu
S

3

Ala → Thr
Gly → Leu → Gly → Leu → Trp → Gly → Asn → Lys → Gly → Cys — Cys
X Xu — Cys ← Ala — Ala ← Gly ← Ile ← Ser
Ala ← Gly ← Ala ← Ile ← Glx Leu → Val → Asp
Asp ← Pro ← Ile ← Pro ← Gly

4

Phe
Lys Ile Pro → Gly
Ile → Ala → DAla Ala → DAbu Ala → Ala → Lys → Dhb → Gly
S S
S
HN — Ala DAla
Tyr Phe
S DAla ← Asn

5

Ala → Val → Gln → DAla → Ala → Dha → Phe → Gly
S S S — DAbu ← Leu ← Pro
Lys ← DAbu ← Asn ← Gly ← Asp ← Ala — Ser ← Trp

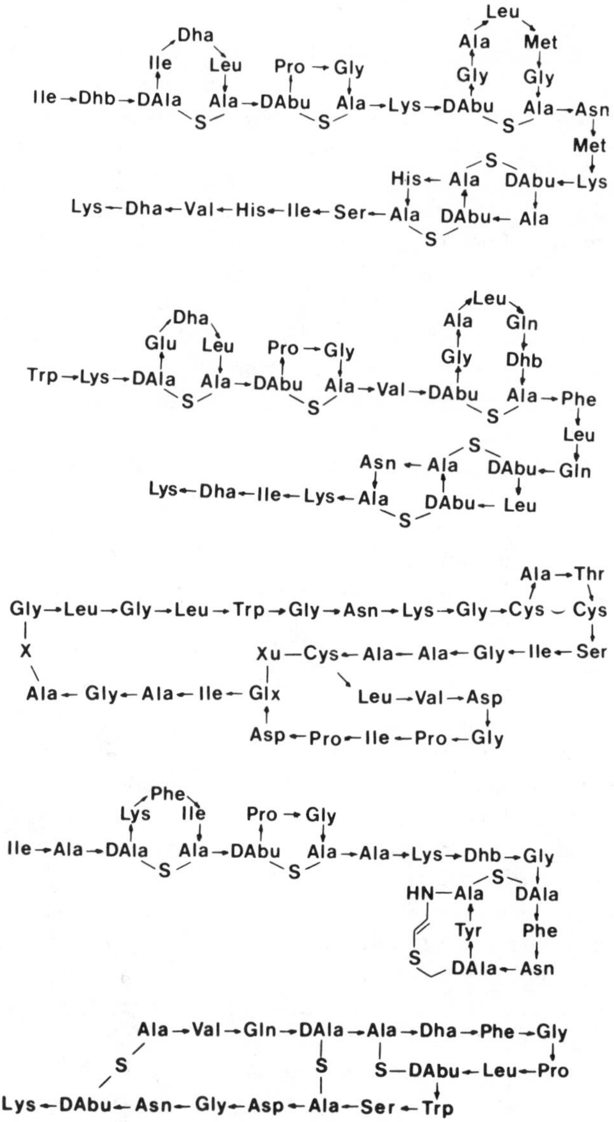

Figure 1 Structures of prokaryotic antibiotic polypeptides of ribosomal origin. *1,* nisin, used as a food preservative; *2,* subtilin and *3,* subtilosin, antibiotics produced by *Bacillus subtilis* (*X* and *Xu* represent unknown compounds); *4,* epidermin, a peptide found effective against *Propionebacterium acne;* and *5,* ancovenin, an inhibitor of angiotensin-converting enzyme.

nonproducers containing plasmids have been found (22), and no structural genes have been mapped so far. In *S. lactis,* sucrose-fermenting ability can be transferred together with nisin resistance and production in a conjugationlike process involving the pSN plasmid (named for sucrose and nisin) (34, 44, 169). A cDNA approach is currently being used to identify possible precursor structures (G. Jung, personal communication).

Another unusual antibiotic structure from a ribosomal origin, subtilosin A, has been found in *Bacillus subtilis* 168 (6). This cyclic 32-peptide with two unidentified residues is formed at the end of vegetative growth and is one of the major antibiotics of at least ten produced by this strain in sporulation medium. Similar compounds have been detected in the subtilin-producing strain *B. subtilis* ATCC 6633 and a strain of *B. natto.*

Peptides and Polypeptides Formed by Enzymes

Enzymatic polypeptide synthesis has often been referred to as template synthesis, in the sense of a protein template encoding the amino acid sequence; such a polyenzyme model was proposed in 1954 by Lipmann (reviewed in 87). In this model a quasilinear or spatial arrangement of activated amino acid substrates directs the peptide sequence. As it was evident that there is no direct complementation between amino acids as found in nucleic acid base pairs, the question arose of whether a complementing protein surface is indeed realized, or whether a sort of timing device (as proposed by Lipmann) guides the sequence of events. From descriptions of the various types of compounds that have now been examined, a more complicated picture emerges. Peptides with repeated sequences require a different enzymic machinery from peptides with a unique sequence or small peptides with up to five reaction steps. Examples of peptides with repeated sequences are poly-γ-DGlu compounds with an average of 750–10^4 residues formed by *Bacillus licheniformis* strains (180, 181), branched multi-Arg-poly(Asp) structures found in *Anabaena cylindrica* (167), and γ-Glu or β-Lys chains (123, 159) from streptothricin-type antibiotics. In the formation of repeated sequences the template may be used repeatedly, and thus may actually be smaller than the final product.

Peptides with a unique sequence have an average length of about seven residues, although some have more than 20 amino acids, and often have cyclic structures (average ring size of about six amino acids). The formation of unique sequences is catalyzed by the following general mechanism. Amino acid–activating units are lined up in one or several multienzymes, depending on the size of the product. Intermediates remain linked to the enzymes, and are transported by a CoA-like cofactor. The multienzyme structures are directly encoded at the gene level. This type of mechanism has been termed the "multienzyme thiotemplate mechanism" (107), since all activated intermediates formed are thiolesters. This mechanism may not be the only type of organized reaction sequence operating, although it has been verified in linear peptides (edeine, linear gramicidin, alamethicin), cyclopeptides (tyrocidine, gramicidin S, cyclosporin), branched cyclopeptides (bacitracin, polymyxin), and depsipeptides (enniatin, beauvericin). Peptidolactone systems (actinomycin, surfactin, destruxin) are currently under investigation.

While all compounds generated with the thiol/multienzyme mechanism contain α,α-peptide bonds in their chains (except for the final ring-closure reactions in bacitracin and polymyxin and corresponding ester bonds in depsipeptides), some structures do contain γ,α-peptide links (e.g. mycobacillin), or in cyclization use β-hydroxy groups (e.g. surfactin) or the β-alanyl carboxyl (e.g. destruxin). For at least mycobacillin a different type of mechanism has been proposed. This type of reaction sequence resembles reaction types found in the formation of glutathione and pantetheine in the early 1950s. The starting compound with the first carboxyl group is activated by a distinct enzyme, which also catalyzes the addition of an amino-group acceptor. Thus a free soluble intermediate is formed, which is the substrate in the following activation reaction. Such a system seemed to be suitable for small peptides only, but the recent studies of Bose's group (41, 42, 117, 133, 134) demonstrate that it functions in the formation of the cyclic 13-peptide mycobacillin. They detected no activation of the constituent amino acids except for the starter compound proline, no enzyme-bound intermediates, and no participation of a pantetheine cofactor. Instead they found the production of soluble intermediates and the activation of intermediate peptides, although a thiotemplatelike organization into multienzymes was observed. It remains to be established whether multifunctional protein structures are involved as in the thiotemplate route. The finding, however, that intermediate peptides can be elongated by a single-step enzymic reaction implies that recognition sites are used for the addition of peptides and amino acids at each elongation site.

Organization of Protein Templates

The presently available information on protein templates is schematically shown in Figure 2. So far, only systems forming tyrocidines, gramicidin S, bacitracin, enniatin, beauvericin, and mycobacillin have been characterized, and the system for linear gramicidin has been partially identified. The organization of the unique multifunctional structure of these protein templates cannot yet be predicted, but must be considered in the interpretation of genetic studies. Defective multifunctional enzymes can usually only be complemented efficiently by integral structures. Consider the gramicidin S synthetase, which consists of two multifunctional enzymes, GS1 and GS2, of 120 and 280 kd, respectively. Both multienzymes catalyze the formation of the enzyme-bound peptide D-Phe-Pro-Val-Orn-Leu, which is followed by head-to-tail cyclization. While GS1 activates and epimerizes phenylalanine, analysis of fragments carrying partial activities has shown that GS2 contains activating units in the peptide sequence Pro-Val-Orn-Leu (5, 97). Defective enzymes isolated from nonproducing mutants retain their size despite defective activation of either proline, valine, or leucine or absence of the cofactor 4'-phosphopantetheine (57, 59). Apparent early-termination mutants with reduced size that activate proline (110 kd) and proline, valine, and ornithine (260 kd) have also been isolated and characterized (15, 67). On the other hand, a nonproducing mutant of bacitracin synthetase, which is a three-enzyme system (see Figure 2), contains an associated or fused structure of the multienzymes BA1 and BA3 that could not be complemented by BA2 or the addition of BA2 and BA3 (52). The arrangement of amino acid–activating units in GS2 and the sequential distribution of the peptide sequence between the multienzymes suggest that the enzymatic code is linked to the sequence of activating centers, which are fixed to multifunctional structures by principles that have not yet been deduced. Does this code permit mutations of the peptides synthesized? At least a partial answer can be drawn from recent studies of the enniatin/beauvericin systems and from various structural studies of peptides.

The cyclohexadepsipeptide enniatin is formed by a strain of *Fusarium*

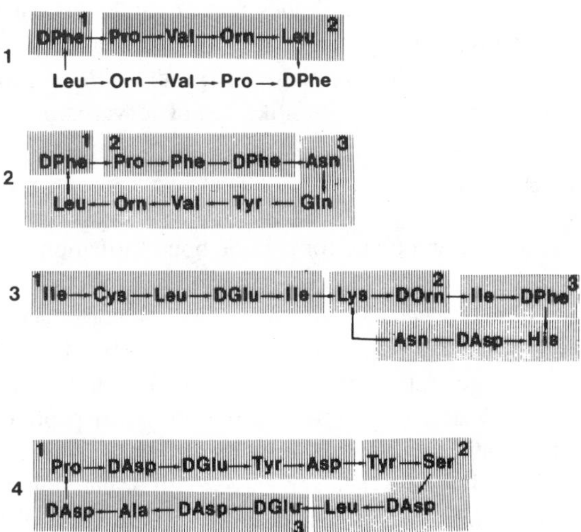

Figure 2 Structures of peptide antibiotics and their multienzymic templates, with sizes of enzymes, where known, as follows. *1,* Gramicidin S (GS1, 120 kd; GS2, 280 kd). *2,* Tyrocidine (TY1, 120 kd; TY2, 230 kd; TY3, 450 kd). *3,* Bacitracin (BA1, 330 kd; BA2, 210 kd; BA3, 380 kd). *4,* mycobacillin.

oxysporum on a 250-kd multienzyme (200–202) by trimerization of activated *N*-methyl-L-valyl-D-hydroxyisovalerate fragments formed by the thiotemplate mechanism. Beauvericin, an analog containing phenylalanine instead of valine, is produced by several fungi, including *Beauveria bassiana*. Attempts to produce beauvericin with enniatin synthetase have failed, although the multienzyme accepted a variety of branched-chain and linear aliphatic amino acids (113, 166). However, a similar multienzyme from *B. bassiana* has been characterized (144–146). This multienzyme has broad specificity for aromatic amino acids, but also accepts saturated hydrophobic amino acids at a reduced rate (144). This replacement of aromatic by aliphatic amino acids has been found in numerous cases. Polymyxin, which is produced by strains of *Bacillus polymyxa* and *Aerobacter aerogenes* by the thiotemplate mechanism (99–101), is an example. Various strains of *Bacillus* have been isolated that produce analogs of this type (178). While minor variations such as Leu/Ile/Val/Thr or Phe/Tyr/Trp can be attributed to overlapping specificity of activation reactions, a Leu/Val-Phe exchange implies variation of the template.

Other structural variations that presumably can be traced to changes in template structure are depicted in Figure 3. A series of cyclotetrapeptides containing the unusual amino acid 2-amino-8-oxoepoxidecanoate (Aoe), produced by various fungi, show intriguing structural homologies that suggest partial homologies of templates. Iturin, mycosubtilin, and bacillomycin, cyclooctapeptides that contain an aliphatic β-amino acid, produced by strains of *B. subtilis,* have structural homologies that imply the exchange of activating units in the corresponding templates. Gratisin, a cyclododecapeptide isolated from a mutant of the gramicidin S producer *B. brevis* (136, 176, 177), may originate from an insertionlike template variation.

Activation Reactions

Carboxyl activation is prerequisite for peptide bond formation, and is the most sensitive enzymatic reaction for the detection of enzymes and multienzymes. Activation generally proceeds by the formation of phosphate or adenylate intermediates with cleavage of β-γ- or α-β-phosphate bonds of triphosphates. We can compare the activation of simple carboxyl groups (acylation), β- or γ-carboxyls of amino acids or peptides, α-carboxyls of peptides, and, most importantly, amino acid carboxyls. Currently known activation reactions are summarized in Table 1. These data can serve as an orientation in the analysis of pathways. The table indicates that α-carboxyls are generally activated as adenylates, as are most simple acids, while the enzymes that activate β- and γ-carboxyls tend to use phosphate.

How can these amino acid activations be differentiated from those of aminoacyl-tRNA synthetases? There have been several attempts to detect similarities between ribosomal and enzymatic systems. Indeed, the study of

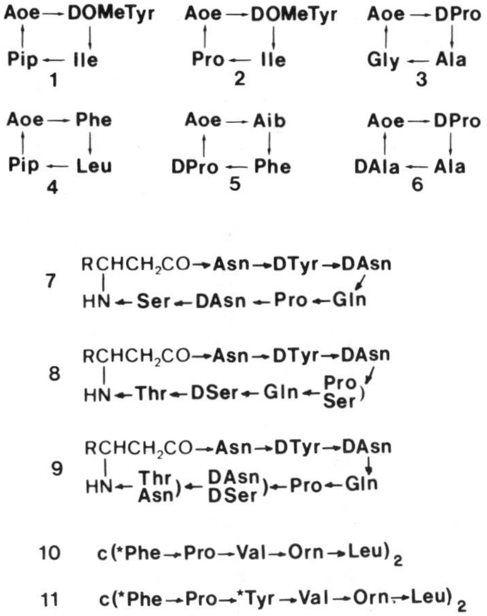

Aoe → DOMeTyr Aoe → DOMeTyr Aoe → DPro
Pip ← Ile Pro ← Ile Gly ← Ala
1 2 3

Aoe → Phe Aoe → Aib Aoe → DPro
Pip ← Leu DPro ← Phe DAla ← Ala
4 5 6

7 RCHCH$_2$CO → Asn → DTyr → DAsn
HN ← Ser ← DAsn ← Pro ← Gln

8 RCHCH$_2$CO → Asn → DTyr → DAsn
HN ← Thr ← DSer ← Gln ← (Pro/Ser)

9 RCHCH$_2$CO → Asn → DTyr → DAsn
HN ← (Thr/Asn) ← (DAsn/DSer) ← Pro ← Gln

10 c(*Phe → Pro → Val → Orn → Leu)$_2$

11 c(*Phe → Pro → *Tyr → Val → Orn → Leu)$_2$

Figure 3 Structures of some peptide antibiotics with variations in enzymic templates indicated. The fungal cyclotetrapeptides Cyl-1 and -2 [*1,2,* from *Cylindrocladium scoparum* (175)], HC-toxins [*3,6,* from *Helminthosporium carbonum* (75, 82)], chlamydocin [*5,* from *Diheterospora chlamydosporia* (18, 157)], and WF-3161 (*4,* from *Petriella guttulata*) all contain the unusual amino acid Aoe (2-amino-8-oxo-9,10-epoxidecanoic acid). The cyclopeptides iturin *(7),* bacillomycin *(8),* and mycosubtiline *(9),* isolated from strains of *Bacillus subtilis,* all have the same LDDLLDL sequence (148), a constant region of residues 1–4, and a variable region at residues 5–8. In gratisin *(11),* an additional Tyr appears to be inserted into the gramicidin S sequence *(10).* Note that configurations (*) have not been determined in gratisin, but Phe is D in gramicidin S and is likely to be D in gratisin, as concluded from synthetic studies (176).

enzymic mechanisms employs identical procedures, such as ATP-[^{32}P]PP$_i$ exchange or γ-[^{32}P]P$_i$ release from ATP. Certain enzymic reactions utilize aminoacyl-tRNA charged by the corresponding synthetase, such as aminoacyl-tRNA transferases or enzymes forming interpeptide bridges in bacterial cell walls. In the search for procedures to selectively study single activation reactions within multienzymes, analogs of ATP have been studied. Aminoacyl-tRNA synthetases of *E. coli* and yeast can be grouped into families according to their use of ATP analogs (27, 39). Unexpectedly, all activation reactions in multienzymes studied so far have shown similar specificity patterns, which differ from those of all types of tRNA synthetases (93; H. von Döhren, unpublished results). Some analogs have been useful in enzymic studies. For example, 2'-dATP is accepted by all thiotemplate activation sites on multienzymes, but not by several tRNA synthetases including Val/Ile/Leu. Such a reaction can be of use in the detection of expression of cloned antibiotic synthetase activities, e.g. in *E. coli* (103).

Table 1 Types of carboxyl activation reactions[a]

Enzyme	Type I (AMP)[b]	Type II (NDP)[c]
Acid-CoA ligases	Acetate, medium (C_4–C_{12}), intermediate (C_3–C_7), and long chains (C_{12}–C_{18}), benzoate, cholate, luciferin, phenylacetate, oxalate, biotin	Succinate, succinate (GDP), medium chains (C_4–C_{12}) (GDP), glutarate
Aminoacyl-tRNA ligases	Ala, Arg, Asn, Asp, Cys, Glu, Gln, Gly, His, Ile, Leu, Lys, Met, Phe, Pro, Ser, Thr, Trp, Tyr, Val	
Peptide synthetases		
Amino acids—α-carboxyl	AcLeu, Aib, Ala, D-Ala, Asn, Cys, Dab, Dap, Glu, Gln, Gly, His, Ile, Leu, D-Leu, Lys, MeBmt, Orn, Phe, D-Phe, Pro, Thr, Trp, Tyr, Val	Pro, D-Ala
Amino acids—other carboxyls	Dha, Ise, β-Tyr	β-Ala, Glu(γ), D-Glu(γ), D-Asp(β)
Peptides—α-caroboxyl	AcLeu-Leu	Glu-Cys, UDPMurNAcGly(Ala), UDPMurNAc-A_1-D-Glu-A_3 (A_3 = Lys, Dpm, Dab, Orn, Hsr), mycobacillin intermediates(α)
Peptides—other carboxyls		UDPMurNAc-A_1-D-Glu(γ), mycobacillin intermediates(γ)
Carboxylic acids	Dhb, pantoate, D-Hiv	UDP-GNAc-lactate

[a]Abbreviations: Aib, aminoadipic acid; Dab, diaminobutyric acid; Dap, diaminopropionic acid; Dha, dihydroxyamino acid; Dhb, dihydroxybenzoic acid; Dpm, diaminopimelic acid; Hiv, hydroxyisovaleric acid; Hsr, homoserine; Ise, isoserine; MeBmt, (4*R*)-4-[(*E*)-2-butenyl]-4, *N*-dimethyl-L-threonine.
[b]Carboxyl activation by cleavage of ATP to AMP and PP_i.
[c]Carboxyl activation by cleavage of NTP to NDP and P_i; ATP if indicated differently.

Reactive Intermediates

An outstanding property of the thiotemplate multienzyme mechanism is the participation of stable thiolesters of amino acids and peptides. These actually resemble the aminoacyl-tRNA/peptidyl-tRNA in the ribosomal system or the acyl-CoA/acyl-ACP intermediates of lipid biosynthesis. Preceding these intermediates are several less stable intermediates: phosphoryl and adenyl enzymes, enzyme-stabilized aminoacyl adenylates, and even aminoacyl enzymes. These intermediates have been discussed in general (168) and for amino acid–activating enzymes (86, 88), so we just point out recent data on the stability of thiol-bound intermediates.

Thiolesters are stable at acid pH and thus decompose by hydrolysis. This hydrolysis is enhanced by thiol reducers, such as dithiothreitol, which are usually included in preparation buffers (33). Stability of active esters is also related to the amino acid peptide structures. For example, ornithyl GS2, an intermediate in gramicidin S formation, is cyclized to 3-amino-2-piperidon, but if ornithine is replaced by lysine the stability of the thiolester ($t_{1/2}2 = 0.5$ hr at 3°C) is increased ninefold, since the reaction is sterically less favored. The dipeptide D-Phe-Pro-GS2, with a half-life of only 0.9 hr, is lost by cyclization to the piperazine-dione, which is favored by the Pro-dipeptide structure. Addition of the third amino acid, valine, and subsequent formation of the intermediate D-Phe-Pro-Val-GS2, increases the half-life hundredfold to 90 hr; however, further elongation again decreases the stability to 0.4 hr owing to the abortion of the tetrapeptide by cyclization of the activated ornithine residue (33).

Transfer of Intermediates

Intermediates can be transferred in intermolecular reactions between multienzymes. Transfers in the thiotemplate system include a D-amino acid (gramicidin S), a tetrapeptide (tyrocidine) that terminates with D-configuration, a pentapeptide and a heptapeptide (bacitracin system), and an acyl dipeptide and a heptapeptide (linear gramicidin system) (see Figure 2). There is no obvious structural correlation between donor and acceptor compounds. Phosphopantetheine may have some role in transfer reactions, since all acceptor sites are located on multienzymes that contain this cofactor. A mutant GS2 enzyme devoid of pantetheine was not able to accept D-phenylalanine from GS1 (58). Some evidence for a soluble cofactor has been reported for the intermediate multienzyme of bacitracin biosynthesis (151). Stable complexes of multienzymes have so far not been detected. Peptide formation by the individual enzymes GS1 and GS2 in the gramicidin S system appears to be linearly dependent on concentration (98; H. von Döhren, unpublished results); thus in the case of stoichiometric amounts of enzymes a second-order relation to protein concentration is found (195). This is highly significant in estimating the total activity of a system in crude preparations. For enniatin synthetase, where cyclohexadepsipeptide formation is catalyzed by a single multienzyme, the linear concentration dependence indicates that the complete process may proceed within one enzyme, not on dimers or trimers (R. Zocher, unpublished data). Bacitracin synthetase multienzymes have been found in a

3:1:2 ratio (BA1:BA2:BA3) by electrophoretic procedures (H. von Döhren & Ø. Frøyshov, unpublished results).

The mechanism of elongation with the swinging cofactor pantetheine is in accordance with the observation that no more than six amino acids are assembled on a single multienzyme; this has been interpreted as a limit to the number of active centers that can be reached by a swinging 20-Å long cofactor. Electron microscopic images obtained for GS2 and enniatin synthetase (94, 96; J. Wecke, P. Giesbrecht, H. Schmiady, B. Tesche & J. Vater, unpublished results) and ferrichrome synthetase (H. Diekmann, personal communication) show flat circular structures without obvious subunit organization, which contain an inner structure 40 Å in diameter. So far the lability of the structures has hampered orientation and statistical evaluation of images, which are required for a detailed structural model.

MODIFICATION REACTIONS

D-Amino Acids

D-Amino acids are widely distributed among peptide antibiotics and contribute to the activity and stability of these compounds. They may be substrates of enzymes or may not be accepted at all. Numerous in vivo studies have employed feeding with D-amino acids to establish precursor configuration. Enzyme studies have shown that some sites with missing stereoselectivity accept both isomers, but they utilize only the D-form in further reactions. Most systems select the L-form and should possess an epimerase function. Sometimes, when the L-configuration is substrate and is not epimerized, the enzyme may bind the D-form with inhibitory effects (gramicidin S system, valine and leucine) (156).

The classical case is the formation of D-Ala-D-Ala in murein biosynthesis. D-Alanine is provided by the alanine racemase and is activated by the synthetase as a phosphate intermediate. D-Ala-D-Ala has also been detected in plants (118). Other dipeptides containing D-amino acids are also known (70). More complex peptides are apparently formed on templates with integrated epimerase functions. Spector (168) has discussed tyrocidine synthetase in detail. In an analysis of mutants with altered structures of GS1, epimerization has been shown to proceed at the thiolester stage (68). If an essential thiol is missing, activation stops with the formation of aminoacyl adenylate in either L- or D-configuration. Other types of isomerizations are known as well. In etamycin, *cis*-4-hydroxy-D-Pro originates from *trans*-4-hydroxy-L-Pro (55, 56, 74). Feeding of L-proline leads to the D-proline–containing analog neoviridogrisein, a more effective antibiotic (142, 143).

Modification of enzyme systems relies on the stereochemistry of peptide precursors. Boente et al (13), using cyclo(L-[^{15}N]Phe-L-[1-^{13}C]Phe, unambiguously showed the in vivo incorporation of an intact cyclodipeptide in bisdethiobis-(methylthio)acetylaranotin. The corresponding LD- or DD-piperazine-2,5-diones were not accepted. The ring systems formed were similar to those of gliotoxins (63, 84) and bipolaramide (114), as shown in Figure 4. Possible epimerizations via dehydroamino acids have been mentioned in the section on ribosomally formed antibiotics.

Dehydroamino Acids

Dehydroamino acids have been detected in various peptides including those of the ribosomal nisin/subtilin type; linear peptides such as antrimycin (131), cirratiomycins (161–163), lavendomycin (183), and celeneamides (170); small cyclopeptides such as alternarolide, AM-toxin, and A-2315; cyclopeptides such as virginiamycin, capreomycin, telomycin, and stendomycin; and complex structures such as thiostrepton, siomycin, berninamycin, and nosiheptide. Dehydroalanine, dehydroaminobutyrate, and Cys-derived thiazoline and thiazole structures are predominant, which indicates that their formation proceeds via β-elimination from serine or threonine. This reaction is well known from *S*-methyl-Cys, *O*-tosyl-Ser, or *O*-phospho-Ser within peptide chains (158). The only stereochemical studies have been carried out with mycelianamide (85) and cryoechinulin (17), and show *cis*-dehydrogenation. An enzyme that may form α,β-dehydro-Trp from Trp-containing peptides (138) is a hemoprotein from *Pseudomonas,* indolyl-3-alkane-α-hydroxylase (150, 174).

The bis-thiazole structure in bleomycine is of special importance in the peptide's DNA-binding properties, but also in its toxicity. Phleomycine, a thiazoline/thiazole structure, proved to be highly toxic, and screens were

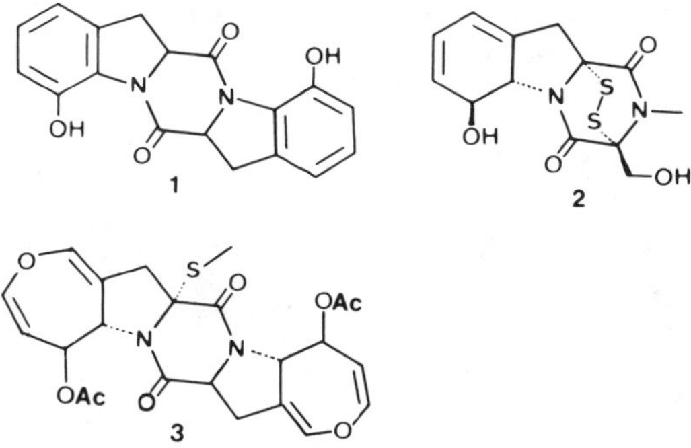

Figure 4 Structures of modified piperazine-2,5-diones. Gliotoxin *(2)* and bisdethiobis-(methylthio)acetyl-aranotin *(3)* can be derived from Phe and cyclo(Phe)$_2$, respectively, but not from other isomers or from Tyr or Tyr peptides. Ring expansion of bipolaramide *(1)* to the oxepin is thought to proceed via arene oxide and epoxidation.

extended to search for analogs with the bithiazole moiety. Fujii (31) has found evidence for an enzyme that transforms phleomycine into bleomycine. A similar bithiazole structure has been detected in myxothiazole, which is produced by *Myxococcus fulvus* (40, 179) (Figure 5). Less common dehydroamino acids are 3,4-dehydro-Val, 3,4-dehydro-Pro, and 2,3-dehydro-Asp, which have been detected in phomopsin A (20).

Unusual Amino Acids and Related Compounds

Various unusual amino acids generally indicate a nonribosomal origin of a peptide. Quite often β-alanine is a constituent. In a simple dipeptide, such as carnosine, the activation proceeds not by adenylate formation, as was reported in the first studies, but by β,γ-phosphate cleavage (9, 10). In a β-alanine peptide, such as 4'-phosphopantothenate, the carboxyl is again phosphate activated. β-Alanine is also a constituent of sarcophagine, an insect peptide with the structure β-Ala-Tyr. A corresponding enzyme system has been found in larvae of *Sarcophaga peregrina* (69). The complete peptide

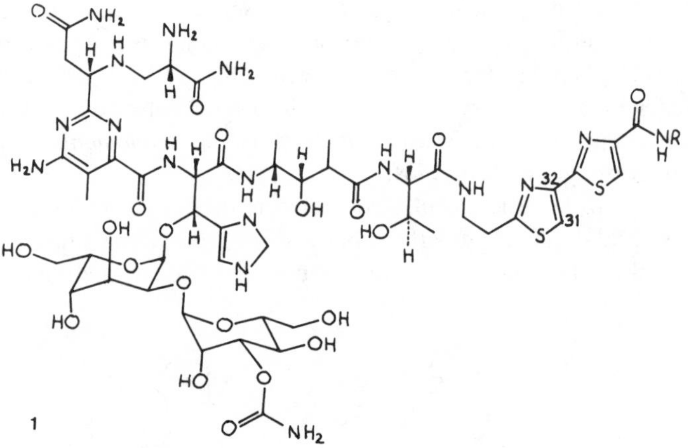

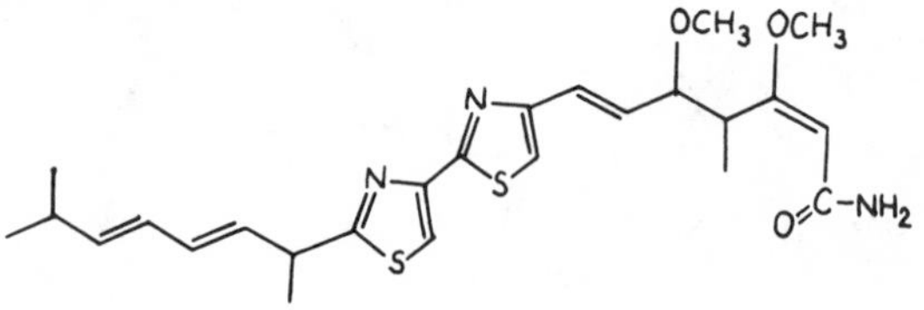

Figure 5 Structures of peptides containing bithiazole moieties derived from adjacent Cys residues. *1*, Bleomycins; *2*, myxothiazole. Various bleomycins have been produced by directed biosynthesis with amine compounds (R).

moiety of CoA has recently been detected as a side chain of the carbapenem OA-6129 (32), and thus provides a link between the so-called primary and secondary metabolism. Acylases that catalyze depantothenylation have been isolated from several strains of *Streptomyces* (104). This type of enzyme also catalyzes acyl exchange of carbapenems and deacylation of various *N*-methylamino acids.

The enzymic reactions that provide the direct link between amino acid and acetate polymerization have not yet been characterized. In a scheme proposed in 1974, Morishima et al (132) postulated that the formation of β-hydroxy-β-amino acids proceeds from activated amino acids by acetate addition via malonyl-CoA. Such β-hydroxy acids are found in bleomycine-type peptides (phleomycine, tallysomycine, zorbamycine), proteinase inhibitors (e.g. pepstatine) produced by streptomycetes, and edeines, which are produced by *Bacillus brevis* V. Thus consecutive peptide and C-C bonds could be assembled in one multienzyme. In at least edeine formation, however, the β-hydroxy acid 2,6-diamino-7-hydroxyazaleic acid is directly activated by the thiotemplate system (108, 109), and is thus generated in advance.

An even more obvious interplay of biosynthetic mechanisms could be that in virginiamycin M, which has been studied by stable isotope techniques (83, 110, 111, 198). The common ring system (Figure 5) originates from amino acids and acetate units [see also A-2315-A from *Actinoplanes* sp. (110, 111)].

Finally, a direct link of the so-called secondary metabolites to the well-characterized glutathione pathway (124), which has been opened up to genetic manipulation by the impressive work of Kimura's laboratory (50, 51, 135, 196), is obvious. Such a link, as suggested by the large number of γ-glutamyl compounds known (16, 21, 71), has now been confirmed by the recent fascinating discovery of heavy metal–complexing peptides in yeast, *Schizosaccharomyces pombe, Neurospora* (47, 54), and higher plants (46). These consist of two to eight repetitions of γ-Glu-Cys, terminating with Gly, and are especially active in Cd^{2+}-binding. Grill et al (47) proposed that these peptides are formed from glutathion or its precursor, γ-Glu-Cys.

Acylations

Amino-group acylations are well-known modifications performed on technical scales with a variety of acylases. These reaction types are not discussed here. (For recent reviews see 72, 115.) We only consider reactions that are essential parts of the peptide antibiotic pathway. As in the ribosomal process, *N*-formyl or *N*-acyl residues are frequently required or formed during initiation. In vitro biosynthetic systems have utilized acetyl-CoA for the 19-peptide alamethicin (129, 130) and formyltetrahydrofolate for linear gramicidin (1, 2). The formation of leupeptin, acetyl-Leu-Leu-argininal, even requires acetyleucine for initiation. A 27-kd acyltransferase that uses acetyl-CoA has been partially purified from the producer *Streptomyces roseus*, but it is not specific for leucine; it also accepts unrelated amino acids and peptides (171). The leupeptin acid synthetase, a 260-kd multienzyme, activates acetyl-Leu and acetyl-Leu-Leu, and then adds arginine, forming acetyl-Leu-Leu-Arg (172). Biosynthesis of polymyxins also requires an acyltransferase to form the initiating *N*-octanoyl-diaminobutyrate (99, 100). Thus a polyenzyme fraction catalyzes peptide formation from octanoyl-CoA and amino acids only if a transferase fraction has been added. In the initiation reaction, polyenzyme-attached diaminobutyrate is acylated by octanoyl-CoA or isomers (101).

In peptidolactone systems currently being studied, the acyl moiety is a dominating structural factor. Enzymes that activate 4-methyl-3-hydroxyanthranilate (4MHA) (actinomycin) and lysergate (ergot peptides) have been characterized (76–79). Amino acids or peptides are thought to be transferred to these enzyme-stabilized acyl adenylates. Specificity studies of the 4MHA-activating enzyme and feeding experiments have led to analogs and even new types of compounds. The final step in actinomycin formation is thought to be the oxidative condensation of two 4MHA pentapeptides to the phenoxazinone chromophore. This reaction either proceeds spontaneously or is catalyzed by phenoxazinone synthetase. Defunctionalized 4MHA analogs may then lead to the production of actinomycin halves (78).

In quinomycin-type antibiotics, which are acylated tetrapeptides linked head-to-tail by ester bonds and bridged by a disulfide or thioether formed by opposite cysteines, various quinoxaline analogs have been introduced by directed biosynthesis (19, 35–37, 155, 197).

N-*Methylation*

Methylated peptide bonds are most commonly found in cyclic peptides. *N*-Methylation is thought to stabilize peptide bonds against proteolytic cleavage, and may influence conformational stabilities (191). A well-studied case is enniatin synthetase (201). The *N*-methyl transferase function is an integral part of the multienzyme. *S*-Adenosylmethionine is used to methylate activated thiolester-bound valine. In the absence of adenosylmethionine, small amounts of unmethylated depsipeptides are formed (200). A 28-dk fragment involved in the transferase reaction has been isolated by affinity labeling; it could serve as a structural probe to identify similar sites in other systems (R. Zocher & A. Billig, unpublished results).

Terminal Modifications

It is well known that the cysteamine moiety of CoA originates from incorporation of cysteine, followed by decarboxylation. A similar mechanism has been considered for the terminal ethanolamine (EA) of linear gramicidin, although it has been shown that EA is accepted by the nascent 15-peptide in vitro (11).

Phosphatidyl-EA, for example, is formed in bacterial systems by decarboxylation of phosphatidylserine but in mammalian and plant systems by direct EA addition (24). Kubota (105) suggested cytidine-diphospho-EA as a suitable donor, and proposed an active intermediate. Recently, a new gramicidin with an alkali-labile lipid component, presumably attached to the EA terminal, has been detected (95).

Amination of carboxyls is apparently not very specific, as has become obvious from various terminal substitutions carried out in vivo and in vitro with bleomycin in the development of new anticancer drugs (31, 185, 186). Several hundred compounds have been synthesized, and efforts are now being directed to the introduction of modifications in other positions (see legend to Figure 5). Spermidine, one of these terminal amines, is also the terminal acceptor for edeines (108, 109) and is contained in glycocinnamoyl antibiotics (182), B-1008 (81), and laterosporamine (165).

Cyclization Reactions

The various types of cyclopeptide structures and some of the reactions involved have been discussed in general (88). No detailed studies on mechanisms are available, but we summarize some data. Stereochemically favored abortive cyclizations of intermediates, which also occur in solid-phase peptide synthesis, have been mentioned in the section on reactive intermediates. We can distinguish many types of reactions, such as thiazolidine formation, peptide or ester bond formation, or oxidative couplings of phenolic rings. No biosynthetic studies are available on the latter type of reaction, which is presently of some interest, but an increasing number of peptides with this feature have been reported. In addition to numerous compounds of the vancomycin group, which have recently been reviewed in this series (8), bouvardin *(Bouvardia terniflora),* piperazinomycin *(Streptoverticillium olivoreticuli),* and RA-peptides *(Rubiae radix)* (62) are examples. With regard to cyclization by peptide and ester bonds, it is evident from structural considerations that protein surfaces direct donor and acceptor sites, since several amino or hydroxy functions can usually react with the respective activated peptide carboxyl. In comparative analysis certain ring sizes dominate, e.g. seven–amino acid rings in branched peptides or six members in lactones. Outstanding examples of size selection are esperin and surfactin; two acyl heptapeptides with the perhaps identical sequence acyl(β-hydroxy)-Glu-Leu-Leu-Val-Asp-D-Leu-Leu form a lactone with either the β-aspartic acid or the terminal Leu-carboxyl. The configuration of the ^{6}Leu, however, has not been determined for esperin (surfactin D) (60, 88, 164).

Significant advances in this field have been made for β-lactam compounds, which are treated elsewhere in this volume (139).

CHARACTERIZATION OF PATHWAYS

Regulation of Biosynthesis

In the study of peptide antibiotics one has to keep in mind that certain conditions are favorable for production, while others are not. Since almost no functional studies of the peptides in their producer organisms have been conclusive, the regulatory factors remain quite uncertain. Studies of Fe^{3+}-transport compounds and Cd^{2+} complexers are on the way to providing an understanding of mechanisms at the molecular level (49). For many peptides, however, nonproducing mutants have been known for a long time but have not led to unambiguous elucidation of the roles of these compounds (for reviews see 73, 89, 91, 123). Most so-called secondary metabolites, or idiolites, are not essential for growth or survival. To study regulatory events and mechanisms a molecular approach is clearly required, with efforts directed toward intramolecular effectors, regulatory molecules, and transcriptional events. The current approach is to purify enzymes, to obtain either antibodies or sequence information, and to attempt to clone the corresponding genes using cDNA or expression systems. Recent activities have focused on aerobactin (a siderophore containing acetylhydroxylysine), actinomycin, gramicidin S, tyrocidine, bacitracin (L. Korsnes, personal communication), and various penicillin- and cephalosporium-producing systems (139).

Bacterial Fe^{3+} uptake is a relatively simple system. In aerobactin it consists of a Fe^{3+}-receptor protein, Fe^{2+}-binding repressor, and three enzymes forming the carrier: *N*-hydroxylase, *N*-acetyltransferase, and synthetase (23, 26, 27). Much more complex are the typical idiolite networks, embedded, for example, in catabolite repression. It is doubtful that other systems besides the well-studied *E. coli* or *B. subtilis* will presently permit a reasonable evaluation (93). The phenoxazinone-synthase gene of *Streptomyces antibioticus* has been cloned (65), but has since been shown to be present in *S. lividans* in a silent form. Upon introduction of certain *S. antibioticus* DNA fragments, expression is triggered (66). Regulatory sequences are now being studied to understand transcriptional activation by 5-fluorouracil (64). In *Bacillus brevis* strains, producers of gramicidin S and tyrocidines, difficulties in protoplasting and transformation have favored the use of *lacZ* expression vectors in the cloning of multienzymes. While part of the GS2 multienzyme has been expressed in *E. coli* as a *lacZ* fusion protein (102, 103), the complete tyrocidine synthetase 1 (TY1) gene *tycA* has been isolated with the help of heterologous antibody probes of the partially homologous gramicidin S multienzymes (122). This DNA fragment contains a transcribed 330-bp leader sequence. It is constitutively expressed in *E. coli,* but in *B. subtilis* 168, as in the producer strain *B. brevis* ATCC 8185, it is expressed at the end of vegetative growth. Marahiel et al have constructed an SPβ *tycA-lacZ* fusion with the transducing phage SPβ to study regulatory events in *B. subtilis* with relation to *SpoO* gene products (121; M. Marahiel, P. Zuber, G. Czekay, R. Losick, unpublished manuscript). Dependence of expression on the *SpoOA* product could be bypassed by an *AbrB* suppressor mutation, causing constitutive expression. Thus the *AbrB* gene product(s) may promote negative regulation of *tycA* expression. This shows that, at least in *Bacillus,* some insight may be gained from analogies with regulatory systems.

Characterization of Enzyme Systems

The general features of enzyme systems have been summarized above and have been extensively reviewed (86, 88, 89, 91, 107). We present an overview of available systems in Table 2 and add a few technical comments. One of the first things to ascertain is whether cells are actually synthesizing the peptide, although synthesis may not coincide with the highest levels of enzyme(s). In a thiotemplate system, high-molecular-weight polypeptides are generally observed in addition to amino acid activation reactions. These can be distinguished in many cases from aminoacyl-tRNA synthetases by the use of ATP analogs such as 2'd-ATP. Between such multifunctional enzymes or polyenzymes certain homologies of protein structures can be detected with heterologous antibodies (14, 192). DNA probes can also be of use in the detection of such homologies. In some cases it may prove difficult to obtain cell-free preparations. In following published procedures care should be taken to use similar organisms. Many membrane-bound forms of enzymes have been reported (7, 29, 89, 188, 189). In at least one case, valinomycin production of *Streptomyces* sp., active cell-free preparations have not been obtained by any of the current procedures (116).

Genetic Approaches

There have been several attempts to characterize nonproducing mutants in order to understand the functioning and organization of multienzyme systems. By far the most thorough studies have been the investigations of Saito et al on the gramicidin S synthetase (57–59, 68, 153). These studies clearly support the concept of multifunctional enzymes, since enzymes with a single inactive site were not changed in size. Other studies have not included characterization of gene products (52, 119, 173, 189), and are therefore only of limited value.

Several years ago many antibiotic genes were expected to have an extrachromosomal location. Curing agents were effective in the conversion of producers to nonproducers, sometimes with a very high frequency (93).

Table 2 Current knowledge of enzymology of peptides

Peptide	Organisms	Structure[a]	Thiotem-plate	Remarks[b]	References
Linear—short					
Carnosin	Rat/chick	β-Ala-His	–	p	9, 10
Sarcophagine	*Sarcophaga peregrina*	β-Ala-Tyr	–	pp	69
Glutathion	Diverse	γ-Glu-Cys-Gly	–	p	50, 124, 135
4'-P-Pantetheine	Diverse	P-pantoic-β-Ala-Cys($-CO_2$)	–	p	89, 90
Leupeptin	*Streptomyces roseus*	AcLeu-Leu-Argal	–	p, me	171, 172
ACV	*Cephalosporium acremonium*	Aad-Cys-D-Val	–	pp	139
Linear—medium					
Edeine	*Bacillus brevis*	β-Tyr-Ise-Dpr-Dahaa-Gly-spermidine	+	pp, 3 me	108, 109
Linear—long					
Gramicidin	*Bacillus brevis*	f-Val-Gly-Ala-D-Leu-Ala-D-Val-Val-D-Val-Trp(D-Leu-Trp)$_3$-ethanolamine	+	pp, 4 me (?)	1, 2, 11, 105
Alamethicin	*Trichoderma viride*	AcAib-Pro-Aib-Ala-Aib-Ala-Gln-Aib-Val-Aib-Gly-Leu-Aib-Pro-Val-Aib-Aib-Glu-Gln-Pheol	+	pp, me	129, 130
Linear polymers					
Multi-Arg-poly(Asp)	*Anabaena cylindrica*	Multi-Arg-poly(Asp)	–	pp	167
γ-D-Glu-capsule	*Bacillus licheniformis*	Poly-γ-D-Glu(n = 750. . .10^4)	–	pp	180, 181

Cyclic					
Cyclopeptin	*Penicillium cyclopium*	c(Anthranilate-Phe)	+	pp	38, 112
Enterochelin	*Escherichia coli*	c(Dhb-Ser)$_3$	−	pp	45
Ferrichrome	*Aspergillus quadricinctus*	c(Gly$_3$OH-Orn$_3$)	+	p, me	61
Gramicidin S	*Bacillus brevis*	c(D-Phe-Pro-Val-Orn-Leu)$_2$	+	p, 2 me	91
Tyrocidine	*Bacillus brevis*	c(D-Phe-Pro-Phe-D-Phe-Asn-Gln-Tyr-Val-Orn-Leu)	+	p, 3 me	91
Cyclosporin	*Tolypocladium inflatum*	c(MeBmt-Abu-Sar-NMe-Leu-Val-NMe-Leu-Ala-D-Ala-NMe-Leu-NMe-Leu-NMe-Val)	+	pp, me	
Mycobacillin	*Bacillus subtilis*	c(Pro-D-Asp-D-Glu-(γ)-Tyr-Asp-Tyr-Ser-D-Asp-Leu-D-Glu-(γ)-D-Asp-Ala-D-Asp)	−	p, 3 me	41, 42, 117, 133
Branched cyclic					
Polymyxin	*Bacillus polymyxa*	Oct-Dab-Thr-Dab-Dab-Dab-Leu-Leu └Thr-Dab-Dab	+	pp, me	7, 99–101, 189
Bacitracin	*Bacillus licheniformis*	Ile-Cys-Leu-D-Glu-Ile-Lys-D-Orn-Ile-Phe └Asn-D-Asp-His	+	p, 3 me	29, 151
Depsipeptides					
Enniatin	*Fusarium oxysporum*	c(NMe-Val-D-Hiv)$_3$	+	p, me	200–202
Beauvericin	*Beauveria bassiana*	c(NMe-Phe-D-Hiv)$_3$	+	pp, me	144, 146

[a]Aad, aminoadipic acid; Abu, aminobutyric acid; Ac, acetyl-; Aib, aminobutyric acid; Argal, argininal; c, cyclo-; Dab, diaminobutyric acid; Dahaa, 2,6-diamino-7-hydroxy acid; Dhb, dihydroxybenzoic acid; Dpr, diaminopropionic acid; f, formyl-; Hiv, hydroxyisovaleric acid; Ise, isoserine; Oct, octanoyl; MeBmt, (4*R*)-4-[(*E*)-2-butenyl]-4-*N*-dimethyl-L-threonine; NMe, *N*-methyl-; Pheol, phenylalaninol; Sar, Sarcosin.

[b]pp, partially purified; p, purified enzyme fractions; me, multienzymes, preceded by the number of multienzymes if known.

Although since then plasmids, cryptic membrane-bound plasmids, and defective phages have been detected in bacterial strains (93, 140, 141), these cannot be related exclusively to the production of peptides. Instead, the chromosomal location of genes has been established for gramicidin S, tyrocidine, and bacitracin, and at least in part for actinomycin (53). Some evidence favors a clustered type of organization of genes of multifunctional enzymes in *Bacillus* (M. Marahiel, M. Krause, G. Czekay, A. Mittenhuber, personal communication). The application of standard molecular genetic procedures in the analysis of mutants is expected to aid in the isolation of useful mutants for the future analysis of these enzyme systems.

FUTURE PROSPECTS

Peptides are derived by the controlled assembly of activated amino acids. Assembly is controlled either by the ribosomal system directed by mRNA, or by complex enzymic systems. The multienzyme thiotemplate system has been the dominant enzymic system studied so far. In this type of assembly multifunctional proteins form templates that activate the constituent amino acids and direct their polymerization from thiolester intermediates with the aid of a pantetheine carrier. We hope that this sequential mechanism will be understood if structures of these multienzymes become available at the molecular level. Recent advances in the cloning of multienzyme structures may eventually lead to an understanding of the genetic organization of such nonribosomal templates, and may permit their exploitation in peptide synthesis. Equally fascinating are the diverse types of modification reactions found either integrated in multienzymes or on separate enzyme sites. Future studies of peptide-modifying enzymes will be of considerable use in the explanation of structure-function relationships, and will find many applications. Finally, considering the functional aspects of peptides within their producer organisms, molecular genetic investigation of the regulatory controls of production should be of considerable help in defining their role in nature.

Acknowledgments

We wish to express our thanks to all our colleagues who have provided published and unpublished materials, and to those who have helped through numerous discussions to shape this overview. Because of space limitations we have not attempted complete coverage or an exhaustive bibliography. Work in the authors' laboratory was mainly supported by grants of the Deutsche Forschungsgemeinschaft (Soderforschungsbereich 9) and the Bundesministerium für Forschung und Technologie.

Literature Cited

1. Akashi, K., Kurahshi, K. 1977. Formylation of enzyme-bound valine and stepwise elongation of intermediate peptides of gramicidin A by a cell-free enzyme system. *Biochem. Biophys. Res. Commun.* 77:259–67
2. Akashi, K., Kurahshi, K. 1978. Enzyme-bound formylvaline and formylvalylglycine, an initiation complex for gramicidin A biosynthesis. *J. Biochem. Tokyo* 83:1219–29
3. Allgaier, H., Jung, G., Werner, R. G., Schneider, U., Zähner, H. 1985. Elucidation of the structure of epidermin, a ribosomally synthesized tetracyclic heterodetic polypeptide antibiotic. *Angew. Chem.* 24:1051–53
4. Allgaier, H., Jung, G., Werner, R. G., Schneider, U., Zähner, H. 1986. Epidermin: sequencing of a heterodetic tetracyclic 21-peptide amide antibiotic. *Eur. J. Biochem.* 160:9–22
5. Altmann, M., von Döhren, H., El-Samaraie, A., Kittelberger, R., Por, M. S., Kleinkauf, H. 1982. Limited proteolysis: studies on the multienzyme GS2 of gramicidin S-synthetase. See Ref. 89, pp. 243–52
6. Babasaki, K., Takao, T., Shimonishi, Y., Kurahashi, K. 1985. Subtilosin A, a new antibiotic peptide produced by *Bacillus subtilis* 168: isolation, structural analysis, and biogenesis. *J. Biochem. Tokyo* 98:585–603
7. Balakrishnan, R., Kaur, S., Goel, A. K., Padmavathi, S., Jayaraman, K. 1980. Biosynthesis of polymyxin by *Bacillus polymyxa*. II. On the nature and interaction of the multienzyme complex with the end product polymyxin. *Arch. Biochem. Biophys.* 200:45–54
8. Barna, J. C. J., Williams, D. H. 1984. The structure and mode of action of glycopeptide antibiotics of the vancomycin group. *Ann. Rev. Microbiol.* 38:339–57
9. Bauer, K., Hallermayer, K., Salnikow, J., Kleinkauf, H., Hamprecht, B. 1982. Biosynthesis of carnosine and related peptides by glial cells in primary cultures. *J. Biol. Chem.* 257:3593–97
10. Bauer, K., Jungblut, P., Kleinkauf, H. 1982. Biosynthesis of carnosine and related peptides. See Ref. 89, pp. 337–46
11. Bauer, K., Roskoski, R. Jr., Kleinkauf, H., Lipmann, F. 1972. Synthesis of a linear gramicidin by a combination of biosynthetic and organic methods. *Biochemistry* 11:3266–71
12. Bérdy, J. 1980. *Handbook of Antibiotic Compounds*. Vol. IV, Pt. 1, 2. Boca Raton, Fla: CRC. 558 pp., 356 pp.
13. Boente, M. I. P., Kirby, G. W., Robins, D. J. 1981. Cyclo-(L-[^{15}N]Phe-L-[1-^{13}C]Phe) is incorporated intact into bisdethiobis-(methylthio)acetylaranotin by *Aspergillus terreus*. *J. Chem. Soc. Chem. Commun.* 1981:619–21
14. Bothe, D. 1986. *Immunologische Untersuchungen an multifunktionellen Enzymen der Peptid- und Depsipeptid-Biosynthese aus* Bacillus brevis *und* Fusarium oxysporum. PhD thesis. Tech. Univ. Berlin. 91 pp.
15. Bothe, D., von Döhren, H., Zschiedrich, H., El-Samaraie, A., Krause, M., Kleinkauf, H. 1982. Further characterization of multienzyme fragments of gramicidin S-synthetase obtained from gramicidin S nonproducer mutants. See Ref. 89, pp. 233–41
16. Campos, L., Marlier, M., Dardenne, G., Casimir, J. 1983. γ-Glutamylpeptides from *Philadelphus coronarius*. *Phytochemistry* 22:2507–8
17. Cardillo, R., Fuganti, C., Ghiringhelli, D., Grasselli, P. 1975. Stereochemical course of the α,β-desaturation of L-tryptophan in the biosynthesis of cryoechinuline A in *Aspergillus amstelodami*. *J. Chem. Soc. Chem. Commun.* 1975:778–79
18. Closse, A., Huguenin, R. 1974. Isolierung und Strukturaufklärung von Chlamydomycin. *Helv. Chim. Acta* 57:533–45
19. Cornish, A., Fox, K. R., Waring, M. J. 1983. Preparation and DNA-binding properties of substituted triostin antibiotics. *Antimicrob. Agents Chemother.* 23:221–31
20. Culvenor, C. C. J., Cockrum, P. A., Edgar, J. A., Frahn, J. L., Gorst-Allman, C. P., et al. 1983. Structure elucidation of phomopsin A, a novel cyclic hexapeptide mycotoxin, produced by *Phomopsis leptostromiformis*. *J. Chem. Soc. Chem. Commun.* 1983: 1259–62
21. Daloze, D., Braekman, J. C., Pasteels, J. M. 1986. A toxic dipeptide from the defense glands of the Colorado beetle. *Science* 233:221–23
22. Davey, G. P., Pearce, L. E. 1982. Production of diplococcin by *Streptomyces cremoris* and its transfer to nonproducing group streptococci. In *Microbiology 1982*, ed. D. Schlessinger, pp. 221–24. Washington DC: Am. Soc. Microbiol.

23. DeLorenzo, V., Bindereif, A., Paw, B. H., Neilands, J. B. 1986. Aerobactin biosynthesis and transport genes of plasmid ColV-K 30 in *Escherichia coli* K-12. *J. Bacteriol.* 165:570–78
24. Dutt, A., Dowhan, W. 1985. Purification and characterization of a membrane-associated phosphatidylserine synthase from *Bacillus licheniformis*. *Biochemistry* 24:1073–79
25. Edo, K., Ishida, N. 1982. Chemistry of neocarzinostatin (NCS) and its biological activity. *Kagaku To Seibutsu* 20:213–15
26. Engelbrecht, F., Braun, V. 1986. Inhibition of microbial growth by interference with siderophore biosynthesis. Oxidation of primary amino groups in aerobactin synthesis by *Escherichia coli*. *FEMS Microbiol. Lett.* 33:223–27
27. Ford, S., Cooper, R. A., Williams, P. H. 1986. Biochemical genetics of aerobactin biosynthesis in *Escherichia coli*. *FEMS Microbiol. Lett.* 36:281–85
28. Freist, W., Sternbach, H., Cramer, F. 1981. Survey of substrate specificity with regard to ATP analogs of aminoacyl-tRNA synthetases from *E. coli* and from bakers' yeast. *Hoppe-Seyler's Z. Physiol. Chem.* 362:1247–54
29. Frøyshov, Ø. 1977. The production of bacitracin synthetase by *Bacillus licheniformis* ATCC 10716. *FEBS Lett.* 81:315–18
30. Frøyshov, Ø., Zimmer, T.-L., Laland, S. G. 1978. Biosynthesis of microbial peptides by the thiotemplate mechanism. In *International Review of Biochemistry, Amino Acid and Protein Biosynthesis II*, Vol. 18, ed. H. R. V. Arnstein, pp. 49–78. Baltimore: Univ. Park Press
31. Fujii, A. 1978. Biogenetic aspects of bleomycin-phleomycin group antibiotics. In *Bleomycin*, ed. S. M. Hecht, pp. 75–91. New York: Springer
32. Fukagawa, Y., Okabe, M., Azuma, S., Kojima, I., Ishikura, T., Kubo, K. 1984. Studies on the biosynthesis of carbapenem antibiotics. I. Biosynthetic significance of the OA-6129 group of carbapenem compounds as the direct precursors for PS-5, epithienamycin A, and C and MM 17880. *J. Antibiot.* 37:1388–93
33. Gadow, A., Vater, J., Schlumbohm, W., Palacz, Z., Salnikow, J., Kleinkauf, H. 1983. Gramicidin S synthetase. Stability of reactive thioester intermediates and formation of 3-amino-2-piperidone. *Eur. J. Biochem.* 132:229–34
34. Gasson, M. J. 1984. Transfer of sucrose fermenting ability, nisin resistance, and nisin production into *Streptococcus lactis* 712. *FEMS Microbiol. Lett.* 21:7–10
35. Gauvreau, D., Waring, M. J. 1984. Directed biosynthesis of novel derivatives of echinomycin by *Streptomyces echinatus*. I. Effect of exogenous analogs of quinoxaline-2-carboxylic acid on the fermentation. *Can. J. Microbiol.* 30: 439–50
36. Gauvreau, D., Waring, M. J. 1984. Studies on antibiotic biosynthesis by protoplasts and resting cells of *Streptomyces echinatus*. I. The synthesis of echinomycin. *Can. J. Microbiol.* 30: 721–29
37. Gauvreau, D., Waring, M. J. 1984. Directed biosynthesis of novel derivatives of echinomycin. II. Purification and structure elucidation. *Can. J. Microbiol.* 30:730–38
38. Gerlach, M., Schwelle, N., Lerbs, W., Luckner, M. 1985. Enzymatic synthesis of cyclopeptine intermediates in *Penicillium cyclopium*. *Phytochemistry* 24: 1935–39
39. Gerlo, E., Freist, W., Charlier, J. 1982. Arginyl-tRNA synthetase from *Escherichia coli* K-12: Specificity with regard to ATP analogs and their magnesium complexes. *Hoppe-Seyler's Z. Physiol. Chem.* 363:365–73
40. Gerth, K., Irschik, H., Reichenbach, H., Trowitzsch, W. 1980. Myxothazol, an antibiotic from *Myxococcus fulvus* (Myxobacteriales). I. Cultivation, isolation, physicochemical and biological properties. *J. Antibiot.* 33:1474–79
41. Ghosh, S. K., Mukhopadyay, N. K., Majumder, S., Bose, S. K. 1983. Fractionation of the mycobacillin synthesizing enzyme system. *Biochem. J.* 215:539–43
42. Ghosh, S. K., Mukhopadyay, N. K., Majumder, S., Bose, S. K. 1985. Functional characterization of constituent enzyme fractions of mycobacillin synthetase. *Biochem. J.* 230:785–89
43. Gibson, B. W., Herlihy, W. C., Samy, T. S. A., Hahm, K.-S., Maedo, H., et al. 1984. A revised primary structure for neocarzinostatin based on fast atom bombardment and gas chromatographic mass spectrometry. *J. Biol. Chem.* 259:10801–6
44. Gonzalez, C. F., Kunka, B. S. 1985. Transfer of sucrose fermenting ability and nisin production phenotype among lactic streptococci. *Appl. Environ. Microbiol.* 49:627–33
45. Greenwood, K. T., Luke, R. K. J. 1980. Studies on the enzymatic synthesis of enterochelin in *Escherichia coli*

K-12, *Salmonella typhimurium* and *Klebsiella pneumoniae*. Physical association of enterochelin synthetase components in vitro. *Biochim. Biophys. Acta* 614:185–95

46. Grill, E., Winnacker, E.-L., Zenk, M. H. 1986. Phytochelatins: The principal heavy-metal complexing peptides in higher plants. *Science* 230:674–75
47. Grill, E., Winnacker, E.-L., Zenk, M. H. 1986. Synthesis of seven different homologous phytochelatins in metal-exposed *Schizosaccharomyces pombe* cells. *FEBS Lett.* 197:115–20
48. Gross, E. 1978. Peptide antibiotics. In *Antibiotics—Isolation, Separation and Purification,* ed. M. J. Weinstein, pp. 415–62. Amsterdam: Elsevier
49. Gross, R., Engelbrecht, F., Braun, V. 1986. Identification of the genes and their polypeptide products responsible for aerobactin synthesis by pColV plasmid. *Mol. Gen. Genet.* 201:204–12
50. Gushima, H., Miya, T., Murata, K., Kimura, A. 1983. γ-Glutamylcysteine production by *Escherichia coli* cells dosed with the γ-glutamylcysteine synthetase gene. *Agric. Biol. Chem.* 47: 1927–28
51. Gushima, H., Yasuda, S., Soeda, E., Yokoto, M., Kondo, M., Kimura, A. 1984. Complete nucleotide sequence of the *E. coli* glutathione synthetase gsh-II. *Nucleic Acids Res.* 12:9299–305
52. Haavik, H. I., Frøyshov, Ø. 1975. Function of peptide antibiotics in producer organisms. *Nature* 254:79–82
53. Haese, A. 1987. *Genetik der Actinomycin C-Biosynthese in* Streptomyces chrysomallus. PhD thesis. Tech. Univ. Berlin
54. Hayashi, Y., Nakagawa, C. W., Murasugi, A. 1986. Unique properties of Cd-binding induced in fission yeast *Schizosaccharomyces pombe*. *Environ. Health Perspect.* 65:13–19
55. Hook, D. J., Vining, L. C. 1973. Biosynthesis of the peptide antibiotic etamycin. Origin of the 3-hydroxypicolinyl and amino acid fractions. *J. Chem. Soc. Chem. Commun.* 1973:185–86
56. Hook, D. J., Vining, L. C. 1973. Biosynthetic precursors of etamycin, a peptidolactone antibiotic from *Streptomyces griseus*. *Can. J. Biochem.* 51: 1630–37
57. Hori, K., Kanda, M., Kurotsu, T., Miura, S., Yamada, Y., Saito, Y. 1981. Absence of pantothenic acid in gramicidin S synthetase 2 obtained from some mutants of *Bacillus brevis*. *J. Biochem. Tokyo* 90:439–47
58. Hori, K., Kanda, M., Miura, S., Yamada, Y., Saito, Y. 1983. Transfer of D-phenylalanine from gramicidin S synthetase 1 to gramicidin S synthetase 2 in gramicidin S synthesis. *J. Biochem. Tokyo* 93:177–88
59. Hori, K., Kurotsu, T., Kanda, M., Miura, S., Yamada, Y., Saito, Y. 1982. Evidence for a single multifunctional polypeptide chain of gramicidin S synthetase obtained from a wild strain and mutant strains of *Bacillus brevis*. *J. Biochem. Tokyo* 91:369–79
60. Hosono, K., Suzuki, H. 1983. Acylpeptides, the inhibitors of cyclic adenosine-3',5'-monophosphate phosphodiesterase. III. Inhibition of cyclic AMP phosphodiesterase. *J. Antibiot.* 36: 679–83
61. Hummel, W., Diekmann, H. 1981. Preliminary characterization of ferrichrome synthetase from *Aspergillus quadricinctus*. *Biochim. Biophys. Acta* 617: 313–20
62. Itokawa, H., Takeya, K., Mori, N., Hamanaka, T., Sonobe, T. 1984. Isolation and antitumor activity of cyclic hexapeptides isolated from *Rubiae radix*. *Chem. Pharm. Bull.* 32:284–90
63. Johns, N., Kirby, G. W. 1985. The biosynthetic incorporation of [phenyl-^{3}H]phenylalanine into gliotoxin. *J. Chem. Soc. Perkin Trans. 1* 1985:1487–90
64. Jones, G. H. 1985. Regulation of phenoxazinone synthase expression in *Streptomyces antibioticus*. *J. Bacteriol.* 163:1215–21
65. Jones, G. H., Hopwood, D. A. 1984. Molecular cloning and expression of the phenoxazinone synthase gene from *Streptomyces antibioticus*. *J. Biol. Chem.* 259:14151–57
66. Jones, G. H., Hopwood, D. A. 1984. Activation of phenoxazinone synthase expression in *Streptomyces lividans* by cloned DNA sequences from *Streptomyces antibioticus*. *J. Biol. Chem.* 259: 14158–64
67. Kambe, M., Imae, Y., Kurahshi, K. 1974. Biochemical studies on gramicidin S non-producing mutants of *Bacillus brevis* ATCC 9999. *J. Biochem. Tokyo* 75:481–93
68. Kanda, M., Hori, K., Kurotsu, T., Miura, S., Nozoe, A., Saito, Y. 1978. Studies on gramicidin S-synthetase. Purification and properties of the light enzyme obtained from some mutants of *Bacillus brevis*. *J. Biochem. Tokyo* 84:435–41
69. Kano, Y., Natori, S. 1984. Sarcophagine (β-alanyl-L-tyrosine) synthesis in

the fat body of *Sarcophaga peregrina* larvae. *J. Biochem.* 95:1041–46

70. Kasai, T., Larsen, P. O. 1980. Chemistry and biochemistry of γ-glutamyl derivatives from plants including mushrooms (basidiomycetes). *Fortschr. Chem. Org. Naturst.* 39:173–286
71. Kasai, T., Shiroshita, Y., Sakamura, S. 1986. γ-Glutamyl peptides of *Vigna radiata* seeds. *Phytochemistry* 25:679–82
72. Kasche, V. 1986. Mechanism and yields in enzyme catalysed equilibrium and kinetically controlled synthesis of β-lactam antibiotics, peptides and other condensation products. *Enzyme Microb. Technol.* 8:2–14
73. Katz, E., Demain, A. L. 1977. The peptide antibiotics of *Bacillus:* biogenesis and possible functions. *Bacteriol. Rev.* 41:449–74
74. Katz, E., Kamal, F., Mason, K. 1979. Biosynthesis of trans-4-hydroxyl-L-proline by *Streptomyces griseoviridus*. *J. Biol. Chem.* 254:6684–90
75. Kawai, M., Rich, D. H., Walton, J. D. 1983. The structure and conformation of HC-toxin. *Biochem. Biophys. Res. Commun.* 111:398–403
76. Keller, U. 1986. Ergot peptide alkaloid synthesis in *Claviceps purpurea*. See Ref. 93, pp. 157–72
77. Keller, U. 1986. High purification and characterization of D-lysergic acid activating enzyme from *Claviceps purpurea*. *Biol. Chem. Hoppe-Seyler* 367:159(Suppl.)
78. Keller, U., Kleinkauf, H., Zocher, R. 1984. 4-Methyl-3-hydroxy-anthranilic acid (4-MHA) activating enzyme from actinomycin producing *Streptomyces chrysomallus*. *Biochemistry* 23:1479–84
79. Keller, U., Zocher, R., Krengel, U., Kleinkauf, H. 1984. D-Lysergic acid activating enzyme from *Claviceps purpurea*. *Biochem. J.* 218:857–62
80. Khoklov, A. S., Reshetov, P. D., Chupova, L. A., Cherches, B. Z., Zhigis, L. S., Stoyachenko, I. A. 1976. Chemical studies on actinoxanthin. *J. Antibiot.* 29:1026–34
81. Kido, Y., Harada, T., Yoshida, T., Umemoto, J., Motoki, Y. 1980. B-1008, a new antibiotic of bacterial origin containing a spermidine moiety. Production, isolation, characterization and biological property. *J. Antibiot.* 33:791–95
82. Kim, S. D., Knoche, H. W., Dunkle, L. D., McCrery, D. A., Tomer, K. B. 1985. Structure of an amino acid analog of the host-specific toxin from *Helminthosporium carbonum*. *Tetrahedron Lett.* 26:969–72
83. Kingston, D. G. I., Kolpak, M. X. 1980. Biosynthesis of antibiotics of the virginiamycin family. 1. Biosynthesis of virginiamycin M_1: determination of the labelling pattern by the use of stable isotope techniques. *J. Am. Chem. Soc.* 102:5964–66
84. Kirby, G. W., Lösel, W., Rao, P. S., Robins, D. J., Sefton, M. A., Talekar, R. R. 1983. Biosynthesis and synthesis of 3-benzyl-6-ethyl-3,6-bis(methylthio)piperazine-2,5-dione, an "unnatural" metabolite of *Gliocladium deliquescens*. *J. Chem. Soc. Chem. Commun.* 1983:810–12
85. Kirby, G. W., Narayanaswami, S. 1976. Stereochemical studies on the biosynthesis of α,β-dehydro-amino acid units of mycelianamide, cyclopenin, and cyclopenol. *J. Chem. Soc. Perkin Trans. 1* 1976:1564–67
86. Kleinkauf, H., Koischwitz, H. 1978. Peptide bond formation in non-ribosomal systems. *Prog. Mol. Subcell. Biol.* 6:59–112
87. Kleinkauf, H., Koischwitz, H. 1980. Gramicidin S-synthetase: On the structure of a polyenzyme template in polypeptide biosynthesis. *Mol. Biol. Biochem. Biophys.* 32:205–16
88. Kleinkauf, H., von Döhren, H. 1981. Nucleic acid independent synthesis of peptides. *Curr. Top. Microbiol. Immunol.* 91:129–77
89. Kleinkauf, H., von Döhren, H., eds. 1982. *Peptide Antibiotics—Biosynthesis and Functions*. Berlin: De Gruyter. 479 pp.
90. Kleinkauf, H., von Döhren, H. 1982. A survey of enzymatic biosynthesis of peptide antibiotics. In *Trends in Antibiotic Research,* ed. H. Umezawa, A. L. Demain, T. Hata, C. R. Hutchinson, pp. 220–32. Tokyo: Jpn. Antibiot. Res. Assoc. 268 pp.
91. Kleinkauf, H., von Döhren, H. 1983. Peptides. In *Biochemistry and Genetic Regulation of Commercially Important Antibiotics,* ed. L. C. Vining, pp. 95–145. Reading, Mass: Addison-Wesley
92. Kleinkauf, H., von Döhren, H. 1985. Peptide antibiotics. In *Comprehensive Biotechnology,* ed. M. Moo-Young, 2:95–135. Oxford: Pergamon
93. Kleinkauf, H., von Döhren, H., Dornauer, H., Nesemann, G., eds. 1986. *Regulation of Secondary Metabolite Formation*. Weinheim: Verlag Chemie. 402 pp.
94. Kleinkauf, H., von Döhren, H., Tesche, B., Zocher, R. 1981. Structures of mul-

tifunctional enzyme templates involved in peptide antibiotic biosynthesis. *FEBS Proc. Meet., 14th, Edinburgh* Abstr. Tue-S14-22

95. Koeppe, R. E. II, Paczkowski, J. A., Whaley, W. L. 1985. Gramicidin K, a new linear channel-forming gramicidin from *Bacillus brevis*. *Biochemistry* 24:2822–26
96. Koischwitz, H. 1978. *Zu Struktur und Funktion des Multienzyms GS2 der Gramicidin S-Synthetase*. PhD thesis. Tech. Univ. Berlin. 48 pp.
97. Koischwitz, H. 1979. Zur Struktur der Proteinmatrize von Gramicidin S. *Hoppe-Seyler's Z. Physiol. Chem.* 360:307
98. Koischwitz, H., Kleinkauf, H. 1976. Gramicidin S-synthetase. Preparation of the multienzyme with high specific activity. *Biochim. Biophys. Acta* 429:1041–51
99. Komura, S., Kurahashi, K. 1980. Biosynthesis of polymyxin E by a cell-free system. *J. Biochem. Tokyo* 88:285–88
100. Komura, S., Kurahashi, K. 1980. Biosynthesis of polymyxin E. III. Total synthesis of polymyxin by a cell-free enzyme system. *Biochem. Biophys. Res. Commun.* 95:1145–51
101. Komura, S., Kurahashi, K. 1985. Biosynthesis of polymyxin E by a cell-free enzyme system. IV. Acylation of enzyme-bound L-2,4-diaminobutyric acid. *J. Biochem. Tokyo* 97:1409–17
102. Krause, M. 1985. *Antibiotika-Synthetase-Gene aus* Bacillus brevis: *Klonierung und Expression des Genfragments der Ornithin-aktivierenden Domäne der Gramicidin S-Synthetase 2 in* Escherichia coli. PhD thesis. Tech. Univ. Berlin. 115 pp.
103. Krause, M., Marahiel, M. A., von Döhren, H., Kleinkauf, H. 1985. Molecular cloning of an ornithine activating fragment of the gramicidin S-synthetase 2 gene from *Bacillus brevis* and its expression in *Escherichia coli*. *J. Bacteriol.* 162:1120–25
104. Kubo, K., Ishikura, T., Fukagawa, Y. 1984. Studies on the biosynthesis of carbapenem antibiotics. II. Isolation and functions of a specific acylase involved in depantothenylation of the OA-6129 compounds. *J. Antibiot.* 37:1394–402
105. Kubota, K. 1982. Generation of formic acid and ethanolamine from serine in biosynthesis of linear gramicidin by a cell-free preparation of *Bacillus brevis* (ATCC 8185). *Biochem. Biophys. Res. Commun.* 105:688–97
106. Kurahashi, K. 1974. Biosynthesis of small peptides. *Ann. Rev. Biochem.* 43: 445–59
107. Kurahashi, K. 1981. Biosynthesis of peptide antibiotics. In *Antibiotics*, Vol. IV, *Biosynthesis*, ed. J. W. Corcoran, pp. 325–52. Berlin: Springer
108. Kurylo-Borowska, Z. 1974. Biosynthesis of edeine. I. Fractionation and characterization of enzymes responsible for biosynthesis of edeines A and B. *Biochim. Biophys. Acta* 351:42–56
109. Kurylo-Borowska, Z. 1975. Edeine synthetase. *Methods Enzymol.* 43:559–67
110. LeFevre, J. W., Glass, T. E., Kolpak, M. X., Kingston, D. G. I. 1983. Biosynthesis of antibiotics of the virginiamycin family. 2. Assignment of the ^{13}C-NMR spectra of virginiamycin M_1 and antibiotic A2315A. *J. Nat. Prod.—Lloydia* 46:475–80
111. LeFevre, J. W., Kingston, D. G. I. 1984. Biosynthesis of antibiotics of the virginiamycin family. 4. Biosynthesis of A2315A. *J. Org. Chem.* 49:2588–93
112. Lerbs, W., Luckner, M. 1985. Cyclopeptine synthetase activity in surface cultures of *Penicillium cyclopium*. *J. Basic Microbiol.* 25:387–91
113. Madry, N., Zocher, R., Grodzki, K., Kleinkauf, H. 1984. Selective synthesis of depsipeptides by the immobilized multienzyme enniatin synthetase. *Appl. Microbiol. Biotechnol.* 20:83–86
114. Maes, C. M., Steyn, P. S., Van Rooyen, P. H., Rabie, C. J. 1982. The structure of bipolaramide, a novel dioxopiperazine from *Bipolaris sorokiniana*. *J. Chem. Soc. Chem. Commun.* 1982:350
115. Mahajan, P. B. 1984. Penicillin acylases. An update. *Appl. Biochem. Biotechnol.* 9:537–54
116. Mahmutoğlu, I. 1981. *Untersuchungen zur Biosynthese des Valinomycins*. PhD thesis. Tech. Univ. Berlin. 96 pp.
117. Majumder, S., Ghosh, S. K., Mukhopadhyay, N. K., Bose, S. K. 1985. Accumulation of peptides by mycobacillin-negative mutants of *Bacillus subtilis* B_3. *J. Gen. Microbiol.* 131:119–27
118. Manabe, H. 1985. Occurence of D-alanyl-D-alanine in *Oryza australiensis*. *Agric. Biol. Chem.* 49:1203–4
119. Marahiel, M. A., Danders, W., Krause, M., Kleinkauf, H. 1979. Biological role of gramicidin S in spore functions. Studies on gramicidin S-negative mutants of *Bacillus brevis* ATCC 9999. *Eur. J. Biochem.* 99:49–55
120. Marahiel, M. A., Krause, M., Czekay, G. 1986. Cloning and expression of the tyrocidine synthetase 1 gene. *Biol. Chem. Hoppe-Seyler* 367:158(Suppl.)

121. Marahiel, M. A., Krause, M., Skarpeid, H.-J. 1985. Cloning of the tyrocidine synthetase 1 gene from *Bacillus brevis* and its expression in *Escherichia coli*. *Mol. Gen. Genet.* 201:231–36
122. Marahiel, M. A., von Döhren, H. 1982. A survey of possible functions of peptide antibiotics in the producer organisms. See Ref. 89, pp. 375–80
123. McGuire, J. J., Bertino, J. R. 1981. Enzymatic synthesis and function of folylpolyglutamates. *Mol. Cell. Biochem.* 38:19–48
124. Meister, A., Tate, S. S. 1976. Glutathione and related γ-glutamyl compounds: biosynthesis and utilization. *Ann. Rev. Biochem.* 45:559–604
125. Miyashiro, S., Kida, T., Shibai, H., Shiio, T., Udaka, S. 1983. Screening and some properties of new macromolecular peptide antibiotics. *J. Antibiot.* 36:1136–43
126. Miyashiro, S., Kida, T., Shibai, H., Shiio, T., Udaka, S. 1984. The fermentation, isolation and characterization of macromolecular peptide antibiotics AN-7A, -7B and -7D. *J. Antibiot.* 37:20–26
127. Miyashiro, S., Kida, T., Shibai, H., Shiio, T., Udaka, S. 1984. The fermentation, isolation and characterization of a macromolecular peptide antibiotic: AN-1. *J. Antibiot.* 37:27–32
128. Miyashiro, S., Udaka, S. 1982. Screening and some properties of new macromolecular peptide antibiotics. *J. Antibiot.* 35:1319–25
129. Mohr, H. 1977. *Biosynthese von Alamethicin*. PhD thesis. Tech. Univ. Berlin. 104 pp.
130. Mohr, H., Kleinkauf, H. 1978. Alamethicin biosynthesis. Acetylation of the alamethicin terminus and attachment of phenylalaninol. *Biochim. Biophys. Acta* 526:375–86
131. Morimoto, K., Shimada, N., Naganawa, H., Takita, T., Umezawa, H. 1981. the structure of antrimycin. *J. Antibiot.* 34:1615–18
132. Morishima, H., Sawa, T., Takita, T., Aoyagi, T., Takeuchi, T., Umezawa, H. 1974. Biosynthetic studies on pepstatin. Biosynthesis of (3*S*,4*S*)-4-amino-3-hydroxy-6-methylheptanoic acid moiety. *J. Antibiot.* 27:267–73
133. Mukhopadhyay, N. K., Ghosh, S. K., Majumder, S., Bose, S. K. 1985. Translocation of mycobacillin synthetase in *Bacillus subtilis*. *Biochem J.* 225:639–43
134. Mukhopadhyay, N. K., Majumder, S., Ghosh, S. K., Bose, S. K. 1986. Characterization of three-fraction mycobacillin synthetase. *Biochem. J.* 240: 265–68
135. Murata, K., Kimura, A. 1982. Cloning of a gene responsible for the biosynthesis of glutathione in *Escherichia coli* B. *Appl. Environ. Microbiol.* 44:1444–48
136. Myaskovskaya, S. P., Zharikova, G. G., Silaev, A. B. 1973. Amino acid composition and some of the physicochemical properties of gratisin. *Vestn. Mosk. Univ. Biol. Pochvoved.* 28:123–25
137. Nishio, C., Komura, S., Kurahashi, K. 1983. Peptide antibiotic subtilin is synthesized via precursor proteins. *Biochem. Biophys. Res. Commun.* 116:751–58
138. Noda, Y., Takai, K., Tokuyama, T., Narumiya, S., Ushiro, H., Hayashi, O. 1977. Enzymatic oxidation of acetyltryptophan-amide and tryptophan-containing peptides. Formation of dehydrotryptophan. *J. Biol. Chem.* 252:4413–15
139. Nüesch, J., Heim, J., Treichler, H. J. 1987. Biosynthesis of beta-lactam antibiotics. *Ann. Rev. Microbiol.* 41:
140. Ochi, K. 1982. Protoplast fusion permits high-frequency transfer of a *Streptomyces* determinant which mediates actinomycin synthesis. *J. Bacteriol.* 150: 592–97
141. Ochi, K. 1982. Control of actinomycin biosynthetic pathway in and actinomycin resistance of *Streptomyces* spp. *J. Bacteriol.* 150:598–603
142. Okumura, Y., Okamura, K., Takei, T., Kouno, K., Lein, J., et al. 1979. Controlled biosynthesis of neoviridogriseins, new homologues of viridogriseins. I. Taxonomy and fermentation. *J. Antibiot.* 32:575–83
143. Okumura, Y., Takei, T., Sakamoto, M., Ishikura, T., Fukagawa, Y. 1979. Controlled biosynthesis of neoviridogriseins, new homologues of viridogriseins. II. Production, biological properties and structure of neoviridogrisein II. *J. Antibiot.* 32:584–92
144. Peeters, H. 1987. *Biosynthese von Beauvericin*. PhD thesis. Tech. Univ. Berlin
145. Peeters, H., Zocher, R., Madry, N., Kleinkauf, H. 1983. Incorporation of radioactive precursors into beauvericin produced by *Paecilomyces fumosoroseus*. *Phytochemistry* 22:1719–20
146. Peeters, H., Zocher, R., Madry, N., Oelrichs, P. B., Kleinkauf, H., Kraepelin, G. 1983. Cell-free synthesis of the

depsipeptide beauvericin. *J. Antibiot.* 36:1762–66

147. Perlman, D. 1978. Antibiotics (peptides). In *Encyclopedia of Chemical Technology,* Vol. 2, ed. R. E. Kirk, D. F. Othmer, M. Grayson, D. Eckroth, pp. 991–1036. New York: Wiley. 3rd ed.
148. Peypoux, F., Pommier, M. T., Marion, D., Ptak, M., Das, B. C., Michel, G. 1986. Revised structure of mycosubtilin, a peptidolipid antibiotic from *Bacillus subtilis*. *J. Antibiot.* 39:636–41
149. Rayman, K., Hurst, A. 1984. Nisin: properties, biosynthesis, and fermentation. In *Biotechnology of Industrial Antibiotics,* ed. E. J. Vandamme, pp. 607–28. New York: Dekker
150. Roberts, J., Rosenfeld, H. J. 1977. Isolation, crystallisation, and properties of indolyl-3-alkane-α-hydroxylase. *J. Biol. Chem.* 252:2640–47
151. Roland, I., Frøyshov, Ø., Laland, S. G. 1975. On the presence of pantothenic acid in the three complementary enzymes of bacitracin synthetase. *FEBS Lett.* 60:305–8
152. Sahl, H.-G., Grossgarten, M., Widger, W. R., Cramer, W. A., Brandis, H. 1985. Structural similarities of the staphylococcin-like peptide pep-5 to the peptide antibiotic nisin. *Antimicrob. Agents Chemother.* 27:836–40
153. Saito, Y. 1982. Some characteristics of gramicidin S-synthetase obtained from mutants of *Bacillus brevis* which could not form D-phenylalanyl-L-prolyl-diketopiperazine. See Ref. 89, pp. 195–208
154. Samy, T. S. A., Hahm, K.-S., Modest, E. J., Lampman, G. W., Keutmann, H. T., et al. 1983. Primary structure of macromomycin, an antitumor antibiotic protein. *J. Biol. Chem.* 258:183–91
155. Santikarn, S., Hammond, S. J., Williams, D. H. 1983. Characterization of novel antibiotics of the triostin group by fast atom bombardment mass spectrometry. *J. Antibiot.* 36:362–64
156. Saxholm, H., Zimmer, T.-L., Laland, S. G. 1972. The mechanism of the inhibition of gramicidin S synthesis by D-leucine. *Eur. J. Biochem.* 30:138–44
157. Schmidt, U., Beuttler, T., Lieberknecht, A., Griesser, H. 1983. Aminosäuren und Peptide. XXXXII. Synthese von Chlamydocin und epi-Chlamydocin. *Tetrahedron Lett.* 24:3573–76
158. Schmidt, U., Häusler, J., Öhler, E., Poisel, H. 1979. Dehydroamino acids, α-hydroxy-α-amino acids and α-mercapto-α-amino acids. *Progr. Chem. Org. Nat. Prod.* 37:252–327
159. Shane, B. 1982. *Corynebacterium* species folylpoly-γ-glutamate synthetase. See Ref. 89, pp. 353–68
160. Shiba, T., Wakamiya, T. 1985. Lanthionine peptide. *Kagaku Kyoto* 40:416–17
161. Shiroza, T., Ebisawa, N., Furihata, K., Endō, T., Seto, H., Ōtake, N. 1982. Isolation and structures of new peptide antibiotics, cirratiomycin A and B. *Agric. Biol. Chem.* 46:865–67
162. Shiroza, T., Ebisawa, N., Furihata, K., Endō, T., Seto, H., Ōtake, N. 1982. The structures of cirratiomycin A and B, the new peptide antibiotics. *Agric. Biol. Chem.* 46:1891–98
163. Shiroza, T., Ebisawa, N., Kojima, A., Furihata, K., Shimazu, A., et al. 1982. Taxonomy of producing organism and production, isolation, physico-chemical properties and biological activities of cirratiomycin A and B. *Agric. Biol. Chem.* 46:1885–90
164. Shoji, J. 1978. Recent chemical studies on peptide antibiotics from the genus *Bacillus*. *Adv. Appl. Microbiol.* 24:187–214
165. Shoji, J., Sakazaki, R., Wakisaka, Y., Koizumi, K., Mayama, M., et al. 1976. Studies on antibiotics from the genus *Bacillus*. XIII. Isolation of a new peptide antibiotic, laterasporamine *J. Antibiot.* 29:390–93
166. Siegbahn, N., Mosbach, K., Grodzki, K., Zocher, R., Madry, N., Kleinkauf, H. 1985. Covalent immobilization of the multienzyme enniatin synthetase. *Biotechnol. Lett.* 7:297–302
167. Simon, R. D. 1976. The biosynthesis of multi-L-arginyl-poly(L-aspartic acid) in the filamentous cyanobacterium *Anabaena cylindrica*. *Biochim. Biophys. Acta* 422:407–18
168. Spector, L. B. 1982. *Covalent Catalysis by Enzymes*. New York: Springer. 276 pp.
169. Steele, J. L., McKay, L. L. 1986. Partial characterization of the genetic basis for sucrose metabolism and nisin production in *Streptococcus lactis*. *Appl. Environ. Microbiol.* 51:57–64
170. Stonard, R. J., Andersen, R. J. 1980. Celeneamides A and B, linear peptide alkaloids from the sponge *Cliona celata*. *J. Org. Chem.* 45:3687–91
171. Suzukake, K., Hori, M., Hayashi, H., Umezawa, H. 1982. Biosynthesis of leupeptin. See Ref. 89, pp. 325–36
172. Suzukake, K., Hori, M., Tamemasa, O., Umezawa, H. 1981. Purification

and properties of an enzyme reducing leupeptin acid to leupeptin. *Biochim. Biophys. Acta* 661:175–81
173. Symons, D. C., Hodgson, B. 1982. Isolation and properties of *Bacillus brevis* mutants unable to produce tyrocidine. *J. Bacteriol.* 151:580–90
174. Takai, K., Ushiro, H., Noda, Y., Narumiya, S., Tokuyama, T., Hayaishi, O. 1977. Crystalline hemoprotein from *Pseudomonas* that catalyzes oxidation of side chains of tryptophan and other indole derivatives. *J. Biol. Chem.* 252:2648–56
175. Takayama, S., Isogai, A., Nakata, M., Suzuki, H., Suzuki, A. 1984. Structure of Cyl-1, a novel cyclotetrapeptide from *Cylindrocladium scoparium. Agric. Biol. Chem.* 48:839–42
176. Tamaki, M. 1984. Studies on peptide antibiotic "gratisin." *Bull. Chem. Soc. Jpn.* 57:3210–20
177. Tamaki, M., Ogata, M., Takimoto, M., Sofuku, S., Muramatsu, I. 1985. Studies on peptides related to gratisin and gramicidin S. *Proc. 22nd Symp. Peptide Chem., Osaka,* 1984, pp. 339–42
178. Terabe, S., Konaka, R., Shoji, J. 1979. Separation of polymyxins and octapeptins by high-performance liquid chromatography. *J. Chromatogr.* 173: 313–20
179. Trowitzsch, W., Reifenstahl, G., Wray, V., Gerth, K. 1980. Myxothiazol, an antibiotic from *Myxococcus fulvus* (Myxobacterales). II. Structure elucidation. *J. Antibiot.* 33:1480–90
180. Troy, F. A. II. 1979. The chemistry and biosynthesis of selected bacterial capsular polymers. *Ann. Rev. Microbiol.* 33:519–60
181. Troy, F. A. II. 1982. Chemistry and biosynthesis of the poly(γ-D-glutamyl) capsule in *Bacillus licheniformis*. See Ref. 89, pp. 49–84
182. Tsou, H.-R., Fiala, R. R., Mowery, P. C., Bullock, M. W. 1984. Biosynthesis of the spermidine and guanidino units in the glycocinnamoyl spermidine antibiotic cinodine. *J. Antibiot.* 37:1382–87
183. Uchida, I., Shigematsu, N., Ezaki, M., Hashimoto, M. 1985. Structure of lavendomycin, a new peptide antibiotic. *Chem. Pharm. Bull.* 33:3053–56
184. Udaka, S., Miyashiro, S. 1982. A new test system for screening macromolecular antitumor antibiotics and its application to culture fluids of actinomycetes. *J. Antibiot.* 35:1312–18
185. Umezawa, H. 1983. Studies on microbial products in rising to the challenge of curing cancer. *Proc. R. Soc. Lond. Ser. B* 217:357–76
186. Umezawa, H., Takita, T., Saito, S., Muraoka, Y., Takahashi, K., et al. 1985. New analogs and derivatives of bleomycin. In *Bleomycin Chemotherapy,* ed. B. I. Sikic, M. Rozencweig, S. K. Carter, pp. 289–301. Academic
187. Umezawa, H., Takita, T., Shiba, T., eds. 1978. *Bioactive Peptides Produced by Microorganisms*. Tokyo/New York: Kodansha/Wiley. 275 pp.
188. Vandamme, E. J. 1981. Properties, biogenesis and fermentation of the cyclic decapeptide antibiotic gramicidin S. In *Topics in Enzyme and Fermentation Biotechnology,* ed. A. Wiseman, pp. 187–261. Chichester, England: Horwood
189. Vasantha, N., Balakrishnan, R., Kaur, S., Jayaraman, K. 1980. Biosynthesis of polymyxin by *Bacillus polymyxa*. I. The status of the biosynthetic multienzyme complex during active antibiotic synthesis and sporulation. *Arch. Biochem. Biophys.* 200:40–44
190. Vining, L. C., Wright, J. L. C. 1977. Biosynthesis of oligopeptides. *Biosynthesis* 5:240–305
191. Vitoux, B., Aubry, A., Cong, M. T., Marrand, M. 1986. *N*-methyl peptides. VIII. Conformational perturbations induced by *N*-methylation of model dipeptides. *Int. J. Pept. Protein Res.* 27:617–32
192. Von Döhren, H. 1986. Structural homologies of multienzymes involved in non-ribosomal formation of peptide antibiotics. *Biol. Chem. Hoppe-Seyler* 367: 77(Suppl.)
193. Wakamiya, T., Shimbo, K., Sano, A., Fukase, K., Shiba, T. 1983. Synthetic study on peptide antibiotic nisin. I. The synthesis of ring A. *Bull. Chem. Soc. Jpn.* 56:2044–49
194. Wakamiya, T., Veki, Y., Shiba, T., Kido, Y., Motoki, Y. 1985. The structure of ancovenin, a new peptide inhibitor of angiotensin I converting enzyme. *Tetrahedron Lett.* 26:665–68
195. Wang, D. I. C., Hamilton, B. K. 1977. Kinetics of enzymic synthesis of peptide antibiotics. *Biotechnol. Bioeng.* 19: 1225–32
196. Watanabe, K., Yamano, Y., Murata, K., Kimura, A. 1986. The nucleotide sequence of the gene for γ-glutamylcysteine synthetase of *Escherichia coli. Nucleic Acids Res.* 14:4393–400
197. Williamson, M. P., Gavreau, D., Williams, D. H., Waring, M. J. 1982. Structure and conformation of fourteen antibiotics of the quinoxaline group de-

termined by ^{1}H NMR. *J. Antibiot.* 35:62–66

198. Wood, R. B., Ganem, B. 1982. Virginiamycin M: absolute configuration and synthetic studies. *Tetrahedron Lett.* 23: 707–10
199. Zimmer, T.-L., Frøyshov, Ø., Laland, S. G. 1979. Peptide antibiotics. In *Economic Microbiology,* Vol. III, *Secondary Products of Metabolism,* ed. A. H. Rose, pp. 123–50. New York: Academic
200. Zocher, R. 1986. The methyltransferase function of the multienzyme enniatin synthetase. *Biol. Chem. Hoppe-Seyler* 367:159(Suppl.)
201. Zocher, R., Keller, U., Kleinkauf, H. 1982. Enniatin synthetase, a novel type of multifunctional enzyme catalyzing depsipeptide synthesis in *Fusarium oxysporum. Biochemistry* 21:43–48
202. Zocher, R., Keller, U., Kleinkauf, H. 1983. Mechanism of depsipeptide formation catalyzed by enniatin synthetase. *Biochem. Biophys. Res. Commun.* 110:292–99

Ann. Rev. Microbiol. 1987. 41:291–300

USE OF ORGAN CULTURES IN MICROBIOLOGICAL RESEARCH

Zell A. McGee and Marion L. Woods, Jr.

Center for Infectious Diseases, Diagnostic Microbiology and Immunology, University of Utah School of Medicine, Salt Lake City, Utah 84132

CONTENTS

INTRODUCTION

Any experimental model is only a valid model of the naturally occurring disease when the molecular mechanisms of pathogenicity in the model are the same as those as in the naturally occurring disease. Organ cultures can be extremely useful and appropriate models, especially when the infection in question occurs in humans, when it has an element of species specificity, and when ethical or practical considerations preclude using humans as the experimental model. The assumption is that the tissues from the host species can be maintained in organ culture and that these tissues maintain receptors or other macromolecules that are responsible for the species specificity of the infection. Organ cultures are most useful and are better models when the infectious agent and the pathologic events that occur in the infection do not

0066-4227/87/1001-0291$02.00

depend on host defenses or serum factors, as may be true of infections in which damage is done by host macrophages or polymorphonuclear leukocytes as they elaborate tissue-destructive enzymes into the environment (e.g. cavity formation by *Mycobacterium tuberculosis*). Thus, organ cultures are ideal for investigation of the various components of the pathologic process when microorganisms produce damage by using attachment to mucosal cells and elaboration of toxic moieties. Once the molecular mechanisms whereby microorganisms cause disease are elucidated, efforts can be made through immunologic or other means to intervene in what appear to be critical steps in the pathologic process, in an effort to protect the host from disease. In experimental animals it may be impossible to determine the microbial molecules responsible for the ability of the microbe to cause disease if serum factors and products elaborated by degenerating professional phagocytes contribute to tissue damage; such contributions confuse the question of which elements of the damage are caused by the microorganism and which are caused by the host response to the infection. In organ cultures there are no systemic humoral or cellular responses to infection to confuse the picture. The next sections deal with specific types of organ cultures and the use to which they have been put for studies of various microbe-host interactions, generally involving microbes that are pathogenic for humans.

NASOPHARYNGEAL ORGAN CULTURES

Stephens et al (25) have reported the characteristics of human nasopharyngeal tissue established and maintained in organ culture. Nasopharyngeal epithelium is comprised primarily of ciliated and nonciliated cells [almost identical to the histological characteristics of the human fallopian tube (12)]. It is possible by monitoring the degree of ciliary activity and ciliary vigor to assess damage to human nasopharyngeal or fallopian tube mucosa in organ culture (12, 13, 27). Similarly, by infecting the organ cultures and removing pieces of tissue at varying intervals thereafter, one can get a sense of the temporal sequence of events in the infectious process. Also, the relative specificity of piliated meningococci for cells from the nasopharynx or the posterior pharynx (26) is accomodated by the model, and apparently specific attachment of meningococci to nonciliated cells in preference to ciliated cells is observed. McGee & Stephens (14) also observed a striking interaction of the nonciliated cells with the meningococci: Microvili of the nonciliated cells crept up over the sides of the meningococci and entrapped the organisms, pulling them against the surface of the nonciliated cells, which then invaginated, pulling the meningococci inside phagocytic vacuoles. There appeared to be either multiplication of meningococci inside the phagocytic vacuoles or coalescence of multiple vacuoles, because large vacuoles containing many meningococci were seen. At a subsequent time meningococci were observed under the epithelium; they may have been transported from the apex to the base of the nonciliated cells and exocytosed there, although there is no direct evidence that this occurred.

In humans, colonization of the nasopharyngeal mucosa by *Haemophilus influenzae* appears to precede invasive, possibly fatal disease. There are no firm data on how *H. influenzae* crosses the nasopharyngeal epithelium to enter the blood stream, although the organisms do appear inside the nasopharyngeal cells in experimental animals when *H. influenzae* is sprayed into their noses (23). Farley et al (5) have studied the various phases of the interaction of human nasopharyngeal mucosa in organ culture with two strains of *H. influenzae,* one having and one lacking the ability to make IgA1 protease. Both strains decreased ciliary function and caused sloughing of ciliated cells. *H. influenzae* cells were observed inside phagocytic vacuoles in nonciliated cells in the nasopharyngeal mucosal epithelium prior to appearing in the subepithelial tissues. The striking similarities between the interaction of *Neisseria gonorrhoeae* with human fallopian tube mucosa and the interaction of *Neisseria meningitidis* and *H. influenzae* with human nasopharyngeal mucosa suggest convergent evolution of these microorganisms as they accommodated to life on human mucosal surfaces.

TRACHEAL ORGAN CULTURES

The pathogenic mechanisms of a variety of agents that are responsible for acute or chronic respiratory disease in humans have been studied to good advantage in organ cultures. Denny (4) has described the interaction of *H. influenzae* with tracheal organ cultures and the resultant sloughing of ciliated cells from the mucosa. He has postulated that this sloughing was caused by the elaboration by *H. influenzae* of lipopolysaccharide (endotoxin) into the organ culture medium with subsequent damage to the ciliated cells. This type of damage to ciliated cells, with sloughing and consequent loss of a toilet mechanism, has been studied in detail in human fallopian tube organ cultures infected with *N. gonorrhoeae* (as described below). It is not clear whether the lipopolysaccharide damages the ciliated cells directly in these systems or whether it acts indirectly by inducing the nonciliated cells to produce a bioactive compound that in turn has activity against the ciliated cells.

Influenza virus is also capable of finding receptors on cells of tracheal organ cultures and multiplying within cells of the tracheal epithelium (22).

Organ cultures have also been used to study the interaction of *Bordetella pertussis* with tracheal epithelium. Collier and his colleagues (2) showed that organisms of *B. pertussis* attached specifically to the cilia of ciliated cells in the tracheal epithelium and that these cells sloughed from the surface. The cells appeared to have suffered damage to their outer membranes. There will be further opportunities to use organ cultures in efforts to determine the molecular mechanisms of pathogenicity of *B. pertussis* more completely.

Studies of the mechanisms of pathogenicity of *Mycoplasma pneumoniae* (15) have been greatly aided by the use of tracheal organ cultures; the localization of the organisms on the surface of ciliated cells in organ culture completely parallels the findings in expectorated sputum of patients with *M. pneumoniae* pneumonia. In this disease the organisms appear to attach to target ciliated cells by means of a polar attachment organelle on each organism. Thus, the distribution of attachment of the organism, and therefore the distribution of the disease process, is determined by the distribution of the receptor for *M. pneumoniae,* which has been determined, by use of tracheal organ cultures, to be neuraminic acid. This substance is distributed primarily in the trachea and bronchi. The distribution of the receptor and thus the disease process is probably responsible for the characteristic substernal localization of pain that is associated with coughing in patients infected with *M. pneumoniae*. Although some of the molecular mechanisms of pathogenicity of *M. pneumoniae* have been elucidated by the use of organ cultures, organ cultures do not completely reproduce the disease, because part of the pathological process is a peribronchial infiltration with round cells that do not appear if experimental animals have been neonatally thymectomized.

Ramphal & Pyle (20) have used tracheal organ cultures to provide evidence that mucin and sialic acid serve as receptors for attachment of *Pseudomonas aeruginosa* in the lower respiratory tract. This attachment is probably an antecedent step to pseudomonas pneumonia.

OVIDUCT OR FALLOPIAN TUBE ORGAN CULTURES

Stalheim & Gallagher (24) have shown that *Ureaplasma urealyticum* isolated from the human genital tract or the bovine genital tract stops ciliary activity of oviduct epithelium in organ culture and causes severe sloughing of ciliated cells. The authors were able to reproduce the effects of the infection by adding ammonia to the organ cultures in concentrations as low as 0.001 M or by adding jack-bean urease. Thus, it would appear that ureaplasmas may induce damage in human tissues by producing ammonia as a result of their elaboration of urease. The same investigators also examined the effect of *Trichomonas fetus* and *Campylobacter fetus* on the ciliary activity of bovine oviduct organ cultures. They found that both of these organisms stopped ciliary activity by inducing loss of cilia from ciliated cells. Since sterile filtrates of *C. fetus* cultures had no effect, there was no compelling reason to believe that elaboration of a toxin or other product led to toxic damage to the ciliated cells.

Hutchinson et al (9) studied the growth and effect of chlamydiae in human and bovine oviduct organ cultures. *Chlamydia trachomatis* appeared to multiply in the organ cultures. The yield was increased by centrifuging the organisms onto the tissues. However, recovery of chlamydiae after the sixth day was rare, which suggests a self-limiting infection. However, no evidence of extensive epithelial cell damage was observed, and there was no loss of ciliary activity. The authors concluded that oviduct damage in acute chlamydial salpingitis may be immunologically mediated.

In studies of the interaction of *N. gonorrhoeae* with human fallopian tube organ cultures, McGee and colleagues (12) found that the various stages in the interaction of the organisms with the mucosa were highly species specific. Human fallopian tubes could not be damaged by commensal *Neisseria* species, even when the organisms were piliated. Organ cultures established with fallopian tube tissue from various animals other than humans were not affected by *N. gonorrhoeae* (10). Thus, the host species specificity of gonococcal infection that is observed in vivo (8) was also observed in the organ culture system (10). Experiments with human volunteers had indicated that piliated gonococci were less pathogenic than nonpiliated gonococci (11). In an experimental model in which loss of ciliary activity of human fallopian tube organ cultures was monitored as an index of damage, McGee and colleagues (13) also found that piliated gonococci were much more virulent than nonpiliated gonococci. The major features of the infection were attachment of gonococci to the microvilli of nonciliated cells (see Figure 1), followed by sloughing of ciliated cells from the epithelium (see Figure 2), followed by apparent invasion of the nonciliated cells by gonococci. [The latter effect actually proved to be phagocytosis and transport of gonococci across the epithelial barrier by the nonciliated cells (13).] This process, called "parasite-directed endocytosis," has features that parallel those described in early pathologic reports of gonococcal infection by Pollock & Harrison (19).

Melly and colleagues (17) found that filter-sterilized fluid from gonococcus-infected organ cultures was capable of reproducing the sloughing of ciliated cells and ciliostasis that was noted in actual gonococcus-infected organ cultures. This suggested that a toxin or toxic factor was involved in the damage. A series of systematic studies by these investigators employing human fallopian tube organ cultures indicated that the toxic factor was heat stable and nondialyzable (17). The most likely candidate to have these characteristics was lipopolysaccharide (LPS). Indeed, LPS (endotoxin) purified from *N. gonorrhoeae* was capable in very small amounts (0.0015 μg ml^{-1}) of damaging human fallopian tube mucosa (7); this damage was similar to that produced by gonococci and was blocked by polymyxin B. The fact that this same preparation of LPS failed to damage the oviducts of a number of animal species (6) suggests that the LPS molecule has at least two functional

Figure 1 Scanning eletron micrograph of normal human fallopian tube mucosa in organ culture. Note ciliated cells (*left* and *top right*) and microvilli on nonciliated cells *(center)*. Magnified about 3500×.

areas. One of these areas is a counterpart of Ehrlich's haptophore; it mediates recognition of a site where the molecule attaches to human cells. This seems most likely to be the carbohydrate portion of the LPS molecule. The experiments in which damage to the fallopian tube organ cultures was blocked by polymyxin B (which binds to and interferes with the action of lipid A) suggest that the lipid A moiety of the LPS is, in Ehrlich's terms, the

Figure 2 Scanning electron micrograph of human fallopian tube mucosa in organ culture, 35 hr after inoculation with a piliated gonococcus. Note damaged, sloughing ciliated cell *(center)* and gonococci attached to microvilli just above and to the right of the sloughing ciliated cell *(arrow)*. Magnified about 2500×.

toxophore portion of the molecule, i.e. the part of the molecule that mediates damage to the receptor cell. Thus, the organ cultures have been of tremendous value in elucidating the molecular mechanisms of pathogenicity used by gonococci in their interaction with human fallopian tube mucosa. The identification of the microbial macromolecules that are critical to the disease process facilitates the identification of candidate molecules for inclusion in a vaccine, because antibody against a molecule that is critical to a step in the disease process may interfere in that step and block the disease process. The organ cultures supply a means of testing antibody against a particular molecule. If an antibody can block the disease process, the antigen used to elicit the antibody may be a good candidate vaccine.

USE OF ORGAN CULTURES FOR IMMUNOLOGIC STUDIES

Information that may be important for developing effective immunization against influenza was reported by Schmidt & Maassab (22). These investigators immunized chickens systemically or via the respiratory route with one of two antigenically distinct influenza A viruses and then attempted infection of tracheal organ cultures with one or the other influenza strain. They found that the tracheal organ cultures from chickens immunized by the intranasal or intratracheal route were more resistant to the viral strain used to immunize the chickens, but were susceptible to the heterotypic strain (i.e. the protection was specific). Five to ten times more virus was required to infect 50% of the tracheal organ cultures from chickens immunized intranasally and intratracheally with active virus than to infect 50% of organ cultures from control, nonimmunized chickens or tracheal organ cultures from chickens with serum antibody induced by intramuscular or intraperitoneal inoculation of inactivated virus. The authors demonstrated that the protection afforded locally immunized mucosa was not due to the presence of serum antibodies in the organ cultures. Thus, some form of protection against influenza, which appears to be highly effective, appears to reside in the tracheal epithelium or the underlying tissues (22). Clearly, more work is needed to establish the nature of this type of protection and predictable means of eliciting immune responses at respiratory and possibly other mucosal surfaces.

Studies with the human fallopian tube have also been instructive about the immune system at mucosal surfaces. Cooper et al (3) have reported that gonococci genetically altered so as not to produce IgA protease were nonetheless capable of engaging in the major stages of the pathologic process, i.e. attachment, damage, invasion, and apparent survival within mucosal cells. Thus, these functions were not reliant upon production of IgA protease. A notable finding in these studies was that the organ cultures infected with gonococci elaborated substantially more IgA than uninfected organ cultures. This suggests that some molecules from gonococci are transported across the epithelium and are capable of stimulating mucosal production of IgA, release of IgA, or both. This means of promoting production of IgA could be extremely useful in certain clinical situations, and the organ cultures should prove useful in further studies of precisely how this response of a mucosal surface might be elicited.

Recently, McQueen et al (16) used intestinal mucosal organ cultures established after intestinal infection with an enteroadherent *Escherichia coli* to assess the immune response of the intestinal mucosa. They concluded that a mucosal response to a natural enteropathogen could be detected and serially followed by this technique. Such use of organ cultures has the potential to provide new information critical for designing vaccines that can elicit an effective immune response at selected mucosal surfaces.

USE OF ORGAN CULTURES IN ADDRESSING SPECIAL PROBLEMS OF INFECTIOUS DISEASES

Organ Cultures of Skin

It is possible that microbial agents that grow in native or maturing skin (e.g. papilloma viruses, herpesviruses, and certain fungi) could be grown and studied in organ cultures that contain keratinocytes in various states of maturation (1).

Organ cultures of blood vessels could prove extremely useful in studying how microorganisms in the blood transgress endothelial barriers to enter the spinal fluid to cause meningitis (18, 21).

ACKNOWLEDGMENTS

This review and much of the work reviewed was supported by Research Grant AI-20265 from the National Institute of Allergy and Infectious Diseases, US Department of Health and Human Services. The authors are grateful to Dr. Edward N. Robinson, Jr. for his helpful suggestions, especially in the early phases of preparation of the manuscript, and to Dr. Harry Ennis, Dr. Howard G. McQuarrie, and Dr. A. Hamer Reiser III for their help in obtaining human fallopian tubes. ZAM is especially grateful to Ms. Jan Aprin for her help in making possible his participation in this project. The authors are also grateful to Dr. Carrolee Barlow and Dr. Loren Hoffman for their help with the photomicrographs. We also thank Ms. Karen Lauridsen for her excellent help in preparation of the manuscript.

Literature Cited

1. Asselineau, D., Bernhard, B., Bailly, C., Darmon, M. 1985. Epidermal morphogenesis and induction of the 67 kD keratin polypeptide by culture of human keratinocytes at the liquid-air interface. *Exp. Cell Res.* 159:536–39
2. Collier, A. M., Peterson, L. P., Baseman, J. B. 1977. Pathogenesis of infection with *Bordetella pertussis* in hamster tracheal organ culture. *J. Infect. Dis.* 136:S196–203
3. Cooper, M. D., McGee, Z. A., Mulks, M. H., Koomey, J. M., Hindman, T. L. 1984. Studies on attachment and invasion of human fallopian tube mucosa by a gonococcal IgA1 protease deficient mutant and its wild type parent. *J. Infect. Dis.* 150:737–44
4. Denny, F. W. 1974. Effect of a toxin produced by *Haemophilus influenzae* on ciliated respiratory epithelium. *J. Infect. Dis.* 129:93–100
5. Farley, M. M., Stephens, D. S., Mulks, M. H., Cooper, M. D., Bricker, J. V., et al. 1986. Pathogenesis of IgA1 protease-producing and -nonproducing *Haemophilus influenzae* in human nasopharyngeal organ cultures. *J. Infect. Dis.* 154:752–59
6. Gregg, C. R., Johnson, A. P., Taylor-Robinson, D., Melly, M. A., McGee, Z. A. 1981. Host species-specific damage to oviduct mucosa by *Neisseria gonorrhoeae* lipopolysaccharide. *Infect. Immun.* 34:1056–58
7. Gregg, C. R., Melly, M. A., Heller-

qvist, C. G., Coniglio, J. G., McGee, Z. A. 1981. Toxic activity of purified lipopolysaccharide of *Neisseria gonorrhoeae* for human fallopian tube mucosa. *J. Infect. Dis.* 143:432–39

8. Hill, J. H. 1944. Experimental infection with *Neisseria gonorrhoeae*. II. Animal inoculations. *Am. J. Syph. Gonorrhea Vener. Dis.* 28:334–78, 471–510
9. Hutchinson, G. R., Taylor-Robinson, D., Dourmashkin, R. R. 1979. Growth and effect of chlamydiae in human and bovine oviduct organ cultures. *Br. J. Vener. Dis.* 55:194–202
10. Johnson, A. P., Taylor-Robinson, D., McGee, Z. A. 1977. Species specificity of attachment and damage to oviduct mucosa by *Neisseria gonorrhoeae*. *Infect. Immun.* 18:833–39
11. Kellogg, D. S. Jr., Peacock, W. L. Jr., Deacon, W. E., Brown, L., Pirkle, C. I. 1963. *Neisseria gonorrhoeae*. I. Virulence genetically linked to clonal variation. *J. Bacteriol.* 85:1274–79
12. McGee, Z. A., Johnson, A. P., Taylor-Robinson, D. 1976. Human fallopian tubes in organ culture: preparation, maintenance, and quantitation of damage by pathogenic microorganisms. *Infect. Immun.* 13:608–18
13. McGee, Z. A., Johnson, A. P., Taylor-Robinson, D. 1981. Pathogenic mechanisms of *Neisseria gonorrhoeae:* observations on damage to human fallopian tubes in organ culture by gonococci of colony type 1 or type 4. *J. Infect. Dis.* 143:413–22
14. McGee, Z. A., Stephens, D. S. 1984. Common pathways of invasion of mucosal barriers by *Neisseria gonorrhoeae* and *Neisseria meningitidis*. *Surv. Synth. Pathol. Res.* 3:1–10
15. McGee, Z. A., Taylor-Robinson, D. 1986. Mycoplasmas. In *Infectious Diseases and Medical Microbiology,* ed. A. I. Braude, pp. 455–60. Philadelphia: Saunders
16. McQueen, C., Dinari, G., Boedeker, E. 1986. Serial measurement of IgA response to an *E. coli* infection (RDEC-1) using organ culture of rectal biopsies. *Gastroenterology* 90:1547
17. Melly, M. A., Gregg, C. R., McGee, Z. A. 1981. Studies of toxicity of *Neisseria gonorrhoeae* for human fallopian tube mucosa. *J. Infect. Dis.* 143:423–31
18. Ogawa, S. K., Yurberg, E. R., Hatcher, V. B., Levitt, M. A., Lowry, F. D. 1985. Bacterial adherence to human endothelial cells in vitro. *Infect. Immun.* 50:218–24
19. Pollock, C. E., Harrison, L. W. 1912. *Gonococcal Infections*. London: Oxford Univ. Press
20. Ramphal, R., Pyle, M. 1983. Evidence for mucins and sialic acid as receptors for *Pseudomonas aeruginosa* in the lower respiratory tract. *Infect. Immun.* 41:339–44
21. Scheld, W. M. 1985. Pathogenesis and pathophysiology of pneumococcal meningitis. In *Bacterial Meningitis,* ed. M. A. Sande, A. L. Smith, R. K. Root, p. 50. New York: Churchill Livingstone
22. Schmidt, R. C., Maassab, H. F. 1974. Local immunity to influenza virus in chicken tracheal organ cultures. *J. Infect. Dis.* 129:637–42
23. Smith, A. L., Roberts, M. C., Haas, J. E., Stull, T. L., Mendelman, P. M. 1985. Mechanisms of *Haemophilus influenzae* type b meningitis. See Ref. 21, pp. 11–21
24. Stalheim, O. H. V., Gallagher, J. E. 1977. Ureaplasmal epithelial lesions related to ammonia. *Infect. Immun.* 15: 995–96
25. Stephens, D. S., Hoffman, L. H., McGee, Z. A. 1983. Interaction of *Neisseria meningitidis* with human nasopharyngeal mucosa: attachment and entry into columnar epithelial cells. *J. Infect. Dis.* 148:369–76
26. Stephens, D. S., McGee, Z. A. 1981. Attachment of *Neisseria meningitidis* to human mucosal surfaces: influence of pili and type of receptor cell. *J. Infect. Dis.* 143:525–32
27. Stephens, D. S., McGee, Z. A., Cooper, M. D. 1987. Cytopathic effects of the pathogenic *Neisseria:* studies using human fallopian tube organ cultures and human nasopharyngeal organ cultures. *Proc. 5th Int. Pathog. Neisseria Conf., Noordwijkerhout, Netherlands, 1986*. In press

Ann. Rev. Microbiol. 1987. 41:301–33

PROKARYOTIC HOPANOIDS AND OTHER POLYTERPENOID STEROL SURROGATES

Guy Ourisson

Centre de Neurochimie, CNRS, 5 rue Blaise Pascal, 67084 Strasbourg, France

Michel Rohmer

Ecole Nationale Supérieure de Chimie, 3 rue Alfred Werner, 68093 Mulhouse, France

Karl Poralla

Institut für Biologie I, Universität Tübingen, An der Morgenstelle, Tübingen, Federal Republic of Germany

---

CONTENTS

0066-4227/87/1001-0301$02.00

INTRODUCTION

Prokaryotic lipids are quite complex, and they have often been discussed. The classes considered have usually been those characterized by long fatty acid chains: either neutral, saponifiable, reserve, or cell-wall lipids, or membrane phospholipids, glycolipids, and the like. In prokaryotes, sterols are usually absent or present in minute amounts, and they have usually been considered insignificant. However, in recent years, new classes of microbial lipids have been discovered. Although they were first thought to be specific to particular microorganisms, subsequent studies have shown that these lipids—hopanoids, carotenoids, archaebacterial phytanyl and bisphytanyl ethers, tricyclopolyprenols, isoarborinol, cycloartenol, and other triterpenes—are in fact more widespread. These novel lipids have the same role in prokaryotic membranes that cholesterol has in eukaryotic ones. We present these lipid classes, show why they can be considered as cholesterol surrogates, and present the hypothesis that they form a phylogenetic series, having led by reasonable evolutionary steps to the sterols of eukaryotes.

These lipids are all derived from the terpene metabolism; their structures therefore bear methyl groups. They are branched and often cyclic, and are thus not easily biodegradable. Therefore they have often been preserved, with only small structural changes, in the organic matter of sediments. Their molecular fossils are often extremely abundant; in the case of hopanoids, the fossils were recognized earlier than the corresponding living lipids. In several cases such fossils have led us to postulate the existence of other still unknown series of microbial lipids. These classes of lipids are biologically essential and extremely widespread, and they are involved in many industrial fermentations. Their most extraordinary character is the fact that they were not found earlier. They probably form the most abundant class of moderately complex organic substances on Earth, if we include their fossils. The fossil hopanoids alone contain at least as much organic carbon as all the living animals, plants, and microorganisms (about 10^{12} t of reduced carbon) (86).

THE ABSENCE OF STEROLS IN PROKARYOTES

Despite repeated reports of the presence of sterols in bacteria and cyanobacteria, it is clear that, except in the two cases reported below, these metabolites are normally not biosynthesized in prokaryotes (2, 13, 38, 105). The amounts reported are usually several orders of magnitude smaller than those found in eukaryotes. At this level, contamination cannot be excluded. For instance, in *Anacystis nidulans* and *Azotobacter chroococcum,* which reportedly contain trace amounts of sterols (99, 112), it has been shown that the stricter the precautions, the lower the sterol concentration found (13). Furthermore, except in two exceptional cases, there is no incontrovertible evidence for de novo sterol biosynthesis in prokaryotes. Incorporation of radioactive precursors has led to very low activities, and the radiochemical purity of the presumed sterols has usually been checked only by chromatography on silica gel. Other authors, however, have presented a more positive view (74 and references therein).

Apparent exceptions are the mycoplasmas and the spirochete *Treponema hyodysenteriae*. These parasitic organisms do incorporate exogenous sterols from their eukaryotic hosts or from the cultural medium but they do not synthesize them (66, 98, 109).

Genuine exceptions are *Methylococcus capsulatus* (7, 8, 15, 104) and *Nannocystis exedens* (57). These organisms contain sterols in amounts comparable to those present in eukaryotes. (This suggests the insignificance of diminutive traces.) These sterols are not the usual mixtures of unspecific phytosterols (which excludes contamination), and they are unambiguously biosynthesized from labeled precursors.

HOPANOIDS

Hopanoids (Figure 1) are derivatives of hopane, which is one of the basic skeletons of pentacyclic triterpenes, although it was not considered important until its discovery in prokaryotes.

Distribution of Hopanoids and Tetrahymanol in Eukaryotes

Hopanoids were initially discovered in eukaryotes, where they can be divided sharply into two distinct families, depending on the presence or absence of an oxygen atom at C-3. The first group derives from the cyclization of squalene epoxide; the second derives from that of squalene itself.

The 3-hydroxy and 3-ketohopanoids have been found occasionally in scattered taxa of higher plants (84, 88). The first of these known, hydroxyhopanone, was isolated from the dammar resin of *Hopea* (Dipterocarpaceae; named in honor of the eighteenth century British botanist John Hope) and is the most easily accessible starting material for hopanoid hemisyntheses. On the other hand, 3-desoxyhopanoids have so far been found only in cryptogams and a few fungi (27, 58, 75, 84, 89). We shall see that they are widespread in prokaryotes; an arresting hypothesis to explain their presence in cryptogams would be a derivation from endosymbiotic cyanobacteria, e.g. from the chloroplasts of the plants (84). This hypothesis has not yet been checked.

Diplopterol (22-hydroxyhopane) is present in small amounts in the protozoon *Tetrahymena pyriformis*, where it accompanies a major isomer, tetrahymanol (a quasi-hopanoid). Tetrahymanol (or diplopterol) has been found so far only in several *Tetrahymena* species (69, 70) and in a fern (122); the

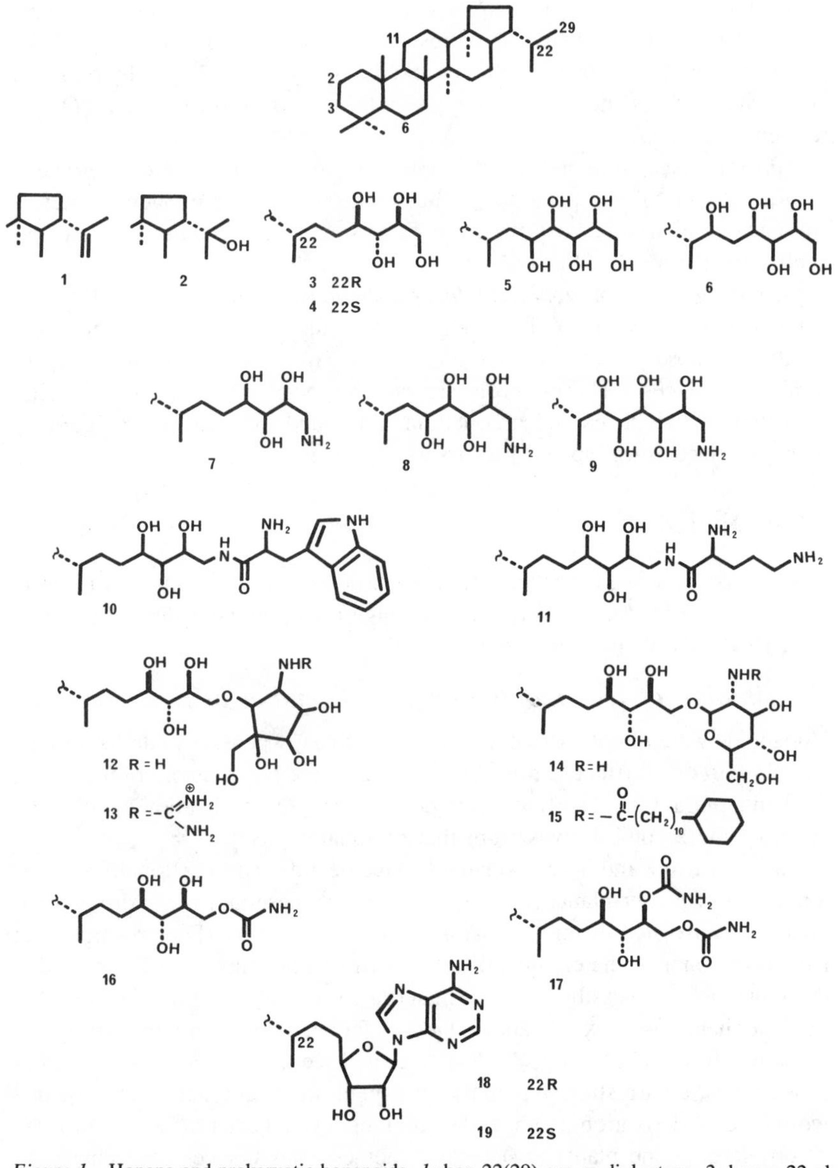

Figure 1 Hopane and prokaryotic hopanoids. *1:* hop-22(29)-ene = diploptene. *2:* hopan-22-ol = diplopterol. *3, 4:* bacteriopanetetrols. *5, 6:* bacteriohopanepentols. *7–9:* bacteriohopaneamino-polyols. *10, 11:* peptidic derivatives of bacteriohopaneaminotriol. *12, 13:* bacteriohopanetetrol ethers of aminocyclopentitols. *14:* glucosamine derivative of bacteriohopanetetrol. *15:* fatty ester of *14*. *16, 17:* carbamoyl derivatives of bacteriohopanetetrol. *18, 19:* adenosylhopanes.

paucity of data on the biochemistry of protozoa and other lower eukaryotes denies any deep significance to this point.

Prokaryotic Hopanoids

C_{30}-HOPANOIDS: DIPLOPTENE AND DIPLOPTEROL The simplest prokaryotic hopanoids are diploptene and diplopterol, which were found first in ferns and then in a few cyanobacteria (34), *M. capsulatus* (8), and *Bacillus acidocaldarius* (28c). Diploptene and diplopterol are in fact present in all hopanoid-containing prokaryotes. In a few cases, minor companions such as hop-21-ene (M. Rohmer, unpublished results) or hop-17(21)-ene (115) have been detected; they may well be isolation artifacts.

MOLECULAR FOSSILS OF BIOHOPANOIDS: GEOHOPANOIDS When organisms die, successive populations of microorganisms process their organic constituents rapidly and nearly quantitatively to carbon dioxide or, anaerobically and in the absence of sulfate, to methane. In sediments, however, huge amounts of organic matter have been accumulated through hundreds of millions of years by a minute leak of the carbon cycle, essentially from those substances most resistant to biodegradation, i.e. insoluble polymers (the so-called kerogens), saturated hydrocarbons, and highly branched polycyclic structures (86). As hopanoids are widespread and abundant in bacteria (see below), we would expect to find their molecular fossils, i.e. substances whose original structural characteristics are preserved enough for us to recognize that the substances are derived from biohopanoids. In fact, these geohopanoids are ubiquitous in sediments, whatever their origin (marine or lacustrine), their age [at least to as old as 1.5×10^9 yr (50)], or their nature (e.g. clay, shale, limestone, evaporite, coal, petroleum). Furthermore, they are abundant (e.g. 50% of the total organic extract of a Lorraine coal) and extremely varied (more than 150 individual structures have been fully identified) (84).

The most complex of these geohopanoids, those that contain up to 35 carbon atoms, were fully identified before any of their bacterial precursors were known (84, 85); thus they are "molecular coelacanths," the molecular analogs of the living fossils of organismic paleontology. The structure of the first bacteriohopane derivative isolated, bacteriohopanetetrol, was in fact established firmly by correlation with some of these previously synthesized molecular fossils (107).

Of course, sedimentary processes are still active today, and C_{30}–C_{35} hopanoids have been found in recent sediments and muds (16, 106).

BACTERIOHOPANE DERIVATIVES: POLYOLS AND AMINOPOLYOLS In typical prokaryotic hopanoids the C_{30} hopane skeleton is linked at C-30 to a

C_5 *n*-alkyl polysubstituted chain, to give C_{35} bacteriohopane derivatives. The most widespread of these derivatives are bacteriohopane-tetrol and -amino-triol.

Bacteriohopanetetrol was first discovered in *Acetobacter aceti* ssp. *xylinum,* where it induces the organization of microfibrils from cellulose micelles (33, 41). Of the two structures proposed, the correct one was selected by a correlation with molecular fossils, which had been correlated with hydroxy-hopanone (107). This structure was independently determined for a lipid isolated from *B. acidocaldarius* (63).

Bacteriohopaneaminotriol was discovered more recently in the Rhodospirillaceae *Rhodomicrobium vannielii* (78) and *Rhodopseudomonas palustris* (S. Neunlist & M. Rohmer, unpublished results) and in methylotrophic bacteria (79, 81).

Structural variations have been found in other bacteria. These involve (*a*) the number and position of the hydroxyl groups in the side chain of the polyols and aminopolyols (10, 33, 79, 81); (*b*) the presence of a supplementary methyl group in ring A, either at C-2β (3, 10) or at C-3β (123); (*c*) the presence of double bonds at C-6, C-11 or both (108); and (*d*) the configuration at C-22 (107) (in most cases only the presumed 22*R*-configuration is encountered).

COMPOSITE BIOHOPANOIDS: PEPTIDES, ETHERS, NUCLEOSIDES In recent years we have carefully studied many strains of prokaryotes. Free polyols have been found only in *Acetobacter aceti* ssp. *xylinum* and *Acetobacter pasteurianus* (M. Rohmer, unpublished results), and free aminopolyols only in methylotrophs (79, 81) and *R. palustris* (S. Neunlist & M. Rohmer, unpublished results). In all other cases, the side chain of the bacteriohopane derivative is linked to polar moieties. For instance, in *B. acidocaldarius,* free bacteriohopanetetrol is accompanied by its glucosamine and *N*-acylglucosamine derivatives (64).

We have recently identified by spectroscopic methods novel composite hopanoids from diverse groups of bacteria. In a composite the bacteriohopane skeleton is linked to moieties that represent some of the major groups of natural products, i.e. amino acids, sugars, and nucleosides. The aminotriol is linked via a peptide bond to tryptophan or ornithine (78), the tetrol is linked by a glycosidic bond to glucosamine derivatives (64), and the tetrol is linked by an ether bond to glucosamine or a novel aminomethylcyclopentitol (100). In addition, the tetrol forms carbamoyl or dicarbamoyl derivatives (S. Neunlist & M. Rohmer, unpublished results), and the hopane skeleton is linked by a C-30/C-5' bond to adenosine (in the 22*R* and the 22*S* series) (80).

The absolute configuration of the latter C-adenosyl-hopane has been established, and the substance has been correlated with bacteriohopanetetrol by two independent routes (S. Neunlist, P. Bisseret & M. Rohmer, unpublished results).

DISTRIBUTION OF HOPANOIDS IN PROKARYOTES Tables 1 and 2 list the bacteria (more than 100 strains) systematically analyzed for the presence of hopanoids (105 and references in the tables). Hopanoids have been found at levels similar to those of sterols in eukaryotes (0.1–2 mg g^{-1} dry weight) in strains widely scattered through numerous taxonomic groups. For instance, they have been found in most cyanobacteria, methylotrophs, Rhodospirillaceae, and acetic acid bacteria, and in various gram-negative and gram-positive bacteria. They have never been found in archaebacteria, purple sulfur bacteria, or enterobacteria. Surprisingly, even though their biosynthesis is independent of molecular oxygen, they have never been found in strict anaerobes. We stress, however, that any of these negative results might later be negated by a change of growth conditions or by an improvement in the analytical procedure followed. There is no foolproof method for the isolation of such highly amphiphilic substances, and no general procedure for the estimation of their amounts. We have usually extracted them with a mixture of chloroform and methanol, and oxidized the crude extract with periodic acid; the polyols thus gave aldehydes, which were reduced with borohydride into primary alcohols, which are easily analyzed (105). This method works for substances that have at least two free vicinal hydroxyl groups that permit cleavage of the side chain; whenever this condition is not met, the method fails. For instance, it would not have permitted the detection of C-adenosylhopane (80), and it does not permit the distinction of bacteriohopanetetrol from aminotriol, as both are cleaved to the same C_{32} alcohol. Isolation, however difficult, is required in these cases.

The frequent presence of hopanoids in prokaryotes is now an accepted fact, and they will probably be found in more strains; we hope that the taxonomic significance of their presence or absence will eventually be understood. This is not yet the case. Recent claims of regularities (115) have been shown by later analyses (105) not to be general.

OTHER PROKARYOTIC POLYTERPENOIDS

All living organisms appear to contain some acyclic polyterpenoids such as polyprenols (dolichols), ubiquinones, or squalene. In the present section the rich polyterpene metabolism of archaebacteria is at the forefront; however, the rate of discovery in this field is such that we must restrict ourselves to a sketchy outline of the presently known facts and a few predictions. We only

Table 1 Prokaryotic strains found to contain hopanoids[a]

Cyanobacteria	Purple nonsulfur bacteria	Methylotrophs	Gram-negative chemoautotrophs	Gram-negative chemoheterotrophs	Gram-positive chemoheterotrophs
Anabaena sp.	*Rhodomicrobium vannielii [7, 10, 11]* (78)	Type I	*Nitrosomonas europea*	*Aerobacter* sp. (11 strains)	*Bacillus acidocaldarius [3, 14, 15]* (63, 64)
Calothrix sp.	*Rhodopseudomonas acidophila* (2 strains) *[12, 16–19]*[b]	"*Methylomonas albus*"		*Azotobacter* sp. (2 strains)	*Corynebacterium* sp. *[3]*
Fisherella sp.	*Rhodopseudomonas palustris [7]*	"*Methylomonas methanica*" *[8, 9]* (80)		*Gluconobacter oxydans*	*Eubacterium limosum*
Lyngbia/Phormidium/Plectonema group (2 strains)	*Rhodopseudomonas sphaeroides*	*Methylomonas* sp.		*Hyphomicrobium* sp.	*Streptomyces* sp. (3 strains)
Nostoc sp. *[7, 9]* (10)	*Rhodospirillum rubrum*	Type II		*Methylobacterium organophilum [3, 12–14]* (100)	
Scytonema sp.		"*Methylocystis parvus*"		"*Pseudomonas cepacia*"	
Synechocystis sp. (2 strains)		"*Methylosinus sporium*"		"*Pseudomonas syringae*"	
		"*Methylosinus trichosporium*" *[7, 8]* (79)		*Zymomonas mobilis [3, 12, 14]*[c]	
		Type III			
		"*Methylococcus capsulatus*" *[7–9]* (81)			

[a] Unless otherwise indicated, all analyses by M. Rohmer and coworkers (105 and unpublished results). Structures, when known, are indicated in brackets and refer to Figure 1. In the other cases, the presence of bacteriohopane derivatives was discovered after degradation (see text).
[b] S. Neunlist & M. Rohmer, unpublished results.
[c] G. Flesch & M. Rohmer, unpublished results.

Table 2 Strains of prokaryotes where no hopanoid could be detected[a]

Cyanobacteria	Purple sulfur and green sulfur bacteria	Gram-negative chemoautotrophs and chemoheterotrophs		Gram-positive bacteria	Archaebacteria
Spirulina sp.	*Chlorobium limicola* (2 strains)	*Agrobacterium tumefaciens*	*Neisseria* sp. (2 strains)	*Actinoplanes brasiliensis*	*Halobacterium cutirubrum*
Synechococcus sp. (2 strains)	*Chromatium vinosum*	*Branhamella catarrhalis*	*Paracoccus denitrificans*	*Arthrobacter globiformis*	*Methanobacterium thermoautotrophicum*
	Ectothiorhodospira sp. (2 strains)	*Brucella abortus*	*Photobacterium phosphoreum*	*Bacillus subtilis*	"*Sarcina littoralis*"
	Thiocapsa roseopersicina	*Brucella melitensis*	*Proteus mirabilis*	*Bacillus stearothermophilus*	*Sulfolobus acidocaldarius*
	Thiocystis violacea	*Caulobacter crescentus*	*Pseudomonas acidovorans*	*Brevibacterium linens*	*Thermoplasma acidophilum*
		Citrobacter freundii	*Pseudomonas aeruginosa*	*Clostridium paraputrificum*	
		Erwinia herbicola	*Pseudomonas chlororaphis*	*Corynebacterium callunae*	
		Escherichia coli	*Pseudomonas diminuta*	*Corynebacterium lilium*	
		Flexithrix sp.	*Pseudomonas fluorescens*	*Corynebacterium michiganense*	
		Klebsiella oxytoca	"*Pseudomonas maltophila*"	*Desulfovibrio desulfuricans*	
		"*Methylomonas clara*"	*Pseudomonas stutzeri*	"*Micrococcus flavus*"	
		"*Methylophilus methylotrophus*"	*Rhizobium lupini*	*Micromonospora* sp.	
		"*Moraxella*" sp. (2 strains)	*Thiobacillus* sp. (2 strains)	*Nocardia coeliaca*	
		Morganella morganii	*Xanthomonas campestris*	*Streptococcus faecalis*	
		Nannocystis exedens		"*Sporosarcina lutea*"	

[a] All analyses performed using the method described in text, with the limitations mentioned (105; G. Flesch & M. Rohmer, unpublished results).

very briefly present bacterial carotenoids, and we present general trends and refer to recent reviews rather than compiling exhaustive inventories.

Squalene and Other Nonpolar Polyprenoid Lipids

Squalene is widely distributed in bacteria (115). In *Staphylococcus, Streptococcus* (115), and streptomycetes (39) it is accompanied by a dehydrosqualene, probably diapophytoene, the precursor of the C_{30} carotenoids of several bacteria (42). Apparently the same dehydrosqualene has been isolated from the halophilic archaebacterium *Halobacterium*. However, archaebacteria are mostly characterized by the presence of dihydro- and polyhydrosqualenes (117), which are rarely if at all present in eubacteria (115) and which rarely culminate in squalane (115, 117). In methanogens at least, squalane is accompanied by other saturated polyterpene hydrocarbons (47, 117), and indirect evidence from marine sediments suggests strongly that many more, indicated by their molecular fossils, will be found in other strains (16). Similarly, several types of quinones (menaquinones, phylloquinones, *Caldariella* quinones) are often found in archaebacteria with partially (or, in a few cases, fully) hydrogenated isoprenoid side chains (116).

Polar Lipids of Archaebacteria

The polar lipids (phospholipids and more complex lipids) of archaebacteria (Figure 2) are characterized by specific structural traits (31, 61, 65). They are not esters of *sn*-1,2-glycerol, but ethers of the antipodal *sn*-2,3-glycerol or of more complex polyols such as the branched nonitol or calditol from "*Caldariella*" (28a). Their long hydrocarbon chains are not the *n*-alkyl chains usually present in eukaryotes nor the minimally branched ones of eubacteria, but C_{20} (phytanyl) or C_{25} regular polyprenoid chains, sometimes partly unsaturated. In groups other than halophiles, these lipids are accompanied by the dimeric C_{40} bis-phytanyl ethers, which are sometimes partly cyclized into strings of cyclopentyl units (23, 28d, 28e, 44, 60). In sediments and petroleums the corresponding hydrocarbons can be found (71), as well as the polar lipids themselves (21). Moreover, other hydrocarbons have been identified that so strikingly resemble known archebacterial lipids that the existence of novel structures in strains not yet analyzed can be predicted with confidence (22, 85). In addition, some isoalkyl glyceryl ethers have been found in nonarchaebacterial thermophilic anaerobes (62).

The presence of isoprenoid ethers appears to be a reliable archaebacterial character. Their absence has been used to confirm the eubacterial nature of a thermophilic organism, *Thermomicrobium roseum,* despite its lack of a peptidoglycan cell wall (another archaebacterial trait, previously considered to be reliable) but in agreement with 16S ribosomal RNA analyses (91).

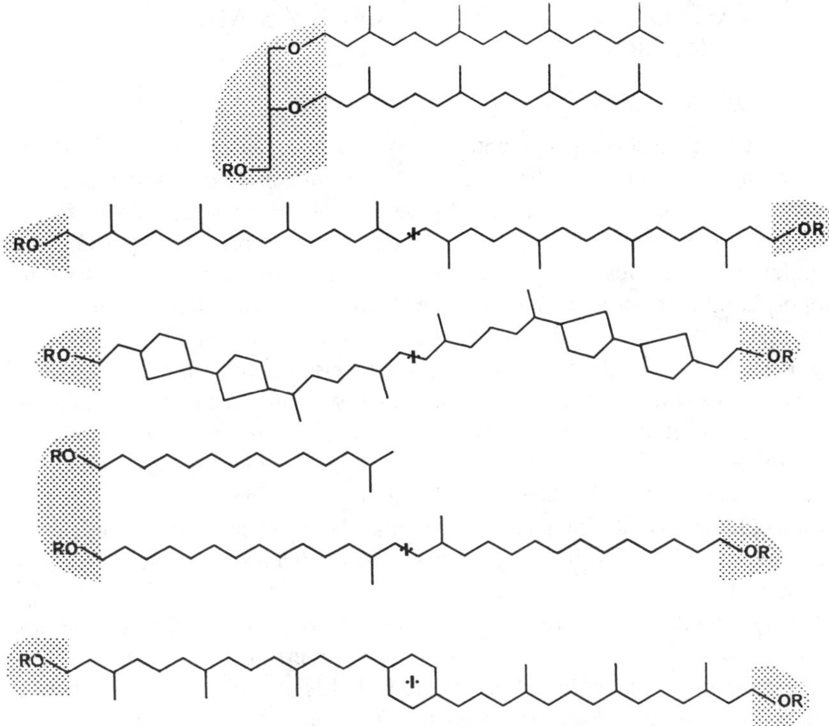

Figure 2 Polar lipids of archaebacteria. *Shaded* portions: polar heads. *Nonshaded* structures: related sedimentary hydrocarbons. The two bottom structures are known only as their molecular fossils; their derivation from polar lipids is assumed by analogy.

Prokaryotic Carotenoids

The structural variety of bacterial carotenoids is impressive; not only do they contain diverse end groups not known elsewhere, but they furthermore break the definition of carotenoids as C_{40} substances, as they comprise also C_{30} or C_{50} analogs. Fortunately, excellent reviews have been written on their structures and biosyntheses (18, 42, 67, 115).

Carotenoids in general, with their extensive chromophore, absorb UV/visible light and have a major role in photoprotection and in energy transfer reactions of photosynthesis. It is also very probable that at least some of them (those that carry highly polar groups at one or both ends) are incorporated in bacterial cell membranes, where they have a mechanical role similar to that of cholesterol in eukaryotes (103). This concept was initially misinterpreted on the basis of wrong structures in which *cis* double bonds were postulated for the carotenoids (74, 76, 110); it has now been substantiated by in vivo data (49, 110) (for in vitro evidence, see the next section).

PROKARYOTIC POLYTERPENOIDS AS MEMBRANE REINFORCERS

Role of Sterols in Phospholipid Bilayers

In water, phospholipids spontaneously form organized systems. If the hydrophobic chains are about 15–20 Å long and the cross sections of the chains and of the hydrophilic head group are of similar size, bilayers can be formed as planar assemblies (e.g. black lipid membranes), liposomes, or unilamellar vesicles. At low temperatures the hydrocarbon chains of the phospholipid are ordered and gel-like. Above a transition temperature, T_m, a second phase, locally disordered like a liquid, appears. T_m is the two-dimensional equivalent of the fusion temperature of a three-dimensional crystal. However, above T_m the long-range structure of the bilayer is maintained (it is often said to be a liquid crystal). The conformational disorder of the hydrocarbon chains in the hydrophobic core becomes distinctly enhanced. As a result of this increased fluidity, passive diffusion of small hydrophilic molecules across the bilayer (in particular diffusion of water molecules) is also increased. At a still higher temperature, the integrity and barrier function of the membrane fade.

Cholesterol and other related sterols are always present in the phospholipid membranes of eukaryotes. Their function is understood on the basis of their effect on pure phospholipid bilayers (11, 28, 121). Addition of cholesterol to phospholipids has no influence on the formation of vesicles. However, as cholesterol concentration increases, the transition between the gel and liquid-crystalline states becomes progressively less sharp, until it is altogether abolished. This effect is partly similar to that when an impurity is added to a solid substance and its sharp melting point becomes a melting range. However, cholesterol in a bilayer is not just an impurity. Below the transition temperature it tends to increase disorder, particularly in the hydrocarbon chains; however, at temperatures above T_m it reduces disorder. Its molecules remain oriented (the 3β-OH group near the aqueous phase); as their tetracyclic ring system is rigid, and as their molecular dimensions (length and cross section) (Figure 3) fit well with those of the hydrophobic part of the phospholipid, cooperative van der Waals forces impart some additional rigidity to the hydrocarbon chains. The system, above T_m, is more ordered than without cholesterol. The molecules are more tightly assembled (condensing effect), so the bilayer is less permeable to small hydrophilic molecules and ions (28).

Sterols also have other, more specific biological effects. In particular, proper sterols (i.e. those devoid of methyl groups at C-4 and C-14) also influence the incorporation of unsaturated fatty acids into phospholipids in *Mycoplasma* (26) and yeast (90, 97). The structural requirements are more strict for discharging this function than for the bulk membrane effect.

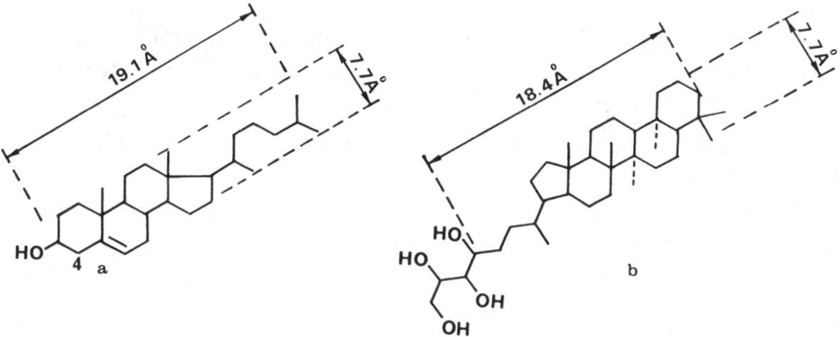

Figure 3 Cholesterol (*a*) and bacteriohopanetetrol (*b*), presented to emphasize their structural analogy. 4-Methylsterols bear a supplementary methyl group at position 4 in *a*.

Functional Equivalence of Structurally Related Polyterpenoids and Sterols

Until a few years ago, sterols were the sole naturally occurring lipids known to stabilize phospholipid bilayers (and by extension biomembranes) by the above condensing effect. This view was subsequently expanded by the discovery of tetrahymanol as a structural equivalent of sterols in the ciliate *Tetrahymena pyriformis* (24, 46), the hypothesis that some triterpenoids played the same role (74, 76), the limitation of this hypothesis to the class of biohopanoids (103), and most directly the demonstration of the condensing effect of two naturally occurring hopanoids (56, 94). The discovery of the tetraether lipids of archaebacteria (60) and the recognition of the mechanical role of the highly polar bacterial carotenoids (the carotenols) (49) have extended the field; the existence of novel polyterpenoids that behave as cholesterol surrogates is now predicted from their fossils (85).

PHYSICOCHEMICAL EVIDENCE Through the use of various physicochemical methods it has been clearly demonstrated that hopanoids influence bilayer properties much as sterols do. The most direct results have been obtained by the study of the mechanical properties of monolayers, which constitute the membrane system best defined physically since that of Langmuir (59). Addition of bacteriohopanetetrol (56) or the derived glycolipid (94), both isolated from the thermoacidophilic bacterium *Bacillus acidocaldarius,* to a phospholipid monolayer led to the expected condensing effect and decrease in the sharpness of the transition. The comparison to cholesterol was extended with phosphatidylcholines containing different fatty acid residues, i.e. straight and branched chains (53, 54) and ω-cyclohexyl fatty acids (51, 52). The condensing effect thus measured is in general larger with cholesterol, but bacteriohopanetetrol was equally efficient when incorporated into *anteiso*-acylphosphatidylcholine monolayers (51). This observation is of particular interest, as at least some bacteria, such as *Bacillus* and *Streptomyces,* are known to contain phospholipids of this type and also hopanoids. However, no wide-range simultaneous studies of phospholipids and hopanoids in the same strains have been made. Furthermore, the largest condensing effect is observed with a combination of *n*-acyl phosphatidylcholines and cholesterol (51).

Differential microcalorimetry is another method appropriate for studying the phase transition by observing its thermal effect. At a molar concentration of 40%, bacteriohopanetetrol completely abolishes the transition (53). Benz et al (6), using black lipid membranes (bilayers of a particular type that permit the evaluation of permeability), showed that both 2-methyl-diplopterol from *Methylobacterium* and the hopane glycolipid improve the barrier properties. The same hopanoids decrease the permeability of liposomes to glycerol (6) and the permeability of unilamellar vesicles to water (9). These results clearly indicate that hopanoids reinforce membranes similarly to cholesterol; the reinforcement is quantitatively of the same order of magnitude, although in most cases it is not as strong.

Bacterial carotenoids, which have been postulated to act as rivets across the bilayers (103), can probably not be investigated by the direct study of monolayers. In a few cases they have been studied in bilayers, as membrane models (9, 49, 72, 73, 120). Most recently, a complete study has been carried out using a naturally occurring bacterial carotenoid, bacterioruberin, and phospholipids of the same origin (T. Lazrak, Y. Nakatani, G. Wolff, M. Kates & G. Ourisson, unpublished results). Previously, only models of bacterial carotenoids had been studied.

Coelution of *n*-acyl phospholipidic vesicles from a Sephadex column showed that they are incorporated into the bilayer, and UV/visible spectroscopy showed that they are oriented across the bilayer (73). The water permeability of these carotenoid-containing vesicles is lower than that of vesicles made up of pure phospholipids (72). ESR studies have shown that the acyl chains are more fluid in the absence of carotenoids (49).

The tetraterpanic tetraether lipids of two strains of archaebacteria have been used for model experiments. The head groups of the tetraether lipids of *Thermoplasma* were glycerol (60), whereas the *Sulfolobus* lipids were characterized by two head groups, one glycerol and one nonitol (28a). However, the experiments with the *Thermoplasma* lipids were carried out on the native fractions (12), whereas those with the *Sulfolobus* lipids were carried out after hydrolysis of the phosphate substituents (37). Thus, direct comparison of the results is not possible.

The results concerning the occurrence of a phase transition in the physiological range of temperature are also conflicting for the two lipids. The glycophospholipid of *Thermoplasma* displayed no phase transition between 0 and 80°C, as indicated by differential microcalorimetry (12); this indicates a liquid-crystalline phase over an unusually wide temperature range. On the other hand, a phase transition was observed at about 40°C in the lipids of *Sulfolobus;* this transition temperature increased progressively with the number of additional cyclopentane rings in the alkyl chain and hence with local rigidity (37).

Small-angle X-ray scattering on the lipids of *Sulfolobus* has shown the occurrence of a variety of phases in the physiological range of temperatures (40, 68). Moreover, the transition between the phases is slower than in conventional monolayer lipids. This is an indication that bipolar lipids tend to stabilize at different phases owing to the limited diffusion of the head groups. Gliozzi et al (36) have compared these results with those obtained by other methods.

Simple molecular considerations suggest that the tetraether lipids probably influence the stability or permeability of membranes composed of their halves, phytanyl ethers, although this has not been directly demonstrated. One can predict that sealing the two halves of a bilayer should make it much more resistant and less permeable.

Liposomes of the tetraether lipid fraction extracted from *Thermoplasma* had much lower permeability for carboxyfluorescein than normal phospholipid liposomes, even at 70°C. These liposomes also proved very resistant against detergents (K. Ring, unpublished observations).

In summary, all the physicochemical studies carried out so far confirm that cholesterol and its polyterpene surrogates are functionally equivalent as membrane strengtheners. These studies should now concentrate on more natural models and on more strict comparisons.

BIOLOGICAL EVIDENCE The first evidence of equivalence between sterols and hopanoids came from early studies of the protozoon *Tetrahymena pyriformis*. When grown in the presence of sterols, this organism incorporates them into its membranes after slight processing (i.e. unsaturation and alkylation). However, when grown in the absence of sterols it biosynthesizes the hopanoid diplopterol and especially its quasi-hopanoid isomer tetrahymanol, and incorporates them into its membranes (24).

For *Mycoplasma,* which is normally cholesterol dependent, the hopanoid diplopterol is an acceptable substitute, but its growth-promoting effect is only about 20% that of cholesterol. The bacteriohopanetetrol glycolipid, however, does not promote growth (55). No attempt was made to test a possible synergistic effect between diplopterol and cholesterol (26). The pentacyclic triterpenoid β-amyrin also supports growth of *Mycoplasma* (82).

The sum of the concentrations of the extended (C_{35}) hopanoids of *B. acidocaldarius,* bacteriohopanetetrol and the derived glycolipids, *N*-acyl-glucosaminyl- and glucosaminyl-bacteriohopanetetrols, can be measured by gas-liquid chromatography after derivatization (93). This total concentration increases from 4 to 16% (w/w) of the total lipids when the temperature is raised from 50 to 65°C. The influence of the pH of the growth medium is weak. The hopanoid content of *Zymomonas mobilis,* which has very different fatty acids (*n*-acyl, saturated and unsaturated, instead of ω-cyclohexyl-acyl) (4), also increases when the growth temperature is raised (111).

Z. mobilis is an anaerobic ethanol-producing bacterium. In the culture fluid the ethanol concentration can reach 14%, which is much higher than most bacteria can tolerate. In continuous cultures, as the ethanol concentration increased, so did the hopanoid content in the total cellular lipids [from 4 to 20% (w/w)](17). It therefore seems that hopanoids have a direct role in stabilizing bacterial membranes that are subjected to potentially lysis-inducing stress.

The experiments described above provide no understanding of the variety of structures found in biohopanoids. Some hopanoids, notably the most elaborate ones, probably have not only a structural role, but in addition a functional one. More information will no doubt accrue as more strains and structures are studied, and as some taxonomic regularities emerge. It will also be very important to investigate the distribution of biohopanoids in the outer and inner halves of the bilayer membranes, as well as their distribution in the different intracytoplasmic membranes of phototrophic, obligate methylotrophic, and nitrifying bacteria.

Early results with *Acholeplasma* membranes have indicated the probable equivalence of carotenols and cholesterol (114) in much the same way as with *Tetrahymena:* Growth of this mycoplasm was normal in the presence of sterols, but was also possible in their absence, when polar carotenoids were produced and localized in the membranes. It was subsequently shown by electron spin resonance that these carotenoids rigidify the membranes of *Acholeplasma* (49). Also, when fed more fluid fatty acids, *Acholeplasma* responded by raising its production of polar carotenoids (110). These experiments should be repeated, in particular to define fully the molecular structures of these carotenols; but they are sufficient to demonstrate that, apart from having a photoprotective role, bacterial carotenoids are physiological membrane reinforcers.

Few physiological experiments have been performed with archaebacteria containing tetraether lipids. Elevation of the growth temperature increased the intrinsic rigidity of the lipids of the thermoacidophilic *Sulfolobus* (Caldariella) by increasing the number of cyclopentane residues along the chain (28d). The biochemistry involved is as unknown as the mechanism by which the two phytanyl chains are linked into the bis-phytanyl skeleton. It would be very interesting to study the relative content of C_{20} and C_{40} lipids at various temperatures in some archaebacteria containing both types; the obvious prediction is that the content of C_{40} lipids will be higher at higher temperatures. (In *Sulfolobus,* the C_{20}:C_{40} ratio is about 1:10, so changes would not be significant.)

The extreme thermophile *Pyrodictium,* which grows at temperatures up to 110°C, and other thermophilic but nonacidophilic archaebacteria produce a mixture of di- and tetraether lipids free of cyclopentane rings (65). It remains to be seen whether the presence of cyclopentane rings is linked to the acidophilic character or to the aerobic character of these bacteria. It should be very rewarding to look for cyclopentane rings in anaerobically grown *Sulfolobus*.

A Structural Rule Characterizing Membrane Reinforcers

It is possible to summarize the results mentioned above by a remarkably simple rule (85): The membrane reinforcers found in prokaryotes are derivatives of the terpene metabolism and are of one of two structural types. They may either be partially rigid amphiphilic molecules of dimensions adapted for one phospholipid layer (about 6 × 6 × 20 Å), or they may carry two distal polar heads and be adapted to a phospholipid bilayer (about 6 × 6 × 40 Å). This rule, however, should not hide the fact that in many bacteria, e.g. *Escherichia coli* and *Bacillus subtilis,* no specific membrane reinforcer has yet been found. It will be interesting to see whether their membranes are stable enough without reinforcers, perhaps owing to some of their intrinsic proteins, or whether reinforcers can be found upon further study.

BIOSYNTHESIS OF PROKARYOTIC POLYTERPENOIDS

The biosynthesis of isoprenoids in prokaryotes has not been studied as thoroughly as in eukaryotes. This is due to the low concentrations of the polyprenols previously known to be ubiquitous in prokaryotes (bactoprenols, isoprenoid quinones) and to the absence (until the discovery of the hopanoids) of sterols and other triterpenoids. It can, however, be safely assumed that the usual pathway of polyterpenoid biosynthesis, established largely through work on eukaryotes, is applicable also to prokaryotes.

Polyprenols and Archaebacterial Lipids

The general biosynthetic pathway accepted for regular polyprenoids linked head-to-tail has been confirmed for the phytanyl and bisphytanyl chains of the lipids of the extremely thermoacidophilic archaebacteria *Sulfolobus ("Caldariella"),* using various ^{13}C-labeled acetates (109). However, nothing is known about the biochemical steps involved in the dimerization of the phytanyl chains or in the formation of the cyclopentane rings along the bis-phytanyl chains. These reactions involve the removal of two nonactivated hydrogen atoms, and may be linked with the active metabolism of molecular hydrogen displayed by archaebacteria in the formation of methane as well as in the partial or complete saturation of geranyl-geraniol to phytanol or of squalene to squalane.

The biosynthesis of prokaryotic carotenoids has been studied rather extensively. In particular, the steps involved in the formation of the C_{30} or C_{50} carotenoids characteristic of bacteria have been elucidated; these steps correspond fully to those in plants. An excellent summary is found in Reference 42.

Hopanoids

Hopanoids, like other isoprenoids, derive from the general polyprenol pathway. However, detailed studies have revealed some surprising results regarding the mode of synthesis of hydroxymethyl-glutaryl CoA (HMGCoA), the precursor of the isopentenyl units via mevalonic acid (MVA), and regarding the weak specificity of the squalene cyclase that leads to the hopane skeleton.

BIOSYNTHESIS OF HMGCoA *Methylobacterium organophilum* fed [1-^{13}C]acetate and [2-^{13}C]acetate produced labeled bacteriohopanetetrol derivatives, and similarly fed *Rhodopseudomonas palustris* produced labeled aminotriol. From the labeling distribution in the pentacyclic hopane skeleton, the following conclusions were drawn for the formation of the C_6 unit of HMGCoA (G. Flesch & M. Rohmer, unpublished results): (*a*) The labeling pattern is not compatible with the usual isoprenoid biosynthetic pathway, as it appears that acetate undergoes several metabolic changes before entering the isoprenoid pathway. (*b*) More precisely, HMGCoA results from the condensation of an acetoacetate derived from two acetates issued from the Krebs cycle, and of one other acetate, most probably derived from glucose catabolism via the Entner-Doudoroff pathway.

The numerous careful studies of isoprenoid biosynthesis in eukaryotes have never revealed such a dual origin of the acetate units involved in HMGCoA biosynthesis. However, studies using radioactive precursors (e.g. 69) would not have detected it, and this is how most of the parsimonious work with prokaryotes has been done. The biosynthesis of the phytanyl chains of the phospholipids has been thoroughly studied with ^{13}C-labeled acetates only in the halophilic archaebacterium *Halobacterium cutirubrum*. In this case also, HMGCoA was found to originate from two pools of acetate; two molecules came from the acetate added in the culture medium, and the third came from L-lysine (30). These unexpected peculiarities may lead to the discovery of new aspects of the regulation of isoprenoid biosynthesis in prokaryotes.

CYCLIZATION OF SQUALENE Unlike most eukaryotic triterpenes, which are derived from squalene epoxide cyclization, hopanoids are formed by direct cyclization of squalene itself. This has been proposed for fern hopanoids (35) and demonstrated for tetrahymanol, the quasi-hopanoid triterpene of *Tetrahymena* (1, 19).

Direct cyclization of squalene into diploptene and/or diplopterol has been observed in cell-free systems prepared from various bacteria (32, 77, 102, 104, 113). The membrane-bound cyclase of *B. acidocaldarius* has been partially purified and its sensitivity to sulfhydryl and histidyl reagents has been shown (77, 113).

Like the eukaryotic squalene epoxide cyclases (32), the prokaryotic squalene-hopanoid cyclases are strongly inhibited by squalene analogs such as 2,3-dihydro-2-azasqualene derivatives (29) or epiminosqualene ($I_{50} < 1$ μM) (25). These substances inhibited the growth of hopanoid-synthesizing bacteria at ~1 μM. Thus, hopanoid biosynthesis is essential for these bacteria (32).

The eukaryotic squalene epoxide cyclases display a rigorous substrate specificity as they leave intact squalene or its (3*R*)-epoxide, but cyclize specifically the (3*S*)-epoxide (5). In contrast, enzyme preparations obtained from *Acetobacter pasteurianus* and *Methylococcus capsulatus* convert squalene into diploptene and diplopterol as in the living cell, but also convert both enantiomers of squalene epoxide (3*R* and 3*S*, which are unnatural substrates for this enzyme) into a mixture of 3α- and 3β-hydroxy derivatives of diploptene and diplopterol (102, 104). In *M. capsulatus* a pathway leading to lanosterol and thence essentially to 4-methyl sterols by cyclization of squalene (3*S*)-epoxide coexists with the hopanoid biosynthesis. During the above-mentioned experiments with cell-free systems from *M. capsulatus* squalene was not transformed into 3-desoxylanostane derivatives, but a mixture of 3β- and 3α-lanosterols was obtained from racemic squalene epoxide; the second cyclase involved is therefore partly substrate specific in this bacterium (it cannot cyclize squalene), but is not as specific as in yeast or rat liver (104).

In *T. pyriformis* a similar weak substrate specificity is observed. Squalene is, as expected, converted into tetrahymanol, but racemic squalene epoxide gives a mixture of two epimeric gamma-ceranediols (14). Bouvier et al (14) have elucidated the stereochemistry of these cyclizations and have shown that the substrate must have had different conformations in the two epoxides. This conformational change has evidently maintained the substrate and the active site of the enzyme in the same relative positions (14). The substrate versatility of the squalene cyclase of *Tetrahymena* is further illustrated by its capacity to cyclize other unnatural substrates such as all-*trans* pentaprenol and hexaprenol derivatives into a series of polycyclic systems (101). Similarly, the cyclase of *B. acidocaldarius* cyclizes all-*trans* farnesol and geraniol derivatives into tricyclic and bicyclic derivatives, respectively (77).

BIOSYNTHESIS OF THE C_5 BACTERIOHOPANE SIDE CHAIN According to the stereochemistry of the chiral centers in the side chain of bacteriohopanetetrol, it seems likely that this C_5 unit could be derived from a ribose derivative, linked through C-5 to the hopane skeleton. This hypothesis is reinforced by the existence of the adenosyl-hopane, which contains such a C-5'–linked ribose unit (G. Flesch & M. Rohmer, unpublished results). The labeling pattern observed in the side chain of the aminotriol of *R. palustris* and the tetrol of *M. organophilum* and *Rhodopseudomonas acidophila* fed [1-^{13}C]acetate and [2-^{13}C]acetate was in agreement with the idea that a pentose derivative behaved as intermediate. Such a pentose should have arisen from the complete nonoxidative pentose phosphate pathway (via transketolase and transaldolase), and not from the oxidative decarboxylation of a hexose. The precise nature of the precursors corresponding to the hopane and pentose moieties and the nature of the complete reaction are being investigated.

RING METHYLATION OF HOPANOIDS Incorporation of L-[*methyl*-^{14}C; ^{3}H]methionine or L-[*methyl*-$^2H_3$]methionine into 2β-methyldiplopterol in *M. organophilum* and into 3β-methyldiplopterol and 3β-methylbacteriohopanepolyols in *A. pasteurianus* has shown that all three protons of the methyl group transferred are retained (124). These methylations may therefore involve as substrate a Δ^2-hopanoid in the first case and a Δ^2-hopanoid or possibly squalene in the case of the 3-methylhopanoids. Substances possessing a cyclopropane ring or an exomethylene group are excluded as intermediates (124).

MOLECULAR EVOLUTION OF MEMBRANE REINFORCERS

We have shown that derivatives of the same metabolic route, polyterpenoids of very diverse structures, have a similar role in reinforcing biomembranes in very diverse organisms. We now show that these substances can be considered as part of an evolutionary sequence, i.e. that they are related to one another by plausible phylogenetic derivations of the enzymes involved.

A few general comments are required as we begin this section. We must first dismiss any misunderstanding that could possibly stem from our frequent reference to results obtained from the study of sediments. The molecular fossils we have mentioned repeatedly cannot be used to trace molecular evolution in the same way that classical fossils are used to trace the evolution of organisms. In the oldest sediments found to contain recognizable geohopanoids or other polyterpenoids (about 1.5 billion years old) (50), the constituents do not differ markedly from those of very young ones such as lake muds (106), marine sediments (16), or hot springs (19). In fact, we postulate that all these molecular fossils come from substances still present in some extant organisms that have not yet been studied from this point of view.

Another important point is that hypotheses of evolution can never be proved; the best we can hope for is to derive schemes compatible with accepted mechanisms, and to derive consequences that can be checked experimentally in model systems. If these consequences prove wrong, the hypothesis must be abandoned; if they prove right, one can entertain the hypothesis further, but it remains a hypothesis.

There have been remarkably few critical discussions of the general concepts applicable to analysis of the evolution of biochemical pathways. Most of them have been restricted to one aspect or to the propagation of one hypothesis. However, Haslam (43) has recently provided an excellent summary with useful references.

General Concepts of Biochemical Evolutionary Theory

The basic processes of biochemical evolution are accepted to be (*a*) random mutation of genes coding for biosynthetic processes and (*b*) natural selection. An important refinement is linked with gene duplication. If a gene coding for an essential function is first duplicated, one of its copies can freely undergo random mutations. Most of these are probably inept, and would be lethal in a single-copy organism. However, with gene duplication the process can continue in a harmless, if useless, way until one of the mutations leads to a positive evolutionary step. Such a step could end with the acquisition of a new function or simply with the replacement of the earlier effector by another, more efficient one.

GENE MUTATIONS AND PRIMITIVITY If this postulated mechanism is applied to enzymatic reactions, most of the observable mutations should be limited to parts of the enzyme not directly linked with the active site, as this is obviously the most sensitive part of the enzymatic machinery. We therefore postulate that enzymes that perform the same reaction on different substrates, even with different substrate selectivities, are related phylogenetically, either by derivation from one another or through a common ancestry.

It is not easy to adapt Horowitz's view of "retrograde evolution" of biosynthetic processes (48) to the case of the polyterpenoids considered here. Instead we consider as "more primitive" the substances that require fewer discrete enzymatic steps and as "most primitive" those that involve substances potentially formed by abiotic processes. For instance, cholesterol must be a more recent acquisition than tetrahymanol, which requires for its biosynthesis at least half a dozen fewer enzymatic reactions. A simple derivative of a polyprenol, which could conceivably be formed from small molecules and an acidic catalyst such as a clay, is considered an early precursor.

We use progressively increased substrate specificity of enzymatic processes as a criterion of primitivity (20), even though the relationship is not obvious (one could also expect progress to be accompanied by a lesser dependence on specific substrates; this is certainly true of many catabolic processes).

Another generally accepted convention is to consider as more primitive the substances that require no molecular oxygen in their biosynthesis. This is based on the accepted assumptions that molecular oxygen is totally or very largely derived in the atmosphere from the activity of photosynthetic organisms, and that these organisms appeared after the more primitive chemolithotrophs (see however 118). This convention has already been used to show the relative modernity of cholesterol (11, 76, 92).

Finally, we propose that enzymatic reactions for which there is good analogy in totally abiotic processes are primitive. For instance, enzymatic hydration of the terminal trisubstituted double bond of a polyprenol to give the corresponding tertiary alcohol, which could be effected (although not very efficiently) simply by catalysis with an inorganic acid (Markovnikof rule), would be more primitive than the hydration in the opposite direction to give a secondary alcohol. This hypothesis does link up with developments on prebiotic evolution, but we only very briefly allude to this extension.

NATURAL SELECTION The concept of evolution through progressive adaptation to environmental pressure can obviously not be applied in a simplistic way. An organism and its environment form a system, and a system does not usually operate optimally when its parts are separately optimized. We therefore do not put much faith in attempts to demonstrate a progressive improvement of one function along a postulated phylogenetic pathway, e.g. membrane reinforcement along the pathway leading to cholesterol (11).

A Plausible Phylogenetic Derivation of Cholesterol

From the information summarized above and from our conjectures (83, 87, 92) and those of Nes et al (75), it is possible to postulate a complete phylogenetic derivation of cholesterol (Figure 4).

HEAD-TO-TAIL CONDENSATIONS: POLYPRENOLS AND SIMPLE DERIVATIVES The basic *iso*-C_5 building block of polyterpenoids requires for its biosynthesis a number of reactions; it is conceivable, however, that simple clay-catalyzed condensation of ethylene and formaldehyde could produce this unit. Anyway, only one enzymatic reaction would be required for the recur-

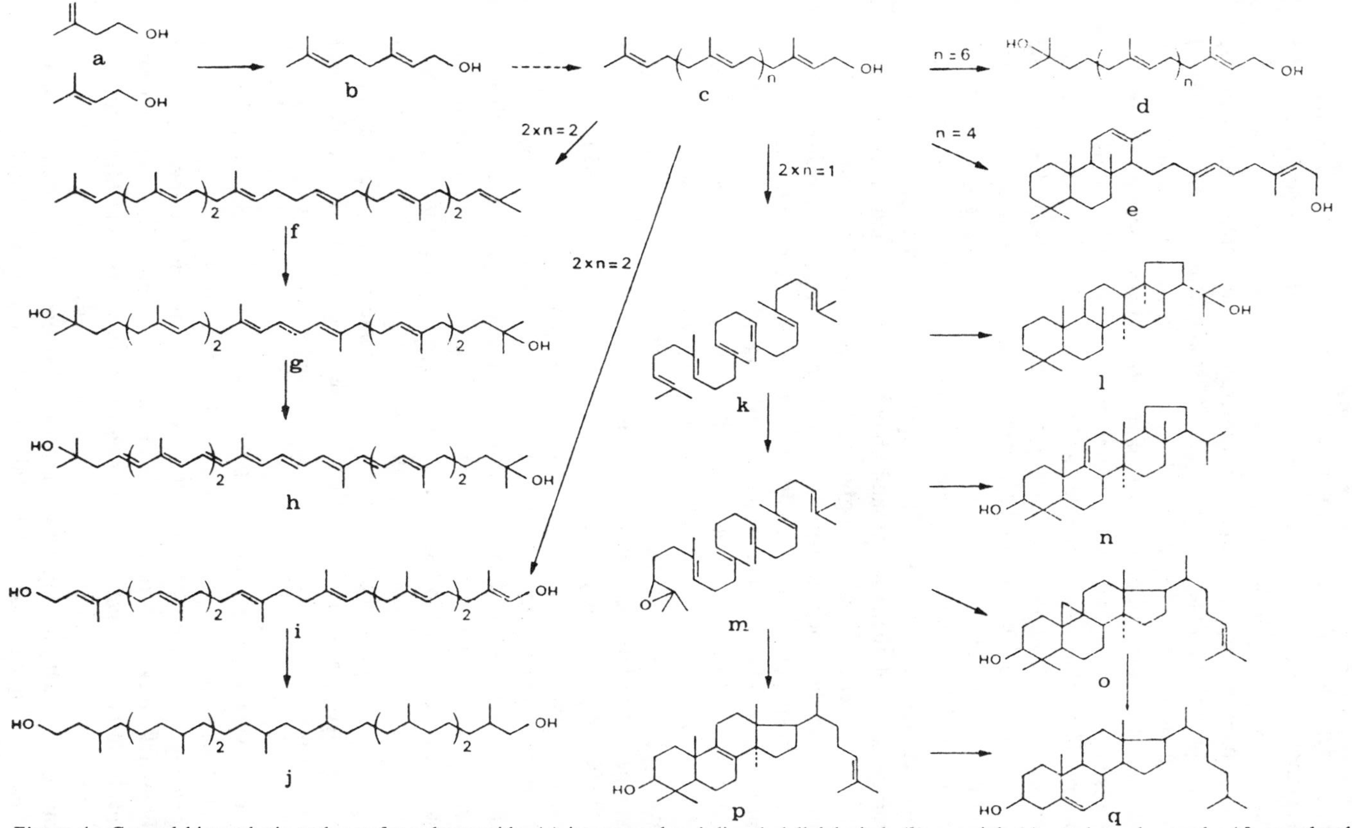

Figure 4 General biosynthetic pathway for polyprenoids: (*a*) isopentenol and dimethylallylalcohol; (*b*) geraniol; (*c*) regular polyprenols; (*d*) postulated hydrated octaprenol; (*e*) postulated tricyclohexaprenol; (*f*) lycopersene; (*g*) postulated hydrated lycopersene or phytoene; (*h*) prototype bacterial carotenoid (spirilloxanthin is the dimethyl ether); (*i*) postulated precursor of bis-phytanol; (*j*) bis-phytanol; (*k*) squalene; (*l*) diplopterol; (*m*) squalene epoxide; (*n*) isoarborinol; (*o*) cycloartenol; (*p*) lanosterol; (*q*) cholesterol.

rent condensation of isopentenol and its isomer dimethylallyl alcohol to polyprenols. Furthermore, this condensation could in principle also be achieved by clay catalysis.

To display the required molecular dimensions and amphiphilic character, the most primitive polyterpenoids capable of strengthening membranes would be derivatives of tetra- or pentaprenol, i.e. the required unsaturated precursors of the C_{20}–C_{25} ethers of archaebacteria. Shorter chains would give micelles, not vesicles, and longer ones would not spontaneously order themselves into bilayers. The polar head of simple alcohols is probably not bulky enough to form membranes; derivatives such as phosphates or glycosides might be formed spontaneously, and might have large enough polar heads. We postulate that these unsaturated chains will be found in archaebacterial lipids.

Such tetra- or pentaprenyl derivatives would be the only types of membrane component accessible from the simple polymerization of isopentenol. Hydrogenation of their unsaturated chains would be required to give, for instance, the saturated phytanyl glycerol ethers of *Halobacterium;* nothing is known about the biochemistry of this addition of molecular hydrogen.

Another simple reaction could give other polyterpenoids capable of reinforcing membranes. The Markovnikof hydration of the terminal double bond of a C_{30}–C_{45} polyprenol would give a bipolar molecule, a dialcohol, which could be oriented across a membrane 30–45 Å thick. These hydrated polyprenols are not known, and their molecular fossils could not be distinguished from those of other polyprenols such as the dolichols. The study of synthetic samples is in progress (B. Chappe, unpublished work), and we predict that these hydroxy polyprenols will be found in nature.

From a chemical point of view, the mechanism leading to Markovnikof hydration could also lead to cyclization. Mono- or polycyclic derivatives of polyprenols would then be formed; those with proper dimensions (about 20 Å long) could be cholesterol surrogates. Indeed, molecular fossils of tricyclic hydrocarbons have often been found in sediments and petroleums, predominantly with up to about 30 carbons but in some cases with as many as 45 (45). We have postulated that they would be found in some widespread microorganism, which could be anaerobic and should be primitive (85).

HEAD-TO-HEAD CONDENSATION: BISPHYTANYL LIPIDS So far, only one condensation has been considered: the head-to-tail condensation of isopentenyl units. New reactions are required to proceed further.

Phytanyl ethers are found alone in halophilic and some methanogenic archaebacteria. The other groups of archaebacteria contain in addition the C_{40} bis-phytanyl dimers formed by head-to-head condensation. Nothing is known about this formation of a new carbon-carbon bond. Even though no intermediate is known, from a chemical point of view this remarkable dimerization would look less unexplainable if it occurred at a stage when the C_{20}

chains still contained double bonds; reactions between saturated chains would be even more unexpected. This argument, however unscientific, links up with our previous postulate of the existence of archaebacterial lipids in their original, unsaturated form; the argument was implicitly accepted in the most recent review of archaebacterial lipid biosynthesis (28b). Again, we predict that the head-to-head dimer of tetraprenol (head-to-head bis-geranyl-geraniol) will be found, either unsubstituted or as a polar derivative of its primary alcohol functions, in some primitive organisms. Such dimers are being synthesized and studied (B. Chappe, unpublished results).

TAIL-TO-TAIL CONDENSATION: SQUALENOIDS AND CAROTENOIDS Head-to-tail and head-to-head dimerizations are not the only ways to condense isoprenoid chains. The tail-to-tail dimerization is the key step in the formation of the C_{30} squalene (tail-to-tail bis-farnesol) and the precursor of the C_{40} carotenoids, prephytoene (tail-to-tail bis-geranyl-geraniol). The chemical steps involved are now well known; they are remarkably complex and practically identical in both series. (Only the last step diverges; squalene results from reduction and prephytoene results from an elimination that produces a double bond.) Even more remarkable is the fact that each enzymatic system carrying out this coupling can accept as substrate, albeit less efficiently, the normal substrate of the other system. There is some substrate selectivity, but it is not very strict with regard to the length of the chains to be linked. It is highly probable that these enzymatic systems, whose structure is still unknown, are related by a minimal mutation that preserves most of the active site.

The biosynthesis of carotenoids involves further steps to dehydrogenate the chain fully. We have postulated that their nonconjugated precursors will be found in some primitive bacterium. Again, the precursors are accessible by synthesis and can be tested (B. Chappe, unpublished results).

ANAEROBIC CYCLIZATIONS: CYCLIC CAROTENOIDS, TRICYCLOPOLYPRENOLS, HOPANOIDS Squalene and the acyclic precarotenoids can undergo the acid-catalyzed, enzyme-mediated reactions described above, to form simple hydrated or cyclic derivatives. The hydrated forms are not known, but they are the precursors of the known acyclic bacterial C_{30} and C_{40} carotenoids. We have postulated that these precursors are membrane "rivet" strengtheners. Monocyclic or distal bicyclic derivatives are the precursors of the more usual carotenoids. However, both the hydration and the cyclization can be initiated, in bacteria, not only by attack of a proton, but by attack of another an isoprenyl cation. This leads to the C_{45} or C_{50} carotenoids, which are known in many forms and which are structurally capable of strengthening membranes (as we have shown with model systems) (10, 72, 73).

Cyclization of squalene to diplopterol is an acid-catalyzed, enzyme-mediated hydration. It is in principle rigorously similar to the cyclization of polyprenols to tricyclopolyprenols postulated earlier. As we have shown above, squalene cyclases of *Tetrahymena pyriformis* (101) and *Bacillus acidocaldarius* (77) can accept regular polyprenols as substrates. However, the cyclases converting squalene to diploptene/diplopterol must be more evolved, because an unfavorable aspect of the cyclization of the third to fifth rings must be surmounted: intermediary secondary instead of tertiary ions. This type of cyclization is termed "anti-Markovnikof."

AEROBIC CYCLIZATIONS: ISOARBORINOL, CYCLOARTENOL, LANOSTEROL The formation of squalene epoxide from squalene requires molecular oxygen. It must therefore be a relatively modern reaction. We have shown that the squalene cyclases of several microorganisms can cyclize both enantiomeric forms of this epoxide. The hopane-diols produced, however, are unfit for any strengthening role in membranes, and squalene epoxide can produce suitable reinforcers only if the cyclization is followed by elimination of a proton and not hydration. Several triterpenes can result. The least modified one, 3β-hydroxyhop-22(29)-ene, has not yet been found in microorganisms or in sediments. However, its isomer, isoarborinol, which is already known in a few higher plants, has been found repeatedly in sediments; we have postulated that it might be a constituent of the membranes of one or more aerobic bacteria (85).

Other isomers with more modified cyclic systems are well known. The tetracyclic isomer of isoarborinol, lanosterol, is the usual precursor of sterols in yeast and other eukaryotes and also of 4-methyl sterols in *Methylococcus capsulatus* (8, 15). In this last organism, which appears to be, from this point of view, a transition with eukaryotes, we have seen that the epoxysqualene-lanosterol cyclase is less substrate selective than that of eukaryotes, as it accepts both (3*R*)- and (3*S*)-epoxides (104); it is therefore probably more primitive. However, its immediate product, lanosterol, is not adapted to membrane reinforcement, probably because of the 14-methyl group protruding on its α-face (11); lanosterol must still be degraded further by a series of catabolic steps.

The immediate product of the cyclization of squalene epoxide in plants, cycloartenol, displays several characteristics that suggest that it is more primitive: It does strengthen membranes, it can replace cholesterol in mycoplasms in vivo (11), it is the precursor of cholesterol in some amebae (95), and finally it is present in other amebae in amounts such that it is probably not only a precursor of sterols, but also a membrane strengthener in its own right (96).

A COMPLETE PHYLOGENY OF STEROLS The sequence of the cyclized derivatives of squalene would then be (Figure 5): tricyclohexaprenol (?),

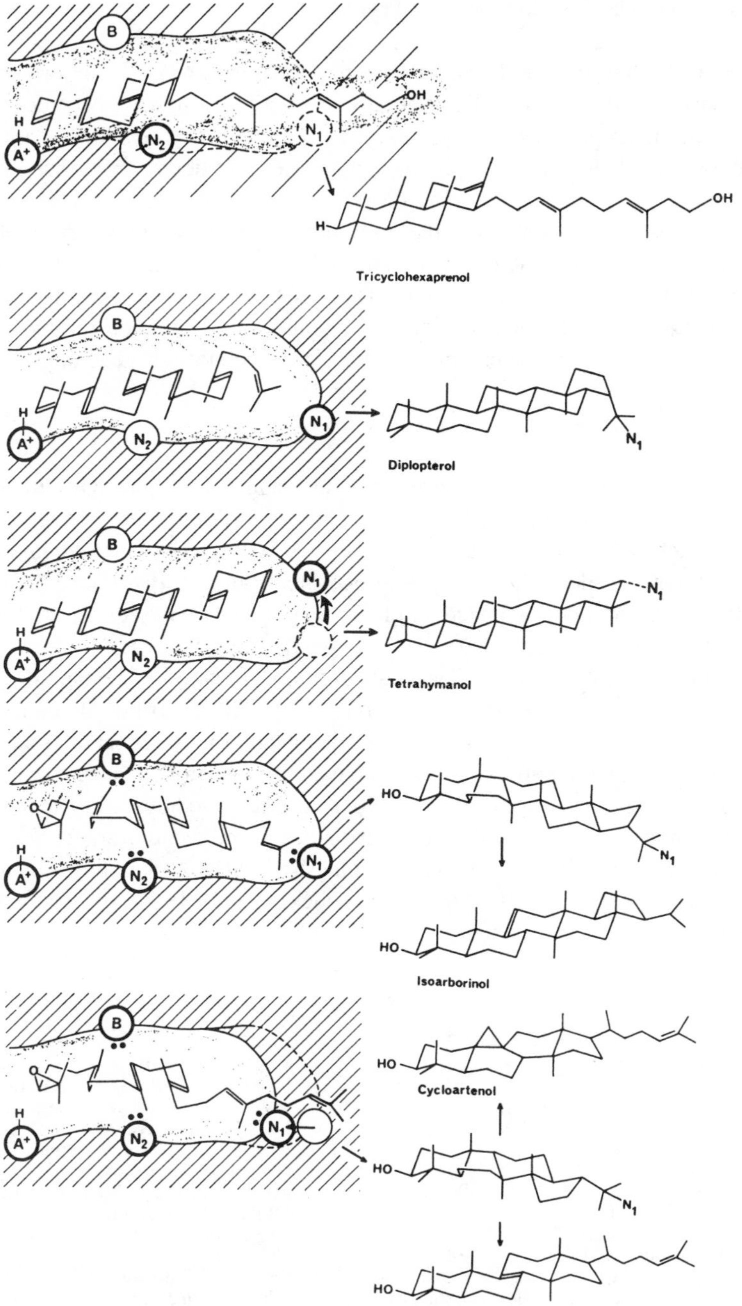

Figure 5 Postulated phylogenetic sequence of the squalene and squalene epoxide cyclases. *A^+–H:* acidic site. *B:* basic site. N_1 and N_2*:* nucleophilic sites. The relative positions of the sites and the general shape of the hydrophobic cavity accomodating the substrate vary slightly from case to case. In the first case N_1 is not involved; in the next two neither *B* nor N_2 is involved.

hopanoids, tetrahymanol, 3β-hydroxyhop-22(29)-ene (?), isoarborinol (?), cycloartenol, lanosterol. The formation of all of these substances from squalene or its epoxide involves very similar chemistry. It is likely that the cyclases involved form one phylogenetic series and derive from one another by minimal changes in the active site, i.e. point mutations. It will be possible to test this hypothesis when the structures of some of these cyclases are known. Furthermore, it is remarkable that in two organisms, two cyclases apparently coexist: *Methylococcus* produces both hopanoids and lanosterol, and *Tetrahymena* produces both tetrahymanol and diplopterol. This is not proof, but is a strong argument in favor of the operation of gene duplication.

CONCLUSION

As we have seen, many questions remain unanswered, and many plausible predictions remain unproved. It is clear, however, that the initial query, "Are there bacterial sterols?" has been superseded. We now want to know (in phylogenetic order): Do nondehydrogenated polyprenol ethers, hydrated polyprenols, nonhydrogenated carotenoids, polycyclopolyprenols, and isoarborinol exist in some prokaryotes? Do polyprenoid derivatives of any kind exist in bacteria that seem to lack standard hopanoids (novel hopanoids so far not detected, or some of the families mentioned above)? If not, do membrane proteins have a sterol-like mechanical role in addition to a functional one? Do some of the most complex hopanoids have other physiological functions? Do squalene cyclases really form a homologous series? Do hopanoids, membrane carotenoids, or other cholesterol surrogates form useful targets for antimicrobial agents? How is their biosynthesis regulated? Are fern hopanoids produced by chloroplasts or other organelles? If so, why do other plants not produce them? The list continues.

Literature Cited

1. Aberhart, D. J., Caspi, E. 1979. Nonoxidative cyclization of squalene by *Tetrahymena pyriformis*. Incorporation of a 3β-hydrogen (deuterium) atom into tetrahymanol. *J. Am. Chem. Soc.* 101: 1013–19
2. Asselineau, J. 1962. In *Les lipides bactériens*. pp. 162, 287. Paris: Hermann. 350 pp.
3. Babadjamian, A., Faure, R., Laget, M., Duménil, G., Padieu, P. 1984. Occurrence of triterpenoids in methanol-oxidizing bacteria. 2-Methyl-22-hydroxyhopane from *Corynebacterium*. *Chem. Commun.* 1984:1657–58
4. Barrow, K. D., Collins, J. G., Rogers, P. L., Smith, G. M. 1983. Lipid composition of an ethanol-tolerant strain of *Zymomonas mobilis*. *Biochim. Biophys. Acta* 753:324–30
5. Barton, D. H. R., Jarman, T. R., Watson, K. C., Widdowson, D. A., Boar, R. B., Damps, K. 1975. Investigations on the biosynthesis of steroids and terpenoids. XII. Biosynthesis of 3β-hydroxy-triterpenoids and -steroids from (3*S*)-2,3-epoxy-2,3-dihydrosqualene. *J. Chem. Soc. Perkin Trans. 1* 1975:1134–38
6. Benz, R., Hallmann, D., Poralla, K., Eibl, H. 1983. Incorporation of hopanoids with phosphatidylcholines containing oleic and ω-cyclohexyldodecanoic acid in lipid bilayer membranes. *Chem. Phys. Lipids* 34:7–24
7. Bird, C. W., Lynch, J. M., Pirt, S. J.,

Reid, W. W. 1971. The identification of hop-22(29)-ene in prokaryotic organisms. *Tetrahedron Lett.* 1971:3189–90

8. Bird, C. W., Lynch, J. M., Pirt, S. J., Reid, W. W., Brooks, C. J. W., Middleditch, B. S. 1971. Steroids and squalene in *Methylococcus capsulatus* grown on methane. *Nature* 230:473–74
9. Bisseret, P., Wolff, G., Albrecht, A. M., Tanaka, T., Nakatani, Y., Ourisson, G. 1983. A direct study of the cohesion of lecithin bilayers: the effect of hopanoids and α,ω-dihydroxycarotenoids. *Biochem. Biophys. Res. Commun.* 110:320–24
10. Bisseret, P., Zundel, M., Rohmer, M. 1985. Prokaryotic triterpenoids. 2. 2β-Methylhopanoids from *Methylobacterium organophilum* and *Nostoc muscorum,* a new series of prokaryotic triterpenoids. *Eur. J. Biochem.* 150:29–34
11. Bloch, K. 1983. Sterol structure and membrane function. *CRC Crit. Rev. Biochem.* 14:47–92
12. Blöcher, D., Gutermann, R., Henkel, B., Ring, K. 1984. Physicochemical characterization of tetraether lipids from *Thermoplasma acidophilum.* Differential scanning calorimetry studies on glycolipids and glycophospholipids. *Biochim. Biophys. Acta* 778:74–80
13. Bouvier, P. 1978. *Biosynthèse de stéroïdes et de triterpénoïdes chez les procaryotes et les eucaryotes.* PhD dissertation. Univ. Louis Pasteur, Strasbourg, France. 145 pp.
14. Bouvier, P., Berger, Y., Rohmer, M., Ourisson, G. 1980. Nonspecific biosynthesis of gammacerane derivatives by a cell-free system from *Tetrahymena pyriformis. Eur. J. Biochem.* 112:549–56
15. Bouvier, P., Rohmer, M., Benveniste, P., Ourisson, G. 1976. Δ 8(14)-steroids in the bacterium *Methylococcus capsulatus. Biochem. J.* 159:267–71
16. Brassell, S. C., Wardroper, A. M. K., Thomson, I. D., Maxwell, J. R., Eglinton, G. 1981. Specific acyclic isoprenoids as biological markers of methanogenic bacteria in marine sediments. *Nature* 290:693–96
17. Bringer, S., Hätner, T., Poralla, K., Sahm, H. 1985. Influence of ethanol on the hopanoid content and the fatty acid pattern in batch and continuous cultures of *Zymomonas mobilis. Arch. Microbiol.* 140:312–16
18. Britton, G. 1984. Carotenoids and polyterpenoids. *Nat. Prod. Rep.* 1:67–85
19. Caspi, E., Zander, J. M., Greig, J. B., Mallory, F. B., Conner, R. L., Landrey, J. R. 1968. Evidence for a non-oxidative cyclization of squalene in the biosynthesis of tetrahymanol. *J. Am. Chem. Soc.* 90:5563–65
20. Chapman, D. J., Ragan, M. A. 1980. Evolution of biochemical pathways. Evidence from comparative biochemistry. *Ann. Rev. Plant Physiol.* 31:639–78
21. Chappe, B., Albrecht, P., Michaelis, W. 1982. Polar lipids of archaebacteria in sediments and petroleums. *Science* 217:65–67
22. Chappe, B. (Chap Sim), Michaelis, W., Albrecht, P., Ourisson, G. 1979. Fossil evidence for a novel series of archaebacterial lipids. *Naturwissenschaften* 66:522–23
23. Comita, P. B., Gagosian, R. B., Pang, H., Costello, C. E. 1985. Structural elucidation of a unique macrocyclic membrane lipid from a new, extremely thermophilic, deep-sea hydrothermal vent archaebacterium, *Methanococcus jannaschii. J. Biol. Chem.* 259:15234–41
24. Conner, R. L., Landrey, J. R., Burns, C. H., Mallory, F. B. 1968. Cholesterol inhibition of pentacyclic triterpenoid biosynthesis in *Tetrahymena pyriformis. J. Protozool.* 15:600–5
25. Corey, E. J., Ortiz de Montellano, P. R., Lin, K., Dean, P. D. G. 1963. 2,3-Iminosqualene, a potent inhibitor of enzymic cyclization of 2,3-oxidosqualene to sterols. *J. Am. Chem. Soc.* 89:2797–98
26. Dahl, J. S., Dahl, C. E. 1983. Coordinate regulation of unsaturated phospholipid, RNA, and protein synthesis in *Mycoplasma capricolum* by cholesterol. *Proc. Natl. Acad. Sci. USA* 80:692–96
27. Das, J. S., Manhoato, S. B. 1983. Triterpenoids. *Phytochemistry* 22:1071–95
28. Demel, R. A., de Kruyff, B. 1976. The function of sterols in membranes. *Biochim. Biophys. Acta* 457:109–32

28a. de Rosa, M., de Rosa, S., Gambacorta A., Bu'Lock J. D. 1980. Structure of calditol, a new branched-chain nonitol, and of the derived tetraether lipids in thermoacidophile archaebacteria of the *Caldariella* group. *Phytochemistry* 19: 249–54

28b. de Rosa, M., Gambacorta, A. 1986. Lipid biogenesis in archaebacteria. *Syst. Appl. Microbiol.* 7:278–85

28c. de Rosa, M., Gambacorta, A., Minale, L., Bu'lock, J. 1971. Bacterial triterpenes. *Chem. Commun.* 1971:619–20

28d. de Rosa, M., Gambacorta, A., Nicolaus, B., Bu'lock, J. D. 1980. Complex lipids of *Caldariella acidophila,* a thermoacidophilic archaebacterium. *Phytochemistry* 19:821–25

28e. de Rosa, M., Gambacorta, A., Nicolaus, B., Ross, H. N. M., Grant, W. D., Bu'Lock, J. D. 1982. An asymmetric archaebacterial diether lipid from alkaliphilic halophiles. *J. Gen. Microbiol.* 128:343–48

29. Duriatti, A., Bouvier-Navé, P., Benveniste, P., Schuber, F., Delprino, L., et al. 1985. *In vitro* inhibition of animal and higher plant 2,3-oxidosqualene-sterol cyclases by 2-aza-2,3-dihydrosqualene and derivatives, and other ammonium-containing molecules. *Biochem. Pharmacol.* 34:2765–77

30. Ekiel, I., Sprott, G. D., Smith, I. C. 1986. Mevalonic acid is partially synthesized from amino acids in *Halobacterium cutirubrum:* a ^{13}C nuclear magnetic resonance study. *J. Bacteriol.* 166:557–64

31. Fewson C. A. 1986. Archaebacteria. *Biochem. Educ.* 14:103–15

32. Flesch, G., Rohmer, M. 1987. Growth inhibition of hopanoid synthesizing bacteria by squalene cyclase inhibitors. *Arch. Microbiol.* 147:100–4

33. Förster, H. J., Biemann, K., Haigh, W. G., Tattrie, N. H., Colvin, J. R. 1973. The structure of novel C_{35} pentacyclic triterpenes from *Acetobacter xylinum*. *Biochem. J.* 135:133–43

34. Gelpi, E., Schneider, H., Mann, J., Oró, J. 1970. Hydrocarbons of geochemical significance in microscopic algae. *Phytochemistry* 9:603–12

35. Ghisalberti, E. L., de Souza, N. J., Rees, H. H., Goodwin, T. W. 1970. Biosynthesis of the triterpene hydrocarbons of *Polypodium vulgare*. *Phytochemistry* 11:1817–23

36. Gliozzi, A., Bruno, S., Basak, T. K., de Rosa, M., Gambacorta, A. 1986. Organization and dynamics of bipolar lipids from *Sulfolobus solfataricus* in bulk phase and in monolayer membranes. *Syst. Appl. Microbiol.* 7:266–71

37. Gliozzi, A., Paoli, G., de Rosa, M., Gambacorta, A. 1983. Effect of isoprenoid cyclization on the transition temperature of lipids in thermophilic archaebacteria. *Biochim. Biophys. Acta* 735:234–42

38. Goldfine, H. 1972. Comparative aspects of bacterial lipids. *Adv. Microb. Physiol.* 8:1–58

39. Gräfe, U., Reinhardt, G., Hänel, F., Schade, W., Gumpert, J. 1985. Occurrence of squalene and dehydrosqualene in streptomycetes. *J. Basic Microbiol.* 25:503–7

40. Gulik, A., Luzzati, V., de Rosa, M., Gambacorta, A., 1986. Structure and polymorphism of tetraether lipids from *Sulfolobus solfataricus*. I. Crystallographic analysis. *Syst. Appl. Microbiol.* 7:258–61

41. Haigh, G., Förster, H. J., Biemann, K., Tattrie, N. H., Colvin, J. R. 1973. Induction of bacterial cellulose microfibrils by a novel terpenoid from *Acetobacter xylinum*. *Biochem. J.* 135:145–49

42. Harrison, D. M. 1986. The biosynthesis of carotenoids. *Nat. Prod. Rep.* 3:205–15

43. Haslam, E. 1986. Secondary metabolism—fact and fiction. *Nat. Prod. Rep.* 3:217–49

44. Heathcock, C. B., Finkelstein, B. L., Aoki, T., Poulter, C. D. 1985. Stereostructure of the archaebacterial C_{40} diol. *Science* 229:862–64

45. Heissler, D., Ocampo, R., Albrecht, P., Riehal, J.-J., Ourisson, G. 1984. Identification of long-chain tricyclic terpene hydrocarbons (C_{21}–C_{30}) in geological samples. *J. Chem. Soc. Chem. Commun.* 1984:496–98

46. Holz, G. G. Jr., Conner, R. L. 1973. The composition, metabolism and roles of lipids in *Tetrahymena*. In *Biology of* Tetrahymena, ed. A. M. Eliott, 4:99–122. Stroudsbury, UK: Dowdon, Hutchinson & Ross. 508 pp.

47. Holzer, G., Oro, J., Tornabene, T. G. 1979. Gas chromatographic–mass spectrometric analysis of neutral lipids from methanogenic and thermoacidophilic bacteria. *J. Chromatogr.* 186: 795–809

48. Horowitz, N. H. 1945. On the evolution of biochemical synthesis. *Proc. Natl. Acad. Sci. USA* 31:153–57

49. Huang, L., Huang, A. 1974. Regulation of membrane lipid fluidity in *Acholeplasma laidlawii:* effect of carotenoid pigment content. *Biochim. Biophys. Acta* 352:361–70

50. Jackson, M. J., Powell, T. G., Summons, R. E., Sweet, I. P. 1986. Hydrocarbon shows and petroleum source rocks in sediments as old as 1.7×10^9 years. *Nature* 322:727–29

51. Kannenberg, E. 1983. Membraneigenschaften von Hopanoiden und Lipiden, die Fettsäuren mit endständigen Verzweigung and ω-Cyclohexanringen enthalten. PhD dissertation. Univ. Tübingen, Germany. 176 pp.

52. Kannenberg, E., Blume, A., Geckeler, K., Poralla, K. 1985. Properties of hopanoids and phosphatidylcholines containing ω-cyclohexane fatty acids in monolayer and liposome experiments. *Biochim. Biophys. Acta* 814:179–85

53. Kannenberg, E., Blume, A., McElhaney, R. N., Poralla, K. 1983. Monolayer and calorimetric studies of phosphatidylcholines containing branched-chain fatty acids and of their interactions with cholesterol and with a bacterial hopanoid in model membranes. *Biochim. Biophys. Acta* 733:111–16
54. Kannenberg, E., Blume, A., McElhaney, R. N., Poralla, K. 1986. Mixed monolayer studies on the interactions of synthetic phosphatidylcholines containing branched-chain fatty acids and a hopane glycolipid isolated from the thermoacidophilic *Bacillus acidocaldarius. Chem. Phys. Lipids* 39:145–53
55. Kannenberg, E., Poralla, K. 1982. The influence of hopanoids on growth of *Mycoplasma mycoides. Arch. Microbiol.* 133:100–2
56. Kannenberg, E., Poralla, K., Blume, A. 1980. A hopanoid from the thermoacidophilic *Bacillus acidocaldarius* condenses membranes. *Naturwissenschaften* 67:458–59
57. Kohl, W., Gloe, A., Reichenbach, H. 1983. Steroids from the myxobacterium *Nannocystis exedens. J. Gen. Microbiol.* 129:1629–35
58. Kulshreshta, M. J., Kulshreshta, D. K., Rastogi, R. P. 1972. The triterpenoids. *Phytochemistry* 11:2369–81
59. Langmuir, I. 1917. The constitution and fundamental properties of solids and liquids. II. Liquids. *J. Am. Chem. Soc.* 39:1848–906
60. Langworthy, T. A. 1977. Long-chain diglycerol tetraether from *Thermoplasma acidophilum. Biochim. Biophys. Acta* 487:37–50
61. Langworthy, T. A. 1985. The lipids of archaebacteria. In *The Bacteria,* Vol. 8, *Archaebacteria,* ed. C. R. Woese, R. S. Wolfe, pp. 459–97. New York: Academic. 582 pp.
62. Langworthy, T. A., Holzer, G., Zeikus, J. G., Tornabene, T. G. 1983. *Iso-* and *anteiso*-branched glycerol diethers of the thermophilic anaerobe *Thermodesulfotobacterium commune. Syst. Appl. Microbiol.* 4:1–17
63. Langworthy, T. A., Mayberry, W. R. 1976. A 1,2,3,4-tetrahydroxypentane-substituted pentacyclic triterpene from *Bacillus acidocaldarius. Biochim. Biophys. Acta* 431:570–77
64. Langworthy, T. A., Mayberry, W. R., Smith, P. F. 1976. A sulfonolipid and novel glucosaminyl glycolipids from the extreme thermoacidophile *Bacillus acidocaldarius. Biochim. Biophys. Acta* 431:550–69
65. Langworthy, T. A., Pond, J. L. 1986. Membranes and lipids in thermophiles. In *Thermophiles. General, Molecular and Applied Microbiology,* ed. T. D. Brock, pp. 107–36. New York: Wiley. 316 pp.
66. Lemcke, R. M., Burrows, M. R. 1980. Sterol requirement for the growth of *Treponema hyodysenteriae. J. Gen. Microbiol.* 116:539–43
67. Liaen-Jensen, S., Andrewes, A. G. 1972. Microbial carotenoids. *Ann. Rev. Microbiol.* 26:225–48
68. Luzzati, V., Gulik, A. 1986. Structure and polymorphism of tetraether lipids from *Sulfolobus solfataricus.* II. Conjectures regarding biological significance. *Syst. Appl. Microbiol.* 7:262–65
69. Mallory, F. B., Conner, R. L., Landrey, J. R., Zander, J. M., Greig, J. B., Caspi, E. 1968. The biosynthesis of tetrahymanol from (4*R*)-(4-^{3}H, 2-^{14}C) mevalonic acid. *J. Am. Chem. Soc.* 90:3564–66
70. Mallory, F. B., Gordon, J. T., Conner, R. L. 1963. The isolation of a pentacyclic triterpenoid alcohol from a protozoan. *J. Am. Chem. Soc.* 85:1362–63
71. Michaelis, W., Albrecht, P. 1979. Molecular fossils of archaebacteria in kerogen. *Naturwissenschaften* 66:420–21
72. Milon, A., Lazrak, T., Albrecht, A.-M., Wolff, G., Weill, G., et al. 1986. Osmotic swelling of unilamellar vesicles by the stopped-flow light scattering method. Influence of vesicle size, solute, temperature, cholesterol and three α,ω-dihydroxycarotenoids. *Biochim. Biophys. Acta* 859:1–9
73. Milon, A., Wolff, G., Ourisson, G., Nakatani, Y. 1986. Organisation of carotenoid-phospholipid bilayer systems. Incorporation of zeaxanthin, astaxanthin, and their C_{50} homologues into dimyristoylphosphatidylcholine vesicles. *Helv. Chim. Acta* 69:12–24
74. Nes, W. D., Heftmann, E. 1981. A comparison of triterpenoids with steroids as membrane components. *J. Nat. Prod.* 44:377–400
75. Nes, W. D., Heupel, R. C., Le, P. H. 1984. A comparison of sterol biosynthesis in fungi and tracheophytes and its phylogenetic and functional implications. In *Structure, Function and Metabolism of Plant Lipids,* ed. P.-A. Segenthaler, W. Eichenberger, 1:207–16. New York: Elsevier. 634 pp.
76. Nes, W. R. 1974. Role of sterols in membranes. *Lipids* 9:596–612
77. Neumann, S., Simon, H. 1986. Purification, partial characterization and

substrate specificity of a squalene cyclase from *Bacillus acidocaldarius*. *Biol. Chem. Hoppe-Seyler* 367:723–29
78. Neunlist, S., Holst, O., Rohmer, M. 1985. Prokaryotic triterpenoids. The hopanoids of the purple non-sulfur bacterium *Rhodomicrobium vannielii:* an aminotriol and its aminoacylderivatives, *N*-tryptophanyl and *N*-ornithyl aminotriol. *Eur. J. Biochem.* 147:561–68
79. Neunlist, S., Rohmer, M. 1985. The hopanoids of *Methylosinus trichosporium:* aminobacteriohopanetriol and aminobacteriohopanetetrol. *J. Gen. Microbiol.* 131:1363–67
80. Neunlist, S., Rohmer, M. 1985. A novel hopanoid, 30-(5'-adenosyl)hopane, from the purple non-sulfur bacterium *Rhodopseudomonas acidophila,* with possible DNA interactions. *Biochem. J.* 228:769–71
81. Neunlist, S., Rohmer, M. 1985. Novel hopanoids from the methylotrophic bacteria *Methylococcus capsulatus* and *Methylomonas methanica:* (22*S*)-35-aminobacteriohopane-30, 31, 32, 33, 34-pentol and (22*S*)-35-amino-3β-methylbacteriohopane-30, 31, 32, 33, 34-pentol. *Biochem. J.* 231:635–39
82. Odriozola, J. M., Waitzkin, E., Smith, T. L., Bloch, K. 1978. Sterol requirements of *Mycoplasma capricolum*. *Proc. Natl. Acad. Sci. USA* 75:4107–9
83. Ourisson, G. 1986. Bacterial lipids disclosed by their fossils: their evolution towards sterols. In *Natural Products and Biological Activities,* ed. H. Imura, T. Goto, T. Murachi, T. Nakajima, pp. 45–57. Tokyo/Amsterdam: Univ. Tokyo Press/Elsevier. 371 pp.
84. Ourisson, G., Albrecht, P., Rohmer, M. 1979. The hopanoids: palaeochemistry and biochemistry of a group of natural products. *Pure Appl. Chem.* 51:709–29
85. Ourisson, G., Albrecht, P., Rohmer, M. 1982. Predictive microbial biochemistry: from molecular fossils to procaryotic membranes. *Trends Biochem. Sci.* 7: 233–39
86. Ourisson, G., Albrecht, P., Rohmer, M. 1984. The microbial origin of fossil fuels. *Sci. Am.* 251(2):44–51
87. Ourisson, G., Rohmer, M. 1982. Prokaryotic polyterpenes: phylogenetic precursors of sterols. *Curr. Top. Membr. Transp.* 17:153–82
88. Ourisson, G., Rohmer, M., Anton, R. 1979. From terpenes to sterols: macroevolution and microevolution. *Recent Adv. Phytochem.* 13:131–62
89. Pant, P., Rastogi, R. P. 1979. The triterpenoids. *Phytochemistry* 18:1095–108
90. Pinto, W. J., Lozano, R., Sekula, B. C., Nes, W. R. 1983. Stereochemical distinct roles for sterol in *Saccharomyces cerevisiae*. *Biochem. Biophys. Res. Commun.* 112:47–54
91. Pond, J. L., Langworthy, T. A., Holzer, G. 1986. Long-chain diols: a new class of membrane lipids from a thermophilic bacterium. *Science* 231:1134–36
92. Poralla, K. 1982. Considerations on the evolution of steroids as membrane components. *FEMS Microbiol. Lett.* 13: 131–35
93. Poralla, K., Härtner, T., Kannenberg, E. 1984. Effect of temperature and pH on the hopanoid content of *Bacillus acidocaldarius*. *FEMS Microbiol. Lett.* 23:253–56
94. Poralla, K., Kannenberg, E., Blume, A. 1980. A glycolipid-containing hopane isolated from the acidophilic, thermophilic *Bacillus acidocaldarius* has a cholesterol-like function in membranes. *FEBS Lett.* 113:107–10
95. Raederstorff, D., Rohmer, M. 1985. Sterol biosynthesis *de novo* via cycloartenol by the soil amoeba *Acanthamoeba polyphaga*. *Biochem. J.* 231:609–15
96. Raederstorff, D., Rohmer, M. 1987. Sterol biosynthesis via cycloartenol and other biochemical features related to photosynthetic phyla in the amoebae *Naegleria lovaniensis* and *Naegleria gruberi*. *Eur. J. Biochem.* In press
97. Ramgopal, M., Bloch, K. 1983. Sterol synergism in yeast. *Proc. Natl. Acad. Sci. USA* 80:712–15
98. Razin, S. 1982. Sterols in mycoplasma membranes. *Curr. Top. Membr. Transp.* 17:183–205
99. Reitz, R. C., Hamilton, J. G. 1968. The isolation and identification of two sterols from two species of blue-green algae. *Comp. Biochem. Physiol.* 25:401–16
100. Renoux, J. M., Rohmer, M. 1985. Prokaryotic triterpenoids. New bacteriohopane cyclitol ethers from the methylotrophic bacterium *Methylobacterium organophilum*. *Eur. J. Biochem.* 151:405–10
101. Renoux, J. M., Rohmer, M. 1986. Enzymatic cyclization of all-*trans* pentaprenyl and hexaprenyl methyl ethers by a cell-free system from the protozoon *Tetrahymena pyriformis*. The biosynthesis of scalarane and polycyclohexaprenyl derivatives. *Eur. J. Biochem.* 155:125–32
102. Rohmer, M., Anding, C., Ourisson, G. 1980. Non-specific biosynthesis of hopane triterpenes by a cell-free system from *Acetobacter pasteurianus*. *Eur. J. Biochem.* 112:541–47

103. Rohmer, M., Bouvier, P., Ourisson, G. 1979. Molecular evolution of biomembranes: structural equivalents and phylogenetic precursors of sterols. *Proc. Natl. Acad. Sci. USA* 76:847–51

104. Rohmer, M., Bouvier, P., Ourisson, G. 1980. Non-specific lanosterol and hopanoid biosynthesis by a cell-free system from the bacterium *Methylococcus capsulatus*. *Eur. J. Biochem*. 112:557–60

105. Rohmer, M., Bouvier-Navé, P., Ourisson, G. 1984. Distribution of hopanoid triterpenes in prokaryotes. *J. Gen. Microbiol*. 130:1137–50

106. Rohmer, M., Dastillung, M., Ourisson, G. 1980. Hopanoids from C_{30} to C_{35} in recent muds. Chemical markers for bacterial activity. *Naturwissenschaften* 67:456–58

107. Rohmer, M., Ourisson, G. 1976. Structure des bactériohopanetétrols d'*Acetobacter xylinum*. *Tetrahedron Lett*. 1976:3633–36

108. Rohmer, M., Ourisson, G. 1986. Unsaturated bacteriohopanepolyols from *Acetobacter aceti* ssp. *xylinum*. *J. Chem. Res*. 1986:S356–57, M3037–59

109. Rottem, S. 1980. Membrane lipids of mycoplasmas. *Biochim. Biophys. Acta* 604:65–90

110. Rottem, S., Markowitz, O. 1979. Carotenoids act as reinforcers of the *Acholeplasma laidlawii* lipid bilayer. *J. Bacteriol*. 140:944–48

111. Schmidt, A., Bringer-Meyer, S., Poralla, K., Sahm, H. 1986. Effect of alcohols and temperature on the hopanoid content of *Zymomonas mobilis*. *Appl. Microbiol. Biotechnol*. 25:32–36

112. Schubert, K. Rose, G., Wachtel, H., Hörhold, C., Ikekawa, N. 1968. Zum Vorkommen von Sterinen in Bakterien. *Eur. J. Biochem*. 5:246–51

113. Seckler, B., Poralla, K. 1986. Characterization and partial purification of squalene-hopene cyclase from *Bacillus acidocaldarius*. *Biochim. Biophys. Acta* 881:356–63

114. Smith, P. F., Henrikson, C. V. 1966. Growth inhibition of *Mycoplasma* by inhibitors of polyterpene biosynthesis and its reversal by cholesterol. *J. Bacteriol*. 91:1854–58

115. Taylor, R. F. 1984. Bacterial triterpenoids. *Microbiol. Rev*. 48:181–98

116. Thurl S., Buhrow, I., Schäfer, W. 1985. New types of menaquinones from the thermophilic archaebacterium *Thermoproteus tenax*. *Biol. Chem. Hoppe-Seyler* 366:1079–83

117. Tornabene, T. G., Langworthy, T. A., Holzer, G., Oro, J. 1979. Squalenes, phytanes and other isoprenoids as major neutral lipids of methanogenic and thermoacidophilic "archaebacteria." *J. Mol. Evol*. 13:73–83

118. Towe, K. M. 1978. Early precambrian oxygen: a case against photosynthesis. *Nature* 274:657–60

119. Ward, D. M., Brassell, S. C., Eglinton, G. 1985. Archaebacterial lipids in hot-spring microbial mats. *Nature* 318:656–59

120. Yamamoto, H. Y., Bangham, A. D. 1978. Carotenoid organization in membranes. Thermal transition and spectral properties of carotenoid-containing liposomes. *Biochim. Biophys. Acta* 507:119–27

121. Yeagle, P. L. 1985. Cholesterol and the cell membrane. *Biochim. Biophys. Acta* 822:267–87

122. Zander, J. M., Caspi, E., Pendley, G. N., Mitra, C. R. 1969. The presence of tetrahymanol in *Oleandra wallichii*. *Phytochemistry* 8:2265–67

123. Zundel, M., Rohmer, M. 1985. Prokaryotic triterpenoids. 1. 3β-Methylhopanoids from *Acetobacter* species and *Methylococcus capsulatus*. *Eur. J. Biochem*. 150:23–27

124. Zundel, M., Rohmer, M. 1985. Prokaryotic triterpenoids. 3. The biosynthesis of 2β-methylhopanoids and 3β-methylhopanoids of *Methylobacterium organophilum* and *Acetobacter pasteurianus* ssp. *pasteurianus*. *Eur. J. Biochem*. 150:35–39

Ann. Rev. Microbiol. 1987. 41:335–61

PHYSIOLOGY, BIOCHEMISTRY, AND GENETICS OF THE UPTAKE HYDROGENASE IN RHIZOBIA

Harold J. Evans, Alan R. Harker, Hans Papen, Sterling A. Russell, F. J. Hanus, and Mohammed Zuber

Laboratory for Nitrogen Fixation Research, Oregon State University, Corvallis, Oregon, 97331

CONTENTS

INTRODUCTION

In 1941 Phelps & Wilson (93) reported the presence of a hydrogenase system in nodules of *Pisum sativum* formed by strain 311 of *Rhizobium leguminosarum*. More than 25 years later, Dixon (38–40) confirmed this report, and also concluded that nodules formed by *R. leguminosarum* with hydrogenase activ-

0066-4227/87/1001-0335$02.00

ity evolved little or no H_2 during N_2 fixation because the H_2 produced as a by-product of N_2 fixation was consumed internally by an H_2 oxidation system. Dixon's research showed for the first time that H_2 recycling involved H_2 evolution associated with the nitrogenase reaction per se and H_2 uptake via an H_2 oxidation process that was not directly related to the nitrogenase system. However, the H_2 that served as a substrate for the H_2 uptake system was derived from the nitrogenase reaction. No H_2 uptake capability was observed in laboratory cultures of any *Rhizobium* species or in bacteroids of *Rhizobium* species other than *R. leguminosarum* (38–40).

Schubert & Evans (110) reported in 1976 that the majority of nodulated legumes lose 30–50% of their nitrogenase electron flux as H_2. This and other information raised many questions about the role of H_2 recycling in the biology of N_2-fixing legumes. What are the biochemical components of the H_2 oxidation process? How widely is H_2 recycling distributed among legume inoculants? What are the advantages and disadvantages of H_2 recycling capacity? Does efficient H_2 recycling capacity result in measurable increases in quantities of N_2 fixed? If H_2 recycling capability results in economic benefits, what are the possibilities of transferring the genetic determinants for H_2 recycling capacity within and among *Rhizobium* species?

During the past ten years partial answers to many of these questions have been provided, but as expected, there are some disagreements among researchers. During the last few years, several reviews relevant to hydrogenase and H_2 recycling biology have been published (1, 8, 21, 22, 46, 81, 92, 100). Here we present our interpretation of recent developments in research on aspects of the physiology, biochemistry, and genetics of H_2 metabolism in *Rhizobium*.

DIHYDROGEN EVOLUTION DURING N_2 FIXATION

Mechanistic Considerations

Since most nodulated legumes expend 25% or more of their nitrogenase electron flux in the evolution of H_2 (22, 52), the basic mechanisms of H_2 evolution during the nitrogenase reaction need to be considered. Figure 1 illustrates general aspects of functioning nitrogenase; Burgess (25) has given more details of the mechanism. As shown, the reductant to support the nitrogenase reaction in *Rhizobium* is provided by reduced ferredoxin (Fd) (30). In *Klebsiella pneumoniae,* reduced flavodoxin is the natural electron donor to nitrogenase (112). Electrons are transferred to the Fe protein of nitrogenase, which in the presence of MgATP is converted to a low potential component with altered conformation. The Fe protein combines with and dissociates from the MoFe protein as one electron is transferred. Two MgATP molecules are hydrolyzed during each cycle of association-dissociation (see 25).

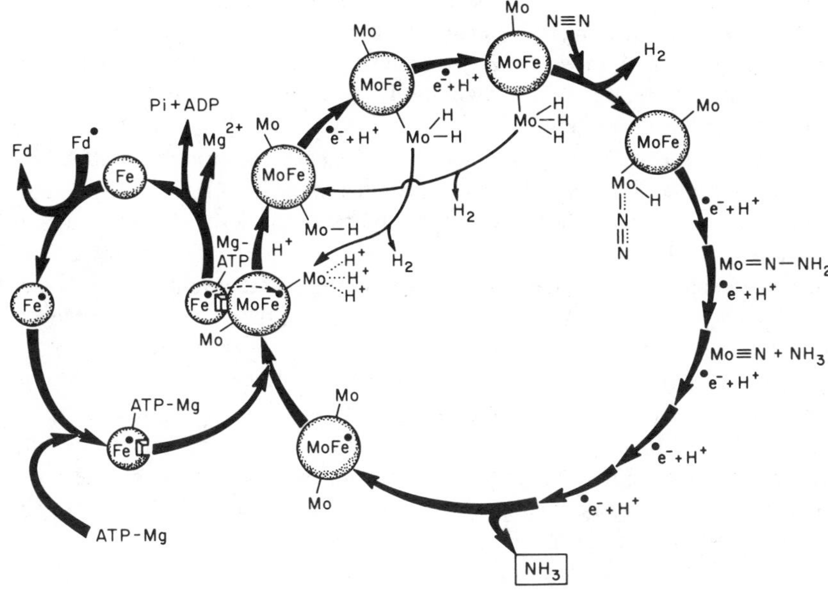

Figure 1 A diagram illustrating the roles of the Fe protein (Fe) and MoFe protein (MoFe) in the fixation of N_2 and evolution of H_2 by the nitrogenase-catalyzed reaction. The scheme has been constructed using information from Chatt (31), Lowe et al (80), and Postgate (97). For details see text.

In Figure 1 three electron transfers from the Fe protein to the MoFe protein generate a trihydride of the MoFe protein, as originally proposed by Chatt (31). According to a scheme based upon kinetic analyses (80, 122), an enzyme trihydride, referred to as E_3H_3, would be formed in three steps, with mono- (E_1H_1) and dihydride (E_2H_2) MoFe protein derivatives as intermediates. According to Chatt (31), N_2 may displace two hydrogen atoms from the MoFe protein trihydride, which results in the binding of N_2 and evolution of H_2. Six additional electrons are required for the formation of two molecules of NH_3. From the chemistry of N_2 fixation by model metal complex, Chatt (31) postulated that the mono- and dihydride MoFe complexes would not function as initial reactants with N_2. The necessity of N_2 for the formation of H_2 from the MoFe protein trihydride would account for the minimum stoichiometry of one mole of H_2 evolved per mole of N_2 reduced. A need for N_2 is also consistent with data of Simpson & Burris (113) that showed that 27% of the nitrogenase electron flux was utilized in H_2 evolution even under 50 atm of N_2. These results strongly support the argument that H_2 evolution is an essential aspect of the N_2 fixation mechanism. To explain the evolution of H_2 when argon replaces N_2 over the nitrogenase reaction, or the evolution of more than one mole of H_2 per mole of N_2 fixed, Lowe et al (80) included in their scheme a possibility for back reactions of either $MoFeH_2$ (E_2H_2) or $MoFeH_3$ (E_3H_3) resulting in N_2-independent H_2 evolution (Figure 1). Furthermore, Thorneley & Lowe (122) proposed that the MoFe protein dihydride (E_2H_2) may be prevented from evolving H_2 by combination with the MgATP complex of the Fe protein.

Dihydrogen evolution by the nitrogenase reaction is strongly inhibited by C_2H_2 and several other alternate substrates (see 25). Presumably, these compounds react with the enzyme at some stage in the oxidation-reduction cycle prior to the appearance of the trihydride form of the MoFe protein. It is of considerable interest that H_2 evolution by the alternate nitrogenase reaction, discovered by Bishop and coworkers (19, 20), was not inhibited by C_2H_2; in addition, about 50% of the nitrogenase electrons were allocated to N_2 and the remainder to H^+ under several conditions. There is no direct evidence for the existence of hydride derivatives of the MoFe protein (25). However, the MoFe protein from *Azotobacter vinelandii,* which is reduced by two electrons per Mo, catalyzes H_2-dependent reduction of methylene blue; this information is not inconsistent with the existence of a MoFe protein hydride derivative.

Partitioning of electrons to N_2 or H^+ during catalysis by purified nitrogenase components is dependent upon the turnover rate of the MoFe component of nitrogenase (26, 58). Reactions containing optimal proportions of the two nitrogenase components and optimal concentrations of MgATP and reductant produced the highest MoFe protein turnover rates, maximum allocation of electrons to N_2, and minimum allocation of electrons to H^+. In contrast, low MoFe turnover rates associated with a suboptimal concentration of the Fe protein or an insufficiency of MgATP or reductant produced a minimum allocation of electrons to N_2 and a maximum allocation to H^+ or C_2H_2. Burris et al (27) have pointed out that "despite the method of varying the electron flux, the control point for distribution of electrons to various substrates is at dinitrogenase (MoFe protein) and its reduction state governs whether electrons go preferentially to protons, to acetylene or to N_2 for their respective reductions."

Recently, Simpson et al (115) used purified nitrogenase components for the catalysis of a reaction in which they monitored the time course of H_2 evolution and N_2 fixation with increasing electron equivalents transferred to the MoFe protein. They also concluded that H_2 is evolved at two different oxidation states of the MoFe protein.

Biological Effects

In vivo the supply of ATP and reductant in nodule bacteroids may sometimes be insufficient to provide the magnitude of electron flux through nitrogenase necessary for generation of the postulated MoFe protein trihydride required for maximum N_2 fixation, while allowing less reduced MoFe protein components to catalyze H_2 evolution. This would explain the high H_2 evolution rates (i.e. greater than one H_2 per N_2 fixed) by intact legume nodules that lack an H_2 recycling system (52). Perhaps environmental conditions where an adequate supply of photosynthate is available to nodules could maximize allocation of electrons to N_2 and minimize H_2 evolution. However, many poorly understood factors influence allocation of nitrogenase electrons under in vivo conditions.

The capacity of nodule bacteroids to recycle H_2 produced by nitrogenase is the major factor determining the extent of H_2 evolution into the atmosphere (see Figure 2). This point is supported by data showing that 72 nodule samples from several legumes inoculated with Hup^- strains lost a mean of 32% of their nitrogenase electron flux as H_2, whereas 54 nodule samples formed by efficient Hup^+ inoculants lost a mean of 3.8% of the electron flux as H_2 (52). The distribution of H_2 recycling capability among rhizobial species used for inoculation of appropriate hosts was presented by Eisbrenner & Evans (46) and Brewin (22). The percentages of Hup^+ strains in the different rhizobial species in the latter tabulation were: *R. leguminosarum,* 9.3% of 171; *Rhizobium meliloti,* 21% of 19 (all activities were weak); *Rhizobium trifolii,* 0% of 7; *Rhizobium phaseoli,* 0% of 10; *Bradyrhizobium japonicum,* 21% of 1432; *Bradyrhizobium* sp. (*Vigna,* cowpea), 85% of 13; *Bradyrhizobium* sp. (*Vigna,* mungbean), 93% of 30. For reasons that are not obvious, a minority of most species of rhizobium express H_2 recycling capability. The lack of H_2 recycling undoubtedly accounts for the estimate that one million or more metric tons of H_2 are produced globally each year by nodulated legumes. This is a large proportion of the estimated 2.4–4.7 million metric tons of H_2 produced annually worldwide by all N_2-fixing organisms (34).

Several reviews have discussed the factors that affect the evolution of H_2 from nodules that do or do not have H_2 recycling capacity (22, 46, 81). Among these are temperature, light intensity, stage in the growth cycle, the genetic constitution of the host legume used in the symbiosis, and the vigor of H_2 recycling capacity of nodule bacteroids. For the most part these in vivo effects have not been systematically examined.

DIHYDROGEN OXIDATION AND CHEMOLITHOTROPHIC GROWTH

Dihydrogen not only is oxidized by the nodule bacteroids formed by Hup^+ strains of *B. japonicum,* but is also utilized by laboratory cultures of the Hup^+ strains of *B. japonicum,* provided that appropriate conditions are maintained that allow derepression of the H_2 oxidation system (78, 82). Research from

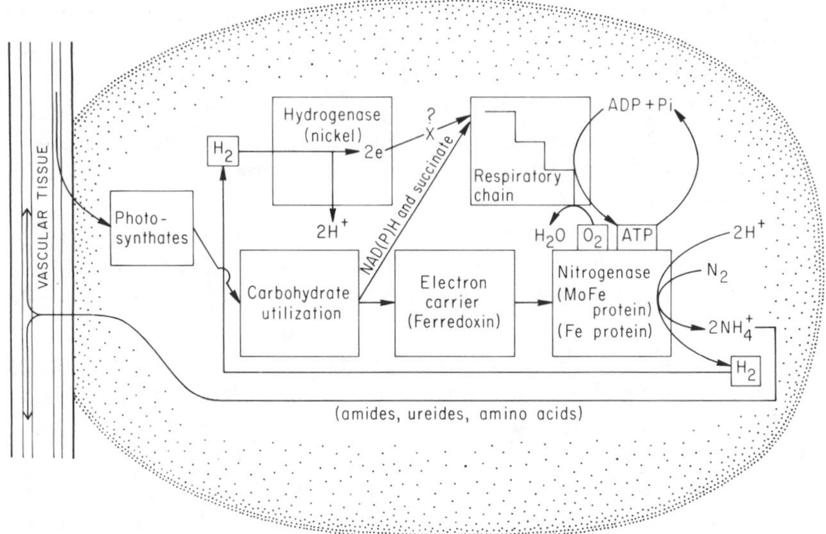

Figure 2 A scheme illustrating the relationship between H_2 evolution during N_2 fixation and H_2 oxidation (i.e. H_2 recycling) in a legume nodule formed by a H_2 uptake positive (Hup^+) strain of *Rhizobium*.

our laboratory (82) demonstrated that H_2 oxidation in cultures of *B. japonicum* strains that possessed the genetic determinants for H_2 oxidation required decreased O_2 tension over the cultures (1–2%), low-level carbon substrates in media, and H_2 in the atmosphere over the cultures. Hup^+ strains of *B. japonicum* not only expressed H_2 oxidation capability under appropriate conditions in laboratory cultures, but also grew as chemolithotrophs, utilizing H_2 and CO_2 as sole sources of energy and carbon, respectively (61, 77). When a Hup^+ strain of *B. japonicum* was cultured under conditions appropriate for hydrogenase expression, coordinate induction of ribulose-1,5-bisphosphate carboxylase was observed (114). Furthermore, cells from large batches of chemolithotrophically cultured *B. japonicum* were used to purify and characterize ribulose-1,5-bisphosphate carboxylase/oxygenase (98).

In our laboratory, ammonium salts and nitrates have served as N sources for chemolithotrophically cultured *B. japonicum,* but so far no N_2 fixation with this or other species of rhizobium grown under chemolithotrophic conditions has been demonstrated by our group (77). A strain of *R. trifolii* and *Rhizobium* sp. NEA4 were reported to grow chemolithotrophically with H_2 as the source of energy and either N_2 or ammonium salts as the source of N (125). However, data supporting growth with N_2 were not presented. *Rhizobium* sp. ORS571 *(Sesbania)* fixed N_2 and expressed hydrogenase activity in a medium containing succinate as the major carbon source (117). For reasons that are not apparent, the wild-type Hup^+ strains of *R. leguminosarum* and *Bradyrhizobium* sp. *(Vigna)* and Hup^+ strains of species other than those discussed above not only fail to grow chemolithotrophically but so far have failed to express hydrogenase activity under laboratory conditions. This has impeded progress toward characterization of the genetic determinants for H_2 oxidation in species other than *B. japonicum.*

CONSEQUENCES OF H_2 OXIDATION

The capacity for utilization of H_2 from any source could have some beneficial consequences (reviewed in 22, 46, 51, 81). Several of the potential advantages of H_2 recycling that were proposed by Dixon (40) are now supported by experimental data. H_2-dependent synthesis of ATP was demonstrated with bacteroid preparations of *R. leguminosarum* (39, 87) and with *B. japonicum* bacteroids (48). However, only 5 of 14 strains of *R. leguminosarum* that were examined (87) efficiently coupled H_2 oxidation to ATP synthesis. Furthermore, many Hup^+ strains of *Rhizobium,* including *R. meliloti* and *R. trifolii,* possess insufficient H_2 recycling capacity to recycle a significant portion of the H_2 evolved during N_2 fixation (101). Any rhizobium strain that is capable of chemolithotrophic growth with H_2 as the sole source of energy is capable of coupling H_2 oxidation to ATP synthesis. For reasons that are not known, chemolithotrophic growth capability is most prominent in the Hup^+ strains of *B. japonicum* or other rhizobia that have acquired *B. japonicum* determinants for H_2 oxidation (74). Even though *Rhizobium* sp. ORS571 *(Sesbania)* is not known to grow chemolithotrophically, the addition of H_2 to a N_2-fixing chemostat culture of this bacterium in a succinate-limited medium increased the molar growth yield (based on succinate) by 27–35% (117). As pointed out by DeVries et al (37), only one site for H_2-dependent ATP synthesis is functional in N_2-fixing *Rhizobium* ORS571. Therefore, ATP synthesis supported by carbon substrates is more efficient than that supported by H_2. In regard to efficiency of energy utilization by nodule bacteroids, it is pertinent to remember that ATP synthesis, which is dependent upon H_2 from the nitrogenase reaction, is essentially free because all H_2 produced via the nitrogenase reaction is lost in the absence of H_2 recycling capacity. The results with *Rhizobium* ORS571 are consistent with those reported for hydrogenase-deficient mutants of *A. vinelandii* (5). When the Hup^- mutants and parent strains of *A. vinelandii* were grown in chemostat cultures in a sucrose-limited medium at high dilution rates, the growth of the Hup^+ parent strain was significantly greater than that of the Hup^- mutant. The information available provides conclusive evidence that many Hup^+ strains of rhizobium, particularly the Hup^+ *B. japonicum,* are capable of using H_2 as a substrate to support life processes, and that this is advantageous when the availability of carbon substrates is limited.

Dixon (40) also postulated that utilization of O_2 during H_2 recycling could contribute toward the protection of the labile nitrogenase from O_2 damage. In experiments with bacteroids from *B. japonicum* (48) and *R. leguminosarum* (87), H_2 oxidation by bacteroid suspensions lowered the concentration of dissolved O_2 and protected nitrogenase from damage. As discussed by Minchin et al (85), N_2 fixation in legume nodules is a function of the supply of both O_2 and carbohydrate. Therefore, Drevon et al (43) have argued that consumption of O_2 by the H_2 recycling system in legume nodules grown in liquid culture might deplete the O_2 supply enough to cause decreased N_2 fixation rates. However, increasing the O_2 partial pressure over legumes in long-term trials has not increased N_2 fixation and growth (see 85). Whether the consumption of O_2 that results from H_2 recycling protects nitrogenase from O_2 damage in vivo remains an open question.

Several researchers have reported that rates of respiratory CO_2 evolution of nodules formed by Hup^+ rhizobia are measurably less than those of nodules formed by Hup^- strains (6, 42, 49, 99). According to Drevon et al (42), nodules formed by a Hup^+ strain of *B. japonicum* lost respiratory CO_2 at about a 10% lower rate than nodules formed by a Hup^- revertible mutant of the Hup^+ strain; the authors postulated that saturation of the respiratory electron transport chain with electrons from H_2 inhibited carbohydrate oxidation. It has been reported that root weights of *Vigna unguiculata* (99) and *Glycine max* (51) nodulated with Hup^+ strains of rhizobium were greater than those of plants nodulated by Hup^- strains. This effect is explained by assuming that H_2 oxidation resulted in the conservation of carbohydrates, which were diverted into increased root growth.

According to Dixon & Blunden (41), the increase in nitrogenase activities resulting from variation in incubation periods, temperatures, and O_2 supply was inversely related to the proportion of nitrogenase electrons allocated to N_2. These authors initially suggested that their results could be explained by increased rates of H_2 evolution associated with high nitrogenase activity, resulting in H_2 concentrations sufficient to inhibit N_2 fixation. However, after further experiments, this interpretation is no longer considered tenable (R. O. D. Dixon, personal communication).

Recent research suggests that the capacity for H_2 recycling within legume nodules affects metabolic processes that so far cannot be related directly to the energetics of N_2 fixation or to the protection or inhibition of nitrogenase. Minamisawa et al (83) found that the rates of transport of products of N_2 fixation from nodules of *G. max* formed by Hup^+ strains of *B. japonicum* were consistently greater than transport rates from nodules of the same host formed by Hup^- *B. japonicum* strains. These authors suggested that the H_2 recycling process influenced the balance of carbon substrate utilization and assimilation of nitrogenous compounds. In addition, the concentration of an unidentified compound believed to be a polyamine was more than 10-fold higher in *G. max* nodules formed by several Hup^- strains of *B. japonicum* than in comparable nodules formed by Hup^+ strains (84). Experimental results that can provide insight into these interesting findings will be welcomed.

BIOCHEMISTRY OF H_2 OXIDATION

Components of the H_2 Oxidation System

There is general agreement that ubiquinone (UQ) is involved in the transfer of electrons from H_2 or NADH to O_2 in *B. japonicum* (44, 81, 91). It is also agreed that the respiratory electron transport chain is branched in bacteroids from soybean nodules (45), Hup^c mutant cells of *B. japonicum*, and heterotrophically grown *B. japonicum* that have been derepressed for hydrogenase (81, 88, 90). Furthermore, cytochromes *o* and aa_3 as terminal oxidases, cytochromes of the *b* and *c* types as electron carriers, and UQ as a hydrogen carrier have been identified in *B. japonicum* cells from various sources. However, the physiological electron acceptor of the membrane-bound hydrogenase of *B. japonicum* bacteroids, hydrogenase-derepressed heterotrophic cells, and chemolithotrophically cultured cells remains unknown. This is also the case for *R. leguminosarum* bacteroids, N_2-fixing *Azotobacter*, and the aerobic H_2-oxidizing bacteria in general. The only exception is the pathway of electron transport from H_2 in *Xanthobacter autotrophicus* (104). From membranes of this organism, a homogeneous preparation of hydrogenase was shown to contain a *b*-type cytochrome in a ratio of one mole per mole of the purified hydrogenase. From this result, Schink (104) concluded that the bound *b*-type cytochrome may function as the proximal electron acceptor for hydrogenase in vivo. This observation provided some support for the postulation (46) of a role of a *b*-type cytochrome as a proximal electron acceptor for the membrane-bound hydrogenase in *B. japonicum*. However, homogeneous hydrogenase preparations from chemolithotrophically cultured cells, nodule bacteroids, and Hup^c mutant cells of *B. japonicum* contain no measurable amounts of cytochromes (7, 63, 118).

Work from our laboratory (44, 45, 47) provided spectral evidence for a cytochrome b_{559}, which was postulated to be specifically involved in H_2 utilization in bacteroids and chemolithotrophically grown *B. japonicum* cells. Furthermore, Eisbrenner & Evans (45) reported a positive correlation (r = 0.98) between hydrogenase activities and estimated concentrations of a cytochrome b_{559} in a series of *B. japonicum* strains. In the calculation of the correlation coefficient, data from Hup^- bacteroids of *B. japonicum* were included. However, Maier (81) justly criticized that a cytochrome b_{559} that was preferentially reduced by H_2 would have been overlooked in hydrogenase-deficient strains of *B. japonicum.* We have recalculated the correlation coefficient, omitting data for all Hup^- *B. japonicum* strains listed in Table 1 of Reference 45; The revised coefficient is 0.997. On the basis of the positive correlation between cytochrome b_{559} content and hydrogenase activities and the apparent specificity of H_2 as a reductant for the cytochrome b_{559}, Eisbrenner & Evans (44–46) proposed a generalized scheme in which a cytochrome b_{559} was located upstream from UQ. To emphasize that the location of the cytochrome b_{559} was a postulation rather than a fact, a question mark was included at the appropriate place in Figure 2 of Reference 46.

Since O'Brian & Maier (88–90) found no evidence for the involvement of a cytochrome b_{559} that could be preferentially reduced by H_2 in membranes prepared from derepressed *B. japonicum* SR and Hup^c mutants SR476 and SR481, and since the initial experiments of Eisbrenner & Evans (45) were performed using cyanide-treated whole cells, F. J. Hanus (unpublished results) has examined membranes from chemolithotrophically grown *B. japonicum* 122DES prepared in the same way as the membranes of O'Brian & Maier (88). He has examined the effects of electron donors and the inhibitor 2-*n*-heptyl-4-hydroxyquinoline-*N*-oxide (HQNO) on the extent of reduction at equilibrium of cytochrome b_{559}. Only H_2 as electron donor reduced membrane-bound cytochrome b_{559} to the maximum extent. The reduction of cytochrome b_{559} after addition of NADH or succinate as electron donors at equilibrium never exceeded 33–40% of the reduction obtained with H_2 as the electron donor. Furthermore, the results were essentially identical whether or not HQNO was present in the assays; this shows that HQNO is not an inhibitor of cytochrome b_{559} reduction in membranes of chemolithotrophically cultured *B. japonicum* 122DES.

Hanus's results are consistent with the earlier finding of Appleby (4), who also observed no inhibition of cytochrome reduction by HQNO in membranes prepared from heterotrophically grown cells of *B. japonicum* 505 (Wisconsin). The results also agree with those from Arp's laboratory (95, 96); these researchers could not detect any effects of HQNO on the reduction of cytochromes in membrane preparations of chemolithotrophically cultured *Alcaligenes eutrophus* H16 and *Alcaligenes latus.* In contrast, O'Brian & Maier (88–90) and Maier (81) have reported that HQNO completely inhibited cytochrome reduction in membranes of *B. japonicum* SR and the heterotrophically grown hydrogenase-constitutive (Hup^c) *B. japonicum* mutants SR476 and SR481. These results are surprising and difficult to interpret, since only 72–80% inhibition by HQNO of O_2-dependent H_2 oxidation was reported (81, 88). Hanus's observations that membranes of chemolithotrophically grown *B. japonicum* 122DES contain a cytochrome b_{559}, which is reduced to a maximum extent by H_2 but not by other added substrates, is consistent with the report of Eisbrenner & Evans (45) that cytochrome b_{559}

was specifically reduced by H_2 in *B. japonicum* bacteroids and chemolithotrophically grown *B. japonicum* 122DES (47). These results, together with repeated observations that the cytochrome b_{559} in chemolithotrophically grown *B. japonicum* is not inhibited by HQNO, are not inconsistent with the possibility that a cytochrome of the *b*-type may be located upstream from UQ as suggested previously (46).

The failure to observe H_2-specific reduction of *b*-type cytochromes in heterotrophically grown *B. japonicum* SR (88–91) agrees with the conclusion of Eisbrenner & Evans (Table 1 in Reference 45), who also found no H_2-dependent reduction of cytochrome b_{559} in the same derepressed strain of *B. japonicum*. The results indicate that the cytochrome composition of hydrogenase-derepressed and chemolithotrophically cultured *B. japonicum* cells is not the same, and that much of the apparent disagreement about the details of pathways of electrons from H_2 to O_2 in *B. japonicum* and other aerobic H_2 oxidizers is associated with the variety of bacterial strains, cultural conditions, and methods used by different laboratories.

In our opinion, the spectral methods that have been used so far by the various laboratories to analyze components of the electron transport chain in *B. japonicum* are not adequate for distinguishing among the various *b*-type cytochromes. No attempts have been made to record spectra at the temperature of liquid N_2, which is well known to greatly improve the resolution of the different cytochrome components. From this point of view the method of van Wielink et al (128), which uses combined spectrum deconvolution and potentiometric analysis for in situ characterization of *b*- and *c*-type cytochromes, appears to be the method of choice. For example, these authors were able to resolve four *b*-type cytochrome components in membranes from *Escherichia coli,* which showed α-band maxima at 77°K of 555.7, 556.7, 558.6, and 563.5 nm and midpoint potentials at pH 7.0 of +46, +174, −75, and +187 mV, respectively. The cytochrome exhibiting an α-band maximum of 558.6 nm had a negative midpoint potential of −75 mV. Application of these modern methods to *B. japonicum* membranes could provide detailed information about the midpoint potentials and absorption maxima of the various cytochromes and could contribute toward a more complete understanding of the pathways of electron transport from H_2 to O_2 in these bacteria.

Purification and Properties of Hydrogenase

The membrane-bound hydrogenase from *B. japonicum* was first purified from bacteroids by Arp & Burris (9), who reported detailed physical and kinetic properties. Recent purifications of hydrogenase from chemolithotrophically grown cells (63) and bacteroids (7) of *B. japonicum* have provided new information concerning the enzyme. The detailed properties of hydrogenases from selected bacteria are presented in Table 1. Although a great deal of information has been compiled concerning membrane-bound hydrogenases in general, differences in properties are obvious. The differing methods of various laboratories have influenced the results.

PHYSICAL PROPERTIES The physical properties of the hydrogenase from *B. japonicum* are similar to those of other membrane-bound hydrogenases. The isoelectric point is about pH 5.5, which indicates a predominance of acidic residues. The pH optimum for uptake activity is low, ranging from 5.5 to 6.0. The half-life of reduced hydrogenase exposed to air is approximately 1 hr (7, 118, 129). For this reason, most purifications have been performed anaerobically (7, 63). Schneider et al (109) purified the O_2-sensitive hydrogenases from *A. eutrophus* by affinity chromatography under aerobic conditions. These techniques have been successfully applied to *B. japonicum* hydrogenase (118). The aerobically purified enzyme had a half-life in air at 25°C of about 3 days. There have been no systematic studies of the effects of aerobic versus anaerobic purification procedures on the long-term storage of *B. japonicum* hydrogenase. Anaerobically purified hydrogenase containing reducing agents, stored at −80° C under H_2, was stable for several months and lost only 20–30% of its original activity over the course of one year (A. R. Harker, unpublished results).

The thermal stability of the hydrogenase from *B. japonicum* has not been thoroughly investigated. Stults et al (118) noted that the aerobically purified enzyme was more stable at 25°C than at 0°C. Hydrogenases are stable to temperatures ranging from 50–70°C (see Table 1).

SUBUNITS There is a present consensus that hydrogenase from *B. japonicum* is composed of two subunits of about 63 and 33 kd (Table 1). Initial purification (9) from *B. japonicum* bacteroids yielded an enzyme with an apparent single subunit of 65.3 kd and a specific activity of 19.6 μmol H_2 oxidized per minute per milligram protein with methylene blue as the electron acceptor. The most purified fraction contained a detectable quantity of a second polypeptide. It exhibited no stoichiometric relationship to the larger polypeptide and therefore was regarded as a contaminant. Stults et al (119) identified hydrogenase from a Hup constitutive strain of *B. japonicum* by use of successive electrophoretic procedures and activity staining. They detected a single polypeptide associated with the active hydrogenase. Since no quantitative estimates of specific activity or extent of purification were reported, a comparison of the results with those of other researchers is not possible.

The hydrogenase purified from chemolithotrophically cultured *B. japonicum* contains two subunits of approximately 60 and 30 kd (63). The two subunits were present in a 1 : 1 molar ratio and were immunologically distinct

Table 1 A comparison of some properties of hydrogenases from representative bacteria

Bacterium	Molecular mass (kd)	Subunit mass (kd)	Ni[a]	Fe[a]	Specific[b] activity	K_m	Reversibility[c]	pI	pH optimum	O_2 lability ($T_{1/2}$)	Heat stability	References
Bradyrhizobium japonicum	100	63, 33	0.6	6.5	75	1	1	5.5	5.5–6.0	70 min		7, 11, 63, 118, 119
Azotobacter vinelandii	98	67, 31	0.7	6.6	124	0.9	3	5.8–6.1	6.0–8.5	20 min	70°C	73, 111
Escherichia coli	200	64, 35	0.64	12	10.6	2.0	15	4.7		12 hr	70°C	1, 13, 103
Alcaligenes eutrophus	99	67, 31	0.9	6.1	170	32		6.5	5.5	stable	52°C	105, 109
Rhodopseudomonas capsulata	65	65	1.0	3.6	7.4		4	5.5		5 d	70°C	32, 33, 56
Thiocapsa roseopersicina	68?	47, 25	present	3.9	1.3		+	4.2		6 d	70°C	56, 57, 124
Desulfovibrio gigas	89	63, 26	1.0	11	500[d]		+		7.5–8.0	stable		17, 65, 76, 120

[a] Gram atoms of metal per mole of enzyme.
[b] Specific activity in μmol H_2 oxidized·min^{-1}·mg protein^{-1} with methylene blue as the electron acceptor.
[c] Activity measured as H_2 evolution and expressed as percentage of uptake activity.
[d] Data only available for H_2 evolution.

from each other. The smaller polypeptide was susceptible to proteolysis, but this condition could be obviated by the addition of protease inhibitors during purification. These results have been confirmed for hydrogenase purified from bacteroids (7) and most recently for hydrogenase purified by an aerobic procedure using Hup^c mutants (118). Specific activities from all the recent purifications are comparable, ranging from 65 to 75 μmol H_2 min^{-1} mg protein^{-1} with methylene blue as the electron acceptor.

Ambiguities have been reported in the subunit composition of hydrogenases from other species. The membrane-bound hydrogenase of *A. vinelandii* was initially reported to be a monomer of 60 kd, which retained activity when electrophoresed anaerobically on SDS polyacrylamide gels (73). Subsequent studies revealed two subunits of 67 and 31 kd (12, 111). It has been confirmed that the purified protein, when treated with SDS and electrophoresed anaerobically, retains some activity and migrates as a single polypeptide of ~55–60 kd. When the protein is excised and treated aerobically with SDS its activity is lost and it migrates as two polypeptides of 67 and 31 kd (L. C. Seefeldt & D. J. Arp, personal communication). Hydrogenases from several sources are reported to be resistant to denaturation and inactivation by urea and, to a lesser extent, by SDS (see references in Table 1).

Problems associated with incomplete denaturation and dissociation of subunits may also have influenced results of investigations of uptake hydrogenases from other sources where single subunits have been reported. Hydrogenase from *E. coli* as originally purified consisted of two identical subunits of 56 kd and a native molecular mass of about 113 kd (1). Polypeptides of 60 and 30 kd have been detected in aerobically and anaerobically cultured *E. coli* that cross-react with antibodies against the 60- and 30-kd subunits of *B. japonicum* (64). Hup activity was correlated directly with the presence of these two polypeptides. Recent purification of the membrane hydrogenase from anaerobically cultured *E. coli* showed the presence of two distinct subunits of 64 and 35 kd (103). An analysis of hydrogenase subunits from *Thiocapsa roseopersicina* revealed three polypeptide bands of 68, 47, and 25 kd (57). Tigyi et al (124) have stated that the 68- and 25-kd polypeptides from *T. roseopersicina* are inactive and that upon activation they form the 47-kd protein. The authors suggested the occurrence of protein rearrangements, although incomplete denaturation prior to electrophoresis might be a more probable explanation.

METAL CONTENT All hydrogenases that have been examined in detail are iron-sulfur proteins (see reviews 2, 92). The hydrogenase from *Desulfovibrio gigas* contains one 3Fe center and two Fe_4S_4 centers (120). The hydrogenase of *B. japonicum* is an iron-sulfur protein (7, 9), but the number and type of Fe-S centers present has not been determined.

Nickel has been shown to be a necessary trace element for chemolithotrophic growth and for expression of hydrogenase activity in H_2-oxidizing bacteria (e.g. 14, 54). Furthermore, nickel is now known to be a component of active hydrogenases from many organisms (e.g. 33, 55). Klucas et al (72) used extensively purified chemicals to demonstrate that Ni is specifically required for expression of hydrogenase activity and chemolithotrophic growth of *B. japonicum* 122DES. Subsequently, nickel has been reported to co-migrate with hydrogenase activity during electrophoresis of extracts from derepressed *B. japonicum* (119) and purified protein from chemolithotrophically cultured *B. japonicum* (63). Arp (7) demonstrated that the homogeneous enzyme from *B. japonicum* bacteroids contained 0.6 mole of nickel per mole of enzyme.

CATALYTIC PROPERTIES AND MECHANISMS Kinetic constants have not been measured for many of the uptake hydrogenases. For those that have been studied, the K_m for H_2 is relatively low ($\sim$1 μM) (e.g. 9, 10, 111). The purified hydrogenase of *Alcaligenes latus,* by contrast, has a K_m for H_2 of 20–25 μM (94). The uptake hydrogenases of the aerobic nitrogen-fixing microorganisms, in general, catalyze the reverse reaction at a low rate (Table 1). Some studies with other organisms only note that the enzyme is reversible, while others indicate that rates of H_2 evolution are comparable to rates of H_2 oxidation.

Relatively little work has been reported regarding the mechanism of action of the hydrogenase from *B. japonicum*. Noncompetitive patterns of inhibition in kinetic studies by Arp & Burris (10) suggested a two-site, ping-pong mechanism. Recent work with hydrogenase from *Desulfovibrio gigas* (59) also indicates separate binding sites for H_2 and electron acceptors. The midpoint potentials of the Fe-S center and the Ni center in hydrogenase from *D. gigas* have been identified using an EPR redox titration method in the presence of dyes as mediators (120). With dithionite reduction of the enzyme, the centers were reduced in order of decreasing midpoint potential. Furthermore, the Ni center was preferentially reduced by H_2. Oxidized enzymes from all sources thus far examined have contained Ni in a +3 redox state (e.g. 3, 76, 120). Upon the addition of H_2, a photosensitive monovalent nickel species was detected in EPR studies (127). The photodissociation of this component occurred six times faster in H_2O than in D_2O, indicating involvement of a Ni-H bond.

Many of the nickel-containing uptake hydrogenases have been purified aerobically. These enzymes require a period of activation in the presence of reducing agents and the absence of O_2 (53, 71, 108). The activation of hydrogenase was assumed to be associated with the displacement and removal of O_2, but the addition of O_2 scavengers alone failed to reactivate the oxidized enzyme. Berlier et al (18) and Lissolo et al (79) have proposed a two-step process comprised of O_2 removal followed by reduction of the redox centers. Teixeira et al (121) have proposed a scheme that incorporates information about the lag and reduction phases of the activation and catalytic cycles associated with enzymatic oxidation of H_2. The scheme is based on extensive EPR data from investigation of the hydrogenase from *D. gigas* and involves Ni(III) states in oxygenated, deoxygenated, and active forms of the enzyme. Dihydrogen binding to the active form yields a Ni(III) hydride intermediate. The nickel center is reduced to a Ni(II) state with the release of a proton. All of these data indicate that Ni is the site of H_2 activation and that electrons are subsequently transferred to a mediator binding site via Fe-S clusters.

IMMUNOLOGICAL COMPARISONS An early study of the immunological properties of hydrogenases demonstrated strong homology between the membrane-bound hydrogenases of *A. eutrophus* and of *Pseudomonas pseudoflava* (106). The antisera against the *Alcaligenes* hydrogenase exhibited modest cross-reaction with hydrogenase from *R. capsulata, A. vinelandii, Desulfovibrio* (several species), *Chromatium vinosum,* and *T. roseopersicina,* but no cross-reacting material was detected in *B. japonicum* 122DES, *E. coli* K12, or *Clostridium pasteurianum*. These data were collected using Ouchterlony double-diffusion assays; most of the assays were performed using cytoplasmic or membrane extracts rather than purified hydrogenase proteins.

More recent studies have used ELISA assays and purified proteins and have thus produced more accurate information regarding cross-reactivity. The periplasmic hydrogenase from *D. desulfuricans* exhibits strong immunological homology to that of *D. vulgaris,* but no homology to the membrane-bound enzyme from *D. desulfuricans* or *D. gigas* (130). There was very limited cross-reaction with the membrane-bound enzyme from *E. coli* H61. Antiserum against hydrogenase from *B. japonicum* cross-reacted with purified membrane-bound hydrogenases from four bacteria with the following order of activity: *B. japonicum* > *A. latus* > *A. eutrophus* > *A. vinelandii* (12). Harker et al (64) demonstrated cross-reaction between affinity-purified antibodies against the subunits of *B. japonicum* hydrogenase and polypeptides of comparable size from several strains of Hup^+ *E. coli.* Although these data are by no means comprehensive, they do indicate significant homology between the Ni-containing uptake hydrogenases.

GENETICS OF H_2 OXIDATION

Isolation of a variety of H_2 uptake-deficient mutants by various investigators has revealed the complexity of the regulation of H_2 oxidation in *B. japonicum*. Readers should refer to a recent review by Maier (81) for a detailed account of Hup^- mutants in *B. japonicum*.

Isolation of Uptake Hydrogenase Genes

Studies initiated to determine whether a correlation existed between plasmid content and Hup^+ phenotype in *B. japonicum* revealed that none of the strains with high Hup activities contained discernible plasmids (29). Consequently, a gene bank of *B. japonicum* DNA was constructed in cosmid vector pLAFR1 and was used to isolate cosmid pHU1, which when transferred to Hup^- point mutants of *B. japonicum,* conferred the capacity for H_2 oxidation and chemolithotrophic growth (28, 67). Haugland et al (66) further characterized pHU1 by generating Tn*5* insertion mutations in the *B. japonicum* insert DNA and subsequently scoring phenotypes conferred by these individual Tn*5* insertion mutations after their marker exchange into the *B. japonicum* USDA 122DES chromosome. From these investigations it was concluded that the *hup*-specific sequences spanned about 15 kb of insert DNA on pHU1 (Figure 3) and that the *hup* genes were organized into at least two, and probably three, transcriptional units. However, transfer of pHU1 into Hup^- field isolates of *B. japonicum* and *R. meliloti* did not confer Hup activity in the free-living state; this suggests that additional DNA sequences are required for complementation. Therefore, Lambert et al (74) isolated an additional Hup-complementing cosmid, pHU52, from the *B. japonicum* gene bank, using one of the Tn*5*-generated Hup^- mutants (Tn*5* insertion mutant 2) as the recipient (Figure 3). The cosmid pHU52 contained insert DNA with restriction fragments similar to those in pHU1 and also an additional 5.5-kb *Eco*RI fragment

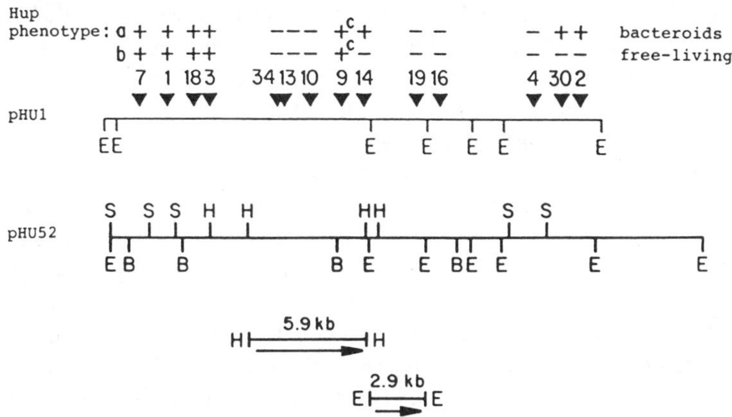

Figure 3 Restriction maps for insert DNA of pHU1 and pHU52. The *triangles* indicate positions of Tn*5* insertions in pHU1 DNA. Positions of insertions in pHU1 are numbered, and + or − below each number indicates whether that insertion resulted in a Hup^+ or Hup^- phenotype in the (*a*) symbiotic state or (*b*) free-living state (66). *B, E, S,* and *H* are the restriction sites recognized by the enzymes *Bgl*II, *Eco*RI, *Sma*I, and *Hin*dIII, respectively. The 5.9-kb *Hin*dIII and 2.9-kb *Eco*RI fragments contain the structural genes for the 60- and 30-kd hydrogenase polypeptide components, respectively (131). *Arrows* indicate the orientation of respective genes. The Hup activity of the mutant resulting from transposon insertion 9 (*c*) was weak.

at the right-hand end (Figure 3). Conjugal transfer of pHU52, in contrast to that of pHU1, conferred Hup activity on all Hup$^-$ *B. japonicum* mutants (except one dominant mutant) and also on free-living Hup$^-$ field isolates of *B. japonicum, R. meliloti, R. leguminosarum,* and *R. trifolii.* Furthermore, pHU52 conferred chemolithotrophic growth capability on these recipient strains (74).

Hom et al (68) recently isolated recombinant cosmids from a *B. japonicum* I-110 gene bank (originally prepared by G. Stacey) that complemented for hydrogenase and nitrogenase activities in a Hup$^-$ Nif$^-$ mutant strain. The *B. japonicum* insert DNA in these *nif/hup* cosmids showed a restriction pattern similar to that of pHU1, although sequence homology must be demonstrated to prove genetic similarity between these two cloned DNA fragments. One of these *nif/hup* recombinant cosmids also complemented several Hup$^-$ Nif$^+$ mutants; this indicates that the cosmid contains a gene involved in both nitrogenase and hydrogenase activities and at least one *hup* gene exclusively involved in H_2 uptake activity. If the *nif/hup* complementing cosmid proves to have homology with pHU1, then the *nif/hup* gene must be located between the Tn*5* insertions (66; R. Haugland, unpublished) because all of the Hup$^-$ Tn*5* insertion mutants were shown to be Nif$^+$.

Hup Determinants in Rhizobium leguminosarum

Hydrogenase activity in *R. leguminosarum,* in contrast to that in *B. japonicum,* is expressed only in symbiotic association with the host. Conditions for derepression of Hup activity have not been defined for free-living *R. leguminosarum.* Consequently, screening for Hup$^-$ mutants and complementation studies must be carried out using nodule bacteroids. The genetic determinants for hydrogenase activity in *R. leguminosarum* 128C53 are located on the nontransmissible plasmid pRL6J1 and are linked to the determinants for nodulation ability residing on the same plasmid (23). Hup$^+$ determinants were transferred to Hup$^-$ strains of *R. leguminosarum* only after recombination of pRL6J1 with pVW5J1, a kanamycin-resistant derivative of a transmissible *R. leguminosarum* plasmid (23, 24, 36). Recently, Sorenson & Wyndaele (116) also reported that the Hup determinants in *R. leguminosarum* MA1 were plasmid-borne and could be mobilized into Hup$^-$ strains only after recombination with the self-transmissible plasmid pVW5J1. Sym-Hup plasmids from Hup$^+$ *R. leguminosarum* strains were successfully transferred into a plasmid-cured strain using the Tn*5*-mob system of the suicide plasmid vector pSUP5011 (102). As bacteroids in pea nodules, two of the three transconjugants containing the pSym-Hup::Tn*5*-mob plasmid exhibited a capacity to oxidize H_2. Interspecies transfer of Hup determinants located on the *R. leguminosarum* recombinant plasmid pIJ1008 into *R. meliloti* and expression of Hup activity in *Medicago sativa* nodules have also been demonstrated (15, 16). Tn*5* insertion mutations that resulted in a Hup^- phenotype in *R. leguminosarum* were successfully complemented by pHU1 (28), a cosmid clone carrying the *hup* genes of *B. japonicum* (70). Furthermore, Tichy et al (123) isolated the plasmid pR1B505 from the gene bank of *R. leguminosarum* by using a subclone of pHU1 as a hybridization probe. It also appears that *hup*-specific DNA sequences are conserved among Hup^+ strains of *R. leguminosarum* (86).

Identification of Uptake Hydrogenase Structural Genes

The *hup*-complementing cosmid pHU52 apparently encodes all essential Hup determinants (74). However, direct evidence for the presence of uptake hydrogenase structural genes on pHU52 has been lacking. Harker et al (62) reported that conjugal transfer of pHU52 into Hup^- rhizobium strains conferred the ability to synthesize both the 60- and 30-kd polypeptide components of the uptake hydrogenase; this suggests that the determinants for hydrogenase subunit synthesis were present on pHU52. Recently, Zuber et al (131) constructed subclones of pHU52 in the plasmid expression vector pMZ545, which generates transcriptional but not translational fusions. These investigators identified two subclones that directed the synthesis of 60-kd and 30-kd proteins in *E. coli* maxicells. Expression of immunologically cross-reactive hydrogenase polypeptides in *E. coli* was also demonstrated when the DNA inserts from these subclones (Figure 3) were placed under the transcriptional control of the lambda P_L promoter. These results provide direct evidence for the presence of uptake hydrogenase structural genes on pHU52.

Symbiotic Expression of hup *Genes*

In view of the potential beneficial effects of H_2 recycling capability in the legume-rhizobium symbiosis, it is desirable to transfer *B. japonicum hup* genes into Hup^- field isolates of rhizobium species to obtain stable expression of Hup activity in nodulated legumes. The conjugal transfer of cosmid pHU52 conferred Hup activity in the free-living state and chemolithotrophic growth capability on several Hup^- wild-type rhizobia (74). Bacteroids from nodules formed by *B. japonicum* USDA138 (pHU52), *R. meliloti* 102F28 (pHU52), and *R. trifolii* SU794 (pHU52) (75) showed low levels of Hup activity consistent with the relative instability of the cosmid in nodule bacteroids. In contrast, acquisition of pHU1 by Hup^- *B. japonicum* and Hup^- *R. meliloti* conferred Hup activity in soybean and alfalfa nodules, respectively, but did not confer activity to these strains in the free-living state. Therefore, Lambert et al (74a) concluded that an extra 5.5-kb *Eco*RI fragment located at the right-hand end of pHU52 (Figure 3) contains a gene or part of a gene required for expression of Hup activity in free-living bacteria but not in nodules. Since the Tn*5* insertions 30 and 2 conferred Hup activity in symbiotic but not in free-living rhizobia (Figure 3), these investigators suggested that the gene required for expression of Hup activity in free-living bacteria probably extends into the proximal 5.0-kb *Eco*RI fragment of pHU1.

Although the Hup-complementing cosmids pHU1 and pHU52 can be transferred within and among species of rhizobium, a major drawback from the viewpoint of agricultural application is the loss of these cosmids at a rapid rate in nodule bacteroids in the absence of selection pressure. It is essential, therefore, to stably integrate the cosmid-borne *hup* genes into the chromosome of the recipient organism if the goal is to construct a strain for practical use under field conditions.

PRACTICAL APPLICATIONS

Reliable conclusions regarding the in vivo benefits from H_2 recycling in legume nodules require evaluation of inoculants in carefully controlled growth trials. The following precautions have been recommended for conducting such experiments (51): (*a*) Trials should use Hup^+ and Hup^- inoculant strains that are isogenic except for the H_2 oxidation characteristic. (*b*) The genetic determinants for H_2 oxidation in the rhizobium strains being evaluated should be stable, and recycling capacity should be sufficiently vigorous that essentially all H_2 evolved during N_2 fixation is consumed. (*c*) The Hup^+ strains being evaluated should efficiently couple H_2 oxidation to ATP synthesis. (*d*) Legume hosts used in strain evaluations should be grown under conditions that permit the nodulated plants to reach maturity normally; this provides the opportunity for compounding any benefits from H_2 recycling during the logarithmic phase of growth. We conclude that most of the evaluations that have been made have not given sufficient attention to the precautions listed above.

Eisbrenner & Evans (46) summarized results of growth trials in which Hup^+ and Hup^- strains were compared and concluded that 9 of 13 trials showed statistically significant increases in the N content of plant materials from the use of Hup^+ strains. More recently, several evaluations have indicated that the Hup characteristic had no significant effect on growth or N_2 fixation. For example, *Pisum sativum* and *Vicia bengalensis* inoculated with a Hup^+ parent strain of *R. leguminosarum* and grown in Leonard jar assemblies produced no more growth than the same hosts inoculated with Hup^- mutants of *R. leguminosarum* produced by Tn*5*-mob insertions (35). However, the strain from which the determinants for Hup were derived in these experiments showed only weak coupling of H_2 oxidation to ATP synthesis. Also, the transfer of Hup determinants from *R. leguminosarum* into *R. meliloti* (16) resulted in increases in H_2 uptake rates of nodules formed by the *R. meliloti* recipient strain, but had no significant effect on dry matter in *Medicago* *sativa*. In these experiments, the nodules formed by the *R. meliloti* that had acquired Hup determinants failed to recycle a majority of the H_2 formed during N_2 fixation. Also, nodulated *M. sativa* was cultured in relatively small plastic containers that may have restricted root growth and necessitated harvest of plants prior to maturity.

Evans et al (51) have described a growth trial with *Glycine max* in which the Hup^- mutant PJ18 was compared as an inoculant with the stable Hup^+ PJ18HR. The Hup^+ strain was isogenic with the Hup^- mutant, with the exception that it contained a functional H_2 recycling system. This experiment included 10 treatment replicates, each in 1-m concrete tiles, provided with all nutrients except fixed nitrogen. Nodules of *G. max* formed by the Hup^+ strain recycled essentially all of the H_2 produced during N_2 fixation throughout the growth period, whereas nodule bacteroids from plants inoculated with the Hup^- strain showed no H_2 uptake capability and therefore recycled little or no H_2. After a growth period of 141 days total N content of plants inoculated with Hup^+ strains had increased 8.6% in seed, 27% in leaves, and 11% in total plants. These differences were significant at the 0.03 or greater confidence level. Significant increases in dry weight of the plant components were also observed, but treatment had no effect on the oil content of the seed. In addition, field trials in which groups of Hup^+ and Hup^- field isolates of *B. japonicum* were compared as inoculants for *G. max* showed an 8.9% ($P < 0.05$) increase in the protein content of grain from plants inoculated with the Hup^+ strain (60). Furthermore, Hungria & Neves (69) have reported that the N content of *Phaseolus vulgaris* pods was positively correlated ($r = 0.87$) with nodule H_2 recycling capability in an experiment involving five cultivars and six *Rhizobium phaseoli* strains.

The weight of evidence from the many growth evaluations and in vitro studies of the potential advantages of H_2 recycling support the conclusion that H_2 recycling capability is a desirable characteristic that may increase the yield and total N content of nodulated legumes under appropriate conditions. However, responses to inoculation with Hup^+ strains may be difficult to demonstrate in the field, especially in soils of the midwestern United States, which are known to contain large populations of highly competitive indigenous strains of *B. japonicum* (see 107). When legumes are grown in soils with relatively high N content, the proportion of required N provided by N_2 fixation is decreased. In this situation responses to different inoculants would be expected to be minimized.

There appears to be no problem in the transfer of the genetic determinants for Hup from one strain or species of rhizobium to another. However, it is desirable to use genetic methods that will facilitate incorporation of Hup determinants into the bacterial chromosome. Although H_2 oxidation genes may be expressed as constituents of extrachromosomal cosmids or plasmids, many of these genetic elements are unstable in the absence of selection pressure, which is the situation that prevails under normal field conditions (see 75). There is a real need for a better understanding of why different leguminous hosts influence Hup expression by rhizobial endophytes (126). More research is also required on the ecological significance of the chemolithotrophic growth capability of Hup^+ rhizobium strains (50, 75).

ACKNOWLEDGMENTS

Research from this laboratory was supported by grants from the National Science Foundation (DBM 85-13846), the Agrigenetics Corporation, and the Oregon Agricultural Experiment Station, from which this is Technical Paper No. 8053. We express appreciation to Mrs. Sheri Woods-Haas for typing the manuscript and to many researchers who provided us with reprints or manuscripts describing research relevant to this review.

Literature Cited

1. Adams, M. W. W., Hall, D. O. 1979. Purification of the membrane-bound hydrogenase of *Escherichia coli*. *Biochem. J.* 183:11–22
2. Adams, M. W. W., Mortenson, L. E., Chen, J. S. 1981. Hydrogenase. *Biochim. Biophys. Acta* 594:105–76
3. Albracht, S. P. J., Kalkman, M. L., Slater, E. C. 1983. Magnetic interaction of nickel (III) and the iron-sulphur cluster in hydrogenase from *Chromatium vinosum*. *Biochim. Biophys. Acta* 724: 309–16
4. Appleby, C. A. 1969. Electron transport systems of *Rhizobium japonicum*. II. *Rhizobium* haemoglobin, cytochromes and oxidases in free-living (cultured) cells. *Biochim. Biophys. Acta* 172:88–105
5. Aquilar, O. M., Yates, M. G., Postgate, J. R. 1985. The beneficial effect of hydrogenase in *Azotobacter chroococcum* under nitrogen-fixing carbon-limiting conditions in continuous and batch cultures. *J. Gen. Microbiol.* 131:3141–45
6. Arima, Y. 1981. Respiration and efficiency of N_2 fixation by nodules formed with a H_2-uptake positive strain of *Rhizobium japonicum*. *Soil Sci. Plant Nutr. Tokyo* 27:115–19
7. Arp, D. J. 1985. *Rhizobium japonicum* hydrogenase: Purification to homogeneity from soybean nodules, and molecular characterization. *Arch. Biochem. Biophys.* 237:504–12
8. Arp, D. J. 1985. H_2-oxidizing hydrogenase of aerobic N_2-fixing microorganisms. In *Nitrogen Fixation and CO_2 Metabolism*, ed. P. W. Ludden, J. F. Burris, pp. 121–31. New York: Elsevier. 445 pp.
9. Arp, D. J., Burris, R. H. 1979. Purification and properties of the particulate hydrogenase from the bacteroids of soybean root nodules. *Biochim. Biophys. Acta* 570:221–30
10. Arp, D. J., Burris, R. H. 1981. Kinetic mechanism of the hydrogen-oxidizing hydrogenase from soybean nodule bacteroids. *Biochemistry* 20:2234–40
11. Arp, D. J., Burris, R. H. 1982. Isotope exchange and discrimination by the H_2-oxidizing hydrogenase from soybean root nodules. *Biochim. Biophys. Acta* 700:7–15
12. Arp, D. J., McCollum, L. C., Seefeldt, L. C. 1985. Molecular and immunological comparison of membrane-bound, H_2-oxidizing hydrogenases of *Bradyrhizobium japonicum, Alcaligenes eutrophus, Alcaligenes latus,* and *Azotobacter vinelandii*. *J. Bacteriol.* 163:15–20
13. Ballantine, S. P., Boxer, D. H. 1985. Nickel-containing hydrogenase isoenzymes from anaerobically grown *Escherichia coli* K-12. *J. Bacteriol.* 163:454–59
14. Bartha, R., Ordal, E. J. 1965. Nickel-dependent chemolithotrophic growth of two *Hydrogenomonas* strains. *J. Bacteriol.* 89:1015–19
15. Bedmar, E. J., Brewin, N. J., Phillips D. A. 1984. Effect of plasmid pIJ1008 from *Rhizobium leguminosarum* on sym-

biotic function of *Rhizobium meliloti*. *Appl. Environ. Microbiol*. 47:876–78

16. Behki, R. M., Selvaraj, G., Iyer, V. N. 1985. Derivatives of *Rhizobium meliloti* strains carrying a plasmid of *Rhizobium leguminosarum* specifying hydrogen uptake and pea-specific symbiotic functions. *Arch. Microbiol*. 140:352–57
17. Bell, G. R., Lee, J. P., Peck, H. D. Jr., LeGall, J. 1978. Reactivity of *Desulfovibrio gigas* hydrogenase toward artificial and natural electron donors or acceptors. *Biochimie* 60:315–20
18. Berlier, Y. M., Fauque, G., Lespinat, P. A., LeGall, J. 1982. Activation, reduction and proton-deuterium exchange reaction of the periplasmid hydrogenase from *Desulfovibrio gigas* in relation with the role of cytochrome c_3. *FEBS Lett*. 140:185–88
19. Bishop, P. E. 1986. A second nitrogen fixation system in *Azotobacter vinelandii*. *Trends Biochem. Sci*. 11:225–27
20. Bishop, P. E., Premakumar, R., Dean, D. R., Jacobson, M. R., Chisnell, J. R., et al. 1986. Nitrogen fixation by *Azotobacter vinelandii* strains having deletions in structural genes for nitrogenase. *Science* 232:92–94
21. Bowien, B., Schlegel, H. G. 1981. Physiology and biochemistry of aerobic hydrogen-oxidizing bacteria. *Ann. Rev. Microbiol*. 35:405–52
22. Brewin, N. J. 1984. Hydrogenase and energy efficiency in nitrogen-fixing symbionts. In *Genes Involved in Plant Microbe-Interactions*, ed. D. P. S. Verma, T. Hohn, pp. 179–203. New York/Vienna: Springer-Verlag. 393 pp.
23. Brewin, N. J., DeJong, T. M., Phillips, D. A., Johnston, A. W. B. 1980. Cotransfer of determinants for hydrogenase activity and nodulation ability in *Rhizobium leguminosarum*. *Nature* 288:77–79
24. Brewin, N. J., Wood, E. A., Johnston, A. W. B., Dibb, N. J., Hombrecher, G. 1982. Recombinant nodulation plasmids in *Rhizobium leguminosarum*. *J. Gen. Microbiol*. 128:1817–27
25. Burgess, B. K. 1985. Nitrogenase mechanism. An overview. In *Nitrogen Fixation Research Progress*, ed. H. J. Evans, P. J. Bottomley, W. E. Newton, pp. 543–49. Dordrecht, Netherlands: Nijhoff. 731 pp.
26. Burris, R. H., Arp, D. J., Benson, D. R., Emerich, D. W., Hageman, R. V., et al. 1980. The biochemistry of nitrogenase. In *Nitrogen Fixation*, ed. W. D. P. Stewart, J. R. Gallon, pp. 37–54. London: Academic. 451 pp.
27. Burris, R. H., Arp., D. J., Hageman, R. V., Houchins, J. P., Sweet, W. J., Tso, M. Y. 1981. Mechanism of nitrogenase action. In *Current Perspectives in Nitrogen Fixation*, ed. A. H. Gibson, W. E. Newton, pp. 56–66. New York: Elsevier/North-Holland. 534 pp.
28. Cantrell, M. A., Haugland, R. A., Evans, H. J. 1983. Construction of a *Rhizobium japonicum* gene bank and use in isolation of a hydrogen uptake gene. *Proc. Natl. Acad. Sci. USA* 80:181–85
29. Cantrell, M. A., Hickok, R. E., Evans, H. J. 1982. Identification and characterization of plasmids in hydrogen uptake positive and hydrogen uptake negative strains of *Rhizobium japonicum*. *Arch. Microbiol*. 131:102–6
30. Carter, K. R., Rawlings, J., Orme-Johnson, W. H., Becker, R. R., Evans, H. J. 1980. Purification and characterization of a ferredoxin from *Rhizobium japonicum* bacteroids. *J. Biol. Chem*. 255:4213–23
31. Chatt, J. 1980. Chemistry relevant to the biological fixation of nitrogen. See Ref. 26, pp. 1–17
32. Colbeau, A., Chabert, J., Vignais, P. M. 1983. Purification, molecular properties and localization in the membrane of the hydrogenase of *Rhodopseudomonas capsulata*. *Biochim. Biophys. Acta* 748:116–27
33. Colbeau, A., Vignais, P. M. 1983. The membrane-bound hydrogenase of *Rhodopseudomonas capsulata* is inducible and contains nickel. *Biochim. Biophys. Acta* 748:128–38
34. Conrad, R., Seiler, W. 1980. Contribution of hydrogen production by biological nitrogen fixation to the global hydrogen budget. *J. Geophys. Res*. 85:5493–98
35. Cunningham, S. D., Kapulnik, Y., Brewin, N. J., Phillips, D. A. 1985. Uptake hydrogenase activity determined by plasmid pRL6JI in *Rhizobium leguminosarum* does not increase symbiotic nitrogen fixation. *Appl. Environ. Microbiol*. 50:791–94
36. DeJong, T. M., Brewin, N. J., Johnston, A. W. B., Phillips, D. A. 1982. Improvement of symbiotic properties in *Rhizobium leguminosarum* by plasmid transfer. *J. Gen. Microbiol*. 128:1829–38
37. DeVries, W., Stam, H., Stouthamer, A. H. 1984. Hydrogen oxidation and nitrogen fixation in rhizobia, with special attention focused on strain ORS 571. *Antonie van Leeuwenhoek J. Microbiol. Serol*. 50:505–24
38. Dixon, R. O. D. 1967. Hydrogen uptake and exchange by pea root nodules. *Ann. Bot. London* 31:179–88

39. Dixon, R. O. D. 1968. Hydrogenase in pea root nodule bacteroids. *Arch. Microbiol.* 62:272–83
40. Dixon, R. O. D. 1972. Hydrogenase in legume root nodule bacteroids: occurrence and properties. *Arch. Microbiol.* 85:193–201
41. Dixon, R. O. D., Blunden, E. A. G. 1983. The relative efficiency of nitrogen fixation in pea root nodules. *Plant Soil* 75:131–38
42. Drevon, J. J., Frazier, L., Russell, S. A., Evans, H. J. 1982. Respiratory and nitrogenase activities of soybean nodules formed by hydrogen uptake negative (Hup^-) mutant and revertant strains of *Rhizobium japonicum* characterized by protein patterns. *Plant Physiol.* 70: 1341–46
43. Drevon, J. J., Kalia, V. C., Heckmann, M. O., Salsac, L. 1985. Influence of hydrogenase on the growth of *Glycine* and *Vigna* spp. See Ref. 25, p. 358
44. Eisbrenner G., Evans, H. J. 1982. Carriers in the electron transport from molecular hydrogen to oxygen in *Rhizobium japonicum* bacteroids. *J. Bacteriol.* 149:1005–12
45. Eisbrenner, G., Evans, H. J. 1982. Spectral evidence for a component involved in hydrogen metabolism of soybean nodule bacteroids. *Plant Physiol.* 70:1667–72
46. Eisbrenner, G., Evans, H. J. 1983. Aspects of hydrogen metabolism in nitrogen-fixing legumes and other plant-microbe associations. *Ann. Rev. Plant Physiol.* 34:105–36
47. Eisbrenner, G., Hickok, R. E., Evans, H. J. 1982. Cytochrome patterns in *Rhizobium japonicum* cells grown under chemolithotrophic conditions. *Arch. Microbiol.* 132:230–35
48. Emerich, D. W., Ruiz-Argueso, T., Ching, T. M., Evans, H. J. 1979. Hydrogen-dependent nitrogenase activity and ATP formation in *Rhizobium japonicum* bacteroids. *J. Bacteriol.* 137:153–60
49. Emerich, D. W., Ruiz-Argueso, T., Russell, S. A., Evans, H. J. 1980. Investigation of the H_2 oxidation system in *Rhizobium japonicum* 122DES nodule bacteroids. *Plant Physiol.* 66:1061–66
50. Evans, H. J., Emerich, D. W., Lepo, J. E., Maier, R. J., Carter, K. R., et al. 1980. The role of hydrogenase in nodule bacteroids and free-living rhizobia. See Ref. 26, pp. 55–81
51. Evans, H. J., Hanus, F. J., Haugland, R. A., Cantrell, M. A., Xu, L. S., et al. 1985. Hydrogen recycling in nodules affects nitrogen fixation and growth of soybeans. *Proc. World Soybean Conf., 3rd, Ames, Iowa,* pp. 935–42. Boulder, Colo: Westview. 1262 pp.
52. Evans, H. J., Purohit, K., Cantrell, M. A., Eisbrenner, G., Russell, S. A. 1981. Hydrogen losses and hydrogenases in nitrogen-fixing organisms. See Ref. 27, pp. 84–96
53. Fernandez, V. M., Aguirre, R., Hatchikian, E. C. 1984. Reductive activation and redox properties of hydrogenase from *Desulfovibrio gigas*. *Biochim. Biophys. Acta* 790:1–7
54. Friedrich, B., Heine, E., Finck, A., Friedrich, C. G. 1981. Nickel requirement for active hydrogenase formation in *Alcaligenes eutrophus*. *J. Bacteriol.* 145:1144–49
55. Friedrich, C. G., Schneider, K., Friedrich, B. 1982. Nickel in the catalytically active hydrogenase of *Alcaligenes eutrophus*. *J. Bacteriol.* 152:42–48
56. Gogotov, I. N. 1986. Hydrogenases of phototrophic microorganisms. *Biochimie* 68:181–87
57. Gogotov, I. N., Zorin, N. A., Serebriakova, L. T., Kondratieva, E. N. 1978. The properties of hydrogenase from *Thiocapsa roseopersicina*. *Biochim. Biophys. Acta* 523:335–43
58. Hageman, R. V., Burris, R. H. 1980. Electron allocation to alternative substrates of *Azotobacter* nitrogenase is controlled by the electron flux through dinitrogenase. *Biochim. Biophys. Acta* 591:63–75
59. Hallahan, D. L., Fernandez, V. M., Hatchikian, E. C., Hall, D. O. 1986. Differential inhibition of catalytic sites in *Desulfovibrio gigas* hydrogenase. *Biochimie* 68:49–54
60. Hanus, F. J., Albrecht, S. L., Zablotowicz, R. M., Emerich, D. W., Russell, S. A., Evans, H. J. 1981. Yield and N content of soybean seed as influenced by *Rhizobium japonicum* inoculants possessing the hydrogenase characteristic. *Agron. J.* 73:368–72
61. Hanus, F. J., Maier, R. J., Evans, H. J. 1979. Autotrophic growth of H_2-uptake positive strains of *R. japonicum* in an atmosphere supplied with hydrogen gas. *Proc. Natl. Acad. Sci. USA* 76:1788–92
62. Harker, A. R., Lambert, G. R., Hanus, F. J., Evans, H. J. 1985. Further evidence that two unique subunits are essential for expression of hydrogenase activity in *Rhizobium japonicum*. *J. Bacteriol.* 164:187–91
63. Harker, A. R., Xu, L. S., Hanus, F. J., Evans, H. J. 1984. Some properties of the nickel-containing hydrogenase of

chemolithotrophically grown *Rhizobium japonicum*. *J. Bacteriol.* 159:850–56

64. Harker, A. R., Zuber, M., Evans, H. J. 1986. Immunological homology between the membrane-bound uptake hydrogenases of *Rhizobium japonicum* and *Escherichia coli*. *J. Bacteriol.* 165:579–84
65. Hatchikian, E. C., Bruschi, M., LeGall, J. 1978. Characterization of the periplasmid hydrogenase from *Desulfovibrio gigas*. *Biochem. Biophys. Res. Commun.* 82:451–61
66. Haugland, R. A., Cantrell, M. A., Beaty, J. S., Hanus, F. J., Russell, S. A., Evans, H. J. 1984. Characterization of *Rhizobium japonicum* hydrogen uptake genes. *J. Bacteriol.* 159:1006–12
67. Haugland, R. A., Hanus, F. J., Cantrell, M. A., Evans, H. J. 1983. Rapid colony screening method for identifying hydrogenase activity in *Rhizobium japonicum*. *Appl. Environ. Microbiol.* 45:892–97
68. Hom, S. S. M., Graham, L. A., Maier, R. J. 1985. Isolation of genes (*nif/hup* cosmids) involved in hydrogenase and nitrogenase activities in *Rhizobium japonicum*. *J. Bacteriol.* 161:882–87
69. Hungria, M., Neves, M. C. P. 1986. Interacao entre cultivares de *Phaseolus vulgaris* e estirpes de *Rhizobium* na fixacao e transporte do nitrogenio. *Pesqui. Agropecu. Bras.* 21:127–40
70. Kagan, S. A., Brewin, N. J. 1985. Mutagenesis of a *Rhizobium* plasmid carrying hydrogenase determinants. *J. Gen. Microbiol.* 131:1141–47
71. Kakuno, T., Kaplan, N. O., Kamen, M. D. 1977. *Chromatium* hydrogenase. *Proc. Natl. Acad. Sci. USA* 74:861–63
72. Klucas, R. V., Hanus, F. J., Russell, S. A., Evans, H. J. 1983. Nickel: a micronutrient element for hydrogen-dependent growth of *Rhizobium japonicum* and for expression of urease activity in soybean leaves. *Proc. Natl. Acad. Sci. USA* 80:2253–57
73. Kow, Y. W., Burris, R. H. 1984. Purification and properties of membrane-bound hydrogenase from *Azotobacter vinelandii*. *J. Bacteriol.* 159:564–69
74. Lambert, G. R., Cantrell, M. A., Hanus, F. J., Russell, S. A., Haddad, K. R., Evans, H. J. 1985. Intra- and interspecies transfer and expression of *Rhizobium japonicum* hydrogen uptake genes and autotrophic growth capability. *Proc. Natl. Acad. Sci. USA* 82:3232–36

74a. Lambert, G. R., Harker, A. R., Cantrell, M. A., Hanus, F. J., Russell, S. A., et al. 1987. Symbiotic expression of cosmid-borne *Rhizobium japonicum* hydrogenase genes. *Appl. Environ. Microbiol.* 53:422–28

75. Lambert, G. R., Harker, A. R., Zuber, M., Dalton, D. A., Hanus, F. J., et al. 1985. Characterization, significance and transfer of hydrogen uptake genes from *Rhizobium japonicum*. See Ref. 25, pp. 209–15
76. LeGall, J., Ljungdahl, P. O., Moura, I., Peck, H. D. Jr., Xavier, A. V., et al. 1982. The presence of redox-sensitive nickel in the periplasmic hydrogenase from *Desulfovibrio gigas*. *Biochem. Biophys. Res. Commun.* 106:610–16
77. Lepo, J. E., Hanus, F. J., Evans, H. J. 1980. Further studies on the chemoautotrophic growth of hydrogen uptake positive strains of *Rhizobium japonicum*. *J. Bacteriol.* 141:664–70
78. Lim, S. T. 1978. Determination of hydrogenase in free-living cultures of *Rhizobium japonicum* and energy efficiency of soybean nodules. *Plant Physiol.* 62:609–11
79. Lissolo, T., Pulvin, S., Thomas, D. 1984. Reactivation of the hydrogenase from *Desulfovibrio gigas* by hydrogen. *J. Biol. Chem.* 259:11725–30
80. Lowe, D. J., Thorneley, R. N. F., Postgate, J. R. 1984. The mechanism of substrate reduction by nitrogenase. In *Advances in Nitrogen Fixation Research*, ed. C. Veeger, W. E. Newton, pp. 133–38. The Hague, Netherlands: Nijhoff. 760 pp.
81. Maier, R. J. 1986. Biochemistry, regulation, and genetics of hydrogen oxidation in *Rhizobium*. *CRC Crit. Rev. Biotechnol.* 3:17–38
82. Maier, R. J., Campbell, N. E. R., Hanus, F. J., Simpson, F. B., Russell, S. A., Evans, H. J. 1978. Expression of hydrogenase activity in free-living *Rhizobium japonicum*. *Proc. Natl. Acad. Sci. USA* 75:3258–62
83. Minamisawa, K., Arima, Y., Kumazawa, K. 1983. Transport of fixed nitrogen from soybean nodules inoculated with H_2-uptake positive and negative *Rhizobium japonicum* strains. *Soil Sci. Plant Nutr. Tokyo* 29:85–92
84. Minamisawa, K., Arima, Y., Kumazawa, K. 1984. Unidentified compound specifically accumulated in soybean nodules formed with H_2-uptake negative *Rhizobium japonicum* strains. *Soil Sci. Plant Nutr. Tokyo* 30:435–44
85. Minchin, F. R., Sheehy, J. E., Witty, J. F. 1985. Factors limiting N_2 fixation by the legume-*Rhizobium* symbiosis. See Ref. 25, pp. 285–91
86. Nelson, L. M., Grosskopf, E., Tichy,

H. V., Lotz, W. 1985. Characterization of *hup*-specific DNA in *Rhizobium leguminosarum* strains of different origin. *FEMS Microbiol. Lett.* 30:53–58

87. Nelson, L. M., Salminen, S. O. 1982. Uptake hydrogenase activity and ATP formation in *Rhizobium leguminosarum* bacteroids. *J. Bacteriol.* 151:989–95
88. O'Brian, M. R., Maier, R. J. 1982. Electron transport components involved in hydrogen oxidation in free-living *Rhizobium japonicum*. *J. Bacteriol.* 152: 422–30
89. O'Brian, M. R., Maier, R. J. 1983. Involvement of cytochromes and a flavoprotein in hydrogen oxidation in free-living *Rhizobium japonicum*. *J. Bacteriol.* 155:481–87
90. O'Brian, M. R., Maier, R. J. 1985. Expression of cytochrome *o* in hydrogen uptake constitutive mutants of *Rhizobium japonicum*. *J. Bacteriol.* 161:507–14
91. O'Brian, M. R., Maier, R. J. 1985. Role of ubiquinone in hydrogen-dependent electron transport in *Rhizobium japonicum*. *J. Bacteriol.* 161:775–77
92. Odom, J. M., Peck, H. D. Jr. 1984. Hydrogenase, electron-transfer proteins, and energy coupling in the sulfate-reducing bacteria *Desulfovibrio*. *Ann. Rev. Microbiol.* 38:551–92
93. Phelps, A. S., Wilson, P. W. 1941. Occurrence of hydrogenase in nitrogen-fixing organisms. *Proc. Soc. Exp. Biol. Med.* 47:473–76
94. Pinkwart, M., Schneider, K., Schlegel, H. G. 1983. Purification and properties of the membrane-bound hydrogenase from N_2-fixing *Alcaligenes latus*. *Biochim. Biophys. Acta* 745:267–78
95. Podzuweit, H. G., Arp, D. J. 1987. Evidence for a branched respiratory system with two cytochrome oxidases in autotrophically grown *Alcaligenes latus*. *Arch. Microbiol.* 146:332–37
96. Podzuweit, H. G., Arp, D. J., Schlegel, H. G., Schneider, K. 1986. Investigation of the H_2-oxidizing activities of *Alcaligenes eutrophus* H16 membranes with artificial electron acceptors, respiratory inhibitors and redox-spectroscopic procedures. *Biochimie* 68:103–11
97. Postgate, J. R. 1982. *The Fundamentals of Nitrogen Fixation*. London: Cambridge Univ. Press. 252 pp.
98. Purohit, K., Becker, R. R., Evans, H. J. 1982. D-Ribulose-1,5-bisphosphate carboxylase/oxygenase from chemolithotrophically grown *Rhizobium japonicum* and inhibition by D-4-phosphoerythronate. *Biochim. Biophys. Acta* 715:320–29
99. Rainbird, R. M., Atkins, C. A., Pate, J. S., Sanford, P. 1983. Significance of hydrogen evolution in the carbon and nitrogen economy of nodulated cowpea. *Plant Physiol.* 71:122–27
100. Robson, R. L., Postgate, J. R. 1980. Oxygen and hydrogen in biological nitrogen fixation. *Ann. Rev. Microbiol.* 34:183–7
101. Ruiz-Argueso, T., Maier, R. J., Evans, H. J. 1979. Hydrogen evolution from alfalfa and clover nodules and hydrogen uptake by free-living *Rhizobium meliloti*. *Appl. Environ. Microbiol.* 37:582–87
102. Ruiz-Argueso, T., Palacios, J. M., Leyva, A., Villa, A. 1985. Mobilization of *sym-hup* plasmids from strains of *Rhizobium leguminosarum*. See Ref. 25, p. 231
103. Sawers, R. G., Boxer, D. H. 1986. Purification and properties of membrane-bound hydrogenase isoenzyme 1 from anaerobically grown *Escherichia coli* K12. *Eur. J. Biochem.* 156:265–75
104. Schink, B. 1982. Isolation of a hydrogenase-cytochrome *b* complex from cytoplasmic membranes of *Xanthobacter autotrophicum* GZ 29. *FEMS Microbiol. Lett.* 13:289–93
105. Schink, B., Schlegel, H. G. 1979. The membrane-bound hydrogenase of *Alcaligenes eutrophus*. I. Solubilization, purification, and biochemical properties. *Biochim. Biophys. Acta* 567:315–24
106. Schink, B., Schlegel, H. G. 1980. The membrane-bound hydrogenase of *Alcaligenes eutrophus*. II. Localization and immunological comparison with other hydrogenase systems. *Antonie van Leeuwenhoek J. Microbiol. Serol.* 46:1–14
107. Schmidt, E. L., Robert, F. M. 1985. Recent advances in the ecology of *Rhizobium*. See Ref. 25, pp. 379–85
108. Schneider, K., Cammack, R., Schlegel, H. G., Hall, D. O. 1979. The iron-sulphur centres of soluble hydrogenase from *Alcaligenes eutrophus*. *Biochim. Biophys. Acta* 578:445–61
109. Schneider, K., Pinkwart, M., Jochim, K. 1983. Purification of hydrogenases by affinity chromatography on Procion Red-agarose. *Biochem. J.* 213:391–98
110. Schubert, K. R., Evans, H. J. 1976. Hydrogen evolution: a major factor affecting the efficiency of nitrogen fixation in nodulated symbionts. *Proc. Natl. Acad. Sci. USA* 73:1207–11
111. Seefeldt, L. C., Arp, D. J. 1986.

Purification to homogeneity of *Azotobacter vinelandii* hydrogenase: a nickel and iron containing dimer. *Biochimie* 68:25–34

112. Shah, V. K., Stacey, G., Brill, W. J. 1983. Electron transport to nitrogenase. *J. Biol. Chem.* 258:12064–68
113. Simpson, F. B., Burris, R. H. 1984. A nitrogen pressure of 50 atmospheres does not prevent evolution of hydrogen by nitrogenase. *Science* 224:1095–97
114. Simpson, F. B., Maier, R. J., Evans, H. J. 1979. Hydrogen-stimulated CO_2 fixation and coordinate induction of hydrogenase and ribulose-bisphosphate carboxylase in a H_2-uptake positive strain of *Rhizobium japonicum*. *Arch. Microbiol.* 123:1–8
115. Simpson, F. B., Thiele, J., Liang, J., Burris, R. H. 1985. Quantitation of the number of electrons stored by dinitrogenase during the lag periods that precede formation of H_2, HD, and uptake of N_2. II. The effect of high pressures of H_2 and D_2 on HD formation by nitrogenase. See Ref. 25, p. 614
116. Sorenson, G. M., Wyndaele, R. 1986. Effect of transfer of symbiotic plasmids and of hydrogenase genes *(hup)* on symbiotic efficiency of *Rhizobium leguminosarum* strains. *J. Gen. Microbiol.* 132:317–24
117. Stam, H., van Verseveld, H. W., de Vries, W., Stouthamer, A. H. 1984. Hydrogen oxidation and efficiency of nitrogen fixation in succinate-limited chemostat cultures of *Rhizobium* ORS 571. *Arch. Microbiol.* 139:53–60
118. Stults, L. W., Moshiri, F., Maier, R. J. 1986. Aerobic purification of hydrogenase from *Rhizobium japonicum* by affinity chromatography. *J. Bacteriol.* 166:795–800
119. Stults, L. W., O'Hara, E. B., Maier, R. J. 1984. Nickel is a component of hydrogenase in *Rhizobium japonicum*. *J. Bacteriol.* 159:153–57
120. Teixeira, M., Moura, I., Xavier, A. V., DerVartanian, D. V., LeGall, J., et al. 1983. *Desulfovibrio gigas* hydrogenase: redox properties of the nickel and iron-sulfur centers. *Eur. J. Biochem.* 130: 481–84
121. Teixeira, M., Moura, I., Xavier, A. V., Huynh, B. H., DerVartanian, D. V., et al. 1985. Electron paramagnetic resonance studies on the mechanism of activation and the catalytic cycle of the nickel-containing hydrogenase from *Desulfovibrio gigas*. *J. Biol. Chem.* 260: 8942–50
122. Thorneley, R. N. F., Lowe, D. J. 1985. Kinetics and mechanism of the nitrogenase enzyme system. In *Molybdenum Enzymes,* ed. T. G. Spiro, pp. 221–84. New York: Wiley. 611 pp.
123. Tichy, H. V., Richter, P., Lotz, W. 1985. Molecular cloning of uptake hydrogenase genes of *Rhizobium leguminosarum* B10. See Ref. 25, p. 32
124. Tigyi, G., Bagyinka, C., Kovacs, K. L. 1986. Protein structural changes associated with active and inactive states of hydrogenase from *Thiocapsa roseopersicina*. *Biochimie* 68:69–74
125. Tilak, K. V. B. R., Schneider, K., Schlegel, H. G. 1984. Autotrophic growth of strains of *Rhizobium* and properties of isolated hydrogenase. *Curr. Microbiol.* 10:49–52
126. Van Berkum, P., Keyser, H. H., Weber, D. F. 1985. Examination of Hup expression by soybean bradyrhizobia belonging to different serogroup phenotypes. See Ref. 25, p. 361
127. Van der Zwaan, J. W., Albracht, S. P. J., Fontijn, R. D., Slater, E. C. 1985. Monovalent nickel in hydrogenase from *Chromatium vinosum*. *FEBS Lett.* 179: 271–77
128. van Wielink, J. E., Oltmann, L. F., Leeuwerik, F. J., De Hollander, J. A., Stouthamer, A. H. 1982. A method for in situ characterization of *b*- and *c*-type cytochromes in *Escherichia coli* and in complex III from beef heart mitochondria by combined spectrum deconvolution and potentiometric analysis. *Biochim. Biophys. Acta* 681:177–90
129. Xu, L. S., Evans, H. J. 1985. Purification and some properties of hydrogenase from autotrophically cultured *Rhizobium japonicum*. *Sci. Sin.* 28:819–27
130. Ziomek, E., Martin, W. G., Williams, R. E. 1984. Immunological and enzymatic properties of the *Desulfovibrio desulfuricans* hydrogenase. *Can. J. Microbiol.* 30:1197–204
131. Zuber, M., Harker, A. R., Sultana, M. A., Evans, H. J. 1986. Cloning and expression of *Bradyrhizobium japonicum* uptake hydrogenase structural genes in *Escherichia coli*. *Proc. Natl. Acad. Sci. USA* 83:7668–72

Ann. Rev. Microbiol. 1987. 41:363–82

THE MITOCHONDRIAL GENOME OF KINETOPLASTID PROTOZOA: Genomic Organization, Transcription, Replication, and Evolution

Larry Simpson

Department of Biology and Molecular Biology Institute, University of California, Los Angeles, California 90024

CONTENTS

INTRODUCTION

The mitochondrial or kinetoplast DNA of the kinetoplastid protozoa is one of the most unusual DNA structures in nature. The purpose of this review is to bring this interesting system to the attention of microbiologists and others

0066-4227/87/1001-0363$02.00

interested in mitochondrial genetic systems and to demonstrate both the extent of our knowledge and the gaps in our understanding. I stress only those studies published in the last two to three years, since comprehensive reviews of previous work have been published (22, 87, 89, 99).

Kinetoplastid protozoa belong to the order Kinetoplastida and comprise monogenetic species in the genera *Crithidia, Leptomonas, Blastocrithidia,* and *Herpetomonas* and digenetic species in the genera *Leishmania, Trypanosoma, Phytomonas,* and *Endotrypanum.* These lower eukaryotic cells are known in the literature as "trypanosomatid protozoa," "kinetoplastid protozoa," "hemoflagellate protozoa," or simply "trypanosomes," although the latter is in the strict sense a genus name. All of these species are parasitic; the monogenetic species are parasites of invertebrates such as *Diptera* and *Hemiptera,* and the digenetic are parasites of both invertebrates and vertebrates such as mammals and lizards. The kinetoplastid protozoa appear to be among the most highly diverged eukaryotic cells known. Perhaps as a consequence of their long separation from the main eukaryotic lineages, they possess many unique biological properties not found in higher organisms: a subplasma membrane exoskeleton of microtubules surrounding the entire cell; a posttranscriptional addition of a strongly conserved 35-nucleotide miniexon sequence at the 5' end of most if not all cytosol mRNAs; a single mitochondrion containing a network of thousands of catenated minicircle DNA molecules and a smaller number of catenated maxicircle molecules; and a highly plastic karyotype, as visualized by orthogonal field gel electrophoresis of chromosomal molecules ranging from 50 kb to approximately 2000 kb. The digenetic species have, in addition, biphasic life cycles that involve substantial changes in morphology and physiology. Such charges include, for example, the complete repression of mitochondrial biosynthetic activity in the bloodstream form of the African *brucei* group trypanosome followed by derepression in the insect form, and the ability of the amastigote forms of mammalian *Leishmania* to withstand lysosomal attack within the intracellular environment of the macrophage (87, 89).

Kinetoplast DNA represents the sole mitochondrial DNA of the kinetoplastid protozoa. It consists of two molecular species, minicircles and maxicircles, catenated together into a single giant network of DNA situated adjacent to the basal body of the flagellum. There are approximately 5×10^3–10^4 minicircles and 20–50 maxicircles per network. In most species there is more than one class of minicircle sequence present in the same cell. The different sequence classes, however, share a strongly conserved region representing approximately 10% of the minicircle length. Maxicircle DNA is apparently homogeneous within a single organism and contains mitochondrial structural genes homologous to those found in other organisms as well as several still unidentified open reading frames.

KINETOPLAST MINICIRCLE DNA

Is There a Gene Product?

There is only one result in the literature that supports the presence of a minicircle-specific transcript. In 1979, Fouts & Wolstenholme (29) reported the existence of a 240-nucleotide transcript from the 2500-nucleotide minicircle DNA of *Crithidia acanthocephali*. This result, however, has never been confirmed with gel technology, and there are several references to negative results in probing Northern RNA blots with minicircle probes in the *Leishmania tarentolae, Crithidia luciliae, Phytomonas davidi,* and *Trypanosoma brucei* systems (39, 41, 89, 99, 100). A small RNA transcript, however, would probably not be detected by the Northern blotting procedure owing to lack of binding to the filter, and an unstable transcript might not be apparent in steady-state RNA. Shlomai & Zadok (84) approached this problem by cloning random minicircle fragments from *Crithidia fasciculata* in bacterial expression vectors and using rabbits to elicit antisera that reacted specifically with *Crithidia* antigens. Their conclusion that kDNA minicircles contain long open reading frames (ORFs) that are transcribed and translated is probably premature for several reasons. First, nonspecific cross-reaction of polyclonal antisera raised against bacterial fusion proteins with parasite proteins could affect the results. Secondly, the published sequence of the predominant minicircle sequence class in *C. fasciculata* does not contain long ORFs (101). Finally, the short ORFs covering the conserved regions derived from the sequences of minicircles from *T. brucei* (15, 48), *Trypanosoma lewisi* (78), *Trypanosoma cruzi* (63), and *L. tarentolae* (51) show no significant similarity on the amino acid level. A final judgment on this important question must await sequencing of the cloned putative minicircle fragments and the putative minicircle transcript, identification of the minicircle ORF, and characterization of the cellular antigens that react with the antiserum.

Sequence Organization in Different Species

The size of kDNA minicircles varies from 465 bp in *Trypanosoma vivax* (10) to 2515 bp in *C. fasciculata* (7, 101). All kDNA minicircles analyzed appear to have the same type of genomic organization, i.e. a small conserved region and a larger variable region. "Conserved" in this sense implies sequence similarity within the different minicircle sequence classes within a single species or strain. The simplest type of organization, found in minicircles of *T. brucei, Trypanosoma equiperdum,* and *L. tarentolae,* is a single conserved region. The 1-kb *T. brucei* minicircles contain a 130-bp conserved region (15, 48), which is quite similar to that found in *T. equiperdum* (3). The three 870-bp *L. tarentolae* minicircles sequenced contain a 160–270-bp conserved region (51). Two conserved regions of 173 and 177 bp situated 180° apart are found in the 2.5-kb minicircles of *C. fasciculata* (101). There are two direct repeats of 95 bp in the 1-kb minicircles of *T. lewisi* (78). Four conserved regions of 118 bp are found in the 1.4-kb *T. cruzi* minicircle, situated at 90° intervals (58, 63, 68).

Oligomer sequences within the conserved regions are also conserved among kinetoplastid species. Kidane et al (51) first recognized a dodecamer present in *L. tarentolae, T. brucei, T. equiperdum,* and *T. cruzi,* and Ponzi et al (78) discovered a pentadecamer present in *T. lewisi, T. brucei,* and *T. equiperdum*. The dodecamer sequence, GGGGTTGGTGTA, is also present in the major minicircle class of *C. fasciculata* strain Cf-C1 (101). This sequence has been termed the "universal" minicircle sequence (74), although adoption of this term may be premature, since sequences from several kinetoplastid genera are not yet available. Ntambi & Englund (74) showed that this sequence represents an origin of replication of the minicircle.

The sequence similarities of the conserved regions from the minicircles of the six species examined correspond with the presumed evolutionary affinities of the species. For example, as stated above, the *T. brucei* and *T. equiperdum* conserved sequences are almost identical, although the other minicircle sequences differ. The *T. cruzi* conserved sequences are more closely related to the *T. lewisi* sequences than to the *T. brucei* sequences and are least closely related to the *L. tarentolae* sequences (63). *T. equiperdum* is thought to be a recent genetic variant of *T. brucei* that has lost the tsetse host. *T. cruzi* and *T. lewisi* are both classified, in terms of developmental localization in the insect host, as the more primitive stecorarian trypanosomes. *T. brucei,* in contrast, is a salivarian trypanosome. These sequence homologies indicate that evolutionary constraints on sequence changes within the conserved region of the minicircle are greater than constraints on changes within the variable regions, as expected from the functional importance of an origin of replication. The origin of changes in the variable region is uncertain, but is probably a combination of a high rate of nucleotide substitutions and segmental rearrangements, possibly involving recombination between minicircles. Segmental rearrangements have been indicated by an electron-microscope heteroduplex analysis of *Crithidia* minicircles (42) and by a partial sequence analysis of a minor minicircle class from *L. tarentolae* kDNA (72). In addition, a high rate (approximately 0.3%) of single nucleotide changes was observed in multiple clones of the same sequence class from *C. fasciculata* (7, 101).

The heterogeneity of the variable region varies from species to species. The most extreme case is *T. brucei,* which has at least 200–300 different minicircle sequence classes within the same cell (98). The frequency of the different sequence classes is also variable. However, frequency has only been measured quantitatively for one cloned minicircle of *T. brucei,* which appeared in 500 copies per cell (100), and for five minicircles of *L. tarentolae,* which comprised 50% of the total minicircle DNA. Of these five minicircles, one class (pKSR1/pLt19) represented 26% of the total population (51). An additional five minicircle sequence classes were detected by partial sequence analysis of cloned fragments (72). The kDNA from the Borst laboratory strain of *C. fasciculata* was estimated to contain 13 different minicircle sequence classes by summation of restriction-fragment gel bands (40, 42, 56). In the case of the Cf-C1 strain of *C. fasciculata,* however, over 90% of the kDNA minicircles consist of a single sequence class, which exhibits approximately 0.3% nucleotide variation in different clones (7, 101). This strain variation was unexpected, since Camargo et al (13) had observed no differences in the kDNA restriction profiles of three *C. fasciculata* strains from different labs. The variation between the Borst strain and the Cf-C1 strain has been confirmed by a side-by-side gel comparison (L. Simpson, unpublished results). It should be noted that visual estimates of frequencies of minicircle classes can be highly inaccurate owing to nonlinear staining of overloaded bands. Furthermore, it should be remembered that kinetic complexity values of kDNA minicircles from the Borst strain of *C. fasciculata* and two other *Crithidia* species are only 1–2.5 times minicircle unit size (28, 42, 56).

The highly complex minicircles from the African trypanosomes appear to possess a sequence adjacent to the conserved region that maintains a characteristic purine versus pyrimidine strand bias and terminates with a series of oligo (A) tracts (48). In addition there is an 18-bp sequence that occurs as 3–4 pairs of inverted repeats throughout the variable region. These inverted repeats flank 102–110-bp segments that show slight sequence conservation among different minicircle classes. Jasmer & Stuart (49) have suggested that these transposonlike sequences represent a mechanism for the development of minicircle sequence diversity in this group of cells. However, such sequences are absent in minicircle DNA from *L. tarentolae, T. lewisi,* and *T. cruzi,* and therefore could not represent a general mechanism for generation of sequence diversity. The oligo (A) tracts are a constant feature of minicircle sequences from all species so far examined.

In addition to the *C. fasciculata* Cf-C1 strain, two other kinetoplastid species, *T. evansi* (24, 31) and *T. equiperdum* (3, 80), possess apparently homogeneous kDNA minicircle populations within a single cell. These species probably represent genetic variants of *T. brucei* that have lost the ability to develop in the insect host and are transmitted contaminatively or venereally. Maxicircle DNA is absent from *T. evansi* (24) and modified in *T. equiperdum* (31). Minicircle DNA from one strain of *T. equiperdum* has been sequenced, and minicircles from six strains of *T. evansi* have been compared by restriction enzyme analysis. Two types of minicircle sequences were found in the six *T. evansi* strains; this suggests that *T. evansi* had at least two separate origins from the presumed ancestral *T. brucei* stock. The level of sequence homogeneity in *T. evansi* and *T. equiperdum* is unknown; sequence analysis of multiple homologous clones is necessary to determine the actual nucleotide sequence variation. The reason for minicircle sequence homogeneity in these two species is unknown, but the minicircle sequence heterogeneity in most species appears to be correlated with the presence of a functional maxicircle. Borst et al (9) have suggested that the recently discovered genetic recombination that occurs in the insect phase of the *T. brucei* life cycle is responsible for creating minicircle sequence heterogeneity, but direct evidence is lacking.

Sequence Heterogeneity as a Diagnostic and Epidemiological Tool

The existence of multiple classes of minicircle sequences within a single cell is mirrored by the existence of different sequence classes in different species and even in different strains of the same species. This diversity implies a rapid rate of sequence change in nature. The rate of minicircle sequence change varies among the different genera. The extensive minicircle sequence complexity within the *T. brucei* group in fact renders this molecule unusable as a marker to distinguish different stocks or strains. However, lesser minicircle complexity renders the sequences of pathogenic *Leishmania* and *T. cruzi* ideal for this purpose. In addition, the high copy number of the minicircle DNA within the network provides an amplified target for visualization of bands in a gel from few cells and for hybridization techniques.

In the case of *T. cruzi,* different stocks exhibited characteristic minicircle restriction fragment profiles in acrylamide gels (69). These patterns were stable over a 2-yr period of continuous culture and did not change during the parasite life cycle. In addition, Southern hybridization analysis showed that differences in minicircle restriction profiles represent differences in sequences. Morel et al (69) proposed the term "schizodeme" to describe subpopulations of *T. cruzi* (or any species of kinetoplastid protozoa) that possess similar minicircle restriction profiles in acrylamide gel electrophoresis. Since similar restriction profiles imply similarities in minicircle DNA sequences, this term is also applicable to subpopulations that exhibit cross-hybridization with kDNA probes as shown by dot blot hybridization. Schizodeme analysis of *T. cruzi* has proved to be a sensitive and reliable method for distinguishing strains and stocks (16, 30, 71).

Schizodeme analysis has been used to show that different strains of *T. cruzi* can simultaneously coexist in the vertebrate host and that the inoculation schedule and the method of reisolation affect the recovery of individual strains (16). In one experiment, mice were doubly infected with the F and Y strains of *T. cruzi*. The parasites were recovered from the blood either directly into culture or through passage in irradiated mice and were identified by schizodeme analysis. The researchers found that both the timing and the method of parasite isolation act as selective factors.

Labeled kDNA and cloned minicircles have also been used as probes in dot blot hybridization experiments with *T. cruzi* strains (32, 62, 83). In addition, oligonucleotide probes synthesized from the known sequence of a cloned *T. cruzi* CL strain minicircle have been used to distinguish different stocks and strains in dot blot experiments with intact cells spotted onto filters (70). Oligonucleotides from different minicircle regions have different specificities: A sequence from the conserved region was specific for the *T. cruzi* species (i.e. versus *Leishmania*), and sequences from the variable region were specific for homologous *T. cruzi* stocks.

In the case of the pathogenic *Leishmania,* schizodeme analysis by comparison of acrylamide gel profiles and direct Southern and dot blot hybridization using kDNA or intact cells probed with labeled kDNA or cloned minicircle DNA have been used with success (1, 37, 44, 45, 50, 60, 61, 97). A clinically useful variant of the dot blot method is the touch blot method; fragments of infected lesions are briefly touched to nitrocellulose filters, leaving parasite amastigotes attached to the filter (103). Owing to the high copy number of kDNA minicircles, a hybridization signal can be seen with as few as 100 cells. This method was used to distinguish *Leishmania brasiliensis* subspecies from *Leishmania mexicana* subspecies and from *Leishmania tropica*. Schizodeme analysis has also been used to distinguish *Leishmania major* from *L. tropica* (50, 97). A cloned kDNA minicircle has been used as a sensitive and specific diagnostic probe for Kenyan visceral leishmaniasis (76).

Cloned minicircle fragments from visceral *Leishmania* species exhibited different taxonomic specificities in hybridization assays (2, 61). One cloned fragment could distinguish visceral and cutaneous species. Another cloned fragment was specific for the strain from which it was derived. Similar results were obtained with cloned minicircle fragments from *Leishmania chagasi*. These results imply either an intermolecular heterogeneity in minicircle sequence specificity, an intramolecular heterogeneity, or both. The intramolecular sequence heterogeneity is clearly due to the presence of both conserved and variable regions. Rogers & Wirth (81) recently used hybridization specificity with minicircle DNA from the same species or group to show the existence of a gradient of sequence conservation within the conserved region of the *Leishmania* minicircle. Based on sequence analysis of two cloned *L. mexicana amazonensis* minicircles, the boundary between the conserved and strain-specific variable region appears to be sharp; this implies the existence of a mechanism that maintains sequence conservation of the conserved region while allowing the variable region to diverge rapidly (D. Wirth, personal communication). The discovery of a gradient of sequence conservation within the conserved region is important, since it indicates that kDNA hybridization can be a precise as well as sensitive method for diagnosis and classification down to the group and strain levels for the leishmanial diseases and probably for other kinetoplastid-caused diseases.

Analysis of Conformational Bend in Minicircle DNA

Cloned or native kDNA minicircles and minicircle fragments from most kinetoplastid species exhibit much slower electrophoretic migration in acrylamide gels than in agarose gels. This was first shown for minicircle DNA from *L. tarentolae* (14, 51, 88, 92) and was also found to be true for minicircle DNA from *T. brucei, T. equiperdum, Herpetomonas muscarum,* and *C. fasciculata* (54, 64, 79), but not for that from *T. cruzi* (63, 69). Such abnormal migration behavior in a cloned minicircle fragment from *L. tarentolae* [which was originally thought to represent a 490-nucleotide fragment and which was later shown by resequencing (66) to be identical to a 410-nt Sau3A fragment of the KSR1/LT19 minicircle (51)] was ascribed (65) to the existence of a conformational sequence-dependent bend in the DNA molecule. The abnormal migration could be eliminated by cutting the minicircle fragment within the putative bend region (51), which was localized at one edge of the region that was conserved among three cloned and sequenced minicircles from *L. tarentolae*. The precise location of the bend was determined by a gel electrophoretic analysis of circularly permuted minicircle DNA fragments to be within a regular repeat of the sequence element $CA_{5-6}T$ with 10-bp periodicity (105) (nucleotide 20 of the KSR1 sequence). A second adjacent bend region (nucleotides 78–137 of the KSR1 sequence) was located by analysis of the gel migration behavior of restriction fragments (20). This region also contains runs of oligo (dA) with 10-bp periodicity.

Two mechanisms have been proposed to account for this conformational bend. In the first model it is assumed that DNA remains in the B conformation and that the dinucleotide ApA produces a wedge by a combination of tilt and roll (102), or that a stereochemical clash of consecutive purine residues on opposite strands deforms the helix (34–36). In the second model it is assumed that the oligo (dA) tracts confer a non-B conformation on the DNA strand and that bending occurs at junctions between regions of B-DNA and oligo (dA) tracts (57, 59). High salt, increased temperature, and binding of distamycin all decrease the bend, as measured by electrophoretic migration in acrylamide (20, 105). There is some evidence against the ApA wedge model (36), but it is not yet possible to decide between the purine clash model and the junction bend model on the basis of existing evidence.

The most striking bend is found in the *C. fasciculata* minicircle, which has 16 successive oligo (dA) tracts at one point in the circle (54, 79, 101). This bent segment can be easily visualized in the electron microscope as 200–300-bp loops within the larger molecule and as looped 300-bp linear fragments (33). It is significant that the single bend in the 2.5-kb *C. fasciculata* minicircle is well separated from the two conserved regions (79) and is not at one edge of the conserved region as in *L. tarentolae* (51) and *T. brucei* (15). This appears to eliminate the possibility that the bend is mechanistically involved in DNA replication. Also, *T. cruzi* minicircles do not contain a bend and replicate quite well.

Muhich & Simpson (72) recently reported a single major cleavage of multiple minicircle sequences from *L. tarentolae* kDNA by mung bean nuclease in the presence of formamide. No specific sequence was common to all minicircle cleavage sites, but the main region of nuclease cleavage was approximately 350 bp from the unique SmaI site in the conserved region. The authors suggested that the nuclease recognized an unusual structural feature of the DNA molecule such as the static bend.

The function of the minicircle bend is unknown, as is the function of the minicircle itself. Perhaps the bend has a role in DNA packaging into the network and formation of the highly organized kDNA nucleoid, or in the targeting of newly replicated networks to a catenation site on the network.

The existence of a sequence-dependent bend in the kDNA minicircle is but one example of local polymorphism of DNA structure. In some cases such polymorphisms are apparently involved with DNA function. One example is sequence-induced DNA curvature at the origins of replication of bacteriophage λ (106) and simian virus 40 (SV40) (82) and at a yeast autonomously replicating sequence (96). Another is protein-induced curvature caused by the binding of catabolite gene–activating protein to its recognition sequence near the promotor of the *Eschericia coli* lac operon (106).

Replication of Minicircle DNA

The decatenation-recatenation model for the replication of kDNA minicircles (21) was based on the early observations that newly replicated kDNA was localized in two loci at the edge of the network; this DNA was chased into a peripheral, widening ring of open circular minicircle DNA in the network, which only became covalently closed after the S phase (86, 94). This model has been substantiated by direct analysis of replicating minicircle intermediates in vivo in *C. fasciculata* and *T. equiperdum* (52, 53, 74, 75) and in an isolated *C. fasciculata* organelle system (6, 8). The minicircle initiates replication with random detachment of the covalently closed circle from the network by a putative topoisomerase II activity, followed by an apparent Cairns-type replication. The newly synthesized heavy strand consists of small oligonucleotides. The light strand, in the case of *T. equiperdum,* contains a single gap of 10 nucleotides with a 5' attached ribonucleotide overlapping the "universal" dodecamer sequence (75). In the case of *C. fasciculata,* the light strand has two gaps overlapping the two "universal" dodecamer sequences (6, 8). The gapped minicircles are then recatenated to the edge of the network. Only after the S phase is completed are all discontinuities filled in and nicks ligated.

KINETOPLAST MAXICIRCLE DNA

Mitochondrial Ribosomal RNA Genes

The 9S and 12S RNAs are the major stable RNA species in purified kinetoplast fractions from several kinetoplastid species (91). They represent the smallest rRNAs yet identified. In spite of radically different sizes and primary sequences as compared to known rRNAs, models of secondary structure can be constructed for the 9S RNA and for portions of the 12S RNA (17, 19, 23, 95). These models conform well to portions of the *E. coli* 16S and 23S rRNA models (12, 104). These models can be verified and extended by the evolutionary method of Woese et al (104), in which compensatory nucleotide substitutions are observed in helical regions of sequences from different kinetoplastid species that preserve base pairing. By comparing the sequences of the 9S and 12S RNAs from *L. tarentolae* and *T. brucei,* several helical regions in the models were verified (17, 19). Similar partial models of secondary structure were constructed from a comparison of 9S and 12S sequences from *T. brucei* and *C. fasciculata* (95). Verification and extension of these models will depend on determination of additional 9S and 12S RNA sequences from other kinetoplastid species and analysis of the alignments.

The major conclusion from the preceding rRNA analyses is that in spite of a high degree of primary sequence divergence and the absence of several regions that are conserved in most other rRNAs, there is a striking conservation of many of the sequences and structures that are implicated in the biosynthetic function of the ribosome, e.g. the peptidyl transferase region in the 12S RNA. To explain the small size of these rRNAs, we have speculated that some of the functions of rRNA have been taken over by proteins and that those that remain are crucial for basic translational functions (19). Thus, an analysis of these unusually small rRNAs may prove to be of general significance in terms of understanding the structure-function relationships of rRNA.

Mitochondrial Structural Genes and Open Reading Frames

The basic organization of the maxicircle genome in all kinetoplastid species examined consists of an actively transcribed informational region containing the rRNA genes and structural genes and a nontranscribed divergent or variable region containing a variety of repeated sequences of unknown function. The complete nucleotide sequences of the informational regions and portions of the divergent regions of the maxicircle DNA from *L. tarentolae* and *T. brucei* and of portions of the maxicircle DNA from *C. fasciculata* are known (4, 18, 25, 38, 46, 77). Simpson et al (90) and de la Cruz et al (84) have compared the maxicircle genomes of *L. tarentolae* and *T. brucei* in detail. They identified specific ORFs as structural genes homologous to those in other organisms by measuring the statistical significance of amino acid alignments and by analyzing the similarity of hydropathic profiles of the translated amino acid sequences. In *L. tarentolae* the following structural genes were identified: cytochrome oxidase subunits I, II, and III; cytochrome *b;* and NAD dehydrogenase subunits 1, 4, and 5. In addition, several ORFs are present that most likely represent protein coding sequences. The criteria used to identify a maxicircle ORF as a functional gene are (*a*) identification of the translated amino acid sequence as a known protein; (*b*) transcription; (*c*) a T:A ratio of approximately 2 on the coding strand; (*d*) conservation of amino acid sequence between *L. tarentolae* and *T. brucei;* and (*e*) conservation of a characteristic codon bias profile between the two species (18, 90).

Appropriate translation initiation codons could not be identified for several of the maxicircle genes (CYb, ND1). Cross-species alignments of translated amino acid sequences of maxicircle genes have aided in assignment of putative initiation codons for COI, COII, ND4, ND5, MURF1, and MURF2. Leucine may represent an alternative initiation codon in the maxicircle genome. Clearly, however, amino acid sequences of the protein products are required for definitive identification of initiation codons of maxicircle genes. In this regard, an antiserum directed against a carboxy-terminal peptide of the putative COII gene product has recently been generated (J. Shaw & L. Simpson, unpublished results). This antiserum detected a polypeptide of the predicted molecular weight in Western blot analysis of a mitochondrial lysate from *L. tarentolae*. Amino acid sequence analysis of this and other kinetoplast gene products should establish unequivocally the functional nature of the kinetoplast transcription-translation system, confirm the specific mitochondrial genetic code employed by this species, and perhaps allow the determination of the specific translation initiation codons used.

Comparison of the Maxicircle Genomes from Two Kinetoplastid Species

The informational portions of the maxicircle genomes of *T. brucei* and *L. tarentolae* are colinear in terms of gene localization and gene polarity except for two regions, one between the 9S RNA gene and the CYb gene and one 3' of the CYb gene (73, 90). The COIII gene and the ORF3–4 and ORF12 genes are unique to *L. tarentolae*. MURF1 and MURF2 represent two unidentified genes that are conserved between these two species.

The absence of the COIII gene in the *T. brucei* maxicircle genome is unprecedented and is due either to a nuclear localization of the COIII gene or to the absence of this subunit from the holoenzyme in this species.

Short regions of DNA sequence homology are present within the nonhomologous regions, but their significance is not understood. In addition, several short GC-rich regions that do not show gene characteristics are nevertheless transcribed, and the pattern of intergenic guanine versus cytosine strand bias is strongly conserved between *T. brucei* and *L. tarentolae* in the absence of nucleotide sequence conservation. Short transcripts off the C-rich strand are found in steady-state RNA (47, 90). The function of these regions is unknown, but it may involve tRNA synthesis, RNA processing, or regulation of transcription.

The COII Frameshift

The COII gene presents a special problem in that there is a −1 frameshift in the DNA sequence that is conserved between *T. brucei* and *L. tarentolae* (18, 38, 77). RNA sequencing of the *T. brucei* COII mRNA has indicated that there is an insertion of several uridines in this region of the transcript; this insertion overcomes the frameshift and allows translation of the complete COII protein (5). This represents a novel posttranscriptional activity, which must be studied for confirmation and for understanding of the mechanism involved. There is one other gene, the MURF2 gene in *L. tarentolae,* which may exhibit the same phenomenon; the DNA sequence gives rise to two overlapping ORFs, yet there is a single transcript, and both ORFs are homologous to a single MURF2 ORF in *T. brucei.*

Divergence of Maxicircle Genes from Mitochondrial Genes of Other Organisms

All identified maxicircle genes have diverged significantly from homologous genes from other organisms. As discussed above, the rRNA genes have diverged so far from rRNA genes of other eukaryotic cells that structural analysis was necessary for definitive identification. The structural genes have diverged to great but differing extents. The most conserved gene is COI, and the least conserved is COIII. In all cases, however, the extent of sequence similarity is less between the kinetoplastid gene and the homologous gene from fungal or animal mitochondrial genomes than between the same genes from fungal and animal lines. The sequence similarity was determined by calculating both the statistical significance of amino acid alignments and the absolute amino acid mismatches.

The Divergent Region of the Maxicircle

The divergent or variable region was first detected in a comparison of maxicircle DNAs from several *T. brucei* strains; this region of the molecule showed up to a 1-kb size variation between strains. In comparisons of the maxicircle DNAs of *L. tarentolae* and *T. brucei,* this region showed no cross-hybridization. A blot hybridization comparison of maxicircle DNAs from *Crithidia oncopelti, Crithidia luciliae, Leptomonas pessoai,* and *Leptomonas gymnodactyli* allowed Maslov et al (67) to construct a general model of the structural organization of maxicircles; the molecule is represented by a

17-kb conservative region common to all kinetoplastids and a divergent region showing both length and sequence variation. Several fragments of the divergent region from *L. tarentolae* and also from *T. brucei* have been cloned and sequenced. Fragments from both species contain tandem repeats of varying length which constitute appropriate substrates for the presumed insertion/deletion and rearrangement events that produce sequence change within this region of the maxicircle. Similar conclusions were reached by a previous electron microscope heteroduplex analysis of maxicircle DNA from *T. brucei* (11). The function of the divergent region is unknown, but a putative origin of replication of the leading strand was roughly localized to a site within the divergent region of the *C. fasciculata* maxicircle. In addition, several fragments of the divergent region of *L. tarentolae* and *C. oncopelti* were found to have autonomous replicating sequence activity in yeast (43, 67).

Transcription of Maxicircle Genes

The conserved region of the maxicircle is actively transcribed, yielding high steady-state levels of the 9S and 12S rRNAs and lower levels of polyadenylated mRNAs and smaller G-rich RNAs of unknown significance (47). Transcripts for six *L. tarentolae* structural genes and several ORFs have been identified, and the locations of the 5' ends have been determined by primer runoff (85). The distance from the 5' end to the putative translation initiation codon (where determined) varies from 20 to 64 nucleotides. This arrangement differs from that in the human mitochondrion and is similar to that in the yeast mitochondrion, where there are multiple promotors and substantial 5' untranslated sequences. Transcription occurs from one strand for all genes but ND1, MURF1, and COI. Single transcripts cover the overlapping ORF3 and ORF4 and the overlapping ORF5 and ORF6 (=MURF2). The number of promotors is unknown. In some cases transcripts of higher molecular weight can be seen in Northern blots; these transcripts may be precursors. A preliminary capping experiment indicated the presence of separate promotors for the 9S and 12S rRNA genes and three to four additional promotors for the structural genes, but these must be confirmed by direct sequence analysis of the capped species (85). In the case of *T. brucei,* low-abundance high–molecular weight putative precursor bands were seen in Northern blots for several of the maxicircle genes (25).

The abundance of the steady-state transcripts from the *L. tarentolae* maxicircle genes showed some variation, which implies either separate transcriptional control or processing. For example, the 1800-nucleotide COI and the 1200-nucleotide CYb transcripts were much more abundant than the 1000-nucleotide MURF2 transcript (85, 93).

Regulation of Maxicircle Transcription During the Life Cycle of Trypanosoma brucei

Extensive Northern blot analysis of maxicircle transcription in *T. brucei* has indicated the presence of double transcripts differing in size by 150–200 bp for all potential protein-coding genes except ND5 (25–27, 46). The size difference is not due to polyadenylation, as both transcripts occur in poly (A)+ RNA (25, 27, 46). In *L. tarentolae,* single RNA species are associated with ND4, ND5, ORF3–4, COI, and MURF2, whereas the COII, COIII, and CYb genes appear to hybridize to at least two transcripts each (85).

The life cycle of *T. brucei* involves dramatic changes in mitochondrial morphology and physiology. The long, slender bloodstream form (LS-BF) in the mammalian bloodstream lacks mitochondrial cytochromes and contains an empty mitochondrial tubule without cristae. The short, stumpy bloodstream form (SS-BF) is fixed the in G_0 phase and possesses some Krebs cycle enzymes and a mitochondrial NADH diaphorase activity. These forms differentiate into the procyclic trypomastigote form (PF) in the insect midgut or in culture. Established procyclic forms contain a complete functional phosphorylating cytochrome chain. The regulation of these biosynthetic changes is complex, involving both nuclear and mitochondrial genes. Modulation of several maxicircle gene transcripts occurs in the different stages of the life cycle. The transition from the LS-BF to the SS-BF appears to be an important developmental step, since the steady-state level of the 9S and 12S mitochondrial ribosomal RNAs increases approximately 30-fold and the originally undetectable levels of the CYb, COI, and COII transcripts increase to levels approaching those in PF cells (68a). On the other hand, Feagin, Jasmer, et al (25–27, 46) have reported that the larger of the two transcripts for the COI, COII, and CYb genes is more abundant in PF cells than in LS-BF cells by 2–9 fold, whereas the smaller transcript is more abundant in SS-BF than in either LS-BF or PF cells. These quantitative differences may reflect parasite strain differences or differences in the purity of the LS-BF cell population.

Regulation appears to be complex; it involves changes in the steady-state levels of one of the two transcripts for each gene as well as changes in the degree of polyadenylation. Usually the bloodstream form transcripts show less polyadenylation than the procyclic form transcripts (25–27, 46). The single ND5 transcript is unusual in that it is either more abundant in BF than in PF cells (25–27, 46) or has the same steady-state levels in both (68a).

One factor involved in the generation of multiple transcripts of the CYb gene in *T. brucei* is an insertion of 36 uridines at the 5' end of the mRNA which are not encoded in the DNA sequence (26a). This is analogous to the insertion of four uridines in the *T. brucei* maxicircle COII transcript, which overcomes a translational frameshift in the DNA sequence (5). The mechanism for this is unknown. It is clear that extensive transcriptional modulation occurs at the mitochondrial level during the life cycle of the African trypanosomes.

CONCLUSIONS

I have reviewed the current status of our understanding of the unusual mitochondrial genome of the parasitic kinetoplastid protozoa and have pointed out the utility of this genome as a model system for the study of several basic biological problems. The very existence of such an unusual biological phenomenon as the kinetoplast DNA is of great inherent interest. I trust that further study of these organisms will remove some of the mystery and add to our understanding of the selective pressures involved in the evolution of mitochondrial genomes in general and of this mitochondrial genome in particular.

ACKNOWLEDGMENTS

I would like to thank the following colleagues who kindly sent preprints and communicated unpublished work for this review: C. Morel, D. Ray, P. Jackson, D. Wirth, K. Stuart, P. Englund, D. Crothers, D. Maslov, D. Barker, and T. Spithill. Work from my laboratory was supported in part by research grant AI 09102 from the National Institutes of Health.

Literature Cited

1. Arnot, D., Barker, D. 1981. Biochemical identification of cutaneous leishmanias by analysis of kinetoplast DNA. II. Sequence homologies in *Leishmania* kinetoplast DNA. *Mol. Biochem. Parasitol.* 3:47–56
2. Barker, D., Gibson, L., Kennedy, W., Nasser, A., Williams, R. 1986. The potential of using recombinant DNA species-specific probes for the identification of tropical *Leishmania*. *Parasitology* 91:8139–74
3. Barrois, M., Riou, G., Galibert, F. 1982. Complete nucleotide sequence of minicircle kinetoplast DNA from *Trypanosoma equiperdum*. *Proc. Natl. Acad. Sci. USA* 78:3323–27
4. Benne, R., DeVries, B., Van den Burg, J., Klaver, B. 1983. The nucleotide sequence of a segment of *Trypanosoma brucei* mitochondrial maxi-circle DNA that contains the gene for apocytochrome *b* and some unusual unassigned reading frames. *Nucleic Acids Res.* 11:6925–41
5. Benne, R., Van den Burg, J., Brakenhoff, J., Sloof, P., Van Boom, J., Tromp, M. 1986. Major transcript of the frameshifted *coxII* gene from trypanosome mitochondria contains four nucleotides that are not encoded in the DNA. *Cell* 46:819–26
6. Birkenmeyer, L., Ray, D. 1986. Replication of kinetoplast DNA in isolated kinetoplasts from *Crithidia fasciculata*. *J. Biol. Chem.* 261:2362–68
7. Birkenmeyer, L., Sugisaki, H., Ray, D. S. 1985. The majority of minicircle DNA in *Crithidia fasciculata* strain CF-C1 is of a single class with nearly homogeneous DNA sequence. *Nucleic Acids Res.* 13:7107–18
8. Birkenmeyer, L., Sugisaki, H., Ray, D. 1987. Structural characterizations of site-specific discontinuities associated with replication origins of minicircle DNA from *Crithidia fasciculata*. *J. Biol. Chem.* 262:2384–92
9. Borst, P., Fase-Fowler, F., Gibson, W. 1987. Kinetoplast DNA of *Trypanosoma*

evansi. *Mol. Biochem. Parasitol.* 23: 31–38

10. Borst, P., Fase-Fowler, F., Weijers, P., Barry, J., Tetley, L., Vickerman, K. 1985. Kinetoplast DNA from *Trypanosoma vivax* and *T. congolense*. *Mol. Biochem. Parasitol.* 15:129–42
11. Borst, P., van der Ploeg, M., van Hoek, J., Tas, J., James, J. 1982. On the DNA content and ploidy of trypanosomes. *Mol. Biochem. Parasitol.* 6:13–23
12. Brimacombe, R., Maly, P., Zwieb, C. 1983. The structure of ribosomal RNA and its organization relative to ribosomal protein. *Prog. Nucleic Acid Res. Mol. Biol.* 28:1–48
13. Camargo, E., Mattei, D., Barbieri, C., Morel, C. 1981. Electrophoretic analysis of endonuclease-generated fragments of kinetoplast DNA, of esterase isoenzymes and of surface proteins as aids for species identification of insect trypanosomatids. *J. Protozool.* 29:251–58
14. Challberg, S., Englund, P. 1980. Heterogeneity of minicircles in kinetoplast DNA of *Leishmania tarentolae*. *J. Mol. Biol.* 138:447–72
15. Chen, K., Donelson, J. 1980. The sequences of two kinetoplast DNA minicircles of *Trypanosoma brucei*. *Proc. Natl. Acad. Sci. USA* 77:2445–49
16. Deane, M., Sousa, M., Pereira, N., Goncalves, A., Momen, H., Morel, C. 1984. *Trypanosoma cruzi:* Inoculation schedules and re-isolation methods select individual strains from doubly infected mice, as demonstrated by schizodeme and zymodeme analyses. *J. Protozool.* 31:276–80
17. de la Cruz, V., Lake, J., Simpson, A., Simpson, L. 1985. A minimal ribosomal RNA: Sequence and secondary structure of the 9S kinetoplast ribosomal RNA from *Leishmania tarentolae*. *Proc. Natl. Acad. Sci. USA* 82:1401–5
18. de la Cruz, V., Neckelmann, N., Simpson, L. 1984. Sequences of six structural genes and several open reading frames in the kinetoplast maxicircle DNA of *Leishmania tarentolae*. *J. Biol. Chem.* 259:15136–47
19. de la Cruz, V., Simpson, A., Lake, J., Simpson, L. 1985. Primary sequence and partial secondary structure of the 12S kinetoplast (mitochondrial) ribosomal RNA from *Leishmania tarentolae:* Conservation of peptidyltransferase structural elements. *Nucleic Acids Res.* 13:2337–56
20. Diekmann, S., Wang, J. 1985. On the sequence determinants and flexibility of the kinetoplast DNA fragment with abnormal gel electrophoretic mobilities. *J. Mol. Biol.* 186:1–11
21. Englund, P. 1979. Free minicircles of kinetoplast DNA in *Crithidia fasciculata*. *J. Biol. Chem.* 254:4895–900
22. Englund, P. 1981. Kinetoplast DNA. In *Biochemistry and Physiology of Protozoa,* ed. M. Levandowsky, S. Hutner, 4:333–81. New York: Academic. 2nd ed.
23. Eperon, I., Janssen, J., Hoeijmakers, J., Borst, P. 1983. The major transcripts of the kinetoplast DNA of *T. brucei* are very small ribosomal RNAs. *Nucleic Acids Res.* 11:105–25
24. Fairlamb, A., Weislogel, P., Hoeijmakers, J., Borst, P. 1978. Isolation and characterization of kinetoplast DNA from bloodstream form of *Trypanosoma brucei*. *J. Cell Biol.* 76:293–309
25. Feagin, J., Jasmer, D., Stuart, K. 1985. Apocytochrome *b* and other mitochondrial DNA sequences are differentially expressed during the life cycle of *Trypanosoma brucei*. *Nucleic Acids Res.* 13:4577–96
26. Feagin, J., Jasmer, D., Stuart, K. 1986. Differential mitochondrial gene expression between slender and stumpy bloodforms of *Trypanosoma brucei*. *Mol. Biochem. Parasitol.* 20:207–14

26a. Feagin, J., Jasmer, D., Stuart, K. 1987. Developmentally regulated addition of nucleotides within apocytochrome *b* transcripts in *Trypanosoma brucei*. *Cell* In press

27. Feagin, J., Stuart, K. 1985. Differential expression of mitochondrial genes between the life cycle stages of *Trypanosoma brucei*. *Proc. Natl. Acad. Sci. USA* 82:3380–84
28. Fouts, D., Manning, J., Wolstenholme, D. 1975. Physicochemical properties of kinetoplast DNA from *Crithidia acanthocephali, Crithidia luciliae* and *Trypanosoma lewisi*. *J. Cell Biol.* 67:378–99
29. Fouts, D., Wolstenholme, D. 1979. Evidence for a partial transcript of the small circular component of kinetoplast DNA of *Crithidia acanthocephali*. *Nucleic Acids Res.* 6:3785–804
30. Frasch, A., Goijman, S., Cazzulo, J., Stoppani, A. 1981. Constant and variable regions in DNA minicircles from *Trypanosoma cruzi* and *Trypanosoma rangeli*: Application to species and stock differentiation. *Mol. Biochem. Parasitol.* 4:163–70
31. Frasch, A., Hajduk, S., Hoeijmakers, J., Borst, P., Brunel, F., Davison, J.

1980. The kDNA of *Trypanosoma equiperdum*. *Biochim. Biophys. Acta* 607:397–410
32. Frasch, A., Sanchez, D., Stoppani, A. 1984. Homogeneous and heterogeneous mini-circle subpopulations in *Trypanosoma cruzi* kinetoplast DNA. *Biochim. Biophys. Acta* 782:26–33
33. Griffith, J., Bleyman, M., Rauch, C., Kitchin, P., Englund, P. 1986. Visualization of the bent helix in kinetoplast DNA by electron microscopy. *Cell* 46:717–24
34. Hagerman, P. J. 1984. Evidence for the existence of stable curvature of DNA in solution. *Proc. Natl. Acad. Sci. USA* 81:4632–36
35. Hagerman, P. 1985. Sequence dependence of the curvature of DNA: A test of the phasing hypothesis. *Biochemistry* 24:7033–37
36. Hagerman, P. 1986. Sequence-directed curvature of DNA. *Nature* 321:449–50
37. Handman, E., Hocking, R., Mitchell, G., Spithill, T. 1983. Isolation and characterization of infective and noninfective clones of *Leishmania tropica*. *Mol. Biochem. Parasitol.* 7:111–26
38. Hensgens, A., Brakenhoff, J., De Vries, B., Sloof, P., Tromp, M., et al. 1984. The sequence of the gene for cytochrome *c* oxidase subunit I, a frameshift containing gene for cytochrome *c* oxidase subunit II and seven unassigned reading frames in *Trypanosoma brucei* mitochondrial maxicircle DNA. *Nucleic Acids Res.* 12:7327–44
39. Hoeijmakers, J., Borst, P. 1982. RNA from the insect trypanosome *Crithidia luciliae* contains transcripts of the maxicircle and not of the minicircle component of kinetoplast DNA. *Biochim. Biophys. Acta* 521:407–11
40. Hoeijmakers, J., Schoutsen, B., Borst, P. 1982. Kinetoplast DNA in the insect trypanosomes *Crithidia luciliae* and *Crithidia fasciculata*. I. Sequence evolution and transcription of the maxicircle. *Plasmid* 7:199–209
41. Hoeijmakers, J., Snijders, A., Janssen, J., Borst, P. 1981. Transcription of kinetoplast DNA in *Trypanosoma brucei* bloodstream and culture forms. *Plasmid* 5:329–50
42. Hoeijmakers, J., Weijers, P., Brakenhoff, C., Borst, P. 1982. Kinetoplast DNA in the insect trypanosomes *Crithidia luciliae* and *Crithidia fasciculata*. III. Heteroduplex analysis of the *Crithidia luciliae* minicircles. *Plasmid* 7:221–29
43. Hughes, D., Simpson, L., Kayne, P., Neckelmann, N. 1984. Autonomous replication sequences in the maxicircle kinetoplast DNA of *Leishmania tarentolae*. *Mol. Biochem. Parasitol.* 13: 263–75
44. Jackson, P. 1987. Differentiation of *Leishmania* species and strains by analysis of restriction endonuclease–produced kinetoplast DNA fragments and DNA-DNA hybridization with 32P-kDNA from type isolates. In *Atomic Energy Applications in Parasitology, A Training Course,* ed. E. Hayunga, M. Steck. Vienna: Int. Atomic Energy Assoc./ Dep. Energy. In press
45. Jackson, P., Wolhieter, J., Jackson, J., Sayles, P., Diggs, C., Hockmeyer, W. 1984. Restriction endonuclease analysis of *Leishmania* kinetoplast DNA characterizes parasites responsible for visceral and cutaneous disease. *Am. J. Trop. Med. Hyg.* 33:808–19
46. Jasmer, D., Feagin, J., Payne, M., Stuart, K. 1985. Diverse patterns of expression of the cytochrome *c* oxidase subunit I gene and unassigned reading frames 4 and 5 during the life cycle of *Trypanosoma brucei*. *Mol. Cell. Biol.* 5:3041–47
47. Jasmer, D., Feagin, J., Payne, M., Stuart, K. 1987. Variation of G-rich mitochondrial transcripts among stocks of African trypanosomes. *Mol. Biochem. Parasitol.* 22:259–72
48. Jasmer, D., Stuart, K. 1986. Conservation of kinetoplastid minicircle characteristics without nucleotide sequence conservation. *Mol. Biochem. Parasitol.* 18:257–70
49. Jasmer, D., Stuart, K. 1986. Sequence organization in African trypanosome minicircles is defined by 18 base pair inverted repeats. *Mol. Biochem. Parasitol.* 18:321–32
50. Kennedy, W. P. 1984. Novel identification of differences in the kinetoplast DNA of *Leishmania* isolates by recombinant DNA techniques and in situ hybridization. *Mol. Biochem. Parasitol.* 12:313–25
51. Kidane, G., Hughes, D., Simpson, L. 1984. Sequence heterogeneity and anomalous electrophoretic mobility of kinetoplast minicircle DNA in *Leishmania tarentolae*. *Gene* 27:265–77
52. Kitchin, P., Klein, V., Englund, P. 1985. Intermediates in the replication of kinetoplast DNA minicircles. *J. Biol. Chem.* 260:3844–51
53. Kitchin, P., Klein, V., Fein, B., Englund, P. 1984. Gapped minicircles. A

novel replication intermediate of kinetoplast DNA. *J. Biol. Chem.* 24:15532–39

54. Kitchin, P., Klein, V., Ryan, K., Gann, K., Rauch, C., et al. 1986. A highly bent fragment of *Crithidia fasciculata* kinetoplast DNA. *J. Biol. Chem.* 261:11302–9
55. Kleisen, C. M., Borst, P. 1975. Are 50% of all cellular proteins synthesized on mitochondrial ribosomes in *Crithidia luciliae? Biochim. Biophys. Acta* 390:78–81
56. Kleisen, C., Borst, P., Weijers, P. 1976. The structure of kinetoplast DNA. I. The mini-circles of *Crithidia luciliae* are heterogeneous in base sequence. *Eur. J. Biochem.* 64:141–51
57. Koo, H., Wu, H., Crothers, D. 1986. DNA bending at adenine-thymine tracts. *Nature* 320:501–6
58. Leon, W., Frank, A., Hoeijmakers, J., Fase-Fowler, F., Borst, P., et al. 1980. Maxi-circles and mini-circles in kinetoplast DNA from *Trypanosoma cruzi. Biochim. Biophys. Acta* 607:221–31
59. Levene, S., Wu, H., Crothers, D. 1986. Bending and flexibility of kinetoplast DNA. *Biochemistry* 25:3988–95
60. Lopes, U., Momen, H., Grimaldi, G., Marzochi, M., Pacheco, R., Morel, C. 1984. Schizodeme and zymodeme characterization of *Leishmania* in the investigation of foci of visceral and cutaneous leishmaniasis. *J. Parasitol.* 70:89–98
61. Lopes, U., Wirth, D. 1986. Identification of visceral *Leishmania* species with cloned sequences of kinetoplast DNA. *Mol. Biochem. Parasitol.* 20:77–84
62. Macina, R., Sanchez, D., Affranchino, J., Engel, J., Frasch, A. 1985. Polymorphism within minicircle sequence classes in the kinetoplast DNA of *Trypanosoma cruzi* clones. *Mol. Biochem. Parasitol.* 16:61–74
63. Macina, R., Sanchez, D., Gluschankof, D., Burrone, O., Frasch, A. 1986. Sequence diversity in the kinetoplast DNA minicircles of *Trypanosoma cruzi. Mol. Biochem. Parasitol.* 21:25–32
64. Marini, J., Effron, P., Goodman, T., Singleton, C., Wells, R., et al. 1984. Physical characterization of a kinetoplast DNA fragment with unusual properties. *J. Biol. Chem.* 259:8974–79
65. Marini, J., Levene, S., Crothers, D., Englund, P. 1982. A bent helix in kinetoplast DNA. *Cold Spring Harbor Symp. Quant. Biol.* 47(Pt. 1):279–83
66. Marini, J., Levene, S., Crothers, D., Englund, P. 1983. Bent helical structure in kinetoplast DNA—Correction. *Proc. Natl. Acad. Sci. USA* 24:7678
67. Maslov, D., Kolesnikov, A., Zaitseva, G. 1984. Conservative and divergent base sequence regions in the maxicircle kinetoplast DNA of several trypanosomatid flagellates. *Mol. Biochem. Parasitol.* 12:351–64
68. Mattei, D., Goldenberg, S., Morel, C., Azevedo, H., Roitman, I. 1977. Biochemical strain characterization of *Trypanosoma cruzi* by restriction endonuclease cleavage of kinetoplast DNA. *FEBS Lett.* 74:264–68

68a. Michelotti, E., Hajduk, S. 1987. Developmental regulation of trypanosome mitochondrial DNA gene expression. *J. Biol. Chem.* 262:927–32

69. Morel, C., Chiari, E., Camargo, E., Mattei, D., Romanha, A., Simpson, L. 1980. Strains and clones of *Trypanosoma cruzi* can be characterized by restriction endonuclease fingerprinting of kinetoplast DNA minicircles. *Proc. Natl. Acad. Sci. USA* 77:6810–14
70. Morel, C., Goncalves, A., Simpson, L., Simpson, A. 1984. Recent advances in the development of DNA hybridization probes for the detection and characterization of *Trypanosoma cruzi. Mem. Inst. Oswaldo Cruz* 79:51–53 (Suppl.)
71. Morel, C., Simpson, L. 1980. Characterization of pathogenic trypanosomatidae by restriction endonuclease fingerprinting of kinetoplast DNA minicircles. *Am. J. Trop. Med. Hyg.* 29:1070–74 (Suppl.)
72. Muhich, M., Simpson, L. 1986. Specific cleavage of kinetoplast minicircle DNA from *Leishmania tarentolae* and identification of several additional minicircle sequence classes. *Nucleic Acids Res.* 14:5531–56
73. Muhich, M., Simpson, L, Simpson, A. 1983. Comparison of the maxicircle DNAs of *Leishmania tarentolae* and *Trypanosoma brucei. Proc. Natl. Acad. Sci. USA* 80:4060–64
74. Ntambi, J., Englund, P. 1985. A gap at a unique location in newly replicated kinetoplast DNA minicircles from *Trypanosoma equiperdum. J. Biol. Chem.* 260:5574–79
75. Ntambi, J., Shapiro, T., Ryan, K., Englund, P. 1986. Ribonucleotides associated with a gap in the newly replicated kinetoplast DNA minicircles from *Trypanosoma equiperdum. J. Biol. Chem.* 261:11890–95
76. Oster, C., Jackson, P., Muigai, R., Wasunna, K., Rashid, J., et al. 1987. Diagnosis of Kenyan visceral leishmaniasis using a kinetoplast DNA probe. *East Afr. Med. J.* In press

77. Payne, M., Rothwell, V., Jasmer, D., Feagin, J., Stuart, K. 1985. Identification of mitochondrial genes in *Trypanosoma brucei* and homology to cytochrome *c* oxidase II in two different reading frames. *Mol. Biochem. Parasitol.* 15:159–70
78. Ponzi, M., Birago, C., Battaglia, P. 1984. Two identical symmetrical regions in the minicircle structure of *Trypanosoma lewisi* kinetoplast DNA. *Mol. Biochem. Parasitol.* 13:111–19
79. Ray, D., Hines, J., Sugisaki, H., Sheline, C. 1986. kDNA minicircles of the major sequence class of *Crithidia fasciculata* contain a single region of bent helix widely separated from the two origins of replication. *Nucleic Acids Res.* 14:7953–65
80. Riou, G., Barrois, M. 1979. Restriction cleavage map of kinetoplast DNA minicircle from *Trypanosoma equiperdum*. *Biochem. Biophys. Res. Commun.* 90:405–9
81. Rogers, D., Wirth, D. 1987. Kinetoplast DNA minicircles: Regions of extensive sequence divergence. *Proc. Natl. Acad. Sci. USA* 84:565–69
82. Ryder, K., Silver, S., DeLucia, A., Fanning, E., Tegtmeyer, P. 1986. An altered DNA conformation in origin region I is a determinant for the binding of SV40 large T antigen. *Cell* 44:719–25
83. Sanchez, D., Madrid, R., Engel, J., Frasch, A. 1984. Rapid identification of *Trypanosoma cruzi* isolates by "dot blot" hybridization. *FEBS Lett.* 168:139–42
84. Shlomai, J., Zadok, A. 1984. Kinetoplast DNA minicircles of trypanosomatids encode for a protein product. *Nucleic Acids Res.* 12:8017–28
85. Simpson, A., Hughes, D., Simpson, L. 1985. Trypanosoma brucei: differentiation of in vitro–grown bloodstream trypomastigotes into procyclic forms. *J. Protozool.* 32:672–77
86. Simpson, A., Simpson, L. 1976. Pulse-labeling of kinetoplast DNA: Localization of two sites of synthesis within the networks and kinetics of labeling of closed minicircles. *J. Protozool.* 23: 583–87
87. Simpson, L. 1972. The kinetoplast of the hemoflagellates. *Int. Rev. Cytol.* 32:139–207
88. Simpson, L. 1979. Isolation of maxicircle component of kinetoplast DNA from hemoflagellate protozoa. *Proc. Natl. Acad. Sci. USA* 76:1585–1588
89. Simpson, L. 1986. Kinetoplast DNA in trypanosomid flagellates. *Int. Rev. Cytol.* 99:119–79
90. Simpson, L., Neckelmann, N., de la Cruz, V., Simpson, A., Feagin, J., et al. 1987. Comparison of the maxicircle (mitochondrial) genomes of *Leishmania tarentolae* and *Trypanosoma brucei* at the level of nucleotide sequence. *J. Biol. Chem.* 262:6182–96
91. Simpson, L., Simpson, A. 1978. Kinetoplast RNA of *Leishmania tarentolae*. *Cell* 14:169–78
92. Simpson, L., Simpson, A., Kidane, G., Livingston, L., Spithill, T. 1980. The kinetoplast DNA of the hemoflagellate protozoa. *Am. J. Trop. Med. Hyg.* 29:1053–63 (Suppl.)
93. Simpson, L., Simpson, A., Livingston, L. 1982. Transcription of the maxicircle kinetoplast DNA of *Leishmania tarentolae*. *Mol. Biochem. Parasitol.* 6:237–57
94. Simpson, L., Simpson, A., Wesley, R. 1974. Replication of the kinetoplast DNA of *Leishmania tarentolae* and *Crithidia fasciculata*. *Biochim. Biophys. Acta* 349:161–72
95. Sloof, P., Van den Burg, J., Voogd, A., Benne, R., Agostinelli, M., et al. 1985. Further characterization of the extremely small mitochondrial ribosomal RNAs from trypanosomes: a detailed comparison of the 9S and the 12S RNAs from *Crithidia fasciculata* and *Trypanosoma brucei* with rRNAs from other organisms. *Nucleic Acids Res.* 13:4171–90
96. Snyder, M., Buchman, A., Davis, R. 1986. Bent DNA at a yeast autonomously replicating sequence. *Nature* 324:87–89
97. Spithill, T., Grumont, R. 1984. Identification of species, strains and clones of *Leishmania* by characterization of kinetoplast DNA minicircles. *Mol. Biochem. Parasitol.* 12:217–36
98. Stuart, K. 1979. Kinetoplast DNA of *Trypanosoma brucei:* Physical map of the maxicircle. *Plasmid* 2:520–28
99. Stuart, K. 1983. Kinetoplast DNA, a mitochondrial DNA with a difference. *Mol. Biochem. Parasitol.* 9:93–104
100. Stuart, K., Gelvin, S. 1982. Localization of kinetoplast DNA maxicircle transcripts in bloodstream and procyclic form *Trypanosoma brucei*. *Mol. Cell. Biol.* 2:845–52
101. Sugisaki, H., Ray, D. 1987. Kinetoplast DNA minicircles of the insect trypanosomatid *Crithidia fasciculata* contain two nearly identical 173 bp and 177 bp sequences involved in DNA replication. *Mol. Biochem. Parasitol.* 23:253–64
102. Trifonov, E., Sussman, J. 1980. The pitch of chromatin DNA is reflected in

its nucleotide sequence. *Proc. Natl. Acad. Sci. USA* 77:3816–820

103. Wirth, D., Pratt, D. 1982. Rapid identification of *Leishmania* species by specific hybridization of kinetoplast DNA in cutaneous lesions. *Proc. Natl. Acad. Sci. USA* 79:6999–7003

104. Woese, C., Gutell, R., Gupta, R., Noller, H. 1983. Detailed analysis of the higher order structure of 16S-like ribosomal ribonucleic acids. *Microbiol. Rev.* 47:621–69

105. Wu, H., Crothers, D. 1984. The locus of sequence-directed and protein-induced DNA bending. *Nature* 308:509–13

106. Zahn, K., Blattner, F. R. 1985. Sequence-induced DNA curvature at the bacteriophage lambda origin of replication. *Nature* 317:451–53

Ann. Rev. Microbiol. 1987. 41:383–407

THE EPIDEMIOLOGIC, CLINICAL, AND MICROBIOLOGIC FEATURES OF HEMORRHAGIC COLITIS

Lee W. Riley

Division of Infectious Diseases, Stanford University Medical Center, Stanford, California 94305

CONTENTS

INTRODUCTION

The term "hemorrhagic colitis" describes a distinct clinical syndrome characterized by abdominal cramping and bloody diarrhea, with radiologic or colonoscopic evidence of colonic mucosal edema, erosion, or hemorrhage, in the absence of conventional enteric pathogens in the stools. Other terms such

0066-4227/87/1001-0383$02.00

as "evanescent colitis" (61), "transient ischemic colitis" (15, 61), and "reversible segmental colitis" (34, 61) have been used to describe a similar syndrome. An illness with these clinical features but distinct from inflammatory bowel diseases, pseudomembranous colitis, or ischemic colitis was described as a "new disease entity" as early as 1971 in five young adults in Philadelphia (61). Etiology could not be determined in these patients. Since then, others have reported sporadic cases of similar illness in the United States (47, 55, 61, 78, 85), Japan (89, 101), and Europe (4, 15, 63, 100) and have suggested a variety of agents to be associated with the disease.

In 1982, the first recognized outbreaks of hemorrhagic colitis occurred in Oregon and Michigan (83, 84). Investigation of these outbreaks revealed that the illness was associated with an *Escherichia coli* serotype that was rare at the time, O157:H7. This *E. coli* did not produce enterotoxins associated with the enterotoxigenic *E. coli* (ETEC), did not, unlike enteroinvasive *E. coli* (EIEC), invade enterocytes, and did not exhibit the characteristic epithelial cell adherence property of the so-called enteropathogenic *E. coli* (EPEC). Additionally, the disease did not resemble the profuse watery diarrheal illness produced by recognized toxin-producing organisms such as ETEC or *Vibrio cholerae,* dysentery produced by invasive organisms such as EIEC, shigellae, or *Entamoeba histolytica,* or the relapsing, protracted, nonbloody diarrhea produced by EPEC. Hence, a fourth group of *E. coli* associated with human diarrhea became established, and another manifestation of an enteric illness became recognized as a distinct clinical entity of infectious etiology.

Since the occurrence of the outbreaks in 1982, continued epidemiologic, clinical, and laboratory investigations have clearly established *E. coli* O157:H7 to be an important etiologic agent of hemorrhagic colitis. This review discusses the results of these investigations.

EPIDEMIOLOGY

Much of the current understanding of the epidemiologic features of hemorrhagic colitis associated with *E. coli* O157:H7 is derived from outbreak investigations and surveillance initiated in the United States and Canada. These investigations have provided new insights into such epidemiologic characteristics as the geographic and seasonal variations in disease prevalence, host factors, modes of transmission, and the reservoir of the etiologic agent.

Outbreak Investigations

In the outbreaks that occurred in Oregon and Michigan, 47 persons were identified whose illness met the outbreak case definition, with (*a*) severe abdominal pain, (*b*) grossly bloody diarrhea, and (*c*) stools that did not yield shigellae, salmonellae, campylobacters, ova, or parasites (84). Case-control studies revealed that the illness was associated with ingestion of a hamburger at a common chain of fast-food restaurants (84). *E. coli* O157:H7 was recovered from 50% of stool specimens of the cases and from none of healthy controls. The same strain of *E. coli* was isolated from a beef patty from a suspected lot of meat in Michigan.

Since these two outbreaks, five other documented outbreaks of hemorrhagic colitis have occurred in North America (Table 1). All of these were associated with *E. coli* O157:H7. Four of these subsequent outbreaks occurred in institutional settings (three nursing homes and one day-care center). At least six more outbreaks (two in nursing homes, two in day-care centers, one on a military base, and one in a community setting) have been reported to the Centers for Disease Control (CDC) through 1986 (Enteric Diseases Branch, CDC, personal communication). The only outbreak of hemorrhagic colitis reported outside of North America was a community-wide outbreak associated with *E. coli* O157 (flagellar antigen not specified) that occurred in England in July, 1985 (17).

Surveillance Studies

Soon after the Oregon outbreak, the CDC initiated an informal national surveillance for hemorrhagic colitis in the United States (13, 82). During 20 months of surveillance, 103 persons with an illness clinically compatible with hemorrhagic colitis were reported from 28 states (82). Stool specimens were examined in 76 (74%) of them, and *E. coli* O157:H7 was recovered from 28 (36%) of the 76 specimens. The high frequency of isolation of the organism in this prospective study further strengthened the association of hemorrhagic colitis with *E. coli* O157:H7.

In addition to initiating this national survey, the CDC retrospectively examined over 3000 *E. coli* cultures serotyped at the reference laboratory

Table 1 Reported outbreaks of hemorrhagic colitis associated with *Escherichia coli* 0157 : H7, 1982–1985

Place	Date month/year	Number of cases	Setting	Suspected vehicle	Reference
Oregon	2–3/82	26	Community	Hamburger	(84)
Michigan	5–6/82	21	Community	Hamburger	(84)
Ontario	11/82	31*	Nursing home	Hamburger	(49)
Quebec	5/83	17	Community	—	(50)
Nebraska	9/84	34*	Nursing home	Hamburger	(87)
North Carolina	9–10/84	36	Day care center	—	(96)
Ontario	9/85	53*	Nursing home	Sandwiches	(37, 51)

* Residents only.

since 1973, and detected only one strain with the O157:H7 serotype. This strain was recovered from a 50-year-old California woman in 1975 during an episode of acute, self-limited, grossly bloody diarrhea (84).

Surveillance for cases of hemorrhagic colitis was also conducted at local levels. In a 12-month hospital-based survey in Chicago, only two strains of *E. coli* O157:H7 were isolated from 2552 stool specimens (36). In contrast, in a hospital-based study in Calgary, Canada, 19 isolates of *E. coli* O157:H7 from 125 patients with bloody diarrhea were recovered during a 6-month period (75). In a later study in the same location the organism was recovered from 116 of 5228 patients with diarrhea (including nonbloody diarrhea) (74). In Calgary, *E. coli* O157:H7 is now recognized as the most common cause of bloody diarrhea of infectious etiology; it accounts for nearly 50% of such cases (C. H. Pai, personal communication). Another hospital-based study in Vancouver, Canada identified *E. coli* O157:H7 from 34 patients during a two-year survey (33). In Vancouver during this period, *E. coli* O157:H7 was the second most frequently isolated enteric bacterial pathogen, after *Campylobacter* (33). In a community-based survey in Newfoundland, the public health laboratories isolated *E. coli* O157:H7 from stool specimens of 7 (15%) of 47 patients who had bloody diarrhea (80).

Reports outside North America of sporadic cases of hemorrhagic colitis associated with *E. coli* O157:H7 are sparse. In the United Kingdom, Day et al (20) reported isolation of only one O157:H7 strain from more than 15,000 strains of *E. coli* recovered during the five years preceding 1983. It is not known if the organism was isolated from a case of hemorrhagic colitis. In Japan, the prefectural public health laboratory in Osaka identified *E. coli* O157:H7 from one of 464 diarrheal stool specimens during a 9-month retrospective study in 1984 (48). It was isolated from a stool of a 34-month-old child with symptoms compatible with hemorrhagic colitis (48).

The preceding surveillance studies suggest that there are geographic differences in disease prevalence. However, because of possible variation in reporting practices and laboratory screening methods, these differences cannot be accurately measured. Nevertheless, comparable studies, such as the hospital-based studies of Chicago, Calgary, and Vancouver, suggest that hemorrhagic colitis associated with *E. coli* O157:H7 is more common in Western Canada. These apparent differences may reflect differences in the proximity of the population at risk to the sources or reservoir of *E. coli* O157:H7. These studies show that in certain areas of North America *E. coli* O157:H7 can no longer be considered rare.

The large number of cases identified in the 21-month Calgary survey made possible determination of seasonal variation in disease incidence (74). *E. coli* O157:H7–associated hemorrhagic colitis was most frequently seen during June–September (75% of the annual total) (C. H. Pai, personal communication). Such a seasonal variation was not observed in the 20-month survey conducted in the United States (82); the number of culture-proven cases may have been too small to detect such a fluctuation in the latter study.

Host Factors, Modes of Transmission, and Reservoir

In the community-wide outbreaks that occurred in the United States, the age of the patients varied from 4 to 76 years and the distribution of the sexes was nearly equal (84). The subsequent outbreaks in nursing homes and a day-care center confirmed that persons in all age groups are at risk for the disease (1, 37, 49, 51, 87, 96). However, these latter outbreaks also suggest that persons at the extreme ends of the age spectrum may be more at risk. In the nursing home outbreak that occurred in London, Ontario, both residents and staff developed the illness, but the attack rate among the residents was more than twice that among the younger staff members (1). In the day-care center outbreak, the diarrhea attack rate was 48% for children less than 4 years of age and 12% for older children (96). However, these differences could have resulted from differences in exposure to the source of infection, and may not reflect true differences in host susceptibility.

The results from investigations of earlier outbreaks suggested that hemorrhagic colitis is primarily a food-borne illness (84, 87). Case-control studies have documented that ingestion of hamburger meat was significantly associated with the disease in the outbreaks in Oregon, Michigan, and Nebraska (84, 87). In the community-wide outbreaks secondary transmission was not apparent.

In contrast, in the two outbreaks in the nursing homes in Ontario both transmission by common food and secondary person-to-person transmission were indicated (1, 37, 49, 51). Hamburgers were ingested shortly before the outbreak in Ottawa (although no case-control study was conducted to implicate them) (49), and sandwiches that included ham, turkey, and cheese were felt to have been the primary vehicles of infection in the London, Ontario outbreak (1). The distribution of the onset of illness was characterized by a cluster within the first 1–3 days of the outbreak, followed by a second, more evenly distributed wave of illness among both residents and staff, which suggests secondary transmission (1, 49).

Person-to-person transmission appeared to be the predominant mode of spread of hemorrhagic colitis in the day-care center outbreak in North Carolina (96). No single food item was shared among most of the children attending the center (96); a case-control study of family members found that 10 (18%) of 56 family members of ill children but only one (2%) of 45 family members of healthy children developed diarrhea; this suggests secondary spread within families.

It is noteworthy that person-to-person transmission of hemorrhagic colitis is suggested only in institutional settings. Although this may be related to the increased susceptibility to *E. coli* O157:H7 of the elderly or very young persons, transmission to the staff of the nursing homes may result from their repeated exposures to the infected patients they care for. It is also possible that infected staff may contribute to the intrainstitutional spread.

Since the initial reports of food-borne hemorrhagic colitis (83, 84), it has become apparent that the disease also spreads person-to-person. The patterns of food-borne transmission in the community and institutions and secondary person-to-person spread within institutions are similar to the known patterns of disease transmission observed with EPEC (77, 99) and EIEC (56, 66). The disproportionate representation of nursing homes (5 of 13) among the recognized outbreaks of hemorrhagic colitis is striking and raises concerns, but is unexplained. The frequency of the disease in institutional settings, the possibility of a high secondary transmission rate, and the high mortality in some outbreaks (36% in London, Ontario) should alert the staff of these institutions to the importance of careful enteric precautions during suspected outbreaks of hemorrhagic colitis.

Hamburger meat was most frequently implicated in the outbreaks examined by case-control studies. In August, 1986, two children in separate family dairy farms in Wisconsin developed hemorrhagic colitis that was complicated by hemolytic-uremic syndrome (see discussion below) shortly after they drank raw milk (59). Their stool cultures yielded *E. coli* O157:H7. The association of the disease with ingestion of raw milk and with hamburger meat suggested that cattle may be a reservoir for *E. coli* O157:H7. Fecal specimens from animals on both of the farms were examined, and *E. coli* O157:H7 was isolated from heifers from both herds. This was the first time this organism was isolated from an animal source. Hence, the organism can be transmitted from animal to human. Established recommendations for avoiding diseases such as salmonellosis by proper handling of food derived from cattle should also apply to the prevention of hemorrhagic colitis.

CLINICAL MANIFESTATIONS

Natural History

Hemorrhagic colitis has a spectrum of clinical manifestations. A typical case begins with sudden onset of severe abdominal cramping, followed within hours by watery diarrhea. Upper gastrointestinal symptoms such as nausea and vomiting occur early. Many women have compared the abdominal cramping to labor during childbirth; others have described the pain as worse than that of appendicitis. Some cases in the outbreaks were indeed mistakenly diagnosed as appendicitis, and the patients were subjected to exploratory laparotomy. Within 1–2 days, the watery diarrhea progresses to bloody diarrhea, with profuse bloody discharge resembling lower gastrointestinal hemorrhage. In contrast to the symptoms of dysentery, mucus is scanty in the bloody stool. More recent observations have noted that *E. coli* O157:H7 can be recovered from patients with nonbloody diarrhea as well as from asymptomatic persons. Hence, infection with this organism does not always lead to the full expression of the syndrome just described.

The physical examination is unremarkable. Abdominal distention can be seen in elderly patients (87). Fever is characteristically absent or low grade. However, Ryan et al (87) noted high fever (>38°C) late in the course of the illness in some of the elderly patients.

The incubation period of hemorrhagic colitis is 3–4 days. This figure is derived from investigations in which the time of exposure to the vehicle of infection was known (84). Ryan et al noted a median incubation period of 8 days (87). They suggested that the longer incubation period may reflect a lower inoculum dose. The average duration of illness is about 8 days. The illness is self-limited in most patients, and most do not require any specific therapy other than supportive care. However, serious complications and deaths do occur.

Laboratory Tests

Routine admission laboratory tests have revealed an elevated white blood cell count (mean of 13–14 $\times$ 10^3 cells μl^{-1}) with slight left shift. Hematocrit, despite the bleeding, is generally not decreased. Barium enema, if performed early, shows a thumbprinting pattern that suggests submucosal edema or hemorrhage, usually in the ascending or transverse colon, sometimes in a segmental distribution. This type of radiologic finding, except for the segmental distribution, is characteristic of patients with ischemic colitis (58, 93); some of the patients in one nursing home outbreak (87) were initially thought to have ischemic colitis. Unnecessary laparotomy was averted only because their illnesses were part of a recognized outbreak of *E. coli* O157:H7 hemorrhagic colitis.

Sigmoidoscopic examination usually reveals normal rectum and sigmoid colon, with blood descending from above. Colonoscopic examination shows erythema, hemorrhage, and edema in the ascending and proximal transverse colon. Occasionally, lesions may extend to the descending colon.

The involvement of the gastrointestinal tract other than the colon is not well documented. Ryan reported adynamic ileus from plain X rays of six patients in the nursing home outbreak in Nebraska (87). In the same outbreak, a technicium red blood cell scan in one patient suggested that hemorrhage was confined to the colon.

Pathology

Post-mortem examinations were performed on three of four fatalities of the Nebraska nursing home outbreak (87). Two who had the typical symptoms of hemorrhagic colitis had dilatation and hyperemia of the cecum, ascending colon, and transverse colon. Histologic examination of these sections revealed focal areas of acute and chronic inflammation, with congestion and hemorrhage in the lamina propria. No ulcerations were noted. The third case had no gross abnormalities, but patchy eosinophilic infiltrates were noted in the microscopic examination of the ascending colon. Light and immunofluorescence microscopic examination showed no areas of bacterial adherence or invasion (87).

Complications

One serious complication of hemorrhagic colitis that is now well documented is the hemolytic-uremic syndrome (HUS). The typical variety of this syndrome, called "idiopathic" HUS, is an illness with an unknown pathogenic mechanism that usually affects children. It is characterized by a prodrome of diarrhea followed by a triad of microangiopathic hemolytic anemia, thrombocytopenia, and acute renal failure (26, 28, 39). Over the years, a variety of infectious agents have been associated with this disease, including other enteric pathogens such as *Shigella dysenteriae* type 1 (79), *Campylobacter jejuni* (14), and *Salmonella typhi* (2). In 1983, Karmali et al (42) reported sporadic cases of HUS associated with cytotoxin-producing *E. coli* in stools. *E. coli* O157:H7 was the most common *E. coli* serotype identified. The United States national surveillance for hemorrhagic colitis detected one culture-confirmed case complicated by HUS in a 25-year-old woman (82). In the day-care center outbreak in North Carolina, 8% of children with diarrhea and 27% of children with bloody diarrhea developed HUS (96). HUS was also reported among adults in the nursing home outbreaks of Nebraska (87) and London, Ontario (51); sporadic cases associated with *E. coli* O157:H7 were reported among adults in the Seattle area (65).

In a previously mentioned two-year study at a children's hospital in Vancouver, Canada, Gransden et al (33) isolated *E. coli* O157:H7 from nine children with HUS. During the same period, the organism was isolated from 25 other children with diarrhea (18 of these had bloody diarrhea). Two of the children with HUS died, while none of children with only diarrhea died.

Although the proportion of patients with hemorrhagic colitis who will develop HUS is difficult to predict accurately from the above data, the day-care center outbreak and the Vancouver study suggest that *E. coli* 0157:H7–associated hemorrhagic colitis can progress to HUS in a substantial proportion of children. Efforts should be directed toward examining the risk factors responsible for the development of HUS in some persons with hemorrhagic colitis.

In addition to HUS, a related syndrome, thrombotic thrombocytopenic purpura (TTP), has been described in two adults following episodes of hemorrhagic colitis (64, 87). TTP is an extension of HUS that includes neurologic symptoms and fever. Other unusual complications include hemorrhagic cystitis and balanitis (32), convulsions, sepsis with another organism, anemia, and iatrogenic upper gastrointestinal bleeding following attempts to relieve abdominal distention by nasogastric suction (87). Most of these latter complications have occurred in elderly patients.

No deaths were reported in the community-wide outbreaks (84). However, it has become clear that residents of nursing homes and patients who develop HUS have an increased risk of death. The first death of a patient with this disease was that of an elderly woman with metastatic cervical cancer (49). Four (12%) of 34 residents with diarrheal illness in the Nebraska nursing home outbreak died (87). In London, Ontario, 19 (36%) of 53 residents with diarrhea died (51). The latter case fatality rate is higher than that of any other enteric illnesses recently reported from institutions.

The immediate cause of death was identified for some of the patients, but was not obvious for many of them. The mean age of the 19 patients who died in the Ontario outbreak was higher than that of the other residents (1). Ryan et al reported that high fever without another identified source of infection preceded death in two of the four patients who died (87). Late-onset high fever during the course of this otherwise afebrile or minimally febrile illness may indicate poor prognosis.

Management

Early outbreak studies in which patients were hospitalized revealed that the illness was self-limited. Drugs, such as antimicrobial agents to which *E. coli* O157:H7 is susceptible, did not alter the duration of the illness or hospitalization (84). Ryan et al (87) did not detect any marked clinical response to antimicrobial therapy among patients who developed complications. The current recommendation for the treatment of hemorrhagic colitis, therefore, is supportive care. Patients should be monitored closely in anticipation of potential complications such as HUS.

Differential Diagnosis

Typical hemorrhagic colitis can be distinguished clinically from bloody diarrhea or dysentery seen in shigellosis, *Campylobacter* and enteroinvasive *E. coli* enteritis, amebiasis, or other enteric illnesses such as necrotizing enterocolitis or pseudomembranous colitis by the lack of prominent fever. Cases of colitis with bloody discharge have been associated with cytomegalovirus (8, 98, 106), but the disease almost always occurs in patients who are immunosuppressed or who have a history of ulcerative colitis. Since *E. coli* O157:H7 can cause diarrhea without blood, the milder form of the illness may be more difficult to differentiate, unless timely microbiologic examination of feces is made.

As mentioned earlier, the bloody diarrhea with thumbprinting pattern seen upon barium enema examination of an elderly patient could be mistaken for ischemic colitis (58). Radiologic appearance may help to distinguish the two diseases, as segmental distribution of the thumbprinting is not usually seen in ischemic colitis (58). A thumbprinting pattern is also occasionally seen in patients with ulcerative colitis and Schonlein-Henoch purpura (34, 93). If bloody diarrhea occurs as part of an outbreak, interventions such as angiography or exploratory laparotomy should be avoided until more information is obtained.

In young adults, hemorrhagic colitis could be mistaken for the first appearance of an inflammatory bowel disease. Fever is usually prominent when bloody diarrhea occurs with ulcerative colitis. Early colonoscopic and microbiologic examinations should help to differentiate these diseases.

As mentioned in the introduction, illness of unknown etiology resembling hemorrhagic colitis had been described before the discovery of *E. coli* O157:H7 as a pathogen. Several cases of an illness termed "transient ischemic colitis" in young adults (15, 61) were reported following a description of "reversible segmental colitis" or "evanescent colitis" in five young adults in Philadelphia in 1971 (61). Sakurai et al (89) reported eight cases of acute, self-limited, right-sided hemorrhagic colitis associated with oral administration of ampicillin. Others have reported a similar illness associated with antimicrobial agents (4, 55, 63, 78, 85, 100). These cases were distinguished from antibiotic-associated pseudomembranous colitis both clinically and by the absence of pseudomembranes in the colon and *Clostridium difficile* toxin in the stool. Kilpatrick et al (47) described two young women with symptoms compatible with hemorrhagic colitis associated with use of oral contraceptives. Totani (101) described cases of hemorrhagic colitis in which *Klebsiella oxytoca* was isolated in pure cultures from stools; however, no control specimens were obtained in this study. In the investigation of hemorrhagic colitis in Oregon *K. oxytoca* was found in fecal specimens of both cases and controls (84).

Many of the above sporadic cases of hemorrhagic colitis reported before 1982 could certainly have been caused by *E. coli* O157:H7. However, this *E. coli* is not isolated from every case of hemorrhagic colitis. Failure to isolate the *E. coli* from some cases is partly explained by the timing of the stool specimen collection. Several investigations have shown that delayed specimen collection (>6 days after onset of illness) can significantly reduce the chance of isolation of the *E. coli* (75, 84, 104). There are probably other causes of hemorrhagic colitis, and since all of the other groups of *E. coli* associated with diarrhea include more than one serotype, it is certainly possible that there are other *E. coli* serotypes associated with hemorrhagic colitis. However, in the United States and Canada, evidence so far indicates that *E. coli* O157:H7 is the single most common cause of this syndrome.

MICROBIOLOGY

Growth and Survival Characteristics of Escherichia coli *0157:H7*

Doyle & Schoeni (22) showed that *E. coli* O157:H7 can survive up to 9 months at −20°C in ground beef. They observed that *E. coli* O157:H7, unlike most *E. coli,* grows poorly at 44–45.5°C, which is the temperature generally used to recover *E. coli* from foods. These characteristics of this *E. coli* strain should be considered during investigations of food-borne outbreaks of hemorrhagic colitis.

Biochemical Features of Escherichia coli *O157:H7*

The strains of *E. coli* O157:H7 isolated during the outbreaks in Oregon and Michigan failed to ferment sorbitol within 7 days (104). Some of the isolates from sporadic cases were found to ferment sorbitol after 24 hr but not earlier (23, 75). Since 95% of the *E. coli* examined at the CDC fermented sorbitol within 24 hr, screening for sorbitol fermentation is useful in selecting *E. coli* strains for serotyping. Exploiting this biochemical feature, Farmer & Davis (23) developed a quick, single-tube screening test using a medium containing D-sorbitol and H7 antiserum to detect *E. coli* O157:H7 from *E. coli* colony picks. Approximately 14% of other motile *E. coli* strains express the H7 flagellar antigen. Only 3 of 300 *E. coli* strains from various animal sources gave a false positive result with this technique.

Haldane et al (35) reported that of 327 sorbitol-negative *E. coli* isolates, 37 (11.3%) were serotype O157:H7. They found that these *E. coli* strains decarboxylated ornithine and lysine; inclusion of such decarboxylation tests would have increased the specificity of the sorbitol-only screen.

Pathogenesis

Until the discovery that *E. coli* O157:H7 was a pathogen, *E. coli* associated with diarrhea was classified into three groups: enterotoxigenic *E. coli* (ETEC), enteroinvasive *E. coli* (EIEC), and enteropathogenic *E. coli* (EPEC) (see Table 2). They were classified on the basis of their distinct clinical manifestations and virulence determinants. Each group belonged to a specific set of O-antigen serogroups. *E. coli* O157:H7 does not elaborate heat-stable (ST) or heat-labile (LT) toxins produced by ETEC (88); it does not invade enterocytes, as EIEC do (102); and it does not attach to HeLa or HEp-2 tissue culture cells in the characteristic pattern termed "localized adherence" that has recently been recognized as a feature of EPEC (3, 16, 19, 54, 86, 91). The organism does not cause the profuse, watery diarrhea seen with ETEC, nor the febrile, dysentery-like illness seen with EIEC. EPEC typically causes watery diarrhea in infants, although volunteer studies have shown that adults can display symptoms when infected (52). EPEC-induced diarrhea may occur in frequent relapses or a prolonged course lasting more than 2 wk (16, 86). Belnap & O'Donnell (7) reported nine fatal cases of EPEC-induced enteritis with bloody discharge in children during a large outbreak in the early 1950s in Virginia. However, these children developed the bloody diarrhea shortly before death, an average of 19 days after the onset of illness. These features are not seen in typical hemorrhagic colitis associated with *E. coli* O157H:H7. Hence, *E. coli* O157:H7 does not appear to cause illness by any of the recognized mechanisms of virulence associated with the traditional groups of *E. coli;* it thus comprises a new group, enterohemorrhagic *E. coli* (EHEC).

The pathogenesis of hemorrhagic colitis associated with *E. coli* O157:H7 is presently unknown. Johnson et al (38a) and O'Brien et al (70) reported in 1983 that this organism produces high levels of cytotoxin active against Vero cells (African green monkey kidney cells). Cytotoxin active against Vero cells, or Vero toxin (VT), was first described in strains of *E. coli* by Konowalchuk et al in 1977 (48a). O'Brien et al (71) found that in *E. coli* O157:H7 this cytotoxin can be neutralized by purified antiserum against Shiga toxin; they named the cytotoxin "Shiga-like toxin" (SLT).

Because Shiga toxin is known to be a potent cytotoxin and an enterotoxin elaborated in large amounts by *Shigella dysenteriae* type 1, Karmali et al (41) and O'Brien et al (70, 71) speculated that SLT may be a virulence determinant of *E. coli* O157:H7. Shiga toxin can cause ascending motor paralysis in rabbits and mice, and hemorrhage, edema, and ulcerations in the gut of a variety of laboratory animals when given intravenously (12). At the molecular level the toxin is composed of two subunits; a single polypeptide comprises the A subunit and multiple copies of a second polypeptide comprise the B subunit (21, 68, 72). The toxin is believed to affect protein synthesis by catalytic inactivation of the 60S ribosome in eukaryotic cells (10, 81). SLT has biological activities (cytotoxicity to Vero and HeLa cells, mice lethality, and rabbit ileal loop enterotoxicity) similar to those of Shiga toxin. O'Brien et

Table 2 Patterns of illness and mechanisms of virulence associated with *Escherichia coli* groups that cause human enteric illness

E. coli group	Pattern of illness	Proposed virulence mechanism
Enterotoxigenic	Watery diarrhea, low fever	Heat-labile and heat-stable toxins
Enteroinvasive	Dysentery-like diarrhea, high fever	Enterocyte invasion
Enteropathogenic	Watery diarrhea, protracted diarrhea, low–moderate fever	Enteroadherence (?)
Enterohemorrhagic	Watery-hemorrhagic diarrhea, no–low fever	Unknown

al (71) reported that in one outbreak strain of *E. coli* O157:H7 (933), SLT production was determined by a converting phage. A cloned DNA segment from the phage (933J) that determines SLT production is homologous to DNA sequences in Shiga toxin–producing strains of shigellae (67). O'Brien et al (69) further characterized the SLT and found that the relative mobility in SDS polyacrylamide gel of the A and B subunits of the purified SLT is indistinguishable from that of the subunits of Shiga toxin purified from a strain of *Shigella dysenteriae* type 1. The amino acid sequence of the B subunit of Shiga toxin was recently determined (95). It is a polypeptide of 69 amino acids containing one disulfide bridge, and it does not have similarity greater than 30% with any other amino acid sequences reported to date (95). The translated amino acid sequence of the B subunit of SLT has been determined to be identical to that of the B subunit of Shiga toxin (38). Hence, SLT and Shiga toxin share in common similar antigenic, genetic, structural, and functional features, and may indeed be the same toxin, although it is still possible that the A subunits are different.

Within the last year, Strockbine et al (97) and Scotland et al (94) reported that some strains of *E. coli* O157:H7 elaborate a second cytotoxin active against Vero or HeLa cells that is antigenically distinct from SLT (VT). Strockbine et al (97) have designated the cytotoxin neutralizable by antiserum against Shiga toxin as Shiga-like toxin 1 (SLT-1) and the second cytotoxin, which is not neutralizable by the antiserum, as Shiga-like toxin 2 (SLT-2). Scotland et al (94) have referred to the two cytotoxins as Vero toxin I (VT1) and Vero toxin II (VT2), respectively. A Southern hybridization study by Willshaw et al (105) suggested that there is some relatedness between the DNA fragments of VT1 and VT2 genes. Evidence suggests that SLT-1 and VT1 are the same cytotoxin, and presumably SLT-2 and VT2 are the same toxin. SLT-2 appears to be found primarily in culture filtrates, whereas SLT-1 predominates in cell lysates (97). SLT-2 appears to have the same biological activities as SLT-1, but its structure and modes of action have not yet been characterized. Padhye et al (73) have described a Vero-cell active, 64-kd polypeptide purified from a strain of *E. coli* O157:H7 that is not neutralized by antisera against Shiga toxin. Whether this is the purified SLT-2 (or VT2) is unknown.

The role of these newly recognized cytotoxins of *E. coli* O157:H7 in the pathogenesis of hemorrhagic colitis remains to be understood. Although the production of a toxin by *Shigella* has been appreciated since the turn of the century (18, 25), the discovery that the Shiga bacillus produces cell-free toxin capable of causing fluid accumulation in the ligated rabbit ileal loop (45, 46) stimulated new debate on the role of Shiga toxin in diarrhea pathogenesis (30, 44, 53). Cantey (11) recently reviewed the role of Shiga toxin in the pathogenesis of various infectious diseases. Therefore, in this review, discussion of the Shiga toxin or SLT (VT) is limited to its role in the pathogenesis of hemorrhagic colitis. It should be noted that most of the studies related to the cytotoxins of *E. coli* O157:H7 were performed prior to the discovery of the second cytotoxin in this strain. Hence, it is not always clear from the reading of the original sources that "SLT" or "VT" represents a homogeneous protein. In the interpretation of the results this uncertainty should therefore be taken into consideration.

Koch's postulates, unfortunately, do not apply to proving the pathogenic capacity of a toxin. Therefore, to date, attempts to link SLT (VT) with the clinical manifestations of hemorrhagic colitis have relied on epidemiologic, clinical, and animal-model investigations. For the remainder of this review I discuss in detail some of these studies.

EPIDEMIOLOGIC EVIDENCE The proposal that *E. coli* O157:H7 is an etiologic agent of hemorrhagic colitis is based on epidemiologic observations derived from several case-control studies of geographically separate disease outbreaks that repeatedly yielded a statistically significant association of this serotype with the disease. Subsequent prospective studies confirmed the original association. To link SLT or VT to hemorrhagic colitis, similar epidemiology should be observed for other organisms that produce SLT (VT). Recently, several investigators have identified other human-infecting isolates of *E. coli* that produce high levels of SLT (40, 42, 57). These include strains belonging to serogroups O26, O111, O113, O121, O128, and O145. To date, none of these strains has been implicated in recognized outbreaks of hemorrhagic colitis; every outbreak of hemorrhagic colitis since 1982 has been associated only with *E. coli* O157:H7. Furthermore, prospective surveys at both the national and local levels have shown that 15–50% of fecal isolates from patients whose illness met the clinical definition of hemorrhagic colitis have been *E. coli* O157:H7. The organism was isolated from these patients in widely separate geographic areas (Florida to Alaska) (82); no other single serotype of *E. coli* with comparable isolation frequency and geographic distribution was identified in the same group of patients. Hence, the epidemiologic observations that linked *E. coli* O157:H7 to hemorrhagic colitis are still lacking for other SLT- or VT-producing strains of *E. coli*.

In a retrospective study, Marques et al (57) examined 418 cultures of *E. coli* obtained mostly from the CDC, isolated over a period of 30 years. They separated these strains according to the levels of cell-associated cytotoxin they produced, i.e. no detectable level, low levels, moderate levels, and high levels. They found that 20 (74%) of 27 persons with hemorrhagic colitis had *E. coli* isolates that produced moderate to high levels of the cytotoxin, whereas only one (2%) of 48 persons without an enteric illness yielded a high level–producing isolate. Hence, it appeared that *E. coli* strains that produced moderate to high levels of cell-associated cytotoxin were significantly associated with hemorrhagic colitis. However, 18 of the 20 isolates from patients with hemorrhagic colitis were *E. coli* O157:H7; if these strains were excluded, the number of isolates (two) from persons with hemorrhagic colitis would become too small to yield any statistically meaningful conclusions.

Marques et al (57) examined the cytotoxin production of *E. coli* strains isolated from persons with HUS. Isolates that produced moderate to high levels of the cytotoxin were recovered from all seven persons with HUS. Statistically significant association of these strains with HUS is clear when these patients are compared to persons without diarrhea ($p<0.001$). However, since this was a retrospective study and precise clinical history may not have been available, the above results must be interpreted with caution. Nevertheless, the study does support Karmali et al's (40, 42) association of SLT (VT)-producing *E. coli* with HUS, but it does not link organisms other than *E. coli* O157:H7 to hemorrhagic colitis. In summary, epidemiologic evidence to date remains insufficient to support or refute the association of other SLT (VT)-producing organisms with hemorrhagic colitis.

CLINICAL EVIDENCE Abdominal pain, bloody diarrhea, and lack of prominent fever are clinical features shared by hemorrhagic colitis and the prodromal phase of some cases of HUS. A radiologic pattern like the thumbprint is also seen in the colitis of HUS. Because of this clinical similarity, Karmali et al (41) suggested that the two diseases are manifestations of the same underlying process. The association of VT-producing *E. coli* with HUS and the association of VT-producing *E. coli* O157:H7 with hemorrhagic colitis and HUS raised the possibility that VT itself is involved in this underlying process.

Although the data are limited, the clinical manifestations are reflected by similarity in some of the histopathologic descriptions of the colonic lesions seen in both diseases. Colonic and rectal biopsy studies of some of the patients with hemorrhagic colitis seen in Calgary, Canada showed normal mucosa to mild and acute inflammation (75). Postmortem examination of three adult fatalities of hemorrhagic colitis (87) showed dilatation and hyperemia of the cecum, ascending colon, and transverse colon; microscopy revealed focal areas of acute and chronic inflammation and congestion and focal hemorrhage in the lamina propria.

The colonic lesions described in cases of HUS span a wider spectrum. Bar-Ziv et al (5) described a 3½-year-old boy with HUS whose rectal biopsy showed nonspecific inflammation without ulcerations. Berman (9) reported on five children with HUS who initially sustained rectal bleeding and whose sigmoidoscopic examination revealed changes compatible with ulcerative colitis; biopsies from three of them showed edema and submucosal hemorrhage. In a study of 52 autopsies of 678 patients with this syndrome, kidneys and colonic wall were the most frequent sites with small vessel thrombosis (31). The colonic lesions in these cases consisted of areas of hemorrhage and necrosis with mucosal ulcerations, especially within the cecum and ascending colon. Others have reported colonic lesions including stricture (90) and gangrene requiring surgery (92) in association with HUS.

Although the colonic lesions of hemorrhagic colitis described are nonspecific and seemingly milder than those described for HUS, most histopathologic descriptions to date have been anecdotal, and the disease could certainly have a wider spectrum. It should also be noted that the HUS cases reported above did not have a specific etiology and therefore are not known to have been associated with SLT (VT)-producing organisms. I know of no systematic study of histopathologic examinations of the colon of patients with HUS specifically associated with other SLT (VT)-producing *E. coli*. An O111 *E. coli* serotype was implicated in a diarrhea outbreak that included nine fatal cases with bloody diarrhea (7, 60). In these patients, whose clinical presentation and autopsy findings were consistent with the diagnosis of HUS, the colon showed extensive damage with hemorrhagic necrosis and ulceration of the mucosa (60). So far such lesions have not been reported in hemorrhagic colitis associated with *E. coli* O157:H7. It is not known if this O111 *E. coli* strain produced a cytotoxin.

The clinical observations do suggest that the symptoms of colitis and the lesions in the colon in hemorrhagic colitis and HUS are produced by the same mechanism. However, whether this mechanism is mediated solely by SLT (VT) is unclear. Since human volunteer studies with the purified cytotoxins of *E. coli* O157:H7 are precluded by the severity of the potential complications, further clinical investigations to provide evidence linking SLT (VT) with hemorrhagic colitis will be difficult.

EVIDENCE FROM ANIMAL STUDIES With an appropriate animal model, several strategies can be used to provide evidence for the role of a toxin (as opposed to an intact organism) in pathogenesis. These include demonstration that (*a*) the toxigenic strain, but not its nontoxigenic isogenic variant, produces the disease, (*b*) a vaccine that elicits a toxin-specific antibody is protective, (*c*) a soluble receptor analog of the natural receptor for the toxin can confer protection by competitive inhibition of the natural receptor, and (*d*) the toxin purified to homogeneity produces the same disease as the organism that produces the toxin. Some of these strategies have already been undertaken to demonstrate the pathogenic capacity of SLT (VT).

In early animal studies the pathogenic potential of *E. coli* O157:H7 was examined in attempts to identify an appropriate animal model. Farmer et al (24) were able to induce diarrhea (nonbloody) in 5–10-day-old rabbits by nasogastrically inoculating them with an outbreak strain of *E. coli* O157:H7. The bacteria failed to produce diarrhea in older rabbits, 2-week-old guinea pigs, 3-week-old mice, and young rhesus monkeys in the same study. Beery et al (6) showed that *E. coli* O157:H7 colonizes the cecae of chicks but fails to induce diarrhea. Francis et al (27) produced watery diarrhea in gnotobiotic pigs with an orally administered strain of *E. coli* O157:H7 and demonstrated that the bacteria can attach to the epithelial surface in the cecum and the colon. The bacteria also attached in small foci to the ileal and rectal mucosal surfaces. The epithelial cells in the areas of adherent bacteria were distorted, with effacement of the microvilli and projections or invaginations of the plasma membrane; these lesions resembled the intestinal mucosal lesions caused by enteropathogenic *E. coli*. None of these animal models, however, reproduced the disease seen in humans, i.e. grossly bloody diarrhea with the colon as the predominant site of disease.

An intriguing study on the role of VT in hemorrhagic colitis was reported by Pai et al (76). They confirmed Farmer et al's observation (24) that the susceptibility of the rabbit to *E. coli* O157:H7 is related to age. They also found that a high-VT-producing strain of *E. coli* O157:H7 produced histologic changes in the colon that were reproduced in infant rabbits given VT alone (76). These changes occurred predominantly in the mid and distal colon, and included apoptosis (individual cell death) in the surface epithelium, increased mitotic activity in the crypts, and a mild to moderate infiltration of neutrophils in the lamina propria and epithelium. Bacteria were seen attached to the luminal surface of the colon, cecum, and gut-associated lymphoid tissue (Peyer's patches, sacculus rotundus, and appendix), but never inside these cells or in the lamina propria. In the small intestine, only scanty foci of bacteria attached to the villi were observed. Similar histologic changes (without bacterial attachment) were seen in rabbits given VT intragastrically, and the major changes were limited to the colon. One rabbit developed bloody diarrhea and extensive mucosal and submucosal congestion in addition to the histologic changes described above.

The histologic changes described in the infant rabbits are similar to the changes described by Keenan et al (43) in the rabbit ligated ileal loop induced by Shiga toxin derived from a strain of *Shigella dysenteriae* type 1 and SLT derived from a strain of *E. coli* O157:H7. Both toxins produced similar changes and appeared to selectively affect the mature columnar absorptive epithelial cells. Ultrastructural studies suggested that the cells underwent a process of apoptosis; this finding was similar to that reported in the study by Pai et al (76).

However, despite the similarity of the histologic changes described in the two preceding studies, the data from these studies are not completely consistent. Pai et al (76) observed that the epithelium of the small intestine was spared the changes seen in the colon despite a high concentration of VT in the small intestine, and concluded that the colonic epithelium may be more specifically susceptible to VT. Keenan et al (43), in contrast, concluded that SLT affects a specific cell population of the ligated segment of the small intestine. If VT and SLT are the same toxin, they should share a common pathway in the pathogenesis of diarrhea, and thus the site of action of the toxins should be the same. Mobassaleh et al (62) reported that Shiga toxin did not bind to microvillus membranes of small intestine from rabbits less than 21 days of age; this suggests that the expression of the Shiga toxin receptors on the mucosal surface of the rabbit small intestine is age dependent. The production of diarrhea by *E. coli* O157:H7 or VT was also shown to be age dependent in 5–10-day-old rabbits (24) and 3–11-day-old rabbits (76). Hence, there seems to be a discrepancy of results between the cytotoxin studies that used the rabbit ligated small intestine and the studies in which diarrhea was induced by bacteria or toxin in intact infant rabbits. Perhaps these discrepancies can be resolved if the VT examined by Pai et al (76) is not the same toxin as the SLT examined by Keenan et al (43). SLT or Shiga toxin may have a specific site of action in the small intestine, whereas VT, a different cytotoxin, may have a specific site of action in the colon. Such an interpretation is certainly possible, as *E. coli* O157:H7 is now known to produce a second cytotoxin active against Vero cells (VT2 or SLT-2), as mentioned above (94, 97).

Tzipori et al (103) examined the role of the two cytotoxins of *E. coli* 0157:H7 in diarrheal pathogenesis in another animal experiment. They examined orally inoculated gnotobiotic pigs with two strains of *E. coli* O157:H7: EDL 931, which produces large quantities of VT neutralizable by antiserum to Shiga toxin, and 3100-85, which produces moderate amounts of VT not neutralizable with the anti-Shiga toxin. Interestingly, they found no difference in the clinical manifestations or histopathologic lesions in the intestine of piglets inoculated with either strain. The lesions described appeared to be more severe than those described in the piglet study of Francis et al (27). Lesions included foci of degenerating epithelial cells; some lesions progressed to complete necrosis and erosion of the surface epithelium in the ileum and the cecum (103). This study used a variation of the strategy of inoculating two isogenic strains of which one strain lacks the toxin-producing capacity of the other. It is not clear what the results mean. It is possible that these lesions were caused by yet another virulence factor common to both strains of *E. coli* O157:H7; the cytotoxins may have had no role in the pathogenesis of diarrhea in the piglets.

Hence, despite some interesting findings from the animal studies, the role of SLT or VT in the pathogenesis of hemorrhagic colitis is unsolved. There are several problems. One is uncertainty concerning the appropriateness of the animal model. It is not clear that the disease elicited in infant rabbits or gnotobiotic pigs is the same disease seen in human cases of hemorrhagic colitis. Pai et al (76) did examine one of the strategies outlined above to provide evidence for the pathogenicity of VT with the infant rabbit model; at least one animal developed bloody diarrhea with administration of VT alone (although not with the *E. coli*). Other strategies could certainly be designed to test this model, although the short period of susceptibility of infant rabbits precludes demonstrating protection by a vaccine that may elicit a cytotoxin-specific antibody. Another problem is the lack of clear definition and biological and antigenic characterization of the cytotoxins used in the experiments, which made comparison and extrapolation from one study to another impossible. The study by Pai et al (76) clearly suggested that a VT produced by *E. coli* O157:H7 had pathogenic potential, but it is not clear if this VT was a homogeneous polypeptide with a specific antigenic characteristic identical to the so-called SLT described by Keenan et al (43).

In conclusion, the pathogenic potential of Vero cell–active cytotoxin(s) of *E. coli* O157:H7 in hemorrhagic colitis is so far not supported by any epidemiologic findings, is difficult to demonstrate by clinical observations, and is inconclusive from animal studies. Although it is tempting to attribute all of the manifestations of a syndrome to a single agent such as a cytotoxin, it is not at all clear that hemorrhagic colitis can be characterized in this manner. Like shigellosis, hemorrhagic colitis clearly has two distinct phases of illness, the initial watery diarrhea phase and the later hemorrhagic phase, which corresponds to the dysentery phase of shigellosis. The production of mucus, blood, and fever that characterizes the dysentery syndrome is believed to result from the inflammation and destruction of the colonic mucosa following invasion of shigella into the colonic epithelium and its subsequent multiplication in the cells (29). Although *E. coli* O157:H7 does not invade the epithelial cells and hence does not cause dysentery, it may attach to the colonic mucosa, as suggested by animal studies, and thereby deliver to the enterocytes a cytotoxin that produces the characteristic syndrome. Like shigella, *E. coli* O157:H7 may possess a complex of virulence determinants that are all necessary to elicit the full spectrum of the syndrome.

CONCLUSIONS

Hemorrhagic colitis is a unique enteric illness that is associated most frequently with *E. coli* serotype O157:H7. It is transmitted primarily by food, but person-to-person spread appears to be an important mode of transmission in certain settings such as day-care centers and nursing homes. Evidence to date suggests that cattle may be an important reservoir of the *E. coli,* and the organism is recognized with higher frequency in certain areas of North America. Clinical features of the disease vary from self-limited, watery, and bloody diarrhea to serious complications such as hemolytic-uremic syndrome or thrombotic thrombocytopenic purpura, and the disease can result in death. Although Vero cell–active cytotoxins elaborated by *E. coli* O157:H7 may have a role in the pathogenesis of hemorrhagic colitis, evidence from epidemiologic, clinical, and animal studies is inconclusive.

Future epidemiologic studies should include examination of risk and host factors for infection, especially in institutional settings; detailed characterization of the modes of secondary spread; determination of the reasons for the apparent endemicity in certain areas of North America; further exploration of the importance of cattle as a reservoir of *E. coli* O157:H7; and implementation of intervention measures based on information obtained from the above studies. Clinical studies may involve systematic comparison of the histopathologic changes in the colon in hemorrhagic colitis and in HUS; identification of risk factors for the progression of hemorrhagic colitis into HUS; examination of the causes of death; and identification of modalities to prevent HUS and death in patients who acquire hemorrhagic colitis. Finally, laboratory studies may involve development of rapid diagnostic methods according to the types of cytotoxin produced by the *E. coli;* structure and function analyses of the cytotoxins; determination of other virulence determinants; exploration of the receptor-ligand interaction of the intact organism or the cytotoxins with the host intestinal epithelium; further testing of the infant rabbit model and identification of additional appropriate animal models; and clarification of the role of other serotypes of *E. coli* in the pathogenesis of hemorrhagic colitis. These studies are already underway in several centers.

Acknowledgments

I thank Dr. Gary K. Schoolnik for his advice and critical review of the manuscript, Dr. Alison D. O'Brien, Dr. C. H. Pai, and the members of the Enteric Diseases Branch of the Centers for Disease Control for providing me with unpublished manuscripts or information, and Nicolas Y. Riley and Jesse Furman for their patience.

Literature Cited

1. Alberta Social Services and Community Health. 1985. *Escherichia coli* O157:H7 in a nursing home. *Epidemiol. Notes Rep.* 9:485–89
2. Baker, N. M., Mills, A. E., Rachman, I., Thomas, J. E. P. 1974. Haemolytic-uraemic syndrome in typhoid fever. *Br. Med. J.* 2:84–87
3. Baldini, M. M., Kaper, J. B., Levine, M. M. 1983. Plasmid-mediated adhesion in enteropathogenic *Escherichia coli*. *J. Pediatr. Gastroenterol. Nutr.* 2:534–38
4. Barrison, I. G., Kane, S. P. 1978. Penicillin-associated colitis (letter). *Lancet* 2:843
5. Bar-Ziv, J., Ayoub, J. I. G., Fletcher, B. D. 1974. Hemolytic-uremic syndrome: a case presenting with acute colitis. *Pediatr. Radiol.* 2:203–6
6. Beery, J. T., Doyle, M. P., Schoeni, J. L. 1985. Colonization of chicken cecae by *Escherichia coli* associated with hemorrhagic colitis. *Appl. Environ. Microbiol.* 49:310–15
7. Belnap, W. D., O'Donnell, J. J. 1955. Epidemic gastroenteritis due to *Escherichia coli* 0111: Review of the literature, with the epidemiology, bacteriology, and clinical findings of a large outbreak. *J. Pediatr.* 47:178–93
8. Bennett, M. R., Fine, A. P., Hanlon, J. T. 1985. Cytomegalovirus hemorrhagic colitis in a nontransplant patient. *Postgrad. Med.* 77:227–29
9. Berman, W. 1972. The hemolytic-uremic syndrome: Initial clinical presentation mimicking ulcerative colitis. *J. Pediatr.* 81:275–78
10. Brown, J. E., Ussery, M. A., Leppla, S. H., Rothman, S. W. 1980. Inhibition of protein synthesis by Shiga toxin. Activation of the toxin and inhibition of peptide elongation. *FEBS Lett.* 117:84–88
11. Cantey, R. J. 1985. Shiga toxin—an expanding role in the pathogenesis of infectious diseases. *J. Infect. Dis.* 151:766–71
12. Cavanagh, J. J. B., Howard, J. G., Whitby, J. L. 1956. The neurotoxin of *Shigella shigae*. A comparative study of the effects produced in various laboratory animals. *Br. J. Exp. Pathol.* 37:272–8
13. Centers for Disease Control. 1982. Isolation of *E. coli* O157:H7 from sporadic cases of hemorrhagic colitis—United States. *Morb. Mortal. Wkly. Rep.* 31:580–85
14. Chamovitz, B. N., Hartstein, A. I., Alexander, S. R., Terry, A. B., Short, P., et al. 1983. *Campylobacter jejuni*–associated hemolytic-uremic syndrome in a mother and daughter. *Pediatrics* 71:253–56
15. Clark, A. W., Lloyd-Mostyn, R. H. L., Sadler, M. R. de C. 1972. "Ischaemic" colitis in young adults. *Br. Med. J.* 4:70–72
16. Clausen, C. R., Christie, D. C. 1983. Chronic diarrhea in infants caused by adherent enteropathogenic *Escherichia coli*. *J. Pediatr.* 100:358–61
17. Communicable Disease Surveillance Centre. 1985. Haemorrhagic colitis: East Anglia. *Commun. Dis. Rep.* 85(28):2
18. Conradi, H. 1903. Über losliche, durch aseptische Autolyse erhaltene Giftstoffe von Ruhr- und *Typhus-bazillen*. *Dtsch. Med. Wochenschr.* 29:26–28
19. Cravioto, A., Gross, R. J., Scotland, S. M., Rowe, B. 1979. An adhesive factor found in strains of *Escherichia coli* belonging to the traditional infantile enteropathogenic serotypes. *Curr. Microbiol.* 3:95–99
20. Day, N. P., Scotland, S. M., Cheasty, T., Rowe, B. 1983. *Escherichia coli* O157:H7 associated with human infections in the United Kingdom (letter). *Lancet* 1:825
21. Donohue-Rolfe, A., Keusch, G. T., Edson C., Thorley-Lawson, D., Jacewicz, M. 1984. Pathogenesis of *Shigella* diarrhea. IX. Simplified high yield purification of *Shigella* toxin and characterization of subunit composition and function by the use of subunit-specific monoclonal and polyclonal antibodies. *J. Exp. Med.* 160:1767–81
22. Doyle, M. P., Schoeni, J. L. 1984. Survival and growth characteristics of *Escherichia coli* associated with hemorrhagic colitis. *Appl. Environ. Microbiol.* 48:855–56
23. Farmer, J. J., Davis, B. R. 1985. H7 antiserum–sorbitol fermentation medium: a single tube screening medium for detecting *Escherichia coli* O157:H7 associated with hemorrhagic colitis. *J. Clin. Microbiol.* 22:620–25
24. Farmer, J. J., Potter, M. E., Riley, L. W., Barrett, T. J., Blake, P. A., et al. 1983. Animal models to study *Escherichia coli* O157:H7 isolated from patients with hemorrhagic colitis (letter). *Lancet* 1:702–3

25. Flexner, S., Sweet, J. E. 1906. The pathogenesis of experimental colitis, and the relation of colitis in animals and man. *J. Exp. Med.* 8:514–35
26. Fong, J. S. C., DeChadarevian, J. P., Kaplan, B. S. 1982. Hemolytic-uremic syndrome. Current concepts and management. *Pediatr. Clin. North Am.* 29:835–56
27. Francis, D. H., Collins, J. E., Duimstra, J. R. 1986. Infection of gnotobiotic pigs with an *Escherichia coli* O157:H7 strain associated with an outbreak of hemorrhagic colitis. *Infect. Immun.* 51:953–56
28. Gasser, C., Gautier, E., Stech, A., Siebenmann, R. E., Dechslin, R. 1955. Hämolytisch-urämische Syndrome: bilatorale Nierenrindennekrosen bei akuten erworbenen hämolytischen Anämien. *Schweiz. Med. Wschr.* 85:905–9
29. Gemski, P., Formal, S. B. 1975. Shigellosis: an invasive infection of the gastrointestinal tract. In *Microbiology*, ed. D. Schlessinger, pp. 165–69. Washington, DC: Am. Soc. Microbiol.
30. Gemski, P., Takeuchi, A., Washington, O., Formal, S. B. 1972. Shigellosis due to *Shigella dysenteriae* 1: Relative importance of mucosal invasion versus toxin production in pathogenesis. *J. Infect. Dis.* 126:523–30
31. Gianantonio, C. A., Vitracco, M., Mendilaharzu, F., Gallo, G. E., Sojo, E. T. 1973. The hemolytic-uremic syndrome. *Nephron* 11:174–92
32. Gransden, W. R., Damm, M. A. S., Anderson, J. D., Carter, J. E., Lior, H. 1985. Haemorrhagic cystitis and balanitis associated with Vero-toxin producing *Escherichia coli* O157:H7. *Lancet* 2:150
33. Gransden, W. R., Damm, M. A. S., Anderson, J. D., Carter, J. E., Lior, H. 1986. Further evidence associating hemolytic uremic syndrome with infection by Verotoxin producing *Escherichia coli* O157:H7. *J. Infect. Dis.* 154:522–24
34. Grossman, H., Berdon, W. E., Baker, D. H. 1965. Reversible gastrointestinal signs of hemorrhage and edema in the pediatric age group. *Radiology* 84:33–39
35. Haldane, D. J. M., Damm, A. S., Anderson, J. D. 1986. Improved biochemical screening procedure for small clinical laboratories for Vero (Shiga-like-)-toxin-producing stains of *Escherichia coli* O157:H7. *J. Clin. Microbiol.* 24:652–53
36. Harris, A. A., Kaplan, L. J., Goodman, L. J., Doyle M., Landau, W., et al. 1985. Results of a screening method used in a 12-month stool survey for *Escherichia coli* O157:H7. *J. Infect. Dis.* 152:775–77
37. Hockin, J. C., Lior, H. 1986. Haemorrhagic colitis due to *Escherichia coli* O157:H7. A rare disease? *Can. Med. Assoc. J.* 134:25–26
38. Jackson, M. P., Newland, J. W., Holmes, R. K., O'Brien, A. D. 1987. Nucleotide sequence analysis of the structural genes for Shiga-like toxin 1 encoded by bacteriophage 933J from *Escherichia coli*. *Microb. Pathog.* 2:147–53
38a. Johnson, W. M., Lior, H., Bezanson, G. S. 1983. Cytotoxic *Escherichia coli* O157:H7 associated with haemorrhagic colitis in Canada (letter). *Lancet* 1:76
39. Kaplan, B. S., Thomson, P. D., De-Chandarevian, J. P. 1976. The hemolytic-uremic syndrome. *Pediatr. Clin. North Am.* 23:761–77
40. Karmali, M. A., Petric, M., Lim, C., Fleming, P. C., Arbus, G. S., et al. 1985. The association between idiopathic hemolytic-uremic syndrome and infection by verotoxin-producing *Escherichia coli*. *J. Infect. Dis.* 151:775–82
41. Karmali, M. A., Petric, M., Lim, C., Fleming, P. C., Steele, B. T. 1983. *Escherichia coli* cytotoxin, haemolytic-uraemic syndrome, haemorrhagic colitis (letter). *Lancet* 2:1299–300
42. Karmali, M. A., Steele, B. T., Petric, M., Lim, C. 1983. Sporadic cases of haemolytic-uraemic syndrome associated with faecal cytotoxin and cytotoxin producing *Escherichia coli* in stools. *Lancet* 1:619–20
43. Keenan, K. P., Sharpnack, D. D., Collins, H., Formal, S. B., O'Brien, A. D. 1986. Morphologic evaluation of the effects of Shiga toxin and *Escherichia coli* Shiga-like toxin on the rabbit intestine. *Am. J. Pathol.* 125:69–80
44. Keusch, G. T., Donohue-Rolfe, A., Jacewicz, M. 1982. Shigella toxin(s): Description and role in diarrhea and dysentery. *Pharmacol. Ther.* 15:403–38
45. Keusch, G. T., Grady, G. F., Mata, L. J., McIver, J. 1972. The pathogenesis of *Shigella* diarrhea. Enterotoxin-production by *Shigella dysenteriae* 1. *J. Clin. Invest.* 51:1212–18
46. Keusch, G. T., Mata, L. J., Grady, G. F. 1970. Shigella enterotoxin: Isolation and characterization. *Clin. Res.* 18: 442
47. Kilpatrick, Z. M., Silverman, J. F., Betancourt, E., Farman, J., Lawson, J. P. 1968. Vascular occlusion of the colon

and oral contraceptives. Possible relation. *N. Engl. J. Med.* 278:438–40

48. Kobayashi, K., Harada, K., Nakatsukasa, M., Kanno, I., Ishii, T., et al. 1986. A retrospective study of hemorrhagic colitis associated with *Escherichia coli* O157:H7. *Kansenshogaku Zasshi* 59:1056–60 (In Japanese)

48a. Konowalchuk, J., Speirs, J. I., Stavric, S. 1977. Vero response to a cytotoxin of *Escherichia coli. Infect. Immun.* 18:775–79

49. Laboratory Centre for Disease Control. 1983. Hemorrhagic colitis in a home for the aged—Ontario. *Can. Dis. Wkly. Rep.* 9:29–32

50. Laboratory Centre for Disease Control. 1983. Hemorrhagic colitis associated with *Escherichia coli* O157:H7—Newfoundland and Labrador. *Can. Dis. Wkly. Rep.* 9:182–84

51. Laboratory Centre for Disease Control. 1986. Hemorrhagic colitis in a nursing home in Ontario. *Can. Med. Assoc. J.* 134:50

52. Levine, M. M., Bergquist, E. J., Nalin, D. R., Waterman, D. H., Hornick, R. B., et al. 1978. *Escherichia coli* strains that cause diarrhea but do not produce heat-labile or heat-stable enterotoxins and are non-invasive. *Lancet* 1:1119–22

53. Levine, M. M., DuPont, H. L., Formal, S. B., Hornick, R. B., Takeuchi, A., et al. 1973. Pathogenesis of *Shigella dysenteriae* 1 (Shiga) dysentery. *J. Infect. Dis.* 127:261–70

54. Levine, M. M., Edelman, R. 1984. Enteropathogenic *Escherichia coli* of classic serotypes associated with infant diarrhea: epidemiology and pathogenesis. *Epidemiol. Rev.* 6:31–51

55. Manashil, G. B., Kern, J. A. 1973. Nonspecific colitis following oral lincomycin therapy. *Am. J. Gastroenterol.* 60:394–99

56. Marier, F., Wells, J. G., Swanson, R. C., Callahan, W., Mehlman, I. J. 1973. An outbreak of enteropathogenic *Escherichia coli* food-borne disease traced to imported French cheese. *Lancet* 2:1376–78

57. Marques, L. R. M., Moore, M. A., Wells, J. G., Wachsmuth, I. K., O'Brien, A. D. 1986. Production of Shiga-like toxin by *Escherichia coli. J. Infect. Dis.* 154:338–41

58. Marston, A., Pheils, M. T., Thomas, M. L., Morson, B. C. 1966. Ischemic colitis. *Gut* 7:1–15

59. Martin, M. L., Shipman, L. D., Wells, J. G., Potter, M. E., Hedberg, K., et al. 1986. Isolation of *Escherichia coli* O157:H7 from dairy cattle associated with two cases of haemolytic-uraemic syndrome (letter). *Lancet* 2:1043

60. McKay, D. G., Wahle, G. H. 1955. Epidemic gastroenteritis due to *Escherichia coli* 0111:B4. *Arch. Pathol.* 60:679–93

61. Miller, W. T., DePoto, D. W., Scholl, H. W., Raffensperger, E. C. 1971. Evanescent colitis in the young adult: a new entity? *Radiology* 100:71–78

62. Mobassaleh, M., Donohue-Rolfe, A. Montgomery, R. K., Keusch, G. T., Grand, R. J. 1985. Postnatal development of *Shigella* toxin (ST) binding sites on rabbit intestinal microvillus membranes. *Intersci. Conf. Antimicrob. Agents Chemother., Minneapolis,* p. 149. Washington, DC: Am. Soc. Microbiol. (Abstr.)

63. Mogg, G. A. G., Keighley, M. R. B., Burdon, D. W., Alexander-Williams, J., Youngs, D., et al. 1979. Antibiotic-associated colitis—a review of 66 cases. *Br. J. Surg.* 66:738–42

64. Morrison, D. M., Tyrrell, D. L. J., Jewell, L. D. 1985. Colonic biopsy in Verotoxin-induced hemorrhagic colitis and thrombotic thrombocytopenic purpura (TTP). *Am. J. Clin. Pathol.* 86:108–12

65. Neill, M. A., Agosti, J., Rosen, H. 1985. Hemorrhagic colitis due to *Escherichia coli* O157:H7 preceding adult hemolytic-uremic syndrome. *Arch. Intern. Med.* 145:2215–17

66. Neubauer, M., Aldova, E., Duben, J. 1982. Outbreak of enterocolitis caused by an enteroinvasive *Escherichia coli* (serotype 0164, synonym serotype 147). *Zentralbl. Bakteriol. Parasitenkd. Infektionskr. Hyg. Abt. Orig. Reihe A* 252:507–13

67. Newland, J. W., Strockbine, N. A., Miller, S. F., O'Brien, A. D., Holmes, R. K. 1985. Cloning of Shiga-like toxin structural genes from a toxin converting phage of *Escherichia coli. Science* 230:179–81

68. O'Brien, A. D., LaVeck, G. D. 1983. Purification and characterization of *Shigella dysenteriae* 1-like toxin produced by *Escherichia coli. Infect. Immun.* 40:675–83

69. O'Brien, A. D., Lively, T. A., Chang, T. W., Gorbach, S. L. 1983. Purification of *Shigella dysenteriae* 1 (Shiga)-like toxin from *Escherichia coli* O157:H7 strains associated with haemorrhagic colitis (letter). *Lancet* 1:573

70. O'Brien, A. D., Lively, T. A., Chen,

M. E., Rothman, S. W., Formal, S. B. 1983. *Escherichia coli* O157:H7 strains associated with haemorrhagic colitis in the United States produce a *Shigella dysenteriae* 1 (Shiga)-like cytotoxin (letter). *Lancet* 1:702

71. O'Brien, A. D., Newland, J. W., Miller, S. F., Holmes, R. K., Smith, H. W., et al. 1984. Shiga-like toxin-converting phages from *Escherichia coli* strains that cause hemorrhagic colitis or infantile diarrhea. *Science* 226:694–6
72. Olsnes, S., Reisbig, R., Eiklid, K. 1981. Subunit structure of *Shigella* cytotoxin. *J. Biol. Chem.* 256:8732–38
73. Padhye, V. V., Kittell, F., Doyle, M. P. 1986. Purification and physicochemical properties of a unique Vero cell cytotoxin from *Escherichia coli* O157:H7. *Biochem. Biophys. Res. Commun.* 139:424–30
74. Pai, C. H., Ahmed, H., Sims, V., Woods, D. E. 1986. Epidemiology of hemorrhagic colitis due to *Escherichia coli* O157:H7 and other Verotoxin (VT) producing *E. coli:* a two-year prospective study. *Intersci. Conf. Antimicrob. Agents Chemother., New Orleans,* p. 130. Washington, DC: Am. Soc. Microbiol. (Abstr.)
75. Pai, C. H., Gordon, R. T., Sims, H. V., Bryan, L. E. 1984. Sporadic cases of hemorrhagic colitis associated with *Escherichia coli* O157:H7. *Ann. Intern. Med.* 101:738–42
76. Pai, C. H., Kelley, J. K., Meyers, G. L. 1986. Experimental infection of infant rabbits with Verotoxin-producing *Escherichia coli*. *Infect. Immun.* 51:16–23
77. Paulozzi, L. J., Johnson, K. E., Kamahele, L. M., Clausen, C. R., Riley, L. W., et al. 1986. Diarrhea associated with adherent enteropathogenic *Escherichia coli* in an infant and toddler center, Seattle, Washington. *Pediatrics* 77:296–300
78. Pittman, F. E., Pittman, J. C., Humphrey, C. D. 1974. Colitis following oral lincomycin therapy. *Arch. Intern. Med.* 134:368–72
79. Raghupathy, P., Date, A., Shastry, J. C. M., Sudarsanam, A., Jadhav, M. 1978. Haemolytic-uraemic syndrome complicating *Shigella* dysentery in south Indian children. *Br. Med. J.* 1:1518–21
80. Ratnam, S., March, S. B. 1986. Sporadic occurrence of hemorrhagic colitis associated with *Escherichia coli* O157:H7 in Newfoundland. *Can. Med. Assoc. J.* 134:34–45
81. Reisbig, R., Olsnes, S., Eiklid, K. 1981. The cytotoxic activity of *Shigella* toxin. Evidence for catalytic inactivation of the 60S ribosomal subunit. *J. Biol. Chem.* 256:8739–44
82. Remis, R. S., MacDonald, K. L., Riley, L. W., Puhr, N. D., Wells, J. G., et al. 1984. Sporadic cases of hemorrhagic colitis associated with *Escherichia coli* O157:H7. *Ann. Intern. Med.* 101:624–26
83. Riley, L. W. 1985. Hemorrhagic colitis—a "new" disease. *Clin. Microbiol. Newsl.* 7:47–49
84. Riley, L. W., Remis, R. S., Helgerson, S. D., McGee, H. B., Wells, J. G., et al. 1983. Hemorrhagic colitis associated with a rare *Escherichia coli* serotype. *N. Engl. J. Med.* 308:681–85
85. Rimmer, M. J., Freeman, A. H., Low, F. M. 1982. The barium enema diagnosis of penicillin-associated colitis. *Clin. Radiol.* 33:529–33
86. Rothbaum, R., McAdams, A. J., Gianella, R., Partin, J. C. 1982. A clinicopathologic study of enterocyte-adherent *Escherichia coli:* a cause of protracted diarrhea in infants. *Gastroenterology* 83:441–54
87. Ryan, C. A., Tauxe, R. V., Hosek, G. W., Wells, J. G., Stoez, P. A., et al. 1986. *Escherichia coli* O157:H7 diarrhea in a nursing home: clinical, epidemiological, and pathological findings. *J. Infect. Dis.* 154:631–38
88. Sack, R. B. 1975. Human diarrheal disease caused by enterotoxigenic *Escherichia coli*. *Ann. Rev. Microbiol.* 29:333–53
89. Sakurai, Y., Tsuchiya, H., Ikegami, F., Funatomi, T., Takasu, S., et al. 1979. Acute right-sided hemorrhagic colitis associated with oral administration of ampicillin. *Dig. Dis. Sci.* 24:910–15
90. Sawaf, H., Sharp, M. J., Youn, K. J., Jewell, P. A., Rabbani, A. 1978. Ischemic colitis and stricture after hemolytic-uremic syndrome. *Pediatrics* 61:315–16
91. Scaletsky, I. C. A., Silva, M. L. M., Trabulsi, L. R. 1984. Plasmid-mediated adhesion in enteropathogenic *Escherichia coli* to HeLa cells. *Infect. Immun.* 45:534–36
92. Schwartz, D. L., Becker, J. M., So, H. B., Schneider, K. M. 1978. Segmental colonic gangrene: a surgical emergency in the hemolytic-uremic syndrome. *Pediatrics* 62:54–56
93. Schwartz, S., Boley, S., Lash, J., Sternhill, V. 1963. Roentgenologic aspects of reversible vascular occlusion of the colon and its relationship to ulcerative colitis. *Radiology* 80:625–35

94. Scotland, S. M., Smith, H. R., Rowe, B. 1985. Two distinct toxins active on Vero cells from *Escherichia coli* O157:H7 (letter). *Lancet* 2:885–86
95. Seidah, N. G., Donohue-Rolfe, A., Lazure, C., Auclair, F., Keusch, G. T., Chretien, M. 1986. Complete amino acid sequence of *Shigella* toxin B-chain. *J. Biol. Chem.* 261:13928–31
96. Spika, J. S., Parsons, J. E., Nordenberg, D., Wells, J. G. Gunn, R. A., et al. 1986. Hemolytic-uremic syndrome and diarrhea associated with *Escherichia coli* O157:H7 in a day care center. *J. Pediatr.* 109:287–91
97. Strockbine, N. A., Marques, L. R. M., Newland, J. W., Smith, H. W., Holmes, R. K., et al. 1986. Two toxin-converting phages from *Escherichia coli* O157:H7 strain 933 encode antigenically distinct toxins with similar biological activities. *Infect. Immun.* 53:135–40
98. Tamura, H. 1973. Acute ulcerative colitis associated with cytomegalic inclusin virus. *Arch. Pathol. Lab. Med.* 96:164–67
99. Thomson, S. 1956. Is infantile gastroenteritis fundamentally a milk-borne infection? *J. Hyg.* 54:311–14
100. Toffler, R. B. 1978. Acute colitis related to penicillin and penicillin derivatives. *Lancet* 2:707–9
101. Totani, T. 1978. Acute hemorrhagic enteritis due to *Klebsiella oxytoca. Nippon Rinsho* 36:1308–9 (In Japanese)
102. Tulloch, E. F., Ryan, K. J., Formal, S. B., Franklin, F. A. 1973. Invasive enteropathic *Escherichia coli* dysentery. *Ann. Intern. Med.* 79:13–17
103. Tzipori, S., Wachsmuth, I. K., Chapman, C., Birner, R., Brittingham, J., et al. 1986. The pathogenesis of hemorrhagic colitis caused by *Escherichia coli* O157:H7 in gnotobiotic pigs. *J. Infect. Dis.* 154:712–16
104. Wells, J. G., Davis, B. R., Wachsmuth, I. K., Riley, L. W., Remis, R. S., et al. 1983. Laboratory investigation of hemorrhagic colitis outbreaks associated with a rare *Escherichia coli* serotype. *J. Clin. Microbiol.* 18:512–20
105. Willshaw, G. A., Smith, H. R., Scotland, S. M., Rowe, B. 1985. Cloning of genes determining the production of Vero cytotoxin by *Escherichia coli. J. Gen. Microbiol.* 131:3047–53
106. Wolfe, B. M., Cherry, J. D. 1971. Hemorrhage from cecal ulcers of cytomegalovirus infection: report of a case. *Ann. Surg.* 177:490–94

Ann. Rev. Microbiol. 1987. 41:409–33

RAPID EVOLUTION OF RNA VIRUSES

D. A. Steinhauer and J. J. Holland

Department of Biology, University of California at San Diego, La Jolla, California 92093

CONTENTS

INTRODUCTION

Viruses with RNA genomes are the most ubiquitous cellular parasites known. They are found intracellularly in nearly all life forms from plants and animals to fungi and prokaryotes. They are diverse in size, structure, genome orga-

0066-4227/87/1001-0409$02.00

nization, and replication strategy, and have been classified into several groups based on these and other criteria (106). RNA viruses have been the object of extensive study because they are responsible for a variety of medically and economically important diseases of man, plants, and animals. Many RNA viral infections, however, are inapparent or asymptomatic, causing little or no obvious detriment to the host. The extraordinary evolutionary success of RNA viruses is attributable to their ability to use different replication strategies and to adapt themselves to the widely varying biological niches encountered during virus spread in a single host or multihost network.

RNA viruses have been recognized as highly mutable since the earliest studies. In the 1940s Burnet & Bull (22) observed phenotypic changes associated with passage of influenza A virus in chick embryos, and Kunkel (95) reported early evidence for the mutability of plant viruses from studies with tobacco mosaic virus (TMV). Subsequent genetic studies on TMV (112), Newcastle disease virus (61), phage Qβ (160), reovirus (53), and vesicular stomatitis virus (126, 127) provided further evidence of RNA genome instability.

In recent years modern techniques in molecular and cellular biology have allowed detailed study of natural virus isolates and of defined laboratory virus strains. Emerging from these studies is a view that RNA virus populations are extremely heterogeneous, which allows for great adaptability and rapid evolution of RNA genomes. In this review we discuss evidence for the heterogeneous (quasispecies) nature of RNA genomes, some factors influencing their rapid evolution, and biological implications of this rapid evolution. The large volume of relevant subject matter and references cannot be covered in detail in this review. For further details and references several other recent reviews are available covering RNA virus evolution (14, 41, 42, 75, 132) as well as evolution of influenza viruses (122, 164), poliovirus (90), and AIDS (HIV) viruses (30).

RNA VERSUS DNA GENOMES

All known cellular organisms use DNA genomes for storage of genetic information. RNA genomes, although they probably preceded DNA as the original information carriers (49), are now found only in host-dependent cellular parasites (e.g. RNA viruses, viroids, yeast killer plasmids). The reason for this is obvious; DNA is more stable than RNA, and owing to the elaborate proofreading and repair mechanisms that have evolved, DNA is replicated with much greater fidelity.

Rates of Evolution

Rates of evolution have now been calculated from available sequence data for several cellular genes in a number of different animal species. The rates of neutral mutation vary among genes and among species, but average less than 10^{-9} substitutions per site per year (19, 100).

Rates of evolution of continuously replicating RNA viruses may exceed by a million fold the rates of evolution of their DNA-based hosts. Nottay et al (116) reported mutation of 1–2% of the poliovirus genome during the 1-yr spread of an outbreak in 1978. Buonagurio et al (21) sequenced the NS gene of influenza A viruses isolated over 50 years. They found a uniform rate of evolution of about 2×10^{-3} substitutions per site per year. The rate of evolution of the AIDS virus has been estimated (66) to be between 10^{-2} and 10^{-3} substitutions per site per year for the *env* gene (that for *gag* is about 10-fold lower).

Gojobori & Yokoyama (59) compared rates of evolution for the v-*mos* gene of Maloney murine sarcoma virus with those for its cellular homolog, c-*mos*. The rate of nucleotide substitution was estimated to be 1.31×10^{-3} per site per year for v-*mos* versus 1.71×10^{-9} for c-*mos;* again, a millionfold difference was demonstrated in rates of evolution of RNA versus DNA.

Polymerase Fidelity

Obvious reasons for the differences in evolution rates of DNA and RNA are the shorter generation times for replicating RNA viruses and the error-prone nature of viral RNA polymerases compared to DNA polymerases.

Mutation rates for DNA genomes have been estimated to be between 10^{-7} and 10^{-11} per base pair per replication (46, 56). Drake (46) observed that in prokaryotes genome size was inversely proportional to mutation rates. He suggested that larger genomes must pay the energy costs of increased fidelity (proofreading) and repair in order to produce viable genome copies, while smaller genomes would not require the same fidelity and would not need to evolve energy-consuming mechanisms which would limit variability. The fidelity of DNA polymerases measured in vitro varies significantly with enzyme, template, and conditions (56) and generally does not reflect estimated in vivo fidelity, perhaps because of differing degrees of proofreading in the in vitro reactions.

Error frequencies have been estimated by different methods for several RNA polymerases (see below). Estimates vary, but most average about 10^{-4} errors per site following a limited number of replications. The limited fidelity of RNA polymerases reflects the lack of effective proofreading mechanisms. Although evidence for 3' →5 ' excision has been reported for influenza virus polymerase (81), and although 3' →5 ' pyrophosphorylytic excision has been documented with T4-modified RNA polymerase (85), corrective proofreading of elongating RNA has not been demonstrated.

Interestingly, DNA mutation rates of immunoglobulin variable region genes can be as high as 10^{-3} per cell generation during some stages of B-cell development (116a, 145a, 162). Perhaps when mutation is desirable to generate variability, proofreading mechanisms are suppressed or bypassed and fidelity of DNA replication becomes comparable to that of RNA.

RECOMBINATION OF GENETIC INFORMATION IN RNA VIRUSES

In addition to stepwise evolution due to the accumulation of point mutations, RNA viruses have available to them mechanisms of reassortment and recombination that allow for large-scale evolutionary jumps. True intermolecular recombination, which is a major means for DNA evolution, occurs frequently with certain RNA viruses but may occur rather infrequently with others.

Genome Reassortment of Segmented RNA Viruses

A number of RNA virus groups have evolved multipartite or segmented genomes. Each segment contains a viral gene (or genes), and the full complement of segments is normally necessary for virus-particle infectivity. Mixed infection of a host cell by different strains of a virus then allows for the rise of progeny viruses with differing subsets of genome segments from each parent.

Keroack & Fields (89) and Tyler et al (159) have used reassortment experiments to identify reovirus gene segments responsible for various viral functions and traits (see also references in 89, 159). Allen & Desselberger (2) demonstrated the utility of reassortant rotaviruses for amplifying segments of poorly growing human rotavirus in bovine rotavirus recombinants that can replicate efficiently. Shope et al (145) used reassortment to correlate bunyavirus pathogenicity in mice with the M (middle size) RNA segment.

In several examples, e.g. influenza A strains (140) and reovirus (136), low-virulence parental viruses have given rise to highly virulent reassortants. Recently Riviere & Oldstone (133) studied the pathogenesis of reassortants from three parental strains of lymphocytic choriomeningitis virus (LCMV), an arenavirus that has two genome segments. Two of the reassortants caused lethal disease in mice, while parental strains and reciprocal reassortants did not. Some virulent pandemic episodes that are caused periodically by influenza A virus result from reassortment of human strains with avian, animal, or other human strains, which generates antigenically novel virus (13, 36, 52, 164, 166).

Recombination in RNA Virus Genomes and Defective Genomes

RETROVIRUSES Retroviruses are unique among the RNA viruses in that they replicate via a DNA intermediate, which can be integrated within the host-cell chromosome. This allows these viruses the luxury of two distinct life-styles: that of a replicating RNA virus with inherent high mutability, and that of a stably integrated provirus that is replicated only infrequently by high-fidelity DNA replication machinery.

Mutations can be generated by a variety of mechanisms during retrovirus replication. Point mutations occur at a high frequency during reverse transcription (12, 34, 60). Deletions, insertions, duplications, and other rearrangements can also occur during reverse transcription, as can recombination with other viral (or cellular) sequences (45, 119, 144). Such recombinations are most likely generated by template switching or by a copy-choice mechanism (29). Recombination can also take place between unintegrated viral DNA and integrated provirus DNA, defective provirus DNA, or cellular DNA with sequence homology. The transcription of retroviral genomic RNA by cellular RNA polymerase can be another major cause of mutation.

Retroviruses, retrotransposons, and related elements, because of their unique replicative mechanisms, may have significant effects on the evolution of eukaryotic genomes (7). Processed pseudogenes and certain highly reiterated DNA sequences, which make up over 5% of the human genome, probably arose by reverse transcription of cellular RNAs (7). The discovery that the transposable Ty element of yeast (15) and copia-like elements in *Drosophila* (55, 138) are related to retroviruses raises the possibility that retroviruslike elements are ubiquitously scattered throughout eukaryotic genomes. For example, representatives of the L1 family of highly repetitive DNA sequences, which are found throughout mammalian genomes, contain sequences with homology to known reverse transcriptases (70, 101). Transposition mediated by these types of elements may profoundly affect host-genome evolution by promoting recombination and insertional mutagenesis, and more importantly by dispersing endogenous promotor/enhancer regions throughout host genomes.

STANDARD RNA VIRUSES Molecular recombination for generation of virus diversity has been demonstrated only for a limited number of nonretroviral RNA viruses.

For many years true intermolecular RNA recombination (in nonretroviruses) could be demonstrated only for picornaviruses (32, 92). Recombination of genome RNA occurs frequently during picornavirus replication, apparently by template switching during negative-strand synthesis (93). In tissue culture, recombinants between isogenic strains of foot-and-mouth disease virus were detected at a frequency of 0.92% (107), and intratypic poliovirus recombinants were seen at a frequency of 0.13% between genetic markers separated by only 190 bases (93). Frequencies of recombination between less closely related strains of virus are much lower (93, 156). However, intertypic poliovirus recombinants have been readily isolated from humans vaccinated with the three Sabin attenuated strains (111) and from close contacts of vaccinees (90). Indeed, Minor et al (111) isolated two distinctly different recombinants from the same vaccinee. These vaccine recombinants appear to represent variants that partially escape the immune responses of the vaccinees.

Lai and coworkers recently demonstrated that homologous recombination of genomic RNA occurs in coronaviruses (96), apparently at a high frequency (105). The discontinuous replication of coronaviruses generates template-free intermediates containing nascent RNA strands complexed with polymerase. These intermediates bind (by specific base pairing) to new templates, generating copy-choice recombinants.

Recombination has also recently been reported for the plant virus brome mosaic virus (20). This virus has a positive-strand tripartite genome with a conserved stem-loop structure at the 3' end of each segment. Deletions in this region of one genome segment were repaired during infection (with restoration of the stem-loop structure) by both homologous and nonhomologous recombination with the 3' ends of unaltered gene segments.

Intermolecular recombination in negative-strand RNA genomes is much less common. One clear example is a mosaic defective interfering RNA of influenza (84), but this and other examples of intermolecular recombination show no base-pairing specificity in the nascent strand–polymerase leaps to new templates (i.e. are generally nonhomologous). Therefore, generation of viable infectious intermolecular recombinant genomes of negative-strand viruses may be rare. However, even extremely rare recombination events may have a significant role in RNA virus evolution. Recombinant viruses of another type, defective interfering (DI) particles, have been demonstrated for nearly every type of animal virus. The generation of nearly all DI particles can be explained by some variation of copy-choice intramolecular recombination (78, 98, 99, 124). The significance of DI particle generation is discussed later in this review.

HETEROGENEITY OF RNA GENOMES

The Quasispecies Concept

Eigen and coworkers developed the quasispecies concept to describe the distributions of self-replicating RNAs, which they proposed were the first genes and the precursors to life on earth (48–50). They noted several reasons to believe that RNAs preceded DNA as information carriers. In cells DNA monomers are synthesized via ribose intermediates, DNA replication is initiated on RNA primers, and information is processed by RNA-protein machinery (49). The fact that RNA can fold into a large variety of three-dimensional structures and the recent findings that RNA can have catalytic properties (63, 168) further strengthen the argument for RNAs as the first genes.

The ability to self-replicate may have been unique to RNA in a prebiotic soup. Inoue & Orgel (80) have demonstrated that under appropriate conditions template-directed RNA synthesis will occur in the absence of enzyme. Prebiotic, nonenzymatic RNA synthesis of this type would have proceeded with very high error frequencies, on the order of 10^{-1} to 10^{-2} (50, 80). The error threshold probably limited the maximum length of early self-replicating RNAs to between 50 and 100 bases (49). (The error threshold of RNA synthesis through evolution has similarly limited the length of present-day RNA genomes and cellular mRNAs.) The distribution of early RNAs was such that sequences that were most fit (i.e. those that were most stable and that had superior replicative ability) would have been present in higher concentrations. However, replicative infidelity would have ensured a large proportion of related nonidentical sequences. Eigen and colleagues (49) stated that "the result of self replication competition had to be the master sequence together with a huge swarm of mutants derived from it and from which it had no way of escape." They termed this mutant distribution a "quasispecies" distribution.

Evidence for a Quasispecies Distribution of Genomes in Present-Day RNA Virus Populations

BIOLOGICAL EVIDENCE There have been numerous documentations of heterogeneity in RNA virus populations. We cite only a few examples of extreme heterogeneity in virus populations of known origin and history.

The first recognition that RNA viruses might have quasispecies distributions was made in Wiessmann's laboratory with phage Qβ (11, 40, 43). Domingo et al (43) showed by T_1 fingerprinting that individual clones isolated from a multiply passaged Qβ stock virus differed from the stock virus by an average of one to two nucleotides, in spite of the fact that the stock-virus fingerprint had remained unchanged over the previous 50 passages. In competition experiments, variant clones were consistently outgrown by parental wild-type virus. Domingo et al concluded "that the phage population was in a dynamic equilibrium with viable mutants arising at a high rate on one hand and being selected against on the other," and added, "the genome of Qβ phage cannot be described as a defined structure, but rather as a weighted average of a large number of individual sequences." They further pointed out that the relative growth rate of a variant must be extremely close to that of the wild type (within 0.01–0.1%) in order to be detected by ordinary sequencing of the equilibrium population.

Several sequencing studies have demonstrated extreme heterogeneity of viral populations. Fiers and coworkers observed sequence heterogeneity when phages Qβ and MS2 were first sequenced (37, 84a). Variability in several virus populations has been suggested by sequence differences among overlapping cDNA clones generated from viral RNA (4, 54, 58, 141). Catteneo et al (24) reported over 1% sequence difference among overlapping cDNA clones of measles virus RNA isolated from brain tissue of a single subacute sclerosing panencephalitis (SSPE) patient. Hahn et al (66) recently sequenced the *env* gene of AIDS virus isolates collected at different times from a maternally infected child. All isolates differed, and the differences suggested that they had not evolved sequentially, but had diverged from a common ancestor.

Heterogeneity has been detected in populations of foot-and-mouth disease virus from single natural isolates (39, 135) and after replication in cell culture (147). Poliovirus shed by vaccinated individuals can be extremely variable (90, 91, 111). Indeed, one study (91) reported evidence that over 100 genetic changes occurred during vaccine virus replication in one or two individuals. Likewise, Sabara et al (137) detected heterogeneity among each of six natural isolates of bovine rotavirus examined. Recently several cocirculating antigenic variants of influenza A were isolated during a single epidemic in a semiclosed community (121); similar variation had been observed earlier with influenza B (120). The investigators suggested that the epidemics were probably initiated by single individuals' excretion of a mixture of antigenic variants, which were differentially selected for in subsequently infected individuals.

QUANTITATIVE EVIDENCE Relatively few experiments have been performed to measure RNA virus polymerase error frequencies (or mutation frequencies) during a limited number of virus replication cycles.

The reversion rate for one extracistronic point mutant of phage Qβ was calculated to be about 10^{-4} per genome doubling (11, 40). The mutation frequency at one site in the Rous sarcoma virus genome was estimated to be about 3×10^{-4} per virus passage (31), which is consistent with high in vitro error frequencies demonstrated with reverse transcriptase of avian myeloblastosis virus (12, 60).

Steinhauer & Holland (153) recently developed a technique for direct measurement of the polymerase error frequencies at selected G residues in viral RNA. Error frequencies at one highly conserved site in vesicular stomatitis virus (VSV) exceeded 10^{-4} in clonal pools of virus replicated in vivo and in transcription products in vitro. Similar error frequencies have been observed repeatedly at two other sites of the VSV genome that have been extensively characterized. Significantly, at one site at which two of the three possible mutations generate lethal stop codons in the N gene a similarly high error frequency was also determined (D. A. Steinhauer & J. J. Holland, unpublished).

Heterogeneity has also been demonstrated in single plaques of influenza A virus. Parvin et al (123) sequenced the NS gene from multiple individual virus plaques that were generated during growth of a single plaque. Of approximately 92×10^3 bases sequenced, seven mutations were detected, all of which were neutral with respect to growth kinetics. Lethal or very debilitating mutations would not, of course, have been scored in this study. Interestingly, when Parvin et al did the same experiment on a segment of the VP1 gene of poliovirus type 1, no mutations were detected in over 95×10^3 bases sequenced. The authors calculated rates of neutral mutation of 1.5×10^{-5} for influenza virus and less than 2.1×10^{-6} for poliovirus at these sites (assuming five replication cycles during growth of each plaque). The reason for the lower rates for poliovirus could be greater restraints on VP1 or differing polymerase error frequencies or purifying selection difference, but it is not possible at present to distinguish among them.

Durbin & Stollar (47) recently reported a remarkably low error frequency (less than 5×10^{-7}) for the reversion of a particular Sindbis virus host-restricted mutant. It would be interesting if certain viral RNA sites are replicated with much higher fidelity than others, and the molecular basis for this would require investigation. Stec et al (152) observed that Sindbis virus variants resistant to monoclonal antibodies arose spontaneously at frequencies between $10^{-3.5}$ and 10^{-5}; thus it is possible that the reversion rate of Durbin & Stollar (47) might have reflected the combined frequencies of two complementing mutations, one of which was not in the sequenced region of the genome. In any case, much more work is needed to quantitate mutation rates and polymerase error frequencies at defined sites for a variety of RNA viruses (including retroviruses and retrotranscribing DNA viruses such as hepatitis B virus).

HIGH MUTATION RATES VERSUS RAPID RATES OF EVOLUTION

Wild-Type and Relatively Stable Equilibrium Populations

In spite of great heterogeneity within virus populations, high mutation rates are not always reflected in rapid evolution. There are conditions under which viruses can replicate efficiently and continuously, and yet accumulate few if any viable, competitive mutations in the virus population. The studies of Domingo et al (43), with phage Qβ first illustrated this. The same wild-type sequence predominated through extensive passage despite the fact that a large portion of the population was shown to be variant at any one time. Once a stable equilibrium population has been reached, virus may be able to replicate for extensive periods with little evolution as long as conditions remain unchanged. This does not mean that the predominating sequence cannot stray to some degree from the master sequence or that a rare event (many mutations at one time, or a recombinational event) could not give rise to a more fit variant (a jump from one fitness peak to another). It simply means that among the distribution of variants generated by the wild-type sequence, none have a competitive advantage over the parental master sequence.

Observations with VSV support this conclusion. Laboratory strains of VSV with different passage histories over many years accumulated few if any nucleotide changes as revealed by T_1 fingerprinting (28). Spindler et al (148) fingerprinted isolates obtained during 232 dilute passages of VSV in BHK_{21} cells. Although several spot changes were seen in intermediate passages, all reverted to wild type by passage 232. Recently, after 529 dilute passages of this virus, T_1 oligonucleotide mapping again revealed the accumulation of not a single oligonucleotide spot change (D. A. Steinhauer & J. J. Holland, unpublished).

Other viruses also demonstrate remarkable stability in some situations. The type 3 Sabin poliovirus vaccine differed from its neurovirulent progenitor at only 10 nucleotide positions after 53 in vitro and 21 in vivo passages in monkey tissue (150). In 1977 H1N1 influenza A virus reappeared in the human population after 27 years of dormancy with sequences nearly identical to those of the 1950 virus (88, 114). It is possible that virus was harbored in a nonhuman host where it remained very stable. Alternatively, virus may have remained viable in a nonreplicating state (frozen or desiccated) before reemergence. It is significant that the H1N1 sequences evolved rapidly and in divergent directions in subsequent human outbreaks (130). Rotavirus genome exhibited only limited heterogeneity after years of continuous passage (4).

Many selective forces may stabilize virus populations. These stabilizing factors may include the need for conservation of protein structure and function, RNA secondary structure, glycosylation sites, and phosphorylation sites. Even third-codon changes can be subject to selective pressures (62). Recently, remarkable conservation of certain protein domain sequences has been observed between completely unrelated RNA viruses (1, 3, 69). These observations argue for a modular theory of evolution (16) that suggests that selective forces may often act at the level of subgenomic functional units rather than solely upon intact virus particles.

Conditions That Promote Disequilibrium Favor Rapid, Random Evolution

Any change in environmental conditions that offers variants in a virus population the opportunity to compete favorably with the predominating virus can shift the equilibrium and drive virus evolution. The countless factors that can contribute to disequilibrium include interference by DI particles, different host or cell types, immune selection, transfer from vectors to hosts or vice versa, and temperature shifts. During VSV infection of BHK cells, where low multiplicity passages favored the establishment of a very stable equilibrium population (see previous section), high multiplicity passages led to rapid and random genome evolution (148). Similarly, rapid evolution of the VSV genome was also seen in persistently infected cells (72). In these instances virus evolution was partly driven by DI particles. DI particles are amplified during high multiplicity passages and interfere with parental virus replication. This forces the evolution of standard virus mutants (Sdi$^-$) that are resistant to interference by the predominating DI particles (76, 77). It appears that continuous coevolution takes place at the termini of virus and DI genomes and the replication-encapsidation proteins with which they specifically interact (35, 117). Poliovirus also demonstrates reasonable genome stability in some situations (51, 150), but can evolve extensively upon replication in the human gut (90, 91, 111). This is reflected in the rapid evolution seen as poliovirus outbreaks spread in human host populations (104, 116). Likewise, Takeda et al (155) documented continuous and divergent evolution of enterovirus 70 genomes from about its time of appearance in Africa in the 1960s through two pandemics of acute hemorrhagic conjunctivitis.

Host immune selection can be strongly involved in driving virus evolution. This is clearly demonstrated with influenza virus. Antigenic variants are regularly selected, so previously immune hosts are no longer protected against newly circulating strains (122, 164). Other factors, including reassortment of independently evolving modular genome segments, are important in the rapid evolution of influenza A viruses. Influenza viruses have been reported to exhibit rapid and continuous evolution in vitro even in conditions in which other mutable viruses such as VSV (18) and poliovirus (123) do not evolve as extensively.

Salinovich et al (139) recently demonstrated strong evidence for immune selection of antigenic variants with the lentivirus equine infectious anemia virus (EIAV). Persistent infection by EIAV is characterized by periodic emergence of disease. Isolates from four consecutive disease episodes of a persistently infected pony revealed a novel predominating variant each time. Although the variants were closely related, as shown by T_1 fingerprinting, each variant was distinguishable antigenically by neutralization assays with the pony's serum and by Western blot analysis using monoclonal antibody to the major surface glycoprotein. Antigenic variants can also be selected during infection by visna virus (27) and probably HIV, the AIDS virus (151). The role of these variants in persistence is unclear (64). Cellular immune selection can also drive rapid evolution of viral RNA genomes. Extensive mutations in the G protein (and fewer in M and NS proteins) were observed in variants of VSV selected in vivo by natural killer cell lysis of persistently infected cells (161).

Acute Versus Persistent Infection

RNA viruses that normally cause acute infection in susceptible hosts can sometimes cause a long-term subacute infection. Persistent infection can be initiated by factors such as multiple mutations, interferon, defective interfering particles, host cell properties which prevent efficient virus replication or maturation, etc. (74, 167). Maintenance of the persistent state requires only that some balance be reached between virus replication and host function, and possible viral etiology is suspected for a number of long term degenerative and autoimmune diseases.

Subacute sclerosing panencephalitis is a rare but fatal degenerative disease of the central nervous system resulting from persistent measles virus infection; the disease becomes manifest years after acute infection. In the CNS neurons of SSPE patients, intracellular replication of nucleocapsids continues, but no mature budding virus particles are detected. Several studies suggest that a defect in M (matrix) protein expression is involved (67, 68). Experiments utilizing cells derived from brain tissue from SSPE patients (23) and direct observation of brain from SSPE patients (5, 65) showed a block at the level of M gene mRNA translation. Other workers (143) observed synthesis and then rapid degradation of M protein in one cell line derived from the brain of an SSPE patient. Another study reported a stop codon within the M gene coding region of RNAs isolated from SSPE-patient brain (24). There is no reason to believe that only defects in the M gene are associated with SSPE. Norrby et al (115) have detected M proteins in the brain of some SSPE patients, and Baczko et al (6) have suggested that SSPE can be correlated with defects in other measles genes or gene products which prevent normal virus particle assembly. Haase et al (65) demonstrated repressed synthesis of all viral RNAs in SSPE-affected brain and suggested that generalized constraints (of an unknown nature) on virus gene expression may be important in establishing persistent CNS infections by a variety of viruses.

Whether virus genome degeneration is a cause or a result of SSPE is not clear, but extreme variability has been demonstrated among viral RNAs present in infected brain. Catteneo et al (24) detected as much as 1% sequence variation in the M gene of overlapping cDNA clones from the same area of the same SSPE-patient's brain. They suggested that extreme variability could be responsible for the observed diversity of measles virus gene expression in SSPE patients. This heterogeneity is not unexpected, since long-term intracellular replication of measles virus RNA probably offers reduced opportunity for selection against mutants in the M, F, and H proteins, which are mainly involved in virus maturation, release, and reentry.

Paget's disease is a chronic bone disorder that affects nearly 4% of the human population aged 70 and over. Viral antigens of a number of paramyxoviruses, including measles virus, respiratory syncytial virus, simian virus 5, and parainfluenza virus type 3, have been detected in bone tissue of patients with this disease (10, 109). Recently measles virus RNA was detected by in situ hybridization to bone tissue of a patient with Paget's disease (9), but any role of these RNA viruses in Paget's disease is obscure as yet.

Several viruses have been loosely linked to autoimmune diseases such as multiple sclerosis, rheumatoid arthritis, and juvenile-onset diabetes, but definitive cause and effect is difficult to assess. Virus infections are known to trigger immune responses which can sometimes cross-react with host tissue (57). Srinivasappa et al (149) demonstrated that cross-reactivity of antiviral antibodies with normal tissues is fairly common by screening 14 organs with 600 monoclonal antibodies raised against 11 viruses. Although molecular mimicry may arise frequently in rapidly evolving viruses, it probably only rarely triggers serious diseases. Onset of disease would depend upon other factors such as strength and duration of autoimmunity, major histocompatibility complex (MHC) types, and type and strain of virus.

BIOLOGICAL IMPLICATIONS OF VIRAL RNA GENOME PLASTICITY

Antigenic Change and Vaccine Problems

Very rapid evolution of RNA viruses can create serious problems for vaccine design. Recent advances in molecular virology including the use of synthetic peptides, recombinant viral vectors, and cDNAs as vectors or stable seed inoculum should ensure the effectiveness (and safety) of future vaccines. The challenge will be to design safe vaccines that can elicit strong immune responses and yet be maximally effective in the wake of antigenic changes during virus evolution.

The large number of monoclonal antibodies now available for RNA viruses has allowed studies of antigenic variation in defined populations as well as among isolates collected at different times and locations and from different hosts. The frequency of resistance to monoclonal antibodies can vary greatly (102), but averages between 10^{-4} and 10^{-5} (41, 75, 125). Most viruses exhibit strongly conserved, moderately conserved, and poorly conserved antigenic sites in their surface protein; an ideal vaccine might target strongly conserved epitopes, but this is not always possible. Multivalent vaccines and frequent modification of vaccines may be needed to deal with rapid virus evolution in many circumstances.

Vaccines that elicit a cellular immune response as well as an antibody response may often prove most effective. Cytotoxic T-cell responses are not limited to subtype specificities as defined by antibody epitopes. Polyclonal murine cytotoxic T cells raised against respiratory syncytial virus showed cross-reactivity with all human virus subtypes tested (8). Subtype cross-reaction of Tc cells was demonstrated earlier with influenza A virus (158, 165). Broad T-cell reactivity is to be expected, because cytotoxic T cells can recognize not only virus surface antigens on influenza A–infected cells, but also fragments of the internal nucleocapsid protein presented at the cell surface together with MHC antigens (157, 165). Similarly, with VSV the N protein was demonstrated to be the major antigen recognized by cytotoxic T cells (128). Variability in T-cell target sites remains to be studied, but it is probably lower than that of surface-antigen epitopes.

Other problems for vaccine design include the masking of antigenic determinants by oligosaccharides on enveloped virus glycoproteins (44, 146) and possibly the protection of important regions on surface proteins by functionless antigenic domains that are free to drift without constraint, as proposed for HIV (AIDS) viruses (30).

Even in well-vaccinated populations RNA virus variants are likely to arise at intervals and cause outbreaks of disease. Novel antigenic variants recently caused a small poliovirus outbreak in Finland, where 97% of the population had been vaccinated (104). Sequencing showed that an isolate from the outbreak was related to type 3 polioviruses but significantly different in areas known to be antigenically important (79). The isolated virus was probably not derived from vaccine, but from circulating wild-type virus.

The long-term success of the poliovirus vaccines, as contrasted with the need for frequent updating of the antigenic composition of influenza A vaccines, illustrates clearly that any RNA virus vaccine program must take account of the relative degree of antigenic plasticity (or constraint) involved in the targeted viral epitopes. When these epitopes tolerate extreme variability, effective vaccine design will pose greater challenges, and continuing modification may sometimes be required.

Host-Cell Specificities and Changes in Disease Patterns and Virulence

A major factor determining host-cell specificity is the recognition by virus capsid or glycoprotein surfaces of a specific protein or other receptor structure on the surface of susceptible cells. For example, HIV (AIDS) viruses have been reported to be specific for cells bearing the T4 (CD4) antigen (33, 94, 103). It was demonstrated long ago that poliovirus can replicate in refractory cells if virus genome is introduced intracellularly and thus bypasses the discriminatory receptor-binding step (73). Recently the human poliovirus receptor gene was cloned and transformed into poliovirus-resistant mouse L cells, whereupon they became sensitive to all three poliovirus types (108). Broad–host range viruses use cell-surface components that are common to many cell types and species.

Mere entrance into a cell does not ensure that virus will generate a productive infection. The cell must also allow for proper transcription, replication, and translation of viral nucleic acids and proper maturation of progeny virus. In the retroviruses, host-range determinants can be defined not only for the *env* gene (17, 45) (the likely receptor-binding site), but also for the U3 region at the 3' end of the genome (25, 26, 97, 154). This site contains enhancer sequences that regulate viral RNA transcription in a tissue-specific manner (25).

Regardless of the molecular factors that determine cell, tissue, and species specificity, it is obvious that extremely rapid evolution of RNA genomes will on rare occasions result in marked changes in virus species specificity or tissue tropism and disease patterns. Sometimes only one or a few mutations can cause profound alterations in virulence phenotype or host or tissue tropism of a virus. A few changes in the U3 region of myeloproliferative sarcoma virus are responsible for its expanded host range and disease specificity relative to those of its progenitor virus, Moloney murine leukemia virus (154). A single nucleotide change in the T3 hemagglutinin gene of reovirus type 3 was demonstrated to alter both its growth and its tropism in the CNS (87). Rogers et al (134) demonstrated that a single amino acid substitution in the influenza hemagglutinin can alter receptor-binding specificity. Naeve et al (113) showed with an avian influenza reassortant virus containing human influenza virus hemagglutinin that only two nucleotide changes were necessary for altered receptor binding and replication in ducks. Reagan et al (131) isolated a mutant of coxsackievirus B3, which acquired an alternate site of attachment to virus receptor on HeLa cells following passage in rhabdomyosarcoma cells.

Similarly, viral pathogenesis can be greatly affected by single nucleotide changes. Only a single T_1 spot difference was detected between fingerprints of diabetogenic and nondiabetogenic variants of encephalomyocarditis virus (129). During the influenza A outbreak that occurred in chickens in the eastern United States in 1983, the virus mutated from an avirulent form to a highly virulent variant, which caused up to 80% mortality. Webster and coworkers (163) showed that a single point mutation was probably responsible for the acquisition of virulence. Seif et al (142) and Dietzschold et al (38) correlated virulence of rabies virus with changes in a single site in the glycoprotein. Stanway et al (150) showed only ten sequence differences between the genomes of Sabin type 3 poliovaccine and its neurovirulent progenitor, and Evans et al (51) demonstrated that only one of those mutations, in the 5' noncoding region of the genome, was necessary for reversion of the vaccine strains to neurovirulence. RNA genomes, with their extreme mutation rates and great genetic plasticity, constantly threaten to undergo major or minor changes in tropisms and/or disease propensity.

Unpredictability of Future RNA Virus Disease Outbreaks

To the extent that we associate certain viruses with certain defined disease syndromes (e.g. measles virus with measles, poliovirus with polio, HIV with AIDS), we also tend to overlook the phenotypic plasticity and the inherent unpredictability of all rapidly evolving RNA virus genomes. It is true that virus populations tend toward defined patterns of transmission and pathogenesis once they occupy a stable ecological niche. Hence, even a continuously drifting RNA genome may produce a specific disease syndrome in one preferred host for very long periods. Even when there is rapid and extensive genomic evolution (as in the case of influenza virus A), there may nevertheless be strong conservation of the overall structure and function of viral proteins. This conservation is to be expected, since it maximizes replication and transmission efficiency, virus stability, and recovery and survival of major host species. Conservation does not, however, mitigate the possibility of the rare appearance of new virus strains with markedly different structures, host ranges, tissue tropisms, replication and transmission strategies, and disease patterns and virulence. Variation will be particularly favored whenever new or expanded ecological niches present themselves to evolving virus populations. The global expansion of the human population during the last century is an obvious example of evolving opportunities for viral RNA genomes, particularly when the impact of human activities on other niches is considered (e.g. deforestation, agricultural development, growth of large cities, and construction of rapid transportation networks). Therefore it should not be surprising that previously unknown human, domestic-animal, and plant-crop diseases will appear at intervals, and that many or most will be due to evolving RNA genomes. Human AIDS is but one example of such a newly recognized human disease. Enterovirus 70 is a newly recognized virus that causes pandemics of acute hemorrhagic conjunctivitis in humans. It has diverged continuously since its appearance in Africa in the late 1960s, as documented by the oligonucleotide mapping studies of Takeda et al (155). Presumably this virus was derived from an animal (or human) virus that had not used this ecological niche or mode of transmission.

It is clear from sequencing data now available that despite the presence of some strongly conserved genomic stretches, well over half of the nucleotide positions in many RNA virus genomes can be substituted during their evolution without loss of virus viability. One consequence of this genomic plasticity is the statistical certainty that completely new RNA genomes will continually be generated. A 10-kb genome would be required to test the viability and competitive characteristics of at least 4^{5000} different sequence permutations to find all fit combinations. This is an extremely conservative number, because the total information content of viral genomes can increase and decrease. However, this "small" astronomical number assures that not even a very minute fraction of all possible RNA viruses can ever be produced and tested. More importantly, it also assures that the large size and rapid evolution of the biosphere's viral RNA genome pools will constantly produce genome permutations that have never before existed. The vast majority of these will be nonviable or noncompetitive and will generally disappear. The vast majority of viable and competitive new viruses will be unremarkable in their biological properties. But humans will have to live with, and try to cope with, new RNA virus variants that have particularly destructive phenotypes. However, when destructive viruses such as the AIDS viruses appear at intervals, it should be remembered that the new virus is not in fact a virus. It is a rapidly evolving quasispecies population of RNA genomes, some of which have greater or lesser virulence. Over longer periods of time, less virulent quasispecies populations will inevitably prevail over those that destroy the hosts upon which they depend.

Modulation of RNA Virus Evolution by Defective RNA Genomes

The rapid evolution of RNA genome populations can be affected not only by viable infectious virus mutants generated during replication, but also by the defective nonviable genomes that also arise during RNA virus replication as a result of viral replicase error (71, 78, 98, 99, 124). DI particles can affect RNA virus evolution in several ways. First, defective genomes can act as helper virus–dependent evolving gene "modules" (16) that are able to tolerate extensive genome rearrangement and base substitution (e.g. 118). If rare recombination events return portions of these extensively altered gene segments to infectious particles they can, when viable, cause profound phenotypic changes. DI particles of RNA viruses are often able to modulate virus lethality for cells, thereby allowing persistent infections in which virus maturation is neither frequent nor required for genome survival. This type of chronic intracellular infection without mature virus production can allow accumulation of mutations in genome segments involved in maturation of virus that might otherwise be nonviable or noncompetitive, as discussed earlier for SSPE. Finally, with a variety of animal RNA viruses the presence of DI particles selects for infectious virus mutants (Sdi^{-} mutants) that are relatively resistant to DI particle interference (76, 82, 83, 86). The repetitive escape of Sdi^{-} mutants from DI particle interference followed by recurring generation of new DI particle types can drive rapid viral genome evolution (77).

SUMMARY

The high error rate inherent in all RNA synthesis provides RNA virus genomes with extremely high mutation rates. Thus nearly all large RNA virus clonal populations are quasispecies collections of differing, related genomes (14, 49). These rapidly mutating populations can remain remarkably stable under certain conditions of replication. Under other conditions, virus-population equilibria become disturbed, and extremely rapid evolution can result. This extreme variability and rapid evolution can cause severe problems with previously unknown virus diseases (such as AIDS). It also presents daunting challenges for the design of effective vaccines for the control of diseases caused by rapidly evolving RNA virus populations.

Literature Cited

1. Ahlquist, P., Strauss, E. G., Rice, C. M., Strauss, J. H., Haseloff, J., Zimmern, D. 1985. Sindbis virus proteins nsP1 and nsP2 contain homology to nonstructural proteins from several RNA plant viruses. *J. Virol.* 53:536–42
2. Allen, A. M., Desselberger, U. 1985. Reassortment of human rotaviruses carrying rearranged genomes with bovine rotavirus. *J. Gen. Virol.* 66: 2703–14
3. Argos, P., Kamer, G., Nicklin, M. J., Wimmer, E. 1984. Similarity in gene organization and homology between proteins of animal picornaviruses and a plant comovirus suggest common ancestry of these virus families. *Nucleic Acids Res.* 12:7251–67
4. Arias, C. F., Lopez, S., Espejo, R. T. 1986. Heterogeneity in base sequence among different DNA clones containing equivalent sequence of rotavirus double-stranded RNA. *J. Virol.* 57:1207–9
5. Baczko, K., Carter, M. J., Billeter, M., ter Meulen, V. 1984. Measles virus gene expression in subacute sclerosing panencephalitis. *Virus Res.* 1:585–95
6. Baczko, K., Liebert, U. G., Billeter, M., Catteneo, R., Budka, H., ter Meulen, V. 1986. Expression of defective measles virus genes in brain tissues of patients with subacute sclerosing panencephalitis. *J. Virol.* 59:472–78
7. Baltimore, D. 1985. Retroviruses and retrotransposons: The role of reverse transcription in shaping the eukaryotic genome. *Cell* 40:481–82
8. Bangham, C. R. M., Askonas, B. A. 1986. Murine cytotoxic T cells specific to respiratory syncytial virus recognize different antigenic subtypes of the virus. *J. Gen. Virol.* 67:623–29
9. Baslé, M. F., Fournier, J. G., Rozenblatt, S., Rebel, A., Bouteille, M. 1986. Measles virus RNA detected in Paget's disease bone tissue by *in situ* hybridization. *J. Gen. Virol.* 67:907–13
10. Baslé, M. F., Russell, W. C., Goswami, K. K. A., Rebel, A., Giraudon, P., et al. 1985. Paramyxovirus antigens in osteoclasts from Paget's bone tissue detected by monoclonal antibodies. *J. Gen. Virol.* 66:2103–10
11. Batschelet, E., Domingo, E., Weismann, C. 1976. The proportion of revertant and mutant phage in a growing population as a function of mutation and growth rate. *Gene* 1:27–32
12. Battula, N., Loeb, L. A. 1974. The infidelity of avian myeloblastosis virus deoxyribonucleic acid polymerase in polynucleotide replication. *J. Biol. Chem.* 294:4086–93
13. Bean, W. J., Cox, N. J., Kendal, A. P. 1980. Recombination of human influenza A viruses in nature. *Nature* 284:638–40
14. Biebricher, C. K. 1983. Darwinian selection of self-replicating RNA molecules. *Evol. Biol.* 16:1–52
15. Boeke, J. D., Garfinkel, D. J., Styles, C. A., Fink, G. R. 1985. Ty elements transpose through an RNA intermediate. *Cell* 40:491–500
16. Botstein, D. 1981. A modular theory of virus evolution. In *Animal Virus Genetics,* ed. B. N. Fields, R. Jaenisch, C. F. Fox, pp. 363–84. New York: Academic
17. Bova, C. A., Manfredi, J. P., Swanstrom, R. 1986. *env* genes of avian retroviruses: Nucleotide sequence and molecular recombinants define host range determinants. *Virology* 152:343–54
18. Brand, C., Palese, P. 1980. Sequential passage of influenza virus in embryonated eggs or tissue culture: Emergence of mutants. *Virology* 107:424–33
19. Britten, R. J. 1986. Rates of DNA sequence evolution differ between taxonomic groups. *Science* 231:1393–98
20. Bujarski, J. J., Kaesberg, P. 1986. Genetic recombination between RNA components of a multipartite plant virus. *Nature* 321:528–31
21. Buonagurio, D. A., Nakada, S., Parvin, J. D., Krystal, M., Palese, P., Fitch, W. M. 1986. Evolution of human influenza A viruses over 50 years: Rapid, uniform rate of change in NS gene. *Science* 232:980–82
22. Burnet, F. M., Bull, D. R. 1943.

Changes in influenza virus associated with adaptation to passage in chick embryo. *Aust. J. Exp. Biol. Med. Sci.* 21:55–69
23. Carter, M. J., Willcocks, M. M., ter Meulen, V. 1983. Defective translation of measles virus matrix protein in a subacute sclerosing panencephalitis cell line. *Nature* 305:153–55
24. Catteneo, R., Schmid, A., Rebmann, G., Baczko, K., ter Meulen, V., et al. 1986. Accumulated measles virus mutations in a case of subacute sclerosing panencephalitis: Interrupted matrix protein reading frame and transcription alteration. *Virology* 154:97–107
25. Celander, D., Haseltine, W. A. 1983. Tissue-specific transcription preference as a determinant of cell tropism and leukaemogenic potential of murine retroviruses. *Nature* 312:159–62
26. Chatis, P. A., Holland, C. A., Hartley, J. W., Rowe, W. P., Hopkins, N. 1983. Role for the 3' end of the genome in determining disease specificity of Friend and Maloney murine leukemia viruses. *Proc. Natl. Acad. Sci. USA* 80:4408–11
27. Clements, J. E., Narayan, O., Griffin, D. E., Johnson, R. T. 1980. Genomic changes associated with antigenic variation of visna virus during persistent infection. *Proc. Natl. Acad. Sci. USA* 77:4454–58
28. Clewley, J. P., Bishop, D. H. L., Yang, C. Y., Coffin, J., Schnitzlein, W. M., Reichmann, M. E. 1977. Oligonucleotide fingerprints of RNA species obtained from rhabdoviruses belonging to the vesicular stomatitis virus subgroup. *J. Virol.* 23:152–66
29. Coffin, J. M. 1979. Structure, replication, and recombination of retrovirus genomes: Some unifying hypotheses. *J. Gen. Virol.* 42:1–46
30. Coffin, J. M. 1986. Genetic variation in AIDS viruses. *Cell* 46:1–4
31. Coffin, J. M., Tsichlis, P. V., Barker, C. S., Voynow, S. 1980. Variation in avian retrovirus genomes. *Ann. NY Acad. Sci.* 354:410–25
32. Cooper, P. D. 1977. Genetics of picornaviruses. *Compr. Virol.* 9:133–207
33. Dalgleish, A. G., Beverly, P. C. L., Clapham, P. R., Crawford, D. H., Greaves, M. F., Weiss, R. A. 1984. The CD4(T4) antigen is an essential component of the receptor for the AIDS retrovirus. *Nature* 312:763–67
34. Darlix, J. L., Spahr, P. F. 1983. High spontaneous mutation rate of Rous sarcoma virus demonstrated by direct sequencing of the RNA genome. *Nucleic Acids Res.* 11:5953–67
35. DePolo, N. J., Giachetti, C., Holland, J. J. 1987. Continuing coevolution of virus and defective interfering particles and of viral genome sequences during undiluted passages: Virus mutants exhibiting nearly complete resistance to formerly dominant defective interfering particles. *J. Virol.* 61:454–64
36. Desselberger, U., Nakajima, K., Alfino, P., Perderson, F. S., Haseltine, W. A., et al. 1978. Biochemical evidence that "new" influenza virus strains in nature may arise by recombination (reassortment). *Proc. Natl. Acad. Sci. USA* 76:3341–45
37. de Wachter, R., Fiers, W. 1969. Sequences at the 5'-terminus of bacteriophage QB RNA. *Nature* 221:233–35
38. Dietzschold, B., Wiktor, T. J., Trojanowski, J. Q., Macfarlan, R. I., Wunner, W. H., et al. 1985. Differences in cell-to-cell spread of pathogenic and apathogenic rabies virus *in vivo* and *in vitro*. *J. Virol.* 56:12–18
39. Domingo, E., Davila, M., Ortin, J. 1980. Nucleotide sequence heterogeneity of the RNA from a natural population of foot-and-mouth disease virus. *Gene* 11:333–46
40. Domingo, E., Flavell, R. A., Weissmann, L. 1976. *In vitro* site directed mutagenesis: Generation and properties of an infectious extracistronic mutant of bacteriophage QB. *Gene* 1:3–25
41. Domingo, E., Holland, J. J. 1987. High error rates, population equilibrium and evolution of RNA replication systems. In *RNA Genetics,* ed. E. Domingo, P. Ahlquist, J. J. Holland. Boca Raton, Fla: CRC. In press
42. Domingo, E., Martínez-Salas, E., Sobrino, F., de la Torre, J. C., Portela, A., et al. 1985. The quasispecies (extremely heterogeneous) nature of viral RNA genome populations: Biological relevance—a review. *Gene* 40:1–8
43. Domingo, E., Sabo, D., Taniguchi, T., Weissmann, C. 1978. Nucleotide sequence heterogeneity of an RNA phage population. *Cell* 13:735–44
44. Donis-Keller, H., Rommelaese, J., Ellis, R. W., Hopkins, N. 1980. Nucleotide sequences associated with differences in electrophoretic mobility of envelope glycoprotein gp 70 and with G_{IX} antigen phenotype of certain murine leukemia viruses. *Proc. Natl. Acad. Sci. USA* 77:1642–45
45. Dorner, A. J., Stoye, J. P., Coffin, J. M. 1985. Molecular basis of host range variation in avian retroviruses. *J. Virol.* 53:32–39

46. Drake, J. W. 1969. Spontaneous mutation. *Nature* 221:1128–32
47. Durbin, R. K., Stollar, V. 1986. Sequence analysis of the E1 gene of a hyperglycosylated, host restricted mutant of Sindbis virus and estimation of mutation rate from frequency of revertants. *Virology* 154:135–43
48. Eigen, M. 1971. Self-organization of matter and the evolution of macromolecules. *Naturwissenschaften* 58: 65–523
49. Eigen, M., Gardiner, W., Schuster, P., Winkler-Oswatitsch, R. 1981. The origin of genetic information. *Sci. Am.* 244:88–118
50. Eigen, M., Schuster, P. 1979. *The Hypercycle. A Principle of Natural Self-Organization*. Berlin: Springer-Verlag
51. Evans, D. M. A., Dunn, G., Minor, P. D., Schild, G. C., Cann, A. J., et al. 1985. Increased neurovirulence associated with a single nucleotide change in a noncoding region of the Sabin type 3 poliovaccine genome. *Nature* 314:548–50
52. Fang, R. X., Jou, W. M., Huylebroek, D., Devos, R., Fiers, W. 1981. Complete structure of A/duck/Ukraine/63 influenza hemagglutinin gene: Animal virus as progenitor of human H3 Hong Kong 1968 influenza hemagglutinin. *Cell* 25:315–23
53. Fields, B. N., Joklik, W. K. 1969. Isolation and preliminary charaterization of temperature sensitive mutants of reovirus. *Virology* 37:335–42
54. Fields, S., Winter, G. 1981. Nucleotide sequence heterogeneity and sequence rearrangements in influenza virus cDNA. *Gene* 15:207–14
55. Flavell, A. J. 1984. Role of reverse transcription in the generation of extrachromosomal copia mobile genetic elements. *Nature* 310:514–16
56. Fry, M., Loeb, L. A. 1986. Fidelity of DNA synthesis. In *Animal Cell DNA Polymerases*, pp. 157–83. Boca Raton, Fla: CRC
57. Fujinami, R. S., Oldstone, M. B. A. 1986. Molecular mimicry and autoimmunity. In *Viruses, Immunity, and Autoimmunity*, ed. A. Szentivanyi, H. Friedman, pp. 183–87. New York: Plenum
58. Goelet, P., Lomonossoff, G. P., Butler, B. J. G., Akam, M. E., Gait, M. J., Karn, J. 1982. Nucleotide sequence of tobacco mosaic virus RNA. *Proc. Natl. Acad. Sci. USA* 79:5818–22
59. Gojobori, T., Yokoyama, S. 1985. Rates of evolution of the retroviral oncogene of Maloney murine sarcoma virus and of its cellular homologues. *Proc. Natl. Acad. Sci. USA* 82:4198–201
60. Gopinathan, K. P., Weymouth, L. A., Kunkel, T. A., Loeb, L. A. 1979. Mutagenesis *in vitro* by DNA polymerase from an RNA tumour virus. *Nature* 278:857–59
61. Granoff, A. 1964. Nature of the Newcastle disease virus population. In *Newcastle Disease Virus, An Evolving Pathogen*, ed. R. P. Hanson, pp. 107–8. Madison, Wis: Univ. Wis. Press
62. Gribskov, M., Devereux, J., Burgess, R. R. 1984. The codon preference plot: Graphic analysis of protein coding sequences and prediction of gene expression. *Nucleic Acids Res.* 12:539–49
63. Guerrier-Takada, C., Gardiner, K., Marsh, T., Pace, N., Altman, S. 1983. The RNA moiety of ribonuclease P is the catalytic subunit of the enzyme. *Cell* 35:849–57
64. Haase, A. T. 1986. Pathogenesis of lentivirus infections. *Nature* 322:130–36
65. Haase, A. T., Gantz, D., Eble, B., Walker, D., Stowring, L., et al. 1985. Natural history of restricted synthesis and expression of measles virus genes in subacute sclerosing panencephalitis. *Proc. Natl. Acad. Sci. USA* 82:3020–24
66. Hahn, B. H., Shaw, G. M., Taylor, M. E., Redfield, R. R., Markham, P. D., et al. 1986. Genetic variation in HTLV-III/LAV over time in patients with AIDS or at risk for AIDS. *Science* 232:1548–53
67. Hall, W. W., Choppin, P. W. 1981. Measles-virus proteins in the brain tissue of patients with subacute sclerosing panencephalitis. Absence of the M protein. *N. Engl. J. Med.* 19:1152–55
68. Hall, W. W., Lamb, R. A., Choppin, P. W. 1979. Measles and subacute sclerosing panencephalitis virus proteins: Lack of antibodies to the M protein in patients with subacute sclerosing panencephalitis. *Proc. Natl. Acad. Sci. USA* 76:2047–51
69. Haseloff, J., Goelet, P., Zimmern, D., Alhquist, P., Dasgupta, R., Kaesberg, P. 1984. Striking similarities in amino acid sequence among nonstructural proteins encoded by RNA viruses that have dissimilar genomic organization. *Proc. Natl. Acad. Sci. USA* 81:4358–62
70. Hattori, M., Kuhara, S., Takenaka, O., Sakaki, Y. 1986. L1 family of repetitive DNA sequences in primates may be derived from a sequence encoding a reverse transcriptase-related protein. *Nature* 321:625–27
71. Holland, J. J. 1987. Defective interfering rhabdoviruses. In *The Rhabdovi-*

ruses, ed. R. Wagner. New York: Plenum. In press
72. Holland, J. J., Grabau, E. A., Jones, C. L., Semler, B. L. 1979. Evolution of multiple genome mutations during long-term persistent infection by vesicular stomatitis virus. *Cell* 16:494–504
73. Holland, J. J., McLaren, J. C., Syverton, J. T. 1959. Infection of naturally insusceptible cells with enterovirus ribonucleic acid. *J. Exp. Med.* 110:65–80
74. Holland, J. J., Spindler, K., Grabau, E., Semler, B., Jones, C., et al. 1980. Viral mutation in persistent infection. See Ref. 16, pp. 695–709
75. Holland, J. J., Spindler, K., Horodyski, F., Grabau, E., Nichol, S., Vandepol, S. 1982. Rapid evolution of RNA genomes. *Science* 215:1577–85
76. Horodyski, F. M., Holland, J. J. 1980. Viruses isolated from cells persistently infected with vesicular stomatitis virus show altered interactions with defective interfering particles. *J. Virol.* 36:627–31
77. Horodyski, F. M., Nichol, S. T., Spindler, K. R., Holland, J. J. 1983. Properties of DI particle resistant mutants of vesicular stomatitis virus isolated from persistent infections and from undiluted passages. *Cell* 33:801–10
78. Huang, A. J., Baltimore, D. 1977. Defective interfering animal viruses. *Compr. Virol.* 10:73–106
79. Hughes, P. J., Evans, D. M. A., Minor, P. D., Schild, G. C., Almond, J. W., Stanway, G. 1986. The nucleotide sequence of a type 3 poliovirus isolated during a recent outbreak of poliomyelitis in Finland. *J. Gen. Virol.* 67:2093–102
80. Inoue, T., Orgel, L. E. 1983. A nonenzymatic RNA polymerase model. *Science* 219:859–62
81. Ishihama, A., Mizumoto, K., Kawakami, K., Koto, A., Honda, A. 1986. Proofreading function associated with the RNA-dependent RNA polymerase from influenza virus. *J. Biol. Chem.* 261:10417–21
82. Jacobsen, S., Pfau, C. J. 1980. Viral pathogenesis and resistance to defective interfering particles. *Nature* 283:311–13
83. Jacobsen, S., Schlesinger, S. 1981. Defective interfering particles of Sindbis virus do not interfere with the homologous virus obtained from persistently infected cells but do interfere with Semliki Forest virus. *J. Virol.* 37:840–44
84. Jennings, P. A., Finch, J. T., Winter, G., Robertson, J. S. 1983. Does the higher order structure of the influenza virus nucleoprotein guide sequence rearrangements in influenza viral RNA? *Cell* 34:619–27
84a. Jou, W. M., Haegeman, G., Ysebaert, M., Fiers, W. 1972. Nucleotide sequence of the gene coding for the bacteriophage MS2 coat protein. *Nature* 237:82–88
85. Kassavetis, G. A., Zentner, P. G., Geiduschek, E. P. 1986. Transcription at bacteriophage T4 variant late promoters: An application of a newly-devised promoter-mapping method. *J. Biol. Chem.* 261:14256–65
86. Kawai, A., Matsumoto, S. 1977. Interefering and non-interfering defective particles generated by a rabies small plaque variant virus. *Virology* 76:60–71
87. Kaye, K. M., Spriggs, D. R., Bassel-Duby, R., Fields, B. N., Tyler, K. L. 1986. Genetic basis for altered pathogenesis of an immune-selected antigenic variant of reovirus type 3 (Dearing). *J. Virol.* 59:90–97
88. Kendal, A. P., Noble, G. R., Skehel, J. J., Dowdle, W. R. 1978. Antigenic similarity of influenza A (H1N1) virus from epidemics in 1977–1978 to Scandinavian strains in epidemics of 1950–1951. *Virology* 89:632–36
89. Keroack, M., Fields, B. N. 1986. Viral shedding and transmission between hosts determined by reovirus L2 gene. *Science* 232:1635–38
90. Kew, O. M., Nottay, B. K. 1984. Evolution of oral poliovirus vaccine strain in humans occurs by both mutation and intermolecular recombination. In *Modern Approaches to Vaccines,* ed. R. Chanock, R. Lerner, pp. 357–62. Cold Spring Harbor, NY: Cold Spring Harbor Lab.
91. Kew, O. M., Nottay, B. K., Hatch, M. H., Nakano, J. H., Obijeski, J. F. 1981. Multiple genetic changes can occur in the oral poliovaccines upon replication in humans. *J. Gen. Virol.* 56:337–47
92. King, A. M. Q., McCahon, D., Slade, W. R., Newman, J. W. I. 1982. Recombination in RNA. *Cell* 29:921–28
93. Kirkegaard, K., Baltimore, D. 1986. The mechanism of RNA recombination in poliovirus. *Cell* 47:433–43
94. Klatzmann, D., Champagne, E., Chamaret, S., Gruest, J., Guetard, D., et al. 1984. T-lymphocyte T4 molecule behaves as the receptor for human retrovirus LAV. *Nature* 312:767–68
95. Kunkel, L. O. 1947. Variation in phytopathogenic viruses. *Ann. Rev. Microbiol.* 1:85–100
96. Lai, M. M. C., Baric, R. S., Makino, S., Keck, J. G., Egbert, J., et al. 1985. Recombination between nonsegmented

RNA genomes of murine coronaviruses. *J. Virol.* 56:449–56

97. Laimins, L. A., Khoury, G., Gorman, L., Howard, B., Gruss, P. 1982. Host-specific activation of transcription by tandem repeats from simian virus 40 and Moloney murine sarcoma virus. *Proc. Natl. Acad. Sci. USA* 79:6453–57
98. Lazzarini, R. A., Keene, J. D., Schubert, M. 1981. The origins of defective interfering particles of the negative-strand RNA viruses. *Cell* 26:145–54
99. Leppert, M., Kort, L., Kolakofsky, D. 1977. Further characterization of Sendai virus DI-RNAs: A model for their generation. *Cell* 12:539–52
100. Li, W.-H., Wu, C.-I., Luo, C.-C. 1985. A new method for estimating synonymous and nonsynonymous rates of nucleotide substitution considering the relative likelihood of nucleotide and codon changes. *Mol. Biol. Evol.* 2:150–74
101. Loeb, D. D., Padgett, R. W., Hardies, S. C., Shehee, W. R., Comer, M. B., et al. 1986. The sequence of a large L1Md element reveals a tandemly repeated 5' end and several features found in retrotransposons. *Mol. Cell. Biol.* 6:168–82
102. Lubeck, M. D., Schulman, J. L., Palese, P. 1980. Antigenic variants of influenza virus: Marked differences in the frequencies of variants selected with different monoclonal antibodies. *Virology* 102:458–62
103. Maddon, P. J., Dalgleish, A. G., McDougal, J. S., Clapham, P. R., Weiss, R. A., Axel, R. 1986. The T4 gene encodes the AIDS virus receptor and is expressed in the immune system and the brain. *Cell* 47:333–48
104. Magrath, D. I., Evans, D. M. A., Ferguson, M., Schild, G. C., Minor, P. D., et al. 1986. Antigenic and molecular properties of type 3 poliovirus responsible for an outbreak of poliomyelitis in a vaccinated population. *J. Gen. Virol.* 67:899–905
105. Makino, S., Keck, J. G., Stohlman, S. A., Lai, M. M. C. 1986. High frequency RNA recombination of murine coronaviruses. *J. Virol.* 57:729–37
106. Matthews, R. E. F. 1983. Classification and nomenclature of viruses. *Intervirology* 17:1–199
107. McCahon, D. D., Slade, W. R., Priston, R. A. J., Lake, J. R. 1977. An extended genetic recombination map of foot-and-mouth disease virus. *J. Gen. Virol.* 35:555–65
108. Mendelsohn, C., Johnson, B., Lionetti, K. A., Nobis, P., Wimmer, E., Racaniello, V. R. 1986. Transformation of a human poliovirus receptor gene into mouse cells. *Proc. Natl. Acad. Sci. USA* 83:7845–59
109. Mills, B. G., Holst, P. A., Stabile, E., Adams, J. S., Rude, R. K., et al. 1985. A viral antigen-bearing cell line derived from culture of Paget's bone cells. *Bone* 6:257–68
110. Deleted in proof
111. Minor, P. D., John, A., Ferguson, M., Icenogle, J. P. 1986. Antigenic and molecular evolution of the vaccine strain of type 3 poliovirus during the period of excretion by a primary vaccine. *J. Gen. Virol.* 67:693
112. Mundry, K. W., Gierer, A. 1958. Die Erseugung von Mutationen des Tabakmosaikvirus durch chemische Behandlung seiner Nucleinsäure *in vitro*. *Z. Vererbungsl.* 89:614–30
113. Naeve, C. W., Hinshaw, V. S., Webster, R. G. 1984. Mutations in the hemagglutinin receptor-binding site can change the biological properties of an influenza virus. *J. Virol.* 51:567–69
114. Nakajima, K., Desselberger, U., Palase, P. 1978. Recent human influenza A (H1N1) viruses are closely related genetically to strains isolated in 1950. *Nature* 274:334–39
115. Norrby, E., Kristensson, K., Brzosko, W. J., Kapsenburg, J. G. 1985. Measles virus matrix protein detected by immune fluorescence with monoclonal antibodies in the brain of patients with subacute sclerosing panencephalitis. *J. Virol.* 56:337–40
116. Nottay, B. K., Kew, O. M., Hatch, M. H., Heyward, J. T., Obijeski, J. F. 1981. Molecular variation of type 1 vaccine-related and wild poliovirus during replication in humans. *Virology* 108:405–23

116a. O'Brien, R. L., Brinster, R. L., Storb, U., 1987. Somatic hypermutation of an immunoglobulin transgene in κ transgenic mice. *Nature* 326:405–9

117. O'Hara, P. J., Horodyski, F. M., Nichol, S. T., Holland, J. J. 1984. Vesicular stomatitis virus mutants resistant to defective interfering particles accummulate stable 5'-terminal and fewer 3'-terminal mutations in a stepwise manner. *J. Virol.* 49:793–98
118. O'Hara, P. J., Nichol, S. T., Horodyski, F. M., Holland, J. J. 1984. Vesicular stomatitis virus defective interfering particles can contain extensive genomic sequence rearrangements and base substitutions. *Cell* 36:915–24
119. O'Rear, J. J., Temin, H. M. 1982. Spontaneous changes in nucleotide se-

quence in proviruses of spleen necrosis virus, an avian retrovirus. *Proc. Natl. Acad. Sci. USA* 79:1230–34

120. Oxford, J. S., Abbo, H., Corcoran, T., Webster, R. G., Smith, A. J., et al. 1983. Antigenic and biochemical analysis of field isolates of influenza B virus: Evidence for intra- and inter- epidemic variation. *J. Gen. Virol.* 64:2367–77
121. Oxford, J. S., Salum, S., Corcoran, T., Smith, A. J., Grilli, E. A., Schild, G. C. 1986. An antigenic analysis using antibodies of influenza A (H3N2) viruses isolated from an epidemic in a semi-closed community. *J. Gen. Virol.* 67:265–74
122. Palese, P., Young, J. F. 1982. Variation of influenza A, B and C viruses. *Science* 215:1468–74
123. Parvin, J. D., Moscona, A., Pan, W. T., Leider, J. M., Palese, P. 1986. Measurement of the mutation rates of animal viruses: Influenza A virus and poliovirus type 1. *J. Virol.* 59:377–83
124. Perrault, J. 1981. Origin and replication of defective interfering particles. *Curr. Top. Microbiol. Immunol.* 93:152–209
125. Portner, A., Webster, R. G., Bean, W. J. 1980. Similar frequency of antigenic variants in Sendai, vesicular stomatitis virus, and influenza A viruses. *Virology* 104:235–38
126. Pringle, C. R. 1970. Genetic characteristics of conditional lethal mutants of vesicular stomatitis virus induced by 5-fluorouracil, 5-azacytidine and ethylmethane sulfonate. *J. Virol.* 5:559–67
127. Pringle, C. R., Duncan, I. B. 1971. Preliminary physiological characterization of temperature sensitive mutants of vesicular stomatitis virus. *J. Virol.* 8:56–61
128. Puddington, L., Bevan, M. J., Rose, J. K., LeFrançois, L. 1986. N protein is the predominant antigen recognized by vesicular stomatitis virus–specific cytotoxic T cells. *J. Virol.* 60:708–17
129. Ray, U. R., Aulakh, G. S., Schubert, M., McClintock, P. R., Yoon, J. W., Notkins, A. L. 1983. Virus-induced diabetes mellitus. XXV. Difference in the RNA fingerprints of diabetogenic and non-diabetogenic variants of encephalomyocarditis virus. *J. Gen. Virol.* 64:947–50
130. Raymond, F. L., Caton, A. J., Cox, N. J., Kendal, A. P., Brownlee, G. G. 1986. The antigenicity and evolution of influenza H1 hemagglutinin from 1950–1957 and 1977–1983: Two pathways from one gene. *Virology* 148:275–87
131. Reagan, K. J., Goldberg, B., Crowell, R. L. 1984. Altered receptor specificity of coxsackievirus B3 after growth in rhabdomyosarcoma cells. *J. Virol.* 49:635–40
132. Reanney, D. C. 1982. The evolution of RNA viruses. *Ann. Rev. Microbiol.* 36:47–73
133. Riviere, Y., Oldstone, M. B. A. 1986. Genetic reassortants of lymphocytic choriomeningitis virus: Unexpected disease and mechanism of pathogenesis. *J. Virol.* 59:363–68
134. Rogers, G. N., Paulson, J. C., Daniels, R. S., Skehel, J. J., Wilson, I. A., Wiley, D. C. 1983. Single amino acid substitutions in the influenza hemagglutinin change receptor binding specificity. *Nature* 304:76–78
135. Rowlands, D. J., Clarke, B. E., Carrol, A. R., Brown, F., Nicholson, B. H., et al. 1983. Chemical basis of antigenic variation in foot-and-mouth disease virus. *Nature* 306:694–97
136. Rubin, D. H., Fields, B. N. 1980. Molecular basis of reovirus virulence. Role of the M2 gene. *J. Exp. Med.* 152:853–57
137. Sabara, M., Deregt, D., Babiuk, L. A., Misra, V. 1982. Genetic heterogeneity within individual bovine rotavirus isolates. *J. Virol.* 44:813–22
138. Saigo, K., Kugimiya, W., Matsuo, Y., Inouye, S., Yoshioka, K., Yuki, S. 1984. Identification of the coding sequence for a reverse transncriptase-like enzyme in a transposable genetic element in *Drosophila melanogaster*. *Nature* 312:659–61
139. Salinovich, O., Payne, S. L., Montelaro, R. C., Hussain, K. A., Issel, C. J., Schnorr, K. L. 1986. Rapid emergence of novel antigenic and genetic variants of equine infectious anemia virus during persistent infection. *J. Virol.* 57:71–80
140. Scholtissek, C., Vallbracht, A., Flehmig, B., Rott, R. 1979. Correlation of pathogenicity and gene constellation of influenza A virus. II. Highly neurovirulent recombinants derived from non-virulent or weakly neurovirulent parent virus strains. *Virology* 95:492–98
141. Schubert, M., Harmison, G. G., Meier, E. 1984. Primary structure of the vesicular stomatitis virus polymerase (L) gene: Evidence for a high frequency of mutations. *J. Virol.* 51:505–14
142. Seif, I., Coulon, P., Rollin, P. E., Flamand, A. 1984. Rabies virulence: Effect on pathogenicity and sequence characterization of rabies virus mutations affecting antigenic site III of the glycoprotein. *J. Virol.* 53:926–34
143. Sheppard, R. D., Raine, C. S., Born-

stein, M. B., Udem, S. A. 1986. Rapid degradation restricts measles virus matrix protein expression in a subacute sclerosing panencephalitis cell line. *Proc. Natl. Acad. Sci. USA* 83:7913–17

144. Shimotohno, K., Temin, H. 1982. Spontaneous variation and synthesis in the U3 region of the long terminal repeat of an avian retrovirus. *J. Virol.* 41:162–71

145. Shope, R. E., Tignor, E. J., Rozhon, E. J., Bishop, D. H. L. 1981. The association of bunyavirus middle sized RNA segment with mouse pathogenicity. In *The Replication of Negative Strand Viruses,* ed. D. H. L. Bishop, R. W. Compans, pp. 147–52. New York: Elsevier/North-Holland

145a. Siekevitz, M., Kocks, C., Rajewsky, K., Dildrop, R. 1987. Analysis of somatic mutation and class switching in naive and memory B cells generating adoptive primary and secondary responses. *Cell* 48:757–70

146. Skehel, J. J., Stevens, D. J., Daniels, R. S., Douglas, A. R., Knossow, M., et al. 1984. A carbohydrate side chain on hemagglutinins of Hong Kong influenza viruses inhibits recognition by a monoclonal antibody. *Proc. Natl. Acad. Sci. USA* 81:1779–83

147. Sobrino, F., Davila, M., Ortin, J., Domingo, E. 1983. Multiple genetic variants arise in the course of replication of foot-and-mouth disease virus in cell culture. *Virology* 123:310–18

148. Spindler, K. R., Horodyski, F. M., Holland, J. J. 1982. High multiplicities of infection favor rapid and random evolution of vesicular stomatitis virus. *Virology* 119:96–108

149. Srinivasappa, J., Saegusa, J., Prabhakar, B. S., Gentry, M. K., Buchmeier, J., et al. 1986. Molecular mimicry: Frequency of reactivity of monoclonal antiviral antibodies with normal tissues. *J. Virol.* 57:397–401

150. Stanway, G., Hughes, P. J., Mountford, R. C., Reeve, P., Minor, P. D., et al. 1984. Comparison of the complete nucleotide sequences of the genomes of the neurovirulent poliovirus P3/Leon/37 and its attenuated Sabin vaccine derivative P3/Leon $12a_1b$. *Proc. Natl. Acad. Sci. USA* 81:1539.

151. Starcich, B. R., Hahn, B. H., Shaw, G. M., McNeely, P. D., Modrow, S., et al. 1986. Identification and characterization of conserved and variable regions in the envelope gene of HTLV-III/LAV, the retrovirus of AIDS. *Cell* 45:637–48

152. Stec, D. S., Waddell, A., Schmaljohn, C. S., Cole, G. A., Schmaljohn, A. L. 1986. Antibody-selected variation and reversion in Sindbis virus neutralization epitopes. *J. Virol.* 57:715–20

153. Steinhauer, D. A., Holland, J. J. 1986. Direct method for quantification of extreme polymerase error frequencies at selected single base sites in viral RNA. *J. Virol.* 57:219–28

154. Stocking, C., Kollek, R., Bergholz, U., Ostertag, W. 1986. Point mutations in the U3 region of the long terminal repeat of Moloney murine leukemia virus determine disease specificity of the myeloproliferative sarcoma virus. *Virology* 153:145–49

155. Takeda, N., Miyamura, K., Ogino, T., Natori, S., Yamazaki, N., et al. 1984. Evolution of enterovirus type 70: oligonucleotide mapping analysis of RNA genome. *Virology* 134:375–88

156. Tolskaya, E. A., Romanova, L. I., Kolesnikova, M. S., Agol, V. I. 1983. Intertypic recombination in poliovirus: Genetic and biochemical studies. *Virology* 124:121–32

157. Townsend, A. R. M., Rothbard, J., Gotch, F. M., Bahadur, G., Wraith, D., McMichael, A. J. 1986. The epitopes of influenza nucleoprotein recognized by cytotoxic T lymphocytes can be defined with short sythetic peptides. *Cell* 44:959–68

158. Townsend, A. R. M., Skehel, J. J. 1984. The influenza A virus nucleoprotein gene controls the induction of both subtype specific and cross reactive cytotoxic T cells. *J. Exp. Med.* 160:552–63

159. Tyler, K. L., McPhee, D. A., Fields, B. N. 1986. Distinct pathways of viral spread in the host determined by reovirus S1 gene segment. *Science* 233:770–74

160. Valentine, R. C., Ward, R., Strand, M. 1969. The replication cycle of RNA bacteriophages. *Adv. Virus Res.* 15:1–59

161. VandePol, S. B., Holland, J. J. 1986. Evolution of vesicular stomatitis virus in athymic nude mice: Mutations associated with natural killer cell selection. *J. Gen. Virol.* 67:441–51

162. Wabl, M., Burrows, P. D., von Gabain, A., Steinberg, C. 1985. Hypermutation at the immunoglobulin heavy chain locus in a pre-B-cell line. *Proc. Natl. Acad. Sci. USA* 82:479–82

163. Webster, R. G., Kawaoka, Y., Bean, W. J. 1986. Molecular changes in A/chicken/Pennsylvania/83 (H5N2) influenza virus associated with acquisition of virulence. *Virology* 149:165–73

164. Webster, R. G., Laver, W. G., Air, G. M., Schild, G. C. 1982. Molecular

mechanisms of variation in influenza viruses. *Nature* 196:115–21

165. Yewdell, J. W., Bennink, J. R., Smith, G. L., Moss, B. 1985. Influenza A virus nucleoprotein is a major target antigen for cross reactive anti-influenza A virus cytotoxic T lymphocytes. *Proc. Natl. Acad. Sci. USA* 82:1785–89
166. Young, J. F., Palese, P. 1979 Evolution of influenza A viruses in nature: Recombination contributes to genetic variation of H1N1 strains. *Proc. Natl. Acad. Sci. USA* 76:6547–51
167. Youngner, J. S., Preble, O. T. 1980. Viral persistence: Evolution of viral populations. *Compr. Virol.* 16:73–135
168. Zaug, A. J., Cech, T. R. 1986. The intervening sequence RNA of *tetrahymena* is an enzyme. *Science* 231:470–75

Ann. Rev. Microbiol. 1987. 41:435–64

BACTERIAL BIOFILMS IN NATURE AND DISEASE

J. William Costerton

Departments of Biology and Infectious Diseases, University of Calgary, Calgary, Alberta, Canada T2N 1N4

K.-J. Cheng

Agriculture Canada Research Station, Lethbridge, Alberta, Canada T1J 4B1

Gill G. Geesey

Department of Microbiology, Long Beach State University, Long Beach, California 90840

Timothy I. Ladd

Department of Biology, St. Mary's University, Halifax, Nova Scotia, Canada B3H 3C3

J. Curtis Nickel

Department of Urology, Queen's University, Kingston, Ontario, Canada K7L 3N6

Mrinal Dasgupta

Department of Nephrology, University of Alberta Hospital, Edmonton, Alberta, Canada T6G 2E5

Thomas J. Marrie

Depatment of Medicine, Dalhousie University, Halifax, Nova Scotia, Canada B3H 2Y9

0066-4227/87/1001-0435$02.00

CONTENTS

INTRODUCTION

The growth of bacteria in pure cultures has been the mainstay of microbiological technique from the time of Pasteur to the present. Solid media techniques have allowed the isolation of individual species from complex natural populations. These pure isolates are intensively studied as they grow in batch cultures in nutrient-rich media. This experimental approach has served well in providing an increasingly accurate understanding of prokaryotic genetics and metabolism and in facilitating the isolation and identification of pathogens in a wide variety of diseases. Further, vaccines and antibiotics developed on the basis of in vitro data and tested on test-tube bacteria have provided a large measure of control of these pathogenic organisms.

During the last two decades microbial ecologists have developed a series of exciting new techniques for the examination of bacteria growing in vivo, and often in situ, in natural environments and in pathogenic relationships with tissues. The data suggest that these organisms differ profoundly from cells of the same species grown in vitro. Brown & Williams (12) have shown that bacteria growing in infected tissues produce cell surface components not found on cells grown in vitro and that a whole spectrum of cell wall structures may be produced in cells of the same species in response to variations in nutrient status, surface growth, and other environmental factors (67). We and others (28) have used direct ecological methods to examine bacterial cells growing in natural and pathogenic ecosystems, and we find that many important populations grow in adherent biofilms and structured consortia that are not seen in pure cultures growing in nutrient-rich media. In fact, it is difficult to imagine actual natural or pathogenic ecosystems in which the bacteria would be as well nourished and as well protected as they are in single-species batch cultures.

In this review we summarize and synthesize the data generated by the new direct methods of studying mixed natural bacterial populations in situ. Generally, morphological data give us a basic concept of community structure, direct biochemical techniques monitor metabolic processes at the whole-community level, and specific probes define cell surface structures in situ. Any in vitro techniques used in these ecological studies are selected to mimic the natural ecosystem as closely as possible. In our estimation, data from studies of bacteria growing in single-species batch cultures continue to be very valuable. However, these data represent a single, and perhaps unrepresentative, point in the broad spectrum of bacterial characteristics expressed in response to altered environmental factors. In retrospect, it may become apparent that the phenotypic plasticity of bacteria (12, 107) and their ability to form structured and cooperative consortia will prove to be their most remarkable characteristics.

STRUCTURE AND DYNAMICS OF BACTERIAL BIOFILMS

Bacteria in natural aquatic populations have a marked tendency to interact with surfaces (120). Recent work has demonstrated that many bacteria associate with surfaces in transient apposition, particularly in oligotrophic marine environments (75). Some of these bacteria adhere to these surfaces, initially in a reversible association and eventually in an irreversible adhesion (76), and initiate the development of adherent bacterial biofilms. We recognize the inaccuracy of simple physical models in which bacteria are represented as smooth 1-μm particles with various surface properties and are placed in computer-simulated association with similarly homogeneous substrate surfaces. The substrate surfaces of importance in natural and medical systems are invariably coated with adsorbed polymers, and the surfaces of bacteria growing in these environments are a forest of protruding linear macromolecules such as pili, lipopolysaccharide O antigen or teichoic acid, and exopolysaccharides (9). Thus, the initial association between bacterial and substrate surfaces in real aquatic environments is difficult to model, simply because it consists of the association of numerous linear polymers. However, Fletcher and associates (39), in a useful and very extensive series of empirical experiments, have shown that the rate of bacterial adhesion to a wide variety of surfaces is responsive to some physical characteristics such as hydrophobicity (40).

A bacterial cell initiates the process of irreversible adhesion by binding to the surface using exopolysaccharide glycocalyx polymers (Figure 1). Cell division then produces sister cells that are bound within the glycocalyx matrix, initiating the development of adherent microcolonies. The eventual production of a continous biofilm on the colonized surface is a function of cell division within microcolonies (69) and new recruitment of bacteria from the planktonic phase (Figure 1). Consequently, the biofilm finally consists of single cells and microcolonies of sister cells all embedded in a highly hydrated, predominantly anionic matrix (110) of bacterial exopolymers and trapped extraneous macromolecules. As the bacterial biofilm gradually occludes the colonized surface, newly recruited bacteria adhere to the biofilm itself (Figure 1, *F*). Differences in the rates of colonization of various surfaces usually disappear in long-term colonization experiments.

PHYSIOLOGY OF BIOFILM BACTERIA

In their very comprehensive review (12), Brown & Williams have provided detailed experimental evidence for Smith's earlier conclusion (107) that the molecular composition of bacterial cell walls is essentially plastic and is remarkably responsive to the cell's growth environment. The iron deprivation inherent in growth within infected tissues profoundly changes the cell wall protein composition of gram-negative bacteria. Thus, the cell walls of bacteria recovered directly from the sputum of cystic fibrosis patients (3) or from infected urine (105) or burn fluids (4) exhibit iron depletion and differ radically from those of cells of the same organism grown in batch culture in vitro. Similar bacterial surface changes are seen in response to alterations in

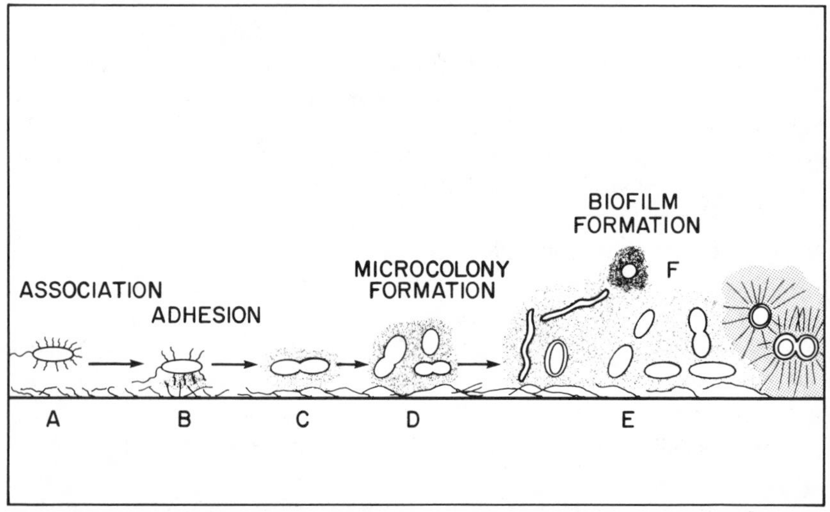

Figure 1 Bacteria may associate reversibly with a polymer-coated surface (*A*), or they may adhere irreversibly (*B*) and divide (*C*) to produce microcolonies (*D*) within an adherent multispecies biofilm (*E*). The biofilm grows by internal replication and by recruitment (*F*) from the bulk fluid phase.

growth rate (67, 118), exposure to subinhibitory concentrations of certain antibiotics (66), and growth on solid surfaces (67). Equally profound differences in enzyme activity have been noted between bacterial cells adherent to surfaces and planktonic cells of the same organism (C. S. Dow, R. Whittenbury & D. Kelly, unpublished). These data suggest that sessile cells have more active reproduction and general metabolism, while planktonic cells are phenotypically committed to motility and to the colonization of new surfaces.

In light of the remarkable phenotypic plasticity documented above and the fact that biofilm bacteria live in specific microenvironments, we must conclude that these cells are structurally and functionally very different from planktonic cells grown in rich media at high growth rates in batch culture. It seems unwise to extrapolate from batch-culture data conclusions about bacteria growing in biofilms in natural and pathogenic environments. Therefore, we have adapted existing methods for planktonic bacteria (60) to study the physiology and resistance to antimicrobial agents of biofilm bacteria in situ.

When microbial biofilms are developed on rock surfaces suspended in flowing streams, the overall metabolic activity of cells within the mixed adherent populations can be assessed by a modification of the heterotrophic potential technique (30, 61). Biofilm cells have a greater capacity to convert C^{14} glutamate to labeled cell components and to $C^{14}O_2$ than cells in the planktonic phase of the same stream or dispersed biofilm cells from equivalent surfaces. The increased metabolic activity of biofilm cells may result from phenotypic changes in response to sessile growth (C. S. Dow, R. Whittenbury & D. Kelly, unpublished). It may also result from nutrient trapping (Figure 2, *A*), which occurs when organic nutrients are bound to the biofilm matrix and are readily dissociated for use by the component organisms. High rates of biofilm development in oligotrophic environments (45) and in distilled water systems argue for the importance of this nutrient trapping strategy. Nutrients produced by component organisms (e.g. by photosynthesis) also enter the biofilm, and microcolonies of cells capable of primary production of nutrients are often surrounded by heterotrophic organisms that are stimulated by the exudates to grow and to produce adjacent microcolonies (Figure 2, *B*). The seasonal death and cell lysis of primary producers often radically stimulates biofilm growth (45), because biofilms tend to trap and recycle cellular components. When biofilms form on the surfaces of insoluble nutrients (e.g. cellulose), the initial events of adhesion (Figure 2, *C*) favor specific bacteria that can digest that substrate (e.g. cellulolytic bacteria). The primary colonizers in such a system produce cell-associated digestive enzymes that attack the insoluble substrate and produce soluble nutrients that stimulate the growth of adjacent heterotrophic organisms until a digestive consortium is formed (Figure 2, *C*). Electron transfer may also take place in these consortia. The conformation and juxtaposition of the component organisms vis-a-vis their insoluble substrate is optimally maintained by the biofilm mode of growth. In modern techniques for the in situ study of biofilms an adherent population is treated almost like a multicellular tissue in matters of nutrient uptake, nutrient cycling, respiration, and overall growth.

Because of the matrix-enclosed mode of growth of biofilm bacteria, a substantial ion exchange matrix arises between the component cells and the liquid phase of their environment (Figure 2). Additionally, the gellike state of the predominantly polysaccharide biofilm matrix limits the access of antibacterial agents (Figure 2, *D*), such as antibodies (6), bacteriophage, and phagocytic eukaryotic cells, to its component bacteria. Therefore, biofilm bacteria are substantially protected from amebae, white blood cells (96, 104, 115), bacteriophage, surfactants (47), biocides (100), and antibiotics (90, 91).

DISTRIBUTION OF BIOFILMS IN NATURAL AND PATHOGENIC ENVIRONMENTS

The formation of biofilms on surfaces can be regarded as a universal bacterial strategy for survival and for optimum positioning with regard to available

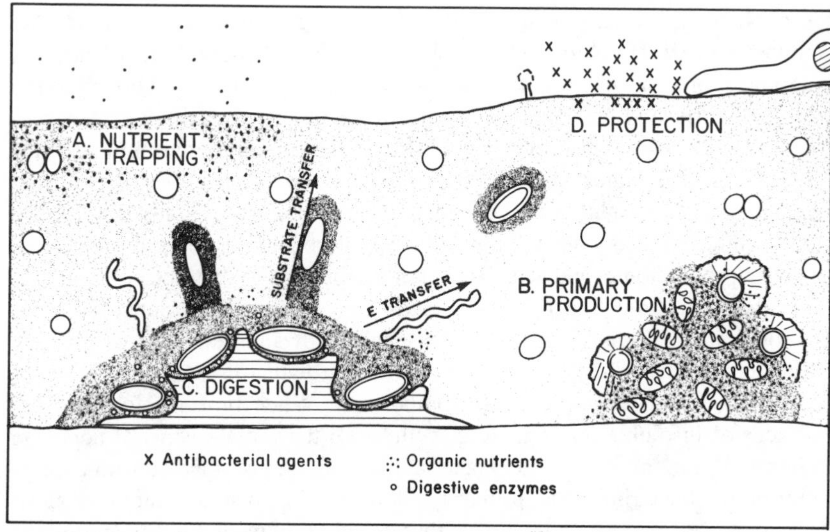

Figure 2 Modes of nutrient acquisition and protection of a bacterial biofilm. (*A*) Nutrient trapping. (*B*) Primary production. (*C*) Formation of a digestive consortium. (*D*) Exclusion of antibacterial substances.

nutrients. Bacterial populations living in this protected mode of growth produce planktonic cells, with much reduced chances of survival. These detached cells may colonize new surfaces or may burgeon to form large planktonic populations in those rare environments where nutrients are plentiful and bacterial antagonists are few.

Natural Aquatic Environments

In an exhaustive survey of the sessile and planktonic bacterial populations of 88 streams and rivers, the sessile population exceeded the planktonic population by 3–4 logarithm units in pristine alpine streams and by 200 fold in sewage effluents (65). A very widespread exception to the general trend occurs in the extremely oligotrophic environments of the deep ocean and deep groundwater, where bacteria are in an advanced stage of starvation (2). These cells alter their cell surfaces (56) and their patterns of peptide synthesis (51) in response to starvation and do not expend their scarce metabolic resources in exopolysaccharide synthesis unless they are revived by nutrient stimulation. Specific insoluble nutrient substrates (e.g. cellulose, solid hydrocarbons) within these aquatic environments are rapidly colonized by bacteria specialized for their digestion; the consistent presence of these materials in an aquatic system stimulates the development of large and vigorous populations of primary and secondary colonizers.

Industrial Aquatic Systems

The relatively high levels of nutrients and the high surface areas in many industrial aquatic systems predispose these systems to biofilm formation. Adherent populations bedevil most industrial systems (24) by plugging filters and injection faces, fouling products, and generating harmful metabolites (e.g. H_2S). Bacteria gradually colonize the water-cooled side of metal surfaces in heat exchangers, and the resultant biofilm insulates against heat exchange so effectively that exchange efficiency is gradually reduced to $<10\%$ of designed values. This costly problem may now be solved by a recently patented biofilm removal technique (23) in which slow freezing cycles are used to produce large ice crystals within the biofilm. Thawing of the crystals then leads to complete removal of the biofilm.

Bacterial corrosion of metals is an economically important consequence of bacterial biofilm formation that illustrates several fascinating aspects of the structure and physiology of these adherent bacterial populations. The bacteria most commonly associated with the corrosion of metals are the anaerobic sulfate-reducing bacteria (SRB), but other sulfur-cycle organisms are also important in this process (92). Geesey et al (44) have found that metal corrosion can occur, even in the absence of living bacterial cells, when two polymers with different metal-binding capacities (Figure 3) are adsorbed to adjacent areas of a metallic surface. Little et al (64) have reported that a measurable corrosion potential is generated between an uncolonized metal surface and a metallic surface colonized by bacteria. Bacteria organized into structural consortia occupy developing corrosion pits (27), and local pH differences as great as 1.5 units can occur in the lower zones of biofilms growing on metallic surfaces. These data have allowed us to construct a conceptual model of a functioning bacterial "corrosion cell" (Figure 3) using established physicochemical concepts. The microcolony of sister cells at site *A* on the metallic surface would produce an exopolysaccharide matrix that would bind metal ions (+), and their metabolic activities would generate a local pH of 7.0. In an adjoining site (*B*) another bacterium (probably an SRB) would establish a consortium with other organisms, and the coordinated metabolic activity of this structured community would generate a lowered local pH (perhaps 5.5) and an exopolysaccharide matrix with a low natural affinity for soluble metal ions. The scenario described so far is hypothetical, but the physicochemical differences between sites *A* and *B* would, by immutable physicochemical laws, cause the mobilization of metal at site *B* and the formation of a corrosion pit. Thus, bacterial corrosion of metals is really an activity of structured bacterial biofilms in which physicochemical differences

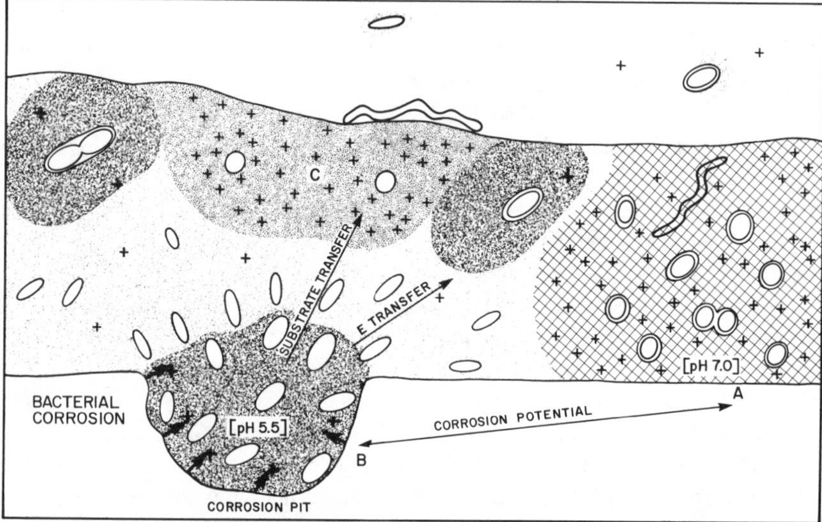

Figure 3 Bacteria within biofilms adherent to metal surfaces grow in the form of discrete microcolonies whose exopolymers may (*A* and *C*) or may not (*B*) bind certain metal ions. Metabolic activities in some microcolonies may produce local differences in pH, abetted by electron and substrate transfer in structured consortia (*B*). These local surface differences can produce an electrochemical "corrosion cell."

between adjacent loci on the metallic surface are created and maintained by differential metal-binding and metabolic activity until deep corrosion pits have been produced. Engineers have long noted that regular scraping (pigging) of pipelines prevents corrosion. We have now shown, in the field and in the laboratory, that pigging disturbs these highly structured bacterial corrosion cells; several days are required to reestablish their structure and activity (J. W. Costerton, unpublished data).

Medical Biomaterials

Medical biomaterials are the plastic, rubber, and metallic materials that are used to construct the myriad of medical devices and prostheses. With the remarkable modern advances in medicine and the concommitant increases in the numbers of elderly and immunocompromised patients, the use of these biomaterials has increased exponentially. An increasing number of bacterial infections is centered on implanted devices (31). These infections have the following unique characteristics. (*a*) They often have indolent pathogenic patterns with alternating quiescent and acute periods. (*b*) There may be an initial response to antibiotic therapy, but relapses are frequent because bacteria in the biofilms are protected from antibiotics and constitute uncontrolled foci that often necessitate the removal of the device. (*c*) While these infections are often polymicrobial, the predominant bacteria are either common members of the autochthonous skin or bowel flora or very common environmental organisms (e.g. *Pseudomonas*) that are often only pathogenic in immunocompromised patients. (*d*) Bacteria may be difficult to recover from adjacent fluids when the device is in place and from the device itself when it is removed.

All of these disease characteristics indicated that bacterial biofilms might be involved in biomaterial-related bacterial infections. Therefore, methods of direct observation and quantitative recovery developed in our environmental and industrial studies have been used to examine a large number of medical devices that were demonstrable foci of infections. Extensive bacterial biofilms were found on transparent dressings, sutures (50), wound drainage tubes, hemodialysis buttons, hemasite access devices, intraarterial and intravenous catheters (94), Hickman and silastic cardiac catheters (112), central venous lines, Swann-Ganz pulmonary artery catheters, dacron vessel repair sections, cardiac pacemakers (70, 72), bioprosthetic and mechanical heart valves (57), Tenckhoff peritoneal catheters (74), Foley urinary catheters (87) and urine collection systems, nephrostomy tubes, ureteral stents, biliary stents, penile prostheses (88), intrauterine contraceptive devices (IUDs) (71), endotracheal tubes, and prosthetic hip joints (48, 49). In all instances of bacterial colonization of medical biomaterials, extensive bacterial biofilms were seen by scanning and transmission electron microscopy (SEM and

TEM). Large numbers of living organisms were recovered by quantitative (scraping-vortex-sonication) recovery techniques (90), but not by routine methods used in clinical laboratories, including the otherwise very useful Maki technique (68). We have now implanted several of these medical devices (e.g. Foley catheters (85), cardiac catheters, Tenckhoff catheters, intrauterine contraceptive devices) in experimental animals to follow the time course of their bacterial colonization. We have noted that autochthonous and environmental organisms commonly form biofilms on accessible biomaterial surfaces. Further, these biofilms spread up to 100 cm along these colonized surfaces in as few as 3 days, in spite of active host defense systems (Figure 4, *A*) and prophylactic doses of antibiotics (Figure 4, *B*). These bacterial biofilms on the biomaterial surfaces do not usually cause overt infection or even detectable inflammation. However, they do constitute a nidus of infection when the host defense system fails to contain them. Consequently, they are able to give rise to disseminating planktonic cells that trigger acutely pathological sequellae (Figure 4, *C*). Conventional therapeutic doses of antibiotics often suffice to kill the disseminated bacteria (Figure 4, *B*) and to control the symptoms of infection, but the biofilm cells are not killed (90); thus, they constitute a continuing nidus for relapse of the infection.

Because it protects component cells from host defense mechanisms and

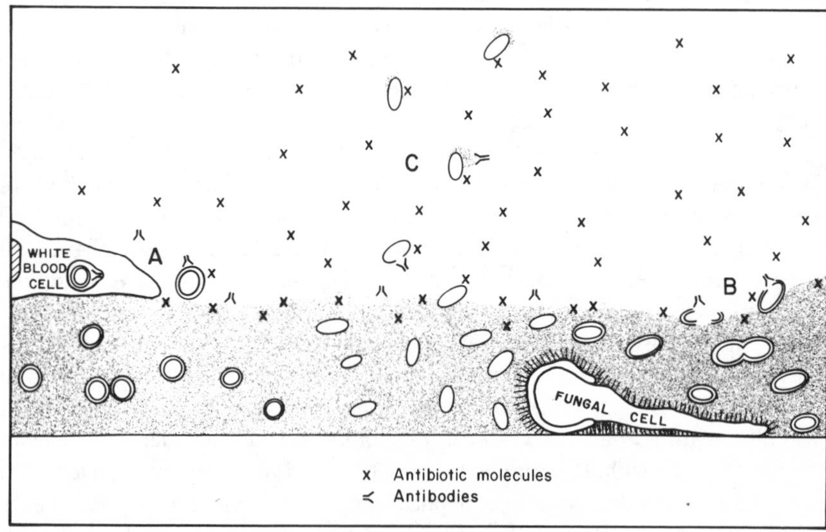

Figure 4 Microbial cells growing in adherent biofilms on medical biomaterials are generally protected from host defense mechanisms and antibacterial agents. Phagocytic white blood cells (*A*) and antibodies (*B*) can kill bacteria at the biofilm surface and planktonic bacteria (*C*) that have left the protection of the biofilm to initiate an acute disseminated infection, but usually cannot kill bacteria within the deeper layers of the biofilm.

from antibiotics, the biofilm mode of growth is of pivotal importance in the progressive colonization of biomaterials. Currently, the most effective strategy for the prevention of biomaterial-associated infections takes advantage of this fact. By following the progress of an auxotrophically marked uropathogenic strain of *Escherichia coli,* Nickel et al (85) established that biofilm development on the plastic and rubber surfaces of the luminal route through the Foley catheter and urine collection system is an important mechanism of bacterial access to the catheterized bladder. Accordingly, they have placed a plastic luminal sleeve impregnated with a powerful industrial biocide (26) in the drainage spout of a new urine collection system. Bacteria are attracted to this biocide-releasing surface, where they are killed; thus biofilm development is not initiated. This strategy delayed bladder colonization for 6–8 days in a rabbit model system (86). The sleeve device is currently being tested in patients.

Digestion and Biodeterioration

Biodeterioration of materials, including the digestion of insoluble nutrients by bacterial populations in the digestive tracts of higher animals, usually involves a focused enzymatic attack on particular loci at the material surface. This focused attack produces the pitting characteristic of the biodeterioration of a variety of insoluble substrates ranging from the digestion of cellulose to the corrosion of stainless steel. Physical attachment is necessary for active cellulose digestion, and the enzymes involved remain in particularly close association with bacterial cells growing in biofilms on the surface of cellulose fibers (Figure 5).

The colonization of cellulose fibers exposed to normal rumen flora is very rapid. Primary cellulose-degrading organisms can be heavily enriched by placing sterile cellulose fibers in rumen fluid, allowing 5 min for the adhesion of the cellulose degraders, and recovering the colonized fibers by centrifugation (83). Monocultures of the three main cellulose-degrading species from ruminants (*Bacteroides succinogenes, Ruminococcus albus,* and *Ruminococcus flavefaciens*) adhere avidly to cellulose fibers, but they degrade this insoluble substrate at a rate much slower than that in natural digestive systems. Cells of *B. succinogenes* are closely apposed to the surface of the cellulose (Figure 5, *A*), and small cell-derived vesicles are produced, which retain adhesive and cellulolytic properties (41). Conversely, cells of the gram-positive coccoid *Ruminococcus* species adhere to the cellulose surface with much greater separation (Figure 5, *B*) and do not usually detach cellulolytic vesicles.

The specific adhesion of bacteria to cellulose is the essential first step in ruminant digestion, but realistic rates of cellulose digestion are not achieved in the laboratory until the organisms are combined with other bacteria (58) or

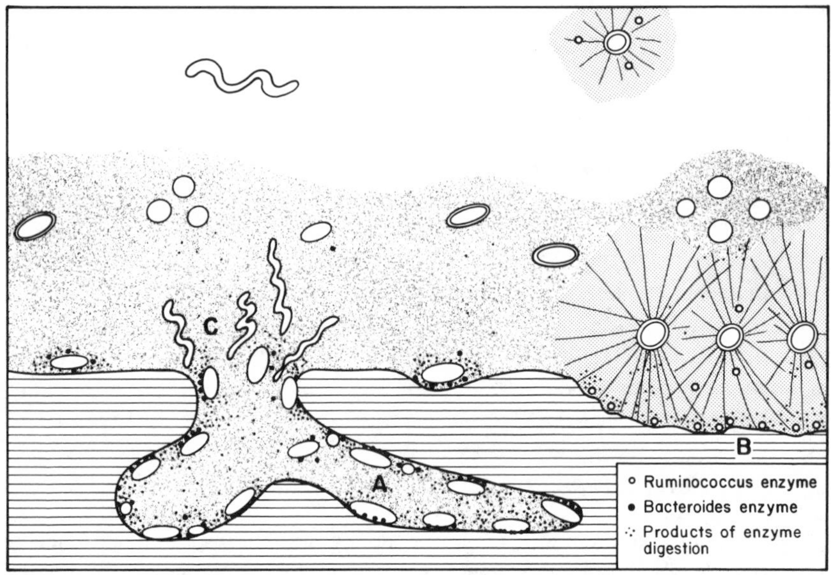

Figure 5 In the bacterial digestion of cellulose, *Ruminococcus* cells remain at some distance from the substrate surface while their enzymes initiate digestion (*B*). Other cellulolytic bacteria (e.g. *Bacteroides*) adhere intimately to the substrate (*A*) and detach vesicles and enzymes that cause a focal pitting attack. This attack is accelerated by the metabolic activity of noncellulolytic bacterial members (*C*) of structured consortia.

with fungi (116) to promote the development of functional microbial consortia (Figure 5). The products of cellulose digestion associate with the biofilm; these soluble nutrients are available to the cellulolytic bacteria themselves and to other heterotrophic organisms that may be attracted by chemotaxis and stimulated to divide and to form structured consortia. Cells of *Treponema bryantii* and *Butyrivibrio fibrisolvens* (Figure 5, *C*) are very commonly found in association with adherent cellulolytic cells of *Bacteroides succinogenes* (Figure 5, *A*); while these spirochaetes and curved rods have no cellulolytic enzymes, they enhance the cellulolytic activity of the *Bacteroides* species (>9%) when grown in mixed biofilms on cellulose fibers (58). These noncellulolytic bacteria may accelerate cellulose digestion by drawing products away from the consortium (Figure 5, *C*). Methane-producing bacteria combine with cellulolytic fungi to produce the most active in vitro cellulotytic consortium reported to date (116). Even though they are not intimately associated with the primary cellulolytic organisms in structured consortia, other heterotrophic bacteria growing in these mixed biofilms may subsist, directly or indirectly, on the soluble products of cellulose digestion and may contribute to the generation of the volatile fatty acids (VFAs) that are the main vehicle of nutrient transfer to the host animal. When the cellulose fibers are completely digested, all of the bacterial components of this digestive biofilm become, perforce, planktonic organisms that await the provision of similar nutrient substrates that will be specifically colonized and rapidly digested by a reconstituted digestive biofilm. Structurally complicated forage materials undergo a complex and sequential bacterial attack (20). This ecological process gradually accelerates as the bacterial populations of the rumen become adapted to new feeds, as when forage-fed range cattle are transferred to feedlots and fed barley concentrates. Some chemical agents (e.g. methyl cellulose) cause the complete dissociation of adherent cells of *B. succinogenes* from cellulose fibers (59, 83), while other agents (e.g. 3-phenyl propanoic acid) enhance adhesion (110). The specificity of the bacterial degradation of cellulose suggests that some manipulation of this important microbial activity may be practicable in the near future.

Natural and Protective Associations with Tissue Surfaces

Direct observation and quantitative recovery have been particularly successful in the definition of bacterial populations on tissue surfaces. In a systematic examination of the bacterial populations at 25 sites in the digestive tracts of more than 100 cattle, Cheng and his colleagues (21) observed many bacterial morphotypes growing as adherent biofilms on food materials and tissue surfaces. Morphological "keys" (e.g. details of cell wall structure) have been used to equate many of these morphotypes to isolates obtained by homogenization and quantitative microbiological examination of the same food materials and tissues. Three distinct bacterial communities have been located in this organ system (19). Small numbers of bacteria live preferentially in the rumen fluid; the largest bacterial population is cellulolytic and is associated with food materials (see the preceding section); and perhaps the most unique population forms an adherent biofilm (Figure 6) on the surface of the stratified squamous epithelium of the rumen wall. This latter bacterial community, which develops during the first three days of life, contains many facultative organisms that consume the oxygen that diffuses from the animal tissue (Figure 6, *A*); it thus protects the fastidious anaerobes within the rumen. Proteolytic bacteria within this tissue surface biofilm (Figure 6, *B*) digest the dead distal cells of the stratified tissue and recycle their cellular components to benefit the host animal. Many of the bacteria in this tissue surface biofilm produce urease, as detected in cultures and as visualized in situ by a newly developed histochemical technique (81). Urease has a vital role in the digestive physiology of the host animal; it converts the urea that diffuses through the rumen wall (Figure 6, *C*) to ammonia, which constitutes an essential nitrogen source for the bacteria within the rumen. The epithelial tissue does not produce urease (19) and thus depends on bacterial

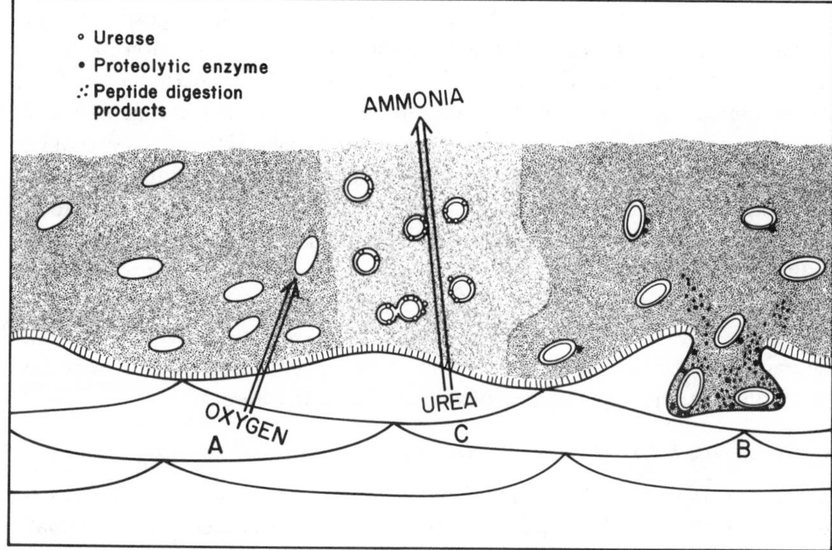

Figure 6 Component bacteria of the complex adherent biofilms found on the epithelial tissues of the bovine rumen facilitate the overall function of this organ by utilizing oxygen (*A*), digesting and recycling dead epithelial cells (*B*), and converting urea to ammonia (*C*).

cells within this specific tissue surface biofilm to produce an enzyme essential to the survival of the host animal. This is a clear demonstration of what we believe to be a general principle of microbial ecology: Stable tissue surface ecosystems attract and maintain adherent bacterial populations whose enzymatic activities are often integrated with those of the tissue itself.

Bacterial relationships with the surface of mucus-covered tissues such as the intestine are much more complex. Rozee et al (99), using new methods that allow the thick ($\sim$400 μm) mucus layer to be retained during preparation for electron microscopy, have visually confirmed Freter et al's (42) notion that most intestinal bacteria inhabit the mucus itself and that very few are actually attached to the tissue surface in normal animals. When the physical integrity of the mucus layer was disturbed by treatment with lectins (7) or by irradiation (114), autochthonous bacteria and protozoa that usually grow in this viscous phase proliferated and formed adherent biofilms on the tissue surface. Thus, mucus-covered tissues are not usually colonized by biofilms per se, but they are covered by dynamic viscous structures within which bacteria live and proliferate and only rarely associate directly with tissue surfaces. These vigorous mucus-associated microbial populations constitute a formidable ecological barrier that prevents the access of extraneous bacterial pathogens to their targets at the tissue surface (1). Direct observation and quantitative recovery have yielded very useful data regarding intestinal diseases in which mobile pathogens (e.g. *Campylobacter jejeuni*) must traverse the mucus layer to invade specific gut cells (38) and regarding ecological diseases in which antibiotic stress leads to an overgrowth of toxin-producing bacteria (e.g. *Clostridium difficile*). These modern techniques are especially useful when tampon-induced changes in the human vagina stimulate the growth of autochthonous staphylococci (36) and the location of these toxin-producing cells, in relation to epithelial tissue, is more important than their total numbers within the affected organ.

The autochthonous bacterial populations of several tissue surfaces have been described by modern ecological methods (19). Well developed bacterial biofilms have been seen on the epithelia of the distal human female urethra (73), on the tissue surfaces of the rabbit vagina (55), and on the epithelia of the human vagina and cervix (8). These adherent organisms have been characterized following quantitative recovery. The same methods have been used to define the adherent bacterial populations on tissue surfaces in surgically constructed structures such as the ileal urinary conduits that connect the ureters to the body surface following surgical removal of the bladder. Cheng et al (21) have suggested that well established tissue surface biofilms composed of autochthonous bacteria act as ecological barriers to upstream colonization by pathogenic organisms (18, 101). This concept is supported by data on bladder infections that follow the disturbance of distal-urethra biofilm populations by broad-spectrum antibiotic therapy. Similarly, the physical bypassing of the urethral surface biofilm population by urinary catheters (87), of the adherent cervical population by IUDs (71), and of the bile duct population by biliary catheters (A. G. Speer, P. B. Cotton & J. W. Costerton, manuscript submitted) leads to rapid bacterial invasion of the upstream organ in each system. Recent successes in the prevention of bladder infections in rats by the colonization of their distal urinary tract tissues with autochthonous strains of *Lactobacillus* (98) encourage some hope that we may be able to reinforce and manipulate these ecological barrier populations to prevent upstream infections. We are presently examining the natural extent of these protective tissue surface biofilms in organ systems that are colonized at their distal extremity but consistently sterile in their proximal organ (e.g. urethra–bladder; cervix–uterus; duodenum–gall bladder). A more complete understanding of the nature of the sustained boundaries of autochthonous bacterial colonization will enable us to predict what procedures and treatments are likely to produce failure of these ecological barriers and consequent upstream colonization by pathogens.

Pathogenic Associations with Tissue Surfaces

Sustained interest in bacterial pathogenic mechanisms throughout the last three decades has resulted in detailed accounts of pili (111) and cell wall proteins that promote adhesion to target tissues, and of toxins (107) that mediate pathogenic effects on host tissues. These elegant molecular mechanisms have appealed to our collective affection for order and simplicity; researchers have even developed model animal infections in which the genetic deletion of a single pathogenic mechanism renders the bacterial cell nonpathogenic. However, these attractive concepts have often fared poorly in the real world. A specific pathogenic mechanism, although real and fully operative, is frequently only one of many factors that facilitate bacterial pathogenicity (32, 84). Pathogenic bacteria often use more than one mechanism to mediate their attachment to target tissues (17), and most pathogens must persist and multiply in the infected system in order to exert deleterious effects on the host. The frequent encounters of healthy individuals with pathogenic bacteria actually lead to overt disease in a minute fraction of instances; the study of the entire etiological process from contact, through persistence, to infection reveals dozens of points (107) at which the pathogenic process may be aborted. In short, pathogenesis is an ecological process in which a particular bacterial species occasionally colonizes and persists in spite of all adverse environmental factors to produce a population sufficiently numerous, active, and well-located to exert a pathological effect upon the host.

Figure 7 is a hypothetical diagram that summarizes some of the postulated events of the early stages of the formation of pathogenic biofilms on tissue surfaces. The bacterial characteristic that is emphasized is the exceptional phenotypic plasticity that allows cells of a given species to change their cell surface characteristics radically in response to changing environmental factors (1, 12). Bacterial pili are avid and specific mediators of adhesion to receptors on target tissues (Figure 7, *A* and *C*). However, specific adhesion is not sufficient to establish colonization if the adherent cells lack the exopolysaccharide glycocalyx that affords protection from surfactants and other important host defense factors (Figure 7, *A*). Pathogenic bacteria often employ less specific and less avid exopolysaccharide-mediated adhesion mechanisms, which may act alone (Figure 7, *B*) or in concert with the pili (Figure 7, *C*). Exopolysaccharide-mediated adhesion is strong and resistant to shear forces (13), while pili are comparatively fragile structures (93). Although the occlusion of a target tissue surface by a confluent autochthonous bacterial biofilm (see the preceding section) would virtually preclude specific interactions between bacterial pili and tissue surface receptors, nonspecific glycocalyx-mediated adhesion could still be operative.

Once established on the tissue surface, the adherent pathogens must compete for iron with remarkably effective host siderophores. The expression of the genes controlling bacterial siderophore production (Figure 7, *D*) is a sine qua non of bacterial growth and persistence on the colonized tissue (108). Because phagocytic host cells are chemotactically attracted to complexes of

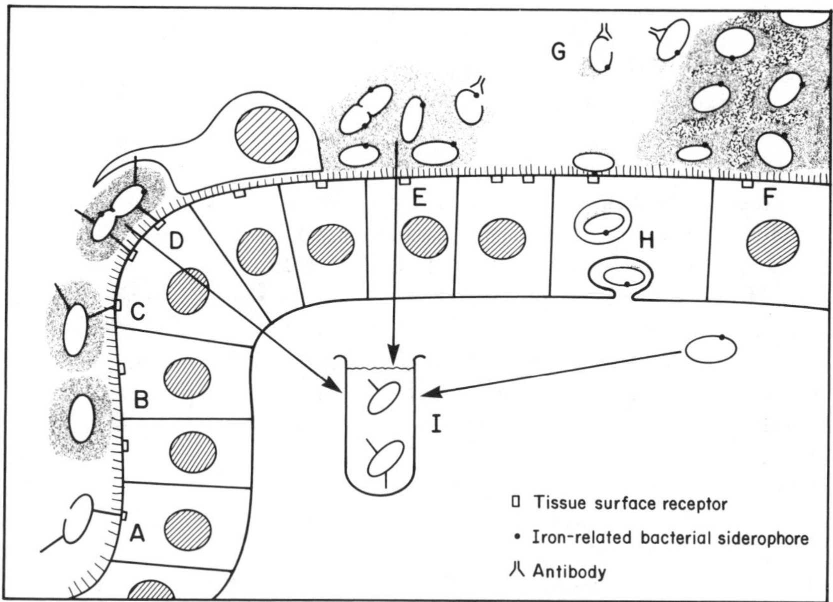

Figure 7 Diagrammatic representation of the variety of phenotypic forms bacteria may assume during the etiology *A-D* of acute (*E, G,* and *H*) or chronic (*F*) disease, contrasted with their phenotype when grown in nutrient-rich batch culture (*I*). See text for detailed explanation.

secretory IgA with the pili of many bacteria (1), bacterial survival may depend on the deletion of these structures from the bacterial surface (Figure 7, *E*). Development of glycocalyx-enclosed microcolonies (Figure 7, *E*) sufficiently large to withstand phagocytosis (52) would also contribute to bacterial persistence on the colonized tissue. We estimate that very few contacts with pathogenic bacteria proceed to this protected microcolony stage, but those organisms that do establish adherent "beachheads" can then proliferate to form a tissue surface biofilm (Figure 7, *F*). Bacteria within these protected tissue surface biofilms are not always overtly pathogenic (50), but their persistence and growth eventually result in massive unresolved bacterial masses resembling those seen in the lung in cystic fibrosis (62) and in the medullary cavity in osteomyelitis (78). Growth within protected glycocalyx-enclosed biofilms imposes several constraints on bacterial pathogenicity, in that bacterial cells and toxins are retained within the biofilm matrix or neutralized by antibodies or phagocytes at the biofilm surface; thus bacteremia and toxemia are rarely seen in these chronic diseases. Bacterial antigens at the biofilm surface stimulate the production of antibodies. These antibodies cannot penetrate the biofilm to resolve the infection, but they do form immune complexes at the biofilm surface (35) (Figure 7, *E* and *G*) and thus often damage the colonized tissue (103). When bacterial biofilms form on vascular surfaces, such as the endothelium of heart valves, they accrete blood components such as fibrin and platelets (102). The resultant vegetations (82) often grow sufficiently large to interfere with the mechanical functions of these organs. When the metabolic activities of bacteria within tissue surface biofilms produce insoluble salts, the biofilm matrices trap the resultant crystals, and "infection stones" develop in the affected organs. Bacterial biofilms form the structural matrix of the struvite calculi that develop in the kidney when urease-producing bacteria produce ammonia that is deposited as $MgNH_4PO_4$ (89). Similarly, the production of deconjugated cholesterol or calcium bilirubinate can lead to the development of thick occlusive biofilms in the biliary tract (A. G. Speer, P. B. Cotton & J. W. Costerton, manuscript submitted). The pivotal role of bacterial biofilms in these occlusive diseases results from the tendency of their exopolysaccharide matrices to trap and accrete fine particles that would otherwise move harmlessly through the system.

Like biofilm-covered biomaterials (Figure 4), the tissue surface biofilms formed in chronic infections are foci of humoral and cellular inflammatory reactions that are usually sufficient to kill individual bacterial cells released at their surfaces (Figure 7, *G*). However, host defense mechanisms sometimes fail, especially in debilitated patients, and disseminated single cells of these pathogenic species may escape to exert their full toxic or invasive potential (Figure 7, *H*). Because of phenotypic plasticity (12), a totally different phenotype is produced when bacterial cells are recovered from any etiological stage of the infection and grown in an iron-rich medium (Figure 7, *I*) in the absence of host defense factors.

RESISTANCE OF BIOFILM BACTERIA TO ANTIMICROBIAL AGENTS

The functional environments of individual bacterial cells growing within biofilms differ radically from those of planktonic cells in the same ecosystem, and even more radically from those of planktonic cells in batch culture in nutrient media (Figures 1–7). We must now accept the unequivocal evidence (12) that bacteria respond to changes in their environment by profound phenotypic variations in enzymatic activity (C. S. Dow, R. Whittenbury, D. Kelly, unpublished), cell wall composition (105), and surface structure (4). These phenotypic changes involve the target molecules for biocides, antibiotics, antibodies, and phagocytes, and also involve the external structures that control the access of these agents to the targets. Therefore, we must expect that biofilm bacteria will show some alterations in susceptibility to these antibacterial agents. Further changes in susceptibility to antibacterial agents are dictated by the structure of biofilms, within which the bacterial cells grow embedded in a thick, highly hydrated anionic matrix that conditions the environment of individual cells and constitutes a different solute phase from the bulk fluids of the system. Biofilms could therefore increase the concentration of a soluble antibacterial agent in the cellular environment by trapping and concentrating its molecules much as it traps and concentrates nutrients (Figure 2, *A*). On the other hand, the numerous charged binding sites on the matrix polymers and on the most distal cells in the biofilm may afford the innermost cells a large measure of protection from these agents. Peterson et al (95) have shown that aminoglycoside and peptide antibiotics are bound to the lipopolysaccharide molecules of *Pseudomonas aeruginosa* in a manner similar to that proposed here. We speculate that the dissociation constants of molecules bound within the biofilm are important because they dictate the concentrations of nutrients and of antibacterial agents available at the surface of individual cells.

An ecological examination of the whole aquatic system is a necessary first step in controlling bacterial problems in industry. Fouling, corrosion, and plugging all involve bacterial biofilms, but traditional monitoring procedures sample the much smaller planktonic populations that are intermittently shed from these adherent communities. Treatment with traditional concentrations of biocides kills these planktonic organisms and gives the appearance of success, but leaves the biofilm populations virtually unaffected (100). Thus millions of dollars have been wasted in ineffectual treatments. Costerton et al (27) and McCoy & Costerton (80) have developed a series of biofilm samplers that constitute parts of a pipe wall which are removable to yield an accurate biofilm sample. When biofilm sampling indicates an important problem, biocides are used at concentrations that completely kill the biofilm bacteria. These sampling and biocide procedures have been very useful in the oil recovery industry (27).

Direct sampling of biofilm populations is seldom possible in human disease, but the biofilms within the patient can be mimicked in vitro. When urine containing a uropathogenic strain of *P. aeruginosa* is passed through a modified Robbins device containing discs of catheter material (90, 91) the biomaterials develop a bacterial biofilm that closely resembles those observed on catheters recovered from patients infected by the same organisms (88). When planktonic cells within this system were treated with 50 $\mu g\ ml^{-1}$ of tobramycin, 8 hr of contact was sufficient to yield a complete kill, but 12-hr contact with 1000 $\mu g\ ml^{-1}$ of tobramycin did not kill the biofilm cells (90). When these resistant biofilms were washed free of tobramycin and dispersed to yield single cells, these essentially planktonic cells were susceptible to 50 $\mu g\ ml^{-1}$ of the antibiotic. This study has now been expanded to include other antibiotics and other organisms (J. C. Nickel & J. W. Costerton, unpublished data), and the inherent resistance of biofilm bacteria to antibiotics has been found to be a general phenomenon. A bacteremic patient with an endocardial pacemaker (70) presumed to have developed a *Staphylococcus aureus* biofilm on its biomaterial surfaces was treated for 6 wk with 12 g cloxacillin and 600 mg rifampin per day (72). When staphylococcal bacteremia recurred after antibiotic therapy was discontinued, the pacemaker was removed. It was covered by a thick biofilm (70) containing living cells of *S. aureus*. Chronic diseases in which the pathogenic bacteria grow in biofilms on tissue surfaces must be treated with high sustained doses of antibiotics to kill any of these protected organisms (77, 97). Complete resolution of these continuing bacterial foci is rarely achieved.

We conclude that bacteria within biofilms are generally much more resistant to biocides and antibiotics than their planktonic counterparts. We are presently examining this resistance to determine whether it derives from physiological changes in these sessile organisms or from the penetration barriers provided by the exopolysaccharide matrix and the distal cells of the biofilm itself. Recent successes in the prevention of biofilm development by the incorporation of biocides (86) or antibiotics (113) into biomaterials during polymerization have shown that antibacterial molecules leaching from a biomaterial surface can influence bacterial adhesion and subsequent biofilm development. While this preemptive strategy may be successful in preventing some of the infections associated with the short-term use of medical devices (86), control of infections during the long-term use of such devices will require an ecological understanding of the microbiology of the system and the judicious use of antibiotics in adequate doses.

When medical or industrial problems involve planktonic cells, as in acute bacteremic diseases or bacterial contamination in microchip production, antibacterial agents should be tested against planktonic cells in menstrua resembling those of the affected systems. Such tests routinely permit successful selection of antibiotics and biocides to solve these problems. Even when medical and industrial problems involve biofilm bacteria, tests against planktonic cells may be very useful; they indicate efficacy against planktonic cells that may detach from the biofilms, and negative planktonic results disqualify an agent for use against biofilm organisms. However, when the objective of treatment is the eradication of bacterial biofilms in either medical or industrial systems, antibacterial agents must be tested against biofilm bacteria. Tests against biofilm bacteria have now led to the identification of penetrating industrial biocides (100), but it is sobering that virtually all antibiotics in current use were selected for their efficacy against planktonic bacteria. The extension of biofilm test methods into the medical area is expected to identify penetrating antibiotics that may be useful in the treatment of the chronic biofilm-related diseases that are increasingly common in modern medical practice.

In vitro experiments have shown that encapsulated bacteria are less susceptible than their unprotected counterparts to the bactericidal and opsonic effects of specific antibodies (6) and to uptake by phagocytic cells (96, 104, 115). However, the most useful perceptions of the sensitivity of biofilm bacteria to antibodies and phagocytes are derived from studies of infections in animals and patients. Ward et al (K. Ward, M. R. W. Brown & J. W. Costerton, unpublished data) initiated growth of biofilm bacteria on biomaterial surfaces in the peritoneum of experimental animals and monitored the host animals' immune responses by crossed immunoelectrophoresis (XIE) and by immune blotting of SDS PAGE gels. They noted that large amounts of antibodies were produced against a small number of surface-located antigens, but that these antibodies failed to resolve the chronic infections. Similarly, large amounts of antibodies were produced against a limited number of bacterial antigens when cells of *P. aeruginosa* grew in biofilmlike masses in the lungs of rats in the chronic lung infection model (15) and in the lungs of cystic fibrosis patients (33). These antibodies did not control the further growth of biofilm bacteria within the lung, and they contributed to the formation of immune complexes (53), which profoundly damaged the infected tissue. Direct examination of biofilm-colonized peritoneal biomaterials (34) and of lung tissue in chronic *Pseudomonas* infections (15) showed that the phagocytic cells characteristic of the inflammatory response are mobilized and activated in response to the presence of these biofilm bacteria. However, the very prolonged course of the chronic infections (>20 yr in many cases) indicates a failure of the combined humoral and cellular defense mechanisms of infected hosts. Thus, while the molecular basis of the resistance of biofilm bacteria to clearance by antibodies and phagocytes remains obscure, it is unequivocally a general phenomenon that biofilms persist in spite of the vigorous immunological and inflammatory reactions of the infected host.

IMPACT OF THE BIOFILM CONCEPT ON EXPERIMENTAL DESIGN

The development of methods for the quantitative recovery of bacteria from biofilm populations was an important first step in the genesis of the biofilm concept (45). Because these biofilms are coherent and continuous structures, they can be removed by scraping with a sterile scalpel blade. Additional sessile cells can then be removed from the colonized surface by irrigation using a Pasteur pipette. The Robbins device (79, 90) provides disclike sample studs whose sides and back are sterile. The scraped material, the irrigation fluid, and the stud itself can be processed together by vortex mixing and gentle sonication (45, 90). Plating of the dispersed bacterial cells from fully disrupted natural biofilms usually yields a count about one logarithmic unit lower than the acridine orange direct count (AODC) of the same preparation

(43). We attribute this consistent discrepancy to the incomplete dispersal of biofilm fragments into single cells and to the presence of some dead bacteria that were either dead within the biofilm or killed during recovery and processing. Some bacteria remain on the colonized surface and some may be killed during dispersal by vortex mixing and sonication, so quantitative recovery data should be expressed as the minimum number of organisms present in the biofilm on each square centimeter of the colonized surface. New methods are now available for determining the viability [2-(*p*-iodophenyl)-3-(*p*-nitrophenyl)-5-phenyltetrazolium chloride (INT) technique (119)] and the metabolic activity [sessile heterotrophic potential (30, 60)] of bacteria within biofilms. The surfaces within a given system should be examined by several of these techniques to determine whether the majority of living cells within the biofilm are recovered by the quantitative recovery techniques used. Recovery media must often be adjusted to obtain isolates corresponding to the important morphological types (e.g. spirochetes) and physiological activities (e.g. urease activity) seen by direct observation of the biofilm in situ.

A preliminary ecological examination of an entire aquatic system is a necessary step in the design of a sampling program that will yield reliable and useful data. In environmental, industrial, and medical systems a simple preliminary statement of the approximate number and type (e.g. gram-positive rods) of bacterial cells on surfaces and in fluids throughout the system allows us to design a rational sampling program. If planktonic bacterial cells constitute the problem, planktonic samples are of paramount importance, and well designed sessile samples may permit detection of biofilms that periodically shed these free-floating cells. Conventional planktonic minimal inhibitory concentration (MIC) data will predict the concentration of antibiotics or biocides necessary to control these planktonic populations, and sessile MIC data (90) will predict the amount of these agents necessary to kill bacteria within the biofilms where they may have originated.

When a problem demonstrably involves biofilm bacteria, it is important that biofilm samples be obtained and that sessile MIC data be used to design biocide and antibiotic treatment strategies. When sessile samples cannot be obtained, planktonic bacteria from the affected system may be induced to produce biofilms outside the system, and potential control agents may be tested against these sessile organisms (90). The most important principle in these basically ecological studies of whole systems is that many different types of samples and many different types of data are useful, but we must not extrapolate from test tube–grown cells to individual planktonic cells or to biofilm cells in determining activities or sensitivity to antibacterial agents.

Because we acknowledge the phenomenal phenotypic plasticity of bacteria, we place an especially high value on direct observations or measurements of the activity of bacterial cells growing in the system of interest. Modern monoclonal antibody probes (16) allow us to detect the bacterial production of specific adhesins (17) or toxins on the infected tissue and thus to assess their role, if any, in the etiology of particular diseases. Alternatively, the immune system of an infected animal or patient constitutes a remarkably sensitive monitor of bacteria growing in affected tissues. When bacteria grow within protected biofilms in chronic bacterial infections, the host's immune system recognizes and produces antibodies against only a limited number of bacterial antigens (54, 63). Detailed XIE (63) or immunoblotting (4) analyses of sera from infected patients reveal which bacterial surface structures are produced in situ and sufficiently accessible to the immune system to stimulate the production of specific antibodies. The emergence of planktonic bacteria from protected biofilm foci can be detected by a sharp rise in the number of different antibodies produced (54), and these instances of bacterial dissemination can often be effectively treated with antibiotics. Special immunological methods can detect both the formation of immune complexes (35, 53, 117) and their resolution by natural increases in leukocyte elastase activity (37) or by therapy with immune suppressants (5). When the formation of immune complexes can be accurately monitored in individual patients, this devastating sequel of chronic bacterial infections due to the biofilm mode of growth may be effectively controlled by selective immune suppression (5). In many chronic biofilm diseases the causative organisms are difficult to recover because of sampling and microbiological problems, but experience leads us to expect the involvement of certain pathogenic bacteria (46). The use of XIE and immunoblotting techniques may permit immunologic identification of these pathogens by the detection of rising titers of antibodies against specific marker antigens characteristic of certain species (e.g. *Bacteroides fragilis*).

The biofilm concept has prompted the development of several new techniques, but perhaps its greatest impact in the area of methodology has been to force us to examine conventional sampling and analytical techniques to determine exactly what each can tell us about the coordinated functioning of whole microbial ecosystems in nature and in disease.

EPILOG

Because the breadth of this review has necessitated some generalizations, it may be useful to summarize major points that are supported by unequivocal evidence. Using methods of light and electron microscopy that can cause their loss but not their acquisition, bacterial biofilms have been found adhered to most surfaces in all except the most oligotrophic environments (22). Detailed examination of biofilms in several of these environments has shown a secondary level of organization in which bacteria form structured consortia with cells of other species (106) or with cells of host tissues (19). Within these consortia, instances of physiological cooperativity have been detected (19, 106). Biofilms have been found on the surfaces of biomaterials (31, 48, 70) and tissues (62, 78, 82) in chronic bacterial diseases characterized by their resistance to antibiotic chemotherapy and clearance by humoral or cellular host defense mechanisms (24). Bacterial biofilms constitute a hydrated viscous phase composed of cells and their exopolysaccharide matrices, within which molecules and ions may occur in concentrations different from those in the bulk fluid phase (29). Bacterial cells respond to changes in their immediate environments by a remarkable phenotypic plasticity (12) involving changes in their physiology (60; C. S. Dow, R. Whittenbury & D. Kelly, unpublished), their cell surface structure (10), and their resistance to antimicrobial agents (11, 90). Compelling evidence shows that bacteria growing in diseased tissue (3, 4, 105) differ from test tube–grown cells of the same organism in several important parameters, and extrapolation from in vitro data to actual bacterial ecosystems is now even more strongly contraindicated.

Techniques have been developed for the direct observation of biofilm bacteria (30) on inert surfaces and tissues, including mucus-covered epithelia (99), and for their quantitative recovery from these colonized surfaces (45, 90). New in situ techniques allow the detection of specific cell surface structures (14, 16), the measurement of general (30) and specific (81) metabolic activities, and the assessment of antibiotic sensitivity (90) in intact biofilms. The demonstrated ubiquity of bacterial biofilms, their established importance in many industrial and medical problems, and this new availability of methods for their detection and analysis should herald an exciting era in which microbiologists will come to recognize and perhaps counteract this basic bacterial strategy for growth and survival in nature and disease.

ACKNOWLEDGMENTS

We are most grateful to many colleagues who have helped us explore the many facets of the biofilm concept: S. D. Acers, J. G. Banwell, T. J. Beveridge, A. W. Bruce, R. C. Y. Chan, A. W. Chow, G. Döring, J. P. Fay, A. G. Gristina, L. Hagberg, N. Høiby, R. T. Irvin, M. Jacques, G. King, H. Kudo, J. Lam, D. W. Lambe, K.-J. Mayberry-Carson, W. F. McCoy, J. Mills, G. Reid, K. Rozee, K. Sadhu, P. Sullam, C. Svanborg-Eden, C. Whitfield, and R. C. Wyndham.

Bacterial biofilms have been defined by a very capable technical team including Cathy Barlow, Ivanka Ruseska, Francine Cusack, Wayne Jansen, Stacey Grant, Ian Lee, Ushi Sabharwal, Liz Middlemiss, Lee Watkins, Katherine Jackober, Barbara Stewart, and Jim Robbins and led by Joyce Nelligan and Kan Lam. We are especially grateful to Kan Lam and to Merle Olson for their tireless work in the design and execution of animal experiments, to M. R. W. Brown and Henry Eisenberg for their support of the biofilm concept during the dark days, and to Zell McGee for stimulating clinical insights into the effects of phenotypic plasticity on the etiology of specific diseases. R. G. E. Murray has been an inspiring example of unbridled scientific curiosity, vast enthusiasm for broad scientific concepts, and generosity to colleagues large and small.

Literature Cited

1. Abraham, S. N., Beachey, E. H. 1985. Host defences against adhesion of bacteria to mucosal surfaces. *Adv. Host Def. Mech.* 4:63–88
2. Amy, P. S., Morita, R. Y. 1984. Starvation-survival patterns of sixteen isolated open ocean bacteria. *Appl. Environ. Microbiol.* 45:1109–15
3. Anwar, H., Brown, M. R. W., Day, A., Weller, P. H. 1984. Outer membrane antigens of mucoid *Pseudomonas aeruginosa* isolated directly from the sputum of a cystic fibrosis patient. *FEMS Microbiol. Lett.* 24:235–39
4. Anwar, H., Shand, G. H., Ward, K. H., Brown, M. R. W., Alpar, K. E., Gowar, J. 1985. Antibody response to acute *Pseudomonas aeruginosa* infection in a burn wound. *FEMS Microbiol. Lett.* 29:225–30
5. Auerbach, H. S., Williams, M., Kirkpatrick, J. A., Colten, H. R. 1985. Alternate day prednisone reduces morbidity and improves pulmonary function in cystic fibrosis. *Lancet* 2:686–88
6. Baltimore, R. S., Mitchell, M. 1980. Immunologic investigations of mucoid strains of *Pseudomonas aeruginosa:* Comparison of susceptibility by opsonic antibody in mucoid and nonmucoid strains. *J. Infect. Dis.* 141:238–47
7. Banwell, J. G., Howard, R., Cooper, D., Costerton, J. W. 1985. Pathogenic sequellae of feeding phytohemagglutinin *(Phaseolus vulgaris)* lectins to conventional rats. *Appl. Environ. Microbiol.* 50:68–80
8. Bartlett, J. G., Moon, N. E., Goldstein, P. R., Goren, B., Onderdonk, A. B., Polk, B. F. 1978. Cervical and vaginal flora: ecological niches in the female lower genital tract. *Am. J. Obstet. Gynecol.* 130:658–61
9. Beveridge, T. J. 1981. Ultrastructure, chemistry and function of the bacterial wall. *Int. Rev. Cytol.* 72:229–317
10. Brook, I., Cole, R., Walker, R. I. 1986. Changes in the cell wall of *Clostridium* species following passage in animals. *Antonie van Leeuwenhoek J. Microbiol. Serol.* 52:273–80
11. Brown, M. R. W., Williams, P. 1985. Influence of substrate limitation and growth phase on sensitivity to antimicrobial agents. *J. Antimicrob. Chemother.* 15(A):7–14
12. Brown, M. R. W., Williams, P. 1985. The influence of environment on envelope properties affecting survival of bacteria in infections. *Ann. Rev. Microbiol.* 39:527–56
13. Bryers, J. D., Characklis, W. G. 1981. Early fouling biofilm formation in a turbulent flow system: overall kinetics. *Water Res.* 15:483–91
14. Caputy, G. G., Costerton, J. W. 1985. Immunological examination of the glycocalyces of *Staphylococcus aureus* strains Wiley and Smith. *Curr. Microbiol.* 11:297–302
15. Cash, H. A., Woods, D. E., McCullough, B., Johanson, W. G., Bass, A. G. 1979. A rat model of chronic respiratory infection with *Pseudomonas aeruginosa*. *Am. Rev. Respir. Dis.* 119:453–59
16. Chan, R., Acres, S. D., Costerton, J. W. 1982. The use of specific antibody to demonstrate glycocalyx, K99 pili, and the spatial relationships of K99^{+} enterotoxigenic *Escherichia coli* in the ileum of colostrum fed calves. *Infect. Immun.* 37:1170–80
17. Chan, R., Lian, C. J., Costerton, J. W., Acres, S. D. 1983. The use of specific antibodies to demonstrate the glycocalyx and spatial relationships of K99^{-}, F41^{-}, entertoxigenic strains of *Escherichia coli* colonizing the ileum of colostrum-deprived calves. *Can. J. Comp. Med.* 47:150–56
18. Chan, R. C. Y., Reid, G., Irvin, R. T., Bruce, A. W., Costerton, J. W. 1985. Competitive exclusion of uropathogens from human uroepithelial cells by lactobacillus whole cells and cell wall fragments. *Infect. Immun.* 47:84–89
19. Cheng, K.-J., Costerton, J. W. 1980. Adherent rumen bacteria—their role in the digestion of plant material, urea and dead epithelial cells. In *Digestion and Metabolism in the Ruminant,* ed. Y. Ruckebush, P. Thivend, pp. 227–50. Lancaster, UK: MTP

20. Cheng, K.-J., Fay, J. P., Howarth, R. E., Costerton, J. W. 1980. Sequence of events in the digestion of fresh legume leaves by rumen bacteria. *Appl. Environ. Microbiol.* 40:613–25
21. Cheng, K.-J., Irvin, R. T., Costerton, J. W. 1981. Autochthonous and pathogenic colonization of animal tissues by bacteria. *Can. J. Microbiol.* 47:461–90
22. Costerton, J. W. 1979. The role of electron microscopy in the elucidation of bacterial structure and function. *Ann. Rev. Microbiol.* 33:459–79
23. Costerton, J. W. 1983. The removal of microbial fouling by ice nucleation. *US Patent No. 4,419,248*
24. Costerton, J. W. 1984. The formation of biocide-resistant biofilms in industrial, natural and medical systems. *Dev. Ind. Microbiol.* 25:363–72
25. Costerton, J. W. 1984. The etiology and persistence of cryptic bacterial infections: a hypothesis. *Rev. Infect. Dis.* 6:S608–16
26. Costerton, J. W. 1985. Biomedical devices containing isothiazalones to control bacterial growth. *US Patent No. 4,542,169*
27. Costerton, J. W., Geesey, G. G., Jones, P. A. 1987. Bacterial biofilms in relation to internal corrosion, monitoring, and biocide strategies. *Corrosion 87, San Francisco,* pp. 1–9. Houston: North Am. Conf. Eng.
28. Costerton, J. W., Irvin, R. T., Cheng, K.-J. 1981. The bacterial glycocalyx in nature and disease. *Ann. Rev. Microbiol.* 35:299–324
29. Costerton, J. W., Marrie, T. J. 1983. The role of the bacterial glycocalyx in resistance to antimicrobial agents. In *Role of the Envelope in the Survival of Bacteria in Infection,* ed. C. S. F. Easmon, J. Jeljaszewicz, M. R. W. Brown, P. A. Lambert, pp. 63–85. London/New York: Academic
30. Costerton, J. W., Nickel, J. C., Ladd, T. I. 1986. Suitable methods for the comparative study of free-living and surface-associated bacterial populations. In *Bacteria in Nature,* ed. J. S. Poindexter, E. R. Leadbetter, pp. 49–84. New York: Plenum
31. Costerton, J. W., Nickel, J. C., Marrie, T. J. 1985. The role of the bacterial glycocalyx and of the biofilm mode of growth in bacterial pathogenesis. *Roche Sem. Bact.* 2:1–25
32. Darfeuille-Michaud, A., Forestier, C., Joly, B., Cluzel, R. 1986. Identification of a nonfimbrial adhesive factor of an enterotoxigenic *Escherichia coli* strain. *Infect. Immun.* 52:468–75
33. Dasgupta, M. K., Lam, J., Döring, G., Harley, F. L., Zubbenbuhler, P., et al. 1987. Prognostic implications of circulating immune complexes and *Pseudomonas aeruginosa*—specific antibodies in cystic fibrosis. *J. Clin. Lab. Invest.* 23:25–30
34. Dasgupta, M. K., Ulan, R. A., Bettcher, K. B., Burns, V., Lam, K., et al. 1986. Effect of exit site infection and peritonitis on the distribution of biofilm encased bacterial microcolonies on Tenckhoff catheters in patients undergoing continuous ambulatory peritoneal dialysis. In *Advances in Continuous Ambulatory Peritoneal Dialysis,* ed. R. Khanna, K. D. Nolph, B. Prowant, Z. J. Twardowski, D. G. Oreopoulos, pp. 102–9. Toronto: Univ. Toronto Press
35. Dasgupta, M. K., Zubbenbuhler, P., Abbi, A., Harley, F. C., Brown, N. E., et al. 1986. Combined evaluation of circulating immune complexes and antibodies to *Pseudomonas aeruginosa* as an immunologic profile in relation to pulmonary function in cystic fibrosis. *J. Clin. Immunol.* 7:51–58
36. Davis, J. P., Chesney, P. J., Wand, P. J., LaVenture, M. 1980. Toxic shock syndrome: epidemiological features, recurrence, risk factors and presentation. *N. Engl. J. Med.* 303:1429–35
37. Döring, G., Goldstein, W., Botzenhart, K., Kharazmi, A., Høiby, N., et al. 1986. Elastase from polymorphonuclear leukocytes—a regulatory enzyme in immune complex disease. *Clin. Exp. Immunol.* 64:597–605
38. Field, L. H., Underwood, J. L., Berry, L. J. 1984. The role of gut flora and animal passage in the colonization of adult mice with *Campylobacter jejuni. J. Med. Microbiol.* 17:59–66
39. Fletcher, M. 1980. The question of passive versus active attachment mechanisms in non-specific bacterial adhesion. In *Microbial Adhesion to Surfaces,* ed. R. C. W. Berkeley, J. M. Lynch, J. Melling, P. R. Rutter, B. Vincent, pp. 197–210. Chichester, UK: Horwood
40. Fletcher, M., Loeb, G. I. 1979. Influence of substratum characteristics on the attachment of a marine pseudomonad to solid surfaces. *Appl. Environ. Microbiol.* 37:67–72
41. Forsberg, C. W., Beveridge, T. J., Hellstrom, A. 1981. Cellulase and xylanase release from *Bacteroides succinogenes* and its importance in the rumen environment. *Appl. Environ. Microbiol.* 42: 886–96
42. Freter, R., O'Brien, P. C. M., Macsai, M. S. 1981. Role of chemotaxis in the

association of motile bacteria with intestinal mucosa: *In vivo* studies. *Infect. Immun.* 34:234–40
43. Geesey, G. G., Costerton, J. W. 1979. Microbiology of a northern river: bacterial distribution and relationship to suspended sediment and organic carbon. *Can. J. Microbiol.* 25:1058–62
44. Geesey, G. G., Iwaoka, T., Griffiths, P. R. 1987. Characterization of interfacial phenomena occurring during exposure of a thin copper film to an aqueous suspension of an acidic polysaccharide. *J. Colloid Interface Sci.* In press
45. Geesey, G. G., Mutch, R., Costerton, J. W., Green, R. B. 1978. Sessile bacteria: an important component of the microbial population in small mountain streams. *Limnol. Oceanogr.* 23:1214–23
46. Gorbach, S. L., Bartlett, J. G. 1974. Anaerobic infections. *N. Engl. J. Med.* 290:1177–84
47. Govan, J. R. W. 1975. Mucoid strains of *Pseudomonas aeruginosa:* the influence of culture medium on the stability of mucus production. *J. Med. Microbiol.* 8:513–22
48. Gristina, A. G., Costerton, J. W. 1984. Bacteria-laden biofilms—a hazard to orthopedic prostheses. *Infect. Surg.* 3:655–62
49. Gristina, A. G., Costerton, J. W. 1985. Bacterial adherence to biomaterials and tissue. *J. Bone Jt. Surg.* 67-A:264–73
50. Gristina, A. G., Price, J. L., Hobgood, C. D., Webb, L. X., Costerton, J. W. 1985. Bacterial colonization of percutaneous sutures. *Surgery* 98:12–19
51. Groat, R. G., Schultz, J. E., Zychlinsky, E., Bockman, A., Matin, A. 1986. Starvation proteins in *Escherichia coli:* Kinetics of synthesis and role in starvation survival. *J. Bacteriol.* 168:486–93
52. Hagberg, L., Lam, J., Svanborg-Edén, C., Costerton, J. W. 1986. Interaction of a pyelonephritogenic *Escherichia coli* strain with the tissue components of the mouse urinary tract. *J. Urol.* 136:165–72
53. Høiby, N., Döring, G., Schiøtz, P. O. 1986. The role of immune complexes in the pathogenesis of bacterial infections. *Ann. Rev. Microbiol.* 40:29–53
54. Høiby, N., Flensborg, E. W., Beck, B., Friis, B., Jacobsen, S. V., Jacobsen, L. 1977. *Pseudomonas aeruginosa* infections in cystic fibrosis. Diagnosis and prognostic significance of *Pseudomonas aeruginosa* precipitins determined by means of immunoelectrophoresis. *Scand. J. Respir. Dis.* 58:65–79
55. Jacques, M., Olson, M. E., Crichlow, A. M., Osborne, A. D., Costerton, J. W. 1986. The normal microflora of the female rabbit's genital tract. *Can. J. Comp. Med.* 50:272–74
56. Kjelleberg, S., Hermansson, M. 1984. Starvation-induced effects on bacterial surface characteristics. *Appl. Environ. Microbiol.* 48:497–503
57. Kluge, R. M. 1982. Infections of prosthetic cardiac valves and arterial grafts. *Heart Lung* 11:146–51
58. Kudo, H., Cheng, K.-J., Costerton, J. W. 1987. Interactions between *Treponema bryantii* and cellulolytic bacteria in the *in vitro* degradation of straw cellulose. *Can. J. Microbiol.* 33:244–48
59. Kudo, H., Cheng, K.-J., Costerton, J. W. 1987. Electron microscopic study of the methyl cellulose–mediated detachment of cellulolytic rumen bacteria from cellulose fibers. *Can. J. Microbiol.* 33:267–72
60. Ladd, T. I., Costerton, J. W., Geesey, G. G. 1979. Determination of the heterotrophic activity of epilithic microbial populations. In *Native Aquatic Bacteria: Enumeration, Activity, and Ecology. ASTM Tech. Publ. No. 695,* ed. J. W. Costerton, R. R. Colwell, pp. 180–95. Philadelphia: *Am. Soc. Test. Mater.*
61. Ladd, T. I., Ventullo, R. M., Wallis, P. M., Costerton, J. W. 1982. Heterotrophic activity and biodegradation of labile and refractory compounds by ground water and stream microbial populations. *Appl. Environ. Microbiol.* 44:321–29
62. Lam, J. S., Chan, R., Lam, K., Costerton, J. W. 1980. The production of mucoid microcolonies by *Pseudomonas aeruginosa* within infected lungs in cystic fibrosis. *Infect. Immun.* 28:546–56
63. Lam, J. S., Mutharia, L. M., Hancock, R. E. W., Høiby, N., Lam, K., et al. 1983. Immunogenicity of *Pseudomonas aeruginosa* outer membrane antigens examined by crossed immunoelectrophoresis. *Infect. Immun.* 42:88–98
64. Little, B., Wagner, P., Gerchakov, S. M., Walsh, M., Mitchell, R. 1987. Involvement of a thermophilic bacterium in corrosion processes. *Corrosion* 42: 533–36
65. Lock, M. A., Wallace, R. R., Costerton, J. W., Ventullo, R. M., Charlton, S. E. 1984. River epilithon: towards a structural-functional model. *Oikos* 44:10–22
66. Lorian, V., Atkinson, B., Waluschka, A., Kim, Y. 1982. Ultrastructure, in vitro and in vivo, of staphylococci exposed to antibiotics. *Curr. Microbiol.* 7:301–4
67. Lorian, V., Zak, O., Suter, J.,

Bruecher, C. 1985. Staphylococci, in vitro and in vivo. *Diagn. Microbiol. Infect. Dis.* 3:433–44
68. Maki, D. G., Weise, C. E., Sarafin, H. W. 1977. A semi-quantitative method of identifying intravenous catheter-related infection. *N. Engl. J. Med.* 296:1305–9
69. Malone, J. S., Caldwell, D. E. 1983. Evaluation of surface colonization kinetics in continuous culture. *Microb. Ecol.* 9:299–306
70. Marrie, T. J., Costerton, J. W. 1982. A scanning and transmission electron microscopic study of an infected endocardial pacemaker lead. Circulation 66:1339–43
71. Marrie, T. J., Costerton, J. W. 1983. A scanning and transmission electron microscopic study of the surfaces of intrauterine contraceptive devices. *Am. J. Obstet. Gynecol.* 146:384–94
72. Marrie, T. J., Costerton, J. W. 1984. Morphology of bacterial attachment to cardiac pacemaker leads and power packs. *J. Clin. Microbiol.* 19:911–14
73. Marrie, T. J., Harding, G. K. M., Ronald, A. R., Dikkema, J., Lam, J., et al. 1979. Influence of mucoidy on antibody coating of *Pseudomonas aeruginosa*. *J. Infect. Dis.* 139:357–61
74. Marrie, T. J., Noble, M. A., Costerton, J. W. 1983. Examination of the morphology of bacteria adhering to intraperitoneal dialysis catheters by scanning and transmission electron microscopy. *J. Clin. Microbiol.* 18:1388–98
75. Marshall, K. C. 1985. Mechanisms of bacterial adhesion to solid-water interfaces. In *Bacterial Adhesion, Mechanisms and Physiological Significance*, ed. D. C. Savage, M. Fletcher, pp. 133–61. New York: Plenum
76. Marshall, K. C., Stout, R., Mitchell, R. 1971. Mechanisms of the initial events in the sorption of marine bacteria to surfaces. *J. Gen. Microbiol.* 68:337–48
77. Mayberry-Carson, K. J., Tober-Meyer, B., Lambe, D. W. Jr., Costerton, J. W. 1986. An electron microscopic study of the effect of clindamycin therapy on bacterial adherence and glycocalyx formation in experimental *Staphylococcus aureus* osteomyelitis. *Microbios* 48:189–206
78. Mayberry-Carson, K. J., Tober-Meyer, B., Smith, J. K., Lambe, D. W., Costerton, J. W. 1984. Bacterial adherence and glycocalyx formation in osteomyelitis experimentally induced with *Staphylococcus aureus*. *Infect. Immun.* 43: 825–33
79. McCoy, W. F., Bryers, J. D., Robbins, J., Costerton, J. W. 1981. Observations in fouling biofilm formation. *Can. J. Microbiol.* 27:910–17
80. McCoy, W. F., Costerton, J. W. 1982. Growth of *Spaerothilus natans* in a tubular reactor system. *Appl. Environ. Microbiol.* 43:1490–94
81. McLean, R. J. C., Cheng, K.-J., Gould, W. D., Costerton, J. W. 1985. Cytochemical localization of urease in a rumen *Staphylococcus* sp. by electron microscopy. *Appl. Environ. Microbiol.* 49: 253–55
82. Mills, J., Pulliam, L., Dall, D., Marzouk, J., Wilson, W., Costerton, J. W. 1984. Exopolysaccharide production by viridans streptococci in experimental endocarditis. *Infect. Immun.* 43:359–67
83. Minato, H., Suto, T. 1978. Technique for fractionation of bacteria in rumen microbial ecosystem. II. Attachment of bacteria isolated from bovine rumen to cellulose powder *in vitro* and elution of attached bacteria therefrom. *J. Gen. Appl. Microbiol.* 24:1–16
84. Nagy, B., Moon, H. W., Isaacson, R. E. 1976. Colonization of porcine small intestine by *Escherichia coli:* ileal colonization and adhesion by pig enteropathogens that lack K88 antigen and by some acapsular mutants. *Infect. Immun.* 13:1214–20
85. Nickel, J. C., Grant, S. K., Costerton, J. W. 1985. Catheter-associated bacteriuria: an experimental study. *Urology* 26:369–75
86. Nickel, J. C., Grant, S. K., Lam, K., Olson, M. E., Costerton, J. W. 1987. Evaluation in a bacteriologically-stressed animal model of a new closed catheter drainage system incorporating a microbicidal outlet tube. *Urology* In press
87. Nickel, J. C., Gristina, A. G., Costerton, J. W. 1985. Electron microscopic study of an infected Foley catheter. *Can. J. Surg.* 28:50–54
88. Nickel, J. C., Heaton, J., Morales, A., Costerton, J. W. 1986. Bacterial biofilm in persistent penile prosthesis–associated infection. *J. Urol.* 135:586–88
89. Nickel, J. C., Olson, M. E., McLean, R. J. C., Grant, S. K., Costerton, J. W. 1987. An ecological study of infected urinary stone genesis in an animal model. *Br. J. Urol.* 59:1–10
90. Nickel, J. C., Ruseska, I., Costerton, J. W. 1985. Tobramycin resistance of cells of *Pseudomonas aeruginosa* growing as a biofilm on urinary catheter material. *Antimicrob. Agents Chemother.* 27:619–24
91. Nickel, J. C., Ruseska, I., Whitfield, C., Marrie, T. J., Costerton, J. W.

1985. Antibiotic resistance of *Pseudomonas aeruginosa* colonizing a urinary catheter *in vivo*. *Eur. J. Clin. Microbiol.* 4:213–18
92. Obuekwe, C. O., Westlake, D. W. S., Cook, F. D., Costerton, J. W. 1981. Surface changes in mild steel coupons from the action of corrosion-causing bacteria. *Appl. Environ. Microbiol.* 41:766–74
93. Pearce, W. A., Buchanan, T. M. 1980. Structures and cell membrane–binding properties of bacterial fimbriae. In *Receptors and Recognition,* Ser. B, Vol. 6, *Bacterial Adherence,* ed. E. H. Beachey, pp. 289–344. New York: Chapman & Hall
94. Peters, G., Locci, R., Pulverer, G. 1981. Microbial colonization of prosthetic devices. II. Scanning electron microscopy of naturally infected intravenous catheters. *Zentralbl. Bakteriol. Mikrobiol. Hyg. I Abt. Orig. B* 173:293–99
95. Peterson, A. A., Hancock, R. E. W., McGroarty, E. J. 1985. Binding of polycationic antibiotics and polyamines to lipopolysaccharides of *Pseudomonas aeruginosa*. *J. Bacteriol.* 164:1256–61
96. Peterson, P. K., Wilkinson, B. J., Kim, Y., Schmeling, D., Quie, P. G. 1978. Influence of encapsulation on staphylococcal opsonization and phagocytosis by human polymorphonuclear leukocytes. *Infect. Immun.* 19:943–49
97. Rabin, H. R., Harley, F. L., Bryan, L. E., Elfring, G. L. 1980. Evaluation of a high dose tobramycin and ticarcillin treatment protocol in cystic fibrosis based on improved susceptibility criteria and antibiotic pharmacokinetics. In *Perspectives in Cystic Fibrosis,* ed. J. M. Sturgess, pp. 370–75. Toronto: Can. Cystic Fibrosis Found.
98. Reid, G., Chan, R. C. Y., Bruce, A. W., Costerton, J. W. 1985. Prevention of urinary tract infection in rats with an indigenous *Lactobacillus casei* strain. *Infect. Immun.* 49:320–24
99. Rozee, K. R., Cooper, D., Lam, K., Costerton, J. W. 1982. Microbial flora of the mouse ileum mucous layer and epithelial surface. *Appl. Environ. Microbiol.* 43:1451–63
100. Ruseska, I., Robbins, J., Lashen, E. S., Costerton, J. W. 1982. Biocide testing against corrosion-causing oilfield bacteria helps control plugging. *Oil Gas J.* 1982:253–64
101. Saigh, J. H., Sanders, C. C., Sanders, W. E. 1978. Inhibition of *Neisseria gonorrhoeae* by aerobic and facultative anaerobic components of the endocervical flora: Evidence for a protective affect against infection. *Infect. Immun.* 19:704–10
102. Scheld, W. M., Vaone, J. A., Sande, M. A. 1978. Bacterial adherence in the pathogenesis of endocarditis. Interaction of bacterial dextran, platelets, and fibrin. *J. Clin. Invest.* 61:1394–404
103. Schiøtz, P. O., Jørgenson, M., Flensborg, E. W., Faerø, O., Husby, S., et al. 1983. Chronic *Pseudomonas aeruginosa* lung infection in cystic fibrosis. *Acta Paediatr. Scand.* 72:283–87
104. Schwarzmann, S., Boring, J. R. III. 1971. Antiphagocytic effect of slime from a mucoid strain of *Pseudomonas aeruginosa*. *Infect. Immun.* 3:762–67
105. Shand, G. H., Anwar, H., Kadurugamuwa, J., Brown, M. R. W., Silverman, S. H., Melling, J. 1985. *In vivo* evidence that bacteria in urinary tract infections grow under iron-restricted conditions. *Infect. Immun.* 48:35–39
106. Shelton, D. R., Tiedje, J. M. 1984. Isolation and partial characterization of bacteria in an anaerobic consortium that mineralizes 3-chlorobenzoic acid. *Appl. Environ. Microbiol.* 48:840–48
107. Smith, H. 1977. Microbial surfaces in relation to pathogenicity. *Bacteriol. Rev.* 41:475–500
108. Sokol, P. A., Woods, D. E. 1984. Relationship of iron and extracellular virulence factors to *Pseudomonas aeruginosa* lung infections. *J. Med. Microbiol.* 18:125–33
109. Stack, R. J., Cotta, M. A. 1986. Effect of 3-phenylpropanoic acid on growth of and cellulose utilization by cellulolytic ruminal bacteria. *Appl. Environ. Microbiol.* 52:209–10
110. Sutherland, I. W. 1977. Bacterial exopolysaccharides—their nature and production. In *Surface Carbohydrates of the Prokaryotic Cell,* ed. I. W. Sutherland, pp. 27–96. London: Academic
111. Svanborg-Edén, C., Hansson, H. A. 1978. *Escherichia coli* pili as possible mediators of attachment to human urinary tract epithelial cells. *Infect. Immun.* 21:229–37
112. Tenney, J. H., Moody, M. R., Newman, K. A., Schimpff, S. C., Wade, J. C., et al. 1986. Adherent microorganisms on lumenal surfaces of long-term intravenous catheters: Importance of *Staphylococcus epidermidis* in patients with cancer. *Arch. Intern. Med.* 146:1949–54
113. Trooskin, S. Z., Donetz, A. P., Harvey, R. A., Greco, R. S. 1985. Prevention of catheter sepsis by antibiotic bonding. *Surgery* 97:547–51

114. Walker, R. I., Brook, I., Costerton, J. W., MacVittie, T., Mybul, M. L. 1986. Possible association of mucous blanket integrity with postirradiation colonization resistance. *Radiat. Res.* 104:346–57
115. Whitnak, E., Bisno, A. L., Beachey, E. H. 1981. Hyaluronate capsule prevents attachment of Group A streptococci to mouse peritoneal macrophages. *Infect. Immun.* 31:985–91
116. Wood, T. M., Wilson, C. A., McCrae, S. I., Joblin, K. N. 1986. A highly active extracellular cellulase from the anaerobic rumen fungus *Neocallimastix frontalis*. *FEMS Microbiol. Lett.* 34:37–40
117. Woods, D. E., Bryan, L. E. 1985. Studies on the ability of alginate to act as a protective immunogen against *Pseudomonas aeruginosa* in animals. *J. Infect. Dis.* 151:581–88
118. Zak, O., Sande, M. A. 1982. Correlation of *in vitro* activity of antibiotics with results of treatment in experimental animal models and human infection. In *Action of Antibiotics in Patients*, ed. L. D. Sabath, pp. 55–67. Berne, Switzerland: Huber
119. Zimmermann, R. R., Iturriagu, R., Becker-Birck, J. 1978. Simultaneous determination of the total number of aquatic bacteria and the number thereof involved in respiration. *Appl. Environ. Microbiol.* 36:926–35
120. Zobell, C. E. 1943. The effect of solid surfaces on bacterial activity. *J. Bacteriol.* 46:39–56

Ann. Rev. Microbiol. 1987. 41:465–505

ENZYMATIC "COMBUSTION": THE MICROBIAL DEGRADATION OF LIGNIN[1,2]

T. Kent Kirk

Forest Products Laboratory, Forest Service, United States Department of Agriculture, One Gifford Pinchot Drive, Madison, Wisconsin 53705

Roberta L. Farrell

Repligen Corporation, One Kendall Square, Building 700, Cambridge, Massachusetts 02139

CONTENTS

INTRODUCTION

Lignin is the most abundant renewable aromatic material on earth. Growing evidence indicates that the complex plant polymer is biodegraded by a unique

[1]Dedicated to the memory of Professor Erich Adler.

[2]The US Government has the right to retain a nonexclusive, royalty-free license in and to any copyright covering this paper.

enzymatic "combustion," i.e. a nonspecific enzyme-catalyzed burning. Lignin is degraded by a narrower array of microbes than the other major biopolymers. Lignin biodegradation is central to the earth's carbon cycle because lignin is second only to cellulose in abundance and, perhaps more significantly, because lignin physically protects most of the world's cellulose and hemicelluloses from enzymatic hydrolysis.

Research on lignin biodegradation has accelerated greatly during the past 10 years, mainly because of the substantial potential applications of bioligninolytic systems in pulping, bleaching, converting lignins to useful products, and treating wastes (52, 58, 59, 103, 126, 130, 158, 270). Rapid progress in lignin biodegradation research is reflected in the number of reviews published in the last 5–6 years on the subject in general (25a, 34, 37, 110, 111, 126, 154, 179) or selected aspects of it (26, 103, 113, 114, 146, 155, 162, 168, 274). In addition, two international meetings have focused on lignin biodegradation (115, 163).

Our review encapsulates the major findings of the past 5–6 years, emphasizing the research that immediately preceded and followed the discovery in 1982 of the first lignin-degrading enzyme. That discovery projected the field into the realm of biochemistry and molecular biology and opened up new prospects for application.

LIGNIN AS A SUBSTRATE

Lignin is found in higher plants, including ferns, but not in liverworts, mosses, or plants of lower taxonomic ranking. Wood and other vascular tissues generally are 20–30% lignin. Most lignin is found within the cell walls, where it is intimately interspersed with the hemicelluloses, forming a matrix that surrounds the orderly cellulose microfibrils. In wood, lignin in high concentration is the glue that binds contiguous cells, forming the middle lamella.

Biosynthetically, lignin arises from three precursor alcohols: *p*-hydroxycinnamyl (coumaryl) alcohol, which gives rise to *p*-hydroxyphenyl units in the polymer; 4-hydroxy-3-methoxycinnamyl (coniferyl) alcohol, the guaiacyl units; and 3,5-dimethoxy-4-hydroxycinnamyl (sinapyl) alcohol, the syringyl units. Free radical copolymerization of these alcohols produces the heterogeneous, optically inactive, cross-linked, and highly polydisperse polymer. Most gymnosperm lignins contain primarily guaiacyl units. Angiosperm lignins contain approximately equal amounts of guaiacyl and syringyl units. Both types of lignin generally contain only small amounts of *p*-hydroxyphenyl units. For convenience we use the term "lignin" for the family of related polymers.

Figure 1 is a schematic formula for an angiosperm lignin, depicting several important features; note the nomenclature of the carbons (Unit 1). Over 10 interphenylpropane linkage types occur, including four that predominate. The schematic illustrates only the major linkages; the dominant linkage (>50%) is the β-O-4 type, seen between units 1 and 2, 2 and 3, 4 and 5, 6 and 7, 7 and 8, and 13 and 14. In the polymerization process, secondary reactions lead to cross-linking between lignin and hemicelluloses (Unit 5 in Figure 1). In addition, the lignins of grasses and certain woods contain aromatic or cinnamic acids esterified through side-chain hydroxyl groups of lignin (Unit 12 in Figure 1). The major features of lignin structure are now well understood (2, 61, 70, 240); although additional details are still being clarified, it is not likely that they will be of consequence to biodegradation research.

The structural features of lignin dictate unusual constraints on biodegradative systems responsible for initial attack: They must be extracellular, nonspecific, and nonhydrolytic. Analogies with other biopolymer-degrading systems, which are hydrolytic and specific, and with the intracellular systems

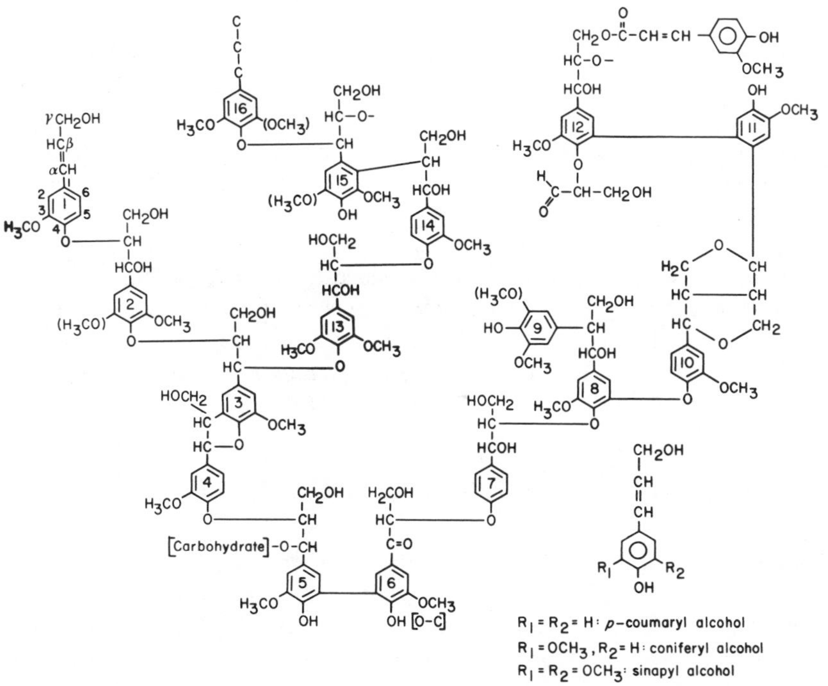

$R_1 = R_2 = H$: *p*-coumaryl alcohol
$R_1 = OCH_3, R_2 = H$: coniferyl alcohol
$R_1 = R_2 = OCH_3$: sinapyl alcohol

Figure 1 Schematic structural formula for lignin, adapted from Adler (2). The structure illustrates major interunit linkages and other features described in the text; it is not a quantitatively accurate depiction of the various substructures. The three precursor alcohols are shown at the *lower right;* their polymerization, following one-electron oxidation, produces lignin.

that degrade low–molecular weight aromatics cannot readily be drawn. We define lignin as a 600–1000-kd molecule, i.e. too big to enter cells. We have chosen not to treat the biodegradation of low–molecular weight lignin-related compounds, except where these compounds have been used as models to elucidate specific reactions in the polymer.

The structural features of lignin also cause problems in methodology, mainly in quantitative determination. Research has been greatly aided by the introduction of ^{14}C-labeled lignins, produced either in vivo (36, 38, 95) or in vitro (95, 157, 159). Conversion of labeled lignins to $^{14}CO_2$ has served as a useful assay for biodegradation, even though it does not indicate the full extent of degradation because lignin is biodegraded in part to water-soluble intermediates (e.g. 187, 231). Care must be taken in preparing labeled lignins. Incorporation of precursors by plants into nonlignin components (proteins, aromatic acids esterified to hemicelluloses or to lignin, and other aromatics such as lignans) must be avoided or circumvented (15, 34). One must also be sure that the lignin is polymeric. This is a potential problem with in vitro lignins [often termed "dehydrogenative polymerizates" (DHPs)] and possibly also with lignins labeled in vivo.

Another important point for microbiologists is that some of the linkages of lignin are unstable, particularly those involving C_α (as between Units 3 and 13, and 15 and 16 in Figure 1). Slow abiotic degradation, favored by high-temperature, acidic, or alkaline environments, releases small fragments. We suspect that the facile biodegradation of such fragments has led to erroneous conclusions concerning polymer biodegradation.

Crawford (34), Janshekar & Fiechter (126), and Buswell & Odier (25a) have discussed the merits of various isolated lignins used in biodegradation studies. In most recent work synthetic lignins or lignin-labeled plant tissues have been used. Some studies, however, have been done with kraft lignins and lignosulfonates, both of which are structurally modified by-products of commercial pulping operations (see 25a, 34, 126, 210, 240).

MICROBIOLOGY OF LIGNIN BIODEGRADATION

Despite numerous studies, it is not entirely clear which microbes, other than certain fungi, degrade the lignin polymer. The uncertainty reflects the experimental difficulties mentioned above and insufficient comprehensive study with selected species. Moreover, the apparent inability of microorganisms to use lignin as sole carbon/energy source for growth precludes the isolation of lignin-degraders by standard enrichment procedures and the use of growth on lignin as a criterion for degradative ability.

Anaerobic Conditions

Lignin is apparently not biodegraded anaerobically. Zeikus et al (275) studied the decomposition in anaerobic lake sediments of synthetic lignins, an alkali-degraded synthetic lignin, a dimeric lignin model compound, and lignin-related phenols, all labeled with ^{14}C. Both $^{14}CO_2$ and $^{14}CH_4$ were monitored. Degradation during 110 days was limited to the low–molecular weight materials (<600 daltons). Colberg & Young (32) obtained similar results in studies with Douglas fir wood labeled by [^{14}C]phenylalanine feeding. In a study by Holt & Jones (117), beech wood buried in anaerobic seawater, freshwater, or brackish muds was only slightly degraded after 18 mo. Similarly, [^{14}C]lignin-labeled aspen wood was not degraded significantly during a 6-mo anaerobic incubation in seven soils (212). We suggest that very limited anaerobic metabolism of lignin-labeled plant tissues by various microflora during extensive incubations (14, 16) can be attributed to nonlignin components or to low–molecular weight materials freed abiotically.

Aerobic Conditions

Neither rapid nor extensive bacterial degradation, even under highly aerobic conditions, has been reported. The most rapid and extensive degradation described to date is caused by certain fungi, particularly the white-rot fungi, in highly aerobic environments.

BACTERIA Bacterial lignin degradation has been most extensively studied in actinomycetes, particularly *Streptomyces* spp. *Streptomyces viridosporus* and *Streptomyces setonii* caused losses of 32–44% of the lignin in spruce, maple, and *Agropyron* lignocelluloses, as determined by chemical analyses of the insoluble residues (9). Characterization of lignin isolated from *S. viridosporus*–degraded spruce phloem indicated oxidative alterations similar to those reported for white-rot fungi (35). Nevertheless, in all studies of degradation of [^{14}C]lignin-labeled lignocelluloses by the streptomycetes, a maximum of about 20% of the ^{14}C has been converted to $^{14}CO_2$, and attack on fully lignified xylem tissues has been minimal. In recent studies, maximum degradation of [^{14}C]lignin in wheat straw to $^{14}CO_2$ was 8% in 14 days (199, 200).

Degradation of grass tissues by the actinomycetes produces a water-soluble residue termed "acid-precipitable polymeric lignin," or APPL (39, 201, 221, 222), which contains varying amounts of carbohydrate. APPLs, therefore, might result from polysaccharide rather than lignin degradation (201). The lignin component is similar to sound lignin (39, 201) and is resistant to further degradation by the actinomycetes that produced it (201, 222).

Recent studies with other bacteria have failed to demonstrate extensive degradation. Experiments that indicated high rates and extents of degradation of a kraft lignin by various bacteria did not employ ^{14}C-labeling, and cell adsorption was not ruled out (42). Janshekar & Fiechter (125) isolated strains of *Nocardia, Pseudomonas,* and *Corynebacterium* for ability to grow on lignin-related phenols, but none was able to degrade any of four different lignins. A bacterium reported to degrade lignin released only 3% of the ^{14}C from [^{14}C]lignin-labeled *Spartina* lignocellulose, whereas pepsin solubilized 19% of the label (150). Although a *Xanthomonas* strain mineralized over 30% of synthetic [^{14}C]lignin in 20 days, the lignin also underwent abiotic depolymerization (149); a very recent study indicated that the low–molecular weight components are mineralized (H. Kern & T. K. Kirk, unpublished). Benner et al (17), using ^{14}C-labeling, found that bacteria were primarily responsible for the degradation of plant material in certain environments, but mineralization of the lignin was minimal.

Nilsson & Holt (208) described a new type of wood decay caused by bacterial consortia that appear to belong to the Myxobacteriales or Cytophagales (G. Daniel & T. Nilsson, personal communication). Analyses of woods after 3-mo exposure to such a consortium indicated extensive lignin loss from birch and pine woods.

To date, however, no study has shown that lignin is mineralized rapidly or extensively by aerobic bacteria. As in the case of the anaerobes, a limiting factor might be the size of the lignin polymer. Lignin-related dimeric compounds (72, 85, 86, 142, 220, 238) and even tetrameric (638- and 666-dalton) compounds (129) are metabolized by numerous species of bacteria, presumably intracellularly, and apparently with little specificity.

FUNGI Soft-rot wood decay caused by various species of ascomycetes and fungi imperfecti involves lignin degradation, although wood polysaccharides are preferentially degraded (see 154, 160). *Chaetomium piluliferum* converted 20–30% of ^{14}C-labeled synthetic lignins to $^{14}CO_2$ in 50 days (96). Unfortunately, the soft-rot fungi have received very little attention; virtually nothing is known about the enzymology of their degradation of lignin.

Several ascomycetes, fungi imperfecti, and phycomycetes that are not associated with soft rot of wood, including 12 marine fungi (245), 18 *Trichoderma* strains (65), and *Trichoderma harzianum* (147), failed to degrade lignin significantly. A supposed kraft lignin degradation by a *Candida* sp. (31) could have been cell adsorption. A report of degradation of ^{14}C-labeled kraft lignin by a strain of *Aspergillus fumigatus* (131) did not include the molecular size of the lignin components, so it is not clear that high–molecular weight material was degraded. Several recent studies have treated *Fusarium* species, and especially their degradation of phenolic-lignin model compounds. These species do not, however, appear to degrade lignin. Thus

Norris (209) reported that an isolate of *Fusarium solani* released only 4–5% of the ^{14}C in labeled synthetic lignins as $^{14}CO_2$ in 30 days, and Sutherland et al (246), using 18 *Fusarium* strains, found a maximum of 5% conversion of ^{14}C-labeled lignin in spruce wood to $^{14}CO_2$ in 60 days.

A few species of ascomycetes (e.g. species of *Xylaria, Libertella,* and *Hypoxylon*) cause white-rot wood decay accompanied by substantial lignin loss. None of these fungi have been studied extensively, but recent work has shown that they fail to degrade gymnosperm wood (guaiacyl lignin) and that they preferentially attack the syringyl units of angiosperm lignin (T. Nilsson, J. R. Obst & T. K. Kirk, unpublished). This finding has interesting implications for the ligninolytic enzymes of these fungi, which remain to be studied.

Lignin degradation by non–white-rotting basidiomycetes has been reported. The gasteromycete *Cyathus stercoreus,* which is associated with litter decomposition, degrades lignin in wheat straw (269); the degradation is as extensive as that by various white-rot fungi (3). Several other *Cyathus* species degrade grass lignin, but members of the related genera *Nidula* and *Crucibulum* do not (1). Ectomycorrhizal fungi (*Cenococcum, Amanita, Tricholoma,* and *Rhizopogon*) only slowly mineralized ^{14}C-labeled synthetic lignins and corn lignin (253); however, poor degradation by two white-rot fungi included for comparison suggests that further studies with different culture conditions are needed.

Basidiomycetes that cause the important brown-rot wood decay also partially decompose lignin. They are closely related to the white-rot fungi; many were formerly assigned to the same genera. They invade the lumens of wood cells, where they secrete enzymes that decompose and remove the polysaccharides, leaving behind a brown, modified lignin residue. Studies show that the lignin undergoes limited aromatic hydroxylation and ring cleavage, but that the major effect is demethylation of aromatic methoxyl groups (154, 156). In accord, brown-rot fungi converted substantial percentages of $[O^{14}CH_3]$lignins to $^{14}CO_2$, whereas conversion of side-chain carbon to CO_2 was significantly lower, and conversion of aromatic carbons lower still (96, 159). Ability of the brown-rot fungi to mineralize the backbone lignin polymer thus seems to be limited. Some brown-rot fungi degrade dimeric model compounds, but to different, as yet unidentified, products from those produced by white-rot fungi (48, 49). Overall, however, lignin degradation by this group of fungi has been studied very little.

LIGNIN DEGRADATION BY WHITE-ROT FUNGI

The white-rot basidiomycetes degrade lignin more rapidly and extensively than other studied microbial groups. Like the brown-rot fungi, they invade the lumens of wood cells, where they secrete enzymes that degrade lignin and the other wood components. The electron-microscopic features of wood decay by white-rot fungi, including species that selectively degrade the lignin plus hemicelluloses, have been studied recently (19–21, 215, 216, 237). These studies reveal, among other things, that lignin is degraded at some distance from the hyphae and is removed progressively from the lumens toward the middle lamella.

During its mineralization by white-rot fungi, lignin undergoes a number of oxidative changes, including aromatic ring cleavage (see 26). Recent studies have shown that a progressive depolymerization occurs and releases a wide array of low–molecular weight fragments (27, 28, 230, 231, 247, 248). Fragments of <1 kd seem to predominate (57, 187).

One species of white-rot fungus, *Phanerochaete chrysosporium* [=*Sporotrichum pulverulentum* (23, 223)] has been studied widely. Culture conditions for lignin degradation have been optimized, and *P. chrysosporium* exhibits the highest reported rates of lignin degradation. Yang et al (272) reported degradation of 2.9 g lignin per gram of fungal cell protein per day in a wood pulp; assuming 15% protein in the mycelium (63), this rate is about 200 mg lignin per gram of mycelium per day. Ulmer et al (255) reported rates three times higher for a lignin from wheat straw.

In the following sections, we briefly review recent progress in the study of the physiology, biochemistry, genetics, and molecular biology of degradation of lignin by *P. chrysosporium*. Where available, data for other white-rot fungi are included.

Physiology

Research during the late 1970s demonstrated that several nutritional and cultural parameters are important for lignin degradation by *P. chrysosporium:* (*a*) presence of a cometabolizable substrate, (*b*) high oxygen tension, (*c*) growth as mycelial mats rather than as submerged pellets in agitated cultures, (*d*) correct choice of buffer, (*e*) correct levels of certain minerals and trace elements, and (*f*) growth-limiting amounts of nutrient nitrogen (reviewed in 25a, 154, 168). Through subsequent research this list has been refined and physiological features have been elucidated.

Recent studies have added to the evidence that lignin is not a growth substrate for white-rot fungi. *P. chrysosporium* and *Lentinula edodes* metabolize various lignin preparations only when an alternate carbon/energy source is present (176, 255). Several studies have confirmed earlier observations that hemicelluloses and cellulose, or added carbohydrates, are always metabolized with the lignin in lignocelluloses (e.g. 4, 19, 186, 189, 215, 230, 269, 272). Nevertheless, the balance between energy-producing and energy-yielding reactions in lignin mineralization is not known.

Earlier observations (167, 185) that molecular oxygen can be crucial in determining the rate of lignin degradation by *P. chrysosporium* as well as by certain other white-rot fungi have been confirmed (13, 108, 232, 273). Increasing the O_2 level in the medium has a multiple effect (13); it leads to an increase in the titer of the ligninolytic system, including ligninase (see section on biochemistry, below), and the H_2O_2-producing system(s) (54, 55), and it also increases the activity of the existing lignin-degrading system, evidently by increasing the supply of O_2 for degradative reactions and for H_2O_2 production. Extracellular H_2O_2 production can be rate limiting in *P. chrysosporium* (170).

Culture agitation, usually used to increase oxygen tension, almost completely suppressed lignin degradation (167) as well as metabolism of dimeric models (45), synthesis of veratryl alcohol (244), and formation of ligninase (55). More recently, however, degradation of lignin by submerged pellets in agitated cultures has been achieved by using a mutant strain (82) or by adding detergent (12, 123), veratryl alcohol (180), or benzyl alcohol (T. K. Kirk & S. C. Croan, unpublished) to cultures of wild-type strains. The agitation-induced suppression and its alleviation have not been explained.

The choice of buffer in the culture medium can affect lignin degradation significantly. Kern (148), for example, showed that cultures buffered with a polymer, polyacrylic acid, mineralize lignosulfonate more rapidly than those buffered with 2,2-dimethylsuccinate, which is widely used.

A medium formulated in initial studies (167) was later found to contain about the correct amounts and balance of several inorganic nutrients (127). Some stimulation was noted when Zn, Fe, and Mo concentrations were decreased 10-fold over the basal level (127). In addition, Ca concentration is important; 1 mM is more favorable than either 0.1 or 10 mM (176). Ligninase production was increased by increasing either Cu or Mn (161). Growing evidence implies that Mn is important in lignin degradation: Not only does increased Mn lead to increased ligninase production, but this element accumulates as MnO_2 deposits during degradation of lignin in wood by several white-rot fungi (18). Also, Mn concentration has a marked influence on lignin degradation by *L. edodes* (176). Finally, a Mn-dependent peroxidase has recently been discovered in *P. chrysosporium* (see section on biochemistry, below).

In *P. chrysosporium,* lignin is degraded only during secondary (idiophasic) metabolism, which is triggered by limiting cultures for nutrient N (127, 153, 167, 228, 229), C, or S (127). Lignin degradation by several other species (105, 106, 177, 213), but not by all white-rot fungi (69, 176, 177), is stimulated by N-limitation. N-limited conditions are natural for the white-rot fungi because wood is N-poor (33). For practical purposes N, rather than C or S, is usually limited in experiments with *P. chrysosporium;* triggering of lignin degradation by S-limitation is not easily demonstrated, and C-limitation leads to autolysis and only transient lignin degradation (127).

In addition to lignin degradation, other features of secondary metabolism triggered by N-limitation in *P. chrysosporium* have been studied. In glucose-grown cultures, appearance of new hyphal outgrowths (124, 187), formation of an extracellular glucan (162, 178, 191), and de novo synthesis of the secondary metabolite veratryl (3,4-dimethoxybenzyl) alcohol (191, 193, 244) are manifestations of secondary metabolism. The N-regulated transitions from primary to secondary metabolism and vice versa are associated with increased and decreased levels of cyclic AMP, respectively; interestingly, cyclic AMP levels are controlled in part by secretion (194, 195).

The transition from primary to secondary metabolism is associated with a transient increase in intracellular glutamate (63); addition of exogenous glutamate or other nitrogen sources sharply suppresses secondary metabolism, including activity of enzymes involved in lignin degradation (62). The titer of ligninase (see section on biochemistry) is lowered sharply by adding glutamate to cultures (55). Variation in glutamate content is associated with changes in the levels of glutamate-synthesizing and glutamate-degrading enzymes (24, 62; reviewed in 211).

The association of veratryl alcohol with lignin degradation has been studied in N-limited cultures of *P. chrysosporium*. Veratryl alcohol is synthesized from phenylalanine via 3,4-dimethoxycinnamyl alcohol, which is oxidized to 1-(3,4-dimethoxyphenyl)glycerol; this is then oxidized to veratraldehyde (244). Interestingly, these oxidations are catalyzed by ligninase in vitro, but their catalyst in vivo is not yet known. In vivo, the veratraldehyde is reduced to the alcohol (244). Veratryl alcohol is a substrate for ligninase; it is oxidized to veratraldehyde and other products (see section on biochemistry). Another possible contribution of veratryl alcohol is that it stimulates the oxidation of other compounds by ligninase (93, 104). In addition, exogenous veratryl alcohol hastens the appearance of the ligninolytic system ($[^{14}C]$lignin $\rightarrow$ $^{14}CO_2$) as a part of secondary metabolism; in the absence of added veratryl alcohol, the ligninolytic system and the biosynthesized metabolite appear simultaneously (188). Added veratryl alcohol also increases the titer of ligninase (55, 161, 184), probably via induction (56). Leisola et al (188) suggested that veratryl alcohol is the normal inducer of the ligninolytic system. Recent studies have shown that veratryl alcohol is also synthesized de novo by the white-rot fungi *Coriolus versicolor* (141), *Pycnoporus cinnabarinus, Phlebia radiata* (105), and a *Trametes* species and by four other unidentified white-rot fungi (H. Silva & E. Agosin, personal communication). However, whether it is synthesized by all white-rot species is not yet known. In any case, the association of veratryl alcohol with lignin degradation and secondary metabolism deserves further study. Studies with mutants might help elucidate its importance, but have so far provided only inconclusive data (191).

Recent studies have shown that added lignin also increases the titers of ligninase and the complete lignin-degrading system (55, 254). This probably explains the failure to saturate the lignin-degrading system of *P. chrysosporium* with lignin; the more lignin is added to cultures, the more is degraded (30, 176). The ligninolytic system of *L. edodes,* by contrast, is easily saturated (176).

The role of the extracellular glucan in lignin degradation by *P. chrysosporium,* if any, remains speculative. The glucan, which is of undetermined structure, seems to occur both as a hyphal sheath (161) and free in culture fluid (251). Reportedly it is remetabolized when carbon becomes limiting (53). Leisola et al (178) presented evidence that the glucan inhibits lignin degradation. Recent studies, however, have linked degradation of wood by white-rot fungi with the presence of mucopolysaccharide hyphal sheaths and extracellular tripartite membranes (66, 217). Palmer et al (217) suggested that wood-degrading enzymes are embedded in the sheaths. The tight, transient binding of lignin to mycelia during degradation (30, 124) might involve the glucan. Although extracellular ligninase is easily separated from the glucan (251), recently reported particulate bodies with ligninase (?) activity (88) might have been formed by sedimenting the activity with the glucan at high centrifuge speeds.

Biochemistry

The recent discovery of several enzymes that are thought to have roles has projected lignin biodegradation research into the realm of biochemistry. These enzymes include ligninases, Mn peroxidases, phenol-oxidizing enzymes, and H_2O_2-producing enzymes.

LIGNINASE (LIGNIN PEROXIDASE) In 1983, two groups announced discovery in *P. chrysosporium* of an extracellular H_2O_2-requiring enzyme activity that catalyzes several of the reactions formerly seen with intact cultures (81, 169). A third group (243) reported an extracellular membrane-bound enzyme activity which in retrospect might have been the same. Papers published by the first two groups later in 1983 described reactions catalyzed by the crude enzyme (75) and by the isolated enzyme, which, importantly, partially depolymerized methylated lignins (250).

Ligninase activity has recently been detected in other white-rot fungi, including *Phlebia radiata* (107; A. Hatakka, M.-L. Niku-Paavola, personal communication), *Panus tigrinus* (196), *Coriolus versicolor, Pleurotus ostreatus,* and *Bjerkandera adusta* (R. Waldner, M. Leisola & A. Fiechter, personal communication). Failure to detect the enzyme in *Fomes lignosus,*

Trametes cingulata (R. Waldner, M. Leisola & A. Fiechter, personal communication), and *L. edodes* (176) might reflect assay insensitivity or failure to solubilize the enzyme. Oki et al (214) recently detected an enzymatic activity in *L. edodes* that cleaves model compounds in the same manner as ligninase.

Nature and properties Ligninase from *P. chrysosporium* was initially isolated by various chromatographic procedures and was shown to contain one mole of protoheme IX per mole of enzyme, to have molecular mass of 41–42 kd, and to be glycosylated (82, 174, 251). In reactions of ligninase with $H_2{}^{16}O_2$, ${}^{18}O_2$, and lignin model compounds, ${}^{18}O$ was incorporated into C_α–C_β cleavage products; therefore the enzyme was referred to as an H_2O_2-requiring oxygenase (82, 251).

Subsequent spectroscopic studies have shown that the ligninase is distinct from P_{450} oxygenases, shares some properties with oxygen-carrying heme proteins, and is a true peroxidase. ESR spectral studies (8, 219) showed that the iron is high-spin Fe(III). Detailed ESR and electronic spectral studies indicated that the heme environment resembles those of other peroxidases (8, 172). Raman resonance spectroscopy confirmed the high-spin ferric nature of the heme iron and indicated that the fifth ligand in the pentacoordinate heme is histidine (8, 172). From their resonance Raman studies, Andersson et al (8) concluded that the active site in ligninase is hexacoordinate at low temperatures. Kuila et al (172) showed that ligninase is pentacoordinate at room temperature. Recent work has confirmed the temperature dependence of axial ligation (M. H. Gold, personal communication).

Ligninases from other fungi are only beginning to be characterized. Two ligninases from *Phlebia radiata* are apparently very similar in size to those from *P. chrysosporium* (M.-L. Niku-Paavola, personal communication). A somewhat larger ligninase (50 kd) has also been isolated from *C. versicolor* (J. Palmer & C. Evans, personal communication).

Enzyme mechanism Renganathan & Gold (234) characterized the H_2O_2-oxidized forms of ligninase by electronic absorption spectroscopy. They demonstrated formation of compound I (the two-electron oxidized form) and compound II (the one-electron oxidized form). Compound I was converted to compound II by one equivalent of a one-electron substrate such as a phenol, or by 0.5 equivalent of the two-electron substrate, veratryl alcohol. Ligninase therefore resembles HRP in many of its properties, although it has a higher oxidation potential than HRP (98; P. J. Kersten, K. E. Hammel, B. Kalyanaraman & T. K. Kirk, unpublished).

Hammel et al (99) showed that O_2 uptake during cleavage of a hydrobenzoin lignin model by ligninase results from addition of O_2 to carbon-centered radicals. These authors, as well as Renganathan et al (236), also showed that ligninase cleaves β-1 model substrates under anaerobic conditions using peroxide alone as the oxidant. These studies proved that the essential mechanism of ligninase is peroxidative.

Tien et al (252) made a detailed study of the oxidation of veratryl alcohol by ligninase. With excess veratryl alcohol they observed a stoichiometry of one mole of veratraldehyde per H_2O_2 and estimated a maximum turnover number of 7.8 sec^{-1}. Steady-state kinetic studies indicated a ping-pong mechanism ($K_m = 29\ \mu M$ for H_2O_2 and 72 μM for veratryl alcohol) in which H_2O_2 first reacts with the enzyme, and the oxidized enzyme then reacts with veratryl alcohol. Attempts to detect intermediate substrate-free radicals were not successful. The results indicated that veratryl alcohol is oxidized at the active site by direct oxygenation or via two rapid one-electron oxidations.

The oxidation of the normal secondary metabolite veratryl alcohol by ligninase might have special significance. Harvey et al (104) showed that veratryl alcohol increases the rate of oxidation of anisyl substrates by ligninase/H_2O_2; in its absence these substrates are only incompletely oxidized. Haemmerli et al (93) similarly reported stimulation by veratryl alcohol of ligninase oxidation of benzo(*a*)pyrene. Harvey et al (104) suggested that veratryl alcohol is oxidized to a cation radical (see below), which acts as a diffusible one-electron oxidant to interact with other substrates. However, the results of Tien et al (252) seem to discount this postulate. It seems more likely that the veratryl alcohol simply protects the enzyme from inactivation by anisyl substrates. Alternatively, it might act as an electron relay at the enzyme active site, or it might alter enzyme conformation.

The reactions of aromatic substrates on ligninase oxidation were puzzling at first in their diversity and complexity, but can now be understood. Investigations by Kersten et al (152) and Hammel et al (97, 99) established the basic simplicity of ligninase oxidation: Susceptible aromatic nuclei are oxidized by one electron, and this produces unstable cation radicals, which undergo a variety of nonenzymatic reactions. Based on direct ESR spectroscopic observations of the cation radicals produced by purified ligninase, Kersten et al (152) proposed that this is the basic mechanism that accounts for the various reactions. At about the same time, Schoemaker et al (241) suggested a cation-radical mechanism based on some of the reactions catalyzed, and Harvey et al (102) showed that chemical one-electron oxidation of certain compounds gives the same products as ligninase. Hammel et al (97) proved the one-electron mechanism of ligninase action by showing that radical coupling dimers are produced stoichiometrically on anaerobic ligninase cleavage of special model compounds, that carbon-centered radical products can be trapped under anaerobic cleavage, and that peroxyl radicals are produced by addition of O_2 to carbon-centered radicals in aerobic ligninase reactions.

The key reaction of ligninase with lignin model compounds therefore is one-electron oxidation. With certain substrates, such as with veratryl alcohol, a second electron apparently can be removed from the substrate before it leaves the enzyme active site (252). Phenolic substrates are oxidized to phenoxy radicals, as evidenced by dimerization of 4-*tert*-butylguaiacol (251) and by direct ESR detection (E. Odier, M. D. Mozuch, B. Kalyanaraman & T. K. Kirk, unpublished). Reactions of phenoxy radicals are discussed below in connection with phenol-oxidizing enzymes. The following discussion deals with nonphenolic substrates.

Whether an aromatic nucleus is a substrate for ligninase depends in part on its oxidation potential (98). Strong electron-withdrawing substituents such as C_α-carbonyl groups tend to inactivate aromatic nuclei to oxidation by ligninase, whereas alkoxyl groups activate it. The positions of the latter groups also affect oxidizability by ligninase. In lignin, the positions of the alkoxyl groups are set, but the number varies from one to three (Figure 1); the oxidation rates are expected, therefore, to be in the order syringyl > guaiacyl > *p*-hydroxyphenyl. The nature and pattern of the substituents also affect the subsequent reactions of the cation (171), including nucleophilic attack by water or an internal hydroxyl group, loss of the acidic proton at C_α, and C_α–C_β cleavage.

A variety of sequential reactions can follow these initial reactions. Included are addition of molecular oxygen to carbon-centered radicals, one-electron oxidation or reduction, and (in the absence of O_2, which scavenges radicals) radical-radical coupling. The many reactions that the cation radicals undergo lead to many different products; this explains the surprisingly large number of degradation intermediates formed from lignin as it is degraded by white-rot fungi (26–28, 247, 248).

It is this nonspecific oxidation of lignin, which leads to a variety of subsequent reactions and products determined only by the kinetics of reaction intermediates, that leads us to conclude that the initial process is essentially enzymatic "combustion" (see final section, below).

In the following we illustrate the diversity of reactions and products with model compounds of the β-O-4 type, which is the dominant lignin substructure. Figure 2 is based on studies from four different laboratories with isolated ligninase and many model compounds. Higuchi (114) has summarized the reactions leading to most of these products and the mechanistic information gained with specially synthesized models and various isotopes. Ligninase can oxidize ring *A* and ring *B*. In lignin the two rings might have the same substitution and be equally susceptible; the more accessible ring would be expected to be oxidized first. In Figure 2, we have labeled the arrows *A* or *B* to denote which ring was oxidized.

Products *1–3* in Figure 2 are formed on C_α–C_β cleavage (75, 82, 91, 171,

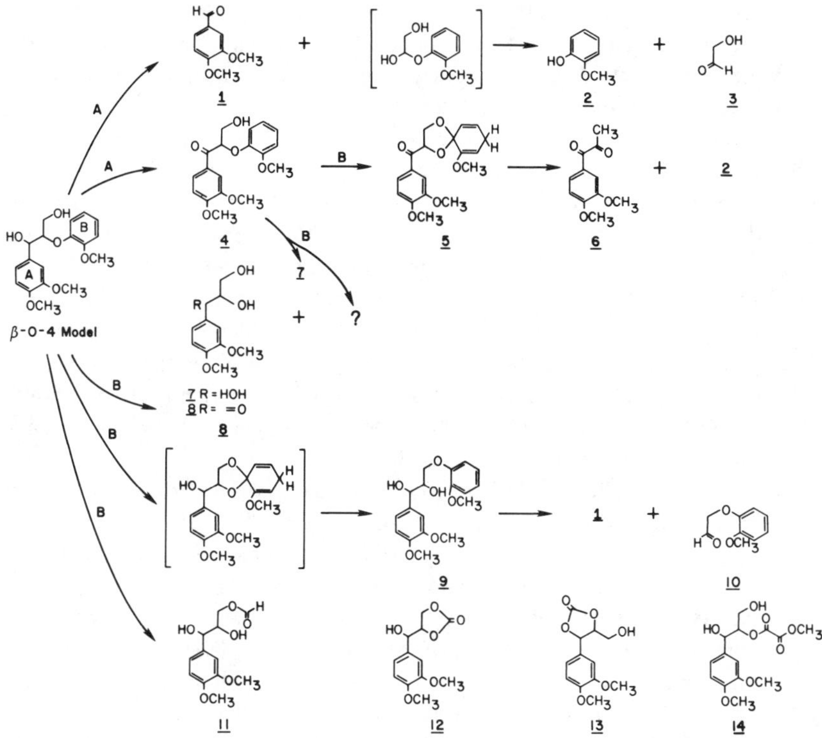

Figure 2 Products of oxidation of β-O-4 model compounds by ligninase/H_2O_2. Oxidation can be in ring *A* or ring *B*, as indicated (see text).

202, 203, 236, 251, 252). By analogy with other types of models, it is clear that the new hydroxyl oxygen atom at C_β comes from O_2 (97, 99, 251). Studies have shown that C_α–C_β cleavage is the major reaction in intact cultures of *P. chrysosporium* (45, 47, 64, 116). Importantly, studies of polymeric lignin that had been partially degraded by white-rot fungi also showed that C_α–C_β cleavage is prominent (26).

Product *4* is formed by C_α oxidation; this is analogous to veratryl alcohol oxidation to veratraldehyde (75, 82, 91, 171, 236, 250, 251). C_α oxidation in model compounds is also seen in intact cultures (64), and polymeric lignin contains substantial numbers of C_α-carbonyl groups after partial degradation by white-rot fungi (26, 29).

Product *5* formation involves an intramolecular nucleophilic attack by the C_γ-hydroxyl group at C_4 of the ring *B* cation and subsequent reduction (171). Spontaneous decomposition of product *5* yields *6* and *2* (171).

Products *7* and *8* (75, 82, 91, 171, 202, 203) result from nucleophilic attack on the ring *B* cation, again at C_4, but this time by water. Product *8* arises from C_α-carbonyl structures such as that in product *4* (171), or perhaps by C_α-oxidation of product *7*. However, product *7* is more likely to undergo C_α–C_β cleavage to products *1* and *3*. Formation of product *7* in intact cultures has been studied extensively (46, 47, 256, 260, 262).

Demethoxylation products (not shown), reflecting attack by water at C_3 of the ring *B* cation, are prominent in anaerobic ligninase reactions (203). Their importance in aerobic reactions is unclear, but small amounts of methanol are formed from β-O-4 models (171). Ligninase-initiated demethoxylation could be responsible for the formation of methanol from lignin and related low–molecular weight aromatics by *P. chrysosporium* (7). It is unclear, however, whether ligninase oxidation alone accounts for the substantial methoxyl deficiency in lignin after partial degradation by white-rot fungi (26).

Product *9* is presumably formed via an intermediate analogous to *5*, (shown in brackets), with subsequent opening of the heterocyclic ring. C_α–C_β cleavage produces *10* and *1* (121, 202, 258).

Products *11–14* (114, 259, 261, 263) involve cleavage of ring *B* following its oxidation. Umezawa et al (261) suggested that the initial ring cleavage product might be a methylmuconate formed by oxygenative cleavage between C_3 and C_4; this has now been proven (T. Higuchi, personal communication). Demethoxylation is not involved (114). Products *11–14* are closely related to each other. They are substrates for further ligninase oxidation, and have only been found in trace amounts. Even so, all have been identified in intact cultures (114). Ligninase oxidation probably accounts for ring cleavage fragments found in the lignin polymer following partial degradation by white-rot fungi (26, 29, 94) and for various low–molecular weight ring cleavage products formed during polymer degradation (26–28, 247, 248).

The first ring cleavage product identified in cultures of *P. chrysosporium* was a cyclic carbonate formed from a β-O-4 model compound (257). Shortly thereafter, Leisola et al (183) identified two isomeric lactones formed from veratryl alcohol via oxygenative cleavage between the methoxyl-bearing carbons. It was demonstrated that ligninase is responsible for their formation (114, 183).

Model β-1 compounds (Units 9 and 10, Figure 1) undergo reactions analogous to those involved in the formation of products *1–4* from the β-O-4 models (Figure 3); the relative simplicity of their reactions has provided much insight into the mechanism of ligninase action (82, 91, 97, 99, 236, 250, 251). The reactions shown in Figure 3 were first found in intact cultures (44, 80, 165, 207), and arise from ring *A* oxidation. Reactions initiated by ligninase oxidation of ring *B* have not been studied per se; it is possible that the products would be the same as those from oxidation of ring *A*. Also, ring cleavage products from ligninase oxidation of β-1 models have not yet been

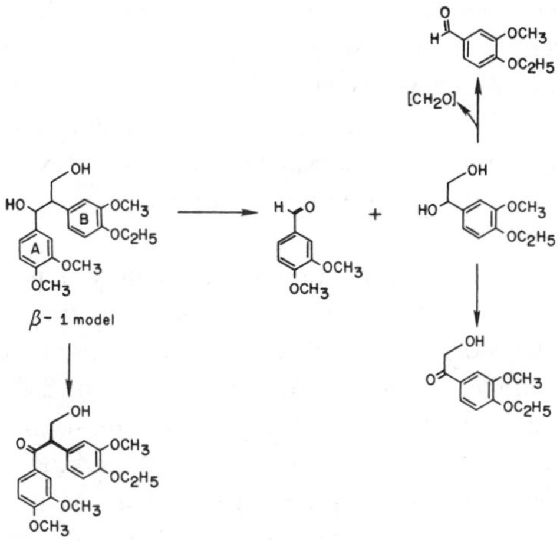

Figure 3 Oxidation of β-1 model compounds by ligninase/H_2O_2.

observed, but such products were tentatively identified among low–molecular weight compounds formed during white-rot of wood (28).

Interestingly, one-electron oxidants, including isolated hemes with appropriate oxidants, convert β-1 models to the same products as ligninase (90, 118, 242), and should have utility in basic and applied studies.

Most of the products in Figures 2 and 3 have been identified in intact cultures of *C. versicolor* (135, 138–140) as well as *P. chrysosporium*. Initial studies with a number of other white-rot fungi indicate degradation of β-O-4 and β-1 models via the same reactions (48, 49). Thus it seems clear that ligninase is primarily responsible for the initial degradation of nonphenolic β-O-4 and β-1 model compounds and presumably also for the initial degradation of the same structures in lignin.

Ligninase oxidation of lignin model compounds representing other substructures has not yet been examined. However, some (not all) of the reactions in the degradation of a nonphenolic β-5 model compound (substructure 3–4, Figure 1) in intact cultures of *P. chrysosporium* (206) can probably be attributed to ligninase. Other spontaneous reactions that follow ligninase oxidation have been seen with single-ring compounds; these include dihydroxylation of styryl structures and hydroxylation of C_α-methylene groups (75, 82, 236, 251).

In addition to lignin-related aromatics, polycyclic aromatics and certain dibenzodioxins with appropriate oxidation potentials are oxidized by ligninase to quinones and unidentified products (93, 98, 239). The reactions are consistent with a cation-radical mechanism: The cation radical from dibenzodioxin oxidation was directly demonstrated, and $H_2{}^{18}O$-labeling studies were consistent with a cation-radical intermediate from ligninase oxidation of pyrene to pyrenediones (98).

Intact cultures of *P. chrysosporium* degrade a variety of chlorinated aromatics (10, 22, 43, 119); this indicates that the degrading system is nonspecific. The role of ligninase is not yet known, but it oxidizes several chlorophenols (K. E. Hammel, personal communication).

Multiplicity In one of the early studies electrophoresis indicated that ligninase activity in *P. chrysosporium* might be associated with more than one protein (250). Later work verified this multiplicity (59, 123, 161, 181, 182, 219, 235). The number of isoenzymes reported (from 2 to 15) reflects not only separation efficiency, but probably postsynthesis modifications, and perhaps differences in strains, culture conditions, and ages. The number of structural genes involved is not yet known. All of the isoenzymes oxidize veratryl alcohol. Those examined further (59, 181, 219, 235; R. L. Farrell, K. Murtagh, M. Tien & T. K. Kirk, unpublished) have molecular mass of 39–43 kd and similar spectral and catalytic properties, and they are glycoproteins. Polyclonal antibody reactions and protease digestion patterns indicate considerable homology. The isoenzymes differ in isoelectric points, in specific activity by at least fivefold, and slightly in K_m for veratryl alcohol oxidation and for H_2O_2 reduction. The important question of whether they differ in oxidation potential has not yet been answered.

Tien et al (252) provided evidence that each preparation of ligninase isoenzyme H-8 (161) consists of a mixture of active and inactive enzyme. The two could not be separated. Once isolated, ligninase is stable (219). Paszczyński et al (219) found that the specific activity of the ligninases varied with culture age, but this might reflect differences in isoenzyme mixtures.

Production Ligninase production by *P. chrysosporium* has been increased substantially through improvement of strains and culture parameters. Originally the activities, measured in N-starved stationary cultures, were low (~5 units per liter) based on veratryl alcohol oxidation (251). Buswell et al (25) later measured over 400 U liter^{-1} in N-sufficient stationary cultures of a new strain, INA-12, grown on glycerol. Faison & Kirk (55) showed that the various culture parameters that favor complete degradation of lignin similarly influence ligninase titer in cultures. Ligninase titer is increased by adding lignins or related low–molecular weight aromatics (55), including veratryl alcohol (55, 184), to cultures.

Volume scale-up was achieved with a mutant strain, SC-26, in rotating disc fermenters (161) and with a wild-type strain immobilized on the interior wall of a rotating plastic carboy (219). Linko et al (190) achieved continuous and repeated batch production of up to 245 U liter^{-1} of ligninase with wild-type mycelium immobilized in agar or agarose gels.

Further improvements in ligninase production have been obtained with agitated submerged cultures. Gold et al (82) first reported ligninase production in such cultures by a mutant strain, and Leisola et al (184) later demonstrated production of 60 U liter^{-1} in wild-type cultures to which veratryl alcohol had been added. Activities of up to 670 U liter^{-1} (specific activity = 36 U mg^{-1}) were obtained in carbon-limited cultures with concentrated mycelial pellets. Jäger et al (123) found that detergent in the medium permits wild type strains to produce ligninase, and to degrade lignin to CO_2, in agitated submerged cultures, and Asther et al (12) measured activity of over 1250 U liter^{-1} (specific activity = 29 U mg^{-1}) with strain INA-12 growing in a glycerol medium to which oleic acid emulsified with Tween 80 had been added. This is the highest activity reported to date, but it represents production of less than 50 mg liter^{-1} of ligninase protein, which suggests that considerable improvements might still be made. We have obtained activity of over 1000 U liter^{-1} with strain SC-26 in 2-liter stirred tank fermentors, and have found that culture additives, pellet size, stirring rate, and aeration are critical parameters (T. K. Kirk & S. Croan, unpublished).

MANGANESE PEROXIDASE Kuwahara et al (174) discovered a peroxidase activity different from ligninase in extracellular growth fluid of ligninolytic cultures of *P. chrysosporium*. The isolated 46-kd enzyme exhibited a requirement for H_2O_2, Mn(II), and lactate. Like horseradish peroxidase, it oxidized phenol red, *o*-dianisidine, and polymeric dyes. A similar enzyme of 45–47 kd isolated from ligninolytic cultures (120) also required Mn and H_2O_2, but did not require lactate (possibly because tartrate was present as buffer) or oxidize phenol red. It contained an easily dissociable heme.

Both Mn peroxidases were later purified and characterized (74, 218). The enzymes now seem to be either identical or isoenzymic. Both contain a single protoheme IX, with high-spin ferric iron. The enzyme oxidizes Mn(II) to Mn(III), which in turn oxidizes the organic substrates (73, 74, 218, 219). Glenn & Gold (74) reported 3–20× stimulation of oxidation rates by lactate or related compounds, which probably reflects stabilization of the Mn(III). The enzyme described by Paszczyński et al (219) contained 17% neutral carbohydrate and a high proportion of acidic amino acids. Leisola et al (181) separated six Mn peroxidases from the extracellular fluid of *P. chrysosporium,* and we separated four (T. K. Kirk & R. L. Farrell, unpublished). Mn peroxidases appear to function as phenol-oxidizing enzymes and perhaps participate in H_2O_2 production.

LACCASE AND OTHER PHENOL-OXIDIZING ENZYMES Most white-rot fungi produce extracellular laccase (EC 1.10.3.2). This blue copper oxidase catalyzes the one-electron oxidation of phenols to phenoxy radicals, eventually transferring four electrons to O_2 (233). The effect on the substrate phenols is the same as that of horseradish peroxidase, despite fundamental differences in enzyme mechanism.

Laccase apparently has a role in sexual fruiting (271) and in lignin degradation. Work with various phenolic model compounds and isolated laccase or HRP shows that certain degradative reactions occur, particularly with syringyl models (114, 168). Among the consequences of the one-electron oxidation in lignin-related phenols are C_α-oxidation, limited demethoxylation, and aryl-C_α cleavage (168). Phenol-oxidizing enzymes account for many of the degradative reactions in phenolic models seen in intact cultures of lignin-degrading fungi (84, 114, 132, 133) and *Fusarium solani* (113, 122, 136, 137).

Recent work shows that laccase can also cause C_α–C_β cleavage in phenolic syringyl structures (T. Higuchi, personal communication). Earlier work (134) had shown that HRP as well as whole cultures of *P. chrysosporium* caused the same cleavage in a phenolic syringyl dimer. Presumably all of the enzymes that catalyze one-electron oxidation of phenols catalyze similar reactions in lignin.

Coupling/polymerization is a major consequence of one-electron oxidation of lignin-related phenols and isolated lignins (e.g. 132, 192; see 168). Polymerization of phenols is also a consequence of ligninase oxidation (92); as discussed below, polymerization by ligninase and the other phenol-oxidizing enzymes must be limited in vivo by mechanism(s) yet to be elucidated.

Phanerochaete chrysosporium belongs to a minority of white-rot fungi that produce no detectable laccase. But apparently all white-rot fungi secrete enzymes capable of oxidizing phenols. Simple color tests are used to determine whether these are of the laccase type or the peroxidase type (which includes ligninase and Mn peroxidase) (100).

H_2O_2-PRODUCING ENZYMES The discovery that H_2O_2 is required for ligninase activity prompted several investigations into its origin. Crude cell extracts from ligninolytic cultures produce H_2O_2 in the presence of added glucose (68). Peroxisomelike structures just beneath the cell walls in hyphae that stain for catalase/H_2O_2 might be the in vivo site of the activity (67). Glucose oxidase activity in ligninolytic mycelia (226, 227) was characterized as a glucose-1-oxidase (gox) (144). Because the activity is produced during growth on various sugars and is associated only with ligninolytic cultures, Kelley & Reddy (145, 227) concluded that gox is the primary source of H_2O_2. Gox mutants had lost their ability to degrade [^{14}C]lignin to $^{14}CO_2$ while retaining other idiophasic functions, and Gox^+ revertants regained their ligninolytic capability (143); exogenous gox, however, did not restore ligninolytic activity to a Gox mutant. In related work, Eriksson et al (53) purified and characterized an intracellular glucose-2-oxidase that was also idiophasic. None of these studies with glucose oxidases demonstrated that the enzyme's action results in production of extracellular H_2O_2. Maltseva et al (196), however, reported extracellular idiophasic glucose oxidase activity in *Panus tigrinus*.

Other work has focused on other possible sources of extracellular H_2O_2. Greene & Gould (87) observed its production by washed, starved mycelia in the absence of added substrates, and attributed H_2O_2 production to fatty acyl CoA oxidase. Paszczyński et al (218) showed that the above-described Mn peroxidase oxidizes various reduced substrates, including glutathione, NADPH, and dihydroxymaleic acid, with the coupled reduction of O_2 to H_2O_2 (see also 11). Surprisingly, extracellular NAD(H) and NADP(H) are present in N-starved ligninolytic cultures (175). Very recently, Kersten & Kirk (151) demonstrated a new extracellular, idiophasic enzyme, glyoxal oxidase, in ligninolytic cultures. Glyoxal, methyl glyoxal, and several other α-hydroxy carbonyl and dicarbonyl compounds serve as substrates; their oxidation is coupled to the reduction of O_2 to H_2O_2. Both glyoxal and methyl glyoxal were identified in the extracellular culture fluid of idiophasic cultures (151). In summary, it appears that the H_2O_2 required for ligninase activity may be supplied by several different oxidases; supply by intracellular enzymes, however, has not been shown directly.

OTHER ENZYMES ACTING ON THE LIGNIN POLYMER As wood is degraded by white-rot fungi, the lignin exposed on the interior microsurfaces within the wood is oxidized. The partially degraded lignin polymer has been produced under controlled laboratory conditions, isolated by solvent extraction, and purified. It has been characterized chemically and physically, most recently by ^{13}C-NMR spectroscopy (26). These studies revealed that fungal attack decreases the methoxyl, phenolic, and aliphatic hydroxyl contents, cleaves aromatic nuclei to aliphatic carboxyl-containing residues, creates new C_α-carbonyl and carboxyl groups, and forms alkoxyacetic acid, phenoxyacetic acid, and phenoxyethanol structures (reviewed in 26). Many of the same effects were observed in specifically ^{13}C -labeled synthetic lignins following incubation with white-rot fungi (94). Surprisingly, ^{13}C spectroscopic studies also revealed significant amounts of aliphatic hydrocarbon structures (26, 29, 95).

It is now apparent that many of these degradative changes in the lignin polymer can be attributed to ligninase; however, some cannot, which indicates that other enzymes participate in polymer breakdown. For example, formation of aromatic carboxyl groups has not been observed in any of the studies with ligninase and model compounds. This suggests that there is an undiscovered extracellular aromatic aldehyde oxidase. Similarly, enzymes other than ligninase are evidently involved in producing the alkoxyacetic acid, phenoxyacetic acid, and phenoxyethanol structures. The aliphatic hydrocarbon structures are perhaps best attributed to covalently bound lipids. As mentioned earlier, it is not known whether the demethoxylation caused by ligninase and phenol-oxidizing enzymes accounts for the methoxyl loss in lignin (7, 26). Brown-rot fungi demethylate methoxyl groups in the polymer (156) by an unknown biochemical mechanism; it is possible that the white-rot fungi possess the same system. Huynh & Crawford (120) reported an extracellular H_2O_2-dependent enzyme activity in *P. chrysosporium* that demethylates 2-hydroxy-3-phenylbenzoic acid.

One of the puzzles of lignin degradation is the mechanism that prevents polymerization of lignin and phenolic products by ligninase and phenol-oxidizing enzymes. Polymerization of lignin is not prominent in vivo (30, 57, 231). This suggests that phenols are rapidly oxidized past the phenoxy radical step or that the radicals are reduced back to the phenols by an undiscovered enzyme and/or mechanism that prevents polymerization.

Westermark & Eriksson (268) suggested some time ago that phenoxy radicals might be reduced back to phenols by the enzyme cellobiose:quinone oxidoreductase (CBQase). CBQase, discovered in *P. chrysosporium* (268), oxidizes cellobiose, transferring the electrons to various quinones. We recently found that CBQase apparently does not prevent polymerization of phenols by ligninase or horseradish peroxidase (E. Odier, M. Mozuch, B. Kalyanaraman & T. K. Kirk, unpublished).

ACTIVATED OXYGEN SPECIES The requirement of H_2O_2 for lignin degradation was discovered through investigations into the possible involvement of activated oxygen species: H_2O_2, superoxide radical ($O_2\cdot^-$), hydroxyl radical ($\cdot OH$), and singlet oxygen ($^1O_2$). All of these species have been reported in biological systems. Only $^1O_2$ and $\cdot OH$ are reactive enough to be considered lignin oxidants. Initial studies by several groups suggested involvement of both of these species (reviewed in 168), but subsequent investigations have discounted involvement of both $^1O_2$ (166, 173) and $\cdot OH$ (164).

Genetics

Genetic approaches have been applied to the study of bioligninolytic organisms, particularly *P. chrysosporium,* to fully elucidate fundamental principles and manipulate the system. The methods of classical genetics such as the use of good selectable markers and the formation of recombinant strains (in the classic sense) are paramount to the basic studies. Comparisons with better-understood prokaryotic secondary metabolic systems cannot be readily drawn, unfortunately, since the enzymes produced in lignin degradation are unlike those in most metabolic pathways (see section on enzymatic "combustion").

Phanerochaete chrysosporium apparently has a classical three-stage life cycle (homokaryotic haploid, multikaryotic haploid, and homozygotic diploid), and it is probably homothallic or self-fertile (H. H. Burdsall & K. Nakasone, personal communication; M. Alic, C. Letzring & M. H. Gold, personal communication). Prolific conidiation, which is unusual among white-rot fungi, gives *P. chrysosporium* distinct advantages over other species for mutant production. Ultraviolet irradiation of asexual spores (conidia) was originally used to induce mutations in *Polyporus adustus* (50). Gold & Cheng (76) facilitated mutant analysis by inducing colonial growth on solid-agar plates and developing replica-plating techniques. They also elucidated the physiological conditions for fruit body formation, which permit genetic recombination (77). They showed that fruiting is controlled by glucose and nitrogen metabolite repression and that adenosine-3',5'-cyclic monophosphate reverses the effects of glucose repression (77).

Table 1 lists major described genetic mutants of *P. chrysosporium*. Various auxotrophic mutants of *P. chrysosporium*, of which most are auxotrophic for amino acids or cofactors (Table 1), have been isolated and used in complementation studies. Gold et al (79, 83) isolated UV and X-ray mutants; some phenotypically similar mutations apparently occurred at different loci. Protoplasts were fused and mycelia were regenerated in various strains (78).

Table 1 Some mutants of *Phanerochaete chrysosporium*

Organism (ATCC number)	Strain or mutation	Phenotype[a]	Reference
32629	44	C^-; X^+; POx^-	6
32629	44-2	C^-; X^-; POx^+	51
32629	63-2	C^-; X^-; POx^+	51
32629	31	C^+; L^+	128
32629	3113	C^-; L^+	128
32629	85118	C^-; L^+	128
24725	SC26	L-enhanced	161
34571	*leu3,1,2*	Leucine auxotroph	204
34571	*ade1,2*	Adenine auxotroph	204
34571	*rib1*	Riboflavin auxotroph	204
34571	LMT 320	Riboflavin auxotroph	191
34571	*nic1*	Nicotinamide auxotroph	204
34571	*arg1,3,4*	Arginine auxotroph	204
34571	*his1,2*	Histidine auxotroph	204
34571	LMT 31	Histidine auxotroph	191
34571	*met1*	Methionine auxotroph	204
34571	*leu1*	Leucine auxotroph	204
34571	LMT 30	Leucine auxotroph	191
34571	LMT 26	POx^-; L-enhanced	191

[a] C = cellulase; X = xylan degradation; POx = phenol oxidase activity; L = lignin degradation; + = positive.

Marker strains carrying multiple mutations were recovered by fruiting heterokaryons of *P. chrysosporium* (5). In further complementation studies certain gene mutations and enzyme deficiencies were identified (204).

Eriksson and coworkers (51) developed various Cel$^-$ as well as phenol oxidase positive (POx$^+$) strains in attempts to improve selective lignin degradation (Table 1). Cross-breeding of homokaryotic Cel$^-$ strains has improved lignin-degrading ability (128). Unfortunately, highly lignin-degrading strains derived from cross-breeding may not be intercrossable, since 75% of the strains did not fruit. Cel$^-$ strains 3113 and 85118 degraded lignin in wood (128) and synthetic lignins (170). In a study with several strains, there was no correlation between lignin degradation and ligninase production (170). SC-26, a mutant selected for enhanced decolorization of a lignin by-product, had about six times the amount of accumulated ligninase as the parent (161, 170).

Gold et al (83) isolated a POx mutant that was pleiotropically lacking in several secondary metabolic functions; a revertant regained all of the functions. Recently, Liwicki et al (191) isolated a number of POx mutants and found that they shared the following idiophasic traits: increased intracellular cAMP levels, sporulation, extracellular glucan, veratryl alcohol synthesis, and lignin-degrading ability. The authors concluded that mutations that result in the loss of lignin-degrading ability are not necessarily pleiotropic with other idiophasic functions.

As pointed out above in the section on physiology, idiophasic metabolism is initiated by N-limitation. Marzluf (198) has extensively studied nitrogen metabolism and its control in *Neurospora crassa, Aspergillus nidulans,* and *Saccharomyces cerevisiae*. Analogous studies with *P. chrysosporium* should contribute to an understanding of N-regulation of lignin degradation. A new selection procedure that uses lignin model compound–amino acid adducts has been developed: ligninase cleaves the compounds, releasing the amino acids, which can serve as growth nitrogen (249). The compounds are now being used to select N-deregulated mutants for lignin degradation (M. Tien, personal communication).

Genetic study of *P. chrysosporium* can be used on the one hand for acquiring more complete basic understanding of the organism, its primary and secondary metabolism, and regulation of ligninolytic activity, and on the other for manipulation in industrial applications. As described in the previous section, a single enzyme, even ligninase, does not completely degrade lignin; therefore, although molecular approaches are warranted for isolation and expression of pertinent ligninolytic enzyme gene(s), fundamental knowledge of the total system is still required. The genetic approach, including manipulation, selection for desired metabolic characteristics, and development of new recombinant strains, is probably the most accessible and clearest approach available for elucidation of bioligninolytic systems.

Molecular Biology

Following the isolation and characterization of ligninase, several laboratories began to apply recombinant DNA research to lignin degradation. The large-scale production of recombinant ligninolytic enzyme(s) in genetically engineered organisms is on the horizon.

CLONING Two similar, simple methods for isolating high-quality DNA and/or RNA that appear general for filamentous fungi have been developed with basidiomycetes, including *P. chrysosporium* (224, 225). Translatable mRNA from *P. chrysosporium* was first isolated (109) by a procedure adapted from that of Garber & Yoder (71).

Comparison of ligninolytic and nonligninolytic mycelial in vitro–translated mRNAs showed several differences in 40–50 species of polypeptides resolved in one-dimensional gels (109). Differences were also observed between in vitro–translated products from 3-day-old and 6-day-old nonligninolytic cultures; this points out the difficulty in interpreting these types of results.

In vitro–translated mRNAs isolated from ligninolytic (N-limited) cultures of *P. chrysosporium* strain BKM (ATCC 24725) contained several translation products in the 25- and 40-kd regions that were not found in mRNAs isolated from nonlignolytic cultures, including a unique 40-kd protein that reacted with affinity-purified polyclonal ligninase antibodies (Y. Devash & A. Anilionis, personal communication).

Ligninase gene(s) encoding isoenzymes of *P. chrysosporium* have been cloned by two groups (60, 252a). Tien & Tu (252a) very recently reported the cloning (in *E. coli*) and sequencing of a cDNA for a ligninase isoenzyme. They showed that synthesis of ligninase is regulated by N-limitation at the mRNA level. Sequence analysis revealed that mature ligninase is preceded by a 28-residue leader, and that the mature protein contains 345 amino acids. Sequence homology around histidine residues of ligninase and turnip, cytochrome *c*, and horseradish peroxidase suggested a similarity in catalytic mechanism (252a). Lignin model compound assays were used to characterize the activities of recombinant ligninase H8, cloned by Farrell (60). With tetramethoxybenzene, veratryl alcohol, and a β-O-4 model (see Figure 2) as substrates, recombinant ligninase catalyzed aromatic demethoxylation, C_α-oxidation, and C_α–C_β cleavage, respectively. The specific activities of recombinant ligninase (rH8) in these assays were essentially equivalent to those of *P. chrysosporium* H8. Oxidation of veratryl alcohol (VA) to veratraldehyde occurred at a specific activity of 25 μmol VA oxidized min^{-1} mg^{-1} enzyme with rH8 compared to 22 μmol VA oxidized min^{-1} mg^{-1} with *P. chrysosporium* H8.

In the presence of kraft lignin (Indulin AT from Westvaco) and milled wood (loblolly pine) lignin the recombinant ligninase exhibited substrate-dependent peroxide uptake. The extent and rate of this reaction were comparable to those of the reaction catalyzed by *P. chrysosporium* ligninase. Analysis of lignin treated with recombinant ligninase by ionization difference spectroscopy indicates that the recombinant enzyme probably catalyzes C_α-oxidation of these polymers, as well as other functional group changes (T. Maione & R. Farrell, unpublished).

Besides the ligninase H8 gene, other genes have been cloned from *P. chrysosporium*. Zhang et al (276) used synthetic oligonucleotide probes to identify putative ligninase cDNAs. P. Broda and coworkers (personal communication) isolated a *trpC* gene by complementation of the corresponding *Escherichia coli* auxotroph. Further, they isolated *P. chrysosporium* acetyl CoA mutants by selecting fluoracetate-resistant strains. This work is aimed at developing a transformation system in *P. chrysosporium*. Rao & Reddy (225) reported several DNA sequences capable of promoting autonomous replication of plasmid Ylp5 in *S. cerevisiae;* these sequences may be useful for vector construction for such a transformation system.

EXPRESSION AND PRODUCTION OF RECOMBINANT ENZYMES The utility of any clone lies in the expression of its gene product. To achieve expression one must consider basic properties of the host organism and their relevance to the gene product of interest. Such properties include fermentability, expression level (percentage of recombinant gene product in the total protein produced), ability to secrete and/or glycosylate, and presence of proteases, which decrease recombinant protein stability.

The major hosts used for expression of recombinant proteins are *Escherichia coli, Bacillus subtilis,* and *Saccharomyces cerevisiae*. Other host organisms that are currently being studied and developed include *Pseudomonas* sp., other yeasts, *Aspergillus nidulans, Aspergillus niger,* mammalian systems, and to a lesser extent plant and insect systems and *Trichoderma reesei*. The highest expression levels have been demonstrated in *E. coli,* which has the disadvantage of inability to glycosylate. In addition, proteins that are normally soluble are often expressed in *E. coli* in an insoluble form in inclusion bodies (101). Solubilization requires denaturation and renaturation, and has been achieved for some proteins (89, 197). Vectors and hosts have been developed for *S. cerevisiae,* and fermentation is probably superior to that by strains of any other species, in large part owing to brewing-industry research (265). The drawbacks of yeast are relatively low expression level and inappropriate glycosylation patterns. The development of transformation systems for *A. nidulans* has permitted investigation for heterologous gene expression in this organism (reviewed in 40, 41, 266). Basic studies will probably require different host organisms (e.g. *A. nidulans*) from those used as industrial expression systems (e.g. *A. niger*). Owing to physiological similarities to fungi such as *P. chrysosporium,* the ability to recognize heterologous promoters and secretion signals, and a high capacity for accumulating protein in the extracellular medium, *Aspergillus* is an excellent choice as a host organism for expression of recombinant ligninolytic organisms.

For recombinant ligninase production, heme incorporation is required in vivo and in vitro. Heme-containing proteins have been expressed in prokaryotic hosts, but usually as the apoprotein. Human myoglobin cDNA expressed as a gene fusion product made up about 10% of *E. coli* total cellular protein. The insoluble recombinant Mb was solubilized from inclusion bodies by denaturation, and apoMb was reconstituted in vitro with heme (267). Electronic absorption spectra were identical to those of native Mb. Nagai et al (205) demonstrated that the oxygen-binding properties of reconstituted recombinant hemoglobin were essentially the same as those of human native hemoglobin. Both of these studies indicate that in vitro incorporation of heme in the recombinant protein resulted in physical characteristics comparable to those of the native protein. P-450, the most similar to ligninase of all cloned proteins, has been successfully expressed in prokaryotic hosts (264). A small percentage of P-450 molecules expressed in *E. coli* appeared to incorporate heme appropriately, but the majority of *E. coli* recombinant P-450 was apoprotein. A greater percentage of P-450 expressed in *S. cerevisiae* had appropriate in vivo incorporation of heme.

Future molecular work will include the cloning of more ligninolytic genes and comparison of their expression and gene products in a variety of hosts. It is likely that ligninase is expressed by a multigenic family, and the structure of the DNAs may give important clues to how they arose (e.g. by gene duplication) and how they are regulated (R. Farrell, unpublished). For eventual application we must develop the most efficient expression systems and maximize the ease of processing active material.

CONCLUSIONS AND RECOMMENDATIONS

Our review leads us to certain general conclusions which in turn point to needs for further research.

1. The literature does not contain convincing evidence that polymeric lignin is biodegraded under anaerobic conditions. We suggest that the slight degradation that has been reported can be attributed to low–molecular weight fragments already present in the samples, freed through the action of hemicellulases or esterases or released abiotically during long incubations.

2. Many aerobic bacteria, including actinomycetes, can degrade certain lignocellulosic materials, and they probably degrade lignin fragments produced abiotically and by other microbes. Some bacteria can apparently degrade a wide range of lignin-related models, which suggests that they possess nonspecific intracellular systems. Further research is needed to determine whether bacteria produce extracellular, polymer-degrading enzymes, i.e. to

determine the ability to attack the polymeric lignin backbone. We suggest that comprehensive studies aimed at maximizing rates and extents of lignin degradation and at identifying the responsible enzymes be undertaken with selected species, with carefully characterized ^{14}C-labeled synthetic lignins, and with dimeric model compounds.

3. Certain ascomycetes and fungi imperfecti, particularly those that cause soft rot and the few that cause a white rot of wood, can degrade lignin. Because of their abundance in soils and plant debris, many of these fungi would seem to be of prime importance in lignin mineralization. As with aerobic bacteria, we suggest that selected species be studied in detail to maximize their rates and extents of lignin degradation and to identify the responsible enzymes.

4. Basidiomycetes other than white-rot fungi have been studied little, but members of other groups clearly degrade lignin. Selected species of litter-degraders, mycorrhizal fungi, brown-rot fungi, and other groups should be investigated. Of particular interest from a biochemical standpoint is the extracellular methoxyl-demethylating system of the brown-rot fungi.

5. Good progress is being made in describing the lignin-degrading enzyme system of white-rot fungi. We suggest that forthcoming biochemical and molecular biological investigations continue to concentrate on *P. chrysosporium;* recent evidence suggests that other white-rot fungi possess a similar enzyme system.

Growing evidence indicates that ligninase (lignin peroxidase) is the key lignin-degrading enzyme of white-rot fungi. Its basic mechanism, involving cation-radical intermediates, has been established, and can explain almost all of the degradation intermediates of model compounds seen in intact whole cultures. Further studies are needed to clarify the mechanism and the relative importance of aromatic ring cleavage by ligninase. Also, the possible role of veratryl alcohol in ligninase catalysis needs to be clarified.

Evidence indicates that enzymes other than ligninase are also involved. What are the roles of the multiple Mn peroxidases and other phenol-oxidizing enzymes? How is repolymerization of reacted lignin and phenolic degradation products controlled in vivo? We suggest that new assays be developed, some based on phenols and polymeric lignin, to facilitate identification of remaining enzymes.

6. Certain physiological questions need to be answered for the *P. chrysosporium* system. What are the roles of the extracellular glucan, Mn, and glutamate? What are the details of induction and regulation of the lignin-degrading enzymes? What is the relationship between energy-consuming and energy-yielding reactions during lignin degradation?

7. We suggest that major emphasis be placed on molecular biology to answer important basic questions about ligninase and other enzymes involved in lignin degradation: (*a*) What is the molecular basis for multiple isoen-

zymes? What purpose does the multiplicity serve? (*b*) What is the role of glycosylation in activity and stability? (*c*) What is the maximal expression of the recombinant ligninolytic enzymes in appropriate vehicles? (*d*) Will heme incorporation into recombinant protein be facilitated in vivo or in vitro?

ENZYMATIC "COMBUSTION"

The size, nonhydrolyzability, heterogeneity, and molecular complexity of lignin dictate that its initial biodegradation be oxidative and nonspecific, and that it be mediated by an extracellular system. It is clear from observations of burning lignin that its conversion to CO_2 and H_2O is thermodynamically favored. In biological systems the oxidative mineralization of organics, unlike combustion, normally takes a channeled route (i.e. metabolic pathway) to thermodynamically stable end products. The reactions are catalyzed by enzymes with a fidelity that is rarely matched even by the best organic chemist. The initial steps in the depolymerization of lignin by white-rot fungi are also catalyzed by enzymes, but with a nonspecificity that leads to a potpourri of diverging reactions that is probably unsurpassed by any other enzyme system. Depolymerization is kinetically favored because ligninases oxidize their substrates by one electron; the diversity of subsequent reactions of the unstable intermediates is a function of their structures. It is this nonspecific subsequent oxidation of lignin that leads us to refer to the process as enzymatic combustion. It is likely that similar types of systems are responsible for the initial reactions in lignin polymer degradation by other groups of microbes. We speculate that enzymatic combustion, as opposed to oxidation catalyzed by specific oxygenases and dehydrogenases, could be important in the microbial degradation of other complex aromatics as well as compounds produced wholly or partly abiotically.

ACKNOWLEDGMENTS

We thank various investigators for sharing papers awaiting publication, as well as unpublished results; this has allowed us to extend coverage somewhat beyond late 1986. We also thank several colleagues for critically reading this manuscript.

Literature Cited

1. Abbott, T. P., Wicklow, D. T. 1984. Degradation of lignin by *Cyathus* species. *Appl. Environ. Microbiol.* 47:585–87
2. Adler, E. 1977. Lignin chemistry. Past, present and future. *Wood Sci. Technol.* 11:169–218
3. Agosin, E., Daudin, J.-J., Odier, E. 1985. Screening of white-rot fungi on [^{14}C]lignin-labelled and [^{14}C]whole-labelled wheat straw. *Appl. Microbiol. Biotechnol.* 22:132–38
4. Agosin, E., Odier, E. 1985. Solid-state fermentation, lignin degradation and resulting digestibility of wheat straw fermented by selected white-rot fungi. *Appl. Microbiol. Biotechnol.* 21:397–403
5. Alic, M., Gold, M. H. 1985. Genetic

recombination in the lignin-degrading basidiomycete *Phanerochaete chrysosporium*. *Appl. Environ. Microbiol.* 50:27–30

6. Ander, P., Eriksson, K.-E. 1976. The importance of phenoloxidase activity in lignin degradation by the white-rot fungus *Sporotrichum pulverulentum*. *Arch. Microbiol.* 109:1–8
7. Ander, P., Eriksson, K.-E. 1985. Methanol formation during lignin degradation by *Phanerochaete chrysosporium*. *Appl. Microbiol. Biotechnol.* 21: 96–102
8. Andersson, L. A., Renganathan, V., Chiu, A. A., Loehr, T. M., Gold, M. H. 1985. Spectral characterization of diarylpropane oxygenase, a novel peroxide-dependent, lignin-degrading heme enzyme. *J. Biol. Chem.* 260:6080–87
9. Antai, S. P., Crawford, D. L. 1981. Degradation of softwood, hardwood, and grass lignocelluloses by two *Streptomyces* strains. *Appl. Environ. Microbiol.* 42:378–80
10. Arjmand, M., Sandermann, H. Jr. 1986. Plant biochemistry of xenobiotics. Mineralization of chloroaniline/lignin metabolites from wheat by the white-rot fungus, *Phanerochaete chrysosporium*. *Z. Naturforsch.* 41c:206–14
11. Asada, Y., Miyabe, M., Kikkawa, M., Kuwahara, M. 1986. Oxidation of NADH by a peroxidase of a lignin-degrading basidiomycete, *Phanerochaete chrysosporium* and its involvement in the degradation of a lignin model compound. *Agric. Biol. Chem.* 50: 525–29
12. Asther, M., Corrieu, G., Drapron, R., Odier, E. 1987. Effect of Tween 80 and oleic acid on ligninase production by *Phanerochaete chrysosporium* INA-12. *Enzyme Microb. Technol.* 9:245–49
13. Bar-Lev, S. S., Kirk, T. K. 1981. Effects of molecular oxygen on lignin degradation by *Phanerochaete chrysosporium*. *Biochem. Biophys. Res. Commun.* 99:373–78
14. Benner, R., Hodson, R. E. 1985. Thermophilic anaerobic biodegradation of [^{14}C]lignin, [^{14}C]cellulose, and [^{14}C]lignocellulose preparations. *Appl. Environ. Microbiol.* 50:971–76
15. Benner, R., MacCubbin, A. E., Hodson, R. E. 1984. Preparation, characterization, and microbial degradation of specifically radiolabelled [^{14}C]lignocelluloses from marine and freshwater macrophytes. *Appl. Environ. Microbiol.* 47:381–89
16. Benner, R., MacCubbin, A. E., Hodson, R. E. 1984. Anaerobic biodegradation of the lignin and polysaccharide components of lignocellulose and synthetic lignin by sediment microflora. *Appl. Environ. Microbiol.* 47:998–1004
17. Benner, R., Moran, M. A., Hodson, R. E. 1986. Biogeochemical cycling of lignocellulosic carbon in marine and freshwater ecosystems: Relative contributions of procaryotes and eucaryotes. *Limnol. Oceanogr.* 31:89–100
18. Blanchette, R. A. 1984. Manganese accumulation in wood decayed by white-rot fungi. *Phytopathology* 74:725–30
19. Blanchette, R. A. 1984. Screening wood decayed by white-rot fungi for preferential lignin degradation. *Appl. Environ. Microbiol.* 48:647–53
20. Blanchette, R. A., Otjen, L., Carlson, M. C. 1987. Lignin distribution in cell walls of birch wood decayed by white rot basidiomycetes. *Phytopathology* In press
21. Blanchette, R. A., Reid, I. D. 1986. Ultrastructural aspects of wood delignification by *Phlebia (Merulius) tremellosus*. *Appl. Environ. Microbiol.* 52:239–45
22. Bumpus, J. A., Tien, M., Wright, D., Aust, S. D. 1985. Oxidation of persistent environmental pollutants by a white rot fungus. *Science* 228:1434–36
23. Burdsall, H. H., Eslyn, W. E. 1974. A new *Phanerochaete* with a *chrysosporium* imperfect state. *Mycotaxon* 1:123–33
24. Buswell, J. A., Ander, P., Eriksson, K.-E. 1982. Ligninolytic activity and levels of ammonia assimilating enzymes in *Sporotrichum pulverulentum*. *Arch. Microbiol.* 133:165–71
25. Buswell, J. A., Mollet, B., Odier, E. 1984. Ligninolytic enzyme production by *Phanerochaete chrysosporium* under conditions of nitrogen sufficiency. *FEMS Microbiol. Lett.* 25:295–99

25a. Buswell, J., Odier, E. 1987. Lignin biodegradation. *CRC Crit. Rev. Biotechnol.* In press

26. Chen, C.-L., Chang, H.-M. 1985. Chemistry of lignin biodegradation. In *Biosynthesis and Biodegradation of Wood*, ed. T. Higuchi, pp. 535–56. San Diego, Calif: Academic. 679 pp.
27. Chen, C.-L., Chang, H.-M., Kirk, T. K. 1982. Aromatic acids produced during degradation of lignin in spruce wood by *Phanerochaete chrysosporium*. *Holzforschung* 36:3–9
28. Chen, C.-L., Chang, H.-M., Kirk, T. K. 1983. Carboxylic acids produced through oxidative cleavage of aromatic rings during degradation of lignin in spruce wood by *Phanerochaete chryso-*

sporium. *J. Wood Chem. Technol.* 3:35–57

29. Chua, M. G. S., Chen, C.-L., Chang, H.-M., Kirk, T. K. 1982. ^{13}C NMR spectroscopic study of spruce lignin degraded by *Phanerochaete chrysosporium*. I. New structures. *Holzforschung* 36:165–72
30. Chua, M. G. S., Choi, S., Kirk, T. K. 1983. Mycelium binding and depolymerization of synthetic ^{14}C-labeled lignin during decomposition by *Phanerochaete chrysosporium*. *Holzforschung* 37:55–61
31. Clayton, N. E., Srinivasan, V. R. 1981. Biodegradation of lignin by *Candida* sp. *Naturwissenschaften* 68:97–98
32. Colberg, P. J., Young, L. Y. 1985. Anaerobic degradation of soluble fractions of (^{14}C)lignocellulose. *Appl. Environ. Microbiol.* 49:345–49
33. Cowling, E. B., Merrill, W. 1966. Nitrogen in wood and its role in wood deterioration. *Can. J. Bot.* 44:1539–54
34. Crawford, R. L. 1981. *Lignin Biodegradation and Transformation*. New York: Wiley. 154 pp.
35. Crawford, D. L., Barder, M. J., Pometto, A. L. III, Crawford, R. L. 1982. Chemistry of softwood lignin degradation by *Streptomyces viridosporus*. *Arch. Microbiol.* 131:140–45
36. Crawford, D. L., Crawford, R. L. 1976. Microbial degradation of lignocellulose: the lignin component. *Appl. Environ. Microbiol.* 31:714–17
37. Crawford, R. L., Crawford, D. L. 1984. Recent advances in studies of the mechanisms of microbial degradation of lignins. *Enzyme Microb. Technol.* 6:434–42
38. Crawford, R. L., Crawford, D. L. 1988. ^{14}C lignins. *Methods Enzymol.* In press
39. Crawford, D. L., Pometto, A. L. III, Crawford, R. L. 1983. Lignin degradation by *Streptomyces viridosporus:* Isolation and characterization of a new polymeric lignin degradation intermediate. *Appl. Environ. Microbiol.* 45:898–904
40. Cullen, D., Gray, G. L., Berka, R. M. 1987. Molecular cloning vectors for *Aspergillus* and *Neurospora*. In *Vectors: A Survey of Molecular Cloning Vectors and Their Uses,* ed. R. L. Rodriguez, D. T. Denhardt. London/Stoneham, Mass: Butterworth. In press
41. Cullen, D., Leong, S. 1986. Recent advances in the molecular genetics of industrial filamentous fungi. *Trends Biotechnol.* 4:285–88
42. Deschamps, A. M., Mahoudeau, G., Lebeault, J. M. 1980. Fast degradation of kraft lignin by bacteria. *Eur. J. Appl. Microbiol. Biotechnol.* 9:45–51
43. Eaton, D. C. 1985. Mineralization of polychlorinated biphenyls by *Phanerochaete chrysosporium*. *Enzyme Microb. Technol.* 7:194–96
44. Enoki, A., Gold, M. H. 1982. Degradation of the diarylpropane lignin model compound 1-(3',4'-diethoxyphenyl)-1, 3-dihydroxy-2-(4''-methoxyphenyl)-propane and derivatives by the basidiomycete *Phanerochaete chrysosporium*. *Arch. Microbiol.* 132:123–30
45. Enoki, A., Goldsby, G. P., Gold, M. H. 1980. Metabolism of the lignin model compounds veratrylglycerol-β-guaiacyl ether and 4-ethoxy-3-methoxyphenylglycerol-β-guaiacyl ether by *Phanerochaete chrysosporium*. *Arch. Microbiol.* 125:227–32
46. Enoki, A., Goldsby, G. P., Gold, M. H. 1981. β-Ether cleavage of the lignin model compound 4-ethoxy-3-methoxyphenylglycerol-β-guaiacyl ether and derivatives by *Phanerochaete chrysosporium*. *Arch. Microbiol.* 129:141–45
47. Enoki, A., Goldsby, G. P., Krishnangkura, K., Gold, M. H. 1981. Degradation of the lignin model compounds 4-ethoxy-3-methoxyphenylglycol β-guaiacyl and vanillic acid ethers by *Phanerochaete chrysosporium*. *FEMS Microbiol. Lett.* 10:373–77
48. Enoki, A., Takahashi, M. 1983. Metabolism of lignin-related compounds by various wood-decomposing fungi. See Ref. 115, pp. 119–33
49. Enoki, A., Takahashi, M., Tanaka, H., Fuse, G. 1985. Degradation of lignin-related compounds and wood components by white-rot and brown-rot fungi. *Mokuzai Gakkaishi* 31:397–408
50. Eriksson, K.-E., Goodell, E. W. 1974. Pleiotropic mutants of the wood-rotting fungus *Polyporus adustus* lacking cellulase, mannanase, and xylanase. *Can. J. Microbiol.* 20:371–78
51. Eriksson, K.-E., Johnsrud, S. C., Vallander, L. 1983. Degradation of lignin and lignin model compounds by various mutants of the white-rot fungus, *Sporotrichum pulverulentum*. *Arch. Microbiol.* 135:161–68
52. Eriksson, K.-E., Kirk, T. K. 1985. Biopulping, biobleaching and treatment of kraft bleaching effluents with white-rot fungi. In *Comprehensive Biotechnology,* ed. C. L. Cooney, A. E. Humphrey, 3:271–94. Toronto: Pergamon
53. Eriksson, K.-E., Pettersson, B., Volc, J., Musilek, V. 1986. Formation and partial characterization of glucose-2-

oxidase, a H_2O_2 producing enzyme in *Phanerochaete chrysosporium*. *Appl. Microbiol. Biotechnol.* 23:257–62

54. Faison, B. D., Kirk, T. K. 1983. Relationship between lignin degradation and production of reduced oxygen species by *Phanerochaete chrysosporium*. *Appl. Environ. Microbiol.* 46:1140–45
55. Faison, B. D., Kirk, T. K. 1985. Factors involved in the regulation of a ligninase activity in *Phanerochaete chrysosporium*. *Appl. Environ. Microbiol.* 49:299–304
56. Faison, B. D., Kirk, T. K., Farrell, R. L. 1986. Role of veratryl alcohol in regulating ligninase activity in *Phanerochaete chrysosporium*. *Appl. Environ. Microbiol.* 52:251–54
57. Faix, O., Mozuch, M. D., Kirk, T. K. 1985. Degradation of gymnosperm (guaiacyl) vs. angiosperm (syringyl/guaiacyl) lignins by *Phanerochaete chrysosporium*. *Holzforschung* 39:203–8
58. Farrell, R. L. 1984. Biocatalysts hold promise for better pulp quality. *Tappi J.* 67:31–33
59. Farrell, R. L. 1987. A new key enzyme for lignin degradation. *Phil. Trans. R. Soc. London Ser. A.* In press
60. Farrell, R. L. 1987. Industrial applications of ligninolytic enzymes. *Kem. Kemi* In press
61. Fengel, D., Wegener, G. 1984. *Wood: Chemistry, Ultrastructure, Reactions*, pp. 132–81. Berlin: de Gruyter
62. Fenn, P., Choi, S., Kirk, T. K. 1981. Ligninolytic activity of *Phanerochaete chrysosporium:* Physiology of suppression by NH_4^+ and L-glutamate. *Arch. Microbiol.* 130:66–71
63. Fenn, P., Kirk, T. K. 1981. Relationship of nitrogen to the onset and suppression of ligninolytic activity and secondary metabolism in *Phanerochaete chrysosporium*. *Arch. Microbiol.* 130:59–65
64. Fenn, P., Kirk, T. K. 1984. Effects of C_α-oxidation in the fungal metabolism of lignin. *J. Wood Chem. Technol.* 4:131–48
65. Flegel, T. W., Meevootisom, V., Kiatapan, S. 1982. Indications of ligninolysis by *Trichoderma* strains isolated from soil during simultaneous screening for fungi with cellulase and laccase activity. *J. Ferment. Technol.* 60:473–75
66. Foisner, R., Messner, K., Stachelberger, H., Röhr, M. 1985. Wood decay by basidiomycetes: Extracellular tripartite membranous structures. *Trans. Br. Mycol. Soc.* 85:257–66
67. Forney, L. J., Reddy, C. A., Pankratz, H. S. 1982. Ultrastructural localization of hydrogen peroxide production in ligninolytic *Phanerochaete chrysosporium* cells. *Appl. Environ. Microbiol.* 44:732–36
68. Forney, L. J., Reddy, C. A., Tien, M., Aust, S. D. 1982. The involvement of hydroxyl radical derived from hydrogen peroxide in lignin degradation by the white rot fungus *Phanerochaete chrysosporium*. *J. Biol. Chem.* 257:11455–62
69. Freer, S. N., Detroy, R. W. 1982. Biological delignification of ^{14}C-labelled lignocelluloses by basidiomycetes: Degradation and solubilization of the lignin and cellulose components. *Mycologia* 74:943–51
70. Freudenberg, K. 1968. The constitution and biosynthesis of lignin. In *Constitution and Biosynthesis of Lignin*, ed. A. C. Neish, K. Freudenberg, pp. 47–122. New York: Springer-Verlag
71. Garber, R. C., Yoder, O. C. 1983. Isolation of DNA from filamentous fungi and separation with nuclear mitochondrial, ribosomal, and plasmid components. *Anal. Biochem.* 135:416–22
72. Girardin, M., Hauteville, M., Metche, M., Tine, E. 1984. Catabolisme du gaîacylglycérol β-gaîacyléther, un modele moléculaire de type lignine, par *Arthrobacter* sp. *C. R. Seances Acad. Sci. Sér. III* 298:351–54
73. Glenn, J. K., Akileswaran, L., Gold, M. H. 1987. Mn(II) oxidation is the principal function of the extracellular Mn-peroxidase from *Phanerochaete chrysosporium*. *Arch. Biochem. Biophys.* 251:688–96
74. Glenn, J. K., Gold, M. H. 1985. Purification and characterization of an extracellular Mn(II)-dependent peroxidase from the lignin-degrading basidiomycete *Phanerochaete chrysosporium*. *Arch. Biochem. Biophys.* 242:329–41
75. Glenn, J. K., Morgan, M. A., Mayfield, M. B., Kuwahara, M., Gold, M. H. 1983. An extracellular H_2O_2-requiring enzyme preparation involved in lignin biodegradation by the white rot basidiomycete *Phanerochaete chrysosporium*. *Biochem. Biophys. Res. Commun.* 114:1077–83
76. Gold, M. H., Cheng, T. M. 1978. Induction of colonial growth and replica plating of the white-rot basidiomycete *Phanerochaete chrysosporium*. *Appl. Environ. Microbiol.* 35:1223–25
77. Gold, M. H., Cheng, T. M. 1979. Conditions for fruit body formation in the white-rot basidiomycete *Phanerochaete*

chrysosporium. Arch. Microbiol. 121: 37–41

78. Gold, M. H., Cheng, T. M., Alic, M. 1983. Formation, fusion, and regeneration of protoplasts from wild-type and auxotrophic strains of the white-rot basidiomycete *Phanerochaete chrysosporium. Appl. Environ. Microbiol.* 46:260–63
79. Gold, M. H., Cheng, T. M., Mayfield, M. B. 1982. Isolation and complementation studies of auxotrophic mutants of the lignin-degrading basidiomycete *Phanerochaete chrysosporium. Appl. Environ. Microbiol.* 44:996–1000
80. Gold, M. H., Enoki, A., Morgan, M. A., Mayfield, M. B., Tanaka, H. 1984. Degradation of the γ-carboxyl-containing diarylpropane lignin model compound 3-(4'-ethoxy-3'-methoxyphenyl)-2-(4''-methoxyphenyl) propionic acid by the basidiomycete *Phanerochaete chrysosporium. Appl. Environ. Microbiol.* 47:597–600
81. Gold, M. H., Glenn, J. K., Mayfield, M. B., Morgan, M. A., Kutsuki, H. 1983. Biochemical and genetic studies on lignin degradation by *Phanerochaete chrysosporium.* See Ref. 115, pp. 219–32
82. Gold, M. H., Kuwahara, M., Chiu, A. A., Glenn, J. K. 1984. Purification and characterization of an extracellular H_2O_2-requiring diarylpropane oxygenase from the white rot basidiomycete, *Phanerochaete chrysosporium. Arch. Biochem. Biophys.* 234:353–62
83. Gold, M. H., Mayfield, M. B., Cheng, T. M., Krisnangkura, K., Shimida, M., et al. 1982. A *Phanerochaete chrysosporium* mutant defective in lignin degradation as well as several other secondary metabolic functions. *Arch. Microbiol.* 132:115–22
84. Goldsby, G. P., Enoki, A., Gold, M. H. 1980. Alkyl-phenyl cleavage of the lignin model compounds guaiacylglycol and glycerol-β-guaiacyl ether by *Phanerochaete chrysosporium. Arch. Microbiol.* 128:190–95
85. Gonzalez, B., Merino, A., Almeida, M., Vicuña, R. 1987. Comparative growth of natural bacterial isolates on various lignin-related compounds. *Appl. Environ. Microbiol.* 52:In press
86. Goycoolea, M., Seelenfreund, D., Rüttimann, C., González, B., Vicuña, R. 1986. Monitoring bacterial consumption of low molecular weight lignin derivatives by high performance liquid chromatography. *Enzyme Microb. Technol.* 8:213–16
87. Greene, R. V., Gould, J. M. 1984. Fatty acyl-coenzyme A oxidase activity and H_2O_2 production in *Phanerochaete chrysosporium* mycelia. *Biochem. Biophys. Res. Commun.* 118:437–43
88. Greene, R. V., Gould, J. M. 1986. H_2O_2-dependent decolorization of Poly R-481 by particulate fractions from *Phanerochaete chrysosporium. Biochem. Biophys. Res. Commun.* 136: 220–27
89. Gribskov, M., Burgess, R. 1983. Overexpression and purification of the sigma subunit of *E. coli* RNA polymerase. *Gene* 26:108–18
90. Habe, T., Shimada, M., Okamoto, T., Panijpan, B., Higuchi, T. 1985. Incorporation of dioxygen into the hydroxylated product during the C–C single bond cleavage of 1,2-bis(*p*-methoxyphenyl)propane-1,3-diol catalysed by hemin. A novel model system for the hemoprotein ligninase. *J. Chem. Soc. Chem. Commun.* 1985:1323–24
91. Habe, T., Shimada, M., Umezawa, T., Higuchi, T. 1985. Evidence for deuterium retention in the products after enzymatic C–C and ether bond cleavages of deuterated lignin model compounds. *Agric. Biol. Chem.* 49:3505–10
92. Haemmerli, S. D., Leisola, M. S. A., Fiechter, A. 1986. Polymerisation of lignins by ligninases from *Phanerochaete chrysosporium. FEMS Microbiol. Lett.* 35:33–36
93. Haemmerli, S. D., Leisola, M. S. A., Sangland, D., Fiechter, A. 1986. Oxidation of benzo(*a*)pyrene by extracellular ligninases of *Phanerochaete chrysosporium. J. Biol. Chem.* 261:6900–3
94. Haider, K., Kern, H. W., Ernst, L. 1985. Intermediate steps of microbial lignin degradation as elucidated by ^{13}C NMR spectroscopy of specifically ^{13}C-enriched DHP-lignins. *Holzforschung* 39:23–32
95. Haider, K., Trojanowski, J. 1975. Decomposition of specifically ^{14}C-labelled phenols and dehydropolymers of coniferyl alcohol as models for lignin degradation by soft and white rot fungi. *Arch. Microbiol.* 105:33–41
96. Haider, K., Trojanowski, J. 1980. A comparison of the degradation of ^{14}C-labeled DHP and corn stalk lignins by micro- and macrofungi and bacteria. See Ref. 163, 1:111–34
97. Hammel, K. E., Kalyanaraman, B., Kirk, T. K. 1986. Substrate free radicals are intermediates in ligninase catalysis. *Proc. Natl. Acad. Sci. USA* 83:3708–12
98. Hammel, K. E., Kalyanaraman, B., Kirk, T. K. 1986. Oxidation of polycyclic aromatic hydrocarbons and

dibenzo[*p*]-dioxins by *Phanerochaete chrysosporium* ligninase. *J. Biol. Chem.* 261:16948–52

99. Hammel, K. E., Tien, M., Kalyanaraman, B., Kirk, T. K. 1985. Mechanism of oxidative C_α-C_β cleavage of a lignin model dimer by *Phanerochaete chrysosporium* ligninase: Stoichiometry and involvement of free radicals. *J. Biol. Chem.* 260:8348–53
100. Harkin, J. M., Obst, J. R. 1973. Syringaldazine, an effective reagent for detecting laccase and peroxidase in fungi. *Experientia* 29:381–87
101. Harris, T. J. R. 1983. Expression of eukaryotic genes in *E. coli*. In *Genetic Engineering,* ed. R. Williamson, 4:127–83. London: Academic
102. Harvey, P. J., Schoemaker, H. E., Bowen, R. M., Palmer, J. M. 1985. Single-electron transfer processes and the reaction mechanism of enzymic degradation of lignin. *FEBS Lett.* 183:13–16
103. Harvey, P. J., Schoemaker, H. E., Palmer, J. M. 1985. Enzymic degradation of lignin and its potential to supply chemicals. In *Plant Products and the New Technology. Annu. Proc. Phytochem. Soc. Eur.* 26:249–66
104. Harvey, P. J., Schoemaker, H. E., Palmer, J. M. 1986. Veratryl alcohol as a mediator and the role of radical cations in lignin biodegradation by *Phanerochaete chrysosporium. FEBS Lett.* 195:242–46
105. Hatakka, A. 1986. *Degradation and conversion of lignin, lignin-related aromatic compounds and lignocellulose by selected white-rot fungi*. PhD thesis. Univ. Helsinki, Finland. 97 pp.
106. Hatakka, A. I., Buswell, J. A., Pirhonen, T. I., Uusi-Rauva, A. K. 1983. Degradation of ^{14}C-labelled lignins by white-rot fungi. See Ref. 115, pp. 176–85
107. Hatakka, A., Tervilä-wilo, A. 1986. Ligninases of white-rot fungi. *Proc. Sov. Finn. Sem. Microb. Degradation Lignocellul. Raw Mater., 1985,* Tbilisi, Georgia, USSR, pp. 65–74. Pushchino, USSR: USSR Acad. Sci. 117 pp.
108. Hatakka, A. I., Uusi-Rauva, A. K. 1983. Degradation of ^{14}C-labelled poplar wood lignin by selected white-rot fungi. *Appl. Microbiol. Biotechnol.* 17:235–42
109. Haylock, R. A., Liwicki, R., Broda, P. 1985. The isolation of mRNA from the basidiomycete fungi *Phanerochaete chrysosporium* and *Coprinus cinerus* and its *in vitro* translation. *J. Microbiol. Methods* 41:55–62
110. Higuchi, T. 1983. Biosynthesis and microbial degradation of lignin. In *The New Frontiers in Plant Biochemistry,* ed. T. Akazawa, T. Asahi, H. Imaseki, pp. 23–46. Tokyo: Jpn. Sci. Soc.
111. Higuchi, T. 1984. Mechanism of microbial degradation of lignin. *Mokuzai Gakkaishi* 30:613–27
112. Deleted in proof
113. Higuchi, T. 1985. Degradative pathways of lignin model compounds. See Ref. 26, pp. 557–78
114. Higuchi, T. 1987. Catabolic pathways and role of ligninase for the degradation of lignin substructure model compounds by white-rot fungi. *Wood Res.* 73:58–81
115. Higuchi, T., Chang, H.-M., Kirk, T. K., eds. 1983. *Recent Advances in Lignin Biodegradation Research.* Tokyo: Uni. 279 pp.
116. Higuchi, T., Nakatsubo, F., Kamaya, Y., Umezawa, T. 1983. Mechanism of β-aryl ether cleavage by *Phanerochaete chrysosporium,* and the role of peroxidase in lignin biodegradation. See Ref. 115, pp. 209–18
117. Holt, D. M., Jones, E. B. G. 1983. Bacterial degradation of lignified wood cell walls in anaerobic aquatic habitats. *Appl. Environ. Microbiol.* 46:722–27
118. Huynh, V.-B. 1986. Biomimetic oxidation of lignin model compounds by simple inorganic complexes. *Biochem. Biophys. Res. Commun.* 139:1104–10
119. Huynh, V.-B., Chang, H.-M., Joyce, T. W. 1985. Dechlorination of chloroorganics by a white-rot fungus. *Tappi J.* 68:98–102
120. Huynh, V.-B., Crawford, R. L. 1985. Novel extracellular enzymes (ligninases) of *Phanerochaete chrysosporium. FEMS Microbiol. Lett.* 28:119–23
121. Huynh, V.-B., Paszczyñski, A., Olson, P., Crawford, R. 1987. Transformations of arylpropane lignin model compounds by a lignin peroxidase of the white-rot fungus *Phanerochaete chrysosporium. Arch. Biochem. Biophys.* 250:139–46
122. Iwahara, S. 1983. Metabolism of lignin-related aromatic compounds by *Fusarium* species. See Ref. 115, pp. 96–111
123. Jäger, A., Croan, S., Kirk, T. K. 1985. Production of ligninases and degradation of lignin in agitated submerged cultures of *Phanerochaete chrysosporium. Appl. Environ. Microbiol.* 50:1274–78
124. Janshekar, H., Brown, C., Haltmeier, T., Leisola, M., Fiechter, A. 1982. Bioalteration of kraft pine lignin by *Phanerochaete chrysosporium. Arch. Microbiol.* 132:14–21
125. Janshekar, H., Fiechter, A. 1982. On

the bacterial degradation of lignin. *Appl. Microbiol. Biotechnol.* 14:47–50

126. Janshekar, H., Fiechter, A. 1983. Lignin: Biosynthesis, applications, and biodegradation. In *Pentoses and Lignin. Adv. Biochem. Eng. Biotechnol.* 27: 119–78
127. Jeffries, T. W., Choi, S., Kirk, T. K. 1981. Nutritional regulation of lignin degradation by *Phanerochaete chrysosporium. Appl. Environ. Microbiol.* 42:290–96
128. Johnsrud, S. C., Eriksson, K.-E. 1985. Cross-breeding of selected and mutated homokaryotic strains of *Phanerochaete chrysosporium* K-3: New cellulase-deficient strains with increased ability to degrade lignin. *Appl. Microbiol. Biotechnol.* 21:320–27
129. Jokela, J., Pellinen, J., Salkinoja-Salonen, M., Brunow, G. 1985. Biodegradation of two tetrameric lignin model compounds by a mixed bacterial culture. *Appl. Microbiol. Biotechnol.* 23:38–46
130. Jurasek, L., Paice, M. 1986. Pulp, paper, and biotechnology. *Chemtech* 16:361–65
131. Kadam, K. L., Drew, S. W. 1986. Study of lignin biotransformation by *Aspergillus fumigatus* and white-rot fungi using ^{14}C-labeled and unlabeled kraft lignins. *Biotechnol. Bioeng.* 28:394–404
132. Kamaya, Y., Higuchi, T. 1983. Degradation of d,l-syringaresinol and its derivatives, β,β'-linked lignin substructure models, by *Phanerochaete chrysosporium. Mokuzai Gakkaishi* 29: 789–94
133. Kamaya, Y., Higuchi, T. 1984. Degradation of lignin substructure models with biphenyl linkage by *Phanerochaete chrysosporium* Burds. *Wood Res.* 70: 25–28
134. Kamaya, Y., Higuchi, T. 1984. Metabolism of 1,2-disyringylpropane-1,3-diol by *Phanerochaete chrysosporium. Mokuzai Gakkaishi* 30:237–39
135. Kamaya, Y., Higuchi, T. 1984. Metabolism of non-phenolic diarylpropane lignin substructure model compound by *Coriolus versicolor. FEMS Microbiol. Lett.* 22:89–92
136. Kamaya, Y., Nakatsubo, F., Higuchi, T. 1983. Degradation of trimeric lignin model compounds, arylglycerol-β-syringaresinol ethers, by *Fusarium solani* M-13-1. *Agric. Biol. Chem.* 47:299–308
137. Katayama, T., Nakatsubo, F., Higuchi, T. 1986. Degradation of a phenylcoumaran, a lignin substructure model, by *Fusarium solani* M-13-1. *Mokuzai Gakkaishi* 32:535–44
138. Kawai, S., Umezawa, T., Higuchi, T. 1985. Metabolism of a non-phenolic β-O-4 lignin substructure model compound by *Coriolus versicolor. Agric. Biol. Chem.* 49:2325–30
139. Kawai, S., Umezawa, T., Higuchi, T. 1985. Arylglycerol-γ-formyl ester as an aromatic ring cleavage product of nonphenolic β-O-4 lignin substructure model compounds degraded by *Coriolus versicolor. Appl. Environ. Microbiol.* 50:1505–8
140. Kawai, S., Umezawa, T., Higuchi, T. 1986. *p*-Benzoquinone monoketals, novel degradation products of β-*O*-4 lignin model compounds by *Coriolus versicolor* and lignin peroxidase of *Phanerochaete chrysosporium. FEBS Lett.* 210:61–65
141. Kawai, S., Umezawa, T., Higuchi, T. 1987. *De novo* synthesis of veratryl alcohol by *Coriolus versicolor. Wood Res.* 73:18–21
142. Kawakami, H., Shumiya, Y. 1983. Degradation of lignin-related compounds and lignins by alkalophilic bacteria. See Ref. 115, pp. 64–77
143. Kelley, R. L., Ramasamy, K., Reddy, C. A. 1986. Characterization of glucose oxidase-negative mutants of a lignin degrading basidiomycete *Phanerochaete chrysosporium. Arch. Microbiol.* 144: 254–57
144. Kelley, R. L., Reddy, C. A. 1986. Purification and characterization of glucose oxidase from ligninolytic cultures of *Phanerochaete chrysosporium. J. Bacteriol.* 166:269–74
145. Kelley, R. L., Reddy, C. A. 1986. Identification of glucose oxidase activity as the primary source of hydrogen peroxide production in ligninolytic cultures of *Phanerochaete chrysosporium. Arch. Microbiol.* 144:248–53
146. Kern, H. W. 1981. Microbial degradation of lignosulfonates. In *Microbial Degradation of Xenobiotics and Recalcitrant Compounds. FEMS Symp.* 12:299–324
147. Kern, H. W. 1983. Transformation of lignosulfonates by *Trichoderma harzianum. Holzforschung* 37:109–15
148. Kern, H. W. 1983. Increased biooxidation of lignosulfonates by *Sporotrichum pulverulentum* in the presence of polyacrylic acid. *Appl. Microbiol. Biotechnol.* 17:182–86
149. Kern, H. W. 1984. Bacterial degradation of dehydropolymers of coniferyl alcohol. *Arch. Microbiol.* 138:18–25
150. Kerr, T. J., Kerr, R. D., Benner, R.

1983. Isolation of a bacterium capable of degrading peanut hull lignin. *Appl. Environ. Microbiol.* 46:1201–6

151. Kersten, P. J., Kirk, T. K. 1987. Involvement of a new enzyme, glyoxal oxidase, in extracellular H_2O_2 production by *Phanerochaete chrysosporium. J. Bacteriol.* 169:2195–202

152. Kersten, P. J., Tien, M., Kalyanaraman, B., Kirk, T. K. 1985. The ligninase of *Phanerochaete chrysosporium* generates cation radicals from methoxybenzenes. *J. Biol. Chem.* 260:2609–12

153. Keyser, P., Kirk, T. K., Zeikus, J. G. 1978. Ligninolytic enzyme system of *Phanerochaete chrysosporium:* Synthesized in the absence of lignin in response to nitrogen starvation. *J. Bacteriol.* 135:790–97

154. Kirk, T. K. 1984. Degradation of lignin. In *Microbial Degradation of Organic Compounds,* ed. D. T. Gibson, pp. 399–437. New York: Dekker

155. Kirk, T. K. 1987. Lignin-degrading enzymes. *Phil. Trans. R. Soc. London Ser. A* In press

156. Kirk, T. K., Adler, E. 1970. Methoxyl-deficient structural elements in lignin of sweetgum decayed by a brown-rot fungus. *Acta Chem. Scand.* 24:3379–90

157. Kirk, T. K., Brunow, G. 1988. Synthetic ^{14}C-lignins. *Methods Enzymol.* In press

158. Kirk, T. K., Chang, H.-M. 1981. Potential applications of bioligninolytic systems. *Enzyme Microb. Technol.* 3:189–96

159. Kirk, T. K., Connors, W. J., Bleam, R. D., Hackett, W. F., Zeikus, J. G. 1975. Preparation and microbial decomposition of synthetic ^{14}C-lignins. *Proc. Natl. Acad. Sci. USA* 72:2515–19

160. Kirk, T. K., Cowling, E. B. 1984. Biological decomposition of solid wood. *Adv. Chem. Ser.* 207:455–87

161. Kirk, T. K., Croan, S., Tien, M., Murtagh, K. E., Farrell, R. L. 1986. Production of multiple ligninases by *Phanerochaete chrysosporium:* Effect of selected growth conditions and use of a mutant strain. *Enzyme Microb. Technol.* 8:27–32

162. Kirk, T. K., Fenn, P. 1982. Formation and action of the ligninolytic system in basidiomycetes. In *Decomposer Basidiomycetes. Br. Mycol. Soc. Symp.* 4, ed. M. J. Swift, J. Frankland, J. N. Hedger, pp. 67–90. Cambridge: Cambridge Univ. Press

163. Kirk, T. K., Higuchi, T., Chang, H.-M., eds. 1980. *Lignin Biodegradation: Microbiology, Chemistry, and Potential Applications,* Vols. 1, 2. Boca Raton, Fla: CRC. 241 pp., 255 pp.

164. Kirk, T. K., Mozuch, M. D., Tien, M. 1985. Free hydroxyl radical is not involved in an important reaction of lignin degradation by *Phanerochaete chrysosporium* Burds. *Biochem. J.* 226:455–60

165. Kirk, T. K., Nakatsubo, F. 1983. Chemical mechanism of an important cleavage reaction in the fungal degradation of lignin. *Biochim. Biophys. Acta* 756:376–84

166. Kirk, T. K., Nakatsubo, F., Reid, I. D. 1983. Further study discounts role for singlet oxygen in fungal degradation of lignin model compounds. *Biochem. Biophys. Res. Commun.* 111:200–4

167. Kirk, T. K., Schultz, E., Connors, W. J., Lorenz, L. F., Zeikus, J. G. 1978. Influence of culture parameters on lignin metabolism by *Phanerochaete chrysosporium. Arch. Microbiol.* 117:277–85

168. Kirk, T. K., Shimada, M. 1985. Lignin biodegradation: The microorganisms involved and the physiology and biochemistry of degradation by white-rot fungi. See Ref. 26, pp. 579–605

169. Kirk, T. K., Tien, M. 1983. Biochemistry of lignin degradation by *Phanerochaete chrysosporium:* Investigations with non-phenolic model compounds. See Ref. 115, pp. 233–45

170. Kirk, T. K., Tien, M., Johnsrud, S. C., Eriksson, K.-E. 1986. Lignin-degrading activity of *Phanerochaete chrysosporium* Burds.: Comparisons of cellulase-negative and other strains. *Enzyme Microb. Technol.* 8:75–80

171. Kirk, T. K., Tien, M., Kersten, P. J., Mozuch, M. D., Kalyanaraman, B. 1986. Ligninase of *Phanerochaete chrysosporium.* Mechanism of its degradation of the non-phenolic arylglycerol β-aryl ether substructure of lignin. *Biochem. J.* 236:279–87

172. Kuila, D., Tien, M., Fee, J. A., Ondrias, M. R. 1985. Resonance raman spectra of extracellular ligninase: Evidence for a heme active site similar to those of peroxidases. *Biochemistry* 24:3394–97

173. Kutsuki, H., Enoki, A., Gold, M. H. 1983. Riboflavin-photosensitized oxidative degradation of a variety of lignin model compounds. *Photochem. Photobiol.* 37:1–7

174. Kuwahara, M., Glenn, J. K., Morgan, M. A., Gold, M. H. 1984. Separation and characterization of two extracellular H_2O_2-dependent oxidases from ligninolytic cultures of *Phanerochaete chrysosporium. FEBS Lett.* 169:247–50

175. Kuwahara, M., Ishida, Y., Miyagawa, Y., Kawakami, C. 1984. Production of extracellular NAD and NADP by a lignin-degrading fungus, *Phanerochaete chrysosporium*. *J. Ferment. Technol.* 62:237–42
176. Leatham, G. F. 1986. The ligninolytic activities of *Lentinus edodes* and *Phanerochaete chrysosporium*. *Appl. Microbiol. Biotechnol.* 24:51–58
177. Leatham, G. F., Kirk, T. K. 1983. Regulation of ligninolytic activity by nutrient nitrogen in white-rot basidiomycetes. *FEMS Microbiol. Lett.* 16:65–67
178. Leisola, M. S. A., Brown, C., Laurila, M., Ulmer, D., Fiechter, A. 1982. Polysaccharide synthesis by *Phanerochaete chrysosporium* during degradation of kraft lignin. *Appl. Microbiol. Biotechnol.* 15:180–84
179. Leisola, M. S. A., Fiechter, A. 1985. New trends in lignin biodegradation. *Adv. Biotechnol. Processes* 5:59–89
180. Leisola, M. S. A., Fiechter, A. 1985. Ligninase production in agitated conditions by *Phanerochaete chrysosporium*. *FEMS Microbiol. Lett.* 29:33–36
181. Leisola, M. S. A., Kozulic, B., Meusdoerffer, F., Fiechter, A. 1987. Homology among multiple extracellular peroxidases from *Phanerochaete chrysosporium*. *J. Biol. Chem.* 262:419–24
182. Leisola, M. S. A., Muessdoerffer, F., Waldner, R., Fiechter, A. 1985. Production and identification of extracellular oxidases of *Phanerochaete chrysosporium*. *J. Biotechnol.* 2:379–82
183. Leisola, M. S. A., Schmidt, B., Thanei-Wyss, U., Fiechter, A. 1985. Aromatic ring cleavage of veratryl alcohol by *Phanerochaete chrysosporium*. *FEBS Lett.* 189:267–70
184. Leisola, M. S. A., Thanei-Wyss, U., Fiechter, A. 1985. Strategies for production of high ligninase activities by *Phanerochaete chrysosporium*. *J. Biotechnol.* 3:97–107
185. Leisola, M. S. A., Ulmer, D., Fiechter, A. 1983. Problem of oxygen transfer during degradation of lignin by *Phanerochaete chrysosporium*. *Appl. Microbiol. Biotechnol.* 17:113–16
186. Leisola, M. S. A., Ulmer, D. C., Fiechter, A. 1984. Factors affecting lignin degradation in lignocellulose by *Phanerochaete chrysosporium*. *Arch. Microbiol.* 137:171–75
187. Leisola, M., Ulmer, D., Haltmeier, T., Fiechter, A. 1983. Rapid solubilization and depolymerization of purified kraft lignin by thin layers of *Phanerochaete chrysosporium*. *Appl. Microbiol. Biotechnol.* 17:117–20
188. Leisola, M. S. A., Ulmer, D. C., Waldner, R., Fiechter, A. 1984. Role of veratryl alcohol in lignin degradation by *Phanerochaete chrysosporium*. *J. Biotechnol.* 1:331–39
189. Levonen-Munoz, E., Bone, D. H., Daugulis, A. J. 1983. Solid state fermentation and fractionation of oat straw by basidiomycetes. *Appl. Microbiol. Biotechnol.* 18:120–23
190. Linko, Y.-Y., Leisola, M., Lindholm, N., Troller, J., Linko, P., Fiechter, A. 1986. Continuous production of lignin peroxidase by *Phanerochaete chrysosporium*. *J. Biotechnol.* 4:283–91
191. Liwicki, R., Paterson, A., MacDonald, M. J., Broda, P. 1985. Phenotypic classes of phenoloxidase-negative mutants of the lignin-degrading fungus, *Phanerochaete chrysosporium*. *J. Bacteriol.* 162:641–44
192. Lobarzewski, J., Trojanowski, J., Wojtas-Wasilewska, M. 1982. The effects of fungal peroxidase on Na-lignosulfonates. *Holzforschung* 36:173–76
193. Lundquist, K., Kirk, T. K. 1978. *De novo* synthesis and decomposition of veratryl alcohol by a lignin-degrading basidiomycete. *Phytochemistry* 17:1676
194. MacDonald, M. J., Ambler, R., Broda, P. 1985. Regulation of intracellular cyclic AMP levels in the white-rot fungus *Phanerochaete chrysosporium* during the onset of idiophasic metabolism. *Arch. Microbiol.* 142:152–56
195. MacDonald, M. J., Paterson, A., Broda, P. 1984. Possible relationship between cyclic AMP and idiophasic metabolism in the white-rot fungus *Phanerochaete chrysosporium*. *J. Bacteriol.* 160:470–72
196. Maltseva, O. V., Myasoedowa, N. M., Leontievsky, A. A., Golovleva, L. A. 1986. Characteristics of the ligninolytic system of *Panus tigrinus* See Ref. 107, pp. 74–82
197. Marston, F. A. O., Lowe, P. A., Doel, M. T., Schoemaker, J. M., White, S., Angal, S. 1984. Purification of calf prochymosin (prorennin) synthesized in *Escherichia coli*. *Biotechnology* 2: 800–4
198. Marzluf, G. A. 1981. Regulation of nitrogen metabolism and gene expression in fungi. *Microbiol. Rev.* 45:437–61
199. McCarthy, A. J., Broda, P. 1984. Screening for lignin-degrading actinomycetes and characterization of their activity against [^{14}C]-lignin-labelled

wheat lignocellulose. *J. Gen. Microbiol.* 130:2905–13

200. McCarthy, A. J., MacDonald, M. J., Paterson, A., Broda, P. 1984. Degradation of [^{14}C]-lignin-labelled wheat lignocellulose by white-rot fungi. *J. Gen. Microbiol.* 130:1023–30
201. McCarthy, A. J., Paterson, A., Broda, P. 1986. Lignin solubilisation by *Thermomonospora mesophila. Appl. Microbiol. Biotechnol.* 24:347–52
202. Miki, K., Renganathan, V., Gold, M. H. 1986. Novel aryl ether rearrangement catalyzed by lignin peroxidase of *Phanerochaete chrysosporium. FEBS Lett.* 203:235–38
203. Miki, K., Renganathan, V., Gold, M. H. 1986. Mechanism of β-aryl ether dimeric lignin model compound oxidation by lignin peroxidase of *Phanerochaete chrysosporium. Biochemistry* 25:4790–96
204. Molskness, T. A., Alic, M., Gold, M. H. 1986. Characterization of leucine auxotrophs of the white-rot basidiomycete *Phanerochaete chrysosporium. Appl. Environ. Microbiol.* 51:1170–73
205. Nagai, K., Perutz, M. F., Payart, C. 1985. Oxygen binding properties of human mutant hemoglobins synthesized in *Escherichia coli. Proc. Natl. Acad. Sci. USA* 82:7252–55
206. Nakatsubo, F., Kirk, T. K., Shimada, M., Higuchi, T. 1981. Metabolism of a phenylcoumaran substructure lignin model compound in ligninolytic cultures of *Phanerochaete chrysosporium. Arch. Microbiol.* 128:416–20
207. Nakatsubo, F., Reid, I. D., Kirk, T. K. 1982. Incorporation of $^{18}O_2$ and absence of stereospecificity in primary product formation during fungal metabolism of a lignin model compound. *Biochim. Biophys. Acta* 719:284–91
208. Nilsson, T., Holt, E. 1983. Bacterial attack occurring in the S_2 layer of wood fibres. *Holzforschung* 37:107–8
209. Norris, D. M. 1980. Degradation of ^{14}C-labeled aromatic acids by *Fusarium solani. Appl. Environ. Microbiol.* 40:376–80
210. Obst, J. R., Kirk, T. K. 1988. Isolation of lignin. *Methods Enzymol.* In press
211. Deleted in proof
212. Odier, E., Monties, B. 1983. Absence of microbial mineralization of lignin in anaerobic enrichment cultures. *Appl. Environ. Microbiol.* 46:661–65
213. Odier, E., Roch, P. 1983. Factors controlling biodegradation of lignin in wood by various white-rot fungi. See Ref. 115, pp. 188–94
214. Oki, T., Shinmoto, M., Ishikawa, H. 1986. Enzymatic degradation of guaiacylglycerol-β-guaiacyl ether. *Mokuzai Gakkaishi* 32:448–56
215. Otjen, L., Blanchette, R. A. 1985. Selective delignification of aspen wood blocks in vitro by three white rot basidiomycetes. *Appl. Environ. Microbiol.* 50:568–72
216. Otjen, L., Blanchette, R. A. 1986. A discussion of microstructural changes in wood during decomposition by white rot basidiomycetes. *Can. J. Bot.* 64:905–11
217. Palmer, J. G., Murmanis, L. L., Highley, T. L. 1983. Visualization of hyphal sheath in wood-decay hymenomycetes. II. White-rotters. *Mycologia* 75:1005–10
218. Paszczyński, A., Huynh, V.-B., Crawford, R. L. 1985. Enzymatic activities of an extracellular, manganese-dependent peroxidase from *Phanerochaete chrysosporium. FEMS Microbiol. Lett.* 29:37–41
219. Paszczyński, A., Huynh, V.-B., Crawford, R. L. 1986. Comparison of ligninase-I and peroxidase-M2 from the white-rot fungus *Phanerochaete chrysosporium. Arch. Biochem. Biophys.* 244:750–65
220. Pellinen, J., Vaisanen, E., Salkinoja-Salonen, M., Brunow, G. 1984. Utilization of dimeric lignin model compounds by mixed bacterial cultures. *Appl. Microbiol. Biotechnol.* 20:77–82
221. Pettey, T. M., Crawford, D. L. 1985. Characterization of acid-precipitable, polymeric lignin (APPL) produced by *Streptomyces viridosporus* and protoplast fusion recombinant *Streptomyces* strains. *Biotechnol. Bioeng. Symp.* 15: 179–90
222. Pometto, A. L. III, Crawford, D. L. 1986. Catabolic fate of *Streptomyces viridosporus* T7A-produced, acid-precipitable polymeric lignin upon incubation with ligninolytic *Streptomyces* species and *Phanerochaete chrysosporium. Appl. Environ. Microbiol.* 51:171–79
223. Raeder, U., Broda, P. 1984. Comparison of the lignin-degrading white-rot fungi *Phanerochaete chrysosporium* and *Sporotrichum pulverulentum* at the DNA level. *Curr. Genet.* 8:499–506
224. Raeder, U., Broda, P. 1985. Rapid preparation of DNA from filamentous fungi. *Lett. Appl. Microbiol.* 1:17–20
225. Rao, T. R., Reddy, C. A. 1984. DNA sequences from a ligninolytic filamentous fungus *Phanerochaete chrysosporium* capable of autonomous

replication in yeast. *Biochem. Biophys. Res. Commun.* 118:821–27
226. Reddy, C. A., Forney, L. J., Kelley, R. L. 1983. Involvement of hydrogen peroxide–derived hydroxyl radical in lignin degradation by the white-rot fungus *Phanerochaete chrysosporium*. See Ref. 115, pp. 153–63
227. Reddy, C. A., Kelley, R. L. 1986. The central role of hydrogen peroxide in lignin degradation by *Phanerochaete chrysosporium*, In *Biodeterioration 6*, ed. S. Barry, D. R. Houghton, D. C. Llewellyn, C. E. O'Rear. pp. 535–42. Slough, UK: Commonw. Agric. Bur.
228. Reid, I. D. 1983. Effects of nitrogen supplements on degradation of aspen wood lignin and carbohydrate components by *Phanerochaete chrysosporium*. *Appl. Environ. Microbiol.* 45:830–37
229. Reid, I. D. 1983. Effects of nitrogen sources on cellulose and synthetic lignin degradation by *Phanerochaete chrysosporium*. *Appl. Environ. Microbiol.* 45: 838–42
230. Reid, I. D. 1985. Biological delignification of aspen wood by solid-state fermentation with the white-rot fungus *Merulius tremellosus*. *Appl. Environ. Microbiol.* 50:133–39
231. Reid, I. D., Abrams, G. D., Pepper, J. M. 1982. Water-soluble products from the degradation of aspen lignin by *Phanerochaete chrysosporium*. *Can. J. Bot.* 60:2357–64
232. Reid, I. D., Seifert, K. A. 1982. Effect of an atmosphere of oxygen on growth, respiration, and lignin degradation by white-rot fungi. *Can. J. Bot.* 60:252–60
233. Reinhammer, B. 1984. Laccase. In *Copper Proteins and Copper Enzymes*, ed. R. Lontie, pp. 1–35. Boca Raton, Fla: CRC
234. Renganathan, V., Gold, M. H. 1986. Spectral characterization of the oxidized states of lignin peroxidase, an extracellular enzyme from the white rot basidiomycete *Phanerochaete chrysosporium*. *Biochemistry* 25:1626–31
235. Renganathan, V., Miki, K., Gold, M. H. 1985. Multiple molecular forms of diarylpropane oxygenase, an H_2O_2-requiring, lignin-degrading enzyme from *Phanerochaete chrysosporium*. *Arch. Biochem. Biophys.* 342:304–14
236. Renganathan, V., Miki, K., Gold, M. H. 1986. Role of molecular oxygen in lignin peroxidase reactions. *Arch. Biochem. Biophys.* 246:155–61
237. Ruel, K., Barnoud, F. 1985. Degradation of wood by microorganisms. See Ref. 26, pp. 441–67
238. Samejima, M., Saburi, Y., Yoshimoto, T., Fukuzumi, T., Nakazawa, T. 1985. Catabolic pathway of guaiacylglycerol-β-guaiacyl ether by *Pseudomonas* sp. TMY1009. *Mokuzai Gakkaishi* 31:956–58
239. Sanglard, D., Leisola, M. S. A., Fiechter, A. 1986. Role of extracellular ligninases in biodegradation of benzo(*a*)pyrene by *Phanerochaete chrysosporium*. *Enzyme Microb. Technol.* 8:209–12
240. Sarkanen, K. V., Ludwig, C. H. 1971. *Lignins. Occurrence, Formation, Structure and Reactions*. New York: Wiley-Interscience. 916 pp.
241. Schoemaker, H. E., Harvey, P. J., Bowen, R. M., Palmer, J. M. 1985. On the mechanism of enzymatic lignin breakdown. *FEBS Lett.* 183:7–12
242. Shimada, M., Habe, T., Umezawa, T., Higuchi, T., Okamoto, T. 1984. The C–C bond cleavage of a lignin model compound 1,2-diarylpropane-1,3-diol, with a heme-enzyme model catalyst tetraphenylprophyrinatoiron(III) chloride in the presence of *tert*-butylhydroperoxide. *Biochem. Biophys. Res. Commun.* 122:1247–52
243. Shimada, M., Higuchi, T. 1983. Biochemical aspects of the secondary metabolism of xenobiotic lignin and veratryl alcohol biosynthesis in *Phanerochaete chrysosporium*. See Ref. 115, pp. 195–208
244. Shimada, M., Nakatsubo, F., Kirk, T. K., and Higuchi, T. 1981. Biosynthesis of the secondary metabolite veratryl alcohol in relation to lignin degradation in *Phanerochaete chrysosporium*. *Arch. Microbiol.* 129:321–24
245. Sutherland, J. B., Crawford, D. L., Speedie, M. K. 1982. Decomposition of ^{14}C-labeled maple and spruce lignin by marine fungi. *Mycologia* 74:511–13
246. Sutherland, J. B., Pometto, A. L. III, Crawford, D. L. 1983. Lignocellulose degradation by *Fusarium* species. *Can. J. Bot.* 61:1194–98
247. Tai, D., Terasawa, M., Chen, C.-L., Chang, H.-M., Kirk, T. K. 1983. Biodegradation of guaiacyl and guaiacyl-syringyl lignins in wood by *Phanerochaete chrysosporium*. See Ref. 115, pp. 44–63
248. Terazawa, M., Tai, D., Chen, C.-L., Chang, H.-M., Kirk, T. K. 1983. Identification of the constituents of low molecular weight fractions obtained from birch wood degraded by *Pha-*

nerochaete chrysosporium. *Proc. Int. Symp. Wood Pulping Chem., Tokyo,* 4:150–55

249. Tien, M., Kersten, P. J., Kirk, T. K. 1987. Selection and improvement of lignin-degrading microorganisms: Potential strategy based on lignin model–amino acid adducts. *Appl. Environ. Microbiol.* 53:242–45
250. Tien, M., Kirk, T. K. 1983. Lignin-degrading enzyme from the hymenomycete *Phanerochaete chrysosporium* Burds. *Science* 221:661–63
251. Tien, M., Kirk, T. K. 1984. Lignin-degrading enzyme from *Phanerochaete chrysosporium*. Purification, characterization, and catalytic properties of a unique H_2O_2-requiring oxygenase. *Proc. Natl. Acad. Sci. USA* 81:2280–84
252. Tien, M., Kirk, T. K., Bull, C., Fee, J. A. 1986. Steady-state and transient-state kinetic studies on the oxidation of 3,4-dimethoxylbenzyl alcohol catalyzed by the ligninase of *Phanerochaete chrysosporium. J. Biol. Chem.* 261:1687–93

252a. Tien, M., Tu, C.-P. D. 1987. Cloning and sequencing of a cDNA for a ligninase from *Phanerochaete chrysosporium. Nature* 326:520–23

253. Trojanowski, J., Haider, K., Hütterman, A. 1984. Decomposition of ^{14}C-labelled lignin, holocellulose and lignocellulose by mycorrhizal fungi. *Arch. Microbiol.* 139:202–6
254. Ulmer, D. C., Leisola, M. S. A., Fiechter, A. 1984. Possible induction of the lignolytic system of *Phanerochaete chrysosporium. J. Biotechnol.* 1:13–24
255. Ulmer, D. C., Leisola, M. S. A., Schmidt, B. H., Fiechter, A. 1983. Rapid degradation of isolated lignins by *Phanerochaete chrysosporium. Appl. Environ. Microbiol.* 45:1795–801
256. Umezawa, T., Higuchi, T. 1984. Incorporation of $H_2^{18}O$ into the C_α but not the C_β position in degradation of a β-O-4 lignin substructure model by *Phanerochaete chrysosporium. Agric. Biol. Chem.* 48:1917–21
257. Umezawa, T., Higuchi, T. 1985. Aromatic ring cleavage in degradation of β-O-4 lignin substructure by *Phanerochaete chrysosporium. FEBS Lett.* 182:257–59
258. Umezawa, T., Higuchi, T. 1985. A novel C_α–C_β cleavage of a β-O-4 lignin model dimer with rearrangement of the β-aryl group by *Phanerochaete chrysosporium. FEBS Lett.* 192:147–50
259. Umezawa, T., Higuchi, T. 1986. Aromatic ring cleavage of β-O-4 lignin model dimers without prior demeth(ox)ylation by lignin peroxidase. *FEBS Lett.* 205:293–98
260. Umezawa, T., Higuchi, T., Nakatsubo, F. 1983. Difference in $^{18}O_2$ incorporation in oxygenative degradation of β-O-4 and β-1 lignin substructures by *Phanerochaete chrysosporium. Agric. Biol. Chem.* 47:2945–48
261. Umezawa, T., Kawai, S., Yokota, S., Higuchi, T. 1986. Aromatic ring cleavage of various β-O-4 lignin model dimers by *Phanerochaete chrysosporium. Wood Res.* 73:8–17
262. Umezawa, T., Nakatsubo, F., Higuchi, T. 1983. Degradation pathway of arylglycerol-β-aryl ethers by *Phanerochaete chrysosporium. Agric. Biol. Chem.* 47:2677–81
263. Umezawa, T., Shimada, M., Higuchi, T., Kusai, K. 1986. Aromatic ring cleavage of β-O-4 lignin substructure model dimers by lignin peroxidase of *Phanerochaete chrysosporium. FEBS Lett.* 205:287–92
264. Unger, B., Sligar, S. G., Gunsalus, I. C. 1986. *Pseudomonas* P-450 cytochromes. In *The Bacteria,* Vol. 9, Pseudomonas, ed. J. Sokatch, pp. 557–89. New York: Academic
265. Van Brunt, J. 1986. Fermentation economics. *Biotechnology* 4:395–401
266. Van Brunt, J. 1986. Fungi: The perfect host? *Biotechnology* 4:1057–62
267. Varadarajan, R., Szabo, A., Boxer, S. G. 1985. Cloning, expression in *Escherichia coli,* and reconstitution of human myoglobin. *Proc. Natl. Acad. Sci. USA* 82:5681–84
268. Westermark, U., Eriksson, K.-E. 1974. Cellobiose:quinone oxidoreductase, a new wood-degrading enzyme from white-rot fungi. *Acta Chem. Scand. Ser. B* 28:209–14
269. Wicklow, D. T., Langie, R., Crabtree, S., Detroy, R. W. 1984. Degradation of lignocellulose in wheat straw versus hardwood by *Cyathus* and related species (Nidulariaceae). *Can. J. Microbiol.* 30:632–36
270. Wood, D. A. 1985. Useful biodegradation of lignocellulose. *Annu. Proc. Phytochem. Soc. Eur.* 26:295–309
271. Wood, D. A. 1985. Production and roles of extracellular enzymes during morphogenesis of basidiomycete fungi. In *Developmental Biology of Higher Fungi,* ed. D. Moore, L. A. Casselton, D. A. Woodand, J. C. Frankland, pp. 375–87. Cambridge, UK: Cambridge Univ. Press
272. Yang, H.-H., Effland, M., Kirk, T. K. 1980. Factors influencing fungal de-

composition of lignin in a representative lignocellulosic, thermomechanical pulp. *Biotechnol. Bioeng.* 22:65–77

273. Yu, H.-s., Eriksson, K.-E. 1985. Influence of oxygen on the degradation of wood and straw by white-rot fungi. *Sven. Papperstidn.* 88:R57–60
274. Zeikus, J. G. 1981. Lignin metabolism and the carbon cycle: Polymer biosynthesis, biodegradation, and environmental recalcitrance. In *Adv. Microb. Ecol.* 5:211–43
275. Zeikus, J. G., Wellstein, A. L., Kirk, T. K. 1982. Molecular basis for the biodegradative recalcitrance of lignin in anaerobic environments. *FEMS Microbiol. Lett.* 15:193–97
276. Zhang, Y.-Z., Zylstra, G. J., Olsen, R. H., Reddy, C. A. 1986. Identification of cDNA clones for ligninase from *Phanerochaete chrysosporium* using synthetic oligonucleotide probes. *Biochem. Biophys. Res. Commun.* 137:649–56

Ann. Rev. Microbiol. 1987. 41:507–41

EXPORT OF PROTEIN: A Biochemical View

L. L. Randall

Biochemistry/Biophysics Program, Washington State University, Pullman, Washington 99164-4660

S. J. S. Hardy

Department of Biology, University of York, Heslington, York Y01 5DD, England

Julia R. Thom

Biochemistry/Biophysics Program, Washington State University, Pullman, Washington 99164-4660

CONTENTS

0066-4227/87/1001-0507$02.00

PROLOGUE

Translocation of specific, newly synthesized polypeptides across or into biological membranes is common to eukaryotic secretion, organelle biogenesis, and prokaryotic protein export. Although there are obvious differences among the systems, it seems likely that underlying principles are shared, and there may be similarities in the molecular mechanisms. Thus, we include some discussion of eukaryotic systems. Since there are available several current and excellent reviews of genetic work on bacterial export (8, 11, 15, 76, 108), the emphasis of this review is on biochemical investigations.

We begin with a brief survey of the gene products that make up the bacterial export machinery and then structure our discussion around the temporal pathway of export, taking in turn each of the stages: recognition, translocation, proteolytic maturation, and release of the final product from the export apparatus. This pathway is common to the export of protein to the periplasm and to the outer membrane. At least two separate mechanisms exist for secretion of protein into the extracellular medium: Some proteins are transferred through the outer membrane from a periplasmic pool, while others are secreted directly from the cytoplasm, apparently utilizing a completely separate pathway. For a comprehensive review of secretion of protein to the extracellular milieu the reader is referred to Pugsley & Schwartz (117).

Cast of Characters

The star of the entire drama of export is of course the exported protein. This is made initially as a precursor containing an amino-terminal extension, the leader sequence,[1] which is removed during the course of export to generate the mature protein. A thoroughly convincing identification of all other cellular components that together mediate export will emerge only from a combination of biochemical and genetic analyses. The genetic approach is made difficult by the requirement for isolation of mutants that are simultaneously defective in export of a number of proteins, many of which are individually essential for viability. A lucid description of the elegant genetic techniques devised to identify genes that encode components of the export apparatus can be found in Bankaitis et al (8). Biochemical analysis of the effects of mutations in several genes has established a role in export for the products of those genes without identifying that role precisely. Temperature-sensitive mutations in two genes, *secA* and *secY (prlA),* cause accumulation of precursors of a number of exported proteins (81, 109, 132). Prior evidence for a direct role of *secY* in export was provided by identification of alleles *(prlA)*

[1]The leader sequence is often called the signal or signal sequence; however, since its precise function is unknown, throughout this review we use the more neutral term "leader."

that suppressed defects in export of individual proteins caused by alterations of their leader sequences (35). Other mutations in the same gene suppressed defects caused by mutations in *secA*, a different component of the apparatus (20). Another locus that has been identified through selection for defects in export is *secD* (47). Mutations at the *prlD* locus have been shown to suppress defects in export of maltose-binding protein that are caused by changes in the leader (6, 125). No periplasmic or outer membrane protein [with the possible exception of the cysteine-binding protein (94)] has yet been identified that is exported independently of the pathway defined by mutations in *secA*, *secY*, or *secD* (see Table 1). This observation suggests that the products of these genes are components common to the export of all proteins. However, the identification of mutations in the nonessential gene *secB* (90, 91) that cause defects in localization of only a subset of proteins implies that there may be parallel pathways or at least loops or branches in the pathway.

Although selection of strains carrying mutations that suppressed export defects was used successfully to identify *secY*, the extension of this approach in an attempt to identify other components has not yielded the anticipated results. Of the many additional loci implicated in this way (15, 108), all of those so far characterized that suppress defects caused by mutations in *secY* and *secA* are genes already known to code for elements involved in protein synthesis, such as ribosomal proteins, including S1, S15, and L34; initiation factor 2; asparaginyl-tRNA synthetase; and seryl-tRNA (92, 131). Thus it appears that perturbations of protein synthesis can indirectly affect protein export. Since protein synthesis requires the products of more than 100 genes, many of the uncharacterized suppressor mutations also may lie in these genes. Sadly, interpretation of genetic analyses is all too often complicated by such nonspecific interactions.

Since extracts of *Escherichia coli* accurately process precursors of exported proteins, it was possible to use a biochemical approach to identify the enzymes involved in removal of leader sequences (161). Two endopeptidases with different specificities have been identified. One, leader peptidase (sp I), has a broad substrate specificity and processes the precursors of a large number of exported proteins (152). The other, lipoprotein signal peptidase (sp II), is specific for lipoproteins (65, 155). When synthesis of leader peptidase, the product of the essential gene *lep*, was repressed in a strain constructed so that *lep* was under the control of the *araB* promoter, precursors to a number of exported proteins accumulated (28). Therefore, it appears that leader peptidase mediates the processing step for the majority of exported proteins. The lipoproteins, a special class of exported proteins, differ from this majority in that their precursors require modification by glyceryl transferase and *O*-acyl transferase before they can be cleaved by lipoprotein signal peptidase (137). Following cleavage, an acyl group is linked to the newly generated amino terminus by an *N*-acyl transferase. The activity in vitro of lipoprotein signal peptidase is inhibited by globomycin, an antibiotic that causes precursors of lipoproteins to accumulate in vivo in the bacterial cell envelope but that has no effect on the export of other proteins (64, 67). The sensitivity of the enzyme to globomycin was exploited to identify *lsp*, the gene encoding lipoprotein signal peptidase (71, 122, 156, 157, 159).

Once leader peptides have been proteolytically removed from precursors of exported proteins, it is assumed that they are degraded. A number of enzymic activities that catalyze degradation of leader peptides in vitro have been described and characterized (65, 66, 69, 107). Unless a mutant which accumulates leader peptides in vivo is isolated, it will remain difficult to assign a role in export to any of these activities.

In summary, the proteins thought to be directly and exclusively involved in export of the majority of proteins from bacteria include SecA, SecD, SecY, PrlD, leader peptidase (Lep), and at least one proteolytic enzyme for degrading the cleaved leader peptide. Thus it is probable that a minimum of six proteins participate in the export of every outer membrane and periplasmic protein. Biochemical analyses have identified proteins involved in export, but those proteins have not been designated as the products of the genes defined. Transport of precursors across the membranes of inverted vesicles in one prokaryotic cell-free system was shown to depend on the presence of a cellular component that sedimented with a coefficient of 12S (104). Studies in vivo with *E. coli* showed that precursor maltose-binding protein was found in association with a cytoplasmic complex that had a sedimentation coefficient of about 16S (J. R. Thom & L. L. Randall, unpublished). It is not clear how these factors are related to one another, or how many separate components they represent. Two laboratories have identified cytoplasmic protein complexes in gram-positive bacteria that may be involved in export (1, 24).

Cell-free systems developed from eukaryotes have been used to identify and partially characterize the function of the macromolecular entities necessary for translocation of proteins across the membrane of the rough endoplasmic reticulum, a process analogous to the export of proteins from bacteria. It is interesting to compare the lists of eukaryotic and prokaryotic components. In eukaryotes, a nucleoprotein complex, the signal recognition particle, which contains a 7S RNA molecule and six polypeptides (144, 145), has been shown to recognize the secretory protein (146) and mediate its interaction with the membrane via an integral membrane protein, the docking protein (52, 100). Two of the polypeptides of this particle are dispensible for translocation in vitro (133, 134) but cause a translational arrest peculiar to the cell-free system derived from wheat germ (99). Two additional microsomal membrane proteins have been implicated in secretion, based on the loss of the capacity of vesicles to translocate precursors in vitro after treatment with

N-ethylmaleimide or trypsin (60). In the case of trypsin treatment the loss of translocation activity was correlated with the loss of ribosome-binding activity, even though two membrane proteins previously implicated in ribosome binding, the ribophorins (87, 88), remained intact. Treatment with *N*-ethylmaleimide blocked productive translocation but did not interfere with ribosome binding (60); thus the affected polypeptide is likely to be involved in a different step, perhaps translocation. As in prokaryotes, the precursor is matured by a protease, signal peptidase (82), and presumably the cleaved leader is degraded. The eukaryotic microsomal signal peptidase has been purified as a complex comprising at least four but not more than six polypeptides (40). These polypeptides are different from the docking protein and from the components of the signal recognition particle and thus may be part of a translocator. (For reviews of eukaryotic secretion see 61, 121, 150.)

The functional correspondence between the eukaryotic and prokaryotic components of the respective export apparatuses is presently obscure and must await a fuller understanding of the mechanism of export and of the roles of all components.

Location of Components

Export of protein from *E. coli* is a vectorial process. In order to describe it fully it is necessary to know the location of the protein being exported at each stage of the process, and also the location of each component of the export apparatus. Thus it is not surprising that many investigations are explicitly concerned with these issues. It is particularly crucial to know whether an intermediate in export or a component of the apparatus is soluble or membrane bound. If it is soluble, is it periplasmic or cytoplasmic? If membrane bound, to which membrane is it bound and in what orientation?

Unfortunately, the experimental methods used to address these questions often yield results that are liable to misinterpretation because the operational definition of a fraction, validated for the normal and permanent components of that fraction, may be inappropriate for defining the location of abnormal or transient components in the export pathway (see 138 for another discussion). The problem can be exacerbated when the bacteria are in an abnormal physiological state, as happens in studies where export is blocked by dissipation of proton-motive force, by induction of a fusion protein, or by a mutation in a component of the export apparatus. Even induction of a MalE-LacZ hybrid that does not block export triggers the heat-shock response (77).

These difficulties are most severe in defining and characterizing the location of membrane components. Centrifugation is almost always the first step in separating membranes from the soluble fraction of a cell lysate. In order that all the membrane material should be found in the pellet, high-speed centrifugation (10^5 G for 2 hr) is required. The pellet generated by this procedure also contains large cytoplasmic structures such as ribosomes, multisubunit proteins, and aggregated proteins, a fact that has on occasion led to components being assigned to the wrong cellular compartment (for discussion of this problem see 57). Further fractionation of the proteins of the pellet into components of the cytoplasmic or outer membranes is accomplished either by exploiting the difference in buoyant density between the two types of membrane through the use of isopycnic density gradient centrifugation in sucrose (102, 110), or by using the differential solubilization of proteins in nonionic detergents (44, 130). The latter method relies on the finding that normal components of the outer membrane are not solubilized by some detergents such as Sarkosyl or Triton X-100 unless the magnesium ions that stabilize the structure through interaction with lipopolysaccharide are removed by a chelating agent, usually EDTA. In contrast, normal components of the cytoplasmic membrane are solubilized by detergent in the absence of EDTA. This technique, developed for differential solubilization of normal stable components of the two membranes, must be applied with caution when initially defining the location of protein.

The method of isopycnic density gradient centrifugation in sucrose has the severe limitation that cofractionation of a component with a membrane is not compelling evidence of association. Indeed, if the sample is applied to the top of the gradient, cofractionation is not even proof that the component and the membrane have the same buoyant density. Any protein sufficiently large to be found in the pellet from the first high-speed centrifugation will sediment through the sucrose, and provided that it is discrete (rather than a nonspecific aggregate) it will be found at a distinct location in the gradient, which may coincide with the equilibrium position of membrane. This phenomenon can be prevented by applying the sample in a dense solution of sucrose to the bottom of a gradient. Since protein uncomplexed with lipid is denser than the most dense sucrose solution, any component that floats away from the bottom of the gradient is membrane bound. A disadvantage of applying the sample to the bottom is that attainment of equilibrium is slower and for small fragments of membrane may take an impossibly long time.

Even if a component were shown by flotation to have the density characteristic of one membrane, it might be located in the other. Fragments of cytoplasmic membrane particularly rich in protein would be significantly denser than bulk cytoplasmic membrane, while fragments of outer membrane partially deficient in lipopolysaccharide would be significantly less dense than bulk outer membrane. There is no reason to suppose that cytoplasmic membrane fragments that contain the export apparatus are typical of the cytoplasmic membrane. Indeed, we found that processed nascent chains of maltose-binding protein, which must span the cytoplasmic membrane in vivo, were associated with membrane that had a buoyant density the same as that of the less dense part of the outer membrane (J. R. Thom & L. L. Randall, unpublished data). It is unlikely that these unfinished fragments of a periplasmic protein are associated with bulk outer membrane. It is more likely that they are associated with domains that are either rich in protein relative to bulk cytoplasmic membrane or located at zones of adhesion that have an intermediate density.

A completely different method for locating the components involved in export is immunoelectron microscopy using colloidal gold (59). This technique does not suffer from the same difficulties in interpretation as the other techniques but has significant limitations of its own. First, it can only be easily used to locate proteins that are abundant. Second, it can only be used to locate protein that is in an antigenic state, i.e. protein that can be recognized by the antiserum used. It is not possible to conclude from use of this technique that all of any given protein has been detected. Forms of the protein may exist that do not react with the antibody and are in a location that differs from that of the species that do react.

That the considerations discussed above are important in identifying the location of aberrant exported proteins such as fusion proteins is shown by a study of the location of a fusion of PhoE with LacZ (139). The amino-terminal 300 amino acid residues of PhoE, an inducible 40-kd component of the outer membrane normally attached to the peptidoglycan, were fused to LacZ, and cells containing the fusion protein were fractionated. All of the PhoE-LacZ hybrid was found in the membrane pellet and was insoluble in Triton X-100 and Sarkosyl. After sucrose density gradient centrifugation with the sample applied at the top of the gradient, the hybrid was found in the fraction with the density characteristic of outer membrane. These findings would indicate that the fusion protein is exported to the outer membrane. However, it was not associated with the peptidoglycan, and immunoelectron microscopy showed that the great majority of detectable PhoE was in the cytoplasm. In another study, various portions of the outer membrane protein OmpA were deleted (46). Immunoelectron microscopy indicated that the mutated proteins were not localized in the outer membrane but were found in large aggregates either in the cytoplasm or in the periplasm. It is not known how these aggregates would have behaved in cell fractionation experiments, but we can speculate that they would most likely have been found, as were PhoE proteins that had portions deleted (19), in the pellet after high-speed centrifugation and thus would have been characterized as membrane-associated proteins. These studies emphasize that techniques designed for the fractionation of normal components of the cell may give misleading results when applied to abnormal and overproduced components.

It should by now be clear that the task of locating the components of the export apparatus and the intermediates in the export pathway is fraught with difficulty. Only when several experimental methods lead to the same conclusion should the location be considered established. This caveat should be kept in mind when evaluating many of the experimental results described in this review.

INITIAL INTERACTIONS

Proteins to be exported from the cytoplasm must specifically interact with the export apparatus to be transferred across the membrane. The defining and essential feature of proteins destined for export is that they are initially synthesized as precursors containing amino-terminal extensions, the leader sequences. It seems likely that one role of the leaders would be to mediate interaction of precursors with proteins that are involved in early steps in export, although alternative or additional roles are possible.

Consideration of the features common to leaders may provide insight concerning their mode of action. Through analysis of 118 eukaryotic and 32 prokaryotic leader sequences, von Heijne (142) found that even though the lengths and amino acid sequences were extremely variable, all leaders comprised three regions: a positively charged (mean net charge +1.7) amino-terminal region, n, of variable length; a central hydrophobic region, h, with minimal length of 7 residues in eukaryotes and 8 residues in prokaryotes; and a polar carboxyl-terminal region, c, which includes the cleavage site. The c region (amino acids −1 to −5 in eukaryotes and −1 to −6 in prokaryotes, numbered with respect to the cleavage site) does not vary in length with the overall length of the leader peptide. Thus, variation in length of the entire leader is accounted for by variation in the h and n regions.

Interaction of Leaders with Other Proteins

First let us consider the possibility that leaders mediate interaction of precursors with a proteinaceous component of the export apparatus. Generally, a specific interaction between two proteins involves complementary steric fit of the surfaces at the interface and thus requires particular amino acid residues at precise positions in the structure of both proteins. Even though there is no sequence homology among leader peptides, it has been suggested that four specific residues in the hydrophobic core of the leaders of maltose-binding protein (−8, −9, −11, −13) and LamB (−7, −10, −11, −12) constitute a recognition site within the leader. Selection for mutants drastically defective in export of these proteins repeatedly yielded mutational substitutions of a charged residue at one of these sites (12, 36). In addition, substitution of either a positively or negatively charged residue at position −9 in the LamB leader (i.e. not one of the four sites) had a negligible effect on export (37). The interpretation that the four critical residues are specifically recognized seems improbable if precursors of both LamB and maltose-binding protein are to interact with the same putative factor. Not only are the four residues not the same in the two proteins, but they do not assume corresponding positions within the structures whether the functional conformation is considered to be α-helix or extended β-strand.

Other genetic data have been interpreted as identifying the product of *secY* as the protein with which leaders directly interact because mutations in leaders of maltose-binding protein, LamB, OmpF, and alkaline phosphatase causing defects in export can be suppressed by extragenic mutations in *prlA (secY)* (15, 34, 35). To eliminate the possibility that suppression is indirect, it is necessary that the pairs of altered proteins be shown to interact in an allele-specific fashion to restore export. However, allele specificity has, in fact, not been demonstrated, as can be seen by comparing the data for the suppression of leader sequence mutations mediated by *prlA* mutations with the data obtained to establish the specific interaction of the products of two genes, *cheC* and *cheZ,* involved in chemotaxis in *E. coli* (114). If an interaction is allele specific, then the effect of a suppressor with any given mutation should not be predictable from its effect with any other mutated allele. Figure 1*A* shows that this is the case for the pair of genes involved in chemotaxis, whereas the order of effectiveness of suppression of signal sequence mutations by the *prlA* alleles is predictably the same for all *prlA* alleles, with the exception of *prlA2* and *prlA3* acting on *malE19-1* (Figure 1*B*). In addition, an alteration in a suppressor that restores interaction with the product of a mutated allele would be expected to have a detrimental effect on interaction with the wild-type product. Half of the *cheZ* mutations isolated as suppressors showed impaired chemotaxis in the presence of wild-type *cheC*. None of the *prlA* mutations impair export (6, 34). The argument in support of direct interaction of SecY and the hydrophobic core of leaders is further weakened by the fact that *prlA* alleles proposed to suppress in an allele-specific manner mutations that change residues within the hydrophobic core were originally isolated as suppressors of the mutations that lack all those residues (*lamBS60 Δ10–21*) (35) or most of those residues *(malE Δ12–18)* (6).

In the absence of direct evidence in support of a stereospecific interaction of the leader with SecY that is made defective by the mutational change and restored by the suppressor, it seems equally likely that the *prlA* alleles suppress defects by increasing the effectiveness of an interaction that does not directly involve binding of the leader. For example, SecY might be the protein that mediates translocation, and the *prlA* mutation might increase the affinity of SecY for the mature portion of the precursor, thus increasing the efficiency of transfer and overcoming a decrease in efficiency of a different interaction that involved the leader.

Although given the evidence available we must conclude that the primary

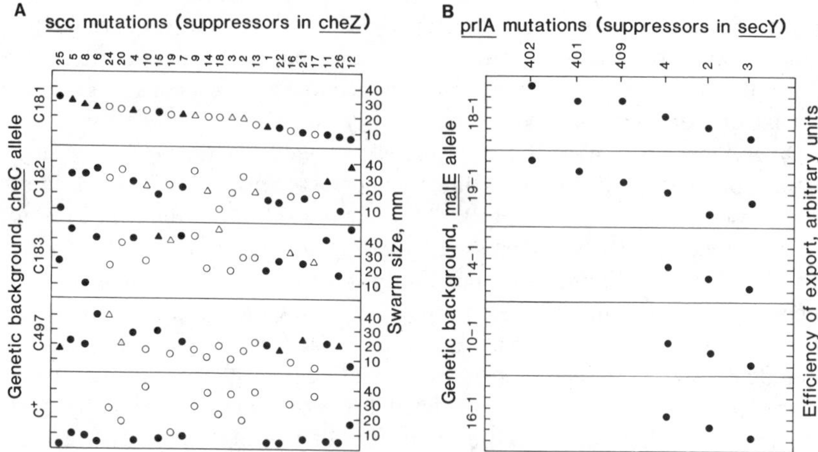

Figure 1 Allele specificity of mutations and their suppressors. *A:* Suppression of chemotaxis-defective mutations in *cheC* by *cheZ* mutations. The data are from Reference 114. *B:* Suppression of *malE* leader sequence mutations by *prlA* alleles. The data are from References 6, 34, and 125. The six *prlA* alleles were ranked according to the order of the efficiency of their suppression of each *malE* allele. The efficiency was plotted in six arbitrarily equivalent increments.

structure is not directly involved in the role of leader peptides, we do not mean to imply that the leaders do not mediate the interaction of precursors with a protein or that the precursor does not interact with SecY. However, it may be that the interaction is mediated in an unconventional manner. Consider the possibility that a factor might bind precursors via the backbone, irrespective of the side chains, and that the role of the leader might be to prevent folding so that the backbone would remain exposed. Although such interaction might seem improbable, a precedent for this proposal is the binding by the protease La (product of the *lon* gene) of its denatured substrate (148).

Interaction of Leaders with the Lipid Bilayer

An alternative or additional role of leaders involving their insertion into the lipid bilayer is indicated by the presence of the hydrophobic core. Bedouelle & Hofnung (13) proposed that a crucial feature of leader peptides is that the hydrophobic core exists in a periodic structure, either α-helix or extended β-strand, that has a critical minimal length of 18 Å. This length was arrived at by comparing the secondary structure, predicted by the guidelines of Chou & Fasman (27), of six wild-type signal sequences and ten different mutated sequences containing single amino acid substitutions that decreased the efficiency of export. The residues substituted in the mutated sequences did not change the prediction of secondary structure, but rather interrupted the length of a continuous hydrophobic segment by insertion of a hydrophilic residue. Two mutational substitutions (a point mutation in *malE* and a deletion in *lamB*) altered a portion of the predicted secondary structure to random coil.

Whereas the mutations of leader peptides described for the earliest isolates (7, 13) were consistent with the proposal of a requirement for a critical length of hydrophobic core, later analyses of ribose-binding protein (70) and maltose-binding protein (126) in which mutations of leaders and the corresponding revertants were characterized indicated that it is not the length of an uninterrupted hydrophobic core that is critical, but rather the overall hydrophobicity.

The predictions of importance of secondary structure of the leader were tested in an experimental system using chemically synthesized peptides with the same sequences as the hydrophobic core of LamB leader peptides of the wild type, a mutant, and a pseudorevertant precursor. Using circular dichroism to assess conformation in helix-promoting environments in vitro, a correlation was established between α-helicity and the ability of the sequence to function in export in vivo (22). Later studies with the same peptides established a correlation of function in vivo with the ability to interact with phospholipids in vitro (23). The wild-type leader peptide that was shown to be unstructured in bulk aqueous solution adopted β-structure when associated electrostatically with the surface of a phospholipid monolayer and underwent a conformational change to α-helix upon insertion (21). During the process of export in vivo when the precursor arrives at the membrane, the leader sequence may interact directly with the bilayer and the conformational change demonstrated for the peptides in vitro may play a role.

TRANSLOCATION

Transfer Independent of Elongation

A key feature of export is the mechanism of transfer of the polypeptide through the bilayer. Knowledge not only of the temporal relationship of synthesis and translocation but also of whether they are mechanistically coupled provides constraints on the mechanisms that can be reasonably proposed. In both eukaryotic and prokaryotic systems, transfer of proteins can occur either cotranslationally, i.e. during synthesis, or posttranslationally, i.e. after synthesis has been completed. In *E. coli* several proteins are exported both cotranslationally and posttranslationally (83). There is no evidence to indicate that the mechanism is different between these two modes of export. In the case of posttranslational transfer (31, 85, 118), the temporal separation eliminates the possibility of mechanistic coupling. Although such coupling could exist when export is cotranslational, in both prokaryotes and eukaryotes transfer seems to be independent of elongation. For some time after biochemical studies in vivo had established this in prokaryotes (118, 160), the conclusion was not considered relevant to eukaryotic systems. It was widely believed that translocation of protein during secretion in eukaryotes was obligatorily coupled to elongation (17, 98). This belief was based on the finding that translocation of protein into vesicles derived from rough endoplasmic reticulum would occur only cotranslationally, a temporal correlation that was misinterpreted as a mechanistic coupling. Now, however, several laboratories have shown that translocation in eukaryotic cell-free systems can occur posttranslationally (53, 96, 123, 147), and even that which occurs during synthesis can be independent of elongation (2, 115). These findings reinforce our contention that temporal differences do not necessarily reflect mechanistic differences.

Source of Energy

Since transfer of polypeptides through the membrane can occur in the absence of protein synthesis, we can eliminate from consideration previously popular models that depended on the energy of elongation in combination with tightly bound ribosomes to force the growing chain either directly through the bilayer (143) or through a proteinaceous channel (51). The question of the source of energy for transfer is thus reopened.

The first indication that export requires energy in addition to that of protein synthesis came from studies in vivo with *E. coli* (29, 30, 32, 38, 112). This conclusion was later extended to gram-positive bacteria (105, 140). Dissipation of proton-motive force by uncouplers or ionophores resulted in accumulation within the cell of precursors of exported proteins. Since the intracellular levels of ATP were normal (4, 30, 32, 38, 105, 160), the blockage in export was not the result of a secondary depletion of ATP. These results demonstrated a role for proton-motive force in export, but did not eliminate the possibility that ATP was also involved. In fact, a requirement for ATP could be demonstrated when a cell-free system from *E. coli* was established allowing posttranslational insertion of proteins into vesicles so that the ATP requirement for protein synthesis could be separated from that for translocation. Chen & Tai (26) concluded that export of alkaline phosphatase and OmpA in vitro was absolutely dependent on ATP and stimulated by proton-motive force. Wickner and his colleagues (50), using a similar system, could show that in the presence of both ATP and proton-motive force translocation of OmpA was efficient, whereas there was insignificant translocation if either was absent. In addition, Mizushima and his coworkers (158) demonstrated that both ATP and proton-motive force were required for translocation in vitro of an OmpF-lipoprotein chimeric protein into inverted vesicles. The fact that proton-motive force was only stimulatory in the cell-free system established by Chen & Tai, whereas there is a clear requirement for it in vivo, may reflect subtle differences between test-tubes and cells. Two such aspects that must be considered are the kinetics of translocation and the structure of isolated vesicles as compared to that of the gram-negative envelope.

The roles of ATP and proton-motive force in translocation have not yet been characterized. Either form of energy might be directly involved in movement of the polypeptides across the bilayer. Alternatively, they might have other roles, such as the cycling of cytosolic factors, as occurs, for example, in protein synthesis with EF-Tu by hydrolysis of GTP, or maintenance of the structure of some component of the membrane or of the translocation apparatus. Any function proposed for proton-motive force should incorporate the finding that the electrical potential ($\Delta\psi$) and the chemical potential of protons (ΔpH) across the membranes function interchangeably in export (4, 158). Thus it would appear that proposals involving electrophoresis are untenable.

The energy inherent in conformational changes of polypeptides has been invoked as the source of energy for insertion of polypeptides into and translocation through membranes (149). Differences in conformation between membrane proteins before and after insertion into membranes have been documented for proteins in the cytoplasmic membrane of *E. coli,* such as the coat protein of phage M13 (80) and leader peptidase (153), and for proteins in the mitochondrial membranes (45, 49, 128). One conformational change might be sufficient to insert a protein from the cytoplasm into the membrane; however, if this mechanism were to be applied to export, a second conformational change would be required for the protein to emerge from the membrane on the other side.

State of Polypeptides Competent for Transfer

Proteins can be translocated after synthesis of substantial stretches of the polypeptide chain; thus the structural state of the polypeptide is an important facet of the mechanism. Investigations of export of maltose-binding protein to the periplasm of *E. coli* established a correlation between the competence of polypeptides for export and the lack of a stably folded tertiary structure (120). Using sensitivity to proteolytic degradation to assess conformation, the kinetics of folding for mutated precursor maltose-binding proteins carrying defective leader peptides and the kinetics of folding of the wild-type precursor were compared. When the newly synthesized wild-type precursor was allowed to enter the normal pathway of export before a blockage was imposed, it was maintained in a state sensitive to proteolytic degradation. In contrast, both wild-type precursor synthesized after a blockage in export had been imposed and precursors with defective leaders folded into a stable conformation, resistant to proteolytic degradation and incompatible with export. These data indicate that the presence of the leader peptide alone is not sufficient to maintain the competent state. However, the leader might directly mediate binding of the precursor to a cellular component, either a soluble factor, membrane protein, or lipid bilayer, that does prevent folding. Alternatively, the leader might retard folding sufficiently to allow interaction of a factor with another region of the unfolded precursor.

Another investigation of maltose-binding protein (125) showed that precursors carrying mutated leaders existed in two states; one fraction was competent for export while the other was retained in the cytoplasm. When a different periplasmic protein, PhoS, was overproduced, a portion of the precursor population remained in the cytoplasm in a folded state (111). These observations are consistent with a kinetic partitioning between the pathway of productive export and the nonproductive intracellular folding of precursors into a stable conformation that cannot be translocated. Such a kinetic partitioning could provide an explanation for the unexpected observation that general defects in export could be suppressed by mutations in components of the protein-synthesis apparatus and by antibiotics that decrease the rate of protein synthesis (92, 131). If, as seems likely, attainment of a stable tertiary conformation requires the presence of almost all of the polypeptide chain, then decreasing the rate of elongation of the polypeptide would increase the time during which the chain is competent for export and might thus overcome defects in export caused by mutations. Apparently at odds with this explanation is the finding that one mutation that suppresses defects in *secY* maps in *infB*, the gene encoding initiation factor 2 (131). The primary effect of such mutations should be to decrease the number of polypeptides synthesized but not the rate of synthesis of individual polypeptides. However, alterations of initiation factor 2 might exert secondary effects on the rate of elongation, e.g. by perturbing the cytoplasmic pool of GTP.

The requirement that polypeptides be devoid of tertiary structure during transfer may apply widely to translocation of protein through membranes. Schatz and coworkers (62, 63) showed that a normally cytosolic enzyme was imported in vitro by yeast mitochondria when it was fused to the amino terminus of an imported mitochondrial protein. It seems that the protein would cross the mitochondrial membrane only if it were unfolded, since addition of a competitive inhibitor of the enzyme that would be expected to stabilize the mature conformation eliminated import (33).

Furthermore, polypeptides that are normally cotranslationally translocated through the membrane of the rough endoplasmic reticulum during secretion might remain competent for posttranslational transfer if they are maintained in a state devoid of stable tertiary structure. This is indicated by the results of Maher & Singer (96). Preprolactin, a protein normally imported by vesicles strictly cotranslationally, was rendered competent for posttranslational transfer when sufficient dithiothreitol was included in the cell-free system to prevent formation of disulfide bonds, which normally stabilize the folded structure of the precursor.

Although it seems clear that the common state of polypeptides that is compatible with transfer through membranes is the absence of stable tertiary structure, the competent state has not yet been completely defined. The polypeptides might have elements of secondary structure (α-helix or β-sheet) that could be involved in transfer; however, at this time there are not data available that address this point. While it seems likely that proteins are at least partially unfolded when they traverse membranes during bacterial export, eukaryotic secretion, and mitochondrial biogenesis, the precise mechanisms of obtaining that state may differ in detail. Whereas in bacteria transfer seems to occur before the polypeptides have folded, it has been proposed that previously established tertiary structure can be disrupted during both mitochondrial biogenesis and eukaryotic secretion (33, 96).

Involvement of Membrane Proteins

Although it seems likely that proteins in addition to the exported ones are involved in translocation, their presence is not a priori a necessity. No proteinaceous apparatus is required in the mechanism proposed in the helical hairpin hypothesis (39; for an alternative view see 135). It is envisaged that a segment of polypeptide undergoing export, even though it might contain polar amino acids, is spontaneously transferred through the membranes by pairing with a hydrophobic helix formed by the leader so that the thermodynamics of the system favor insertion. In this hypothesis, the observed coupling of secretion and protein synthesis was explained by the need to prevent the folding of the polypeptide into a structure that would interfere with the formation of the paired helices; thus the hypothesis can be applied to post-translational transfer, provided that folding is blocked. Nevertheless, integral membrane proteins with roles as yet unspecified have been implicated in both bacterial export and eukaryotic secretion and might function as translocators. The only candidate in *E. coli* at present is the product of *secY (prlA),* which was predicted to be a membrane protein from its nucleotide sequence (25) and which was later directly shown to be an integral component of the cytoplasmic membrane (3). A likely candidate for a eukaryotic translocator would be the component affected by treatment of vesicles derived from endoplasmic reticulum with the sulfhydryl reagent *N*-ethyl maleimide, since the treatment eliminated translocation activity without inhibiting binding of ribosomes and the affected component was shown to be distinct from the docking protein (60).

SITE OF EXPORT IN BACTERIA

Several lines of evidence indicate that protein export in bacteria occurs at specialized sites on the cytoplasmic membrane. Two fractions of the membrane of the gram-positive bacterium *Bacillus subtilis,* which could be separated based on a difference in density due to the presence in one of bound ribosomes, had distinct complements of proteins (97). Thus, it seems that the cytoplasmic membrane might contain specific structural or functional domains involved in export. In *E. coli,* zones of adhesion (9), where the cytoplasmic and outer membranes come into contact, are involved in export of envelope components such as lipopolysaccharide (103) and capsular antigens (10). These adhesions have also been implicated in export of protein. The outer membrane porin was shown by immunoelectron microscopy to emerge at approximately 200 distinct sites corresponding to zones of adhesion on the bacterial surface (136). The site where the receptor for phage λ, LamB, appeared following induction was located on the surface of *E. coli* by using adsorbed phage as a marker. Electron micrographs showed phage clustered at the poles and near the midline of cells (127). Other studies (18) using immunoelectron microscopy showed that maltose-binding protein in the periplasm was found mostly at the poles of the cells. A specialized continuous zone of adhesion that completely encircles the cell, the periseptal annulus (95), might be involved in events that are confined to the septal region.

The site of export of maltose-binding protein within the envelope of *E. coli* was characterized (J. R. Thom & L. L. Randall, unpublished results) using as markers for that site polypeptides that were still nascent but already proteolytically processed and that had previously been shown to span the membrane (118). Isopycnic equilibrium gradient centrifugation using both sedimentation and flotation indicated that the marker nascent maltose-binding protein was equally distributed between two fractions, one with a density characteristic of cytoplasmic membrane (1.16 g ml^{-1}) and the other, unexpectedly, with a density (1.21 g ml^{-1}) close to that characteristic of bulk outer membrane (1.23 g ml^{-1}). The explanation for the unexpected density was suggested by results of immunoelectron microscopy (J. R. Thom, L. L. Randall & M. Bayer, unpublished results) that showed maltose-binding protein from the denser fraction associated with dumbbell-shaped structures, of which half appeared to be outer membrane and half cytoplasmic membrane. The maltose-binding protein was located on the half of the structure that had the appearance characteristic of cytoplasmic membrane vesicles. Therefore, it is tempting to speculate that export sites exist at or are clustered near zones of adhesion.

MATURATION AND RELEASE

Ever since the discovery that exported proteins contain leader peptides, it has seemed likely that processing occurs after translocation across the membrane. However, during normal export, processing is so rapid that determination of the temporal order of events had to await construction of a strain in which leader peptidase was made limiting and the rate of processing was drastically decreased. In this strain precursors were accumulated transiently and were shown to be translocated before they were matured, since they could be degraded by addition of protease that had access to the outer surface of the cytoplasmic membrane (28). This experiment also provides evidence that processing is not required for translocation, a conclusion strengthened by identification of mutations that alter leader sequences to render precursors uncleavable without preventing translocation (84, 86, 89, 93). It is not known if during translocation of bacterial proteins movement of the amino-terminal portion, including the processing site, precedes transfer of the remainder of the polypeptide. However, Schleyer & Neupert (129) have shown that the precursors of two mitochondrial proteins (the F_1 β subunit of ATPase and cytochrome c_1) are partially translocated into the mitochondrial matrix through contact sites between the inner and outer membranes in a step that requires proton-motive force. The presequences are removed and then translocation is completed by a mechanism that is independent of the membrane potential.

Leader peptidase is an integral membrane protein of 36 kd (152, 154). Accessibility of the polypeptide to protease was used to determine its disposition in the membrane. The bulk of the polypeptide is accessible from the periplasmic side of the membrane, with only an amino-terminal fragment of approximately 5 kd exposed on the cytoplasmic face of the membrane (154). Since processing follows translocation (28), the periplasmic domain must contain the catalytic site.

The conclusion that removal of the leader peptide is not necessary for translocation across the membrane is also valid for lipoproteins (68), which are processed by the other identified membrane-bound enzyme, the lipoprotein signal peptidase. Evidence supporting this conclusion was obtained by use of a specific inhibitor of lipoprotein signal peptidase, the antibiotic globomycin, which causes precursors of lipoproteins to accumulate. The major lipoprotein of the cell, the product of the *lpp* gene, is a constituent of the cellular envelope that has glyceride and fatty acid moieties, which are covalently linked to the amino terminus, embedded in the outer membrane. Approximately one third of the molecules are also covalently bound to the peptidoglycan layer through the carboxyl-terminal lysine residue. A fraction of the precursors of lipoprotein that accumulated in cells as a result of treatment with globomycin were found to be covalently attached to the peptidoglycan in the periplasm. The simplest interpretation is that translocation of the lipoprotein can occur without proteolytic processing. Further support for this conclusion is provided by the existence of leader sequence mutations that allow translocation but not maturation (74, 93).

Specific Amino Acids Involved in Recognition of the Processing Site

A survey of all known bacterial leader peptides has highlighted certain regularities in amino acid sequence (141). It is likely that those regularities close to the cleavage site are imposed by the specificity of the leader peptidase. Such regularities include either alanyl, glycyl, or seryl residues on the amino-terminal side of the cleavage site (residue −1) and two residues away either glycyl, alanyl, valyl, leucyl, isoleucyl, or seryl residues (residue −3) (141). In addition, either glycyl or prolyl residues are often found in the short sequence, commonly six or seven residues long, between the cleavage site and the hydrophobic part of the leader peptide. Evidence indicating that these residues and others can be important for leader peptidase activity has been obtained by the isolation or creation of mutants with particular amino acids changed and the concomitant demonstration that the altered precursors are translocated but not processed. Mutations that confer this property are known in the leader peptides of M13 coat protein, β-lactamase, and maltose-binding proteins. In the case of M13 coat protein (89), mutations that meet these criteria occur at positions −1 (Ala to Thr), −3 (Ser to Phe), and −6 (Pro to Ser). In the case of TEM β-lactamase, one change at position −4 (Pro to Ser) caused partial inhibition of processing, while another change at the same position (Pro to Leu) caused complete inhibition (86). Another study using the same protein but a less direct method to demonstrate that the uncleaved precursor was translocated showed that a change at position −6 (Cys to Tyr) caused partial inhibition of processing (84). The importance of the prolyl residue at −4 was confirmed, since a change to tyrosyl prevented detectable processing. It is interesting that an alteration at position −5 (Leu to Phe) had no effect on export or processing. For maltose-binding protein, a mutation that inhibited processing but not translocation of a somewhat aberrant but functional leader sequence occurred at position −3 (Ala to Asp) (42, 43). Thus it is clear that the activity of leader peptidase can be affected by the sequence of the leader peptide up to a distance of at least six residues away from the cleavage site. The residues at positions −1 and −3 and the prolyl residues between −4 and −6 seem to be particularly important.

Specificity for cleavage of precursors of lipoproteins by lipoprotein signal peptidase is determined primarily by glyceride modification of the cysteinyl residue at position +1 and the amino acid residue immediately before the cleavage site, at position −1. Demonstration that the covalent glyceride modification is a prerequisite for recognition by lipoprotein signal peptidase was dependent on establishment of an in vitro assay (137). Envelopes containing labeled lipid prepared from a strain lacking lipoprotein served as donor of the glyceride moiety, and unmodified prolipoprotein accumulated in the envelope of a strain in which protein export was inhibited by induction of a fusion protein served as the acceptor. At pH 5 glyceride was not transferred to prolipoprotein, nor was the precursor cleaved. However, if the precursor was first modified at pH 9, where glyceryl transferase was active, and if the pH was then changed to 5, the modified precursor was cleaved; this indicates that lipoprotein signal peptidase was active at pH 5 but that only modified precursor could serve as substrate. Thus it appears that modification of the cysteinyl residue is required for recognition by the signal peptidase. This conclusion is reinforced by the finding that precursor lipoprotein was not cleaved when the cysteinyl residue was replaced with a glycyl residue so that modification with glyceride was not possible (73). However, glyceride modification of the cysteinyl residue is probably not by itself sufficient for recognition by the signal peptidase. This was indicated by characterization of a precursor of lipoprotein with an alteration in the leader peptide (Gly to Thr at position −1) that was modified but not cleaved (116). Provided that the precursor was translocated and therefore accessible to lipoprotein signal peptidase, the effect of the amino acid change must have been to alter the enzyme-substrate interaction. In addition, substitution of alanyl or seryl residues at position −1 had no effect. Taken together, these results indicate a similar requirement for small amino acid residues at the cleavage site for both signal peptidases. Other changes [Gly to Val or Leu (116) or deletion of Gly (74) at position −1, or Gly to Asp at position −7 (93)] eliminate the modification, and since modification is a prerequisite for cleavage the specific effect of these mutations on recognition by lipoprotein signal peptidase cannot be assessed.

Role of Cleavage

The accumulation of translocated precursors under certain conditions not only shows that cleavage of leader peptides is not necessary for translocation, but also indicates that in these conditions translocation is not reversible. This is not to say however, that translocation would not be reversible in other circumstances, e.g. in the absence of an energized membrane or in the absence of folding of the exported protein. In any event, it is unlikely that the primary function of cleavage is to render export irreversible. Thus other roles must be considered.

Removal of leader sequences might be necessary to render exported proteins functional. However, in all cases examined, isolated precursors exhibited the activity of the mature species. Precursor alkaline phosphatase synthesized in vitro dimerizes to form active enzyme (72). Precursor of maltose-binding protein, whether synthesized in vitro (41) or in vivo (75), was shown to be active in binding ligand by affinity chromatography using a cross-linked amylose resin. A strain of bacteria producing mutated unprocessed precursor of β-lactamase was resistant to a low level of ampicillin (84). Since no mature

β-lactamase was detectable, the precursor species must have been active. These examples indicate that presence of the leader sequence does not by itself prevent attainment of the active conformation of a polypeptide. Nevertheless, it is possible that the leader sequence prevents folding by mediating association of the protein with the bilayer or a component of the export apparatus. In fact, all translocated precursors that have been characterized are membrane associated in vivo [with the single exception of a bizarre precursor of maltose-binding protein with a genetically altered leader sequence (42, 43)]. However, in the one example studied (maltose-binding protein), translocated, membrane-bound precursor seemed to have assumed the stable, mature conformation, since added protease did not degrade the polypeptide entirely but only removed the leader sequence (28). Therefore, even though the conformation of other translocated membrane-associated precursors has not been assessed, it seems unlikely that the primary role of cleavage is to release polypeptides from the membrane so that they can fold, but rather it is to release them so that they can be properly localized. It is clear that many periplasmic proteins such as the amino acid– and sugar-binding proteins must dissociate from the membrane in order to carry out their functions. Outer membrane proteins probably also require cleavage of the leader to gain access to their functional location, since precursor OmpA (28) and the glyceride-modified form of precursor lipoprotein (68) accumulated in the cytoplasmic membrane in the absence of their respective signal peptidase activities. The pathway for localization of lipoprotein is complicated by several sequential processing steps. In the absence of any processing, mutated precursor was found at the density characteristic of outer membrane after isopycnic centrifugation through sucrose gradients (93). Although isopycnic centrifugation is a well-established and reliable technique for separation of bulk cytoplasmic and outer membrane, when it is applied to analysis of distribution of abnormal components within the envelope, interpretation must be made with caution (see section on location of components, above).

Although leader sequences do not seem to block folding directly as discussed above, they might retard folding. In the periplasm such retardation of folding might allow degradation of exported proteins by rendering them sensitive to proteolysis for a significant period. In this scenario, removal of leader sequences would allow rapid folding into a stable state that is resistant to degradation.

Release

A distinct step involving a conformational change to release matured polypeptides from the membrane has been postulated, based on observations that under some conditions cleavage by leader peptidase will not by itself mediate release of periplasmic proteins. Prematurely terminated fragments of three periplasmic proteins (β-lactamase, maltose-binding protein, and glycerol phosphate phosphodiesterase) are found processed but associated with membranes and accessible to externally added proteases (55, 79, 85). In addition, pulse-labeling of wild-type β-lactamase at low temperature (15°C) allowed detection of the completed and processed protein associated with membrane in a state that was sensitive to degradation by added trypsin (101). In contrast, the mature form that was released and found free in the periplasm was resistant to degradation by trypsin. The sensitivity to degradation of the membrane-associated species indicates that it had not attained the mature conformation. It should be noted that interaction with the membrane of the processed species probably differs from membrane association of the precursor that is likely to be anchored by the leader, since precursor maltose-binding protein associated with membrane was shown to have the stable mature conformation (28; see section on role of cleavage, above). The temporal sequence of the cleavage of the signal and the folding of the polypeptide into the final active conformation during normal export has not been defined.

SORTING OF PROTEINS BETWEEN PERIPLASM AND OUTER MEMBRANE

Since inspection of leader sequences reveals no identifiable difference between those of periplasmic and outer membrane proteins that might direct the proteins to their respective compartments, it seems likely that information specific for sorting is contained in the mature polypeptide. Confirmation of this supposition was obtained using chimeric precursors comprising the signal sequence of a protein destined for one compartment and the mature portion of a protein destined for the other. In all cases the mature protein was localized to the proper compartment (11).

Silhavy and colleagues have proposed that specific sequences of the mature portion direct the LamB protein to the outer membrane. Hybrid LamB-LacZ fusions containing increasing lengths of LamB have been used in an attempt to determine which sequences are necessary for proper localization of LamB (14). Induction of certain of these fusions by maltose is lethal to the cell. This maltose-sensitive (Mal^s) phenotype results from interference with the normal pathway of export, as indicated by the accumulation of precursors of exported proteins. Fusions containing the leader sequence and 27 amino acids of the mature protein were resistant to maltose, while fusions containing the leader sequence and 39 or 43 amino acids were maltose sensitive. Thus it is clear that all or part of the sequence between amino acid 27 and amino acid 39 is important for initiating export of LamB. Under the conditions used in the experiments reported (14), 90% of these chimeric proteins were recovered in the soluble fraction after high-speed centrifugation of a lysate. However, it had been demonstrated in previous work (78) that during fractionation other hybrid proteins could be found either in the soluble fraction or in the membrane pellet, depending on the ionic conditions. Therefore Silhavy and coworkers (14) favored the interpretation that the fusions containing 39 amino acyl residues of LamB were associated with the cytoplasmic membrane, based on their Mal^s phenotype. A fusion protein containing the leader sequence and 49 amino acids of the mature protein not only conferred the maltose-sensitive phenotype, but also was partly recovered (60%) in a pellet after high-speed centrifugation. One third of this was found at the density of outer membrane after sedimentation through a sucrose gradient.

Silhavy and his colleagues are aware of the difficulties in interpreting experiments involving cellular fractionation and localization of fusion proteins (see section on location of components for a discussion of the complications) but feel that the difference in behavior during fractionation between the fusion containing 39 amino acids of LamB and that containing 49 amino acids supports the idea that the amino acyl residues between positions 39 and 49 are used by the cell for diverting the LamB protein to the outer membrane (14). This interpretation appears wildly in the literature (8, 108, 121). Studies of LamB fused to a periplasmic protein, alkaline phosphatase, gave contradictory results (58). LamB-PhoA chimeras containing up to 115 amino acid residues of the mature LamB sequence were found in the periplasm rather than in the outer membrane as would be predicted from the results obtained with LamB-LacZ hybrid proteins. Nikaido & Wu (106) independently predicted that the amino acid sequence of LamB between residues 39 and 49 might be important for directing the protein to the outer membrane by comparing the amino acid sequences of the three outer membrane proteins OmpA, LamB, and OmpF. Statistically significant homologies of sequence were found among these proteins; one of the most striking overlapped with residues 39–49 of LamB. Since sequences with some homology to this region were found in three other outer membrane proteins but none were found among representatives of the periplasmic, cytoplasmic, or cytoplasmic membrane proteins, it was suggested (106) that this short sequence was used by the cell to identify outer membrane proteins. However, evidence that in OmpA the homologous sequence spanning residues 1–14 is not essential for localization in the outer membrane was obtained (46) when a mutant form of the protein having residues 4–45 deleted was shown by immunoelectron microscopy to be in the outer membrane. The locations of other deleted forms of OmpA were also characterized. OmpA missing residues 86–227 or 160–325 was located in the periplasm, whereas that without residues 43–84 was located in the outer membrane. A similar analysis of LamB in which immunofluorescence was used to locate the mutant proteins indicated that residues 70–220 could be deleted without affecting its location in the outer membrane (16). The most plausible interpretation reconciling all these data is that proteins are localized in the outer membrane by virtue of their conformation rather than by a unique sorting sequence.

ONE PATHWAY OR MANY?

Evidence that two or more final products of metabolism share a common pathway can be obtained by imposing a blockage of or bottleneck in some step and showing the accumulation of precursors of both products. Such accumulation does not necessarily prove that the step blocked is on a common pathway, since the accumulation of a precursor to one product might saturate a component involved in an earlier common step. Consider the hypothetical situation that each exported protein used a different translocator. Deletion of any one translocator could result in accumulation of all precursor species if there were a common factor used to direct them all to their respective translocators and if this factor were saturated by accumulation of the one precursor.

This caveat is relevant to analysis of the effects of perturbations that inhibit export (Table 1). Induction of expression of a fusion protein, MalE-LacZ, and the conditionally lethal mutation *secA* caused the accumulation of precursors of the seven well-characterized proteins of the cell envelope that were examined. Similar results were obtained for five exported proteins examined in strains carrying a conditionally lethal mutation in *secY*. The rigorous interpretation of these results is that each exported protein must use at least one component of the apparatus that is also used by at least one other exported protein. However, it is likely that the export apparatus comprises a small number of components that mediate export of overlapping subsets of proteins. It is even possible that most components of the apparatus are utilized by all exported proteins. Comparison of the time courses of onset of accumulation of different precursors in response to a given perturbation of export can be helpful in distinguishing among these models. Accumulation of lipoprotein and OmpA precursors began at the same time after the shift of a temperature-sensitive *secA* mutant to the nonpermissive temperature, whereas in a *secY* mutant, accumulation of the precursor of lipoprotein was delayed relative to that of the precursor of OmpA (54). This suggests that the effect of the mutation in *secY* on lipoprotein may be secondary, although other explanations of these data are possible.

The two well-characterized enzymes lipoprotein signal peptidase and leader peptidase define a branch point late in the export pathway. That earlier branches may exist is indicated by the characterization of a mutation, *secB*, which has pleiotropic effects on export but does not affect all proteins of the

Table 1 Perturbations that affect export[a]

Exported protein	Mutation in secA	Mutation in secY	Mutation in secD	Mutation in secB	Dissipation of $\Delta\tilde{\mu}_{H^+}$	Expression of MalE-LacZ	Interference by export-defective MalE
Maltose-binding protein[b]	X (48, 109, 151)	X (132, 151)	X (47)	X (91)	X (38, 112)	X (48, 78)	
Ribose-binding protein[b]	X (48, 90)	ND	X (47)	0 (90, 91)	X (c)	X (48)	0 (d)
Alkaline phosphatase[b]	X (109)	X (e)	X (47)	0 (91)	X (113)	X (78)	X (d)
β-lactamase[b]	X (94)	ND	ND	ND	X (4, 29, 112)	X (94)	0 (d)
OmpA[f]	X (51, 54)	X (51, 54, 132)	X (47)	X (91)	X (38, 112, 160)	X (78)	X (5)
OmpF[f]	X (109)	X (132)	X (47)	X (91)	X (38, 112)	X (78)	ND
LamB[f]	X (109)	ND	X (47)	X (91)	X (38, 112, 113)	X (78)	X (d)
Lipoprotein[f]	X (54, 151)	X (54, 132, 151)	ND	0 (54)	X (124, g)	X (78)	0 (d)
Procoat[h]	0 (151)	0 (151)	ND	ND	X (32)	X/0 (g)	0 (d)

[a] X, effect on export; 0, no effect; X/0, weak effect; ND, not determined. Numbers in parentheses refer to references in *Literature Cited*.
[b] Periplasmic protein.
[c] S. Park & L. Randall, personal communication.
[d] V. Bankaitis & P. Bassford, Jr., personal communication.
[e] K. Strauch & J. Beckwith, personal communication.
[f] Outer membrane protein.
[g] K. Ito, personal communication.
[h] Cytoplasmic membrane protein.

envelope (Table 1). It is particularly interesting that export of maltose-binding protein depends on the product of *secB*, whereas export of ribose-binding protein occurs independently of the *secB* function. Another indication that not all components involved in export of maltose-binding protein and ribose-binding protein are identical is obtained from studying the interference with normal export that occurs when export-defective proteins are produced by the cell. Maltose-binding protein with a defective leader peptide enters the export pathway sufficiently to have a kinetic effect on the export of some proteins, such as OmpA (5), but there is no effect on the export of ribose-binding protein (see Table 1). This kinetic effect requires the presence of a region in the mature sequence of maltose-binding protein (8). Perhaps a component of the apparatus, e.g. the product of *secB*, interacts with the sequence of the mature export-defective precursor to such an extent that it becomes limiting for export of other proteins. In this scheme, ribose-binding protein would not be affected because it would not use the limiting component.

A SPECULATIVE WORKING MODEL

We have tried to propose a model for export that accounts for as much of the experimental data as possible. The model is expected to evolve as new data accumulate, and in this light it is of interest to examine the changes that have occurred in our working model since the previous presentation (119). In that model, based on the similarity between bacterial leader sequences and eukaryotic signals, we made the assumption that the initial steps in export were analogous to those proposed for secretion in eukaryotes. The eukaryotic signal sequence was postulated to mediate association of a cytosolic particle, the signal recognition particle, with the growing nascent chain in such a way as to arrest protein synthesis. The complex, comprising the signal recognition particle, the nascent polypeptide, and the protein-synthesis apparatus, was postulated to bind specifically to the membrane of the endoplasmic reticulum through the affinity of the signal recognition particle for its receptor, the docking protein, located in that membrane.

At the time of our previous proposal, the initial steps of export in bacteria had not been characterized biochemically; however, genetic evidence indicated participation of a proteinaceous apparatus. Thus, our working model included a cytoplasmic factor and a receptor on the cytoplasmic membrane, in analogy to the eukaryotic models. Now there is direct evidence that a cytosolic complex is required for export (104) and that the precursor of at least one exported protein, the maltose-binding protein, associates with such a complex (J. R. Thom & L. L. Randall, unpublished). However, there is no evidence that the cytosolic factor in prokaryotes mediates an arrest of synthesis of precursors. Furthermore, in eukaryotes such an arrest does not seem crucial to secretion (133, 134). We propose a different function for the bacterial particle. The factor may bind the precursor and hinder the polypeptide from folding into a stable tertiary conformation, thus maintaining the polypeptide in a state that is competent for translocation even if it is fully elongated. This might account for the observation that in bacteria export can occur both posttranslationally and cotranslationally. In eukaryotes transfer seems to occur cotranslationally, even though the transfer is not mechanistically coupled to elongation. This may be because many proteins secreted by eukaryotes, such as immunoglobulins, are large polypeptides with structural domains that fold independently. It might be difficult to hinder folding of all the domains simultaneously by the binding of a factor near the amino terminus.

We propose that the bacterial factor not only maintains the competent state of the polypeptide, but also, in analogy to the postulated function for the signal recognition particle, mediates binding of the precursor, whether it be fully elongated or still on ribosomes, to the export apparatus at the membrane (Figure 2). The membranous export apparatus should comprise, at the minimum, leader peptidase and the SecY protein. Either protein might serve as a receptor for the cytosolic factor. The factor would then transfer the precursor to the apparatus that mediates translocation. This transfer would involve a change in the affinity of the factor for the polypeptide. ATP, shown to be necessary for export, might be required at this step. In a manner analogous to the hydrolysis of guanosine triphosphate either by G-proteins or by EF-Tu during a functional cycle, the export factor might hydrolyze ATP and undergo

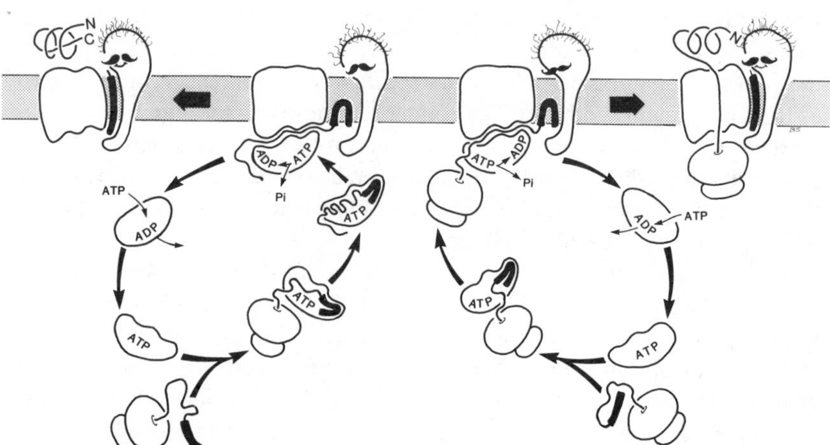

Figure 2 A speculative working model. Cotranslational export is illustrated on the *right* and posttranslational export on the *left*. The character in the membrane with the moustache is leader peptidase, and the other character represents the translocator. A cytosolic particle is included in the pathway and is discussed in the text.

a conformational change that would allow the transfer of the precursor from the factor to the translocator and the release of the unoccupied factor to the cytosol. Although we have included such a cycle in our working model as shown in Figure 2, presently this is purely speculative. An equally possible role for hydrolysis of ATP would be to provide the energy for translocation of the polypeptide.

There are no data available that provide information related to the molecular mechanism of the translocation event. Singer and colleagues (135) have proposed a speculative model detailing the precise mechanism of transfer. However, one should keep in mind that transfer might proceed directly through the bilayer and that the integral membrane protein SecY, which is clearly involved in export, might function solely as a receptor for the cytosolic factor. Nevertheless, with these caveats in mind, we have included a proteinaceous translocator in our working model.

In the scenario proposed here, it should be noted that elongation of the polypeptide to be exported does not necessarily occur at the membrane. It is not known when during synthesis nascent polypeptides interact with the membrane. In the previous model (119), we assumed that the interaction was early and that elongation at the membrane would serve to maintain the polypeptide in a competent state. In light of the recent observation that fully elongated precursor maltose-binding protein is associated with a cytosolic factor, it seems likely that elongation of protein by membrane-bound polysomes results from the relative relationship of the rate of synthesis and the rate of association with membrane. When the association occurs before the polypeptide has been completed, elongation will continue on membrane-bound polysomes and transfer may be initiated cotranslationally. There need not be any mechanistic difference between this type of transfer and that which occurs entirely posttranslationally. However, proteins may exist that must be extruded before they are completed, necessitating a temporal coupling between synthesis and transfer similar to that observed in eukaryotes. If a protein contained domains that could fold into stable structures independently of one another, then interaction of the nascent protein with the putative cytosolic factor might keep the first domain unstructured, but would have little effect on domains comprising stretches of amino acids later in the sequence. If these subsequent portions of the polypeptide were elongated at the export site on the membrane they might be kept unstructured by binding directly to the translocator.

The polypeptides to be exported that interact with proteinaceous components of the export apparatus, whether those components be cytosolic or membrane-bound, must do so in a rather unconventional manner. There is no homology among leader sequences or among the mature portions of exported polypeptides that could be assumed to mediate the usual type of stereospecific interaction between two proteins. It is possible that the export apparatus recognizes the protein to be exported through interaction with the polypeptide backbone. Precedents for this type of recognition might be found in the binding of denatured protein by the protease La (148) or in the transport of peptides by the oligopeptide transport system (encoded by the *opp* genes), which shows no specificity for amino acyl side chains (56). The putative factor involved in export might bind directly to the polypeptide backbone of the leader and might block folding by interacting with critical amino acyl residues near the amino terminus of the mature protein. Alternatively, the leader might mediate binding indirectly by hindering folding to allow the factor to bind the backbone of the mature portion of the polypeptide. Either mode of action would be consistent with the diversity of the amino acid sequences that function as leader peptides and with the conservation of a hydrophobic core among those sequences. The binding site on protease La has a preference for hydrophobic substrates (148); likewise, the putative factor might prefer hydrophobic ligands. Alternatively, the hydrophobic leader peptide might associate with the first hydrophobic stretch of amino acids in the mature protein as it emerges from the ribosome and might thus effectively retard proper folding. In addition to mediating initial interactions, the leader peptide may play a role in the translocation step and, as indicated in Figure 2, may partition into the bilayer when the polypeptide reaches the export site on the membrane.

EPILOGUE

In our previous review in 1984 (119), we suggested that the mechanism of transfer of proteins across membranes was likely to be very similar in eukaryotes and prokaryotes, even though there appeared to be a difference in the degree to which synthesis of proteins was coupled with transfer through the membrane. As discussed here, this apparent difference may reflect a common underlying principle: Polypeptides must be devoid of stable tertiary structure in order to be transferred across the membrane. This principle seems to apply not only to bacterial export and eukaryotic secretion, but also to assembly of protein into mitochondria. The details of the mechanisms may differ; for example, mitochondria may have the capacity to unfold proteins, whereas the proteins may be delivered to the endoplasmic reticulum or the bacterial cytoplasmic membrane in a state devoid of stable tertiary structure. Furthermore, within each system the stability of individual proteins may dictate the degree to which synthesis of the polypeptides must be temporally coupled to transfer.

It is clear from the striking similarities in export of proteins from the cytosol of all living cells that as investigators attempt to further elucidate the molecular mechanism of protein localization, it will be important to keep in mind the results obtained in each of the model systems.

ACKNOWLEDGMENTS

We thank all of our colleagues who made comments on this review. We are particularly grateful to (in alphabetical order) G. Hazelbauer, J. Knowles, and W. Wickner. LLR is supported by NIH grant GM29798.

Literature Cited

1. Adler, L.-Å., Arvidson, S. 1984. Detection of a membrane-associated protein on detached membrane ribosomes in *Staphylococcus aureus*. *J. Gen. Microbiol.* 130:1673–82
2. Ainger, K. J., Meyer, D. I. 1986. Translocation of nascent secretory proteins across membranes can occur late in translation. *EMBO J.* 5:951–55
3. Akiyama, Y., Ito, K. 1985. The SecY membrane component of the bacterial protein export machinery: analysis by new electrophoretic methods for integral membrane proteins. *EMBO J.* 4:3351–56
4. Bakker, E. P., Randall, L. L. 1984. The requirement for energy during export of β-lactamase in *Escherichia coli* is fulfilled by the total protonmotive force. *EMBO J.* 3:895–900
5. Bankaitis, V. A., Bassford, P. J. Jr. 1984. The synthesis of export-defective proteins can interfere with normal protein export in *Escherichia coli*. *J. Biol. Chem.* 259:12193–200
6. Bankaitis, V. A., Bassford, P. J. Jr. 1985. Proper interaction between at least two components is required for efficient export of proteins to the *Escherichia coli* cell envelope. *J. Bacteriol.* 161:169–78
7. Bankaitis, V. A., Rasmussen, B. A., Bassford, P. J. Jr. 1984. Intragenic suppressor mutations that restore export of maltose binding protein with a truncated signal peptide. *Cell* 37:243–52
8. Bankaitis, V. A., Ryan, J. P., Rasmussen, B. A., Bassford, P. J. Jr. 1985. The use of genetic techniques to analyze protein export in *Escherichia coli*. *Curr. Top. Membr. Transp.* 24:105–50
9. Bayer, M. E. 1974. Ultrastructure and organization of the bacterial envelope. *Ann. NY Acad. Sci.* 235:6–28
10. Bayer, M. E., Thurow, H. 1977. Polysaccharide capsule of *Escherichia coli:* microscope study of its size, structure and sites of synthesis. *J. Bacteriol.* 130:911–36
11. Beckwith, J., Ferro-Novick, S. 1986. Genetic studies on protein export in bacteria. *Curr. Top. Microbiol. Immunol.* 125:5–27
12. Bedouelle, H., Bassford, P. J. Jr., Fowler, A. V., Zabin, I., Beckwith, J., Hofnung, M. 1980. Mutations which alter the function of the signal sequence of the maltose binding protein of *Escherichia coli*. *Nature* 285:78–81
13. Bedouelle, H., Hofnung, M. 1981. Functional implications of secondary structure analysis of wild-type and mutant bacterial signal peptides. In *Membrane Transport and Neuroreceptors,* ed. D. Oxender, A. Blume, I. Diamond, C. F. Fox, pp. 399–403. New York: Liss. 450 pp.
14. Benson, S. A., Bremer, E., Silhavy, T. J. 1984. Intragenic regions required for LamB export. *Proc. Natl. Acad. Sci. USA* 81:3830–34
15. Benson, S. A., Hall, M. N., Silhavy, T. J. 1985. Genetic analysis of protein export in *Escherichia coli* K12. *Ann. Rev. Biochem.* 54:101–34
16. Benson, S. A., Silhavy, T. J. 1983. Information within the mature LamB protein necessary for localization to the outer membrane of *E. coli* K12. *Cell* 32:1325–35
17. Blobel, G., Dobberstein, B. 1975. Transfer of proteins across membranes: presence of proteolytically processed and unprocessed nascent immunoglobulin light chains on membrane-bound ribosomes of murine myeloma. *J. Cell Biol.* 67:835–51
18. Boos, W., Staehelin, A. L. 1981. Ultrastructural localization of the maltose-binding protein within the cell envelope of *Escherichia coli*. *Arch. Microbiol.* 129:240–46
19. Bosch, D., Leunissen, J., Verbakel, J., de Jong, M., van Erp, H., Tommassen, J. 1986. Periplasmic accumulation of

truncated forms of outer-membrane PhoE protein of *Escherchia coli* K-12. *J. Mol. Biol.* 189:449–55

20. Brickman, E. R., Oliver, D. B., Garwin, J. L., Kumamoto, C., Beckwith, J. 1984. The use of extragenic suppressors to define genes involved in protein export in *Escherichia coli*. *Mol. Gen. Genet.* 196:24–27
21. Briggs, M., Cornell, D., Dluhy, R., Gierasch, L. 1986. Conformations of signal peptides induced by lipids suggests initial steps in protein export. *Science* 233:206–8
22. Briggs, M., Gierasch, L. 1984. Exploring the conformational role of signal sequences: synthesis and conformational analysis of λ-receptor protein wild-type and mutant signal peptides. *Biochemistry* 23:3111–14
23. Briggs, M., Gierasch, L., Zlotnick, A., Lear, J., DeGrado, W. 1985. *In vivo* function and membrane binding properties are correlated for *Escherichia coli* LamB signal peptides. *Science* 228: 1096–99
24. Caulfield, M. P., Furlong, D., Tai, P. C., Davis, B. D. 1985. Secretory S complex of *Bacillus subtilis* forms a large, organized structure when released from ribosomes. *Proc. Natl. Acad. Sci. USA* 82:4031–35
25. Cerretti, D. P., Dean, D., Davis, G. R., Bedwell, D. M., Nomura, M. 1983. The *spc* ribosomal protein operon of *Escherichia coli:* sequence and cotranscription of the ribosomal protein genes and a protein export gene. *Nucleic Acids Res.* 11:2599–616
26. Chen, L., Tai, P. C. 1985. ATP is essential for protein translocation into *Escherichia coli* membrane vesicles. *Proc. Natl. Acad. Sci. USA* 82:4384–88
27. Chou, P. Y., Fasman, G. D. 1978. Prediction of the secondary structure of proteins from their amino acid sequence. *Adv. Enzymol.* 47:45–148
28. Dalbey, R. E., Wickner, W. 1985. Leader peptidase catalyzes the release of exported proteins from the outer surface of the *Escherichia coli* plasma membrane. *J. Biol. Chem.* 260:15925–31
29. Daniels, C. J., Bole, D. G., Quay, S. C., Oxender, D. L. 1981. Role for membrane potential in the secretion of protein into the periplasm of *Escherichia coli*. *Proc. Natl. Acad. Sci. USA* 78:5396–400
30. Date, T., Goodman, J. M., Wickner, W. T. 1980. Procoat, the precursor of M13 coat protein, requires an electrochemical potential for membrane insertion. *Proc. Natl. Acad. Sci. USA* 77: 4669–73
31. Date, T., Wickner, W. 1981. Procoat, the precursor of M13 coat protein, inserts post-translationally into the membrane of cells infected by wild-type virus. *J. Virol.* 37:1087–89
32. Date, T., Zwizinski, C., Ludmerer, S., Wickner, W. 1980. Mechanisms of membrane assembly: effects of energy poisons on the conversion of soluble M13 coliphage procoat to membrane-bound coat protein. *Proc. Natl. Acad. Sci. USA* 77:827–31
33. Eilers, M., Schatz, G. 1986. Binding of a specific ligand inhibits import of a purified precursor protein into mitochondria. *Nature* 322:228–32
34. Emr, S. D., Bassford, P. J. Jr. 1982. Localization and processing of outer membrane and periplasmic proteins in *Escherichia coli* strains harboring export-specific suppressor mutations. *J. Biol. Chem.* 257:5852–60
35. Emr, S. D., Hanley-Way, S., Silhavy, T. J. 1981. Suppressor mutations that restore export of a protein with a defective signal sequence. *Cell* 23:79–88
36. Emr, S. D., Hedgpeth, J., Clement, J., Silhavy, T. J., Hofnung, M. 1980. Sequence analysis of mutations that prevent export of λ receptor, an *Escherichia coli* outer membrane protein. *Nature* 285:82–85
37. Emr, S. D., Silhavy, T. J. 1982. Molecular components of the signal sequence that function in the initiation of protein export. *J. Cell Biol.* 95:689–96
38. Enequist, H. G., Hirst, T. R., Harayama, S., Hardy, S. J. S., Randall, L. L. 1981. Energy is required for maturation of exported proteins in *Escherichia coli*. *Eur. J. Biochem.* 116:227–33
39. Engelman, D. M., Steitz, T. A. 1981. The spontaneous insertion of proteins into and across membranes: the helical hairpin hypothesis. *Cell* 23:411–22
40. Evans, E. A., Gilmore, R., Blobel, G. 1986. Purification of microsomal signal peptidase as a complex. *Proc. Natl. Acad. Sci. USA* 83:581–85
41. Ferenci, T., Randall, L. L. 1979. Precursor maltose-binding protein is active in binding substrate. *J. Biol. Chem.* 254:9979–81
42. Fikes, J. D., Bankaitis, V. A., Ryan, J. P., Bassford, P. J. Jr. 1987. Mutational alterations affecting the export competence of a truncated but fully functional maltose-binding protein signal peptide. *J. Bacteriol.* In press

43. Fikes, J. D., Bassford, P. J. Jr. 1987. Export of unprocessed precursor maltose-binding protein to the periplasm of cells of *Escherichia coli*. *J. Bacteriol.* In press
44. Filip, C., Fletcher, G., Wulff, J. L., Earhart, C. F. 1973. Solubilisation of the cytoplasmic membrane of *Escherichia coli* by the ionic detergent sodium-lauryl sarcosinate. *J. Bacteriol.* 115:717–22
45. Freitag, H., Janes, M., Neupert, W. 1982. Biosynthesis of mitochondrial porin and insertion into the outer mitochondrial membrane of *Neurospora crassa*. *Eur. J. Biochem.* 126:197–202
46. Freudl, F., Schwarz, H., Klose, M., Movva, N. R., Henning, U. 1985. The nature of information required for export and sorting, present within the outer membrane protein OmpA of *Escherichia coli* K-12. *EMBO J.* 4:3593–98
47. Gardel, C., Benson, S., Hunt, J., Michaelis, S., Beckwith, J. 1987. *secD*, a new gene involved in protein export in *E. coli*. *J. Bacteriol.* 169:1286–90
48. Garwin, J. L., Beckwith, J. 1982. Secretion and processing of ribose-binding protein in *Escherichia coli*. *J. Bacteriol.* 149:789–92
49. Gasser, S. M., Schatz, G. 1983. Import of proteins into mitochondria: *in vitro* studies on the biogenesis of the outer membrane. *J. Biol. Chem.* 258:3427–30
50. Geller, B. L., Movva, N. R., Wickner, W. 1986. Both ATP and the electrochemical potential are required for optimal assembly of pro-OmpA into *Escherichia coli* inner membrane vesicles. *Proc. Natl. Acad. Sci. USA* 83:4219–22
51. Gilmore, R., Blobel, G. 1985. Translocation of secretory proteins across the microsomal membrane occurs through an environment accessible to aqueous perturbants. *Cell* 42:497–505
52. Gilmore, R., Walter, P., Blobel, G. 1982. Protein translocation across the endoplasmic reticulum. II. Isolation and characterization of the signal recognition particle receptor. *J. Cell Biol.* 96:470–77
53. Hansen, W., Garcia, P. D., Walter, P. 1986. *In vitro* protein translocation across the yeast endoplasmic reticulum: ATP dependent post-translational translocation of the prepro-α-factor. *Cell* 45:397–406
54. Hayashi, S., Wu, H. C. 1985. Accumulation of prolipoprotein in *Escherichia coli* mutants defective in protein secretion. *J. Bacteriol.* 161:949–54
55. Hengge, R., Boos, W. 1985. Defective secretion of maltose- and ribose-binding proteins caused by a truncated periplasmic protein in *Escherichia coli*. *J. Bacteriol.* 162:972–78
56. Higgins, C. F. 1984. Peptide transport systems of *Escherichia coli* and *Salmonella typhimurium*. In *Microbiology—1984*, ed. D. Schlessinger, L. Leive, pp. 17–20. Washington, DC: Am. Soc. Microbiol. 441 pp.
57. Hirst, T. R., Randall, L. L., Hardy, S. J. S. 1984. Cellular location of heat-labile enterotoxin in *Escherichia coli*. *J. Bacteriol.* 157:637–42
58. Hoffman, C. S., Wright, A. 1985. Fusions of secreted proteins to alkaline phosphatase: an approach for studying protein secretion. *Proc. Natl. Acad. Sci. USA* 82:5107–11
59. Horisberger, M. 1979. Evaluation of colloidal gold as a cytochemical marker for transmission and scanning electron microscopy. *Biol. Cell* 36:253–58
60. Hortsch, M., Avossa, D., Meyer, D. I. 1986. Characterization of secretory protein translocation: ribosome-membrane interaction in endoplasmic reticulum. *J. Cell Biol.* 103:241–53
61. Hortsch, M., Meyer, D. I. 1986. Transfer of secretory proteins through the membrane of the endoplasmic reticulum. *Int. Rev. Cytol.* 102:215–42
62. Hurt, E. C., Pesold-Hurt, B., Schatz, G. 1984. The amino-terminal region of an imported mitochondrial precursor polypeptide can direct cytoplasmic dihydrofolate reductase into the mitochondrial matrix. *EMBO J.* 3:3149–56
63. Hurt, E. C., Pesold-Hurt, B., Schatz, G. 1984. The cleavable prepiece of an imported mitochondrial protein is sufficient to direct cytosolic dihydrofolate reductase into the mitochondrial matrix. *FEBS Lett.* 178:306–10
64. Hussain, M., Ichihara, S., Mizushima, S. 1980. Accumulation of glyceride-containing precursor of the outer membrane lipoprotein in the cytoplasmic membrane of *Escherichia coli* treated with globomycin. *J. Biol. Chem.* 255: 3707–12
65. Hussain, M., Ichihara, S., Mizushima, S. 1982. Mechanism of signal peptide cleavage in the biosynthesis of the major lipoprotein of the *Escherichia coli* outer membrane. *J. Biol. Chem.* 257:5177–82
66. Ichihara, S., Beppu, N., Mizushima, S. 1984. Protease IV, a cytoplasmic membrane protein of *Escherichia coli* has signal peptide peptidase activity. *J. Biol. Chem.* 259:9853–57

67. Ichihara, S., Hussain, M., Mizushima, S. 1981. Characterization of new membrane lipoproteins and their precursors in *Escherichia coli*. *J. Biol. Chem.* 256: 3125–29
68. Ichihara, S., Hussain, M., Mizushima, S. 1982. Mechanism of export of outer membrane lipoproteins through the cytoplasmic membrane in *Escherichia coli*. *J. Biol. Chem.* 257:495–500
69. Ichihara, S., Suzuki, T., Suzuki, M., Mizushima, S. 1986. Molecular cloning and sequencing of the *sppA* gene and characterization of the encoded protease IV, a signal peptide peptidase, of *Escherichia coli*. *J. Biol. Chem.* 261: 9405–11
70. Iida, A., Groarke, J. M., Park, S., Thom, J., Zabicky, J. H., et al. 1985. A signal sequence mutant defective in export of ribose-binding protein and a corresponding pseudorevertant isolated without imposed selection. *EMBO J.* 4:1875–80
71. Innis, M. A., Tokunaga, M., Williams, M. E., Loranger, J. M., Chang, S.-Y., et al. 1984. Nucleotide sequence of the *Escherichia coli* prolipoprotein signal peptidase (lsp) gene. *Proc. Natl. Acad. Sci. USA* 81:3708–12
72. Inouye, H., Beckwith, J. 1977. Synthesis and processing of alkaline phosphatase precursor *in vitro*. *Proc. Natl. Acad. Sci. USA* 74:1440–44
73. Inouye, S., Franceschi, T., Sato, M., Itakura, K., Inouye, M. 1983. Prolipoprotein signal peptidase of *Escherichia coli* requires a cysteine residue at the cleavage site. *EMBO J.* 2:87–91
74. Inouye, S., Hsu, C.-P. S., Itakura, K., Inouye, M. 1983. Requirement for signal peptide cleavage of *Escherichia coli* prolipoprotein. *Science* 221:59–61
75. Ito, K. 1982. Purification of the precursor form of maltose-binding protein, a periplasmic protein of *Escherichia coli*. *J. Biol. Chem.* 257:9895–97
76. Ito, K. 1986. Genetic control of protein secretion and localisation. *Adv. Biophys.* 21:267–80
77. Ito, K., Akiyama, Y., Yura, T., Shiba, K. 1986. Diverse effects of the MalE-LacZ hybrid protein on *Escherichia coli* cell physiology. *J. Bacteriol.* 167:201–4
78. Ito, K., Bassford, P. J. Jr., Beckwith, J. 1981. Protein localization in *E. coli:* is there a common step in the secretion of periplasmic and outer-membrane proteins? *Cell* 24:707–17
79. Ito, K., Beckwith, J. R. 1981. Role of the mature protein sequence of maltose-binding protein in its secretion across the *E. coli* cytoplasmic membrane. *Cell* 25:143–50
80. Ito, K., Date, T., Wickner, W. T. 1980. Synthesis, assembly into the cytoplasmic membrane and proteolytic processing of the precursor of coliphage M13 coat protein. *J. Biol. Chem.* 255:2123–30
81. Ito, K., Wittekind, M., Nomura, M., Shiba, K., Yura, T., et al. 1983. A temperature-sensitive mutant of *E. coli* exhibiting slow processing of exported proteins. *Cell* 32:789–97
82. Jackson, R. C., Blobel, G. 1977. Post-translational cleavage of presecretory proteins with an extract of rough microsomes from dog pancreas containing signal peptidase activity. *Proc. Natl. Acad. Sci. USA* 74:5598–602
83. Josefsson, L.-G., Randall, L. L. 1981. Different exported proteins in *E. coli* show differences in the temporal mode of processing *in vivo*. *Cell* 25:151–57
84. Kadonaga, J. T., Plückthun, A., Knowles, J. 1985. Signal sequence mutants of β-lactamase. *J. Biol. Chem.* 260:16192–99
85. Koshland, D., Botstein, D. 1982. Evidence for post-translational translocation of β-lactamase across the bacterial inner membrane. *Cell* 30:893–902
86. Koshland, D., Sauer, R. T., Botstein, D. 1982. Diverse effects of mutations in the signal sequence on the secretion of β-lactamase in *Salmonella typhimurium*. *Cell* 30:903–14
87. Kreibich, G., Freienstein, C. M., Pereyra, B. N., Ulrich, B. L., Sabatini, D. D. 1978. Proteins of rough microsomal membranes related to ribosome binding. II. Cross-linking of bound ribosomes to specific membrane proteins exposed at the binding sites. *J. Cell Biol.* 77:488–506
88. Kreibich, G., Ulrich, B. L., Sabatini, D. D. 1978. Proteins of rough microsomal membranes related to ribosome binding. I. Identification of ribophorins I and II, membrane proteins characteristic of rough microsomes. *J. Cell Biol.* 77:464–87
89. Kuhn, A., Wickner, W. 1985. Conserved residues of the leader peptide are essential for cleavage by leader peptidase. *J. Biol. Chem.* 260:15914–18
90. Kumamoto, C. A., Beckwith, J. 1983. Mutations in a new gene, *secB*, cause defective protein localisation in *Escherichia coli*. *J. Bacteriol.* 154:253–60
91. Kumamoto, C., Beckwith, J. 1985. Evidence for specificity at an early step in protein export in *Escherichia coli*. *J. Bacteriol.* 163:267–74

92. Lee, C. A., Beckwith, J. 1986. Suppression of growth and protein secretion defects in *Escherichia coli secA* mutants by decreasing protein synthesis. *J. Bacteriol.* 166:878–83
93. Lin, J. J. C., Kanazawa, H., Ozols, J., Wu, H. C. 1978. An *Escherichia coli* mutant with an amino acid alteration within the signal sequence of outer membrane prolipoprotein. *Proc. Natl. Acad. Sci. USA* 75:4891–95
94. Liss, L. R., Oliver, D. B. 1986. Effects of *secA* mutations on the synthesis and secretion of proteins in *Escherichia coli. J. Biol. Chem.* 261:2299–303
95. MacAlister, T. J., MacDonald, B., Rothfield, L. I. 1983. The periseptal annulus: An organelle associated with cell division in gram-negative bacteria. *Proc. Natl. Acad. Sci. USA* 80:1372–76
96. Maher, P. A., Singer, S. J. 1986. Disulfide bonds and the translocation of proteins across membranes. *Proc. Natl. Acad. Sci. USA* 83:9001–5
97. Marty-Mazars, D., Horiuchi, S., Tai, P. C., Davis, B. D. 1983. Proteins of ribosome-bearing and free membrane domains in *Bacillus subtilis. J. Bacteriol.* 154:1381–88
98. Meyer, D. 1982. The signal hypothesis—a working model. *Trends Biochem. Sci.* 7:320–21
99. Meyer, D. I. 1985. Signal recognition particle (SRP) does not mediate a translational arrest of nascent secretory proteins in mammalian cell-free systems. *EMBO J.* 4:2031–33
100. Meyer, D. I., Krause, E., Dobberstein, B. 1982. Secretory protein translocation across membranes—the role of the "docking protein." *Nature* 297:647–50
101. Minsky, A., Summers, R. G., Knowles, J. R. 1986. Secretion of β-lactamase into the periplasm of *Escherichia coli:* evidence for a distinct release step associated with a conformation change. *Proc. Natl. Acad. Sci. USA* 83:4180–84
102. Miura, T., Mizushima, S. 1968. Separation by density gradient centrifugation of two types of membrane from spheroplast membrane of *Escherichia coli* K12. *Biochim. Biophys. Acta* 150:159–61
103. Mühlradt, P. F., Menzel, J., Golecki, J. R., Speth, V. 1973. Outer membrane of *Salmonella.* Sites of export of newly synthesised lipopolysaccharide on the bacterial surface. *Eur. J. Biochem.* 35:471–81
104. Müller, M., Blobel, G. 1984. Protein export in *Escherichia coli* requires a soluble activity. *Proc. Natl. Acad. Sci.* USA 81:7737–41
105. Murén, E. M., Randall, L. L. 1985. Export of α-amylase by *Bacillus amyloliquefaciens* requires proton motive force. *J. Bacteriol.* 164:712–16
106. Nikaido, H., Wu, H. C. P. 1984. Amino acid sequence homology among the major outer membrane proteins of *Escherichia coli. Proc. Natl. Acad. Sci. USA* 81:1048–52
107. Novak, P., Paul, H. R., Inderjit, K. D. 1986. Localization and purification of two signal peptide hydrolases from *Escherichia coli. J. Biol. Chem.* 261:420–27
108. Oliver, D. 1985. Protein secretion in *Escherichia coli. Ann. Rev. Microbiol.* 39:615–48
109. Oliver, D. B., Beckwith, J. 1981. *E. coli* mutant pleiotropically defective in the export of secreted proteins. *Cell* 25:765–72
110. Osborn, M. J., Munson, R. 1974. Separation of the inner (cytoplasmic) and outer membranes of gram-negative bacteria. *Methods Enzymol.* 31:642–53
111. Pagès, J.-M., Anba, J., Bernadac, A., Shinagawa, H., Nakata, A., Lazdunski, C. 1984. Normal precursors of periplasmic proteins accumulated in the cytoplasm are not exported post-translationally in *Escherichia coli. Eur. J. Biochem.* 143:499–505
112. Pagès, J. M., Lazdunski, C. 1982. Maturation of exported proteins in *Escherichia coli:* requirement for energy, site and kinetics of processing. *Eur. J. Biochem.* 124:561–66
113. Pagès, J.-M., Lazdunski, C. 1982. Membrane potential depolarizing agents inhibit maturation. *FEBS Lett.* 149:51–54
114. Parkinson, J. S., Parker, S. R. 1979. Interaction of the *cheC* and *cheZ* gene products is required for chemotactic behavior in *Escherichia coli. Proc. Natl. Acad. Sci. USA* 76:2390–94
115. Perara, E., Rothman, R. E., Lingappa, V. R. 1986. Uncoupling translocation from translation: implications for transport of proteins across membranes. *Science* 232:348–52
116. Pollitt, S., Inouye, S., Inouye, M. 1986. Effect of amino acid substitutions at the signal peptide cleavage site of the *Escherichia coli* major outer membrane lipoprotein. *J. Biol. Chem.* 261:1835–37
117. Pugsley, A. P., Schwartz, M. 1985. Export and secretion of proteins by bacteria. *FEMS Microbiol. Rev.* 32:3–38
118. Randall, L. L. 1983. Translocation of domains of nascent periplasmic proteins

across the cytoplasmic membrane is independent of elongation. *Cell* 33:231–40
119. Randall, L. L., Hardy, S. J. S. 1984. Export of protein in bacteria: dogma and data. In *Modern Cell Biology*, ed. B. H. Satir, 3:1–20. New York: Liss. 314 pp.
120. Randall, L. L., Hardy, S. J. S. 1986. Correlation of competence for export with lack of tertiary structure of the mature species: a study *in vivo* of maltose-binding protein. *Cell* 46:921–28
121. Rapoport, T. A. 1986. Protein translocation across and integration into membranes. *CRC Crit. Rev. Biochem.* 20: 73–137
122. Regue, M., Remenick, J., Tokunaga, M., Mackie, G. A., Wu, H. C. 1984. Mapping of lipoprotein signal peptidase gene (lsp). *J. Bacteriol.* 158:632–35
123. Rothblatt, J. A., Meyer, D. I., 1986. Secretion in yeast: reconstitution of the translocation and glycosylation of α-factor and invertase in a homologous cell free system. *Cell* 44:619–28
124. Russel, M., Model, P., 1982. Filamentous phage pre-coat is an integral membrane protein: analysis by a new method of membrane preparation. *Cell* 28:177–84
125. Ryan, J. P., Bassford, P. J. Jr. 1985. Post-translational export of maltose-binding protein in *Escherichia coli* strains harboring *malE* signal sequence mutations and either prl^+ or *prl* suppressor alleles. *J. Biol. Chem.* 260:14832–37
126. Ryan, J. P., Duncan, M. C., Bankaitis, V. A., Bassford, P. J. Jr. 1986. Intragenic reversion mutations that improve export of maltose-binding protein in *Escherichia coli malE* signal sequence mutants. *J. Biol. Chem.* 261:3389–95
127. Ryter, A., Shuman, H., Schwartz, M. 1975. Integration of the receptor for bacteriophage lambda in the outer membrane of *Escherichia coli:* coupling with cell division. *J. Bacteriol.* 122:295–301
128. Schleyer, M., Neupert, W. 1984. Transport of ADP/ATP carrier into mitochondria: precursor imported *in vitro* acquires functional properties of the mature protein. *J. Biol. Chem.* 259:3487–91
129. Schleyer, M., Neupert, W. 1985. Transport of proteins into mitochondria: translocation intermediates spanning contact sites between outer and inner membranes. *Cell* 43:339–50
130. Schnaitman, C. A. 1971. Solubilization of the cytoplasmic membrane of *Escherichia coli* by Triton X-100. *J. Bacteriol.* 108:545–52
131. Shiba, K., Ito, K., Yura, T. 1986. Suppressors of the *secY24* mutation: identification and characterization of additional *ssy* genes in *Escherichia coli. J. Bacteriol.* 166:849–56
132. Shiba, K., Ito, K., Yura, T., Cerretti, D. P. 1984. A defined mutation in the protein export gene within the *spc* ribosomal protein operon of *Escherichia coli:* isolation and characterization of a new temperature-sensitive *secY* mutant. *EMBO J.* 3:631–35
133. Siegel, V., Walter, P. 1985. Elongation arrest is not a prerequisite for secretory protein translocation across the microsomal membrane. *J. Cell Biol.* 100: 1913–21
134. Siegel, V., Walter, P. 1986. Removal of the *Alu* structural domain from signal recognition particle leaves its protein translocation activity intact. *Nature* 320:81–84
135. Singer, S. J., Maher, P. A., Yaffe, M. P. 1987. On the translocation of proteins across membranes. *Proc. Natl. Acad. Sci. USA* 84:1015–19
136. Smit, J., Nikaido, H. 1978. Outer membrane of gram-negative bacteria. XVIII. Electron microscopic studies on porin insertion sites and growth of cell surface of *Salmonella typhimurium. J. Bacteriol.* 135:687–702
137. Tokunaga, M., Tokunaga, H., Wu, H. C. 1982. Post-translational modification and processing of *Escherichia coli* prelipoprotein *in vitro. Proc. Natl. Acad. Sci. USA* 79:2255–59
138. Tommassen, J. 1986. Fallacies of *E. coli* cell fractionations and consequences thereof for protein export models. *Microb. Pathog.* 1:225–28
139. Tommassen, J., Leunissen, J., van Damme-Jongsten, M., Overduin, P. 1985. Failure of *E. coli* K-12 to transport PhoE-LacZ hybrid proteins out of the cytoplasm. *EMBO J.* 4:1041–47
140. Tweten, R. K., Iandolo, J. J. 1983. Transport and processing of staphylococcal enterotoxin B. *J. Bacteriol.* 153:297–303
141. von Heijne, G. 1983. Patterns of amino acids near signal-sequence cleavage sites. *Eur. J. Biochem.* 133:17–21
142. von Heijne, G. 1985. Signal sequences: the limits of variation. *J. Mol. Biol.* 184:99–105
143. von Heijne, G., Blomberg, C. 1979. Trans-membrane translocation of proteins, the direct transfer model. *Eur. J. Biochem.* 97:175–81
144. Walter, P., Blobel, G. 1980. Purification of a membrane-associated protein complex required for protein translo-

cation across the endoplasmic reticulum. *Proc. Natl. Acad. Sci. USA* 77:7112–16
145. Walter, P., Blobel, G. 1982. 7SL small cytoplasmic RNA is an integral component of the signal recognition particle. *Nature* 299:691–98
146. Walter, P., Ibrahimi, I., Blobel, G. 1981. Translocation of proteins across the endoplasmic reticulum. I. Signal recognition protein (SRP) binds to *in vitro*–assembled polysomes synthesizing secretory protein. *J. Cell Biol.* 91:545–50
147. Waters, M. G., Blobel, G. 1986. Secretory protein translocation in a yeast cell-free system can occur posttranslationally and requires ATP hydrolysis. *J. Cell Biol.* 102:1543–50
148. Waxman, L., Goldberg, A. L. 1986. Selectivity of intracellular proteolysis: protein substrates activate the ATP-dependent protease (La). *Science* 232: 500–3
149. Wickner, W. 1979. The assembly of proteins into biological membranes: the membrane trigger hypothesis. *Ann. Rev. Biochem.* 48:23–45
150. Wickner, W. T., Lodish, H. F. 1985. Multiple mechanisms of protein insertion into and across membranes. *Science* 230:400–7
151. Wolfe, P. B., Rice, M., Wickner, W. 1985. Effects of two *sec* genes on protein assembly into the plasma membrane of *Escherichia coli*. *J. Biol. Chem.* 260:1836–41
152. Wolfe, P. B., Silver, P., Wickner, W. 1982. The isolation of homogeneous leader peptidase from a strain of *Escherichia coli* which overproduces the enzyme. *J. Biol. Chem.* 257:7898–902
153. Wolfe, P. B., Wickner, W. 1984. Bacterial leader peptidase, a membrane protein without a leader peptide, uses the same export pathway as pre-secretory proteins. *Cell* 36:1067–72
154. Wolfe, P. B., Wickner, W., Goodman, J. M. 1983. Sequence of the leader peptidase gene of *Escherichia coli* and the orientation of leader peptidase in the bacterial envelope. *J. Biol. Chem.* 258:12073–80
155. Yamada, H., Yamagata, H., Mizushima, S. 1984. The major outer membrane lipoprotein and new lipoproteins share a common signal peptidase that exists in the cytoplasmic membrane of *Escherichia coli*. *FEBS Lett.* 166:179–82
156. Yamagata, H., Ippolito, C., Inukai, M., Inouye, M. 1982. Temperature sensitive processing of outer membrane lipoprotein in an *Escherichia coli* mutant. *J. Bacteriol.* 152:1163–68
157. Yamagata, H., Taguchi, N., Daishima, K., Mizushima, S. 1983. Genetic characterization of a gene for prolipoprotein signal peptidase in *Escherichia coli*. *Mol. Gen. Genet.* 192:10–14
158. Yamane, K., Ichihara, S., Mizushima, S. 1987. *In vitro* translocation of protein across *Escherichia coli* membrane vesicles requires both the protonmotive force and ATP. *J. Biol. Chem.* 262:2358–62
159. Yu, F., Yamada, H., Daishima, K., Mizushima, S. 1984. Nucleotide sequence of the *lspA* gene, the structural gene for lipoprotein signal peptidase of *Escherichia coli*. *FEBS Lett.* 173:264–68
160. Zimmermann, R., Wickner, W. 1983. Energetics and intermediates of the assembly of protein OmpA into the outer membrane of *Escherichia coli*. *J. Biol. Chem.* 258:3920–25
161. Zwizinski, C., Wickner, W. 1980. Purification and characterization of leader (signal) peptidase from *Escherichia coli*. *J. Biol. Chem.* 255:7973–77

Ann. Rev. Microbiol. 1987. 41:543–71

AN INQUIRY INTO THE MECHANISMS OF HERPES SIMPLEX VIRUS LATENCY

Bernard Roizman and Amy E. Sears

Marjorie B. Kovler Viral Oncology Laboratories, University of Chicago, 910 East 58th Street, Chicago, Illinois 60637

CONTENTS

0066-4227/87/1001-0543$02.00

INTRODUCTION

Herpes simplex viruses 1 and 2 (HSV-1 and HSV-2) invariably destroy the cells that they infect and in which they multiply. The destruction begins with the first stages of infection; the virus progressively parasitizes the machinery of the cell and even uses the host DNA as a source of deoxynucleotides for the replication of its own DNA. Yet the viruses are best known for their ability to remain latent in humans for the life of the host. In humans, infection with HSV-1 typically occurs during the first six years of age as a consequence of exposure to an adult or child secreting virus. The disease, usually a mild stomatitis, is frequently undifferentiated from other diseases of childhood. In years to come, a large fraction of those infected exhibit recurrent lesions containing virus near the portal of entry, usually at the mucocutaneous junction of the lip. These recurrences are manifested in the face of a state of immunity indistinguishable from that of individuals who do not have recurrences. In the case of HSV-2, infection occurs either at birth or more commonly after the age of consent. The virus is transmitted by sexual contact, and initial lesions and recrudescences usually occur at or near genital organs but also occur in the mouth. While most studies of herpes simplex viruses center on the events of productive infections which invariably lead to cell death, the question that has attracted many investigators, including the authors, into the field, is: How does a virulent virus, which is capable of massive destruction of the cells it infects, stay latent in humans for life without multiplying or destroying the cells in which it resides?

HSV latency is not a novel discovery. It has been the subject of speculative and experimental publications since at least 1905 (20). The puzzle presented by HSV stemmed from the observations that the recurrences and underlying latency could be transmitted from person to person by inoculation of fluid from the vesicles of lesions, but virus could not be isolated from the sites of recurrent lesions in the interim between recrudescences (2, 30, 101, 107). The failure, also, to reproduce latent infection in its exquisite detail in isolated cells in culture seemed to suggest that HSV latency is very different conceptually and mechanistically from phage lysogeny. Indeed, the early literature on HSV latency contains numerous explanations, and many of them seem to have been supported by the experimental evidence and models of the day. Doerr, reviewing the literature in 1938, stated: "All of these observations converge to indicate that the agent of herpes is no infectious agent which is conserved in the site of infection, but that it originates in the human organism, that is, endogenously." (23).

Latency is at the very least a two-body problem, i.e. it involves the interaction of host and viral gene products. Operationally, latency is more complex. We cannot deliver the viral genome directly to the nuclei of ganglionic cells to see if latency will ensue. The third body, the physiologic environment in which the virus is vectored from the site of inoculation to the neuronal nucleus, strongly influences the outcome of the viral inoculation. To shed what little light there is, we therefore first present our current understanding of the virus and its functions as we know them from studies of productive infection of permissive cells. We then examine latency from the standpoint of available data, pressing issues that remain unresolved, and models that remain to be challenged.

> A system . . . is exactly the opposite of a machine, in which the structure of the product depends crucially on strictly predefined operations of the parts. In the system, the structure of the whole determines the operation of the parts; in the machine, the operation of the parts determines the outcome.
>
> Paul Alfred Weiss, in A. Koestler, V. R. Smithies, eds., *Beyond Reductionism,* 1968. London: Hutchinson

PRODUCTIVE INFECTION OF PERMISSIVE CELLS

The Family Herpesviridae

Herpesviruses are defined on the basis of the architecture of their virions (reviewed in 98). All herpes virions have a core consisting of DNA arranged in the form of a torroid; an icosadeltahedral capsid consisting of 162 capsomeres; an amorphous, occasionally asymmetrically placed structure that surrounds the capsid, defined as tegument; and an envelope studded with glycoprotein spikes. The herpesviruses share many features in their patterns of replication, morphogenesis, and ability to remain latent in their hosts. These common features notwithstanding, herpesviruses exhibit differences of at least twofold in the size of their DNAs (120–250 kbp) and G+C content (33–74 mol%) (98). The variability reflects the capacity of these viruses to adapt through selection to the epidemiological exigencies of their hosts rather than the degeneracy of the genetic code.

Herpes Simplex Virus Gene Functions and Expression

HSV expresses approximately 70 proteins in productive infection (Figure 1). These proteins form at least five groups whose expression is coordinately regulated and sequentially ordered in a cascade fashion (98). The first set expressed comprises the α proteins 0, 4, 22, 27, and 47; they appear to be induced by a component of the tegument designated as the α-*trans*-inducing factor or α-TIF (98). Promoter-regulatory domains of α genes contain *cis*-acting sites required for α-TIF–mediated induction (57), but these sites bind host proteins rather than α-TIF (60). Additional *cis*-acting sites with affinity for host transcriptional factors control basal levels and also on or off gene expression (57, 60).

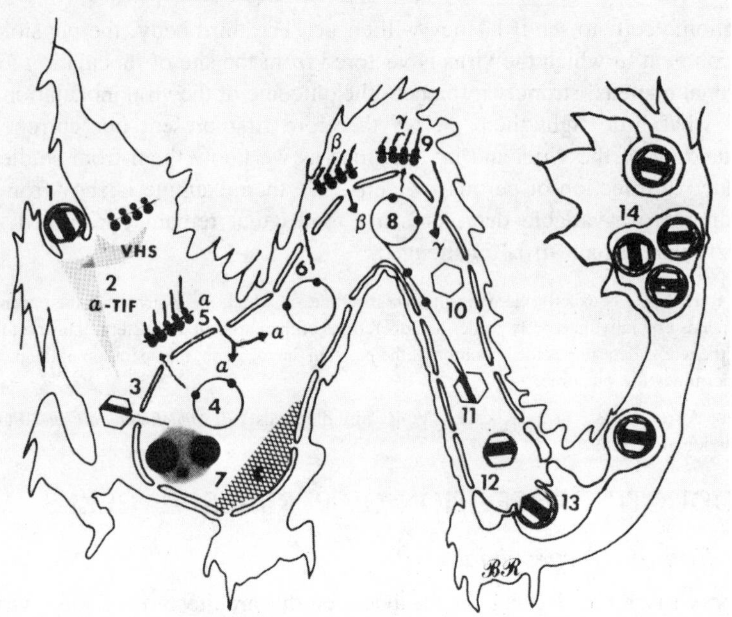

Figure 1 Schematic representation of productive life cycle of HSV. (*1*) Virus adsorbs to the cell and enters by fusion of viral and cell membranes. (*2*) Virion components, including the DNA-containing capsid and tegument proteins (e.g. α-TIF, Vhs) travel through the cytoplasm; the capsid moves to the nuclear pores. (*3*) Viral DNA is ejected from the capsid into the nucleus through the nuclear pores. (*4*) Circularization of the viral DNA occurs immediately after the DNA enters the nucleus. Coordinately regulated and sequentially ordered transcription begins with the α genes, induced by the α-TIF, which has moved into the nucleus. (*5*) The α mRNAs are transported to the cytoplasm and translated; α proteins induce β transcription. (*6*) The β genes are transcribed, and mRNAs are transported to the cytoplasm and translated. (*7*) Chromatin (*c*) is fragmented and marginated to the sides of the nucleus; nucleoli *(dark circles)* are broken down. (*8*) DNA replication begins by a rolling circle mechanism. (*9*) γ genes are transcribed and translated. (*10*) Viral capsids are assembled, containing proteins made mainly during the γ phase of infection. (*11*) Newly formed capsids are filled with viral DNA, which is cleaved to unit length. (*12*) Additional proteins bind to the filled capsids. (*13*) The modified capsids with tegument bud out through the nuclear membrane, acquiring an envelope containing viral glycoproteins. (*14*) Virions move through the membranous structures of the cell; glycoprotein processing is completed. Fully enveloped capsids exit the cell.

The function of most α genes has eluded us. A functional α4 gene product, and possibly other functional α gene products, is required for the expression of subsequent protein groups (98). Independent evidence that the α4 protein regulates gene expression both positively (β and γ) and negatively (the α genes) is consistent with the observation that it binds to promoter-regulatory domains and to the 5' transcribed noncoding sequences of most genes, including the α genes (58–60, 98). Expression of α47 is not essential for viral multiplication in at least some cells in culture (68). The function of the α22 gene, required late in infection, is complemented by the functions expressed by some cells and not others (106). The $\alpha27$ product appears to be required for late gene expression (103). Little is known of the function of $\alpha0$; it seems to act as a general inducer in transient expression systems, but to date no conditional lethal mutants in that gene have been isolated (26, 74, 75).

The β genes, expressed at peak levels between 5 and 7 hr postinfection, are mainly involved in nucleic acid metabolism. Those β genes designated β_1, exemplified by the genes for the major DNA-binding protein ($\beta_1 8$) and ribonucleotide reductase ($\beta_1 6$), are expressed earlier than those in the β_2 class, which includes most β genes (98).

Most γ proteins identified to date are structural components of the virion; they are heterogeneous with respect to the timing of and requirements for their synthesis. Whereas the expression of γ_1 genes (e.g. glycoproteins B and D, α-TIF, and the major capsid protein, $\gamma_1 5$) is only slightly reduced by total inhibition of viral DNA replication, the expression of γ_2 genes (e.g. glycoproteins C and E) stringently requires viral DNA synthesis (98). In addition to the transition from α to β and γ gene expression, posttranscriptional controls determine when the mRNAs are translated (44).

Shut-Off of Viral and Host Gene Expression

The shut-off of host macromolecular metabolism, drastic changes in the structure and integrity of the cell, and sequential shut-off of viral gene expression are characteristic of productively infected cells (98). All are relevant to latency.

Host shut-off is incremental. Immediately after the cell is infected host DNA synthesis is shut off and there is posttranscriptional shut-off of host protein synthesis by one or more structural proteins of the virus designated as virion host shut-off proteins or Vhs. Studies on Vhs mutants indicate that this function is not essential (89).

With onset of synthesis of β and γ proteins, synthesis of host proteins and DNA is completely shut off, and synthesis of host RNA is drastically reduced (98). The mechanism of host shut-off may be different from that caused by Vhs, inasmuch as Vhs mutants also shut off host metabolism late in infection (89). Margination and fragmentation of chromosomes, disaggregation of the nucleoli, distortion of the nucleus with duplication of the nuclear membranes, and insertion of viral glycoproteins into the cellular membranes also occur; these changes provide a target for the host immune response (98). Viral replication without these changes has not been seen.

How far the regulatory cascade can proceed before the infection and cell destruction become irreversible is unclear. The impression gained from studies in cell culture is that expression of α genes alone may not be deleterious to the host (3). It is relevant, therefore, to consider the viral functions that turn off α gene expression.

The first observed inhibition is that of products made later in infection. In the absence of functional β and γ gene products, α protein synthesis continues (98). More recent studies have suggested that the major DNA binding protein ($\beta_1 8$) shuts off α protein synthesis (34, 35). However, in the absence of a functional $\beta_1 8$ protein there is no DNA synthesis or significant synthesis of the γ_1 and γ_2 proteins. The identities of the β or γ proteins which turn off αs are not known.

A second mechanism of α shut-off may be autoregulation by the $\alpha 4$ protein. Evidence is the observation that after the synthesis of a temperature-sensitive $\alpha 4$ protein has declined, shift up to temperature nonpermissive for infected cells results in resumption of α protein synthesis (98). This observation is reinforced by the finding of $\alpha 4$ protein binding sites in the promoter-regulatory domains and 5' transcribed noncoding sequences of α genes (58, 59). The only likely function for these sequences would be to turn off transcription. However, this picture is obscured by the observation that cells that contain a temperature-sensitive $\alpha 4$ gene and that constitutively make the protein do not overexpress $\alpha 4$ when shifted to a nonpermissive temperature (J. Hubenthal-Voss & B. Roizman, unpublished data). The protein is obviously functional since it induces the expression of other viral genes only at permissive temperature. Data from studies of productive infections in permissive cells suggest that the α genes are pivotal in determining the outcome of infection. They are expressed first and are necessary for expression of all other viral genes. Commitment to lytic or latent infection hinges on their activation or deactivation.

The HSV Genome: Sequence and Gene Arrangement

The HSV genome is approximately 140 kbp in size. It consists of two components, L and S, that invert relative to each other; thus viral genomes extracted from virions or infected cells consist of four equimolar isomers that differ solely in the orientation of the L and S components. The isomerization of HSV DNA is due to the internal repetition of the terminal sequences in an inverted orientation. Thus both L and S components consist of unique sequences (U_l and U_S) flanked by inverted repeats. The L component repeat is the terminal 9-kb sequence *ab* and its internal inverted repeat $b'a'$, whereas that of the S component is the 6.5-kb internal inverted repeat $a'c'$ and the terminal *ca* sequences (Figure 2) (98). The presence of these large inverted repeat sequences has evoked numerous speculations concerning their biologic functions. It seems, however, that deletion of the internal inverted repeats ($b'a_n'c'$) does not affect the ability of the virus to replicate in culture or to establish latency (81; B. Meignier & B. Roizman, unpublished data).

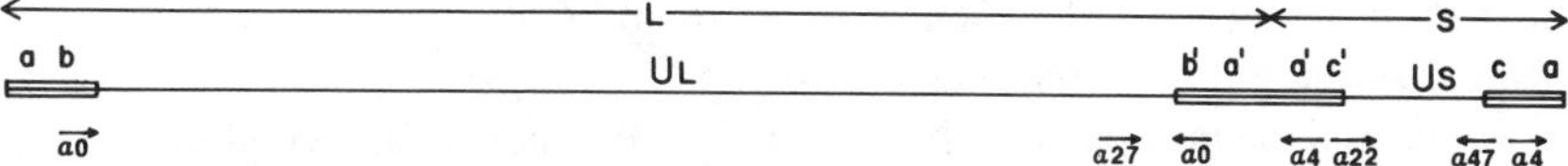

Figure 2 HSV DNA structure. The *top line* indicates the L and S components, which invert relative to one another. The terminal and internal inverted repeated sequences (designated as ab–b'a'a'c'–ca) are shown on the *second line,* along with the locations of the α genes.

The HSV genome contains three origins of viral DNA synthesis, two in the inverted repeats of the S component (*c* sequence) and one in the middle of the L component (98). Viral DNA circularizes immediately after infection without requiring de novo protein synthesis (80), and late in infection the replication occurs via a rolling circle (98). The function of the three origins in replication is not known; at least one origin is dispensible for growth in cell culture and for establishment of latency (63; R. Longnecker, B. Meignier & B. Roizman, unpublished data).

The HSV genome contains small clusters of functionally related genes, but no overall functional or temporal organization. Thus the major DNA-binding protein and DNA polymerase genes flank one origin of DNA replication and there is a cluster of glycoprotein genes in the S component, but at least half of the known glycoprotein genes are scattered in the L component. Only the α genes seem to be organized in a specific cluster. The α4 and α0 genes map in the reiterated sequences of the S and L components, respectively, and both are therefore diploid. The α22 and α47 genes abut, 5' to 5', with the α4 genes, but their structural sequences are in the unique sequences of the S component. The α27 gene is located very close to the right terminus of the L component in the prototype orientation (Figure 2).

Dispensible Genes, Viral Replication, and Latency

Herpes viruses characteristically encode many enzymes and other factors necessary for their replication. Many of these functions may also be expressed by eukaryotic cells. In cells that express these functions, some of the corresponding viral genes are dispensible. The viral gene encoding thymidine kinase was the first gene shown to be dispensible in growing cells in culture (48). However, not all genes with cellular counterparts are dispensible; for example, host DNA polymerases do not substitute for viral DNA polymerase in productively infected cells (98). Surprisingly, the HSV genome encodes structural proteins that are dispensible and that are not complemented by the host. These include, for example, glycoproteins C, G, and E (38, 63, 63a). The glycoprotein C and thymidine kinase genes, all of the genes mapping between α27 and α0, and nearly all of the genes mapping in the unique sequences of the S component (with the exception of that encoding glycoprotein D) have been found dispensible for growth in at least some cultured cells (38, 43, 48, 63, 63a, 68, 82).

In principle, the polymorphism in HSV genomes, dealt with extensively elsewhere (99), indicates that the genome mutates and that mutations in noncritical sites are perpetuated. It is unlikely that nonessential expressed and regulated genes would be perpetuated for millenia. The variability in the size, base composition, and genetic complexity of various genomes of the herpesvirus family suggests that the diversification fulfills the requirements of the viruses for perpetuation in their hosts. Granted that the seemingly dispensible genes are in fact essential, what is their role in the life cycle of the virus?

One obvious explanation is that the virus multiplies in a variety of differentiated cells and that not all of them express all of the functions required for virus growth. For example, neuronal cells that do not divide may lack enzymes necessary for DNA synthesis and would therefore not be permissive for growth of virus mutants defective in these genes. The function of a second set of genes might be to enable the virus to attach to and infect cell types with different surface receptors. A third and perhaps more interesting set would enable the virus to perform functions that are not directly related to multiplication, but that would in effect enable the virus to maintain itself in its host population. There is evidence only for the first set; the thymidine kinase and $\alpha22$ genes are prime examples of genes whose functions can be complemented by host cells (48, 106). The evidence neither excludes nor supports the predicted functions of other dispensible genes, nor do we know at this time just how many viral genes are "dispensible" for growth in cell culture.

> Space travel is utter bilge.
>
> Sir Richard van der Riet Wolley, quoted in Arthur C. Clarke, *Profiles of the Future*, 1973. London: Gollancz

LATENCY

Early Studies and Isolation of Virus From Latently Infected Ganglia

Even before the nature of the infectious agent became known, studies suggested an association of recurrences with nerve trauma and, by extension, with the peripheral nervous system. Cushing's 1905 study (20) and numerous subsequent studies (13, 20, 27, 76, 77, 92) reported the induction of herpetic vesicles in tissues along the track innervated by branches of the trigeminal nerve following section of the posterior root in treatment of trigeminal neuralgia. Concurrently, as noted in the introduction, researchers consistently failed to isolate virus from tissues removed by biopsy from the site of recurrent infections in the interim between recurrences (2, 19, 30, 101, 107). These and other studies led Goodpasture to postulate in 1929 (36) that ". . . it seems quite probable that the virus remains in the ganglia in a latent state after the local lesion has healed." The postulate made anatomical sense; at least in the case of HSV-1 infections of the face, the recurrent lesions usually occur in tissues innervated by the branch of the trigeminal ganglion that innervates the portal of entry (13, 20, 27, 45, 76, 77, 92). Moreover, studies done in the 1930s showed that virus could be reactivated from the brains of experimentally infected rabbits by incubation of the tissue for several days; this indicated that the nervous system was indeed capable of harboring virus during the latent state (78).

Even before demonstration of the presence of latent virus in sensory neurons, the question arose as to whether the virus replicates chronically at a reduced rate (i.e. is maintained in a *dynamic state*) or whether it is maintained in a nonreplicating *static state* (96). The evidence that patients with recurrent infections have high-titer antibody (22, 24, 132); that the lytically infected cell expresses on its surface virus-specific antigens, subsequently identified as viral glycoproteins; and that antibody and complement are capable of lysing the infected cells (98) led to the hypothesis that the virus is retained in the static state (96).

Proof that the trigeminal ganglion is capable of harboring latent virus emerged from early and pioneering studies by Cook, Stevens, and coworkers (17, 109, 110) and others (4, 5, 79, 95) which ultimately led to the isolation of latent virus from trigeminal ganglia removed from human cadavers and explanted in culture (4, 5, 64, 79, 120). Virus has been isolated from sensory ganglia, including those that innervate genital organs, and autonomic ganglia of humans and experimental animals (4, 5, 17, 18, 67, 79, 83, 95, 104, 109, 110, 118–120). Careful analyses indicated that when viral replication is initiated the virus is found predominantly in the neurons (16). Before activation, no evidence of virus replication could be found in neurons or other cells of sensory ganglia.

Perhaps the most important contribution has been the development of a mouse model system; the operational definition of latency developed with this model is applicable to all other model systems (16, 17, 109, 110). In brief, infection of the mouse by inoculation of tissues innervated by sensory ganglia results in local lesions that heal and ultimately disappear. As long as the peripheral lesions at the site of inoculation persist, infectious virus may also be isolated by maceration of the ganglion and inoculation of susceptible cells in culture. After the peripheral lesions have healed, infectious virus may be recovered by explantation of whole ganglia, minced tissue, or dispersed ganglionic cells in cultures of susceptible cells. During this phase, virus cannot be isolated from macerated ganglionic tissue inoculated into cell cultures. The implicit operational definition is that a latent virus is one that must be induced to multiply and that does not exist in infectious form.

According to the model that has emerged, following infection the virus multiplies at the portal of entry, infects sensory nerve endings, and ascends by way of the axons to the nuclei of the sensory neurons, where the latent state is established. When virus multiplication is induced, the virus is transported in the reverse direction by way of the axons to the cells at or near the portal of entry. From the time it enters the neurons until it exits, the virus is shielded from the host immune system.

There is no doubt that sensory ganglia retain latent virus and that the virus can be activated to replicate in cell culture by explantation or in situ in the sensory ganglion and peripheral tissues by localized trauma at the site of initial infection (8, 9, 39, 40). Despite the observation that when the axons that innervate sites of frequent recurrent lesions have been sectioned recurrences have ceased until the tissues have become reinnervated, the role of the latent neuronal virus in human recrudescent lesions has been challenged (102). Several reports indicated that virus could be isolated by cultivating peripheral tissues of experimental animals at the site of inoculation even after extensive enervation of the tissues (41, 105). Challenges of this type in the face of mounting studies on the latent state in neurons are difficult to sustain. The criticism of such systems is intuitive: The isolated virus may reflect chronic multiplication of virus in an environment made susceptible by enervation or by incomplete enervation as suggested by Hill and coworkers (41).

We must accept the notion that the neuron is the most common but perhaps not the only cell capable of harboring HSV in a latent state. The implicit acceptance of the notion that other cells are capable of harboring latent virus does not obviate the basic question of why latency is not established in all human tissue cells capable of being infected. As a corollary, why does the seeding of neuronal cells in culture activate virus? We conclude that establishment of latency requires a cell phenotype that is normally restricted to very few cells in the body and that this phenotype is terminated once the cells are placed in culture. The designation of this phenotype as nonpermissive or restrictive merely restates the problem; it does not in fact illuminate what necessary and specific functions of the host cell render it competent for latency.

Animal Models

Several useful animal model systems have been developed for the study of HSV latency in vivo. The animals used include mice, rats, guinea pigs, rabbits, and monkeys. Sites of inoculation fall into three categories: peripheral (e.g. eye, ear, footpad, vagina), intratissue (cerebrum), or systemic (e.g. peritoneum, circulatory system). To serve as a model, the host must be able to retain the virus in a latent state at a predictable site. The most commonly used model is the mouse, and the highest incidence of latent infections, at least in our hands, has resulted from eye inoculation. In this system, the virus replicates initially at the site of inoculation. Within a few days, infectious virus may also be recovered from macerated ganglia; this suggests that virus replication in the ganglion precedes establishment of latency. The latent virus is not readily reactivated in the mouse; trauma of the ear following inoculation of virus at that site (8, 9, 39, 40) has been reported to cause recurrences in a small fraction of animals. The mouse is therefore a suitable animal for experimental investigations of the requirements for establishment and maintenance of the latent state but not for reactivation. At the other extreme, spontaneous recurrences are common in the guinea pig inoculated vaginally with HSV-2 (104, 108). However, the lesions generated in guinea pigs inoculated with moderate amounts of virus tend to be both common and sterile, and only in animals inoculated with large amounts of virus do the frequent recurrent lesions yield virus (108). The development of sterile lesions is unsettling; it violates the operational definition of latency, which requires emergence of virus at the peripheral site, and raises questions as to whether the recurrences seen in guinea pigs reflect solely virus activation or additional, as yet poorly understood manifestations of the host immune response.

Until the latency-competent phenotype can be reproduced in cells in culture, animal models will provide the sole unambiguous criteria by which both host and viral functions related to the latent state can be defined. The shortcoming of the animal systems is that none exactly match the human situation in all respects. The most economical and easy to handle in large numbers (e.g. mouse, rat, guinea pig) tend also to be less permissive and require relatively high doses of virus for efficient establishment of latency. One explanation of the dosage requirement is that infection of nerve endings is not efficient; to infect nerve endings the virus must either efficiently multiply at the portal of entry or be introduced in sufficient quantity by a route that reduces the dependence on viral multiplication. Highly susceptible animals (e.g. marmosets) require smaller doses of virus, but the animals suffer from high mortality rates even when inoculated by peripheral routes. Because of these shortcomings, no single model can be used to study all aspects of human HSV infection, i.e. primary lesions, latency, and recurrent infection. Table 1 summarizes the advantages and disadvantages of each model.

Fraction of Murine Trigeminal Ganglion Cells Harboring Latent Virus

Data on the number of cells harboring latent virus have largely been obtained by counting infectious centers (plaques) formed after the seeding of dispersed ganglionic cells on top of susceptible cells in culture. This technique is based on the assumptions that the recovery of dispersed cells with and without virus

Table 1 Characteristics of animal models of HSV latency[a]

Animal model and route of innoculation	Advantages	Disadvantages
Mouse eye	Tends to discriminate between ability to multiply and establish latency. Spontaneous induction is negligible.	Not inducible in vivo.
Mouse ear	Inducible with specific HSV strains at low to medium frequency.	Difficult to reproduce.
Rabbit eye	Dependence on peripheral multiplication not well established.	Spontaneous recurrences are common.
Guinea pig vagina	Recurrent genital lesions readily established; model is useful for studies of prevention and therapy.	Biology of recurrent lesions is not well understood, as lesions initiated by low levels of virus tend to be sterile (no detectable virus).
Rat mandibular nerve	Discriminates between multiplication and ability to establish latency inasmuch as rat tissues in situ are nonpermissive; virus can be readily detected by explantation of the ganglia.	No inducible recurrences in vivo.

[a] From References 8, 9, 39, 40, 42, 73, 104, 106, 108, 129.

is approximately the same and that all cells carrying virus in a latent state activate virus replication upon dispersal and cultivation in cell cultures. The data suggest that less than 1% of the neurons in a ganglion harbor virus in the latent form. Neurons constitute 10–20% of the cells in a ganglion; thus, approximately 0.1–0.2% of the cells contain viral genomes that can be reactivated (46, 118).

Form of Viral DNA During Latency in Mice

HSV DNA in virions is for the most part linear; following infection it circularizes, and its replication is initiated from the circular form (80). For HSV, the central question of the state of the viral DNA in neurons becomes particularly complex.

For all linear DNAs that circularize, the transition from the linear to the circular state is marked by disappearance of the terminal fragments and appearance of head-to-tail junction fragments. As described above, in HSV DNA the terminal fragments are repeated internally in an inverted form, and the L and S components invert relative to each other, giving rise to four equimolar isomeric populations of viral DNA (98). These populations of linear molecules contain four terminal fragments and the molar ratio of the terminal to nonterminal fragments is 0.5. The isomeric populations also contain four L-S junction fragments representing head-to-tail fusions of terminal fragments. The molar ratio of these junction fragments to the nonterminal fragments is 0.25. For this reason, a head-to-tail junction fragment of any two termini is present in 25% of the DNA. It follows that if the DNA circularizes, there should be a decrease in the concentration of the terminal fragments and an increase in the concentration of the corresponding head-to-tail L-S junction fragments above the natural background of 0.25 M. We are able to measure at best fourfold differences in DNA fragment concentrations; this accuracy is trivial when the hybridization signals are strong owing to high copy number, but the estimates become less than satisfactory when the copy number is low.

The most extensive published studies on the state of HSV genomes in cells harboring the virus in the latent state were done on the mouse brain stem (93, 94). After eye inoculation in mice, HSV DNA is frequently found in brain-stem tissue; apparently postsynaptic transmission of virus to the brain stem from the trigeminal ganglion is possible as well as ascension to the brain stem by way of the facial nerve (1, 93, 94). Fraser and coworkers have reported that this DNA is nonlinear, and that it is probably either circular or concatomeric; they could not detect the ends of the viral genome by blot hybridizations, and upon scanning the autoradiograms they found that the junction fragments were present in molar excess compared to internal, nonjunction fragments (93, 94). They also investigated the form of the viral genome in the murine trigeminal ganglia harboring latent virus. Again, the HSV-1 DNA was found to be in a nonlinear, endless form, i.e. either integrated or circular. Densitometry scanning was not performed on these samples, so the relative molarity of the junction fragments is not known.

Density gradient centrifugation of DNA samples from latently infected brain-stem tissue identified one fraction of viral DNA at the density of free viral DNA and another at the density of host chromosomal DNA (31). Unfortunately, these results are difficult to interpret because of possible entrapment of viral DNA in the chromosomal band, and neither episomal nor integrated forms could be ruled out.

The striking result of analyses of the brain stem and trigeminal ganglia is the amount of viral DNA recovered. If 1% of neurons harbored virus in a latent state and if only one copy of viral DNA were present in each cell, the expected ratio of viral to host genomes would be of the order of 0.002. However, the actual copy number, apparent from data from several laboratories (12, 87, 93, 94; A. E. Sears & B. Roizman, work in progress), ranges from 0.1 to 1.0, i.e. considerably more than expected. Even if the fraction of cells harboring virus were underestimated by a factor of 10, it would still follow that each neuron harboring virus contains multiple copies of the viral genome; the true number may be as high as several hundred copies per latently infected neuron.

Interpretation of these studies is not straightforward for the following reasons.

1. The possible forms of DNA are (*a*) free linear forms, (*b*) circular episomal forms with one or multiple copies of the viral genome, (*c*) integrated linear forms, and (*d*) integrated concatemeric forms. The studies on the brain stem clearly exclude *a* and *c*, but do not differentiate between *b* and *d*. The problem stems from the fact that in integrated concatemeric forms the ratio of head-to-tail junction fragments to molar nonjunction fragments will approach 2.0 as the size of the concatemer increases. If the integration is random, the junctions between HSV and host DNAs will vary in size and will not be detectable by hybridization.

2. Interpretations of studies done largely if not exclusively on the brain stem are worrisome because activation of latent virus from the brain stem, as compared to that of latent virus from the trigeminal ganglion, is poor (31). Puga and coworkers reported that cloned HSV-1 terminal fragments from DNA extracted from trigeminal ganglia harboring latent virus were for the most part rearranged; the nature of the rearrangements was not determined (86). Results of marker rescue studies in which human ganglia were infected with mutants carrying specific markers also suggest that defective viral genomes may reside in ganglionic cells (11). Given the ease of activating latent virus from trigeminal ganglia, what is the significance of the noninducible virus in the brain stem?

3. The single observation that remains at least for the moment unmolested is that the HSV DNA sequence complexity in ganglionic cells is higher than one genome equivalent per neuronal cell harboring virus in the latent state. Do neurons become infected at a relatively high ratio of virions per cell? Is this DNA the remnant of the DNA replicated early in infection, during the phase of viral replication in the ganglion immediately after infection? Is this DNA replicated after latency is established?

In the case of other DNA viruses that are retained in cells in a latent state, the form of the viral DNA is a major determinant in DNA integrity and gene expression (130). For example, papilloma virus and Epstein-Barr virus (EBV) genomes are maintained as circular episomes in latently infected cells at a copy number in the range of one to a few hundred copies per cell (10, 61). At least in the case of some papilloma viruses, integration of the viral DNA into the host genome is concomitant with cell transformation and usually with loss or rearrangement of the viral sequences (10). The results obtained to date do not refute the hypothesis that HSV-1 DNA is maintained in episomal form in numerous copies per neuron, but as indicated above the conclusions are not compelling. For heuristic reasons, it is useful to consider two hypotheses for the maintenance of multiple copies of the papilloma virus and EBV genomes in an episomal form in cells harboring theses viruses in a latent state. First, the cells harboring these viruses divide or have a limited lifespan. Multiple copies insure that the viruses are transmitted to progeny cells; in addition, each of these viruses is capable of inducing the cells harboring them to multiply, at least in cell culture (47). Second, single copies of integrated viral genomes would encounter difficulties in regenerating termini without specifying as yet unidentified enzymes capable of precisely and specifically replicating or excising the integrated DNA. Neither of these reasons applies to HSV. First, the virus is maintained in a latent form in neurons, which do not divide. Second, HSV genomes differ from those of other DNA viruses capable of remaining in a latent form in that the internal inverted repeats of the terminal sequences could serve as templates for regeneration of the termini (97). The theoretical objections to an integrated state for latent HSV genomes are thus not tenable.

The value of model systems is that they provide a skeleton on which the experimental data can be tried for a fit. The paucity of data renders this fitting premature. But if in spite of the caveats listed above, human papillomaviruses and, more importantly, EBV were to serve as valid models, it would follow that HSV-1 and HSV-2 express functions that cause the copy number of latent viral DNA to be amplified to a discrete level higher than one per cell genome and functions that maintain the latent genomes in an episomal state.

The Search for HSV Functions Related to Latency

METHODS AND RATIONALE The basis for the search for latency-specific functions is conceptually simple. The entry of encapsidated or naked wild-type viral DNA into cells in culture results in viral gene expression. Since viral DNA is transcribed by host enzymes (98), the hypothesis that neurons harboring latent virus lack host factors necessary for the transcription of any viral genes seems farfetched, but it cannot yet be dismissed. Heuristically more attractive is the alternate hypothesis that the expression of a specific viral gene in the environment of the sensory neuron arrests the viral gene cascade and leads to the establishment of latency. The product of this gene, and possibly those of other viral genes, may be required for the maintenance of the latent state. The search for these putative viral genes, referred to collectively as the latency genes, has taken two directions. The first involves analyses of mutants for their ability to establish latency. The second involves analyses of ganglia harboring latent virus for the presence of viral gene products.

STUDIES ON HSV TEMPERATURE-SENSITIVE MUTANTS Fortuitously, the nonpermissive temperature for a large group of temperature-sensitive mutants produced in Glasgow (66) was 38.5°C, the internal temperature of mice. After intracerebral inoculation into mice, several of these mutants failed to establish latent infections (62, 121). In one instance, the wild-type (Ts^+) revertants established latency; others either failed or were not tested (121).

The studies on temperature-sensitive mutants present several important problems. (*a*) The inherent assumption underlying these studies is that the viral genes required for the establishment and maintenance of the latent state also specify a function required for virus replication. (*b*) HSV-1 clinical isolates appear to be temperature sensitive. In fact, to obtain a library of temperature-sensitive viruses, temperature-insensitive mutants capable of growing at the nonpermissive temperature had to be selected by passage of virus stocks at that temperature. An additional problem stems from the fact that mutagenesis of the entire virus genome for selection of temperature-sensitivity mutations also introduces silent mutations in genes not essential for growth in animals (21). Selection of temperature-insensitive revertants does not alleviate the concern that the latency-negative phenotype of the mutants actually reflected mutations in genes not essential for growth. (*c*) A key problem, noted earlier, is that virus multiplication at the site of inoculation and in the ganglion seems to precede establishment of the latent state (50–54, 69, 70, 84, 85, 116–118, 127). Loss of ability to multiply in animal tissues, either because of mutagenesis of the essential gene or because of mutations that are silent in cell culture, may preclude the establishment of latency. The requirement for virus multiplication can be overcome by inoculation of high doses of virus, but data are then ambiguous because wild-type revertants in large inocula could help colonize the ganglion.

In summary, these studies are inconclusive. Mutants derived by site-specific mutagenesis would be more appropriate, but the nonpermissive temperature of most temperature-sensitive mutants produced by site-specific mutagenesis is usually higher than the normal temperature of the mouse.

STUDIES ON THYMIDINE KINASE MUTANTS The gene encoding thymidine kinase (TK) was the first viral gene to be identified that appeared dispensible for growth in cell culture (48). The ease with which TK^- viruses can be isolated and propagated in culture has rendered them useful model viruses for the study of virus pathogenesis and latency. Indeed, early studies have shown that TK^- viruses are less virulent than wild-type viruses (29). TK^- viruses appear to multiply at the site of inoculation, but could not be recovered by explantation of the ganglia (112, 113). For obvious reasons, TK became the bête noire of the latency enterprise. Murine ganglia harboring latent virus were reported to express the TK gene for long intervals after inoculation (128); this led Tenser & Dunstan (112) to hypothesize that TK expression was important for the establishment of latency in sensory neurons. This theory appeared to be supported by a correlation between the extent of the TK activity induced by HSV mutants and their ability to replicate in trigeminal ganglia after infection (114).

The viral thymidine kinase is a promiscuous nucleotide kinase with a reported preference for deoxypyrimidines. In growing cells in culture, normal host enzymes furnish levels of deoxynucleotide monophosphates adequate for viral DNA replication. This apparently is not the case for tissue cells that do not divide. The failure to recover latent virus from animals infected with TK^- viruses may be due to (*a*) a level of replication inadequate for accessing the neurons, (*b*) failure of the latent virus to replicate during reactivation in vitro because of inadequate pools of deoxynucleotide triphosphates, or (*c*) secondary mutations in other genes introduced during mutagenesis procedures that preclude establishment of latency but are not related to the TK gene.

The argument is in any event immaterial. Recent studies were done on a mutant virus that has had 500 bp deleted from the TK gene and that multiplies in TK^- cells in the presence of drugs such as BUdR or Ara T, which depend on TK for their antiviral effect. This virus was able to establish latency in rabbits but not in mice after inoculation by the eye route (B. Meignier, A. E. Sears & B. Roizman, unpublished data); as phosphorylation of thymidine is not likely to be a species-specific event, it appears that at least in the case of this mutant, explanation *c* above is correct.

STUDIES ON MUTANTS WITH DELETIONS IN GENES THAT ARE DISPENSIBLE FOR GROWTH IN CELL CULTURE The discovery that mutants incapable of specifying a major viral glycoprotein (glycoprotein C) are viable (38) has spurred interest in a systematic identification of dispensible viral genes. Unlike the TK gene, which extends the range of HSV to host cells that do not express enzymes with similar functions, the glycoprotein C gene specifies functions that do not appear to be complemented by the cell or to affect the viral host range. Dispensible genes that do not extend the host range may conceivably function only in latency-competent cells. However, a systematic study could not be mounted until techniques for genetic engineering of large genomes became available (82). As discussed earlier, approximately 20% of the genome is now known to be dispensible, including the $\alpha22$ and $\alpha47$ genes. The third body alluded to in the introduction, the influence of the physiologic environment, is most evident in the case of the $\alpha22$ mutant, which failed to establish latency when inoculated into the mouse ear but readily established latency in the mouse trigeminal ganglion following inoculation by the eye route (106).

All the mutants that have been tested (e.g. all except those with deletions in genes currently designated as US2, US3, glycoprotein G, and US5) are capable of establishing latency in the mouse model system. This does not mean that the deleted genes are truly nonessential; rather, their functions in in vivo infection are probably related to replication in peripheral tissues or to reactivation.

The flaw in the virus mutant studies is not in the assumption that HSV encodes a gene dedicated to latency that has no role in lytic infection, but rather in the assumption that the technology being used for its identification will be successful. Systematic deletion studies are not likely to enable us to detect dispensible genes embedded within domains of genes essential for viral replication unless knowledge of these genes comes from another source. Therein rests the value of studies on viral gene expression in neuronal cells harboring latent virus.

VIRAL GENE EXPRESSION IN NEURONS HARBORING LATENT VIRUS Reports on the presence of viral gene products in neurons are sparse but significant, not so much for the findings per se as for the framework in which the analyses were carried out.

In early studies RNA sequences homologous to specific domains of the viral genome were detected in human sensory ganglia harboring latent virus. These domains, however, encode β and γ genes; RNA homologous to α genes was not reported (32, 33). A subsequent report described the detection of the $\alpha 4$ protein but not of late proteins in rabbit ganglia harboring latent virus (37). This list of earlier studies is by no means complete. We expect that as probes become more sensitive other viral transcripts and proteins are likely to be detected. The idea that this type of study might lead to the identification of the genes responsible for at least the maintenance of the latent state is certainly valid; in fact, a recent extensive study of viral transcripts present in latently infected murine ganglia has identified a previously undetected RNA transcribed from the region of the $\alpha 0$ gene, but from the opposite strand (110a).

While these studies are useful for defining the transcriptionally active genes in ganglia during latency, it is more difficult to be certain that these genes are actively involved in the maintenance of the latent state. Early studies on human B lymphocytes carrying latent EBV focused on cells expressing the so-called early antigens (EAs) and viral capsid antigens (VCAs). These studies were important, for they delineated the human immune response to EBV infection. It was not until the discovery of the EBV nuclear antigens (EBNAs) that it became apparent that the hallmark of EBV latency was EBNA and not EA and VCA (90). The significance of EBNA, at a glance, was its presence in all immortalized B lymphocytes, whereas EA and VCA reflected the small fraction of cells committed to viral replication and termination of the latent state. Current studies on HSV gene expression in latently infected ganglia fail to discriminate between neurons harboring the virus in the latent state and either those that have become committed to viral replication or those that harbor defective viral genomes expressing a random fraction of genes that are not lethal to the neuron harboring them.

For a viral gene product to be significantly associated with the maintenance of the latent state, it must be expressed uniformly in all neurons harboring latent virus.

Latency in Cell Culture: A Misnomer?

The operational definition of latency rests on studies of the pathogenesis of human disease and concordant findings in animal studies. The animal models have several major experimental flaws. Foremost, it is difficult to visualize the events taking place in isolated cells from examination of intact animals. Equally importantly, the interposition of the physiologic environment between the inoculated virus and the establishment of latency makes studies on even slightly debilitated virus mutants difficult, costly, and in the case of mutants that fail to establish latency, meaningless. It would be much easier if latency could be established in cells in culture.

Rapp and coworkers pioneered in the establishment of cell culture systems that retain the viral genome in a mode that does not permit the expression of viral lytic functions (14, 15, 100, 122–126). In a typical system, the cells are infected at a supraoptimal temperature (42°C) in the presence of antiviral drugs or inhibitory concentrations of interferon. Once the cultures are established, the cells can be maintained at the elevated temperature indefinitely without the drugs. Lowering the temperature results in virus multiplication. A significant observation is the induction of virus multiplication at the supraoptimal temperature by other viruses such as human cytomegalovirus, varicella zoster, and EBV. This suggests that all of these viruses carry a nonspecific inducer of viral genes distinct from α-TIF. The role of general inducer has been suggested for the product of the $\alpha 0$ gene in HSV-1. The extent of gene expression in these cells has not been reported. Moreover, in contrast to the nonlinear genome seen during in vivo latency, the viral genome appears to persist in a linear conformation (125).

The key question is whether the mechanism by which the viral genome is retained in cells maintained at the supraoptimal temperature is related more than just phenomenologically to the mechanism of latency in neurons. Establishment of viral latency in humans and experimental animals does not depend on supraoptimal temperatures or known antiviral substances. Moreover, in vivo latency is cell specific. If the cell specificity reflects a basic requirement for host factors available only in differentiated latency-competent cells, any cell maintained at supraoptimal temperatures should be capable of expressing these factors.

While latency-competent cells are by definition conditionally restrictive, the description is equally applicable to any temporary antiviral cell state. The mimicry is likely to be insignificant with respect to the mechanism of latency but significant with respect to our understanding of gene expression in nonpermissive cells.

HSV Latency: A Model

In the preceeding sections we examined the available information regarding the state of the viral genome and its expression during the latent state in experimental animal systems. We now present a model of the latent state to define in operational terms the key questions that must be resolved in order to understand this state.

ACCESSION TO THE SENSORY NEURONS It is hypothesized that upon entry of the virus into neuronal cells, the capsid that contains viral DNA is transported to the neuronal nucleus. The key questions are: (*a*) What components of the virion are transported and what are their functions? (*b*) What is the mechanism by which the capsid and associated structural proteins are transported to the nucleus? The answers to these questions are not known even for productively infected cells. Kristensson et al (55) have suggested that the cytoskeleton serves as a conduit; indeed, drugs that affect the integrity of the cytoskeleton and those that affect retrograde flow in axons arrest the transport of virus to the neuronal nucleus. For both productively infected nonneuronal cells and latently infected neuronal cells there must be host proteins that translocate the capsid and viral proteins that specifically interact with the host factors. Neither has been identified in either type of cell, nor is it known whether the same set of proteins serves the same function in both types of cells.

This question is important, for the virion contains *trans*-acting proteins (e.g. α-TIF) that induce α gene expression. These proteins are contained in the tegument, i.e. in the structural component of the virion located between the capsid and the envelope. Failure of the α-TIF and other tegument proteins to be transported to the neuronal nucleus may be the cause of restricted viral gene expression.

ESTABLISHMENT AND MAINTENANCE OF THE LATENT STATE The single key observation that serves as the focal point of our model of latency is that the HSV genome may be present in multiple copies in each neuron that harbors latent virus (12, 87, 93, 94; A. E. Sears & B. Roizman, work in progress). If this observation is correct, several conclusions may follow.

1. By analogy with other models of latent DNA viruses, it is likely that specific viral gene functions maintain both the state of the genome and its copy number, although it is not likely that the genome is amplified by the viral DNA polymerase. We hypothesize that at least one of these two functions is expressed by one or more α proteins, largely because the expression of other viral proteins is dependent on α gene expression and because the α promoters seem to function in a wide variety of cells (56). Of the α proteins, we can exclude α47 and α22 because deletion mutants are latency competent (106;

A. E. Sears & B. Roizman, unpublished data). In addition, $\alpha 27$ appears to regulate late gene expression positively rather than negatively (103).

The two proteins we cannot exclude are $\alpha 4$ and $\alpha 0$. The qualifications of $\alpha 0$ are that it is the earliest α protein expressed and seems to be the earliest turned off (28). In the in vitro transient expression systems, it appears to be a nonspecific *trans*-activating factor (74, 75), but its function in the neuronal cell might be more specific. The qualification of the $\alpha 4$ protein is that it regulates gene expression both positively and negatively. As noted in the section on lytic infections, $\alpha 4$ protein is processed into distinct forms, which bind to specific sequences in the promoter-regulatory domains of all viral genes including α genes (58, 59). Current studies suggest that the different forms of $\alpha 4$ are processed by the cell and that each form binds to specific sequences (N. Michael, D. Spector, T. M. Kristie & B. Roizman, work in progress). It is conceivable, therefore, that the processing of $\alpha 4$ proteins to a unique form precludes viral gene expression of most genes but not of the genes involved in the maintenance of the viral genome.

2. Maintenance of a high copy number implies some mechanism for controlled amplification of the viral genome. We predict that this function is fulfilled by a host polymerase and involves a specific origin of DNA synthesis. The rationale for predicting amplification by a host polymerase stems from two considerations. Firstly, it is not likely that numerous viral genomes enter the neuron and establish latency simultaneously. Secondly, drugs that block the activity of the viral DNA polymerase do not seem to affect the latent state or reduce the frequency of reactivation once the drug is withdrawn (127). The most persuasive argument is that offered by the EBV model of latency. In latently infected B lymphocytes a specific origin of viral DNA synthesis and a set of viral proteins (EBNAs) without known DNA polymerase activity appear to be responsible for the limited amplification and maintenance of the EBV genome in an episomal state (65, 88, 91, 111, 131). The key questions are: When does the HSV DNA amplification occur? And what bearing does it have on the activation of virus multiplication from the latent state?

ACTIVATION OF HSV MULTIPLICATION FROM THE LATENT STATE A model of HSV latency must account for several well-authenticated, key observations (reviewed in 72): (*a*) Not all individuals that carry latent virus exhibit recrudescences. (*b*) The frequency of recurrent lesions is related to the severity and extent of virus multiplication and spread during primary infection. (*c*) A variety of seemingly unrelated stimuli appear to be able to cause recurrence of the lesions. (*d*) Prior colonization of the ganglion with HSV appears to preclude its superinfection by another HSV strain (71, 115). (*e*) Recurrences eventually decrease in frequency and seem to stop altogether.

In the simplest model we wish to examine, the balance between the latent state and viral gene expression, leading to lytic replication and recurrent lesions, rests on the copy number of viral DNA genomes per cell. We hypothesize that each stimulus causes a neuronal perturbation that results in incremental amplification of viral genomes. When in individual cells the increment exceeds the tolerated genome copy number, viral gene expression ensues, and finally the viral progeny are transported along the axon to the periphery. In the context of this model, several observations regarding the activation of virus multiplication can be rationalized as follows.

1. The number of neurons that become infected and harbor virus is proportional to the viral dose to which the nerve endings are exposed and the extent of dissemination of infection. If the extent of amplification of the viral genome is random, the frequency of recurrence will depend on the number of neurons initially infected. Further, each neuron harboring the latent genome will attain the critical copy number at a different time; this would account for the observation that not all latent virus is activated simultaneously.

2. Klein (49) proposed a "round-trip" hypothesis to account for multiple recurrences. According to this hypothesis, the activated virus infects and populates new neurons after each recurrence. The hypothesis is inconsistent with the observation that a second infecting virus does not appear to be able to colonize ganglia that are already latently infected (71). The resistance is probably partly due to host immunity. In addition, the viral genome entering a neuron harboring HSV would become diluted owing to the higher copy number of the preexisting genomes.

3. The random, asynchronous feature of genome amplification during the latent state explains why some individuals suffer no recurrences; the genome copy number does not reach a critical level. The seeming cessation of recrudescences could likewise be due to the exhaustion of neurons with viral genomes that are amplified to a high copy number.

4. The model does not predict the common target of the diverse stimuli that trigger recrudescences, the specific events within the neuron that enable the genome amplification to take place, the role of immune suppression in recrudescences, or the recrudescences seen at very specific intervals after bone-marrow transplants which cannot be explained entirely by the immune suppression that precedes the transplant (25). Sensory neurons and specific lymphocytes have been shown to share receptors for several neuropeptides (6, 7). These neuropeptides stimulate lymphocyte growth in some cases and retard it in others. It is possible, then, that both the local and systemic stimuli that trigger recrudescences are peptides that stimulate the expression of cellular genes involved in DNA synthesis in both lymphocytes and sensory neurons.

We are the products of editing, rather than of authorship.

George Wald, The origin of optical activity, *Ann. NY Acad. Sci.*, Vol. 69, 1957

CONCLUSIONS

The title of this review is identical in part to one published 20 years ago (96); it is being repeated to emphasize the scarcity of explanations of the oldest phenomenon known to the student of human virology. Two points are relevant here.

It is axiomatic that to perpetuate itself, a virus must disseminate in the host population. The role of the latent state in the epidemiology of the virus is important. The latent virus forms a reservoir in the immune host that becomes available for infection of the susceptible population during recrudescences of the herpetic lesions. The localization of the recurrent lesions near or in the mouth and genitals is suitable for efficient transmission. HSV is highly immunogenic and is readily overcome by host defenses; conceivably if HSV were deprived of the capacity to remain latent it could not perpetuate itself in the human population.

To the victim of infection HSV is anathema, an intractable disease that seems to become exacerbated in direct relation to the anxiety it causes. To the student of its structure and manifestations, the virus is an exquisite model of selective adaptation to the exigencies of survival. Above all it is a model of itself; it is also a model of gene expression, regulation and of macromolecular interactions. Models have virtue only if challenged and we hope that ours will be. But even should it not withstand the test of time, much will be learned in the process of knocking it down.

ACKNOWLEDGMENTS

The authors would like to thank Ms. Susanna Rudofsky for translations of papers published in German and quoted in the text. The authors' studies cited in the text were aided by grants from the National Cancer Institute (CA08494 and CA19264) and the American Cancer Society (MV2T).

Literature Cited

1. Anderson, J. R., Field, H. J. 1983. The distribution of herpes simplex type 1 antigen in mouse central nervous system after different routes of inoculation. *J. Neurol. Sci.* 60:181–95
2. Anderson, S. G., Hamilton, J. 1949. The epidemiology of primary herpes simplex infection. *Med. J. Aust.* 36: 308–11
3. Arsenakis, M., Hubenthal-Voss, J., Campadelli-Fiume, G., Pereira, L., Roizman, B. 1986. Construction and properties of a cell line constitutively expressing the herpes simplex virus glycoprotein B dependent on functional $\alpha 4$ synthesis. *J. Virol.* 60:674–82
4. Baringer, J. R., Swoveland, P. 1973. Recovery of herpes simplex virus from human trigeminal ganglia. *N. Engl. J. Med.* 288:648–50
5. Bastian, R. O., Rabson, A. S., Yee, C. L., Tralka, T. S. 1972. Herpes virus hominis: isolation from human trigeminal ganglia. *Science* 178:306–7

6. Blalock, J. E. 1984. Relationships between the neuroendocrine hormones and lymphokines. *Lymphokines* 9:1–13
7. Blalock, J. E. 1984. The immune system as a sensory organ. *J. Immunol.* 132:1067–70
8. Blue, W. T., Winland, R. D., Stobbs, D. G., Kirksey, D. F., Savage, R. E. 1981. Effects of adenosine monophosphate on the reactivation of latent herpes simplex virus type 1 infections in mice. *Antimicrob. Agents Chemother.* 20:547–48
9. Blyth, W. A., Hill, T. J., Field, H. J., Harbour, D. A. 1976. Reactivation of herpes simplex virus infection by ultraviolet light and possible involvement of prostaglandins. *J. Gen. Virol.* 33: 547–50
10. Broker, T. R., Botchan, M. 1986. Papillomaviruses: retrospectives and prospectives. In *Cancer Cells 4/DNA Tumor Viruses,* ed. M. Botchan, T. Grodzicker, P. A. Sharp, pp. 17–36. Cold Spring Harbor, NY: Cold Spring Harbor Lab.
11. Brown, S. M., Subak-Sharpe, J. H., Warren, K. G., Wrobleska, Z., Koprowski, H. 1979. Detection by complementation of defective or uninducible (herpes simplex type 1) virus genomes latent in human ganglia. *Proc. Natl. Acad. Sci. USA* 76:2364–68
12. Cabrera, C. V., Wohlenberg, C., Openshaw, H., Rey-Mendez, M., Puga, A., Notkins, A. L. 1980. Herpes simplex virus DNA sequences in the CNS of latently infected mice. *Nature* 288:288–90
13. Carton, C. A., Kilbourne, E. D. 1952. Activation of latent herpes simplex by trigeminal sensory-root section. *N. Engl. J. Med.* 246:172–76
14. Colberg-Poley, A. M., Isom, H. C., Rapp, F. 1979. Reactivation of herpes simplex virus type 2 from a quiescent state by human cytomegalovirus. *Proc. Natl. Acad. Sci. USA* 76:5948–51
15. Colberg-Poley, A. M., Isom, H. C., Rapp, F. 1981. Involvement of an early cytomegalovirus function in reactivation of quiescent herpes simplex virus type 2. *J. Virol.* 37:1051–59
16. Cook, M. L., Bastone, V. B., Stevens, J. G. 1974. Evidence that neurons harbor latent herpes simplex virus. *Infect. Immun.* 9:946–51
17. Cook, M. L., Stevens, J. G. 1973. Pathogenesis of herpetic neuritis and ganglionitis in mice: Evidence of intra-axonal transport of infection. *Infect. Immun.* 7:272–88
18. Cook, M. L., Stevens, J. G. 1976. Latent herpetic infections following experimental viraemia. *J. Gen. Virol.* 31:75–80
19. Deleted in proof
20. Cushing, H. 1905. Surgical aspects of major neuralgia of trigeminal nerve: Report of 20 cases of operation upon the Gasserian ganglion with anatomic and physiologic notes on the consequence of its removal. *J. Am. Med. Assoc.* 44:1002–8
21. Dargan, D. J., Subak-Sharpe, J. H. 1984. Isolation and characterization of revertants from fourteen herpes simplex virus type 1 (strain 17) temperature sensitive mutants. *J. Gen. Virol.* 65: 477–91
22. Darville, J. M., Blyth, W. A. 1982. Neutralizing antibody in mice with primary and recurrent herpes simplex virus infection. *Arch. Virol.* 71:303–10
23. Doerr, R. 1938. Herpes febrilis. In *Handbuch der Virusforschung,* Vol. 1, ed. R. Doerr, C. Hallaver, pp. 41–45. Vienna: Springer-Verlag
24. Douglas, R. G., Couch, R. B. 1970. A prospective study of chronic herpes virus infection and recurrent herpes labialis in humans. *J. Immunol.* 104:289–95
25. Elfenbein, G. J., Saral, R. 1981. Infectious disease during immune recovery after bone marrow transplantation. In *Infection and the Immune Compromised Host,* ed. J. C. Allen, pp. 157–96. Baltimore: Williams & Wilkins
26. Elkareh, A. A., Murphy, A. J. M., Fichter, T., Efstradiatis, A., Silverstein, S. 1985. "Transactivation" control signals in the promoter of the herpesvirus thymidine kinase gene. *Proc. Natl. Acad. Sci. USA* 82:1002–6
27. Ellison, S. A., Carton, C. A., Rose, H. M. 1959. Studies of recurrent herpes simplex infections following section of the trigeminal nerve. *J. Infect. Dis.* 105:161–67
28. Fenwick, M. L., Clark, J. 1982. Expression of early viral genes: a possible pre-α protein in cells infected with herpes simplex virus. *Biochem. Biophys. Res. Commun.* 108:1454–59
29. Field, H. J., Wildy, P. 1978. The pathogenicity of thymidine kinase–deficient mutants of herpes simplex virus in mice. *J. Hyg.* 81:267–77
30. Findlay, G. M., MacCallum, F. O. 1940. Recurrent traumatic herpes. *Lancet* 1:259–61
31. Fraser, N. W., Muggeridge, M. I., Mellerick, D. M., Rock, D. L. 1984. Molecular biology of HSV-1 latency in a mouse model system. In *Herpesvirus,*

ed. F. Rapp, pp. 159–73. New York: Liss

32. Galloway, D. A., Fenoglio, C. M., McDougall, J. M. 1982. Limited transcription of the herpes simplex virus genome when latent in human sensory ganglia. *J. Virol.* 41:686–91
33. Galloway, D. A., Fenoglio, C., Shevchuk, M., McDougall, J. M. 1979. Detection of herpes simplex RNA in human sensory ganglia. *Virology* 95:265–68
34. Godowski, P. J., Knipe, D. M. 1983. Mutations in the major DNA-binding protein gene of herpes simplex virus type 1 result in increased levels of viral gene expression. *J. Virol.* 47:478–86
35. Godowski, P. J., Knipe, D. M. 1985. Identification of a herpes simplex virus function that represses late gene expression from parental viral genomes. *J. Virol.* 55:357–65
36. Goodpasture, E. W. 1929. Herpetic infections with special reference to involvement of the nervous system. *Medicine* 8:223–43
37. Green, M. T., Courtney, R. J., Dunkel, E. C. 1981. Detection of an immediate early herpes simplex virus type 1 polypeptide in trigeminal ganglia from latently infected animals. *Infect. Immun.* 34:987–92
38. Heine, J. W., Honess, R. W., Cassai, E., Roizman, B. 1974. Proteins specified by herpes simplex virus. XII. The virion polypeptides of type I strains. *J. Virol.* 14:640–51
39. Hill, T. J., Blyth, W. A., Harbour, D. A. 1978. Trauma to the skin causes recurrence of herpes simplex in the mouse. *J. Gen. Virol.* 39:21–28
40. Hill, T. J., Blyth, W. A., Harbour, D. A. 1982. Recurrent herpes simplex in mice: topical treatment with Acyclovir cream. *Antiviral Res.* 2:135–46
41. Hill, T. J., Blyth, W. A., Harbor, D. A. 1983. Recurrence of herpes simplex in the mouse requires an intact nerve supply to the skin. *J. Gen. Virol.* 64:2763–65
42. Iwasaki, Y., Yamamoto, T., Konno, H., Iizuka, H., Kudo, H. 1986. Eradication of herpes simplex virus persistence in rat trigeminal ganglia by retrograde axoplasmic transport. *J. Virol.* 59:242–48
43. Jenkins, F. J., Roizman, B. 1986. Herpes simplex virus recombinants with noninverting genomes frozen in different isomeric arrangements are capable of independent replication. *J. Virol.* 59:494–99
44. Johnson, D. C., Spear, P. G. 1984. Evidence for translational regulation of herpes simplex virus type 1 gD expression. *J. Virol.* 51:389–94
45. Keddie, F. M., Rees, R. B., Epstein, N. N. 1941. Herpes simplex following artificial fever therapy. *J. Am. Med. Assoc.* 117:1327–30
46. Kennedy, P. G. E., Al-Saadi, S. A., Clements, G. B. 1983. Reactivation of latent herpes simplex virus from dissociated identified dorsal root ganglion cells in culture. *J. Gen. Virol.* 64:1629–35
47. Kieff, E., Dambaugh, T., Hummel, M., Heller, M. 1983. Epstein-Barr virus transformation and replication. In *Advances in Viral Oncology,* Vol. 3, ed. G. Klein, pp. 133–82. New York: Raven
48. Kit, S., Dubbs, D. R. 1963. Aquisition of thymidine kinase activity by herpes simplex infected mouse fibroblast cells. *Biochem. Biophys. Res. Commun.* 11: 55–59
49. Klein, R. J. 1976. Pathogenetic mechanisms of recurrent herpes simplex viral infections. *Arch. Virol.* 51:1–13
50. Klein, R. J. 1980. Effect of immune serum on the establishment of herpes simplex virus infection in trigeminal ganglia of hairless mice. *J. Gen. Virol.* 49:401–5
51. Klein, R. J., Friedman-Kien, A. E., Brady, E. 1978. Latent herpes simplex virus infection in ganglia of mice after primary infection and re-inoculation at a distant site. *Arch. Virol.* 57:161–66
52. Klein, R. J., Friedman-Kien, A. E., DeStefano, E. 1979. Latent herpes simplex virus infections in sensory ganglia of hairless mice prevented by acycloguanosine. *Antimicrob. Agents Chemother.* 15:723–29
53. Klein, R. J., Friedman-Kien, A. E., Fondak, A. A., Buimovici-Klein, E. 1977. Immune response and latent infection after topical treatment of herpes simplex virus infection in hairless mice. *Infect. Immun.* 16:842–48
54. Klein, R. J., Friedman-Kien, A. E., Yellin, P. B. 1978. Orofacial herpes simplex virus infection in hairless mice: Latent virus in trigeminal ganglia after topical antiviral treatment. *Infect. Immun.* 20:130–35
55. Kristensson, K., Lycke, E., Roytta, M., Svennerholm, B., Vahlne, A. 1986. Neuritic transport of herpes simplex in rat sensory neurons in vitro. Effects of substances interacting with microtubular function and axonal flow [nocodazole, taxol, and erythro-9-3-(2-hydroxynonyl)-adenine]. *J. Gen. Virol.* 67:2023–28
56. Kristie, T., Batterson, W., Mackem, S.,

Roizman, B. 1983. The regulatory elements in the domains of α genes of herpes simplex virus 1 (HSV-1). In *Proc. Cold Spring Harbor Conf. Enhancers Eukaryotic Gene Expression, Cold Spring Harbor, NY, 1982*, pp. 141–51. Cold Spring Harbor, NY: Cold Spring Harbor Lab.

57. Kristie, T. M., Roizman, B. 1984. Separation of sequences defining basal expression from those conferring α gene regulation within regulatory domains of herpes simplex 1 α genes. *Proc. Natl. Acad. Sci. USA* 81:4065–69
58. Kristie, T. M., Roizman, B. 1986. α4, the major regulatory protein of herpes simplex virus type 1, is stably and specifically associated with promoter-regulatory domains of α genes and of selected other viral genes. *Proc. Natl. Acad. Sci. USA* 83:3218–22
59. Kristie, T. M., Roizman, B. 1986. The binding site of the major regulatory protein α4 specifically associated with the promoter-regulatory domains of α genes of herpes simplex virus type 1. *Proc. Natl. Acad. Sci. USA* 83:4700–4
60. Kristie, T. M., Roizman, B. 1987. Host cell proteins bind to the *cis*-acting site required for virion-mediated induction of herpes simplex virus 1 α genes. *Proc. Natl. Acad. Sci. USA* 84:71–75
61. Lindahl, T., Adams, A., Bjursell, G., Bornkamm, G. W., Kaschaka-Dierich, C., Jehn, U. 1976. Covalently closed circular duplex DNA of Epstein-Barr virus in a human lymphoid cell line. *J. Mol. Biol.* 102:511–30
62. Lofgren, K. W., Stevens, J. G., Marsden, H. S., Subak-Sharpe, J. H. 1977. Temperature sensitive mutants of herpes simplex virus differ in the capacity to establish latent infections in mice. *Virology* 76:440–43
63. Longnecker, R., Roizman, B. 1986. Generation of an inverting herpes simplex virus 1 mutant lacking the L-S junction *a* sequences, an origin of DNA synthesis, and several genes including those specifying glycoprotein E and the α47 gene. *J. Virol.* 58:583–91

63a. Longnecker, R., Roizman, B. 1987. Clustering of genes dispensible for growth in culture in the small component of the HSV-1 genome. *Science* 236:573–79

64. Lonsdale, D. M., Brown, M. S., Subak-Sharpe, J. H., Warren, K. G., Koprowski, H. 1979. The polypeptide and DNA restriction enzyme profiles of spontaneous isolates of herpes simplex virus type 1 from explants of human trigeminal, superior cervical, and vagus ganglia. *J. Gen. Virol.* 43:151–71
65. Lupton, S., Levine, A. 1985. Mapping genetic elements of Epstein-Barr virus that facilitate extrachromosomal persistence of Epstein-Barr virus–derived plasmids in human cells. *Mol. Cell. Biol.* 5:2533–42
66. Marsden, H. S., Crombie, J. K., Subak-Sharpe, J. H. 1976. Control of protein synthesis in herpes virus–infected cells:analysis of the polypeptides induced by wild type and sixteen temperature sensitive mutants of HSV strain 17. *J. Gen. Virol.* 31:347–72
67. Martin, R. G., Dawson, C. R., Jones, P., Togni, B., Lyons, C., Oh, J. O. 1977. Herpes virus in sensory and autonomic ganglia after eye inoculation. *Arch. Ophthalmol.* 95:2053–56
68. Mavromara-Nazos, P., Roizman, B. 1986. Construction and properties of a viable herpes simplex virus 1 recombinant lacking the coding sequences of the α47 gene. *J. Virol.* 60:807–12
69. McKendall, R. R. 1977. Efficacy of herpes simplex virus type 1 immunization in protecting against acute and latent infections by herpes simplex type 2 in mice. *Infect. Immun.* 16:717–19
70. McKendall, R. R., Klassen, T., Baringer, J. R. 1979. Host defenses in herpes simplex infections of the nervous system: Effect of antibody on disease and viral spread. *Infect. Immun.* 23:305–11
71. Meignier, B., Norrild, B., Roizman, B. 1983. Colonization of murine ganglia by a superinfecting strain of herpes simplex virus. *Infect. Immun.* 41:702–8
72. Nahmias, A. J., Roizman, B. 1973. Infection with herpes simplex viruses 1 and 2. *N. Engl. J. Med.* 289:667–74, 719–25, 781–89
73. Nesburn, A. B., Elliott, J. M., Leibowitz, H. M. 1967. Spontaneous reactivation of experimental herpes simplex keratitis in rabbits. *Arch. Ophthalmol.* 78:523–29
74. O'Hare, P., Hayward, G. S. 1984. Expression of recombinant genes containing herpes simplex virus delayed early and immediate early regulatory regions and *trans* activation by herpesvirus infection. *J. Virol.* 52:522–31
75. O'Hare, P., Hayward, G. S. 1985. Evidence for a direct role for both the 175,000 and 110,000 molecular weight immediate-early proteins of herpes simplex virus in the transactivation of delayed-early promoters. *J. Virol.* 53:751–60
76. Pazin, G. J., Armstrong, J. A., Lam,

M. T., Tarr, G. C., Jannetta, P. J., Ho, M. 1979. Prevention of reactivation of herpes simplex virus infection by human leukocyte interferon after operation on the trigeminal root. *N. Engl. J. Med.* 301:225–30
77. Pazin, G. J., Ho, M., Jannetta, P. J. 1978. Herpes simplex virus reactivation after trigeminal nerve root decompression. *J. Infect. Dis.* 138:405–9
78. Pedrau, J. R. 1938. Persistence of the virus of herpes in rabbits immunised with living virus. *J. Pathol. Bacteriol.* 47:447–55
79. Plummer, G. 1973. Isolation of herpesviruses from the trigeminal ganglia of man, monkeys, and cats. *J. Infect. Dis.* 128:345–48
80. Poffenberger, K. L., Roizman, B. 1985. Studies on a non-inverting genome of a viable herpes simplex virus 1. Presence of head-to-tail linkage in packaged genomes and requirements for circularization after infection. *J. Virol.* 53:589–95
81. Poffenberger, K. L., Tabares, E., Roizman, B. 1983. Characterization of a viable, non-inverting herpes simplex virus 1 genome derived by insertion of sequences at the L-S junction. *Proc. Natl. Acad. Sci. USA* 80:2690–94
82. Post, L. E., Roizman, B. 1981. A generalized technique for deletion of specific genes in large genomes: α gene 22 of herpes simplex virus 1 is not essential for growth. *Cell* 25:227–32
83. Price, R. W., Katz, B. J., Notkins, A. L. 1975. Latent infection of the peripheral ANS with herpes simplex virus. *Nature* 257:686–88
84. Price, R. W., Schmitz, J. 1979. Route of infection, systemic host resistance, and integrity of ganglionic axons influence acute and latent herpes simplex virus infection of the superior cervical ganglion. *Infect. Immun.* 23:373–83
85. Price, R. W., Walz, M. A., Wohlenberg, C., Notkins, A. L. 1975. Latent infection of sensory ganglia with herpes simplex virus: Efficacy of immunization. *Science* 188:938–40
86. Puga, A., Cantin, E. M., Wohlenberg, C., Openshaw, H., Notkins, A. L. 1984. Different sizes of restriction endonuclease fragments from the terminal repetitions of the herpes simplex virus type 1 genome latent in trigeminal ganglia of mice. *J. Gen. Virol.* 65:437–44
87. Puga, A., Rosenthal, J. D., Openshaw, H., Notkins, A. L. 1978. Herpes simplex virus DNA and mRNA sequences in acutely and chronically infected trigeminal ganglia of infected mice. *Virology* 89:102–11
88. Rawlins, D., Milman, G., Hayward, S. D., Hayward, G. S. 1985. Sequence specific DNA binding of the Epstein-Barr virus nuclear antigen (EBNA) to clustered sites in the plasmid maintenance region. *Cell* 42:859–68
89. Read, G. S., Frenkel, N. 1983. Herpes simplex virus mutants defective in the virion associated shut-off of host polypeptide synthesis and exhibiting abnormal synthesis of α (immediate early) viral polypeptides. *J. Virol.* 46:498–512
90. Reedman, B. M., Klein, G. 1973. Cellular localization of an Epstein-Barr virus (EBV)–associated complement-fixing antigen in producer and nonproducer lymphoblastoid cell lines. *Int. J. Cancer* 11:499–520
91. Reisman, D., Yates, J., Sugden, B. 1985. A putative origin of replication of plasmids derived from Epstein-Barr virus is composed of two *cis*-acting components. 1985. *Mol. Cell. Biol.* 5:1822–32
92. Richter, R. B. 1944. Observations bearing on the presence of latent herpes simplex virus in the human Gasserian ganglion. *J. Nerv. Ment. Dis.* 99:356–58
93. Rock, D. L., Fraser, N. W. 1983. Detection of HSV-1 genome in central nervous system of latently infected mice. *Nature* 302:523–25
94. Rock, D. L., Fraser, N. W. 1985. Latent herpes simplex virus type 1 DNA contains two copies of the virion DNA joint region. *J. Virol.* 55:849–52
95. Rodda, S., Jack, I., White, D. O. 1973. Herpes simplex virus from the trigeminal ganglion. *Lancet* 1:1395–96
96. Roizman, B. 1966. An inquiry into the mechanisms of recurrent herpes infections of man. In *Perspectives in Virology,* Vol. 4, ed. M. Pollard, pp. 283–304. New York: Harper & Row
97. Roizman, B. 1979. The structure and isomerization of herpes simplex virus genomes. *Cell* 16:481–94
98. Roizman, B., Batterson, W. 1984. The replication of herpesviruses. In *General Virology,* ed. B. Fields, pp. 497–526. New York: Raven
99. Roizman, B., Tognon, M. 1983. Restriction endonuclease patterns of herpes simplex virus DNA: application to diagnosis and molecular epidemiology. *Curr. Top. Microbiol. Immunol.* 104:273–86
100. Russell, J., Preston, C. M. 1986. An *in vitro* latency system for herpes simplex virus type 2. *J. Gen. Virol.* 67:397–403

101. Rustigian, R., Smulou, J. B., Tye, M., Gibson, W. A., Shindell, E. 1966. Studies on latent infection of skin and oral mucosa in individuals with recurrent herpes simplex. *J. Invest. Dermatol.* 47:218–21

102. Sabin, A. B. 1976. Are the herpes simplex-genitalis viruses a cause of certain human cancers? In *Cancer Biology III. Herpes Virus Epidemiology, Molecular Events, Oncogenicity, and Therapy,* ed. C. Borek, D. W. King, pp. 136–50. New York: Stratton

103. Sacks, R., Greene, C. C., Aschman, D. P., Schaffer, P. A. 1985. Herpes simplex virus type 1 ICP 27 is an essential regulatory protein. *J. Virol.* 55:796–805

104. Scriba, M. 1976. Recurrent genital herpes simplex virus (HSV) in guinea pigs. *Med. Microbiol. Immunol.* 162: 201–8

105. Scriba, M. 1981. Persistence of herpes simplex virus (HSV) infection in ganglion and peripheral tissues of guinea pigs. *Med. Microbiol. Immunol.* 169:91–96

106. Sears, A. E., Halliburton, I. W., Meignier, B., Silver, S., Roizman, B. 1985. Herpes simplex virus mutant deleted in the $\alpha 22$ gene: growth and gene expression in permissive and restrictive cells, and establishment of latency in mice. *J. Virol.* 55:338–46

107. Smith, E. B., McLaren, L. C. 1977. Attempts to recover herpes simplex virus from skin sites of recurrent infection. *Int. J. Dermatol.* 16:748–51

108. Stanberry, L. R., Kern, E. R., Richards, J. T., Abott, T. M., Overall, J. C. 1982. Genital herpes in guinea pigs: pathogenesis of the primary infection and description of recurrent disease. *J. Infect. Dis.* 146:397–404

109. Stevens, J. G., Cook, M. L. 1971. Latent herpes simplex virus in spinal ganglia of mice. *Science* 173:843–45

110. Stevens, J. G., Nesburn, A. B., Cook, M. L. 1972. Latent herpes simplex virus from trigeminal ganglia of rabbits with recurrent eye infection. *Nature New Biol.* 235:216–17

110a. Stevens, J. G., Wagner, E. K., Devi-Rao, G. B., Cook, M. L., Feldman, L. T. 1987. RNA complementary to a herpesvirus αmRNA is prominent in latently infected neurons. *Science* 235:1056–59

111. Sugden, B., Marsh, K., Yates, J. 1985. A vector that replicates as a plasmid and can be efficiently selected in B-lymphoblasts transformed by Epstein-Barr virus. *Mol. Cell. Biol.* 5:410–13

112. Tenser, R. B., Dunstan, M. E. 1979. Herpes simplex virus thymidine kinase expression in infection of the trigeminal ganglion. *Virology* 99:417–22

113. Tenser, R. B., Miller, R. L., Rapp, F. 1979. Trigeminal ganglion infection by thymidine kinase negative mutants of herpes simplex virus. *Science* 205:915–17

114. Tenser, R. B., Ressell, S., Dunstan, M. E. 1981. Herpes simplex virus thymidine kinase expression in trigeminal ganglion infection: Correlation of enzyme activity with virus titer and evidence of *in vivo* complementation. *Virology* 112:328–41

115. Thomas, E., Lycke, E., Vahlne, A. 1985. Retrieval of latent herpes simplex virus type 1 genetic information from murine trigeminal ganglia by superinfection with heterotypic virus in vivo. *J. Gen. Virol.* 66:1763–70

116. Tullo, A. B., Shimeld, C., Blyth, W. A., Hill, T. J., Easty, D. L. 1982. Spread of virus and distribution of latent infection following ocular herpes simplex in the non-immune and immune mouse. *J. Gen. Virol.* 63:95–101

117. Tullo, A. B., Shimeld, C., Blyth, W. A., Hill, T. J., Easty, D. L. 1983. Ocular infection with HSV in non-immune and immune mice. *Arch. Ophthalmol.* 101:961–64

118. Walz, M. A., Yamamoto, H., Notkins, A. L. 1976. Immunologic response restricts the number of cells in sensory ganglia infected with herpes simplex. *Nature* 264:554–56

119. Warren, K. G., Brown, S. M., Wroblewska, Z., Gilden, D., Koprowski, H., Subak-Sharpe, J. 1978. Isolation of latent herpes simplex virus from the superior cervical and vagus ganglions of human beings. *N. Engl. J. Med.* 298: 1068–70

120. Warren, K. G., Gilden, D. H., Brown, S. M., Devlin, M., Wroblewska, Z., et al. 1977. Isolation of latent herpes simplex virus from human trigeminal ganglia including ganglia from one patient with multiple sclerosis. *Lancet* 2:637–39

121. Watson, K., Stevens, J. G., Cook, M. L., Subak-Sharpe, J. H. 1980. Latency competence of thirteen HSV-1 temperature sensitive mutants. *J. Gen. Virol.* 49:149–59

122. Wigdahl, B. L., Isom, H. C., de Clerq, E., Rapp, F. 1982. Activation of herpes simplex virus (HSV) type 1 genome by temperature sensitive mutants of HSV type 2. *Virology* 166:468–79

123. Wigdahl, B. L., Isom, H. C., Rapp, F. 1981. Repression and activation of the genome of herpes simplex viruses in hu-

man cells. *Proc. Natl. Acad. Sci. USA* 78:6522–26

124. Wigdahl, B. L., Scheck, A. C., de Clerq, E., Rapp, F. 1982. High efficiency latency and reactivation of herpes simplex virus in human cells. *Science* 217:1145–46
125. Wigdahl, B. L., Scheck, A. C., Ziegler, R. J., de Clerq, E., Rapp, F. 1984. Analysis of the herpes simplex virus genome during *in vitro* latency in human diploid fibroblasts and rat sensory neurons. *J. Virol.* 49:205–13
126. Wigdahl, B. L., Ziegler, R. J., Sneve, M., Rapp, F. 1983. Herpes simplex virus latency and reactivation in isolated rat sensory neurons. *Virology* 127:159–67
127. Wohlenberg, C. R., Walz, M. A., Notkins, A. L. 1976. Efficacy of phosphonoacetic acid on herpes simplex virus infection of sensory ganglia. *Infect. Immun.* 13:1519–21
128. Yamamoto, H., Walz, M. A., Notkins, A. L. 1977. Viral specific thymidine kinase in sensory ganglia of mice infected with herpes simplex virus. *Virology* 76:866–69
129. Yamamoto, T., Iwasaki, Y., Konno, H. 1984. Retrograde axoplasmic transport of Adriamycin: An experimental form of motor neuron disease? *Neurology* 34:1299–304
130. Yates, J., Warren, N., Reisman, D., Sugden, B. 1984. A *cis*-acting element from the Epstein-Barr virus genome that permits stable replication of recombinant plasmids in latently infected cells. *Proc. Natl. Acad. Sci. USA* 81:3806–10
131. Yates, J. L., Warren, N., Sugden, B. 1985. Stable replication of plasmids derived from the Epstein-Barr virus in a variety of mammalian cells. *Nature* 313:812–15
132. Zweerink, H. J., Stanton, L. W. 1981. Immune response to HSV infections: Virus specific antibodies in sera from patients with recurrent facial infections. *Infect. Immun.* 31:624–30

Ann. Rev. Microbiol. 1987. 41:573–93

CHROMOSOMAL CEPHALOSPORINASES RESPONSIBLE FOR MULTIPLE RESISTANCE TO NEWER β-LACTAM ANTIBIOTICS

C. C. Sanders

Department of Medical Microbiology, Creighton University School of Medicine, Omaha, Nebraska 68178

CONTENTS

0066-4227/87/1001-0573$02.00

INTRODUCTION

There are three major mechanisms by which clinical isolates of gram-negative bacteria resist β-lactam antibiotics. These include (*a*) altered outer membrane permeability, (*b*) production of a periplasmic β-lactamase, and (*c*) diminished affinity of the penicillin-binding proteins, the direct targets of β-lactam antibiotics that are located on the cytoplasmic membrane. Of these three mechanisms, production of β-lactamase is by far the most frequently encountered in clinical isolates. Altered permeability is a distant second. Diminished affinity of penicillin-binding proteins is encountered only rarely, primarily among fastidious species such as *Haemophilus influenzae* and *Neisseria gonorrhoeae*. Each of these mechanisms has been the subject of recent reviews (1, 8, 10, 12, 34, 65, 78, 84, 95, 122).

Since β-lactamases are responsible for most of the resistance encountered in clinical isolates of gram-negative bacteria, there has been a major effort in the past 15 years to design new β-lactam antibiotics that are not susceptible to these enzymes. This approach has to date been generally unsuccessful with the penicillin family. However, among the cephalosporins and cephamycins, a number of compounds that resist hydrolysis by many gram-negative β-lactamases have been developed (79, 105). Less conventional β-lactams that are less susceptible to hydrolysis by gram-negative β-lactamases, such as the monobactams, penems, penams, and carbapenems, have also been developed (45, 51, 81, 105, 114). Thus, it was very surprising that when many of these newer β-lactam antibiotics were used clinically, certain gram-negative bacteria rapidly developed resistance to them. This resistance affected multiple β-lactam antibiotics including virtually all penicillins, cephalosporins, cephamycins, and monobactams. It was ultimately found to be mediated by chromosomal cephalosporinases, which after derepression reach extremely high periplasmic concentrations (104).

During therapy with one of the newer β-lactam antibiotics, resistance due to derepression of chromosomal β-lactamases appears to emerge in 10–50% of patients infected with organisms possessing such enzymes (12, 32, 41, 92, 94, 104, 105, 126, 128, 129). Multiresistant mutants are spreading within the hospital environment and have become significant causes of nosocomial infections (7, 66, 83, 91, 104). They are becoming prevalent within burn units, cystic fibrosis centers, intensive care units, and cancer centers, where there is a clustering of patients at risk of infection with organisms that characteristically possess inducible chromosomal cephalosporinases, such as *Enterobacter cloacae* and *Pseudomonas aeruginosa*. This recent increase in the clinical importance of chromosomal cephalosporinases has led to increased interest in their characteristics, the alterations in expression leading to multiple β-lactam resistance, their genetic bases, and the mechanisms responsible for resistance to poorly hydrolyzable β-lactam drugs. These topics are the subjects of this review.

GENERAL CHARACTERISTICS OF CHROMOSOMAL CEPHALOSPORINASES

Nomenclature

Virtually every gram-negative organism possesses a chromosomally mediated cephalosporinase (115). In certain species, such as *Escherichia coli,* the enzyme is constitutively produced, usually at levels too low to confer resistance to β-lactam antibiotics. In other organisms, such as *E. cloacae, Citrobacter freundii,* and *P. aeruginosa,* the enzyme is inducible. Derepression (induction) of the enzyme in these organisms leads to multiple β-lactam resistance. Over the years a number of classification schemes have evolved for β-lactamases, including the chromosomal cephalosporinases (113). These schemes have generated somewhat confusing nomenclature. Richmond & Sykes (100a) originally classified the chromosomal cephalosporinases of gram-negative bacteria as class I enzymes, which were divided into types Ia, Ib, Ic, and Id. Thus the chromosomal cephalosporinases are often referred to as class I or type I β-lactamases. This terminology is easily confused with the Mitsuhashi type 1a and 1b plasmid-mediated β-lactamases and the *Bacillus cereus* type I enzyme (65, 113, 115). This problem with terminology is heightened by the recent classification of the chromosomal cephalosporinases into molecular class C (2, 39, 49). Thus, to avoid further confusion, the chromosomal cephalosporinases of gram-negative bacilli are referred to in this review as R & S I β-lactamases, for the Richmond & Sykes classification.

Physicochemical and Biochemical Characteristics

The R & S I β-lactamases comprise a widely diverse group (Table 1). However, they share many features in common. They are relatively large (30–42 kd) enzymes, usually with alkaline isoelectric points (57, 100a, 115). The active site contains a serine residue, which is acylated by the β-lactam as an intermediate in β-lactam hydrolysis (48, 49, 98). R & S I β-lactamases are commonly referred to as cephalosporinases, since hydrolysis rates (V_{max}) are often much larger for cephalosporins than for penicillins (100a). However, this may be somewhat of a misnomer, since these enzymes are also capable of hydrolyzing certain penicillins with high efficiency (V_{max}/K_m) (8, 49, 100a). They are susceptible to inhibition by cloxacillin but not by parachloromercuribenzoate, clavulanic acid, or sulbactam (80, 100a, 115). Among the well-characterized R & S I β-lactamases (Table 1), the R & S Ic enzyme of *Proteus vulgaris* is the most distinct. Unlike other R & S I enzymes, it readily hydrolyzes the oxyiminocephalosporins and is highly susceptible to inhibition by clavulanic acid (35, 71, 80, 88). Some R & S Ic enzymes also hydrolyze moxalactam and imipenem (88). Enzymes similar to these so-called oxyiminocephalosporinases are also found in *Pseudomonas cepacia, Flavobacterium meningosepticum, Pseudomonas maltophilia, Klebsiella oxytoca,* and *Flavobacterium odoratum* (88). They may be either constitutive or inducible. Since these β-lactamases are quite distinct from the other well-characterized R & S I enzymes and since their role in resistance is not well understood, they are not considered further in this review.

The true substrate profile of R & S I β-lactamases is difficult to ascertain for several reasons. First, very few studies have been performed with purified enzymes and a large number of substrates. Secondly, many of the newer β-lactam antibiotics cannot be accurately examined under the standard conditions usually employed (substrate excess) because they are poor substrates (low V_{max}), for which the enzymes have a very high affinity (low K_m). Thus the K_m for these compounds is usually determined as a K_i; however, the use of different substrates among the various studies makes comparison of data difficult. Finally, the kinetic parameters of the different R & S I enzymes vary among the various genera, species, and even strains of the same species. However, despite all of these problems, several generalizations can be made concerning the interactions of R & S I β-lactamases with various substrates. These are drawn from recent studies of the R & S I enzymes of *E. coli* (67, 93, 109), *E. cloacae* (25, 27, 50, 68, 93, 99, 100, 106, 109), *P. aeruginosa* (54, 76, 106), *C. freundii* (25, 50, 106, 109, 117), *Proteus rettgeri* (64, 87), and *Morganella morganii* (21, 53, 106, 109, 123).

The β-lactam antibiotics can be divided into three groups based upon the affinity of R & S I enzymes for them (K_m) and rates of hydrolysis (V_{max}) (Table 2). Cephalothin is perhaps the best substrate, owing to its high V_{max} and moderate K_m. Many of the older cephalosporins, such as cephaloridine and cefazolin, are also rapidly hydrolyzed, although the R & S I enzymes have a very low affinity for them. Most of the newer cephalosporins are so slowly hydrolyzed despite very high affinity that accurate detection of hy-

Table 1 Classification of chromosomal cephalosporinases

R & S[a] class and type	Expression	Species	Other names
Ia	Inducible[b]	*Enterobacter cloacae*	P99
Ib	Constitutive	*Escherichia coli*	AmpC β-lactamase
Ic	Inducible	*Proteus vulgaris*	Cefuroximase, cefotaximase
Id	Inducible	*Pseudomonas aeruginosa*	Sabath-Abraham enzyme

[a] Based on Reference 100a; see text.
[b] In strain P99 this enzyme was constitutive.

drolysis is not always possible in the in vitro assays. Hydrolysis rates and affinity cannot be used alone to predict correctly the efficiency of the inactivation of each substrate by the R & S I enzyme when it is present in the intact cell (8, 96). Rates of drug penetration into the cell must also be considered. Thus, it is not surprising that resistance to the newer β-lactams mediated by R & S I enzymes was not predicted from an examination of V_{max} and K_m alone, since under conditions of substrate excess the drugs appeared to be very poor substrates (low V_{max}) and in fact good inhibitors (low K_i) of the enzymes.

EXPRESSION OF CHROMOSOMAL CEPHALOSPORINASES LEADING TO RESISTANCE

In gram-negative organisms with inducible R & S I β-lactamases, enhanced expression of enzyme can occur via one of two mechanisms: (*a*) exposure of

Table 2 Kinetic parameters of R & S I enzymes with various β-lactam substrates

Affinity	Rate of hydrolysis
Low ($K_m > 100$ μM)	Low ($V_{max} < 1$ nmol min^{-1} mg^{-1})[a]
Cephaloridine	Cefuroxime
Cefazolin	Cefotaxime
Cefotiam	Moxalactam
Cefpirome	Ceftazidime
BMY-28142	Cefpirome
Moderate ($K_m > 1$ μM)	BMY-28142
Cephalothin	Cefoxitin
Cephalexin	Cefmetazole
Cefamandole	Ampicillin
Cefoperazone	Carbenicillin
Ceftazidime	Cloxacillin
Penicillin G	Aztreonam
High ($K_m < 1$ μM)[b]	Imipenem
Cefuroxime	Moderate ($V_{max} > 1$ nmol min^{-1} mg^{-1})
Cefotaxime	Cephalexin
Moxalactam	Cefotiam
Cefoxitin	Cefamandole
Cefmetazole	Cefoperazone
Ampicillin	Penicillin G
Carbenicillin	High ($V_{max} > 100$ nmol min^{-1} mg^{-1})
Cloxacillin	Cephaloridine
Aztreonam	Cephalothin
Imipenem	Cefazolin

[a] Hydrolysis has not actually been detected for some of these compounds.
[b] K_i usually determined for K_m.

the wild type to an inducer, which causes a temporary increase in enzyme levels, or (*b*) spontaneous mutation of the wild type to the stably derepressed state (104). The ability of the different β-lactam antibiotics to cause derepression of R & S I enzymes is different for the two mechanisms. In general, drugs that are good reversible inducers tend to be inefficient selectors of stably derepressed mutants (106). However, regardless of the mechanism responsible for enhanced expression of R & S I β-lactamases, the end result is multiple drug resistance.

Reversible Induction

A variety of β-lactam antibiotics are capable of inducing R & S I enzymes (8, 11, 26, 37, 69, 70, 73, 74, 82, 89, 106, 127, 133). After the wild-type cell is exposed to such a drug, enzyme levels rise rapidly and usually peak within 2 hr (26, 69, 70, 85). They remain elevated as long as the inducer persists in the environment with its β-lactam ring intact. Physical removal of the inducer or hydrolysis of its β-lactam ring leads to the immediate decline of enzyme levels. This specific induction by β-lactam antibiotics appears to be uninfluenced by (*a*) catabolite repression with glucose, (*b*) anaerobiosis, or (*c*) exogenously supplied cyclic AMP (23, 26, 85). Interestingly, these R & S I β-lactamases can also be induced nonspecifically by tryptophan, thiamine, folic acid, hemin, glycine, and various body fluids (14–16, 23). However, this nonspecific induction may involve a control region distinct from that for specific β-lactam–directed induction, since Gatus et al (23) have recently shown that induction by glycine is abolished by anaerobiosis.

The ability of various β-lactam antibiotics to induce R & S I enzymes varies greatly depending upon the organism examined, the antibiotic, its concentration, and the length of the induction period. Certain β-lactam antibiotics are potent inducers regardless of the species and the assay conditions (8, 69, 70, 74, 89, 106, 119, 133). These include the cephamycins (particularly cefoxitin) and imipenem. These antibiotics induce extremely high levels of R & S I β-lactamases in all organisms examined to date, even at very low concentrations. Certain antibiotics are variable inducers and cause elevated enzyme levels only in particular species or only at high, superinhibitory concentrations (37, 62, 69, 70, 89, 106, 133). Most of the newer cephalosporins fall into this category. In contrast, many of the penicillins appear to be poor inducers, although induction with very high concentrations has been reported (62, 69, 70, 106). Some of this apparent induction with high concentrations may be artificial unless care was taken to prevent killing of the test strain during the induction period. When such killing occurs, total protein levels may fall faster than β-lactamase levels, leading to an artificial increase in enzyme activity per unit protein. This problem is illustrated in Table 3. When a strain of *E. coli* possessing a plasmid-mediated TEM

Table 3 Artificial induction of TEM β-lactamase in *Escherichia coli* owing to killing by 4 μg ml^{-1} cefoxitin during the induction period

Induction time (hr)	Viable count (log_{10})	Protein in sonic extract (mg ml^{-1})	β-Lactamase activity in sonic extract: Uncorrected (units min^{-1})	Corrected (units min^{-1} mg^{-1})
0	8.8	1.0	22	22
2	6.8	0.04	5	124
4	6.6	0.01	3	300
8	7.0	0.02	7	350
30	8.9	0.3	12	36

β-lactamase was killed during the induction period, the enzyme levels suggested incorrectly that this constitutive β-lactamase was inducible.

Stable Derepression

The second mechanism responsible for elevated levels of R & S I β-lactamases involves the spontaneous mutation of the wild type to a stably derepressed state (22, 25, 27, 55). This mutation occurs in one in 10^6 to one in 10^7 cells and probably involves an alteration in the putative repressor protein itself (20, 22, 27). In certain organisms the mutation results in complete derepression, i.e. enzyme is produced constitutively and levels cannot be elevated further by exposure of the mutant to an enzyme inducer (17, 25, 27). However, in *P. aeruginosa* this mutation usually results in only partial derepression, and a second mutational event is required to complete the derepression (6, 17, 22, 33, 47, 59, 63). Mutants derepressed for R & S I β-lactamase are highly stable and persist despite multiple passages in drug-free media.

β-lactam antibiotics vary greatly in their ability to select stably derepressed mutants (106). Older cephalosporins such as cephalothin are poor selectors owing to their moderate inducer activity for some organisms and their high susceptibility to hydrolysis by R & S I β-lactamases. Thus enzyme levels produced by wild-type cells are sufficient to destroy these antibiotics before any selective pressure favoring the stably derepressed mutants is exerted. Cephamycins are also poor selectors. Their high inducer activity for all organisms makes wild-type cells phenotypically identical to the stably derepressed mutants. Thus again there is no selective pressure favoring replication of the mutants. The penems, penams, and carbapenems such as imipenem are poor selectors because they are equally active against wild-type and mutant cells. Replication is not favored for either wild type or mutant in the presence of these antibiotics. The β-lactam antibiotics most capable of exerting a selective pressure favoring growth of stably derepressed mutants are the newer cephalosporins and monobactams. These antibiotics are highly active against the wild-type cells, are generally poor inducers, and are poorly hydrolyzed by R & S I β-lactamases. Thus only mutants with highly elevated enzyme levels will be able to replicate in the presence of these β-lactams.

Products of Derepression

It is unclear whether the β-lactamase produced following derepression is identical to that produced at low levels prior to derepression. Many investigators have compared the characteristics of the enzymes before and after derepression and have found few if any differences (6, 25, 27, 85). Others have reported changes in isoelectric points, substrate profiles, and kinetic parameters (22, 27, 55, 103). The only consistent finding to date has been that production of the enzyme found prior to induction is enhanced whether or not there is evidence for additional enzymes. It is quite possible that in some organisms the apparent production of multiple enzymes following derepression results from the transcription of multiple structural genes (22). In other organisms the multiple enzymes may merely reflect different molecular forms of the same gene product that have undergone different posttranslational processing (97). Both of these possibilities need further assessment.

Derepression and Resistance

The importance of derepression of R & S I enzymes for multiple β-lactam resistance has been readily demonstrated in studies comparing the antibiotic susceptibility of the wild-type organisms to that of the stably derepressed mutants (9, 17, 25, 33, 82, 106, 121). In general, minimal inhibitory concentrations (MICs) of aztreonam and the newer cephalosporins such as cefotaxime, cefoperazone, ceftizoxime, ceftazidime, and moxalactam increase 64 to 256 fold with derepression of β-lactamase. MICs of penicillins such as carbenicillin, mezlocillin, and piperacillin increase 16 to 32 fold, while MICs of amdinocillin, penems, and imipenem are unaffected by derepression of R & S I β-lactamase. These differences are probably due to variations among the β-lactam antibiotics in their (*a*) rate of penetration into the cell, (*b*) interactions with R & S I enzymes, and (*c*) number, location, and type of lethal penicillin-binding proteins. The variation is discussed in detail in a later section.

It is important to note that antibiotics that are good reversible inducers of R & S I β-lactamases are antagonistic for any drug affected by the enzyme including themselves (32, 72, 107, 108). This is because the resistance phenotype expressed by wild-type cells following reversible induction is the same as that of stably derepressed mutants (17). The potential for antagonism has been readily demonstrated in various animal models (24, 46, 52). Thus drugs that are good inducers should never be used in combination with other β-lactam antibiotics.

GENETICS OF CHROMOSOMAL CEPHALOSPORINASES

The genetic organization responsible for inducible R & S I β-lactamases has not been completely delineated for any organism. However, significant progress has been made in determining the organization of genes encoding the enzymes in *C. freundii* (3, 56–58, 77, 131) and *E. cloacae* (30, 42, 43, 110, 111). Much of this progress has been due to similarities with the *ampC* gene responsible for the chromosomal cephalosporinase in *E. coli* K-12 (3, 5, 42, 43, 48, 56). Although the regulation of AmpC β-lactamase in *E. coli* is quite different from that in organisms with inducible enzymes, the similarities between the products of the structural genes suggest a similar evolutionary origin. Thus a description of the genetics of AmpC in *E. coli* as it is now understood is a necessary prelude to delineation of the genetic organization of inducible chromosomal cephalosporinases.

Genetics of AmpC in Escherichia coli

E. coli produces its R & S I β-lactamase constitutively at very low levels. The poor expression of this enzyme results from an inefficient promoter and an attenuator located between the promoter and the structural gene (40). Less than one fourth of the transcripts initiated at the promoter escape attenuation. In *E. coli* K-12 the promoter for *ampC* overlaps the structural gene for fumarate reductase (28). Since this latter enzyme is essential for anaerobic growth (36), mutations affecting the *ampC* promoter that could adversely affect fumarate reductase are highly infrequent (40, 57). Nevertheless, mutants of *E. coli* that hyperproduce AmpC β-lactamase have been described (4, 19, 40, 44, 57, 90). These hyperproducers either possess multiple copies of *ampC* (18, 19, 57, 86) or have an increased rate of transcription owing to mutations affecting the promoter or attenuator (29, 57, 90). Although the 2- to 20-fold increase in enzyme production is not as great as that usually seen with derepression of other R & S I β-lactamases, these hyperproducers do show increased resistance to penicillins, aztreonam, and many of the newer cephalosporins (4, 9, 38, 118).

Inducible R & S I β-Lactamases in Enterobacteriaceae

Among the Enterobacteriaceae possessing inducible R & S I β-lactamases, *C. freundii* and *E. cloacae* have been the most thoroughly studied. Extensive homology between these organisms' β-lactamase and AmpC of *E. coli* has been demonstrated in comparisons of structural genes or gene products (3, 5,

42, 43, 56). However, the regulatory systems governing gene expression are very different from that of *E. coli*. Several laboratories have begun to delineate the genetic organization of the R & S I β-lactamase in *C. freundii* (3, 56, 58, 77, 131). Bergström et al (3) demonstrated that in *C. freundii* the β-lactamase operon is adjacent to the fumarate reductase operon as in *E. coli*. However, whereas in *E. coli* the two regions overlap, in *C. freundii* they are separated by a 1100-bp region designated *ampR*. This region appears to contain regulatory genes for the β-lactamase that have both positive and negative effects on gene expression (58, 77).

Attempts to identify the specific genetic region(s) governing inducibility have been hampered by the need to use *E. coli* as the host for plasmids containing the β-lactamase operon of *C. freundii*. *E. coli* does not efficiently express the *C. freundii* β-lactamase, and expression is often poor or even noninducible (57, 131). Murayama et al (77) have recently shown that a mutation in the chromosome of *E. coli* allows the transformed *C. freundii* β-lactamase to become inducible. However, the induction kinetics are not the same as that seen in the natural host cell. This suggests that the *E. coli* chromosomal gene may not be specifically involved in regulation of β-lactamase. Recently, Lindberg et al (58) have shown that *ampR* is required for inducibility. However, mutations leading to stable derepression appear to occur outside the *ampR-ampC* region cloned by these investigators. This suggests that there is an additional regulatory component that has not yet been identified. Furthermore, stable derepression of the *C. freundii* β-lactamase in *E. coli* transformants appears to result from mutations on the *E. coli* chromosome and not on the transformed genes (58). Thus, these investigators have suggested that stable derepression may not be due to a mutation in a conventional repressor gene (57). Rather, it may be the result of a mutation affecting a metabolic pathway that is present in both *E. coli* and *C. freundii*.

Less detailed information is available for the inducible R & S I β-lactamase of *E. cloacae*. The structural gene for this β-lactamase has been cloned and transferred to *E. coli* (110, 111). Inducibility of the transferred genes was not evaluated in these studies. However, Guerin et al (30) have recently transferred the entire *E. cloacae* β-lactamase operon to *E. coli*. In a minicell system this DNA sequence coded for seven proteins, two of which appeared to be β-lactamase in premature and mature form (30). However, as with *C. freundii*, the expression of the *E. cloacae* β-lactamase was diminished and only weakly inducible in the recipient *E. coli* cells. The problems of gene expression encountered by virtually all investigators using *E. coli* recipients suggest that a more natural host may be needed for precise delineation of the genetic organization responsible for inducible R & S I β-lactamases. Perhaps a more hospitable host for such studies would be a strain of the donor species devoid of part or all of its β-lactamase operon. Clearly, much more work is required before the genetic regulation of inducible R & S I β-lactamases will be completely understood.

MECHANISMS RESPONSIBLE FOR MULTIPLE β-LACTAM RESISTANCE

The precise mechanism responsible for the multiple β-lactam resistance mediated by R & S I enzymes has been the subject of great controversy and is not as yet fully resolved. Once Seeberg et al (110, 111) had cloned the genes responsible for the R & S I β-lactamase in *E. cloacae* and had transferred the genes and thus the resistance to *E. coli,* it was proven conclusively that the enzyme alone was responsible for the resistance. This genetic proof of the mediator of resistance laid to rest all theories concerning other possible mechanisms. It then remained only to explain how a β-lactamase that in vitro did not hydrolyze various antibiotics could in the intact cell mediate high-level resistance to the drugs.

Over the last several years two theories have been developed to explain this apparent paradox. One theory is that the β-lactamases trap poorly hydrolyzable drugs in long-lived, biologically inactive complexes in the periplasmic space, preventing the drugs from reaching their penicillin-binding proteins on the cytoplasmic membrane (31, 104, 120). Such complexes need not involve covalent interaction or enzymic hydrolysis of the drug as long as more enzyme than drug is present. This would make it highly unlikely that free, unbound, and unhydrolyzed drug would ever be present in sufficient concentration to kill the cell. The second, more conventional theory is that within the milieu of the periplasmic space there is slow but efficient hydrolysis of drugs that are poor substrates (60–62, 125, 130). This occurs because enzyme levels are high and drug concentrations are low. Thus, the very low K_m of the newer drugs allows a hydrolysis rate close to $V_{\max}$, which although low, is still faster than the rate of drug penetration into the cell. It now appears that various aspects of both theories help to explain how R & S I β-lactamases cause resistance to many of the newer antibiotics (102).

The Reaction Catalyzed and Methods of Assay

To begin to delineate the mechanism responsible for multiple β-lactam resistance mediated by R & S I β-lactamases, one must consider the reaction catalyzed by β-lactamases in general as well as the assays used to measure the reaction in vitro. The reaction catalyzed can be summarized as

$$E + S \underset{k_{1b}}{\overset{k_{1a}}{\rightleftharpoons}} ES \xrightarrow{k_2} E{-}S \xrightarrow{k_3} E + S_i,$$

where E represents the β-lactamase, S the β-lactam drug, and S_i the hydrolyzed product of the reaction sequence. The β-lactamase initially forms ES, a noncovalent complex with the β-lactam drug. This step is reversible. The enzyme then hydrolyzes the β-lactam ring and becomes acylated by the drug; this form is represented by E–S. In the last step of the reaction, the enzyme cleaves the covalent bond, regenerating the active enzyme E and the hydrolyzed product S_i. Neither of the two intermediates formed in the reaction, ES and E–S, is biologically active nor is the product, S_i. Only the unbound and intact drug, S, is active. Thus, the apparent rate of the reaction sequence will depend not only upon the concentrations of E and S, but also upon the method used to monitor the reaction.

There are numerous methods for measuring β-lactamase activity (101). The iodometric method measures only product, S_i. Thus such methods will underestimate the amounts of biologically inactive drug if k_2 and k_3 are much slower than k_1. UV spectrophotometry assays that measure chemically unaltered drug (S and ES) will also underestimate biologically inactive drug if k_{1a} is greater than k_{1b} or k_2. Only bioassays will correctly measure biologically inactive drug, although they will not distinguish between ES, E–S, or S_i. This is probably why studies involving bioassays, especially those with prolonged incubation, have often shown more inactivation of the newer β-lactam drugs than studies involving other methods (13, 130). Results obtained in tests with drugs that are rapidly hydrolyzed are not greatly influenced by the assay method, since the reaction sequence occurs so quickly. In this case, measuring S or S_i alone will give similar information about the rate of the reaction. Since most of the older cephalosporins are rapidly hydrolyzed by R & S I β-lactamases (high V_{max}; see Table 2), the results obtained in various assays are comparable and generally predict the drugs' lack of activity against organisms that possess such enzymes. However, since the newer cephalosporins are not rapidly hydrolyzed by R & S I β-lactamases (low V_{max}; see Table 2), the results obtained in various assays can be contradictory.

In vitro assays cannot predict the susceptibility of the newer cephalosporins and monobactams to inactivation by R & S I β-lactamases because of the high affinity of the enzymes for these drugs coupled with a very slow rate of hydrolysis. Thus the amount of drug present as ES and E–S at any given time in the reaction sequence will be significant. Assays measuring only S_i or S + ES will thus grossly underestimate the ability of the enzymes to inactivate the drugs, since both ES and E–S are inactive. This problem is compounded further by the necessity to conduct in vitro assays under nonphysiologic conditions of drug excess owing to the inability of detection systems to accurately measure low concentrations of S or S_i. Thus results observed in in vitro assays with the newer β-lactams may not be at all indicative of the actual activity in the intact cell. To date no in vitro method has been developed to accurately measure S alone or ES + E–S + S_i (which represent active and inactive drug, respectively) under the conditions likely to be encountered in the cell following derepression of R & S I β-lactamase. Thus it is not possible to measure amounts of the different forms of the newer β-lactam drugs after they have encountered R & S I β-lactamases intracellularly. Only through such measurements could the precise mechanism responsible for resistance to each drug be delineated. However, this resistance can be partially explained without direct measurement.

Multiform Inactivation of β-Lactam Antibiotics

In the gram-negative cell, the ability of R & S I β-lactamases to protect the cell from any β-lactam drug depends upon (*a*) the ability of the enzyme to interact with the drug to produce ES, E–S, and S_i, (*b*) the rate of drug penetration into the cell, and (*c*) the number, type, and location of penicillin-binding proteins. These factors interdependently determine the fate of the drug.

The number of enzyme molecules in the periplasmic space following derepression of R & S I β-lactamases has been estimated to be as high as 10^5 molecules (8, 125). In addition, the substrate concentration is relatively low owing to the drug's restricted access across the outer membrane (132). Thus the enzyme is in excess within the periplasmic space. Under such conditions the relative amounts of S, ES, E–S, and S_i for each β-lactam can be estimated from the data in Table 2. Drugs with moderate to high affinity (K_m) and high V_{max} (e.g. cephalothin) will exist primarily in the S_i form, since the substrate concentration will not be sufficient to saturate the enzyme. Actual hydrolysis (V_{act}) will be below V_{max} and will depend upon the concentration of drug entering the cell. Drugs with low affinity and moderate to high V_{max} (e.g. cephaloridine) will also exist primarily in the S_i form, providing that the rate of penetration into the cell is sufficient to allow the intracellular concentration of drug to approximate K_m. A very slowly penetrating drug with low affinity might be primarily in the S form, since V_{act} would be much smaller than V_{max}. β-Lactam drugs with low affinity and low V_{max} (e.g. cefpirome) will also comprise more S than S_i. For these drugs, V_{act} will be below V_{max} because of the low substrate concentration. Drugs with high affinity and low V_{max} (e.g. cefuroxime, moxalactam) will exist primarily in the ES and E–S forms, since the intracellular concentration approximates K_m and V_{act} approaches V_{max}. The relative amounts of ES and E–S are difficult to predict. However, drugs that are primarily present in the E–S form will irreversibly inhibit the enzyme in most in vitro assays, since deacylation (k_3) occurs so slowly. Such may be the case for moxalactam and aztreonam. These drugs are deacylated very slowly and once were thought to be suicide inactivators of R & S I β-lactamases (8, 53, 75).

Solely from these considerations of enzyme/drug interactions under conditions of enzyme excess, it is possible to predict correctly, with few exceptions, which β-lactam drugs should be resisted following derepression of R & S I β-lactamase. These include all of the older cephalosporins, the penicillins, the newer cephalosporins, and the monobactams. The ability of R & S I β-lactamases to cause low-level resistance to cefpirome and BMY-28142 (two drugs which should appear primarily in the S form) and their inability to cause high-level resistance to imipenem (which should exist as ES or E–S) is probably due to unique aspects of these drugs' penetration into the cell and interaction with penicillin-binding proteins.

Recent studies by Yoshimura & Nikaido (132) have shown that the zwitterionic β-lactams such as cephaloridine and imipenem penetrate rapidly into the gram-negative cell. Cefpirome and BMY-28142, two other zwitterions, probably also penetrate rapidly. The ability of these compounds to penetrate more rapidly than the monoanionic and dianionic β-lactams probably allows the intracellular concentrations to approach K_m. For cefpirome and BMY-28142 this allows V_{act} to approach V_{max}, and less drug exists in the S form than originally predicted. However, even with greater penetration these drugs are less affected by R & S I β-lactamases than other cephalosporins, since enzyme affinity is still very low and K_m is unlikely to be reached. For imipenem K_m is very small and saturating concentrations may in fact be reached. Thus, more drug may exist in the S form than was originally predicted. Livermore recently demonstrated that mutants of *P. aeruginosa* lacking an inducible R & S I β-lactamase are more susceptible to imipenem than either wild-type cells that have undergone reversible induction or mutants that were stably derepressed for β-lactamase (63a). However, both the reversibly induced wild-type cells and the stably derepressed mutants are still susceptible to relatively low concentrations of imipenem. These results support the prediction that imipenem can rapidly penetrate the cell and saturate the enzyme so that $V_{act} = V_{max}$. Since the rate of drug entry may exceed V_{max}, S accumulates and is free to reach the penicillin-binding proteins in the cell.

Imipenem, like other penams, penems, and carbapenems, binds to penicillin-binding proteins (PBPs) 1 and 2 (124). These are the least numerous essential PBPs in the cell. Saturation of PBP 2 alone, of which the cell contains only 20 molecules, is sufficient to kill the cell (112). The penicillins and cephalosporins, in contrast, bind to the more numerous PBP 3 as well as to PBP 1 and other nonessential PBPs (124). Thus more free drug will be required to kill the cell with these β-lactams than with imipenem. In addition, the three-dimensional location of the different PBPs relative to each other and the β-lactamase is unknown. It is possible that the β-lactamase protects PBP 2 less efficiently than the other targets of β-lactam antibiotics because of its location in the cell.

From a general summary of the various factors that influence the in vitro activity of β-lactam antibiotics (Table 4), it is understandable why imipenem and related compounds are least affected by derepression of R & S I β-lactamases. These drugs have the greatest rate of penetration and require the fewest active drug molecules to bind essential PBPs. Thus, even in the presence of large quantities of β-lactamase the activity of these drugs is least affected. At the other extreme lie the anionic cephalosporins, penicillins, and monobactams. Their rate of penetration into the cell is relatively slow. Thus, the presence of a large quantity of β-lactamase that can interact with these drugs to form ES, E–S, or S_i will significantly diminish the number of bioactive drug molecules that reach their more numerous essential PBPs in the cell. The zwitterionic cephalosporins are less affected than the anionic cephalosporins owing to a lower affinity of the enzyme for the former and a higher rate of penetration into the cell (50, 93). When all of these factors are considered, it is not surprising that high-level resistance can be mediated by a β-lactamase without the formation of S_i at a rate at least equal to that of drug penetration. As long as the enzyme is in sufficient concentration to interact with the drug as it enters the cell, the formation of ES, E–S, and S_i is equally important in inactivating the drug. Since bacterial replication rates greatly exceed the half-lives of ES and E–S (8, 53, 75), it is apparent that with certain of the newer β-lactams, ES and E–S would be the major inactive forms found intracellularly. However, even in this setting S_i would eventually form. When methods are developed that will allow detection of each inactive form specifically it will be possible to determine the relative importance of each form for the resistance mediated by R & S I β-lactamases. However, it is clear that multiform inactivation does occur.

Table 4 Factors governing multiform inactivation of β-lactam antibiotics by R & S I β-lactamases

Interaction with R & S I β-lactamase [affinity *(K_m)*]	Hydrolysis (V_{max})	Penetration into the cell	Primary form in cell	Example
High–moderate	High–moderate	Low	S_i	Cephalothin
Low	High	High	S_i	Cephaloridine
High	Low	Low	ES/E–S	Moxalactam, aztreonam
High	Low	High	S	Imipenem
Low	Low	High	S	Cefpirome, BMY-28142

CONCLUDING REMARKS

With the ever increasing clinical importance of R & S I β-lactamases, it becomes imperative to increase our knowledge about the enzymes themselves, the genetic control responsible for their expression, and the precise mechanisms responsible for the multiple β-lactam resistance mediated by them. Many of the conventional methods routinely used to assess the potential for β-lactamase–mediated resistance are inadequate for the new era of β-lactam antibiotics. Many of the theories regarding β-lactam/β-lactamase interactions also need remodeling to explain how inactive species other than S_i contribute significantly to resistance. Once molecular details of the precise interactions between R & S I β-lactamases and the newer "nonhydrolyzable" antibiotics have been delineated, it should be possible to use a new approach to design enzyme-resistant drugs for gram-negative organisms. Delineation of the control mechanisms responsible for derepression of R & S I β-lactamases should make it possible to design ways to keep these enzymes repressed. If stable repression could be accomplished, the β-lactam antibiotics extant today would be sufficient to solve the clinical problems posed by R & S I β-lactamases.

Acknowledgments

The work on R & S I β-lactamases from my laboratory has been supported over the years by many representatives of the pharmaceutical industry and by the Health Future Foundation. I am greatly indebted to Johan S. Bakken and Robert J. Martig for their help with literature searches, to Ellen Smith-Moland for her excellent technical assistance, and to Carole Sears for typing the manuscript. I am also most appreciative of the constant support and guidance provided to me by my former teacher, respected colleague, closest friend, and husband, W. Eugene Sanders, Jr.

Literature Cited

1. Acar, J. F. 1985. Problems and changing patterns of resistance with gram-negative bacteria. *Rev. Infect. Dis.* 7(Suppl. 4):S545–51
2. Ambler, R. P. 1980. The structure of β-lactamases. *Philos. Trans. R. Soc. London Ser. B* 289:321–31
3. Bergström, S., Lindberg, F. P., Olsson, O., Normark, S. 1983. Comparison of the overlapping *frd* and *ampC* operons of *Escherichia coli* with the corresponding DNA sequence in other gram-negative bacteria. *J. Bacteriol.* 155:1297–305
4. Bergström, S., Normark, S. 1979. β-Lactam resistance in clinical isolates of *Escherichia coli* caused by elevated production of *ampC*-mediated chromosomal β-lactamase. *Antimicrob. Agents Chemother.* 16:427–33
5. Bergström, S., Olsson, O., Normark, S. 1982. Common evolutionary origin of chromosomal beta-lactamase genes in enterobacteria. *J. Bacteriol.* 150:528–34
6. Berks, M., Redhead, K., Abraham, E. P. 1982. Isolation and properties of an inducible and a constitutive β-lactamase from *Pseudomonas aeruginosa*. *J. Gen. Microbiol.* 128:155–59
7. Bryan, C. S., John, J. F. Jr., Pai, S., Austin, T. 1985. Gentamicin vs cefotax-

ime for therapy of neonatal sepsis. Relationship to drug resistance. *Am. J. Dis. Child.* 139:1086–89
8. Bush, K., Sykes, R. B. 1984. Interaction of β-lactam antibiotics with β-lactamases as a cause for resistance. In *Antimicrobial Drug Resistance,* ed. L. E. Bryan, pp. 1–31. New York: Academic. 576 pp.
9. Chabbert, Y. A., Jaffé, A. 1982. Sch 29482: activity against susceptible and β-lactam resistant variants of Enterobacteriaceae. *J. Antimicrob. Chemother.* 9(Suppl. C):203–12
10. Chopra, I. 1984. Antibiotic resistance resulting from decreased drug accumulation. *Br. Med. Bull.* 40:11–17
11. Christenson, J. G., Robertson, T. L. 1986. Interactions of amdinocillin with beta-lactamases. *Abstr. Intersci. Conf. Antimicrob. Agents Chemother. 26th, New Orleans,* Abstr. 1120. Washington DC: Am. Soc. Microbiol.
12. Collatz, E., Gutmann, L., Williamson, R., Acar, J. F. 1984. Development of resistance to β-lactam antibiotics with special reference to third generation cephalosporins. *J. Antimicrob. Chemother.* 14(Suppl. B):13–21
13. Cullmann, W. 1985. Hydrolysis of third generation cephalosporins by *Enterobacter cloacae* beta-lactamase. *Eur. J. Clin. Microbiol.* 4:607–9
14. Cullmann, W., Dalhoff, A., Dick, W. 1984. Nonspecific induction of β-lactamase in *Enterobacter cloacae. J. Gen. Microbiol.* 130:1781–86
15. Cullmann, W., Dick, W. 1985. Evidence of nonspecific induction of beta-lactamase in overproducing variants of *Enterobacter cloacae* and *Citrobacter freundii. Eur. J. Clin. Microbiol.* 4:34–40
16. Cullmann, W., Dick, W., Dalhoff, A. 1983. Nonspecific induction of β-lactamase in *Enterobacter cloacae. J. Infect. Dis.* 148:765
17. Curtis, N. A., Eisenstadt, R. L., Rudd, C., White, A. J. 1986. Inducible type I β-lactamases of gram-negative bacteria and resistance to β-lactam antibiotics. *J. Antimicrob. Chemother.* 17:51–61
18. Edlund, T., Grundström, T., Bjork, G. R., Normark, S. 1980. Tandem duplication induced by an unusual *ampA1-, ampC*-transducing lambda phage: a probe to initiate gene amplification. *Mol. Gen. Genet.* 180:249–57
19. Edlund, T., Grundström, T., Normark, S. 1979. Isolation and characterization of DNA repetitions carrying the chromosomal β-lactamase gene of *Escherichia coli* K-12. *Mol. Gen. Genet.* 173:115–25
20. Findell, C. M., Sherris, J. C. 1976. Susceptibility of *Enterobacter* to cefamandole: evidence for a high mutation rate to resistance. *Antimicrob. Agents Chemother.* 9:970–74
21. Fujii-Kuriyama, Y., Yamamoto, M., Sugawara, S. 1977. Purification and properties of beta-lactamase from *Proteus morganii. J. Bacteriol.* 131:726–34
22. Gates, M. L., Sanders, C. C., Goering, R. V., Sanders, W. E. Jr. 1986. Evidence for multiple forms of type I chromosomal β-lactamase in *Pseudomonas aeruginosa. Antimicrob. Agents Chemother.* 30:453–57
23. Gatus, B. J., Bell, S. M., Jimenez, A. S. 1986. Enhancement of beta-lactamase production by glycine in *Enterobacter cloacae* ATCC 13047. *Pathology* 18: 145–47
24. Goering, R. V., Sanders, C. C., Sanders, W. E. Jr. 1982. Antagonism of carbenicillin and cefamandole by cefoxitin in treatment of experimental infections in mice. *Antimicrob. Agents Chemother.* 21:963–67
25. Gootz, T. D., Jackson, D. B., Sherris, J. C. 1984. Development of resistance to cephalosporins in clinical strains of *Citrobacter* spp. *Antimicrob. Agents Chemother.* 25:591–95
26. Gootz, T. D., Sanders, C. C. 1983. Characterization of β-lactamase induction in *Enterobacter cloacae. Antimicrob. Agents Chemother.* 23:91–97
27. Gootz, T. D., Sanders, C. C., Goering, R. V. 1982. Resistance to cefamandole: derepression of β-lactamases by cefoxitin and mutation in *Enterobacter cloacae. J. Infect. Dis.* 146:34–42
28. Grundström, T., Jaurin, B. 1982. Overlap between *ampC* and *frd* operons on the *Escherichia coli* chromosome. *Proc. Natl. Acad. Sci. USA* 79:1111–15
29. Grundström, T., Normark, S. 1985. Initiation of translation makes attenuation of ampC in *E. coli* dependent on growth rate. *Mol. Gen. Genet.* 198:411–15
30. Guerin, S., Paradis, F., Guay, R. 1985. Cloning and characterization of chromosomally encoded cephalosporinase gene of *Enterobacter cloacae. Can. J. Microbiol.* 32:301–9
31. Gutmann, L., Williamson, R. 1983. A model system to demonstrate that β-lactamase–associated antibiotic trapping could be a potential means of resistance. *J. Infect. Dis.* 8:316–21
32. Gutmann, L., Williamson, R., Kitzis, M.-D., Acar, J. F. 1980. Synergism and antagonism in double beta-lactam antibiotic combinations. *Am. J. Med.* 80(Suppl. 5C):21–29
33. Gwynn, M. N., Rolinson, G. N. 1983.

Selection of variants of gram-negative bacteria with elevated production of type I β-lactamase. *J. Antimicrob. Chemother.* 11:577–81

34. Hamilton-Miller, J. M. T. 1982. β-Lactamases and their clinical significance. *J. Antimicrob. Chemother.* 9(Suppl. B):11–19
35. Hirai, K., Sato, K., Matsubara, N., Katsumata, R., Inoue, M., Mitsuhashi, S. 1981. Immunological properties of beta-lactamases that hydrolyze cefuroxime and cefotaxime. *Antimicrob. Agents Chemother.* 20:262–64
36. Ingledew, W. J., Poole, R. K. 1984. The respiratory chains of *Escherichia coli*. *Microbiol. Rev.* 44:222–71
37. Jacobs, J. Y., Livermore, D. M., Davy, K. W. M. 1984. *Pseudomonas aeruginosa* β-lactamase as a defense against azlocillin, mezlocillin and piperacillin. *J. Antimicrob. Chemother.* 14:221–29
38. Jaffé, A., Chabbert, Y. A., Derlot, E. 1983. Selection and characterization of β-lactam–resistant *Escherichia coli* K-12 mutants. *Antimicrob. Agents Chemother.* 23:622–25
39. Jaurin, B., Grundström, T. 1981. AmpC cephalosporinase of *Escherichia coli* K-12 has a different evolutionary origin from that of β-lactamases of the penicillinase type. *Proc. Natl. Acad. Sci. USA* 78:4897–901
40. Jaurin, B., Grundström, T., Edlund, T., Normark, S. 1981. The *E. coli* beta-lactamase attenuator mediates growth rate–dependent regulation. *Nature* 290: 221–25
41. Jimenez-Lucho, V. E., Saravolatz, L. D., Medeiros, A. A., Pohlod, D. 1986. Failure of therapy in *Pseudomonas* endocarditis: selection of resistant mutants. *J. Infect. Dis.* 154:64–68
42. Joris, B., DeMeester, F., Galleni, M., Reckinger, G., Coyette, J., Frere, J. 1985. The β-lactamase of *Enterobacter cloacae* P99. Chemical properties. N-terminal sequence and interaction with 6-β-halogenopenicillanates. *Biochem. J.* 228:241–48
43. Joris, B., Dusart, J., Frere, J., Van Beeumen, J., Emanuel, E. L., et al. 1984. The active site of the P99 β-lactamase from *Enterobacter cloacae*. *Biochem. J.* 223:271–74
44. Kabins, S. A., Sweeney, H. M., Cohen, S. 1966. Resistance to cephalothin *in vivo* associated with increased cephalosporinase production. *Ann. Intern. Med.* 65:1271–77
45. Kahan, F. M., Kropp, H., Sundeloff, J. G., Birnbaum, J. 1983. Thienamycin: development of imipenem-cilastatin. *J. Antimicrob. Chemother.* 12:1–35
46. Kasai, K. 1986. Antibacterial antagonism of β-lactam antibiotics in experimental infections. *Chemotherapy Basel* 32:148–58
47. King, A., Shannon, K., Eykyn, S., Phillips, I. 1983. Reduced sensitivity to β-lactam antibiotics arising during ceftazidime treatment of *Pseudomonas* infections. *J. Antimicrob. Chemother.* 12:363–70
48. Knott-Hunziker, V., Petursson, S., Jayatilake, G. S., Waley, S. G., Jaurin, B., Grundstrom, T. 1982. Active sites of β-lactamases: the chromosomal β-lactamases of *Pseudomonas aeruginosa* and *Escherichia coli*. *Biochem. J.* 201:621–27
49. Knott-Hunziker, V., Petursson, S., Waley, S. G., Jaurin, S. G., Grundstrom, T. 1982. The acyl-enzyme mechanism of β-lactamase action. *Biochem. J.* 207:315–22
50. Kobayashi, S., Arai, S., Hayashi, S., Kujimoto, K. 1986. β-Lactamase stability of cefpirome (HR 810), a new cephalosporin with a broad antimicrobial spectrum. *Antimicrob. Agents Chemother.* 30:713–18
51. Kropp, H., Gerckens, L., Sundelof, J. G., Kahan, F. M. 1985. Antibacterial activity of imipenem: the first thienamycin antibiotic. *Rev. Infect. Dis.* 7(Suppl. 3):S389–410
52. Kuck, N. A., Testa, R. T., Forbes, M. 1981. *In vitro* and *in vivo* antibacterial effects of combinations of beta-lactam antibiotics. *Antimicrob. Agents Chemother.* 19:634–38
53. Labia, R. 1982. Moxalactam: An oxa-β-lactam antibiotic that inactivates β-lactamases. *Rev. Infect. Dis.* 4(Suppl.): S529–35
54. Labia, R., Beguin-Billecoq, R., Guionie, M. 1981. Behaviour of ceftazidime towards β-lactamases. *J. Antimicrob. Chemother.* 8(Suppl. B):141–46
55. Lampe, M. F., Allan, B. J., Minshew, B. H., Sherris, J. C. 1982. Mutational enzymatic resistance of *Enterobacter* species to beta-lactam antibiotics. *Antimicrob. Agents Chemother.* 21:655–60
56. Lindberg, F., Normark, S. 1986. Sequence of the *Citrobacter freundii* OS60 chromosomal *ampC* β-lactamase gene. *Eur. J. Biochem.* 156:441–45
57. Lindberg, F., Normark, S. 1986. Contribution of chromosomal β-lactamases to β-lactam resistance in enterobacteria. *Rev. Infect. Dis.* 8(Suppl):S292–304
58. Lindberg, F., Westman, L., Normark, S. 1985. Regulatory components in *Citrobacter freundii ampC* β-lactamase

induction. *Proc. Natl. Acad. Sci. USA* 82:4620–24

59. Livermore, D. M. 1983. Kinetics and significance of the activity of the Sabath and Abrahams' β-lactamase of *Pseudomonas aeruginosa* against cefotaxime and cefsulodin. *J. Antimicrob. Chemother.* 11:169–79
60. Livermore, D. M. 1985. Do β-lactamases "trap" cephalosporins? *J. Antimicrob. Chemother.* 15:11–21
61. Livermore, D. M., Riddle, S. J., Davy, K. W. M. 1986. Hydrolytic model for cefotaxime and ceftriaxone resistance in β-lactamase–derepressed *Enterobacter cloacae*. *J. Infect. Dis.* 153:619–22
62. Livermore, D. M., Williams, J. D., Davy, K. W. M. 1985. Cephalosporin resistance in *Pseudomonas aeruginosa,* with special reference to the proposed trapping of antibiotics by beta-lactamase. *Chemioterapia* 4:28–35
63. Livermore, D. M., Williams, R. J., Lindridge, M. A., Slack, R. C. B., Williams, J. D. 1982. *Pseudomonas aeruginosa* isolates with modified β-lactamase inducibility: effects on β-lactam sensitivity. *Lancet* 1:1466–67

63a. Livermore, D. M., Yang, Y. J. 1987. β-Lactamase lability and inducer power of newer β-lactam antibiotics in relation to their activity against β-lactamase inducibility mutants of *Pseudomonas aeruginosa*. *J. Infect. Dis.* 155:775–82

64. Matsuura, M., Nakazawa, H., Inoue, M., Mitsuhashi, S. 1980. Purification and biochemical properties of β-lactamase produced by *Proteus rettgeri*. *Antimicrob. Agents Chemother.* 18:687–90
65. Medeiros, A. A. 1984. β-Lactamases. *Br. Med. Bull.* 40:18–27
66. Medeiros, A. A., Hare, R., Papa, E., Adam, C., Miller, G. H. 1985. Gram-negative bacilli resistant to third generation cephalosporins: β-lactamase characterization and susceptibility to Sch 34343. *J. Antimicrob. Chemother.* 15(Suppl. C):119–32
67. Minami, S., Inoue, M., Mitsuhashi, S. 1980. Purification and properties of cephalosporinase in *Escherichia coli*. *Antimicrob. Agents Chemother.* 18:77–80
68. Minami, S., Inoue, M., Mitsuhashi, S. 1980. Purification and properties of a cephalosporinase from *Enterobacter cloacae*. *Antimicrob. Agents Chemother.* 18:853–57
69. Minami, S., Matsubara, N., Yotsuji, A., Araki, H., Watanabe, Y., et al. 1983. Induction of cephalosporinase production by various penicillins in Enterobacteriaceae. *J. Antibiot.* 36:1387–95
70. Minami, S., Yotsuji, A., Inoue, M., Mitsuhashi, S. 1980. Induction of β-lactamase by various β-lactam antibiotics in *Enterobacter cloacae*. *Antimicrob. Agents Chemother.* 18:382–85
71. Mitsuhashi, S. 1985. Resistance to beta-lactam antibiotics in bacteria. In *Recent Advances in Chemotherapy, Antimicrobial Section 1*, ed. J. Ishigami, pp. 3–9. Tokyo: Univ. Tokyo Press. 874 pp.
72. Moellering, R. C. Jr., Willey, S., Eliopoulos, G. M. 1982. Synergism and antagonism of ceftizoxime and other new cephalosporins. *J. Antimicrob. Chemother.* 10(Suppl. C):69–79
73. Moosdeen, F., Keeble, J., Williams, J. D. 1986. Induction/inhibition of chromosomal β-lactamases by β-lactamase inhibitors. *Rev. Infect. Dis.* 8(Suppl. 5):S562–68
74. Mouton, R. P., van Gestal, M., Bongaerts, G. P. A. 1982. Absence of β-lactamase–inducing effect of moxalactam. *Rev. Infect. Dis.* 4(Suppl):S527–28
75. Murakami, K., Yoshida, T. 1985. Covalent binding of moxalactam to cephalosporinase of *Citrobacter freundii*. *Antimicrob. Agents Chemother.* 27:727–32
76. Murata, T., Minami, S., Yasuda, K., Iyobe, S., Inoue, M., et al. 1981. Purification and properties of cephalosporinase from *Pseudomonas aeruginosa*. *J. Antibiot.* 34:1164–70
77. Murayama, S. Y., Yamamoto, T., Suzuki, I., Sawai, T. 1986. Mutation of *Escherichia coli* capable of expressing gene(s) for β-lactamase production of *Citrobacter freundii*. *Antimicrob. Agents Chemother.* 29:707–9
78. Neu, H. C. 1982. Mechanisms of bacterial resistance to antimicrobial agents, with particular reference to cefotaxime and other β-lactam compounds. *Rev. Infect. Dis.* 4(Suppl):S288–99
79. Neu, H. C. 1982. The new beta-lactamase–stable cephalosporins. *Ann. Intern. Med.* 97:408–19
80. Neu, H. C. 1985. The role of beta-lactamase inhibitors in chemotherapy. *Pharmacol. Ther.* 30:1–18
81. Neu, H. C. 1986. β-Lactam antibiotics: structural relationships affecting *in vitro* activity and pharmacologic properties. *Rev. Infect. Dis.* 8(Suppl. 3):S237–59
82. Neu, H. C., Chin, N. X. 1985. A perspective on the present contribution of beta-lactamases to bacterial resistance with particular reference to induction of beta-lactamase and its clinical significance. *Chemioterapia* 4:63–70
83. Nichols, W. W., Milne, L. M. 1986.

Derepressed β-lactamase synthesis in strains of *Pseudomonas aeruginosa* isolated from patients with cystic fibrosis. *J. Antimicrob. Chemother.* 18:549–50

84. Nikaido, H. 1984. Outer membrane permeability and β-lactam resistance. In *Microbiology—1984,* ed. L. Leive, D. Schlessinger, pp. 381–84. Washington, DC: Am. Soc. Microbiol. 441 pp.
85. Nordström, K., Sykes, R. B. 1974. Induction kinetics of beta-lactamase biosynthesis in *Pseudomonas aeruginosa. Antimicrob. Agents Chemother.* 6:734–40
86. Normark, S., Edlund, T., Grundström, T., Bergström, S., Wolf-Watz, H. 1977. *Escherichia coli* K-12 mutants hyperproducing chromosomal beta-lactamase by gene repetitions. *J. Bacteriol.* 132:912–22
87. Ohya, S., Fujii-Kuriyama, Y., Yamamoto, M., Sugawara, S. 1980. Purification and some properties of β-lactamases from *Proteus rettgeri* and *Proteus inconstans. Microbiol. Immunol.* 24:815–24
88. Okonogi, K., Kuno, M., Higashide, E. 1986. Induction of β-lactamase in *Proteus vulgaris. J. Gen. Microbiol.* 132: 143–50
89. Okonogi, K., Sugiura, A., Kuno, M., Higashide, E., Kondo, M., Imada, A. 1985. Effect of β-lactamase induction on susceptibility to cephalosporins in *Enterobacter cloacae* and *Serratia marcescens. J. Antimicrob. Chemother.* 16:31–42
90. Olsson, O., Bergström, S., Lindberg, F. P., Normark, S. 1983. AmpC β-lactamase hyperproduction in *Escherichia coli:* natural ampicillin resistance generated by horizontal chromosomal DNA transfer from *Shigella. Proc. Natl. Acad. Sci. USA* 80:7556–60
91. Pedersen, S. S., Koch, C., Høiby, N., Rosendal, K. 1986. An epidemic spread of multiresistant *Pseudomonas aeruginosa* in a cystic fibrosis center. *J. Antimicrob. Chemother.* 17:505–16
92. Perronne, C., Régnier, B., Legrand, P., Buré, A., Frottier, J., et al. 1986. Failure of new beta-lactam antibiotics in the treatment of severe infections caused by *Enterobacter cloacae. Presse Med.* 15:1813–18
93. Phelps, D. J., Carlton, D. D., Farrell, C. A., Kessler, R. E. 1986. Affinity of cephalosporins for β-lactamases as a factor in antibacterial efficacy. *Antimicrob. Agents Chemother.* 29:845–48
94. Phillips, I., Shannon, K. 1985. The emergence of resistance and the therapy of septicaemia. *Chemioterapia* 4:90–94
95. Piddock, L. J., Wise, R. 1985. Newer mechanisms of resistance to β-lactam antibiotics in gram-negative bacteria. *J. Antimicrob. Chemother.* 16:279–84
96. Pollock, M. R. 1965. Purification and properties of penicillinase from two strains of *Bacillus licheniformis:* a chemical, physicochemical and physiological comparison. *Biochem. J.* 94:666–75
97. Pollock, M. R. 1968. The range and significance of variations amongst bacterial penicillinases. *Ann. NY Acad. Sci.* 151:502–15
98. Pratt, R. F., Govardhan, C. P. 1984. β-Lactamase–catalyzed hydrolysis of acyclic depsipeptides and acyl transfer to specific amino acid receptors. *Proc. Natl. Acad. Sci. USA* 81:1302–6
99. Richmond, M. H. 1980. β-Lactamase stability of cefotaxime. *J. Antimicrob. Chemother.* 6(Suppl. A):13–17
100. Richmond, M. H. 1982. Susceptibility of moxalacam to β-lactamase. *Rev. Infect. Dis.* 4(Suppl):S522–28

100a. Richmond, M. H., Sykes, R. B. 1973. The β-lactamases of gram-negative bacteria and their possible physiological role. *Adv. Microb. Physiol.* 9:31–88

101. Ross, G. W., O'Callaghan, C. H. 1975. β-Lactamase assays. *Methods Enzymol.* 43:69–85
102. Sanders, C. C. 1984. Inducible β-lactamases and non-hydrolytic resistance mechanisms. *J. Antimicrob. Chemother.* 13:1–3
103. Sanders, C. C., Moellering, R. C. Jr., Martin, R. R., Perkins, R. L., Strike, D. G., et al. 1982. Resistance to cefamandole: a collaborative study of emerging clinical problems. *J. Infect. Dis.* 145:118–25
104. Sanders, C. C., Sanders, W. E. Jr. 1985. Microbial resistance to newer generation β-lactam antibiotics: clinical and laboratory implications. *J. Infect. Dis.* 151:399–405
105. Sanders, C. C., Sanders, W. E. Jr. 1986. The cephalosporins and cephamycins. *Antimicrob. Agents Annu.* 1:66–90
106. Sanders, C. C., Sanders, W. E. Jr. 1986. Type I β-lactamases of gram-negative bacteria: interactions with β-lactam antibiotics. *J. Infect. Dis.* 154:792–800
107. Sanders, C. C., Sanders, W. E. Jr., Goering, R. V. 1982. *In vitro* antagonism of beta-lactam antibiotics by cefoxitin. *Antimicrob. Agents Chemother.* 21:968–75
108. Sanders, C. C., Sanders, W. E. Jr., Goering, R. V. 1983. Influence of clindamycin on derepression of β-

lactamases in *Enterobacter* spp. and *Pseudomonas aeruginosa*. *Antimicrob. Agents Chemother*. 24:48–53

109. Sawai, T., Kanno, M., Tsukamoto, K. 1982. Characterization of eight beta-lactamases of gram-negative bacteria. *J. Bacteriol*. 152:567–71
110. Seeberg, A. H., Tolxdorff-Neutzling, R. M., Wiedemann, B. 1983. Chromosomal β-lactamases of *Enterobacter cloacae* are responsible for resistance to third-generation cephalosporins. *Antimicrob. Agents Chemother*. 23:918–25
111. Seeberg, A. H., Wiedemann, B. 1984. Transfer of the chromosomal *bla* gene from *Enterobacter cloacae* to *Escherichia coli* by RP4::Mini-Mu. *J. Bacteriol*. 157:89–94
112. Spratt, B. G. 1977. The mechanism of action of mecillinam. *J. Antimicrob. Chemother*. 3(Suppl. B):13–19
113. Sykes, R. B. 1982. The classification and terminology of enzymes that hydrolyze β-lactam antibiotics. *J. Infect. Dis*. 45:762–65
114. Sykes, R. B., Bonner, D. P. 1985. Discovery and development of the monobactams. *Rev. Infect. Dis*. 7(Suppl. 4):S579–604
115. Sykes, R. B., Matthew, M. 1976. The β-lactamases of gram-negative bacteria and their role in resistance to β-lactam antibiotics. *J. Antimicrob. Chemother*. 2:115–57
116. Deleted in proof
117. Tajima, M., Takenouchi, Y., Sugawara, S., Inoue, M., Mitsuhashi, S. 1980. Purification and properties of chromosomally mediated β-lactamase from *Citrobacter freundii* GN7391. *J. Gen. Microbiol*. 121:449–56
118. Takahashi, I., Sawai, T., Ando, T., Yamagishi, S. 1980. Cefoxitin resistance by a chromosomal cephalosporinase in *Escherichia coli*. *J. Antibiot*. 33:1037–42
119. Tausk, F., Evans, M. E., Patterson, L. S., Federspiel, C. F., Stratton, C. W. 1985. Imipenem-induced resistance to antipseudomonal β-lactams in *Pseudomonas aeruginosa*. *Antimicrob. Agents Chemother*. 28:41–45
120. Then, R. L., Angehrn, P. 1982. Trapping of nonhydrolyzable cephalosporins by cephalosporinases in *Enterobacter cloacae* and *Pseudomonas aeruginosa* as a possible resistance mechanism. *Antimicrob. Agents Chemother*. 21:711–17
121. Then, R. L., Angehrn, P. 1986. Multiply resistant mutants of *Enterobacter cloacae* selected by β-lactam antibiotics. *Antimicrob. Agents Chemother*. 30:684–88
122. Tipper, D. J. 1985. Mode of action of β-lactam antibiotics. *Pharmacol. Ther*. 27:1–35
123. Toda, M., Inoue, M., Mitsuhashi, S. 1981. Properties of cephalosporinase from *Proteus morganii*. *J. Antibiot*. 34:1469–75
124. Tomasz, A. 1979. From penicillin-binding proteins to the lysis and death of bacteria: a 1979 view. *Rev. Infect. Dis*. 1:434–67
125. Vu, H., Nikaido, H. 1985. Role of β-lactam hydrolysis in the mechanism of resistance of a β-lactamase–constitutive *Enterobacter cloacae* strain to expanded-spectrum β-lactams. *Antimicrob. Agents Chemother*. 27:393–98
126. Vuthien, H., Rolland, M. 1986. *Citrobacter freundii: in vivo* emergence of a mutant resistant to beta-lactam antibiotics during therapy with ceftazidime. *Presse Med*. 15:1241–42
127. Waterworth, P. M., Emmerson, A. M. 1979. Dissociated resistance among cephalosporins. *Antimicrob. Agents Chemother*. 15:497–503
128. Weinbren, M. J., Perinpanayagan, R. M. 1985. Test for β-lactamase production. *Lancet* 2:673–74
129. Weinstein, R. A. 1985. Occurrence of cefotaxime-resistant *Enterobacter* during therapy of cardiac surgery patients. *Chemioterapia* 4:110–12
130. White, A. J., Curtis, N. A. 1985. Hydrolysis of β-lactamase–stable β-lactams by type I "sponge" β-lactamases. *J. Antimicrob. Chemother*. 16:403–5
131. Yamamoto, T., Murayama, S. Y., Sawai, T. 1983. Cloning and expression of the gene(s) for cephalosporinase production of *Citrobacter freundii*. *Mol. Gen. Genet*. 190:85–91
132. Yoshimura, F., Nikaido, H. 1985. Diffusion of β-lactam antibiotics through the porin channels of *Escherichia coli* K-12. *Antimicrob. Agents Chemother*. 27:84–92
133. Yotsuji, A., Minami, S., Araki, Y., Inoue, M., Mitsuhashi, S. 1982. Inducer activity of beta-lactam antibiotics for the beta-lactamases of *Proteus rettgeri* and *Proteus vulgaris*. *J. Antibiot*. 35:1590–93

Ann. Rev. Microbiol. 1987. 41:595–616

HIGH-RESOLUTION NMR STUDIES OF *SACCHAROMYCES CEREVISIAE*

S. L. Campbell-Burk

Department of Biochemistry, Brandeis University, 415 South Street, Waltham, Massachusetts 02254

R. G. Shulman

Department of Molecular Biophysics and Biochemistry, Yale University, 260 Whitney Avenue, New Haven, Connecticut 06511

CONTENTS

INTRODUCTION

In vivo NMR spectroscopy is now widely used to study cellular physiology and energetics in microorganisms, mammalian cell suspensions, perfused organs, intact animals, and humans. The reader is referred to introductory texts (24, 30) for discussions of the fundamental principles of this technique.

0066-4227/87/1001-0595$02.00

High-resolution NMR spectroscopy not only provides real-time information regarding the environment, localization, and levels of intracellular metabolites, but also does so noninvasively. The ability to study cellular events as they naturally occur provides in vivo information that cannot be obtained otherwise and allows one to set up in vitro conditions that duplicate those found in vivo.

In this review we concentrate on NMR studies performed on yeast and yeast spores that have provided detailed explanations of the regulation of metabolic pathways and cellular bioenergetics. Yeast can be grown to and maintained at the high densities necessary for NMR experiments; yeast cell biology and biochemistry are well characterized; and its cell physiology is easily manipulated.

^{31}P AND ^{13}C NMR

Metabolites can be observed by NMR providing they are mobile and concentrated enough to give narrow spectral lines whose intensities are significantly greater than noise. Generally, small free metabolites (MW $\leq 10^3$) whose levels are in the millimolar range satisfy these requirements. In yeast, phosphorus-containing metabolites such as inorganic phosphate (P_i), adenosinetriphosphate (ATP), polyphosphates (PP), concentrated sugar phosphates, nicotinamide adenine dinucleotide phosphate (NAD), and uridine diphosphoglucose (UDPG) can generally be measured by ^{31}P NMR (65). Distinct signals corresponding to various ^{31}P-containing metabolites can often be resolved because nuclei in different chemical environments experience different fields and consequently produce signals at different chemical shifts or frequencies. The chemical shift or frequency differences between various metabolites are measured in parts per million (ppm) relative to a suitable standard. ^{31}P NMR can also be used to monitor intracellular pH, since the chemical shifts of several phosphorylated metabolites are sensitive to pH in the physiological range (31, 32). Because of the involvement of these compounds in high-energy phosphate metabolism and the fact that intracellular pH can be measured, ^{31}P NMR is widely used to follow the bioenergetic status of the cell.

Changes in intracellular pH and phosphorous metabolite levels can often be measured in several minutes or even less than a minute, since the ^{31}P nucleus has a relatively high intrinsic sensitivity and is 100% naturally abundant. However, ^{31}P NMR peaks of metabolites are dispersed over a chemical shift range of ~40 ppm and are fairly broad. Hence, the extent to which metabolites can be monitored and resolved is limited. Linewidths increase at higher field strengths, so resolution does not improve appreciably with field strength, in contrast to that of ^{13}C and proton peaks (22).

The ^{13}C nucleus has a fourfold lower intrinsic sensitivity than the ^{31}P nucleus and furthermore has a natural abundance of only 1.1%. Detection of ^{13}C resonances in metabolites such as sugars, amino acids, and the glycolytic and Krebs cycle intermediates requires ^{13}C enrichment. Selectively labeled precursors make it possible to trace the path of the label through intermediates and end products as a function of time; this permits measurements of fluxes through biochemical pathways (14–16, 19). Similar but less specific information is obtained from tracer methods in which certain atoms are labeled with radioactive isotopes. ^{13}C NMR has several advantages over radioactive ^{14}C labeling. Firstly, the ^{13}C resonance from each carbon in a molecule is different and is generally distinctly resolved, e.g. all five carbons of glutamate are resolved. Secondly, since the specific carbon label in a molecule can be identified, more information about the biosynthetic path is generally obtained. Thirdly, coupling between the ^{1}H and the ^{13}C nuclei (or sometimes just coupling between neighboring carbons) can reveal the amount of unlabeled metabolite as well as the amount labeled (7, 14). Finally, the ^{13}C measurements are made in vivo, so artifacts arising from extract procedures are avoided. The measurements are also made in real time, with spectra taken every few minutes.

The large chemical shift range (~200 ppm) and relatively sharp lines in the ^{13}C spectra allow resolution of a larger number of metabolites than in ^{31}P NMR. Since the information obtained from ^{31}P and ^{13}C spectra is often complementary, it is useful to perform ^{31}P and ^{13}C NMR measurements on a single sample by alternate scan recording using double-tuned probes (71). The energetic state and the pH can then be monitored simultaneously with metabolism.

HOW TO MAKE A NMR MEASUREMENT AND KEEP CELLS VIABLE

NMR is a relatively insensitive spectroscopic technique. While this is not too serious a limitation when large amounts of material are available (as in studies of purified organic compounds), it is one of the main limitations of the technique in vivo. In order to detect physiological levels of many metabolites, it is necessary to perform NMR experiments on dense suspensions of cells, generally 10–50% wet weight/volume. In a wide-bore NMR instrument, large-diameter (20 mm) sample tubes can be used; thus large sample volumes can be measured, but dense suspensions of cells are nearly always needed. It is not trivial to maintain cells in a well oxygenated physiological state at these high concentrations, but the goal is now routinely met. Gillies et al (32) performed many NMR studies on yeast cell suspensions that were oxygenated using a double bubbler apparatus. One set of bubblers of ~100 μm diameter was situated at the bottom of the cell suspension to help oxygenate the cells and prevent them from settling. A set of larger diameter (~0.5 mm) upper bubblers positioned 1–2 cm above the detection coil provided most of the oxygen without producing undue heterogeneity in the detection region. This apparatus was snugly fit (using a teflon holder) into the NMR tube. Substrates could be supplied to the cell suspension by direct addition to the sample tube (3). Gas delivery rates that allowed proper oxygenation of the cell suspension were determined outside the magnet using an O_2 electrode in an apparatus of dimensions similar to those of the NMR tube.

For systems with slower oxygen consumption, an alternative method for oxygenating cell suspensions is to place the suspension in a chamber mixed by a mechanically driven stirrer while oxygen is supplied to the surface above the cell suspension (38).

In studies of *Saccharomyces cerevisiae* carried out using double bubblers, intracellular pH was maintained by removing volatile CO_2 that accumulated in the cell suspension during metabolism. Other metabolic waste products (such as glycerol and ethanol) are either innocuous or nontoxic until high levels accumulate. Thus under many conditions yeast suspensions remained viable for a few hours. However, accumulation of metabolic end products may alter metabolism and must be monitored. A perfused system is preferred because a constant supply of substrates can be maintained while metabolic waste products are simultaneously removed. Consequently, measurement time can be extended by many hours.

NMR experiments have recently been performed on suspensions of yeast cells that were successfully perfused by embedding in an agarose gel matrix (7, 28). This procedure involves mixing cells with agarose, cooling the mixture, and then extruding the gel suspension through a fine capillary to produce threads of ~500 μm diameter. At this diameter, transport of molecules to the cells is fast enough that oxygenated buffer (containing substrate) supplied to the bottom of the NMR tube keeps them well oxygenated. The level of buffer is maintained by removal of effluent buffer containing metabolic wastes to an external reservoir with the aid of a peristaltic pump. The oxygen content in the influent and effluent buffer can be measured with an electrode. The rate of substrate consumption and/or production of intermediates and end products can be determined by sampling the external reservoir as a function of time, either by NMR or chemical assay.

Another approach is to use a hollow-fiber dialysis system (39). The cell suspension is placed in an airtight sample tube, and nutrient-rich medium is pumped through a network of hollow fibers in the sample. These fibers are permeable and allow exchange of small molecules (MW $\lesssim 10^3$ MW) between the cell sample and perfusate. Although this procedure requires the use of a modified NMR sample tube, it does not present the problems associated with the slower diffusion of nutrients to the cells through agar medium.

HOW TO INTERPRET ^{31}P SPECTRA

Figure 1*A* illustrates the ^{31}P spectrum of an anaerobic suspension of *S. cerevisiae* measured after glucose feeding. The cells were grown aerobically in the presence of acetate and were harvested at mid-log phase. The cells were resuspended in the growth medium (lacking the carbon source, acetate) and in enriched medium [50 mM morpholineethanesulfonic acid (MES) with the pH adjusted to 6] and were kept on ice before use (18).

NMR spectra can be readily interpreted once NMR peaks have been identified and pH-sensitive compounds have been calibrated. Resonances in Figure 1 were assigned with the help of spectra taken of cell extracts. The extract spectrum shown at the bottom of Figure 2 was prepared from a suspension of yeast under similar conditions using conventional perchloric acid techniques. Spectra of extracts and intact cells exhibit large differences in linewidths. In the intact cell spectrum, the linewidths are in the range of 30–50 Hz, giving broad overlapping peaks. In contrast, the much narrower lines in the extract spectrum allow resolution of many individual components.

The strategy for the assignment of resonances is as follows. Resonances

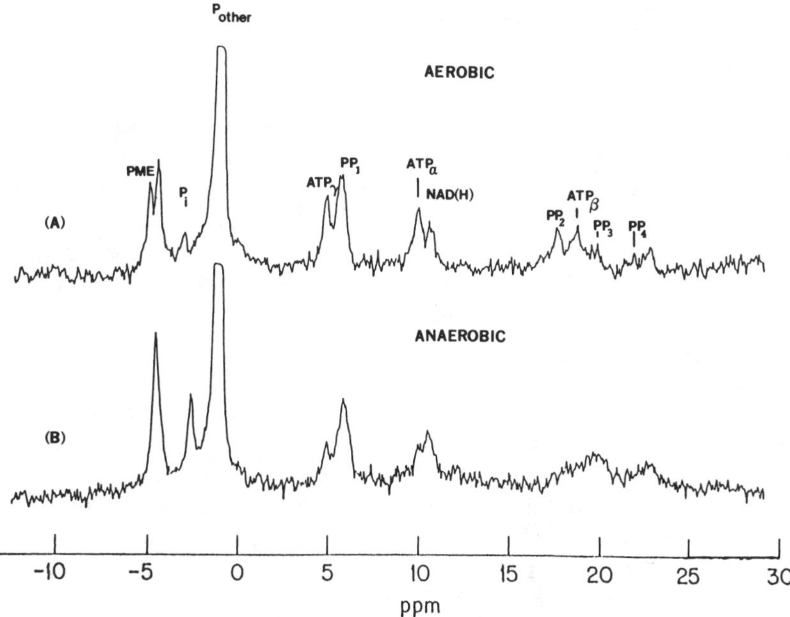

Figure 1 Cells grown with acetate as a carbon source were resuspended in the enriched media at a density of 10% wet weight. Each spectrum is a 16-min accumulation obtained between 10 and 26 min after addition of the carbon source. *A*: Spectrum obtained during aerobic glycolysis after glucose addition. *B*: Spectrum obtained during anaerobic glycolysis after glucose addition.

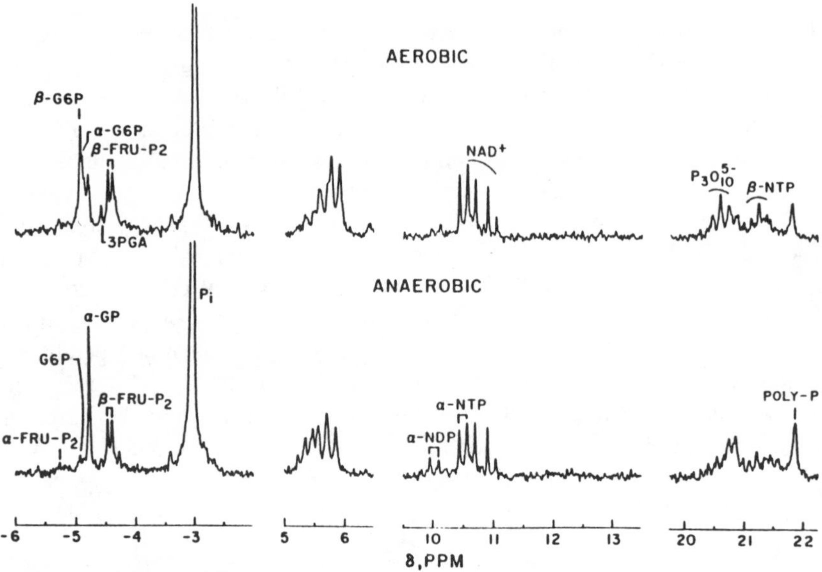

Figure 2 ^{31}P NMR spectra of perchloric acid extracts of *Saccharomyces cerevisiae* obtained during aerobic *(upper)* and anaerobic *(lower)* glycolysis. The cells were grown with acetate as the carbon source, harvested in log phase, and suspended in minimal media prior to extraction.

are tentatively assigned or narrowed down to a few specific metabolites by analyzing chemical shifts and splitting patterns, guided by a general knowledge of the identity and approximate levels of metabolites previously measured in yeast. Peaks are then generally identified in extracts by one or a combination of the following methods: addition of a small amount of the purified compound, titration of the extract, or addition of enzymes(s) that act specifically on the suspected compound.

The spectra of these anaerobic glucose-fed cells show resonances from the α, β, and γ phosphates of ATP and from the terminal, middle, and penultimate phosphates of polyphosphate. Signals are also observed from NAD, UDPG, and P_i. The broad in vivo phosphomonoester (PME) peak is resolved into individual signals from glucose 6-phosphate (G6P) and the α and β anomers of fructose 1,6-bisphosphate (Fru-1,6-P_2) in the extract spectrum.

The pH-sensitive shifts of G6P, Fru-1,6-P_2, and P_i can be used to measure intracellular pH (49). P_i is commonly used to determine intracellular pH since it is better resolved, easily identified, and usually present at detectable levels. At neutral pH, inorganic phosphate exists mainly in the monovalent form ($H_2PO_4^{-1}$, 0.58 ppm) and the divalent form (HPO_4^{2-}, 3.14 ppm, downfield from H_3PO_4). Only one resonance is observed, and it is a weighted average of the chemical shifts of the mono- and divalent forms. Hence, the chemical shift of P_i is determined by the relative amounts of these two species, which are of course pH dependent. Calibration curves have been constructed by plotting the chemical shift of P_i versus pH. The titration curve used to measure intracellular pH should be generated in a medium that reflects the intracellular environment of P_i, since the P_i chemical shift is sensitive to ionic strength (31, 63). Even though it is difficult to determine the intracellular ionic composition exactly, these measurements give accurate values for pH (to better than 0.1 pH unit), since the dependence of P_i chemical shifts upon ionic strength is small at physiological ionic strengths (31).

Yeasts are able to maintain distinct pH environments in vacuoles, mitochondria, and cytosol owing to proton-translocating ATPase associated with the membranes in each of these compartments (56, 66, 79, 82). If the pH difference between any of these compartments is large enough ($\gtrsim$0.3 pH units), phosphate resonances will be resolved. For example, there are two P_i peaks in the in vivo anaerobic spectrum in Figure 1, whereas only one peak is seen in the anaerobic spectrum of the *S. cerevisiae* extract in Figure 2. P_i peaks were assigned to specific compartments by correlating measured pH values and peak intensities to those expected from a given phosphate pool. The higher-field P_i peak was then assigned to P_i originating from the cytosolic pH. This peak (P_i^{in}) may also contain unresolved mitochondrial P_i, although it is uncertain whether mitochondrial P_i is detected by NMR. Intramitochondrial P_i has been observed, however, in isolated mitochondrial suspensions and in hepatocyte suspensions treated with valinomycin (12, 52, 53).

Three distinct P_i peaks were observed in the yeast *Candida utilis;* they were attributed to external, vacuolar, and cytosolic P_i pools (49–51). Nicolay et al (50, 51) made these assignments by observing ^{31}P NMR spectral changes upon separate addition of the protonophore carbonyl *p*-trifluoromethoxyphenylhydrazone (FCCP), the glycolytic inhibitor iodoacetate, and a weak base (NH_4Cl) to these cells.

In asci of the yeast *Pichia pastoris* two P_i peaks were observed. These peaks were assigned to P_i in spore and epiplasmic compartments based on the fact that only one P_i peak was observed for single spores (6).

^{31}P NMR spectra were also obtained for glucose-fed cells under aerobic conditions (upper spectrum, Figure 1). Comparison of anaerobic and aerobic spectra of glucose-fed cells shows distinct differences in the levels and distribution of phosphorus-containing metabolites. In particular, the two differ in their sugar phosphate, P_i, and ATP resonances. The P_i level is much lower with aerobic growth, and the ATP levels are slightly higher. In cell spectra the terminal and α phosphates of ATP and ADP are not resolved. However, the ATP β peak is observed, although it is close to polyphosphate resonances and sometimes obscured. Thus, only the sum of ATP and ADP can be measured from the α and γ peaks of ATP in the intact spectra, but the ATP β peak is an indicator of ATP levels. Differences are also observed between anaerobically and aerobically grown cells in the region of the spectrum (4–5 ppm) containing hexose phosphate and triose phosphate peaks. The clearest difference between aerobic and anearobic spectra is the pH-dependent chemical shift of the intracellular phosphate. This occurs at pH^{in} of 7.5 for aerobically grown cells but at 7.0 for anaerobically grown cells. In addition, the shift occurs at much higher P_i concentrations in the latter.

The extract spectra shown in Figure 2 can be used to measure ATP, ADP, and the various sugar phosphates because the peaks are much better resolved. Hence, it is usually desirable to supplement in vivo measurements of metabolite levels with data from extract spectra. This has been done in the study of glycolytic control, as discussed below.

GLYCOLYTIC CONTROL

NMR studies of glycolytic control in yeast have concentrated on the Pasteur effect (42, 60, 61), which is the decrease in the rate of glucose utilization in respiratory-competent cells transferred from anaerobic to aerobic conditions. The control of enzymatic activities at certain key steps in the glycolytic pathway is believed to be responsible for these small changes in glucose utilization rates. There are three well-established control sites. Glycolytic control at the glucose uptake step is regulated by glucose transport and/or by hexokinase or glucokinase (8, 67). Further down the path, the allosteric enzyme, phosphofructokinase (PFK), catalyzes the phosphorylation of fructose 6-phosphate (F6P) to Fru-1,6-P_2. PFK activity is controlled by a number of allosteric effectors which have been identified by in vitro studies (4, 57, 69). The third known control point is the allosteric enzyme pyruvate kinase, which produces pyruvate from phosphoenolpyruvate (PEP) with the formation of ATP (37).

To study the Pasteur effect in yeast, NMR experiments were conducted under a well-defined set of conditions. Many enzymes are under control of catabolite repression or are inducible (36). For example, when cells are grown in the presence of glucose, levels of enzymes from the tricarboxylic acid cycle, the electron transport chain, and the oxidative phosphorylation pathway are reduced, while glycolytic enzyme levels are elevated. Cells grown under these repressed conditions derive metabolic energy from glycolysis and do not exhibit a Pasteur effect. On the other hand, yeasts grown on nonfermentative substrates such as ethanol or acetate have active respiration and lower levels of glycolytic and gluconeogenic enzymes (29, 45, 58).

Our yeast NMR studies were conducted on acetate-grown derepressed respiratory-competent yeast (15, 17–19). Furthermore, exposure to glucose was kept brief to avoid changes in enzyme levels with time. We investigated the uptake of glucose by measuring the concentration of $[1-{}^{13}C]$glucose with time. V_{max} in the high glucose concentration range under anaerobic conditions was approximately twice the aerobic value (17, 18). A similar relationship between anaerobic and aerobic glucose utilization rates in derepressed yeast cells had been found before by direct assays of glucose levels (44, 70).

Using ^{13}C NMR, den Hollander et al (15) were able to follow the intensities of resolved α and β glucose signals with time. They found that the α anomer was consumed more rapidly than the β anomer. Using a model assuming competition of α and β glucose for the same site, they derived K_m values for uptake of α and β glucose (15).

The enzymatic basis of control of PFK has been studied in detail in a series of NMR experiments. In yeast this enzyme is reported to have the following effectors: ATP, ADP, AMP, P_i, pH, Fru-1,6-P_2, and Fru-2,6-P_2. It is also reported to be sigmoidal with respect to its substrate, F6P. The in vivo metabolic flow through PFK was measured under anaerobic and aerobic conditions in acetate-grown cells (16, 17, 19). The flow is kinetically represented as follows:

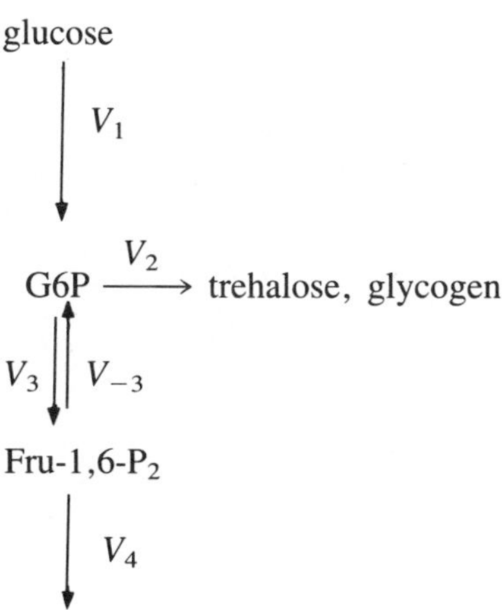

The net glycolytic flow was determined by subtracting the rate of glucose consumption by nonglycolytic pathways (V_2) from the overall glucose utilization rate (V_1). The net glycolytic flux was then corrected for the futile cycling flux through Fru-1,6-P_2ase, i.e. V_{-3}, which was measured as described below. Saturation transfer measurements of the undirectional flow through PFK (V_3), also described below, completed the overall flux determination (11). It is important to note that the two independent rates V_3 and V_{-3} were overdetermined because three measurements were made. As shown below, the results were all consistent.

The overall glucose utilization rate was determined by measuring the time course of [1-^{13}C]glucose consumption using ^{13}C NMR. The rates of nonglycolytic flows such as trehalose and glycogen biosynthesis and the phosphogluconate pathways, which add up to V_2, were evaluated using ^{13}C- and ^{14}C-labeled glucose (17, 19). The ^{13}C-labeled glucose was used to determine amounts of soluble end products such as ethanol, glycerol, and trehalose. The radioisotope ^{14}C data allowed determination of the amounts of the glucose incorporated into glycogen. Flow through the pentose shunt was small and independently evaluated. Hence, it was possible to obtain the difference ($V_1 - V_2$), i.e. the net flow through PFK.

The backward flow through Fru-1,6-P$_2$ase (V_{-3}) was measured by ^{13}C NMR (18). Upon addition of either [1-^{13}C]glucose or [6-^{13}C]glucose to the cells the label distribution in trehalose was used to measure Fru-1,6-P$_2$ase flow (Figure 3). In previous experiments, the label from [6-^{13}C]glucose had been scrambled between the C-1 and C-6 positions of Fru-1,6-P$_2$ by rapid flows through aldolase and triosephosphateisomerase (15). As a result,

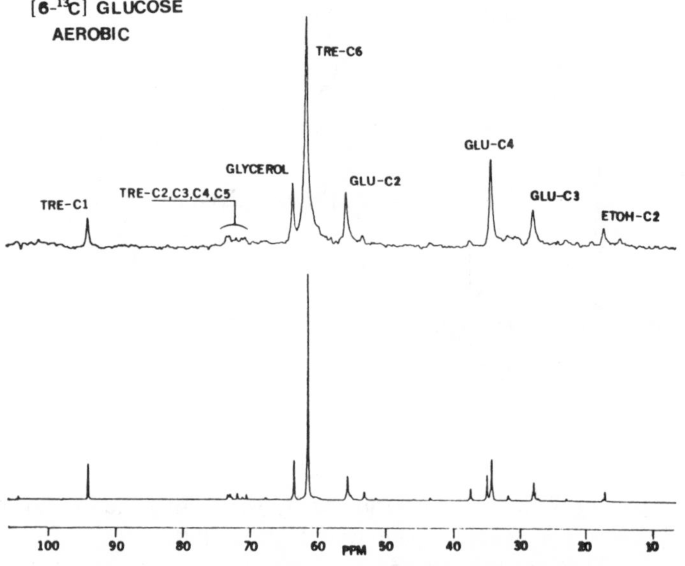

Figure 3 ^{13}C NMR spectra of glucose-grown derepressed cells supplied with [6-^{13}C]glucose under aerobic conditions. The lower trace is the ^{13}C NMR spectrum of an acid extract prepared after aerobic catabolism of [6-^{13}C]glucose. The upper trace is the ^{13}C NMR spectrum obtained approximately 45 min after the addition of [6-^{13}C]glucose. The C-1 : C-6 ratio in trehalose is similar in both the in vivo and extract spectra.

Fru-1,6-P_2 contained ^{13}C at C-1 and C-6, while glucose was only labeled at C-1 (19). We measured the flow through Fru-1,6-P_2ase (V_{-3}) by the amount of C-6 labeling observed in trehalose, since all of the C-6 labeling had to come from Fru-1,6-P_2 (18). The relative amounts of labeling at C-1 and C-6 of trehalose were used to determine V_{-3}/V_1, while V_1 was measured absolutely by ^{13}C NMR. It is interesting to note that futile cycling was extensive ($V_{-3} \approx 20\%\ V_3$) under aerobic conditions but absent under anaerobic conditions. The in vivo rate of flow through PFK was determined from the equation $V_3 = V_1 - V_2 + V_{-3}$, since all three fluxes on the right side of the equation were measured.

Next the concentrations of the eight effectors of PFK, including the substrate F6P, were measured in vivo (17). For six the levels were determined by NMR, while for one they were derived from the NMR data by calculation and for another, Fru-1,6-P_2, they were determined by chemical means (62). It is well established that the response of PFK to any single effector depends on the concentration of other effectors. Therefore, to establish the physiological importance of each effector, in vivo concentrations of all eight effectors of PFK were established in vitro in a partially purified PFK preparation (62). The in vitro rates reproduced the measured in vivo rates of flow through PFK under both aerobic and anaerobic conditions. Then the sensitivity of the PFK rate to individual effectors was investigated by varying the concentration of each effector. These studies demonstrated that the change in PFK activity during the Pasteur effect includes strong and partially compensating contributions from Fru-2,6-P_2 and the substrate F6P. Control by the other six effectors did not change significantly during the switch from anaerobic to aerobic conditions, although some of the concentrations were altered. Hence we not only reproduced the in vivo fluxes in an in vitro experiment, but also identified the controlling effectors that are important during the Pasteur effect.

The accumulation of the PFK product Fru-1,6-P_2 and the extensive scrambling of the ^{13}C label indicate that there is a third control point on the pathway below PFK, presumably at pyruvate kinase.

We feel that these studies, particularly those that have detailed the quantitative understanding of PFK during the Pasteur effect, show that in vivo NMR measurements used in conjunction with other biochemical information have provided knowledge that has quantitatively and qualitatively brought us far beyond the previous state of the field.

CONTROL OF METABOLISM IN YEAST SPORES

When NMR experiments with yeast spores were initiated, the generally accepted explanation of slow metabolism during dormancy was simple: The organism was dry, and therefore its metabolism was slow or nonexistent.

Hydration was considered to be the primary process responsible for the breaking of dormancy (23, 73). However, for the many fungal spores that are not dry the mechanisms responsible for the maintenance and breaking of dormancy had to be more complex (72). The water content of spores showed great variation, and was sometimes higher than in vegetative cells (34, 35). Furthermore, hydration did not always break dormancy. This problem was redirected when ^{31}P NMR spectroscopy showed that the hydration state of yeast ascospores is not significantly different from the hydration state of vegetative yeast cells, since the rotational times for P_i and subsequently for other small molecules were equal in spores and vegetative cells (6). In this particular case the hydration state of the cell did not provide an adequate explanation of the slow metabolism observed during dormancy.

How is dormancy maintained in fungal spores? How is breakdown of the storage products prevented during dormancy? What is the mechanism responsible for the breakdown of storage products when germination is initiated? The answers to these questions are tightly linked. Trehalose is a storage product of most fungal spores and in many species it is the main one (33). During dormancy the trehalose reserve remains fairly stable, while during germination it is usually rapidly depleted. Since protein synthesis and other anabolic processes require energy, Gottlieb (33) argued that no matter what the primary event in breaking of dormancy is it must be translated quickly into an energy-producing event, such as the mobilization of storage products. But what is the role of trehalose? Is it needed during germination to provide energy for the anabolic changes? Or is it stored in order to provide an energy source during dormancy?

^{13}C NMR measurements of exogenous glucose and endogenous trehalose showed that during germination of the yeast ascospore the amount of energy derived from trehalose breakdown was small compared to the amount of energy derived from the medium (5, 76). Hence, for yeast ascospores trehalose breakdown is not the essential energy source for germination, as least not in the rich medium used. It is possible, however, that under different germination conditions trehalose supplies a more substantial fraction of the energy requirements.

Since trehalose was found to be dispensable for energy during germination, the possibility remained that it was needed for metabolism and energy during the dormant period. The metabolism of dormant fungal spores is low but measurable. Hence, any model to explain fungal spore dormancy should be able to explain not only the absence of high metabolic activity but also the presence of low metabolic activity. In vivo NMR experiments (5) showed that in yeast spores the low rate of trehalose consumption is under feedback control by ATP. Deoxyglucose fed to dormant yeast ascospores was shown to be phosphorylated by ATP derived from trehalose breakdown, which during the subsequent glycolysis produced ATP by substrate level phosphorylation.

The amount of deoxyglucose 6-phosphate formed by phosphorylation of deoxyglucose, measured by ^{31}P NMR in vivo, was stoichiometrically consistent with the breakdown of trehalose measured by ^{13}C NMR in the same spores. One trehalose produced two glucoses, each of which produced two ATPs; and indeed, the ratio of trehalose to deoxyglucose 6-phosphate was observed to be 1:4. Hence, the trehalose reserve can be mobilized in response to the metabolic needs of the dormant spore. These experiments indicated that ATP concentration was probably the controlling factor for trehalose mobilization in the dormant spore. This possibility was strengthened by the observation that the ATP concentration in dormant spores (0.2 mM, as measured by in vivo ^{31}P NMR) was the same as the ATP concentration that was shown to inhibit the trehalase extracted from spores (5). Thus the rate of trehalose catabolism during dormancy appears to be controlled by the ATP level via a trehalase feedback control effect. The ATP is clamped at a much lower concentration in spores than in vegetative cells. The low-activity form of trehalase in dormant spores is subject to ATP feedback control, while the activated form found in the germinating spores and vegetative cells is insensitive to ATP and has a much greater V_{max} (76, 81).

Until these NMR experiments, studies of dormancy in fungal spores were mainly descriptive, with few attempts to identify the regulatory metabolic mechanisms involved.

Subsequent work by Thevelein (75) and Uno et al (78) has shown that the trehalase is changed as a result of a cAMP cascade. This agrees with earlier conclusions of van der Plaat (80). They also found that the trehalase changes in yeast are similar to those observed in *Phycomyces* spores after heat activation (75, 78). An excellent review of the regulation of trehalase in fungi has recently appeared (74).

Of more general interest for modern biochemistry is the discovery that the metabolism of yeast spores is profoundly different from that of vegetative cells because of somatic modification of one enzyme, trehalase, by a cAMP cascade. Based upon these NMR results it has not been necessary to postulate any difference in gene expression to describe the metabolic changes observed when the cells differentiate into spores. Just how much of the spores' function can be explained by the changes of trehalase and the consequent lowering of ATP levels from ~3 mM in vegetative cells to ~0.2 mM in ascospores remains to be seen. In some way the reduced levels of ATP may perform the housekeeping functions necessary to keep the spores viable during dormancy while not allowing the rapid metabolism characteristic of vegetative cells.

^{31}P SATURATION TRANSFER STUDIES

^{31}P NMR magnetization transfer techniques have been used to measure steady-state kinetics of certain enzyme-catalyzed reactions in a variety of living systems (2, 41). For yeasts, a particular magnetization transfer technique, termed steady-state saturation transfer (ST), has been used to investigate kinetics of major energy-transducing reactions involving ATP. The basis of the ST technique is that an individual spin will remember a perturbation from equilibrium for a time on the order of its longitudinal relaxation time (T_1) (26, 27, 46, 47). The ST experiment is best illustrated by the case of two species in steady-state exchange, A and B, whose resonances are well resolved.

$$\mathrm{A} \underset{k_{-1}}{\overset{k_1}{\rightleftharpoons}} \mathrm{B}$$

In the above exchange reaction, k_1 and k_{-1} are the first-order rate constants for the interconversion of A and B. Application of an intense radio-frequency field resonant with A will saturate the A resonance and eliminate it from the spectrum. This saturation will be transferred to B and manifested as a reduction in the intensity of B if the conversion of B to A occurs before B relaxes back to equilibrium. Therefore, rate constants can only be measured if they are similar to the longitudinal relaxation rate, ${T_1}^{-1}$. If ${T_1}^{-1}$ is much greater than k, no change in the intensity of B will be observed when A is saturated; if it is much less than k, the B resonance will be obliterated.

The fractional reduction in the intensity of B after the steady state is reached is described by

$$\frac{(M_{\mathrm{B}}^{0} - M_{\mathrm{B}})}{M_{\mathrm{B}}^{0}} = \frac{\Delta M}{M_{\mathrm{B}}^{0}} = \frac{k_{-1}}{k_{-1} + {\mathrm{T}_{1\mathrm{B}}}^{-1}},$$

where M_{B}^{0} is the equilibrium magnetization of B and M_{B} is the magnetization at time T. An independent measurement of $T_{1\mathrm{B}}$ must be obtained before k_{-1} can be determined. Since it is necessary to determine $T_{1\mathrm{B}}$ during exchange between B and A, a specialized T_1 experiment is generally performed to determine the denominator in the right-hand expression above (46). The reverse rate (A → B) can be determined by performing a ST experiment in which the perturbation is applied directly to the B resonance instead of the A. An unusual feature of these experiments is that they measure undirectional flow as opposed to net flow, which is the difference between the forward and reverse rates. One advantage of this kind of measurement is that if forward and reverse rates are individually measured it is possible to determine whether or not an enzyme is operating close to equilibrium in vivo. Unidirectional rates determined by ST are generally derived from experimental measurements of ΔM, but it must be emphasized that they are model dependent. The simple two-state model discussed above is usually used to interpret in vivo results. However, when the species exchange between more than two sites the measurements are more complicated, since additional rate constants must be determined and more experiments are needed. Forsen & Hoffman (27) and Perrin & Johnson (59) discussed such problems for chemical systems, and Ugurbil et al (77) applied the concepts for multiple exchange to in vivo ST results.

Reactions involving $P_i \leftrightarrow$ ATP exchange have been investigated using ^{31}P NMR ST methods with *Saccharomyces cerevisiae* (1, 9, 11). ST experiments were performed by saturating the γ-phosphate of ATP and monitoring the steady reduction in the intracellular intensity of P_i. Interpretation of these experiments gave unidirectional rates of ATP synthesis based upon a model for two-site exchange between ATP and P_i^{in}. For yeast, the reverse rate (ATP $\rightarrow P_i$) could not be measured because T_1 for the ATP γ resonance is too short. Within the simple two-state model it is still possible to have more than one enzymatic reaction for the A $\leftrightarrow$ B exchange. For example, any $P_i \rightarrow$ ATP exchange, such as substrate level phosphorylation or the various ATPase (mitochondrial, plasma membrane, and vacuole) reactions, could give rise to the observed ST effect. To identify the $P_i \rightarrow$ ATP exchange reaction(s) responsible for the observed transfer, specific inhibitors were used and cell growth conditions, feeding, and aeration were varied (11). Results from ST experiments performed on suspensions of *S. cerevisiae* that had been grown to mid-log phase in the presence of acetate indicated that the transfer rates measured only mitochondrial ATPase kinetics (11). The saturation transfer control (M^0) and difference ($M^0 - M$) spectra of aerobically grown ethanol-fed cells are shown in Figure 4*a* and 4*b*, respectively. The P_i^{in} resonance is affected by saturation of the ATP γ peak, as shown by the excursion at 2.8 ppm in the difference spectrum. Unidirectional rates of ATP synthesis were determined from these measurements and compared to O_2-electrode–derived net O_2 consumption rates to obtain an average P:O value of 2.9. The P:O ratio is an index of oxidative phosphorylation; the theoretical value for oxidation of NADH is 3. Since the rates determined by ST are unidirectional and equal to net mitochondrial ATP synthesis rates, the mitochondrial ATPase in these cells appears to be out of equilibrium, operating in the direction of ATP synthesis.

However, the interpretation of ST results is often more complicated, since they depend on yeast growth and feeding conditions. In particular, we have found that P_i-ATP exchange catalyzed by glyceraldehyde 3-phosphate dehydrogenase/phosphoglycerate kinase (GADPH/PGK) will contribute to the observed ST in parallel to contributions from the mitochondrial ATPase under

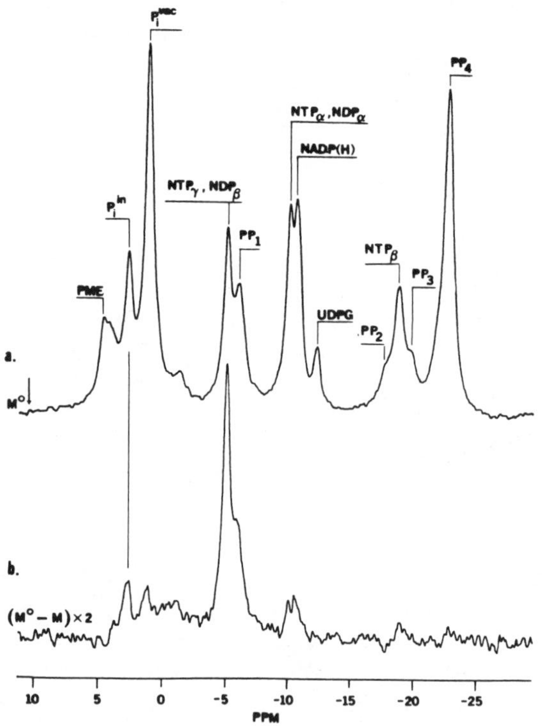

Figure 4 Saturation transfer spectra of aerobic cells supplied with ethanol. Only the M^0 (control) (*a*) and $M^0 - M$ (difference) (*b*) spectra are shown. Spectra were accumulated for 1.5 hr.

conditions where GADPH/PGK activity is high (83, 84). For example, in ST experiments performed on aerobic, glucose-grown, stationary-phase cells supplied with ethanol, a large transfer from γ-ATP to P_i^{in} was observed. This transfer was quantitated and compared to rates of O_2 consumption, and the resultant P:O value was 11. The large P:O value could be explained by GAPDH/PGK contributions to the observed saturation transfer that are not present in acetate-grown cells. In support of this argument, addition of low concentrations of the GAPDH inhibitor iodoacetate to these aerobic, ethanol-fed cells decreased the amount of transfer to P_i^{in}; in this case the P:O value was approximately 3.

Overall, ST results indicate that irreversible synthesis of ATP is catabolized by the mitochondrial ATPase under all conditions examined. Kinetics through GAPDH/PGK were found to depend on metabolic conditions. These measurements illustrate the unique kinds of metabolic information obtained by ST measurements in an in vivo NMR experiment.

^{31}P NMR STUDIES OF PHOSPHATE METABOLISM

The pH-dependent chemical shift of P_i allows phosphates in different pools to be distinguished by ^{31}P NMR. As stated above, under certain conditions it has been possible to monitor levels of P_i and pH associated with the cytosol, vacuole, and external medium simultaneously (50, 51). Moreover, excess phosphate is stored in the yeast vacuole as polyphosphate (PP) (21, 43, 79), and signals from terminal, middle, and penultimate phosphates of PP can be monitored by ^{31}P NMR. Thus, ^{31}P NMR is a useful tool for investigating phosphate metabolism in yeast.

Regulation of P_i utilization during periods of high phosphate demand, such as during nucleic acid synthesis, can be studied by monitoring the time course of P_i and PP. In ^{31}P NMR studies of synchronous *S. cerevisiae* cultures, consumption of phosphate was shown to be rapid during periods of DNA synthesis (32). However, when extracellular P_i (P_i^{ex}) was present in nonlimiting concentrations, transport of P_i^{ex} into the cells occurred rather than mobilization of PP stores. PP degradation was observed only when P_i^{ex} levels were low. Moreover, comparison of PP levels in ^{31}P spectra of cells harvested at exponential and stationary phases of growth showed that rapidly growing cells accumulated less PP than nongrowing stationary-phase cells (18). These results clearly show the utility of PP as a phosphate store.

In ^{31}P NMR studies of aerobic glycolysis in *S. cerevisiae,* levels of cytosolic P_i were shown to be maintained at relatively constant levels by regulation of PP levels. In particular, PP was degraded to vacuolar P_i to replenish levels of cytosolic P_i that had become low owing to phosphorylation of sugars and oxidative phosphorylation. After exhaustion of glucose, excess P_i did not accumulate as cytosolic P_i, but was transported to the vacuole and stored as PP. PP mobilization was also observed during anaerobic fermentation (18, 50).

The possibility that PP is a store for both energy and phosphate has been considered ever since the existence of an enzyme that transfers phosphate from PP to ADP was demonstrated (13, 25). However, the consumption of polyphosphate in vivo is thought to occur by action of polyphosphatase and alkaline phosphatase, which create shorter subunits by sequential hydrolysis, rather than by the production of ATP from PP by polykinase (13, 40). This explanation is consistent with numerous ^{31}P NMR studies of deenergized yeast cells with low levels of ATP and high levels of PP, in which no ATP was regenerated.

Most PP (85%) exists in the form of a single large polymer of ~240 kd. The remaining 15% comprises oligophosphates with an average chain length of 10 phosphate residues (49). Two NMR peaks, one very broad and another narrow, were observed in ^{31}P spectra of yeast. In yeast cell extracts the integrated intensity of the broad resonance corresponded to the fraction with higher molecular weight, while the narrow peak corresponded in intensity to oligophosphate. Consistent with this, the average chain length of oligophosphates in yeast cells has been estimated by ^{31}P NMR analysis of the fraction that gave the narrower peak (49). The relative intensities of the penultimate and middle PP peaks indicated chain lengths of ~13 residues in *S. cerevisiae* (in agreement with the extracted PP) and 20–40 residues in *Candida utilis* (65).

^{23}Na AND ^{39}K NMR STUDIES OF ION TRANSPORT

Intracellular and extracellular pH can be distinguished by ^{31}P NMR, as discussed above. Recently, paramagnetic shift reagents have been used to shift the external ^{23}Na and ^{39}K resonances so that they can be resolved from the resonances of the intracellular ions. The anionic shift reagent used in a series of yeast experiments was $Dy^{3+}(P_3O_{10}^{5-})_2$, which attracted Na^+ and K^+ and shifted their resonances upfield by interactions with the paramagnetic Dy^{3+} (55). NMR measurements of ^{23}Na and ^{39}K first showed that these ions were only 40% visible in NMR spectra because of quadrupolar effects. With this calibration Ogino et al (55) showed that in fresh yeast cells the concentrations of K^+ and Na^+ were ~150 mM and ~2.5 mM, respectively. The H^+ flux was measured by following pH^{in} with ^{31}P NMR. About 1-min resolution was achieved. The H^+ extrusion was correlated with the K^+ uptake. When the cells were deenergized, the K^+ efflux was balanced by H^+ influx with a 1 : 1 stoichiometry, while the Na^+ movements were an order of magnitude smaller. The NMR measurements are very valuable in measuring steady-state ion levels and their changes in response to stimuli.

^{1}H NMR STUDIES OF YEAST

The ^{1}H resonance is intrinsically more sensitive than that of the other nuclei. It has only recently been used to obtain spectra from yeast cells, but for these cells and for intracellular studies in general it has considerable promise. Sillerud et al (68) performed experiments with yeast that had been labeled with ^{13}C by use of enriched acetate. The ^{13}C-labeled lactate was observed by difference NMR spectroscopy of ^{1}H NMR spectra with and without ^{13}C decoupling. The difference spectra had peaks of acetate, glutamate, and aspartate, which showed how the label had been metabolized. The investigators suppressed the intense H_2O peak in these spectra by combining presaturation of H_2O with selective excitation of the metabolite peaks. A more advanced version of this method has recently been used to follow ^{13}C metabolism in the mammalian brain in vivo (64). Other methods of H_2O

suppression and observation of metabolites by ^{1}H NMR in cellular suspensions had previously been reported for *Escherichia coli* (54) and erythrocytes (10).

SUMMARY

High-resolution NMR studies of yeast cells have contributed to our understanding of metabolism and energetics. The above studies of glycolytic control, enzyme kinetics, and metabolism during dormancy have shown how the strengths of NMR investigations can build upon existing knowledge to create a qualitatively different understanding of the processes in yeast.

Literature Cited

1. Alger, J. R., den Hollander, J. A., Shulman, R. G. 1982. In vivo ^{31}P nuclear magnetic resonance saturation transfer studies of adenosinetriphosphate kinetics in *Saccharomyces cerevisiae*. *Biochemistry* 21:2957–63
2. Alger, J. R., Shulman, R. G. 1984. NMR studies of enzymatic rates in vitro and in vivo by magnetization transfer. *Q. Rev. Biophys.* 17:83–124
3. Balaban, R. S., Gadian, D. G., Radda, G. K., Wong, G. G. 1981. An NMR probe for the study of aerobic suspensions of cells and organelles. *Anal. Biochem.* 52:346–49
4. Banuelos, M., Gancedo, C., Gancedo, J. M. 1977. Activation by phosphate of yeast phosphofructokinase. *J. Biol. Chem.* 252:6394–98
5. Barton, J. K., den Hollander, J. A., Hopfield, J. J., Shulman, R. G. 1982. ^{13}C nuclear magnetic resonance study of trehalose mobilization in yeast spores. *J. Bacteriol.* 151:177–85
6. Barton, J. K., den Hollander, J. A., Lee, T. M., McLaughlin, A. C., Shulman, R. G. 1980. Measurement of the internal pH of spores by ^{31}P nuclear magnetic resonance. *Proc. Natl. Acad. Sci. USA* 77:2470–73
7. Baxter, R. L. 1985. Microbiological applications of NMR spectroscopy, part 2. *Microbiol. Sci.* 2:203–211
8. Bisson, L. F., Fraenkel, D. G. 1983. Transport of 6-deoxyglucose in *Saccharomyces cerevisiae*. *J. Bacteriol.* 155:995–1000
9. Brindle, K., Krikler, S. 1985. ^{31}P nuclear magnetic resonance saturation transfer measurements of phosphate consumption in *Saccharomyces cerevisiae*. *Biochim. Biophys. Acta* 847:285–92
10. Brown, F. F., Campbell, I. D., Kuchel, P. W., Rabenstein, D. C. 1977. Human erythrocyte metabolism studied by ^{1}H spin echo NMR. *FEBS Lett.* 82:12–17
11. Campbell, S. L., Jones, K. A., Shulman, R. G. 1985. In vivo ^{31}P nuclear magnetic resonance saturation transfer measurements of phosphate exchange reactions in the yeast *Saccharomyces cerevisiae*. *FEBS Lett.* 193:189–93
12. Cohen, S. M., Ogawa, S., Rottenberg, H., Glynn, P., Yamane, T., et al. 1978. ^{31}P nuclear magnetic resonance studies of isolated rat liver cells. *Nature* 273:554–56
13. Dawes, E. A., Senior, P. J. 1973. The role and regulation of energy reserve polymers in microorganisms. *Adv. Microb. Physiol.* 10:135–43
14. den Hollander, J. A., Behar, K. L., Shulman, R. G. 1981. ^{13}C nuclear magnetic resonance study of transamination during acetate utilization by *Saccharomyces cerevisiae*. *Proc. Natl. Acad. Sci. USA* 78:2693–97
15. den Hollander, J. A., Brown, T. R., Ugurbil, K., Shulman, R. G. 1979. ^{13}C nuclear magnetic resonance studies of anaerobic glycolysis in suspensions of yeast cells. *Proc. Natl. Acad. Sci. USA* 76:6096–100
16. den Hollander, J. A., Shulman, R. G. 1983. ^{13}C NMR studies of in vivo kinetic rates. *Tetrahedron* 39:3529–38
17. den Hollander, J. A., Ugurbil, K., Brown, T. R., Bednar, M., Redfield, C., Shulman, R. G. 1986. Studies of anaerobic and aerobic glycolysis in *Saccharomyces cerevisiae*. *Biochemistry* 25:203–11
18. den Hollander, J. A., Ugurbil, K., Brown, T. R., Shulman, R. G. 1981.

Phosphorous nuclear magnetic resonance studies of the effect of oxygen upon glycolysis in yeast. *Biochemistry* 20:5871–80
19. den Hollander, J. A., Ugurbil, K., Shulman, R. G. 1986. ^{31}P and ^{13}C NMR studies of intermediates of aerobic and anaerobic glycolysis in *Saccharomyces cerevisiae*. *Biochemistry* 25:212–19
20. Deleted in proof
21. Durr, M., Urech, K., Boller, T., Wiemken, A., Schwencke, J., Nagy, M. 1979. Sequestration of arginine by polyphosphate in vacuoles of yeast. *Arch. Microbiol.* 121:169–75
22. Evelhoch, J. L., Ewy, C. S., Siegfried, B. A., Ackerman, J. J. H. 1985. ^{31}P spin-lattice relaxation times and resonance linewidths of rat tissue in vivo: Dependence upon the static magnetic field strength. *Magn. Reson. Med.* 2:410–17
23. Fahey, R. C., Mikolajczyk, S. D., Brody, S. 1978. Correlation of enzymatic activity and thermal resistance with hydration state in ungerminated *Neurospora conidia*. *J. Bacteriol.* 135:868–75
24. Farrar, R., Becker, E. D. 1971. *Pulse and Fourier Transform Nuclear Magnetic Resonance: Introduction to Theory and Methods*. New York: Academic
25. Felter, S., Stahl, A. J. C. 1970. Enzymes du métabolisme des polyphosphates dans la levure. III. Purification et propriétés de la polyphosphate-ADP-phosphotransferase. *Biochimie* 55:245–51
26. Forsen, S., Hoffman, R. A. 1963. Study of moderately rapid chemical exchange reactions by means of nuclear magnetic double resonance. *J. Chem. Phys.* 39:2892–901
27. Forsen, S., Hoffman, R. A. 1964. Exchange rates by nuclear magnetic resonance III: Exchange reactions in systems with several non-equivalent sites. *J. Chem. Phys.* 40:1189–96
28. Foxall, D. L., Cohen, J. S. 1983. Nuclear magnetic resonance studies of perfused cells. *J. Magn. Reson.* 847:285–92
29. Foy, J. J., Bhattacharjee, J. K. 1978. Biosynthesis and regulation of fructose 1,6-biphosphatase and phosphofructokinase in *Saccharomyces cerevisiae* grown in the presence of glucose and gluconeogenic carbon sources. *J. Bacteriol.* 136:647–56
30. Gadian, D. G. 1982. *Nuclear Magnetic Resonance and Its Applications to Living Systems*. Oxford: Clarendon
31. Gillies, R. J., Alger, J. R., den Hollander, J. A., Shulman, R. G. 1982. Intracellular pH measured by NMR: Methods and results. In *Intracellular pH: Its Measurement, Regulation, and Utilization in Cellular Functions*, ed. R. Nuccitelli, D. W. Deamer, pp. 79–104. New York: Liss
32. Gillies, R. J., Ugurbil, K., den Hollander, J. A., Shulman, R. G. 1981. ^{31}P NMR studies of intracellular pH and phosphate metabolism during cell division cycle of *Saccharomyces cerevisiae*. *Proc. Natl. Acad. Sci. USA* 78:2125–29
33. Gottlieb, D. 1976. Carbohydrate metabolism and spore germination. In *The Fungal Spore: Form and Function*, ed. D. J. Weber, W. M. Hess, pp. 141–63. New York: Wiley. 2nd ed.
34. Gottlieb, D. 1978. *The Germination of Fungus Spores*. Shildon, UK: Meadowfield
35. Griffin, D. H. 1981. *Fungal Physiology*. New York: Wiley
36. Hanes, S. D., Bostian, K. A. 1986. Control of cell growth and division in *Saccharomyces cerevisiae*. *CRC Crit. Rev. Biochem.* 21:153–223
37. Hunsley, J. R., Suelter, C. H. 1969. Yeast pyruvate kinase I. Purification and some chemical properties. *J. Biol. Chem.* 244:4815–18
38. Jacobsen, L., Cohen, J. S. 1981. Improved techniques for investigation of cell metabolism by ^{31}P nuclear magnetic resonance spectroscopy. *Biosci. Rep.* 1:141–50
39. Karczmar, G. S., Koretsky, A. P., Bissel, M. J., Klein, M. P., Weiner, M. W. 1983. A device for maintaining viable cells at high densities for nuclear magnetic resonance studies. *J. Magn. Reson.* 53:123–28
40. Katchman, B. J., Fetty, W. O. 1954. Phosphorous metabolism in growing cultures of *Saccharomyces cerevisiae*. *J. Bacteriol.* 69:607–15
41. Koretsky, A. P., Weiner, M. W. 1984. ^{31}P NMR magnetization transfer measurement of phosphorous exchange reactions in vivo. In *Biomedical Magnetic Resonance*, ed. T. L. James, A. R. Margules, pp. 209–30. San Francisco: Radiat. Res. Educ. Found.
42. Krebs, H. A. 1972. The Pasteur effect and the relations between respiration and fermentation. *Essays Biochem.* 8:1
43. Kulaev, I. S. 1975. Biochemistry of inorganic polyphosphates. *Rev. Physiol. Biochem. Pharmacol.* 73:131–207
44. Lynen, F. 1958. Glycolysis in yeast. *Proc. Int. Symp. Enzyme Chem., Tokyo, Kyoto,* pp. 25–34. Tokyo: Marvzen
45. Maitra, P. K., Lobo, Z. 1971. Study of

glycolytic enzyme synthesis in yeast. *J. Biol. Chem.* 246:475–88

46. Mann, B. E. 1977. The application of the Forsen Hoffman spin-saturation transfer method of measuring rates of exchange by means of nuclear magnetic double resonance. *J. Chem. Phys.* 39:2892–901
47. McConnell, H. M. 1958. Reaction rates by nuclear magnetic resonance. *J. Chem. Phys.* 28:430–31
48. Deleted in proof
49. Navon, G., Shulman, R. G., Yamane, T., Eccleshall, T. R., Lam, K. B., et al. 1979. Phosphorus-31 nuclear magnetic resonance studies of wild-type and glycolytic pathway mutants of *Saccharomyces cerevisiae*. *Biochemistry* 18:4487–99
50. Nicolay, K. 1982. *Thesis in NMR Studies of the Bioenergetics and Metabolism of Microorganisms*. PhD thesis. Univ. Groningen, the Netherlands
51. Nicolay, K., Scheffers, W. A., Bruinenberg, P. M., Kaptein, R. 1982. Phosphorus-31 nuclear magnetic resonance studies of intracellular pH, phosphate compartmentation and phosphate transport in yeast. *Arch. Microbiol.* 133:83–89
52. Ogawa, S., Boens, C. C., Lee, T. M. 1981. A ^{31}P nuclear magnetic resonance study of the pH gradient and the inorganic phosphate distribution across the membrane in intact rat liver mitochondria. *Arch. Biochem. Biophys.* 210:740–47
53. Ogawa, S., Rottenberg, H., Brown, T. R., Shulman, R. G., Castillo, C. L., Glynn, P. 1978. High resolution ^{31}P NMR studies of rat liver mitochondria. *Proc. Natl. Acad. Sci. USA* 75:1796–1800
54. Ogino, T., Arata, Y., Fujiwara, S. 1980. Proton correlation nuclear magnetic resonance study of metabolic regulations and pyruvate transport in anaerobic *Escherichia coli* cells. *Biochemistry* 19:3684–91
55. Ogino, T., den Hollander, J. A., Shulman, R. G. 1983. ^{39}K, ^{23}Na and ^{31}P NMR studies of ion transport in *S. cerevisiae*. *Proc. Natl. Acad. Sci. USA* 80:5185–89
56. Oshumi, Y., Anraku, Y. 1981. Active transport of basic amino acids driven by a proton motive force in vacuolar membrane vesicles of *Saccharomyces cerevisiae*. *J. Biol. Chem.* 256:2079–82
57. Passonaeau, J. V., Lowry, O. 1962. Phosphofructokinase and the Pasteur effect. *Biochem. Biophys. Res. Commun.* 7:10–15
58. Perlman, P. S., Mahler, H. R. 1974. Derepression of mitochondria and their enzymes in yeast. *Arch. Biochem. Biophys.* 162:248–71
59. Perrin, C. L., Johnson, E. R. 1979. Saturation transfer studies of three-site exchange kinetics. *J. Magn. Reson.* 33:619–26
60. Racker, E. 1974. History of the Pasteur effect and its pathobiology. *Mol. Cell. Biochem.* 5:17–23
61. Ramaiah, A. 1974. PFK and the control of glycolysis. *Curr. Top. Cell. Regul.* 8:297–345
62. Reibstein, D., den Hollander, J. A., Pilkis, S. J., Shulman, R. G. 1986. Studies on the regulation of yeast phosphfructo-1-kinase: Its role in aerobic and anaerobic glycolysis. *Biochemistry* 25:219–27
63. Roberts, J. K. M., Wade-Jardetsky, N., Jardetsky, O. 1981. Intracellular pH measurements by ^{31}P nuclear magnetic resonance: Influence of factors other than pH on ^{31}P chemical shifts. *Biochemistry* 20:5389–94
64. Rothman, D. L., Behar, K. L., Hetherington, H. P., den Hollander, J. A., Bendall, M. R., et al. 1985. ^{1}H-observe/^{13}C-decouple spectroscopic measurements of lactate and glutamate in the rat brain *in vivo*. *Proc. Natl. Acad. Sci. USA* 82:1633–37
65. Salhany, J. M., Yamane, T., Shulman, R. G., Ogawa, S. 1975. High resolution ^{31}P NMR studies of intact yeast cells. *Proc. Natl. Acad. Sci. USA* 72:4966–970
66. Serrano, R. 1978. Characterization of the plasma membrane ATPase of *Saccharomyces cerevisiae*. *Mol. Cell. Biochem.* 22:51–63
67. Serrano, R., DelaFuente, D. G. 1974. Regulatory properties of the constitutive hexose transport in *Saccharomyces cerevisiae*. *Mol. Cell. Biochem.* 5:161–71
68. Sillerud, L. O., Alger, J. R., Shulman, R. G. 1982. High resolution proton NMR studies of intracellular metabolites in yeast using ^{13}C decoupling. *J. Magn. Reson.* 45:142–50
69. Sols, A. 1976. The Pasteur effect in the allosteric era. In *Reflections in Biochemistry*, ed. A. Kornberg, B. L. Horecker, L. Cornudella, J. Oro, pp. 199–206. Oxford: Pergamon
70. Stickland, L. H. 1956. The Pasteur effect in normal yeast and its inhibition by various agents. *Biochem. J.* 64:503–15
71. Styles, P., Grathwohl, C., Brown, F. F. 1979. Simultaneous multinuclear nuclear magnetic resonance by alternate

scan recording of ^{31}P and ^{13}C spectra. *J. Magn. Reson.* 35:329–36

72. Sussman, A. S. 1966. *The Fungi: An Advanced Treatise,* ed. G. C. Ainsworth, A. S. Sussman, 12:733–64. London: Academic
73. Sussman, A. S. 1969. The prevalence and role of dormancy. In *The Bacterial Spore,* ed. G. W. Gould, A. Hurst, 1:1–38. London: Academic
74. Thevelein, J. M. 1984. Regulation of trehalose mobilization in fungi. *Microbiol. Rev.* 48:42–59
75. Thevelein, J. M. 1984. Cyclic-AMP content and trehalase activation in vegetative cells and ascospores of yeast. *Arch. Microbiol.* 138:64–67
76. Thevelein, J. M., den Hollander, J. A., Shulman, R. G. 1982. Changes in the activity and properties of trehalase during early germination of yeast ascospores: correlation with trehalase breakdown as studied by in vivo ^{13}C nuclear magnetic resonance. *Proc. Natl. Acad. Sci. USA* 79:3503–7
77. Ugurbil, K., Petein, M., Maidan, R., Michurski, S., From, A. H. L. 1986. Measurement of an individual rate constant in the presence of multiple exchanges: application to the myocardial creatine kinase reaction. *Biochemistry* 25:100–7
78. Uno, I., Matsumoto, K., Adachi, K., Ishikawa, T. 1983. Genetic and biochemical evidence that trehalose is a substrate of CAMP-dependent protein kinase in yeast. *J. Biol. Chem.* 258:10867–72
79. Urech, K., Durr, M., Boller, T., Wiemken, A., Schwencke, J. 1978. Localization of polyphosphate in vacuoles of *Saccharomyces cerevisiae. Arch. Microbiol.* 116:275–78
80. van der Plaat, J. B. 1974. Cyclic 3',5'-adenosine monophosphate stimulates trehalose degradation in bakers yeast. *Biochem. Biophys. Res. Commun.* 56:580–87
81. Von Assche, J. A., Carlier, A. R. 1975. Some properties of trehalose from *Phycomyces blakesleeanus. Biochim. Biophys. Acta* 391:154–61
82. Willsky, G. R. 1979. Characterization of the plasma membrane Mg-ATPase from the yeast *Saccharomyces cerevisiae. J. Biol. Chem.* 254:3326–32

REFERENCES ADDED IN PROOF

83. Campbell-Burk, S. L., den Hollander, J. A., Alger, J. R., Shulman, R. G. 1987. ^{31}P saturation transfer and ^{13}C NMR kinetic studies of glycolytic regulation during anaerobic and aerobic glycolysis. *Biochemistry* In press
84. Campbell-Burk, S. L., Jones, K. A., Shulman, R. G. 1987. ^{31}P NMR saturation transfer measurements in *Saccharomyces cerevisiae:* characterization of phosphate exchange reactions by iodoacetate and antimycin A inhibition. *Biochemistry* In press

Ann. Rev. Microbiol. 1987. 41:617–43

MICROBIAL ECOLOGY OF THE COCKROACH GUT

D. L. Cruden and A. J. Markovetz

Department of Microbiology, University of Iowa, Iowa City, Iowa 52242

CONTENTS

INTRODUCTION

The alimentary tract in diverse animal species provides a potential environment for the development of microbial communities. In general, some sections of the tract are more amenable than others to the development of these microbial populations. The best understood microbial gut ecosystem is that of the domestic ruminants and their rumen microbiota (66). Other mammalian systems have been the subject of increased interest from an ecological standpoint [see reviews by Savage (112, 113)]. Work on the gut ecology of insects, with the exception of termites (25), has been fragmentary.

0066-4227/87/1001-0617$02.00

Over 100 different bacteria have been isolated from or passed through cockroaches (100, 109). Many of these bacteria were isolated from the intestine or from feces and thus were regarded as part of the normal gut flora. Such conclusions are open to question since none of these investigations were sufficiently detailed or documented as to the number of times each organism was isolated from different animals, what section of the gut was the source of material, and how many of each type of organism were present. However, these early reports at least serve as a base of information for a more extensive study of the microbial ecology of the gut.

Some 3500 species of cockroaches have been described (102), so our usage of the general term "cockroach" for all such insects is misleading. For the most part, this review is concerned with two cockroaches, *Periplaneta americana* (American cockroach), which belongs to the family Blattidae, and *Eublaberus posticus,* which belongs to the family Blaberidae. Both species are large; adult *P. americana* weigh 1.5–2.5 g, and *E. posticus* weigh 2–4.5 g. Wood-eating cockroaches such as members of the genus *Cryptocercus* are more similar to termites in their physiology and gut microbiota and are not considered in detail.

BIOLOGY OF THE HOST

As a group, cockroaches are distributed over the major land masses of the globe. *P. americana* apparently had its origins in Africa, and through commerce it spread virtually worldwide. This species' preference for warm, moist climates has assured a continued association with humans. Although we tend to associate these insects with indoor places where food is prepared and stored, *P. americana* has also been observed in decaying trees, plants, caves, mines, sewers, latrines, and dumps (59, 106).

The economic impact of cockroach infestations is difficult to assess. Although not considered a major crop pest, *P. americana* has been reported to eat the roots and blossoms of greenhouse plants, particularly orchids (109). Materials eaten or damaged by these opportunistic feeders include paper, bread, fruit, fish, putrid sake, cloth, hides, and hair (106). Much of the food damage occurs by the characteristic cockroach odor imparted to foodstuffs.

The presence of these insects may cause considerable depreciation of property values (59). Generally, numbers vary with the economic status of the property dwellers. In a recent census, an average of 26,400 cockroaches were found per apartment unit in low-cost housing (70).

From a medical standpoint, any insect living in close association with humans represents a potential health problem. Cockroaches do not appear to be major primary vectors for infectious diseases (59). However, because they can feed on human and animal fecal material, they are potential mechanical vectors, able to contaminate food for human consumption with pathogenic microorganisms (108). Cockroach allergens may cause symptoms of hay fever, asthma, and dermatitis in susceptible individuals (107).

Cockroaches are considered omnivorous. Cochran (36) has indicated that such general feeders ingest detritus; this fact is often not appreciated, since most references describe the common pest species of cockroaches, which usually feed on substances associated with humans. Actually, less than 1% of all cockroach species are associated with humans.

Bignell (14) tabulated a variety of digestive enzymes found in *P. americana,* e.g. amylases, invertase, maltase, β-glucosidase, cellulase, esterases, lipases, proteases, and phosphatases. Most are associated with the salivary glands, midgut, and midgut ceca. The cellulase in the midgut but not that in the salivary glands is reported to be of microbial origin (131). Chitinase is distributed throughout the gut and is not of microbial origin (129, 130).

It has been particularly difficult to determine specific nutritional requirements for cockroaches, partly because of the presence of microbial symbiotes (bacteroids) in specialized cells (mycetocytes) in the fat body. Elimination of symbiotes from *P. americana* by antibiotic treatment results in poor growth and diminished reproductive capacity (103). Aposymbiotic *P. americana* are unable to biosynthesize methionine and cysteine from sulfate (60). Such roaches also have reduced levels of ascorbic, folic, and pantothenic acids in the fat body (72). Strong evidence implicates these bacteroids in mobilization of nitrogen from urate deposits in the fat body (29, 83). Inability to cultivate these symbiotes experimentally has left their biochemical function equivocal.

BACTERIA ASSOCIATED WITH GUT REGIONS

The microbes in any gut system have been described by in situ observation and by isolation from the animal. Using media high in sugars and amino acids, incubated aerobically, almost any organism of clinical importance can at one time or another be isolated from the cockroach (47, 59, 106). The definition of autochthonous gut microorganisms (46, modified in 1, 111), the development of more sophisticated methods of culturing strict anaerobes (2, 63), and the beginnings of the development of methods for locating specific types of microbes in situ have made it possible to begin to determine whether microbes are in fact important to the biology of the host.

Foregut

The foregut of most cockroaches comprises close to 50% of the total gut volume (13). Transit time depends on the nature of the food, but some portions of a meal may remain in the foregut for 100 hr (119). Digestive enzymes are introduced into the mass of food in the crop from the salivary glands, and may also reflux back from the midgut (14). The foregut is lined with a thick chitinous wall (22, 85). Spiny projections are flat in the anterior part of the crop and project into the lumen in posterior portions. The surface of the gizzard, just anterior to the midgut, is flat and molarlike. Peristalsis keeps the food in the crop in motion, and the spines aid in mixing.

The chitinous wall of the foregut is less differentiated than that of the exoskeleton (85). The foregut wall is much less permeable to water, ions, and organic compounds than that of the hindgut (74), although Bignell (13) and Hoffman & Downer (61) have presented evidence that lipids may be transported across the foregut wall.

The pH of the foregut is lower and more variable than that in other regions of the gut. Values can be as low as 5.4 (58); variation is partly dependent on the nature of the food.

Few bacteria have been observed attached to the wall of the foregut (14, 22, 85). Those observed by light or electron microscopy appear to be associated with the ingested food. On a volume basis, fewer bacteria have been isolated from the foregut than from other regions of the gut (13). House (65) reported that isolates from the foregut were more qualitatively varied than those from other parts of the gut.

Conditions in the foregut are not unfavorable for microbial activities, and there has been speculation that the variation in pH in the crop could be due to variation in fermentative activity of bacteria ingested with the food (132). However, there have been no reported attempts to count bacteria in food before and after a residence in the crop, although substantial multiplication of organisms should be possible during long transit times.

Midgut

The midgut in the cockroach is a long, straight tube lined with a microvillus brush border. Eight digestive cecae with similar cytology open into the midgut at the anterior end. The insect actively secretes a variety of digestive enzymes into the midgut, and end products are transported back across the midgut epithelium. The food bolus is separated from the epithelium by the peritrophic membrane, a cylindrical structure that is continuously secreted at the anterior end of the midgut. The peritrophic membrane forms a tube that extends the length of the midgut and into the hindgut, where it is apparently shredded by the chitinous spines projecting from the hindgut wall. The peritrophic membrane consists of chitin, protein, and acidic mucopolysaccharide. It forms a porous lattice with openings about 0.15 μm in diameter, which enzymes and other macromolecules, but not bacteria, can pass through (104). The function of the peritrophic membrane is unknown, but the most frequent speculation is that it protects the absorptive surfaces of the insect midgut in the absence of mucous secretions (104).

The pH of the midgut is higher and more stable than that of the crop, partly because of the buffering capacity of the secretions from the Malpighian tubules, which include ammonia, amino acids, K^+, and other ions (73). Midgut pH is 6.6 in *E. posticus* (43) and 6.5–6.7 in *P. americana* (58). These values differ significantly from the high pH (9–>11) found in the pre-cellulolytic midguts of xylophagous insects such as the scarab beetle (6) and cranefly larva (77), or the low pH (2–4) of the midguts of some fly maggots, which is correlated with the digestion of bacteria as the main food source (57). The redox potential is moderately oxidizing, as it is in the midgut of termites (15, 127).

Significant numbers of bacteria have been isolated from the midgut contents of cockroaches. However, no studies have distinguished between organisms located within the tube formed by the peritrophic membrane and the food, and those between the peritrophic membrane and the wall. Bignell (13) reported 7×10^7–10^8 organisms ml^{-1} from the midgut of *Periplaneta* in dilution series on nutrient agar and blood agar, with significantly lower numbers in thioglycollate medium. Cruden & Markovetz (43), using rumen fluid containing medium, obtained counts of approximately 10^8 cells per midgut when samples were incubated aerobically and 3×10^8 cells per midgut when samples were incubated anaerobically. Dilution series on medium with carboxymethyl cellulose as carbon source (40) indicated that the midgut of both *E. posticus* and *P. americana* contained about 10^6 bacteria that could grow aerobically and 10^7 that could grow anaerobically. The most common isolates from midgut contents of both species are the facultatively anaerobic *Enterobacter agglomerans* and *Klebsiella oxytoca* (40; D. L. Cruden, unpublished data). These organisms, along with *Citrobacter freundii,* are the most common facultative isolates from cockroach hindgut contents as well (see below). Ulrich et al (125) isolated only facultatively anaerobic microorganisms, including *Klebsiella* and *Citrobacter,* from the midgut of the cricket.

Electron microscopic examination of the midgut has indicated that a variety of insects harbor a population of bacteria between the peritrophic membrane and the midgut epithelium. Scanning electron micrographs (20) show filamentous microorganisms apparently attached to the peritrophic membrane as well as protruding from between the microvilli of the midgut epithelium of *P. americana*. Occasional protozoa and spirochetes are also present. Spirochetes as well as cuboidal spore-forming bacteria are attached to the midgut epithelium in lower termites (27). Bignell et al (17, 19) have reported the presence of filamentous organisms (possibly actinomycetes) between the peritrophic membrane and midgut wall. Ulrich et al (125) observed bacteria associated with the lumen side of the peritrophic membrane in crickets, but failed to separate the membrane from the rest of the midgut to isolate the organisms. No isolations of these microorganisms have been reported and nothing is known of their metabolic activities. The cells of the microvillus brush border turn over rapidly (14), but the peritrophic membrane may protect microorganisms that enter the space (probably from the hindgut) from being washed out, and enable them to persist and multiply.

Hindgut

The hindgut of the cockroach is approximately as long as the midgut but considerably wider, after a short muscular ileum region where the gut wall again becomes chitinous. In earlier publications we have referred to the saclike anterior two thirds of the hindgut as the paunch, to distinguish it from the posterior black band region discussed below (42, 43). The microbial activity in the cockroach hindgut is considerable (see below). The remnants of the peritrophic membrane channel substances entering from the midgut into the central lumen of the hindgut, from which they can slowly move back to a relatively stable and protected environment along the wall (13).

The hindgut wall is lined with a chitinous cuticle that varies in thickness from 1 to 2 μm in the paunch region. Numerous chitinous spines up to 200 μm long project into the lumen. Analysis of the insect tissue adjacent to the hindgut wall indicated that transport across the chitinous layer is possible (14). Ions, fatty acids, and amino acids (74) and products of bacterial fermentations (24, 62) are rapidly transported into the hemolymph.

The pH of the hindgut is higher than that of the crop or midgut, e.g. 6.4–6.7 in *P. americana* (58) and 7.5 in *E. posticus* (43). Hindgut contents are generally semisolid and opaque, while those of the midgut are less viscous. The hindgut, like that of most termites (15, 127), is anaerobic (59), despite the presence of mitochondria in the insect tissue close to the gut wall.

The anaerobic environment of the hindgut is reflected in the bacteria isolated. Dilution series of rumen fluid–containing media indicated that 10^8 bacteria per *E. posticus* hindgut grow aerobically, while over 10^{10} grow anaerobically (43). Similar results were obtained with *P. americana* (20). Bignell (13) found fewer from *P. americana* growing anaerobically, but he used only thioglycollate medium and nutrient agar for the anaerobic counts.

The most common facultative isolates on a variety of selective and non-selective media are *K. oxytoca, E. agglomerans,* and *C. freundii* (10, 40, 43). *Serratia* and *Streptococcus* species are also frequently isolated (43). Members of one or more of the above species appeared to be present in the hindgut at levels of at least 10^6 per hindgut when dilution series were made in media selective for nitrogen fixing bacteria, chitin degradation, uric acid degradation, and carboxymethyl cellulose degradation (40, 43; D. L. Cruden, unpublished results). In addition, these species all grow on a variety of sugars, sugar alcohols, and amino acids. When the anaerobic microbiota was eliminated by feeding the insects metronidazole, these species became the predominant inhabitants of the hindgut (D. L. Cruden, unpublished results).

These common facultatively anaerobic isolates from the hindgut are similar to those present in large numbers in the cricket (125), crane fly larvae (69), and lower termites (116). We feel they represent a population of adaptable generalists and may scavenge oxygen that diffuses in through the hindgut wall. The removal of almost all of the attached microbiota by administration of metronidazole (21) suggests that these facultative anaerobes are not the main wall-associated organisms, however.

Anaerobic bacteria that have been isolated consistently from cockroaches are more numerous and varied. Isolates from *P. americana* and *E. posticus* are listed in Tables 1 and 2, respectively. These species have been isolated by purification of the most common colony types on nonspecific media or after selection for the ability to degrade chitin or carboxymethyl cellulose. Methanogenic isolates are discussed below. The number of clostridia isolated correlates well with the frequency of sporulating bacteria seen in electron micrographs of the hindgut (42; see below). *Clostridium sporogenes, Clostridium beijerincki,* and *Clostridium bifermentans* have also been reported as part of the normal microbiota in human and mammalian intestinal tracts (63, 64). The most common *Fusobacterium* species isolated is *Fusobacterium necrophorum,* although both *Fusobacterium varium* and *Fusobacterium gonidiaformans* occur. A phage active on a *F. varium* strain from *E. posticus* has been isolated (51), and phage have been observed in electron micrographs of the hindgut of *E. posticus* (39). Fusobacteria are normal inhabitants of both vertebrate and invertebrate intestines, and *F. varium* has been isolated from termites and the oriental cockroach (63, 64).

Table 1 Strictly anaerobic bacteria isolated from the hindgut of *Periplaneta americana*

Consistently isolated[a]	Frequently isolated[b]
Clostridium sporogenes	*Acidaminococcus fermentans*
Fusobacterium varium	*Propionibacterium avidum*
Eubacterium moniliforme	*Bifidobacterium* sp.
Peptococcus variabilis	*Clostridium bifermentans*
Peptostreptococcus productus	*Lactobacillus* sp.
Bacteroides sp.[c]	*Butyrivibrio* sp.
	Coprococcus sp.
	Ruminococcus sp.

[a] Bacteria isolated from the hindgut of every individual dissected over a 2-yr period (20).

[b] Organisms isolated frequently, but not from every individual dissected.

[c] One or more of these species: *B. fragilis, B. putredinis, B. praeacutus, B. ochraceus.*

Table 2 Strictly anaerobic bacteria frequently isolated from the hindgut of *Eublaberus posticus*[a]

Bacteroides praeacutus[b]	*Fusobacterium varium*
Clostridium sporogenes	*Fusobacterium gonidiaformis*[b]
Clostridium carnis	*Fusobacterium necrophorum*
Clostridium bifermentans	*Peptococcus variabilis*
Clostridium mangenoti[b]	*Peptostreptococcus productus*
Clostridium beijerinckii	*Peptostreptococcus anaerobius*
Eubacterium aerofaciens[b]	*Propionibacterium avidum*[b]
Eubacterium tenue	*Propionibacterium freudenreichii*
Eubacterium limosum	*Streptococcus constellatus*
Eubacterium cellulosolvens	*Streptococcus intermedius*
Eubacterium moniliforme	

[a] From References 40, 43, 50 and D. L. Cruden, unpublished data.
[b] Isolated mainly or only from vortexed and washed hindgut wall preparations.

Many isolates from the hindgut cannot rapidly be identified to species or on occasion even to genus, especially when nonconventional enrichments are used. Examples are a *Clostridium* stimulated by inorganic pyrophosphate (38) and a thick-walled organism isolated after enrichment in spent medium (41). Other organisms have been carried in mixed culture for extended periods, often firmly attached to pieces of chitin, but have not been isolated (D. L. Cruden, unpublished results).

Careful electron microscopy can lead to important insights that are applicable to other intestinal systems as well. Microscopic examination reveals that the cockroach hindgut has a dense and extremely varied microbiota (13, 22, 52; Figures 1–3). As in other insect gut systems (5, 7, 27, 125), many of the bacteria are firmly associated with the hindgut wall and resist removal by washing.

The attached population almost entirely disappears when the cockroach is fed metronidazole, which acts against anaerobic bacteria (21). Attachment to the wall appears to be mediated by material that stains with ruthenium red (39; Figure 2). Some rods attached to the chitinous spines of the hindgut wall appear to divide longitudinally, which would insure continued attachment. Other bacteria are firmly attached to bacterial filaments (Figure 1), increasing the weblike array. Many of the morphotypes are similar to those seen in other insect gut systems. Bignell et al (16) have isolated actinomycetelike filamentous bacteria from the chitinous spines of a higher, soil-feeding termite. A lobed coccoid organism (21) resembles one seen in crane fly larvae (69). A darkly stained filamentous rod, which is one of the most common wall-associated organisms in the paunch of both *E. posticus* and *P. americana,* resembles organisms in the hindgut of millipedes (115). A flagellated rod resembles one of the common morphotypes described in a lower termite (27).

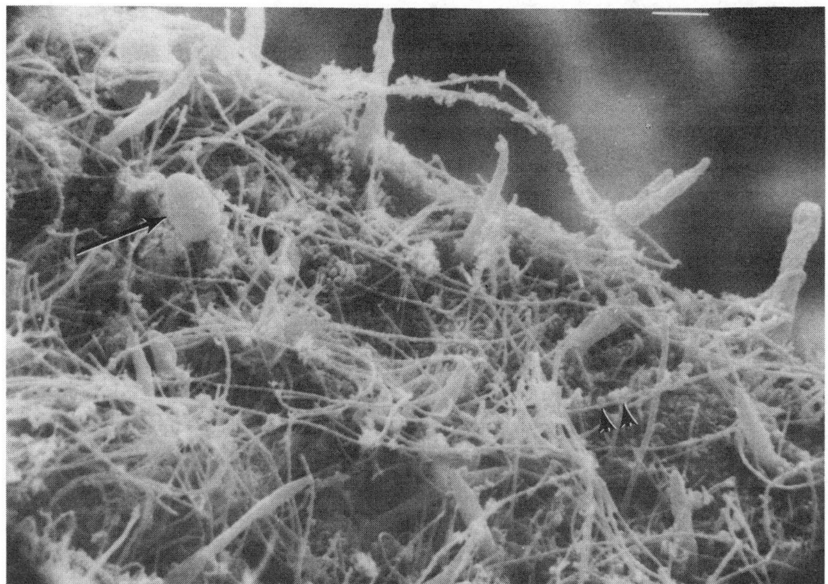

Figure 1 Scanning electron micrograph of paunch region of *Eublaberus posticus* that had been washed to remove all but the most firmly attached microbes. Chitinous spines protrude from the hindgut wall. Bacterial rods and filaments are attached in a dense network to the spines, as well as to the hindgut wall. Coccoid bacteria are attached to bacterial filaments *(arrowheads)*, and a protozoan is present *(arrow)*. *Bar* = 5 μm.

Statistical analysis of the distribution of wall-associated microbes in scanning electron micrographs of the rumen (78) and gut of a higher termite (18) indicated that there are characteristic patterns in the density of colonizing bacteria. This fact, combined with the added resolution and information provided by transmission electron microscopy, enabled us to perform a statistical analysis of the distribution of 14 morphotypes in three wall-associated regions of the hindgut as well as in the lumen (42). All but one of the 14 morphotypes have statistically significant distribution patterns among the various regions. Five morphotypes, when present, constitute a large proportion of the cells (up to 41%) in any micrograph. Six morphotypes rarely make up over 2% of the population, but are consistently present in the region of concentration, with relatively low variance. The remaining three morphotypes are only sporadically present, but where they occur they often constitute 5–10% of the population. Their numbers vary most from animal to animal, but also vary among micrographs from a given animal.

Protozoa (including all types), one of the counted morphotypes, are more often found in the lumen than associated with the wall. In the lumen they

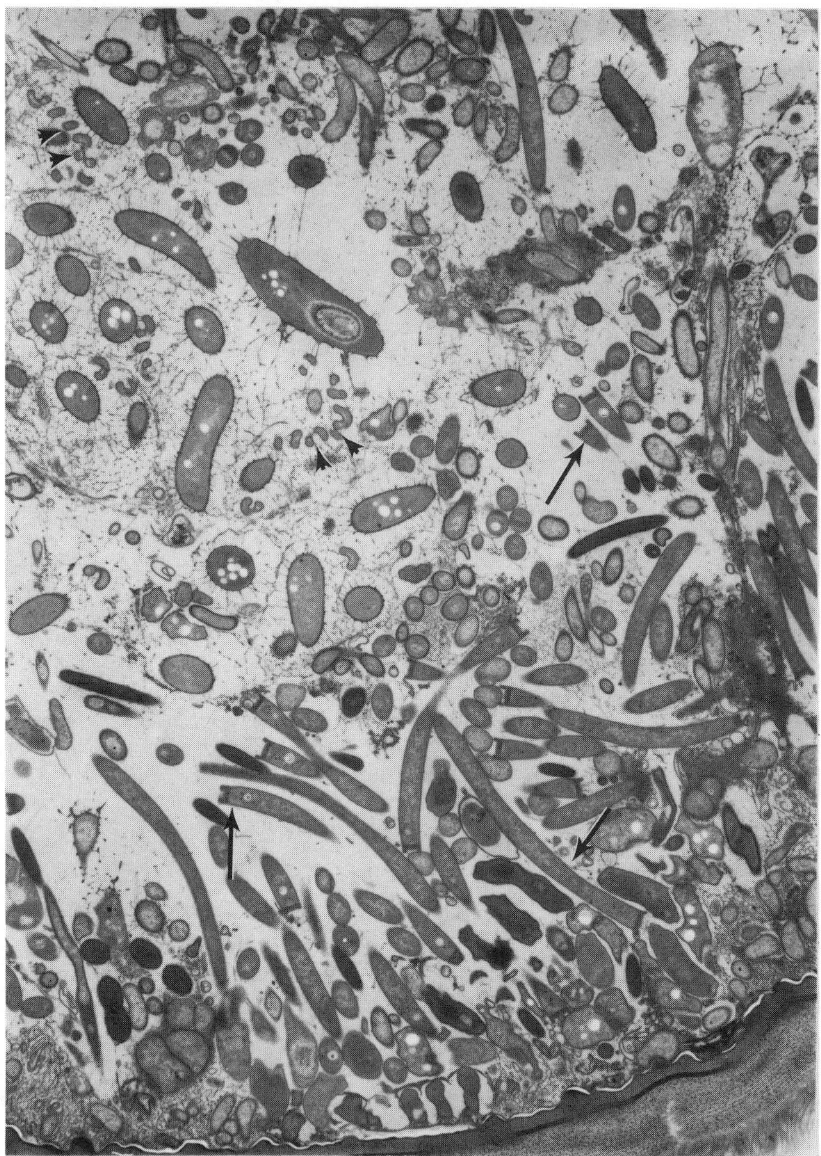

Figure 2 Electron micrograph of the wall-associated microbiota of the anterior paunch of *E. posticus*. The chitinous hindgut wall, which is approximately 2 μm wide, is at the bottom of the micrograph; the epicuticle and laminar endocuticle are visible. This micrograph was prepared with ruthenium red, which stains the characteristic mucopolysaccharide glycocalyces that surround the bacteria in vivo (compare with Figure 3, in which ruthenium red was not used). Bacteria with the distinctive ultrastructure of *Methanospirillum* are present *(arrows)*, as are several types with intracellular inclusions, and a common spore-forming morphotype. Several organisms in the form of tightly coiled helices are also present *(arrowheads*, slightly above and left of center*)*.

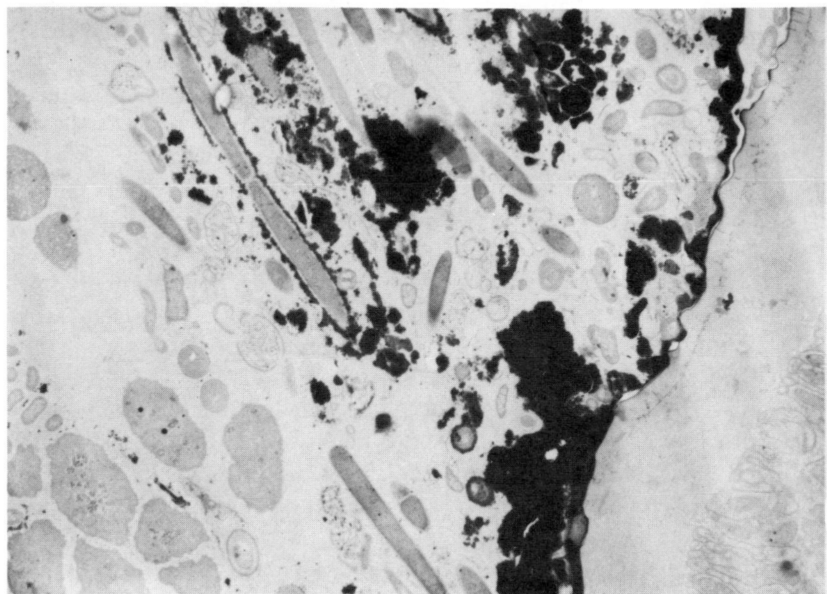

Figure 3 Transmission electron micrograph of the black band region of the hindgut in *E. posticus*, fixed with glutaraldehyde and OsO_4. Electron-dense material accumulates around bacterial cells, as well as in larger deposits at the hindgut wall.

make up approximately 0.2% of the population. This is roughly an order of magnitude more than the number observed in the rumen (66) and an order of magnitude less than that found in the hindguts of lower termites (88). Protozoa are absent from the cricket hindgut (125). Several types of protozoa are usually surrounded by specific types of bacteria (see Figure 2f in Reference 42). Stumm et al (123) have reported that bacteria (most notably methanogens) are specifically associated with protozoa, both intra- and extracellularly, in the rumen. The attachment of methanogens to protozoa, and close interspecific association of microbes in general, may expedite interspecies hydrogen transfer in vivo (126). Spirochetes are firmly attached to specific protozoa in the hindgut of the termite *Mastotermes* (33). Although there are many spirochetes in the cockroach hindgut (42), none appear to be associated with protozoa or other bacteria. In termites, protozoa are attached to the gut wall via specialized attachment organelles (27). We have not observed such attachment in cockroaches; in fact, more protozoa were in the lumen than close to the hindgut wall (42).

Prosthecate organisms constitute about 0.15% of the hindgut population (and up to 1.8% in some micrographs). Their presence in insect guts is surprising, since prosthecate bacteria have most often been isolated from aquatic systems, where their adaptation appeared to be a response to a dilute aqueous environment (45). Poindexter (93) hypothesized that the transport of phosphate, which is growth-limiting in freshwater environments, is maximized in stalked *Caulobacter* cells. In the unbalanced environment of the cockroach hindgut, with a high density of bacterial cells and nutrients available only sporadically, the increased surface area of prosthecate bacteria may be as adaptive as in an aquatic environment, facilitating rapid transport of nutrients (not necessarily only phosphate) as they become available.

In addition to organisms with prosthecae, populations of other bacteria in the cockroach hindgut also have increased surface area. Some form patterned ridges in the cell wall (39; D. L. Cruden, unpublished micrographs), while others form long, tight helices (42; Figure 2). Similar morphotypes have also been reported in vertebrate intestinal systems (68, 92).

Although a significant proportion of the hindgut microbiota is firmly attached to the gut wall and although the hindgut contents are usually semisolid, electron micrographs reveal that many of the normal gut inhabitants are motile. Examples include the large population of spirochetes (42), which constitute over 1% of the population in the black band region (discussed below) of the hindgut; a large rod with a prominent tuft of flagella found only in the black band region (39); and two of the main wall-associated organisms in the paunch region: a filament with the distinctive ultrastructure of *Methanospirillum* and a rod that resembles one in micrographs of termite hindguts (27; Figure 2). The advantages of motility in this environment are unclear, although motile bacteria remaining within the shelter provided by the spines and the network of filamentous attached bacteria are probably not readily eliminated with the fecal pellet. Kaiser & Doetsch (67) showed that translational motility of *Leptospira* increased in a viscous environment. A nonmotile mutant of *Roseburia cecicola,* a known normal inhabitant, colonized the cecum of a germfree mouse but not the cecum of an animal in which a population of microbes was already present (122); this indicates the importance of motility. Even though the cockroach hindgut does not have the mucous coating of the vertebrate intestinal tract, the extracellular bacterial polymers and semisolid consistency may provide a similar environment in which motility is advantageous.

The numbers of cells with the ultrastructure of *Methanospirillum* seen in electron micrographs of the hindgut correlate well with the amount of methane emitted from the insect (discussed below). The ultrastructures of most other bacteria are not as distinctive. However, a preliminary study with an isolate of *F. varium* from the cockroach hindgut showed that it is possible to recognize the organism in hindgut contents using scanning electron microscopy and latex beads coated with specific antiserum to *F. varium* (23).

BLACK BAND The black band was noticed originally in our laboratory as a darker region in the posterior hindgut that is macroscopically visible on dissection. The region varies in size and darkness; both generally increase with age of the insect. It is most noticeable in the large cockroach *E. posticus,* but is also present in laboratory-raised as well as wild *P. americana*. It is also occasionally visible in smaller species (B. A. Stay, personal communication). The macroscopically visible dark color is correlated with electron-dense material in electron micrographs of that region of the hindgut (39; Figure 3). The electron dense material may surround single bacterial cells or microcolonies; this material is apparently complexed with a thin layer of acidic mucopolysaccharides that stain with ruthenium red. This accumulation of electron-dense material resembles the accumulation of iron and manganese in extracellular polymers around *Pedomicrobium*-like organisms described by Ghiorse & Hirsch (55). This material may also accumulate in large masses close to the hindgut wall (Figure 3).

The microbiota of this region is statistically significantly different from that of other wall regions of the hindgut or the lumen (42; see above). In particular, one large irregularly shaped rod with a gram-negative wall structure rarely seen in other fractions is the predominant wall-associated organism; the number of cells with the structure of *Methanospirillum* is drastically lower; and the number of spores is significantly higher. The distinctive microbiota and the electron-dense material appear in electron micrographs of this region even in very small immature stages of *E. posticus* and *P. americana* when the black color is not macroscopically visible (D. L. Cruden, unpublished results). The pH of the black band region (7.7–8.1) is higher than that in any other part of the hindgut (43). In animals fed metronidazole the pH is lower and more variable and the distinctive wall-associated microbiota and the electron-dense material are eliminated (39). Chemical analysis showed that ferrous and sulfide ion concentrations in the black band are higher than in the rest of the hindgut. X-ray microprobe analysis revealed higher concentrations of a variety of elements, including iron and sulfur as well as aluminum and copper, in the electron-opaque material than in any other sites in the hindgut (39, 43). Attempts to isolate sulfate-reducing bacteria from the black band (even from that of insects that were fed Na_2SO_4 in their drinking water for extended periods of time) have failed, but organisms that produce sulfide from organic sulfhydryl compounds are present in high dilutions from the region (39). The presence of ferrous sulfide in the black band indicates that the reduction potential of this region is probably very low. There are no other reports of bacteria specifically isolated from this region.

Deposition of minerals has been reported in various regions of the alimentary tract of a number of insects, including cockroaches. In general, the deposition occurs within the insect tissue, e.g. in the anterior hindgut cells of the German cockroach (3) or in both Malpighian tubules and midgut tissue of the housefly (120, 121). Viewed in the electron microscope these aggregations do not resemble those in the black band region. Moreover, although a variety of elements are present, sulfide has not been reported.

Garnier-Sillam et al (54) have reported organo-mineral aggregates in the feces of termites. Electron micrographs indicate that material similar to that seen in the black band is also present in cockroach feces; this indicates that at least some of the material does not remain in the animal (D. L. Cruden, unpublished micrographs). Bignell et al (19) have reported a black band in a narrow region of the anterior paunch of a soil-feeding termite *(Cubitermes severus),* in which black material is deposited by a dense population of filamentous microorganisms. The pH of the region is high (10.4), even higher than that in the black band region in cockroaches; the authors postulated that alkaline hydrolysis of cellulosic substrates may occur there.

Rectum

Much work has been done on the structure and function of the rectum of a variety of cockroaches (86, 89, 128). In all cases the organ is specialized for very efficient dehydration of fecal pellets and transport of ions, organic compounds, and water into the insect tissue. In *P. americana* the rectum is only 3% of the gut volume (14), but material remains in the rectum for over half the total transit time (59).

Transmission electron micrographs show occasional bacteria associated with the chitinous rectal wall (22, 86). The chitinous spines of the colon end abruptly at the colon-rectum junction, and the wall of the rectum is comparatively smooth. Scanning electron micrographs show a sparse population of rods collected in indentations in the wall (13, 20). The dense, diverse attached microbiota found in the colon is completely missing from this region of the gut. Filamentous organisms, protozoa, and spirochetes appear to be absent. Klug & Kotarski (69) noted a similar abrupt change in microbiota at the colon-rectum junction in crane fly *(Tipula)* larvae. Isolations from rectal contents alone have not been reported.

The amount of feces produced by cockroaches is smaller (on an animal weight basis) than that of other insects (84). This is partly due to the absence of uric acid in feces and the efficient recycling of dietary nitrogen. *E. posticus* has extremely low levels of fecal production, and both *E. posticus* and *P. americana* produce significantly less fecal material than wood-eating cockroaches or termites.

Fecal pellets are very dehydrated, so electron microscopy is difficult. Microbial forms are not as recognizable as in the hindgut because of poor fixation when standard methods are used. However, the pellets appear to consist mostly of undigested plant cell-wall material, large fragments of the peritrophic membrane, and bacterial cells, with a high proportion of spores (D. L. Cruden, unpublished micrographs). Newly hatched insects acquire methanogens, and probably the rest of the anaerobic microbiota, by coprophagy (43). Fecal pellets yield active methanogens at least 72 hr after production when placed in anaerobic medium and gassed with H_2 + CO_2 (D. L. Cruden, unpublished results).

BIOCHEMICAL ACTIVITIES

Cellulose

Cellulase is an enzymatic complex that degrades highly ordered cellulose by an initial reaction with C_1 and C_x enzymes. The degradation of soluble cellulose derivatives, e.g. carboxymethyl cellulose (CMC) or degraded forms of cellulose, is not considered to be true cellulolytic activity (99).

Wharton's group (131) investigated cellulolytic activity in *P. americana* and demonstrated cellulase of both insect origin and bacterial origin in the alimentary tract. The former originated in the salivary glands and the latter was concentrated in the midgut. CMC was the substrate. No attempts were made to isolate cellulolytic bacteria or to estimate the number present. Anaerobic and facultatively anaerobic CMC-degrading bacteria were found in large numbers in the mid- and hindguts of *P. americana* and *E. posticus* (40). Isolates were identified as *K. oxytoca, C. freundii,* and species of *Eubacterium, Clostridium,* and *Serratia*. None of the isolates, nor a cellulolytic reference strain of *Ruminococcus albus* used as a control, digested Whatman No. 1 filter paper in these experiments. When enzyme assays were performed with ball-milled cellulose as substrate using supernatant fluids from cultures grown with CMC, the cellulolytic activity of all strains but one was 30–75% less than when CMC was the substrate (the activity of *Clostridium* remained about the same). Although these bacteria were grown with CMC and the presence of a C_x-cellulase would be expected, the activity seen with ball-milled cellulose indicates that C_1-cellulase activity may also have been present. Alternatively, it could indicate that the ball-milled cellulose also contained polymeric degradation fragments formed during the milling process, which were being degraded by the C_x-enzyme. Bignell (12) showed that [^{14}C]cellulose is degraded in the hindgut of *P. americana*. Feeding of antibiotics sharply reduced $^{14}CO_2$ evolution, which indicates that hindgut bacteria had a role in the degradation of this polysaccharide.

Breznak (25, 26) found no convincing evidence implicating bacteria in cellulose digestion in lower termites that possess cellulolytic protozoa. There is also a lack of evidence supporting bacterial cellulose degradation in the higher termites devoid of protozoa in the hindgut. Fungus-growing higher termites apparently acquire cellulases through the ingestion of fungal enzymes, and acquired enzymes may be involved in other insects that digest cellulose (76). Information pertinent to this point is lacking for cockroaches.

Aerobic and facultative bacteria that could grow on cellulose were found in the mid- and hindgut of desert millipedes, and their role in cellulose digestion was implied when antibiotics were found to decrease [^{14}C]cellulose degradation by approximately 60% (124). Gut microbiota of the scarab beetle, *Oryctes nasicornis,* may be involved in formation of volatile fatty acids from [^{14}C]cellulose (8). Gut microorganisms apparently have no role in cellulose degradation in the silverfish, *Ctenolepisma lineata* (71). No cellulolytic bacteria were found in the larvae of *Costelytra zealandica,* the grass grub beetle, a root feeder on pasture plants (4). The common house cricket, *Acheta domestica,* has a dense and varied hindgut microbiota. Although there are similarities in the gut morphology and diet of crickets and cockroaches, each insect appears to possess different bacterial morphotypes; no cellulolytic bacteria were isolated from *A. domestica* (125). An enlarged hindgut region of the aquatic crane fly larva, *Tipula abdominalis,* has a dense population of morphologically diverse bacteria (69). This insect feeds on detritus, and although no biochemical isolates were reported, it is conceivable that cellulolytic bacteria could be important. Further, Martin (76) suggested that invertebrate detritus feeders could likely benefit from the ingestion of microbial enzymes. Such speculations are being investigated in *Tipula;* Klug (M. J. Klug, personal communication) has indicated that approximately 10% of this insect's nutrition is derived from microorganisms associated with the ingested detritus, while another 10–15% is supplied by microbial activities in the hindgut.

Uric Acid

Uric acid is a major nitrogenous excretory product of numerous insects (30, 34). However, *P. americana* does not excrete uric acid to the exterior; instead, ammonia is present in excreta in relatively large amounts (82). Most cockroach species examined, like *P. americana,* void no urates in their feces (35). However, uric acid is produced and stored in the fat body of the cockroach, and during periods of dietary nitrogen deficiency it can be mobilized (36). There is strong suggestive evidence that symbiotic bacteria in the fat body are the mediators of uric acid degradation (29, 37, 75, 83).

Potrikus & Breznak (95–98) elucidated a more definitive role for bacteria in uric acid degradation and nitrogen recycling in the termite *Reticulitermes flavipes*. The termite synthesizes and stores uric acid in its fat body. Like the cockroach, it excretes little or no uric acid in feces. However, the fat body lacks symbiotes that could degrade urate, and no enzymatic degradation in termite tissue could be demonstrated. The authors found that uricolysis is accomplished anaerobically by hindgut bacteria. Significant numbers of uricolytic bacteria are found in the hindgut; the major isolates are *Bacteroides termitidis, Streptococcus* sp., and *Citrobacter* sp. These isolates use uric acid as an energy source, but only under anaerobic conditions. Ammonia is one of the major products.

The termite and cockroach urate systems appear to be quite similar. However, termites use gut bacteria to degrade uric acid, whereas cockroaches may use the fat-body symbiotes for this purpose. Cochran (36) pointed out that it is not possible to completely rule out the involvement of gut microbes in the degradation of urate by cockroaches.

Bacteria capable of degrading uric acid have been isolated from the hindgut of three cockroaches, *P. americana, E. posticus,* and *Gromphadorhina portentosa* (hissing roach) (10). Six *Clostridium* species were consistently isolated from each cockroach species, but no other uric acid–degrading anaerobes were isolated. *Clostridium scatologenes* and *Clostridium haemolyticum* were the most predominant isolates. *C. freundii* was detected most often in a survey of facultative gut bacteria able to utilize uric acid. It is premature to speculate on whether or not these organisms have any role in the nitrogen recycling of urate. The clostridium isolated on PP_i as an energy source also degrades uric acid (38). *K. oxytoca* has been isolated as N_2-fixer, a CMC degrader, and a uric acid degrader. Such isolations simply demonstrate again that many bacteria are metabolically versatile and can grow on a variety of substrates. It is more difficult to demonstrate that, for example, uric acid is actually degraded in this complex ecosystem. Initially, it must be demonstrated that uric acid actually reaches the hindgut and is available as a potential substrate. Since cockroaches are cannibalistic, any uric acid that reaches the hindgut may have been eaten by the insect.

Crickets *(A. domestica)* differ from cockroaches and termites in that a high percentage of the dry weight of fecal material is uric acid. Significant numbers of uricolytic bacteria are present in the hindgut. *Citrobacter, Klebsiella, Yersina,* one *Streptococcus,* and an unidentified gram-negative bacterium all metabolize uric acid. It is also interesting that no symbiotes are found in the fat bodies of these insects (125).

Bacterial Storage Products

One feature of the cockroach hindgut bacteria that is easily observed in electron micrographs is the large number of cells with inclusions resembling the typical bacterial storage products glycogen, poly-β-hydroxybutyrate (PHB), and polyphosphate (39, 43). The cockroach hindgut contains significantly higher levels of both polyphosphate and PHB than either midgut contents or bovine rumen fluid (43). These levels are decreased when the insects are fed metronidazole. In general, these bacterial storage products are formed in response to an imbalance in nutrients in the environment (44, 118). Cheng et al (32) demonstrated that the rumen microbiota responded to a high-energy ration by accumulating glycogenlike storage granules. There is no information on whether the hindgut bacteria in cockroaches respond similarly to changes in diet, although one might expect that they do not, given the relative independence of diet and methane production (see below).

Anaerobic enrichments in media containing pyrophosphate (PP_i) inoculated with hindgut contents from *Eublaberus* yielded a number of isolates that are stimulated by PP_i (38). Like PP_i-stimulated organisms isolated from aquatic environments (91), all of these are purine degraders. The organism best studied, a *Clostridium* species, contains several enzymes for which PP_i serves as a source of energy, replacing ATP. Tri- and tetrapolyphosphates are also used (38). The presence in the cockroach gut of bacteria that are able to benefit from storage products liberated by other organisms is another indication of the recycling of resources that appears to occur in this environment. Many of the strictly anaerobic bacteria isolated from the cockroach hindgut are members of genera in which PP_i is substituted for ATP [e.g. *Propionibacterium* (87) and *Bacteroides* (101)], but assays for the presence of enzymes in these particular isolates have not been performed.

Methane

Cockroaches emit large amounts of methane. Normal values in laboratory-raised E. *posticus* range from 10 to 25 μmol hr^{-1}; occasionally insects emit as much as 165 μmol hr^{-1} (43). *P. americana,* wild *P. americana,* and *Gromphadorhina* also consistently emit more methane than other insects; scarab beetle larvae (which weigh about 20 g) produce 300–380 nmol hr^{-1} $insect^{-1}$ (6), and among the lower termites, *Reticulitermes tibilias* averages up to 1.3 nmol hr^{-1} $insect^{-1}$ (135) and *R. flavipes* emits approximately 0.4 nmol hr^{-1} $insect^{-1}$ (88). Methane is formed in the cockroach hindgut. Hydrogen, but not methane, is produced in isolated midguts, while no H_2 is detectable in the hindgut (T. E. Gorrell, unpublished experiments). Methane is formed from CO_2 and H_2, and none is made from acetate, which is typical of intestinal systems (81). Many of the hindgut isolates produce large quantities of H_2 [e.g. the fusobacteria (50) and the *Clostridium* species that uses PP_i as an energy source (38)], but it must be consumed very rapidly because it never accumulates to detectable levels. Recently, Breznak & Switzer (28) detected acetogenesis in *P. americana,* as well as in a number of termites and the wood roach, *Cryptocercus punctulatus,* by measuring $^{14}CO_2$ incorporation into acetate. Levels of acetogenesis in termites were up to over 200-fold higher than in *P. americana,* however. In lower termites, acetogenesis may account for up to one third of the acetate formed in the hindgut, which can then cross the gut wall and contribute to the nutrition of the host. In the wood roach, which has hindgut fermentation of cellulose similar to that of the lower termites, levels of acetogenesis are sevenfold higher than in *P. americana*. In *P. americana* most of the $^{14}CO_2$ is incorporated into methane rather than acetate and other nongaseous products, and is thus lost to the host.

The large amount of methane produced by E. *posticus* correlates well with the number of cells with the characteristic morphology of *Methanospirillum* in the hindgut (42). In addition, two other methanogens have been isolated from *E. posticus* (D. L. Cruden, unpublished results). One is a nonmotile short rod that produces methane from H_2 and CO_2 or formate in an inorganic medium but is stimulated by acetate and yeast extract. It is probably a *Methanobrevibacter* species. The other is a coccoid organism isolated from a 10^{-5} dilution of hindgut contents, which produces methane from methanol and H_2. It is antigenically closely related to (if not identical to) *Methanosphaera stadtmaniae,* which was isolated initially from human feces (80) and which has been found in the intestines of a variety of other vertebrates (81).

Methanospirillum was originally isolated from sludge digesters (48, 90). To our knowledge it has not been reported in other gut systems, and we have not seen *Methanospirillum*-like cells in electron micrographs of other intestinal tracts. *Methanobrevibacter* and *Methanosphaera,* on the other hand, appear to be the most common vertebrate intestinal methanogens (79, 81).

Experiments indicate that young cockroaches acquire methanogens by coprophagy (43). The amount of methane produced by an individual remains fairly constant over time, independent of the length of time since the last meal. Methane production by cockroaches is similar to that by humans in that it is apparently the result of endogenous fermentations (81).

There have been claims about the global contributions of termites to recent increases in atmospheric methane (135), as well as counterclaims that the estimates are too high and that the global termite population is relatively unaffected by human activities (117). In view of the large amounts of methane produced by cockroaches (43), the populations of cockroaches in urban environments (70), and the increased urbanization in the developing semitropical and tropical regions where cockroaches thrive, these insects might be likely to contribute to increases in atmospheric methane.

Other Activities

CHITIN Waterhouse & McKellar (130) have reported that chitinase activity is present in the saliva, foregut, midgut, and hindgut of *P. americana*. The activity is purportedly not of microbial origin, and it was not attributed wholly to molting fluid in the cuticle-secreting epithelia of the fore- and hindgut. The authors thought that such activity would have to serve a digestive function, perhaps related to this insect's cannibalism and consumption of shed exoskeletons. They questioned how the gut cuticle and peritrophic membrane escape digestion (130).

A chitin-degrading *Serratia marcescens* was isolated on swollen chitin from the guts of *P. americana* and E. *posticus*. The chitobiase portion of the chitinase complex was found to be periplasmic rather than extracellular (9). The presence of chitinase in the hindgut should have little if any effect on the gut lining, since the chitin-containing layer is covered by a proteinaceous epicuticle which would need to be breached initially. To further complicate the picture, the chitin in the underlying layer is complexed with protein. According to a model for cuticle degradation by endogenous insect enzymes during molt, a trypsinlike protease first unmasks the chitin prior to chitinase activity (31). The chitin-containing peritrophic membrane, which is continually shed into the hindgut where it is shredded in passage, may be subject to microbial degradation. However, the finding of large recognizable fragments of this membrane in the feces indicates that any degradation is probably minor (D. L. Cruden, unpublished results).

Chitinase activity has been demonstrated in extracts of whole termites (129) and in the guts of *Macrotermes ukuzii*, from which a chitinolytic bacterium was isolated (105).

NITROGEN FIXATION A nitrogen-fixing *Klebsiella oxytoca* strain was isolated in our laboratory from the hindgut of *E. posticus* on a nitrogen-free medium incubated anaerobically (D. L. Cruden, unpublished results). It is doubtful that N_2 fixation by gut microorganisms has a significant role in the nitrogen metabolism of this ecosystem. This reaffirms the metabolic versatility of this *Klebsiella* isolate.

The following nitrogen-fixing bacteria have been isolated from termite guts: *C. freundii* from Australian termite species (53) and *Enterobacter agglomerans* from *Coptotermes formosanus* (94). Breznak (25) has discussed the role of N_2 fixation in the termite.

CONTRIBUTIONS TO THE HOST

There is no definitive evidence that the microbiota of the cockroach gut contributes to its host in any major fashion, although most studies to date have used adult insects fed an adequate diet. However, suggestive results lead to speculation on some possible contributions.

Analyses of hindgut contents from *P. americana* reveal typical fermentation products, i.e. acetate, butyrate, and propionate, with caproate, caprylate, isobutyrate, isovalerate, valerate, lactate, and succinate detected in some analyses (20). Such fermentation end products are transported into the hemolymph through the gut wall (24). [^{14}C]Cellulose was also degraded in the hindgut, and this resulted in transport of radioactive end products (12). These results indicate that the insect could derive potentially energy-yielding compounds from the activity of its hindgut microbiota.

The extent of this contribution is inferred from experiments with metronidazole. When this drug was administered orally to adult roaches, essentially all the obligately anaerobic flora was eliminated. The facultative bacteria remaining did not colonize the wall areas vacated by the anaerobes (21). Adult roaches were maintained anaerobe-free for up to 24 mo with no outward adverse effects. However, when nymphs were administered metronidazole, their growth was retarded and their hindguts were smaller; walls were thinner than those of the controls. Their gut contents were translucent and gelatinous, as opposed to the semiliquid, opaque contents of the controls. Metronidazole had no effect on the fat-body symbiotes. Although there have been no studies on the effect of the microbiota on the development of the cockroach gut, Wilson (133) demonstrated that musculature of the lamb rumen depends on the bulk of material in the rumen, whereas mucosal development depends on factors in the diet. Furthermore, mucosal development in the calf rumen is stimulated by fatty acids, i.e. butyrate, propionate, and lactate (49, 110). Perhaps the microbiota have an analogous role in the development and physiology of the cockroach. However, once the host has developed and is fed an adequate diet, the contribution of the bacterial metabolic end products to the nutrition of the host appears to be minimal. The greater amount of methanogenesis than acetogenesis in the hindgut (28) is further indication that the chief end products of microbial fermentations are not of great benefit to the host.

It is possible that the microbiota is an ancillary supplier of vitamins and other nutritional cofactors as in mammalian gut systems (134). These nutrients could become available to the host through direct absorption across the hindgut wall or through adsorption in the midgut following coprophagy. Little is known about coprophagy in cockroaches except that methanogens are acquired via this mechanism (43). Whether cockroaches obtain essential nutrients from ingestion of their own fecal material has not been demonstrated.

An aggregation pheromone that attracts adults and nymphs is elaborated in the feces of *P. americana* (11). This chemical(s) could be produced or modified by the gut microbiota. No statistically significant differences were demonstrated in the aggregation response elicited by extracts of feces from normal and metronidazole-treated adult roaches. Thus it was not ascertained whether obligately anaerobic organisms are involved in production of the aggregation pheromone. The role of facultative bacteria was not ascertained (A. J. Markovetz, unpublished observations).

Scheline (114) has characterized the microbiota of the human colon as an organ with a metabolic potential equal to or greater than that of the liver. In other words, microbial transformations of xenobiotic compounds such as drugs and food constituents may result in biological activity. The relationship between mammalian intestinal microbiota and metabolic transformations has recently been reviewed (56). In cockroaches, the compounds would have to be transformed and then transported through the gut wall or excreted and reingested to affect the host. Clearly, more information is needed on the biochemical potential of this ecosystem as well as on the transport potential of the hindgut. Germfree roaches, i.e. roaches devoid of gastrointestinal microbiota but still retaining their fat-body symbiotes, would be important for the verification of any contributions of the microbiota to the host.

COMMENTS

Savage (113) indicated that gastrointestinal tracts of mammals could be loosely classified into three groups: ruminant, cecal, and "straight tube." The bulk of the information available on gut microbial ecology comes from investigations on the rumen. The ruminant is nutritionally dependent upon microbes and their production of metabolic end products. A mammal possessing a cecum digests a large proportion of its food by intrinsic processes, but a certain portion of nutrients is supplied via microbial activity that occurs in this blind sac. The straight-tube mammals (e.g. the human) are omnivorous, and nutrients are supplied by the action of the gut enzymes. Although activities of microorganisms are not necessary for nutritional processing in the gastrointestinal tract, the large bowel of humans has a significant microbiota. The microbiota of all three types of mammalian gastrointestinal tracts have many similarities.

Insect guts can be loosely classified in a similar manner. The termite is analogous to the ruminant, obtaining its nutrition from microbial processes occurring in the gut. The *Tipula* larva corresponds to the cecum-containing mammal, in that part of its nutrition can be provided by the activity of the gut microbiota. The cockroach is analogous to the human. This insect possesses an extensive hindgut microbiota, which has many similarities to the microbiota of the human colon. The roach is also omnivorous, and our studies have shown that an adult roach deprived of its anaerobic hindgut flora is not nutritionally deprived.

These invertebrates are manageable models for the study of microbe-microbe and microbe-host interactions; perhaps much of the knowledge gained will be applicable to mammalian systems. The effect of diet on the composition and stability of the gut ecosystem is somewhat controversial. The cockroach could well provide an easily manipulated model for studies of this problem.

ACKNOWLEDGMENTS

The authors wish to acknowledge support from the National Institutes of Health (grant AI-13990) and the Rockefeller Foundation.

Literature Cited

1. Alexander, M. 1971. *Microbial Ecology*, pp. 3–21. New York: Wiley. 511 pp.
2. Balch, W. E., Wolfe, R. S. 1976. New approach to the cultivation of methanogenic bacteria: 2-mercaptoethane-sulfonic acid (HS-CoM)–dependent growth of *Methanobacterium ruminantium* in a pressurized atmosphere. *Appl. Environ. Microbiol.* 32:781–91
3. Ballan-Dufrançais, C. 1972. Ultrastructure d'ilèon de *Blatella germanica* (L) (Dictyoptère). Localisation, genèse et composition des concrétions minérales intracytoplasmiques. *Z. Zellforsch. Mikrosk. Anat.* 133:163–79
4. Bauchop, T., Clarke, R. T. J. 1975. Gut microbiology and carbohydrate digestion in the larva of *Costelytra zaelandica*. *NZ J. Zool.* 2:237–43
5. Bayon, C. 1971. La cuticle proctodéale de la larve d'*Oryctes nasicornis* (Coleoptera:Scarabaeidae). Étude au microscope électronique à balayage. *J. Microsc. Paris* 11:353–70
6. Bayon, C. 1980. Volatile fatty acids and methane production in relation to anaerobic carbohydrate fermentation in *Oryctes nasicornis* larvae (Coleoptera:Scarabaeidae). *J. Insect Physiol.* 26:819–28
7. Bayon, C. 1981. Ultrastructure de l'epithelium intestinal et flore parietale chez la larve xylophage d'*Oryctes nasicornis* L (Coleoptera:Scarabaeidae). *Int. J. Insect Morphol. Embryol.* 10:359–71
8. Bayon, C., Mathelin, J. 1980. Carbohydrate fermentation and by-product absorption studied with labeled cellulose in *Oryctes nasicornis* larvae. *J. Insect Physiol.* 26:833–40
9. Becker, G. E. 1980. Chitobiase of *Serratia marcescens*. *Abstr. Annu. Meet. Am. Soc. Microbiol.* 1980:148
10. Becker, G. E., Lozado, E. P., Peterson, T. 1982. Uric acid degrading bacteria in cockroaches. *Abstr. Annu. Meet. Am. Soc. Microbiol.* 1982:103
11. Bell, W. J., Parsons, C., Martinko, E. A. 1972. Cockroach aggregation pheromones: analysis of aggregation tendency and species specificity. *J. Kans. Entomol. Soc.* 45:414–20
12. Bignell, D. E. 1977. An experimental study of cellulose and hemicellulose degradation in the alimentary canal of the American cockroach. *Can. J. Zool.* 55:579–89
13. Bignell, D. E. 1977. Some observations on the distribution of gut flora in the American cockroach, *Periplaneta americana*. *J. Invertebr. Pathol.* 29:338–43
14. Bignell, D. E. 1981. Nutrition and digestion. In *The American Cockroach*, ed. W. J. Bell, K. G. Adiyodi, pp. 57–86. London/New York: Chapman & Hall. 529 pp.
15. Bignell, D. E., Anderson, J. M. 1980. Determination of pH and oxygen status in the guts of lower and higher termites. *J. Insect Physiol.* 26:183–88
16. Bignell, D. E., Oskarsson, H., Anderson, J. M. 1979. Association of actinomycete-like bacteria with soil-feeding termites. *Appl. Environ. Microbiol.* 37:339–42
17. Bignell, D. E., Oskarsson, H., Anderson, J. M. 1980. Colonization of the epithelial face of the peritrophic membrane and the ectoperitrophic space by actinomycetes in a soil-feeding termite. *J. Invertebr. Pathol.* 36:426–28
18. Bignell, D. E., Oskarsson, H., Anderson, J. M. 1980. Distribution and abundance of bacteria in the gut of a soil feeding termite. *J. Gen. Microbiol.* 117:393–403
19. Bignell, D. E., Oskarsson, H., Anderson, J. M., Ineson, P., Wood, T. G. 1983. Structure, microbial associations and function of the so-called "mixed segment" of the gut in two soil-feeding termites, *Procubitermes aburiensis* and *Cubitermes severus* (Termitidae, Termitinae). *J. Zool.* 201:445–80
20. Bracke, J. W. 1977. *Studies on the obligate anaerobic intestinal bacteria of the cockroach,* Periplaneta americana. PhD thesis. Univ. Iowa, Iowa City. 190 pp.
21. Bracke, J. W., Cruden, D. L., Markovetz, A. J. 1978. Effect of metronidazole on the intestinal microflora of the American cockroach, *Periplaneta americana* L. *Antimicrob. Agents Chemother.* 13:115–20
22. Bracke, J. W., Cruden, D. L., Mar-

kovetz, A. J. 1979. Intestinal microbial flora of the American cockroach. *Appl. Environ. Microbiol.* 38:945–55
23. Bracke, J. W., Markovetz, A. J. 1978. Immunolatex localization by scanning electron microscopy of intestinal bacteria from cockroaches. *Appl. Environ. Microbiol.* 35:166–71
24. Bracke, J. W., Markovetz, A. J. 1980. Transport of bacterial end products from the colon of *Periplaneta americana*. *J. Insect Physiol.* 26:85–90
25. Breznak, J. A. 1982. Intestinal microbiota of termites and other xylophagous insects. *Ann. Rev. Microbiol.* 36:323–43
26. Breznak, J. A. 1984. Biochemical aspects of symbiosis between termites and their intestinal microbiota. In *Invertebrate-Microbial Interactions,* ed. J. M. Anderson, A. D. M. Rayner, D. W. H. Walton, pp. 173–203. Cambridge, UK: Cambridge Univ. Press. 349 pp.
27. Breznak, J. A., Pankratz, H. S. 1977. In situ morphology of the gut microbiota of wood-eating termites [*Reticulitermes flavipes* (Kollar) and *Coptotermes formosanus* Shiraki]. *Appl. Environ. Microbiol.* 33:406–26
28. Breznak, J. A., Switzer, J. M. 1986. Acetate synthesis from H_2 and CO_2 by termite gut microbes. *Appl. Environ. Microbiol.* 52:623–30
29. Brooks, M. A. 1970. Comments on the classification of intracellular symbiotes of cockroaches and a description of the species. *J. Invertebr. Pathol.* 16:249–58
30. Bursell, E. 1967. The excretion of nitrogen in insects. *Adv. Insect Physiol.* 4:33–67
31. Charnley, A. K. 1984. Physiological aspects of destructive pathogenesis in insects by fungi: a speculative review. See Ref. 26, pp. 229–70
32. Cheng, K. J., Hironaka, R., Roberts, D. W. A., Costerton, J. W. 1973. Cytoplasmic glycogen in inclusions in cells of anaerobic gram negative rumen bacteria. *Can. J. Microbiol.* 19:1501–6
33. Cleveland, L. R., Grimstone, A. V. 1964. The fine structure of the flagellate *Mixotricha paradoxa* and its associated microorganisms. *Proc. R. Soc. London Ser. B* 159:668–86
34. Cochran, D. G. 1975. Excretion in insects. In *Insect Biochemistry and Function,* ed. D. J. Candy, B. A. Kilby, pp. 177–281. London: Chapman & Hall. 314 pp.
35. Cochran, D. G. 1979. Comparative analysis of excreta and fat body from various cockroach species. *Comp. Biochem. Physiol. A* 64:1–4
36. Cochran, D. G. 1985. Nitrogen excretion in cockroaches. *Ann. Rev. Entomol.* 30:29–49
37. Cochran, D. G., Mullins, D. E. 1982. Physiological processes related to nitrogen excretion in cockroaches. *J. Exp. Zool.* 222:277–85
38. Cruden, D. L., Durbin, W. E., Markovetz, A. J. 1983. Utilization of PP_i as an energy source by a *Clostridium* sp. *Appl. Environ. Microbiol.* 46:1403–8
39. Cruden, D. L., Gorrell, T. E., Markovetz, A. J. 1979. Novel microbial and chemical components of a specific black band region in the cockroach hindgut. *J. Bacteriol.* 140:687–98
40. Cruden, D. L., Markovetz, A. J. 1979. Carboxymethyl cellulose decomposition by intestinal bacteria of cockroaches. *Appl. Environ. Microbiol.* 38:369–72
41. Cruden, D. L., Markovetz, A. J. 1980. A thick walled organism isolated from the cockroach gut by using a spent medium technique. *Appl. Environ. Microbiol.* 39:261–64
42. Cruden, D. L., Markovetz, A. J. 1981. Relative numbers of selected bacterial forms in different regions of the cockroach hindgut. *Arch. Microbiol.* 129:129–34
43. Cruden, D. L., Markovetz, A. J. 1984. Microbial aspects of the cockroach hindgut. *Arch. Microbiol.* 138:131–39
44. Dawes, E. A., Senior, P. J. 1973. The role and regulation of energy reserve polymers in microorganisms. *Adv. Microb. Physiol.* 19:135–278
45. Dow, C. S., Whittenbury, R. 1979. Prosthecate bacteria. In *Developmental Biology of Prokaryotes,* ed. J. H. Parish, pp. 139–65. Berkeley, Calif: Univ. Calif. Press. 297 pp.
46. Dubos, R., Schaedler, R. W., Costello, R., Hoet, P. 1965. Indigenous, normal, and autochthonous flora of the gastrointestinal tract. *J. Exp. Med.* 122:67–76
47. Eaves, G. N., Mundt, J. O. 1960. Distribution and characterization of streptococci from insects. *J. Insect Pathol.* 2:289–98
48. Ferry, J. G., Smith, P. H., Wolfe, R. S. 1974. *Methanospirillum,* a new genus of methanogenic bacteria, and characterization of *Methanospirillum hungatii* sp. nov. *Int. J. Syst. Bacteriol.* 24:465–69
49. Flatt, W. P., Warner, R. G., Loosli, J. K. 1958. Influence of purified materials on the development of the ruminant stomach. *J. Dairy Res.* 41:1593–600
50. Foglesong, M. A., Cruden, D. L., Markovetz, A. J. 1984. Pleomorphism of fusobacteria isolated from the cockroach hindgut. *J. Bacteriol.* 158:474–80

51. Foglesong, M. A., Markovetz, A. J. 1974. Morphology of bacteriophage-like particles from *Fusobacterium symbiosum*. *J. Bacteriol.* 119:325–29
52. Foglesong, M. A., Walker, D. H. Jr., Puffer, J. S., Markovetz, A. J. 1975. Ultrastructural morphology of some procaryotic microorganisms associated with the hindgut of cockroaches. *J. Bacteriol.* 123:336–45
53. French, J. R. J., Turner, G. L., Bradbury, J. F. 1976. Nitrogen fixation by bacteria from the hindgut of termites. *J. Gen. Microbiol.* 95:202–6
54. Garnier-Sillam, E., Villemin, G., Toutain, F., Renoux, J. 1985. Formation of organo-mineral microaggregates in termite faeces. *C. R. Acad. Sci. Ser. III* 301:213–16
55. Ghiorse, W. C., Hirsch, P. 1979. An ultrastructural study of iron and manganese deposition associated with extracellular polymers of *Pedomicrobium*-like budding bacteria. *Arch. Microbiol.* 123:213–26
56. Goldin, B. R. 1986. In situ bacterial metabolism and colon mutagens. *Ann. Rev. Microbiol.* 40:367–93
57. Greenberg, B. 1968. Micro-potentiometric pH determinations of muscoid maggot digestive tracts. *Ann. Entomol. Soc. Am.* 61:365–68
58. Greenberg, B., Kowalski, J., Karpus, J. 1970. Micro-potentiometric pH determination of the gut of *Periplaneta americana* fed three different diets. *J. Econ. Entomol.* 63:1795–97
59. Guthrie, D. M., Tindall, A. R. 1968. *The Biology of the Cockroach*. London: Arnold. 408 pp.
60. Henry, S. M., Block, R. J. 1960. The sulfur metabolism of insects. IV. The role of intracellular symbionts. *Contrib. Boyce Thompson Inst.* 20:317–29
61. Hoffman, A. G. D., Downer, R. G. H. 1976. The crop as an organ of glyceride adsorption in the American cockroach, *Periplaneta americana* L. *Can. J. Zool.* 54:1165–71
62. Hogan, M. E., Slaytor, M., O'Brien, R. W. 1985. Transport of volatile fatty acids across the hindgut of the cockroach *Panesthia cribata* Saussure and the termite *Mastotermes darwiniensis* Froggatt. *J. Insect Physiol.* 31:587–91
63. Holdeman, L. V., Cato, E. P., Moore, W. E. C., eds. 1977. *Anaerobic Laboratory Manual*. Blacksburg, Va: Va. Polytech. Inst. Anaerobe Lab. 4th ed. 152 pp.
64. Holdeman, L. V., Kelley, R. W., Moore, W. E. C. 1983. Family 1. Bacteroidaceae. In *Bergey's Manual of Systematic Bacteriology*, ed. N. R. Krieg, J. G. Holt, 1:602–62. Baltimore, Md: Williams & Wilkins
65. House, H. L. 1949. Nutritional studies with *Blatella germanica* reared under aseptic conditions. II. A chemically defined diet. *Can. Entomol.* 81:105–12
66. Hungate, R. E. 1975. The rumen microbial ecosystem. *Ann. Rev. Ecol. Syst.* 6:38–66
67. Kaiser, G. E., Doetsch, R. N. 1975. Enhanced translational motion of *Leptospira* in viscous environments. *Nature* 255:656–57
68. Kaneuchi, E., Miyazato, T., Shinjo, T., Mitsuoka, T. 1979. Taxonomic study of helically coiled sporeforming anaerobes isolated from the intestines of humans and other animals. *Int. J. Syst. Bacteriol.* 29:1–12
69. Klug, M. J., Kotarski, S. 1980. Bacteria associated with the gut tract of larval stages of the aquatic cranefly *Tipula abdominalis* (Diptera; Tipulidae). *Appl. Environ. Microbiol.* 40:408–16
70. Koehler, P. G., Patterson, R. S., Brenner, K. J. 1987. German cockroach infestations in low-income apartments. *J. Econ. Entomol.* In press
71. Lasher, L., Giese, A. C. 1956. Cellulose digestion by the silverfish *Ctenolepisma lineata*. *J. Exp. Biol.* 33:542–53
72. Ludwig, D., Gallagher, M. R. 1966. Vitamin synthesis by the symbionts in the fat body of the cockroach, *Periplaneta americana* (L). *J. NY Entomol. Soc.* 74:134–39
73. Maddrell, S. H. P. 1977. Insect malpighian tubules. In *Transport of Ions and Water in Animals*, ed. B. L. Gupta, R. B. Moreton, J. G. Oschman, B. H. Wall, pp. 541–69. London: Academic. 721 pp.
74. Maddrell, S. H. P., Gardiner, B. O. C. 1980. The permeability of the cuticular lining of the insect alimentary canal. *J. Exp. Biol.* 85:227–37
75. Malke, H., Schwartz, W. 1966. Untersuchungen über die symbiose von Tieren mit Pilsen und Bakterien. XII. Die Bedeutung der Blattiden Symbioses. *Z. Allg. Mikrobiol.* 6:34–68
76. Martin, M. M. 1984. The role of ingested enzymes in the digestive process of insects. See Ref. 26, pp. 155–72
77. Martin, M. M., Martin, J. S., Kukor, J. J., Merritt, R. W. 1980. The digestion of protein and carbohydrate by the stream detritivore *Tipula abdominalis* (Diptera, Tipulidae). *Oecologia Berlin* 46:360–64
78. McCowan, R. P., Cheng, K. J., Coster-

ton, J. W. 1980. Adherent bacterial populations on the bovine rumen wall: distribution patterns of adherent bacteria. *Appl. Environ. Microbiol.* 39:233–41

79. Miller, T. L., Wolin, M. J. 1982. Enumeration of *Methanobrevibacter smithii* in human feces. *Arch. Microbiol.* 131:14–18
80. Miller, T. L., Wolin, M. J. 1985. *Methanosphaera stadtmaniae* gen. nov., sp. nov.: a species that forms methane by reducing methanol with hydrogen. *Arch. Microbiol.* 141:116–22
81. Miller, T. L., Wolin, M. J. 1986. Methanogens in human and animal intestinal tracts. *Syst. Appl. Microbiol.* 7:223–29
82. Mullins, D. E., Cochran, D. G. 1972. Nitrogen excretion in cockroaches: Uric acid is not a major product. *Science* 177:699–701
83. Mullins, D. E., Cochran, D. G. 1975. Nitrogen metabolism in the American cockroach. II. An examination of negative nitrogen balance with respect to mobilization of uric acid stores. *Comp. Biochem. Physiol. A* 50:501–10
84. Mullins, D. E., Cochran, D. G. 1976. A comparative study of nitrogen excretion in 23 cockroach species. *Comp. Biochem. Physiol. A* 53:393–99
85. Murphy, R. C. 1976. Structure of the foregut cuticle of *Periplaneta americana*. *Experientia* 32:316–17
86. Noirot, C., Noirot-Timothée, C. 1976. Fine structure of the rectum in cockroaches (Dictyoptera): general organization and intracellular junctions. *Tissue Cell* 8:345–68
87. O'Brien, W. E., Bowien, S., Wood, H. G. 1975. Isolation and characterization of a PP_i-dependent phosphofructokinase from *Propionibacterium shermanii*. *J. Biol. Chem.* 250:8690–95
88. Odelson, D. A., Breznak, J. A. 1983. Volatile fatty acid production by the hindgut microbiota of xylophagous termites. *Appl. Environ. Microbiol.* 45:1602–13
89. Oschman, J. L., Wall, B. J. 1969. The structure of the rectal pads of *Periplaneta americana* L. with regard to fluid transport. *J. Morphol.* 127:475–510
90. Patel, G. B., Roth, L. A., van den Berg, L., Clark, D. S. 1976. Characterization of a strain of *Methanospirillum hungatii*. *Can. J. Microbiol.* 22:1404–10
91. Peck, H. D. Jr., Liu, C. L., Varma, A. K., Ljungdahl, L. G., Szulczynski, M., et al. 1983. The utilization of inorganic PP_i, tripolyphosphate, and tetrapolyphosphate as energy sources for the growth of anaerobic bacteria. In *Basic Biology of New Developments in Biotechnology*, ed. A. Hollaender, A. I. Laskin, P. Rogers, pp. 317–48. New York: Plenum. 579 pp.
92. Phillips, M. W., Lee, A. 1983. Isolation and characterization of a spiral bacterium from the crypts of rodent gastrointestinal tracts. *Appl. Environ. Microbiol.* 45:675–83
93. Poindexter, J. S. 1984. Role of prostheca development in oligotrophic aquatic environments. In *Current Perspectives in Microbial Ecology*, ed. M. J. Klug, C. A. Reddy, pp. 33–40. Washington, DC: Am. Soc. Microbiol. 710 pp.
94. Potrikus, C. J., Breznak, J. A. 1977. Nitrogen fixing *Enterobacter agglomerans* isolated from the guts of wood-eating termites. *Appl. Environ. Microbiol.* 33:392–99
95. Potrikus, C. J., Breznak, J. A. 1980. Uric acid–degrading bacteria in guts of termites [*Reticulitermes flavipes* (Kollar)]. *Appl. Environ. Microbiol.* 40:117–24
96. Potrikus, C. J., Breznak, J. A. 1980. Anaerobic degradation of uric acid by gut bacteria of termites. *Appl. Environ. Microbiol.* 40:125–32
97. Potrikus, C. J., Breznak, J. A. 1980. Uric acid in wood-eating termites. *Insect Biochem.* 10:19–27
98. Potrikus, C. J., Breznak, J. A. 1981. Gut bacteria recycle uric acid nitrogen in termites: a strategy for nutrient conservation. *Proc. Natl. Acad. Sci. USA* 78:4601–5
99. Prins, R. A. 1977. Biochemical activities of gut microorganisms. In *Microbial Ecology of the Gut*, ed. R. T. J. Clarke, T. Bauchop, pp. 73–183. London/New York: Academic. 410 pp.
100. Read, H. C. 1933. The cockroach as a possible carrier of tuberculosis. *Am. Rev. Tuberc.* 28:267–72
101. Reeves, R. E., Menzies, R. A., Hsu, D. S. 1968. The pyruvate-phosphate dikinase reaction: the rate of phosphate and the equilibrium. *J. Biol. Chem.* 243:5486–91
102. Rehn, J. W. H. 1951. Classification of the Blattaria as indicated by their wings. *Mem. Am. Entomol. Soc.* 14:1–134
103. Richards, A. G., Brooks, M. A. 1958. Internal symbiosis in insects. *Ann. Rev. Entomol.* 3:37–56
104. Richards, A. G., Richards, P. A. 1977. The peritrophic membrane of insects. *Ann. Rev. Entomol.* 22:219–40
105. Rohrmann, G. F., Rossman, A. Y. 1980. Nutrient strategies of *Macrotermes ukuzii* (Isoptera:Termitidae). *Pedobiologia* 20:61–73

106. Roth, L. M. 1981. Introduction. See Ref. 14, pp. 1–14
107. Roth, L. M., Alsop, D. W. 1978. Toxins of Blattaria. *Handb. Exp. Pharmacol*. 48:465–78
108. Roth, L. M., Willis, E. R. 1957. The medical and veterinary importance of cockroaches. *Smithson. Misc. Collect*. 134:1–147
109. Roth, L. M., Willis, E. R. 1960. The biotic associations of cockroaches. *Smithson. Misc. Collect*. 141:1–468
110. Sanders, E. G., Warner, R. G., Harrison, H. N., Loosli, J. K. 1959. The stimulatory effect of sodium butyrate and sodium propionate on the development of rumen mucosa in the young calf. *J. Dairy Sci*. 42:1600–5
111. Savage, D. C. 1972. Associations and physiological interactions of indigenous microorganisms and gastrointestinal epithelia. *Am. J. Clin. Nutr*. 25:1372–79
112. Savage, D. C. 1977. Interactions between the host and its microbes. See Ref. 99, pp. 277–310
113. Savage, D. C. 1977. Microbial ecology of the gastrointestinal tract. *Ann. Rev. Microbiol*. 31:107–33
114. Scheline, R. R. 1973. Metabolism of foreign compounds by gastrointestinal microorganisms. *Pharmacol. Rev*. 25:451–532
115. Schlüter, U. 1980. Cytopathological alterations in the hindgut of a milliped induced by atypical diet. *J. Invertebr. Pathol*. 36:133–35
116. Schultz, J. E., Breznak, J. A. 1978. Heterotrophic bacteria present in the hindguts of wood-eating termites [*Reticulitermes flavipes* (Kollar)]. *Appl. Environ. Microbiol*. 35:930–36
117. Seiler, W. 1984. Contribution of biological processes to the global budget of CH_4 in the atmosphere. See Ref. 93, pp. 468–77
118. Shively, J. M. 1974. Inclusion bodies of prokaryotes. *Ann. Rev. Microbiol*. 28:167–87
119. Snipes, B. T., Tauber, O. E. 1937. Time requirement for food passage through the alimentary tract of the cockroach *Periplaneta americana*. *Ann. Entomol. Soc. Am*. 30:277–84
120. Sohal, R. S., Peters, P. D., Hall, T. A. 1976. Fine structure and X-ray microanalysis of mineralized concretions in the Malpighian tubules of the housefly, *Musca domestica*. *Tissue Cell* 8:447–58
121. Sohal, R. S., Peters, P. D., Hall, T. A. 1977. Origin, structure, composition and age-dependence of mineralized dense bodies (concretions) in the midgut epithelium of the adult housefly, *Musca domestica*. *Tissue Cell* 9:87–102
122. Stanton, T. B., Savage, D. C. 1984. Motility as a factor in bowel colonization by *Roseburia cecicola*, an obligately anaerobic bacterium from the mouse caecum. *J. Gen. Microbiol*. 130:173–83
123. Stumm, C. K., Gijzen, H. J., Vogels, G. D. 1982. Association of methanogenic bacteria with ovine rumen ciliates. *Br. J. Nutr*. 47:95–99
124. Taylor, E. C. 1982. Role of aerobic microbial populations in cellulose digestion by desert millipedes. *Appl. Environ. Microbiol*. 44:281–91
125. Ulrich, R. G., Buthala, D. A., Klug, M. J. 1981. Microbiota associated with the gastrointestinal tract of the common house cricket, *Acheta domestica*. *Appl. Environ. Microbiol*. 41:246–54
126. Van Bruggen, J. J. A., Zwart, K. B., van Assema, R. M., Stumm, C. K., Vogels, G. D. 1984. *Methanobacterium formicicum*, an endosymbiont of the anaerobic ciliate *Metopus striatus* McMurich. *Arch. Microbiol*. 139:1–7
127. Veivers, P. C., O'Brien, R. W., Slaytor, M. 1980. The redox state of the gut of termites. *J. Insect Physiol*. 26:75–77
128. Wall, B. J., Oschman, J. L. 1973. Structure and function of rectal pads in *Blatella* and *Blaberus* with respect to the mechanism of water uptake. *J. Morphol*. 140:105–18
129. Waterhouse, D. F., Hackman, R. H., McKellar, J. W. 1961. An investigation of chitinase activity in cockroach and termite extracts. *J. Insect Physiol*. 6:96–112
130. Waterhouse, D. F., McKellar, J. W. 1961. The distribution of chitinase activity in the body of the American cockroach. *J. Insect Physiol*. 6:185–95
131. Wharton, D. R. A., Wharton, M. L., Lola, J. E. 1965. Cellulase in the cockroach, with special reference to *Periplaneta americana*. *J. Insect Physiol*. 11:947–59
132. Wigglesworth, V. B. 1972. *The Principles of Insect Physiology*. London: Chapman & Hall. 7th ed. 827 pp.
133. Wilson, A. D. 1963. The influence of diet on the development of parotid salivation and the rumen of the lamb. *Aust. J. Agric. Res*. 14:226–38
134. Wrong, O. M., Edmonds, C. J., Chadwick, V. S. 1981. *The Large Intestine: Its Role in Mammalian Nutrition and Homeostatis*. New York: Wiley. 217 pp.
135. Zimmerman, P. R., Greenberg, J. P., Wandiga, S. O., Crutzen, P. J. 1982. Termites: a potentially large source of atmospheric methane, CO_2 and molecular hydrogen. *Science* 218:563–65

Ann. Rev. Microbiol. 1987. 41:645–75

LEPROSY AND THE LEPROSY BACILLUS: Recent Developments in Characterization of Antigens and Immunology of the Disease

Harvey Gaylord and Patrick J. Brennan

Department of Microbiology, Colorado State University, Fort Collins, Colorado 80523

CONTENTS

INTRODUCTION

Leprosy is among the most prominent severe communicable diseases that may be described as true pestilences and is one that has evoked singular images of horror and fascination throughout history. A unique fear and stigma are associated with the disease because of deformities that occur in less than 30%

0066-4227/87/1001-0645$02.00

of afflicted people. Thirteen million people are now estimated to have leprosy; this partly explains the recent renaissance of scientific interest in the disease. Leprosy also offers exceptional possibilities for gaining insight into immunoregulatory mechanisms in humans. It is not a single clinical entity, but a spectral disease; immune responses to *Mycobacterium leprae* correlate directly with the clinical, bacteriological, and histopathological manifestations of illness. Although the leprosy bacillus was the first bacterial pathogen of humans identified, it remains one of the few that cannot be grown in culture. Nevertheless, the development of the armadillo as a relatively rich source of the bacillus and the publication of a simple protocol for recovery of pure *M. leprae* (28a) have resulted in the availability of unprecedented quantities of the bacillus. The past few years have seen several major developments in leprosy research. The highly antigenic surface glycoconjugates of the bacillus have been defined chemically, particularly phenolic glycolipid I (Figure 1); a revolution in the serodiagnosis of the disease has resulted. The preparation of an array of murine monoclonal antibodies has allowed recognition of at least five highly immunogenic proteins. Recombinant genomic libraries in λ phage, representative of the entire genome of *M. leprae,* have been produced; screening of the gene products with monoclonal antibodies has permitted the identification of recombinant clones expressing the immunogenic proteins of *M. leprae*. The selective anergy of lepromatous patients has been associated with T8 suppressor cells; these cells have been localized in vivo and cloned, and there is evidence of their recognition by the phenolic glycolipid. Finally, T4 helper cells have been cloned; this development promises that the epitopes responsible for cell-mediated and protective immunity will soon be identified.

Other significant events which are not covered in depth in this review are successful vaccination with a mixture of live BCG and killed *M. leprae;* the development of several primate models for leprosy and the recognition of

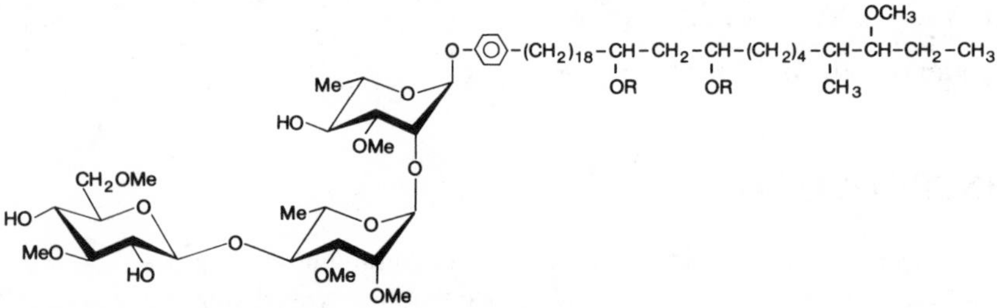

Figure 1 The major phenolic glycolipid of *M. leprae*. *R* = Mixture of C_{30}, C_{32}, and C_{34} mycocerosyl fatty acyl groups.

enzootic leprosy; developments on the issue of genetic susceptibility to leprosy; progress in the treatment of leprosy; and some success in the cultivation of *M. leprae*. A major work (59) has just been issued that covers these and other topics not dealt with here.

THE CARBOHYDRATE- AND LIPID-CONTAINING ENVELOPE OF *MYCOBACTERIUM LEPRAE*

The overt antigens of *Mycobacterium leprae,* like those of all mycobacteria, are carbohydrate based. We now know enough about the composition of *M. leprae* to realize that its carbohydrate-containing antigens are comparatively simple; perhaps in vivo growth ensures only essential anabolism and a modicum of autolysis. The dominant carbohydrate-containing epitopes of *M. leprae* are contained within only three entities: the phenolic glycolipids, lipoarabinomannan, and the arabinogalactan-peptidoglycan complex.

Phenolic Glycolipids

CHEMISTRY The recognition of a distinct class of glycolipid antigens in *M. leprae,* the phenolic glycolipids, arose from a deliberate quest for species-specific glycolipids of the type present in atypical mycobacteria (8). In initial studies a partially purified lipid produced lines of precipitation with antisera from two lepromatous leprosy patients and an infected armadillo; there was no reaction to sera from patients with tuberculosis or *Mycobacterium avium* infection (10). The activity in the lipid fraction was unaffected by mild alkali, which suggests that the lipid was related to one of the mycoside classes. An important development in terms of isolating the glycolipid in quantities sufficient for chemical analysis was the realization that it occurred in enormous quantities in tissue surrounding foci of infection (61). Infrared and NMR spectroscopy showed that the glycolipid was closely related to mycoside A from *Mycobacterium kansasii,* clearly indicating that it was of the oligoglycosylphenolic phthiocerol diester class. However, the product from *M. leprae* differed in that it contained a unique combination of sugars, which were identified as 3,6-di-*O*-methylglucose, 2,3-di-*O*-methylrhamnose, and 3-*O*-methylrhamnose (61). The structure of the oligosaccharide entity was subsequently determined by partial acid hydrolysis, permethylation, ^{1}H NMR, and ^{13}C NMR to be 3,6-di-*O*-methyl-β-D-glucopyranosyl(1→4)2,3-di-*O*-methyl-α-L-rhamnopyranosyl(1→2)3-*O*-methyl-α-L-rhamnopyranosyl→ (63).

Acid hydrolysis of the deacylated glycolipid yielded a phenolic phthiocerol core. Mass spectrometry and proton NMR of the permethylated core established the structure (63):

$$\rightarrow\text{phenol}-CH_2-(CH_2)_{17}-\underset{\displaystyle OH}{\underset{|}{CH}}-CH_2-\underset{\displaystyle OH}{\underset{|}{CH}}-(CH_2)_4-\underset{\displaystyle CH_3}{\underset{|}{CH}}-\overset{\displaystyle OCH_3}{\overset{|}{CH}}-CH_2-CH_3.$$

Combined gas-liquid chromatography and mass spectrometry showed three tetramethyl branched mycocerosic acids esterified to the hydroxyl functions of the branched glycolic chain. Thus the full structural details of the group-specific dimycocerosyl phenolic phthiocerol and its species-specific triglycosyl entity were established (Figure 1).

Two further phenolphthiocerol diester triglycosides have been isolated from *M. leprae* and fully characterized (40, 62): the phenolic glycolipid II (PGL-II), 3,6-di-*O*-methyl-β-D-glucopyranosyl(1→4)3-*O*-methyl-α-L-rhamnopyranosyl(1→2)3-*O*-methyl-α-L-rhamnopyranosyl→, and the phenolic glycolipids III (PGL-III), 6-*O*-methyl-β-D-glucopyranosyl(1→4)2,3-di-*O*-methyl-α-L-rhamnopyranosyl(1→2)3-*O*-methyl-α-L-rhamnopyranosyl →. PGL-II differs from PGL-I in that it lacks an *O*-CH_3 group at C-2 of the penultimate rhamnose, and PGL-III differs in that it lacks an *O*-CH_3 group at C-3 of the distal glucose residue. Independent studies on the structure of the triglycosyl portion of PGL-I (121) are in agreement with the above findings. In addition, nonphenylated, nonglycosylated dimycocerosyl-phthiocerols have been isolated in large quantities from infected armadillo tissue and characterized completely (31, 62) (n = 16, 18; R = mycocerosic acids):

$$CH_3-(CH_2)_n-\underset{\displaystyle OR}{\underset{|}{CH}}-CH_2-\underset{\displaystyle OR}{\underset{|}{CH}}-(CH_2)_4-CH-\overset{\displaystyle OCH_3}{\overset{|}{CH}}-CH_2-CH_3.$$

From the earlier work of Young (130) and later works by Izumi et al (67) and Vemuri et al (126), it is certain the PGL-I and the diacylphthiocerol are also present in large quantities in human lepromas. Seeking further proof for the structure of the unique oligoglycosyl units of the PGLs, Fujiwara et al (38–42) and J. Gigg, R. Gigg, et al (47–49) synthesized the terminal diglycosyl and the entire triglycosyl units or their variations, either as reducing saccharides or combined with various aglycons that make them suitable for subsequent conjugation to carrier proteins.

SUBCELLULAR LOCATION OF PGL-I Early investigators of the microscopic properties of *M. leprae* in situ noted capsular matrices and foamy structures that bound the organism into globi, and decided that they were lipids (55). When the phthiocerol-containing lipids were discovered they were immediately thought to correspond to these bodies. The bulk of the lipids were found in supernatants of homogenized infected tissue after the bacteria had been removed (63), and the quantities of diacylphthiocerol and phenolic glycolipid recovered far exceeded what was expected from the bacillary load; 3.7×10^{10} *M. leprae* within infected tissue (1.44 mg dry weight) produced about 3 mg of cell-dissociated phthiocerol-containing lipids (44, 62, 63). Fukunishi et al (43) observed "small spherical droplets," which obviously correspond to the "peribacillary substances" and "electron-transparent zone" of earlier literature, and concluded that they were composed largely of the phthiocerol-containing lipids, although these bodies were not isolated and analyzed. In recent indirect immunofluorescence experiments, the use of an IgM monoclonal antibody that selectively binds to the distal 3,6-di-*O*-methyl-β-D-glucopyranosyl unit of PGL-I resulted in a fluorescent zone surrounding the bacilli; this clearly demonstrated the surface location of the glycolipid antigen (136).

ANTIGENICITY OF PGL-I AND USE IN SERODIAGNOSIS OF LEPROSY Initially, the antigenicity of PGL-I was merely inferred (10). Payne et al (107) incorporated PGL-I into liposomes that could diffuse in agarose gels and demonstrated precipitation with undiluted sera from patients with active lepromatous leprosy but not with sera from patients with active pulmonary tuberculosis. Several groups simultaneously developed ELISA conditions that were more in accord with the extreme hydrophobicity of native PGL-I. One such assay (25), which provided the basis for most current protocols, involved sonication of the glycolipid in a detergent-free coating buffer to yield small lipid vesicles. The sensitivity of the assay was impressive; 2 μg ml^{-1} buffer (100 ng per well) produced an absorption value of 0.972 ± 0.100 with pooled lepromatous leprosy sera diluted to 1:300. Under these conditions, IgM accounted for about 90% of the total measured immunogluobulins in the antibody response of human sera to PGL-I (25, 132); partially deglycosylated PGL-I, the product devoid of the terminal 3,6-di-*O*-methylglucose, lost most of its ability to recognize the IgM antibodies (40). Likewise, murine monoclonal IgM (136) and monoclonal IgG (7, 79) require the 3,6-di-*O*-methylglucose. Thus this sugar is implicated as the primary epitope of PGL-I.

Young & Buchanan (131) independently developed an ELISA in which the deacylated form of the glycolipid was used. Brett et al (11) used detergent to solubilize the native glycolipid. There was considerable accord in the results reported from these three independent studies: (*a*) Antibodies directed against the glycolipid were seen in sera from leprosy patients but not in sera from uninfected controls or patients infected with mycobacteria other than *M. leprae*. (*b*) Over 90% of clinically diagnosed lepromatous leprosy patients demonstrated antiglycolipid antibodies, compared to less than 50% of tuberculoid leprosy patients. (*c*) Treatment of lepromatous leprosy patients resulted in significantly lower antiglycolipid antibodies. (*d*) The antibody response distinguished between the phenolic glycolipid from *M. leprae* and the structurally related phenolic glycolipids from *Mycobacterium bovis* and *M. kansasii*. The aglycon segment was devoid of activity in one set of studies (25), but exhibited slight cross-reactivity in other later studies (13). All of the investigators agreed that the glycolipid has great potential for the specific diagnosis of lepromatous leprosy. For tuberculoid leprosy, which is much more difficult to diagnose, the assays were disappointingly insensitive.

Several reports have appeared on the application of PGL-I–based serology in the management of leprosy (4, 23, 77) and in gauging the undulations of erythema nodosum leprosum (3, 77). Details of a standardized ELISA, the fruit of collaborative efforts, have been issued (113). The antibody response to PGL-I, at least in immunized animals, is reported to be under genetic control (122). The antigen has been used as a specific tool in demonstrating enzootic leprosy in the armadillo population of parts of the southern United States (124). Simple serological assays for field conditions have been developed (22, 135). In addition, sensitive procedures for the quantitation of PGL-I in serum and urine have emerged; studies using these procedures revealed a rapid decline in the levels of serum PGL-I with the onset of chemotherapy (24, 113, 133, 139). Whether PGL-I can be used for the early detection of leprosy is still moot. Implementation of PGL-I–based immunoassays in leprosy control programs is also hampered by serious issues of cost effectiveness and practicality due to low incidence rates.

NEOGLYCOPROTEINS CONTAINING THE GLYCOSYL UNITS OF PGL-I About the time PGL-I found wide acceptance as a specific antigen for the serodiagnosis of lepromatous leprosy, various semisynthetic relatives, the leprosy-specific neoglycoproteins, became available for widespread use. Fujiwara et al (40) showed in an ELISA inhibition assay that only those synthetic disaccharides with a full complement of methoxyl groups at the nonreducing end were active in binding human leprosy antiglycolipid IgM. Subsequently, it was shown that among the first of the neoglycoproteins [those derived by reductive amination of the synthetic disaccharides (23, 40)], only the *O*-(3,6-di-*O*-methyl-β-D-glucopyranosyl)-(1→4)-(1-deoxy-2,3-di-*O*-methyl-L-rhamnitol-lysyl)-bovine serum albumin was serologically active, whereas those based on the 3-mono-*O*-methyl-β-D-glucopyranoside or the nonmethylated β-D-glucopyranoside were inactive. A variety of disaccharide-containing neoglycoproteins are now available for the worldwide diagnosis of leprosy (9). The second generation neoglycoconjugates synthesized by Chatterjee and colleagues (20, 21) used an 8-methoxycarbonyloctyl linker arm; those produced by Fujiwara et al (41, 42) used a phenylpropionate link; and a third set (14, 47–49) contained the simplest aglycon, based on an allyl group. Extensive interlaboratory testing of these neoglycoproteins (113) indicated that all are comparable, all are highly sensitive and specific for leprosy, and all, being water soluble, are applicable for a wide variety of serological tests (22).

Recently, in a joint project conducted under the auspices of the United States–Japan Cooperative Medical Sciences Program, two forms of the entire trisaccharide unit of PGL-I were synthesized: the *O*-(3,6-di-*O*-methyl-β-D-glucopyranosyl)-(1→4)-*O*-(2,3-di-*O*-methyl-α-L-rhamnopyranosyl)-(1→2)-*O*-(3-*O*-methyl-α-L-rhamnopyranosyl)-(1→9)-oxynonanoyl-bovineserum albumin (natural trisaccharide-octyl-BSA) (NT-O-BSA) (19) (Figure 2*A*) and the *O*-(3,6-di-*O*-methyl-β-D-glucopyranosyl)-(1→4)-*O*-(2,3-di-*O*-methyl-α-L-rhamnopyranosyl)-(1→2)-*O*-(3-*O*-methyl-α-L-rhamnopyranosyl)-phenylpropionyl-BSA (natural trisaccharide-phenylpropionyl-BSA (NT-P-BSA) (41) (Figure 2*B*). The two were compared in extensive studies involving sera from leprosy patients in the United States, the Philippines, Micronesia, and Japan. NT-O-BSA and NT-P-BSA were nearly identical in ability to detect antiglycolipid IgM antibodies in patients with various forms of clinical leprosy. The presence of the phenyl groups in NT-P-BSA did not confer any cross-reactivity against high-reactor tuberculosis sera. In addition, the sensitivity rate of NT-O-BSA and NT-P-BSA was appreciably higher than that of PGL-I or the disaccharide-containing neoglycoproteins; for instance, in one study involving a group of patients from the United States, 106 were found seropositive when NT-O-BSA and NT-P-BSA were used, compared to 97 when PGL-I was used. More importantly, NT-O-BSA and NT-P-BSA showed less propensity to react nonspecifically with sera from an asymptomatic control population; sera from only three (1.8%) of 169 patients in the nonendemic areas of the United States reacted positively to NT-O-BSA, whereas there were seven positive reactions (4.1%) to PGL-I. Clearly, the trisaccharide-containing neoglycoproteins are superior for large-scale serodiagnosis of leprosy.

Lipoarabinomannan and Lipomannan

Several authors (15, 18, 66, 108) noted that upon electrophoresis and immunoblotting, soluble fractions of disrupted *M. leprae* produced a major antigen with an apparent molecular mass of 30–50 kd. Owing to its ready reaction with periodate-Schiff reagents and its resistance to proteolysis, this antigen was thought to be a glycoprotein.

A

O-CH2-CH2-CH2-CH2-CH2-CH2-CH2-CH2-CO-NH-BSA

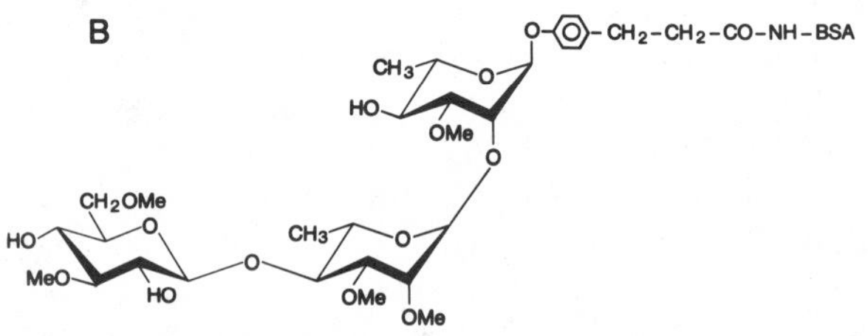

Figure 2 The two trisaccharide-containing neoglycoproteins now available for leprosy serodiagnosis. *A*: The natural trisaccharide-octyl-BSA (NT-O-BSA). *B*: The natural trisaccharide-phenylpropionyl-BSA (NT-P-BSA). Both have 30–40 substitutions per polypeptide chain.

This product has now been identified as the highly anionic, highly immunogenic lipoarabinomannan (LAM) (64). LAM was first recognized as a major soluble, carbohydrate-containing polymer that was extractable from *M. leprae* with ethanol, excluded from gel filtration columns, and highly reactive against lepromatous leprosy sera and a large assortment of monoclonal antibodies. The key to its ultimate purification was anion exchange and gel filtration chromatography in detergent-containing buffers. On polyacrylamide gel electrophoresis (PAGE) it yielded a broad, diffuse band in the vicinity of proteins of 30–35 kd. Besides arabinose and mannose, it contains glycerol, inositol, and phosphate. It is acylated by lactate, succinate, palmitate, and 10-methyloctadecanoate (tuberculostearate). It now appears that the inositol-phosphate is present in the form of phosphatidylinositol. LAM appears to be anchored on phosphatidylinositol or on phosphatidylinositol mannoside, a member of a group of phosphoglycolipids confined to *Mycobacterium* and related genera. Thus, LAM may be an amphipathic multiglycosylated extension of the phosphomannoinositides. There is evidence for subtle structural differences in the LAM from *M. leprae* and *Mycobacterium tuberculosis:* Although many of the monoclonal antibodies reactive to the carbohydrates of *M. leprae* (36, 66) recognize LAM from *M. leprae,* they do not react readily with LAM from *M. tuberculosis* (45). It is now thought that the LAM from *M. tuberculosis* contains a set of alkali-labile inositol-phosphates (64) in addition to the phosphatidylinositol, and they may affect the binding of some antibodies (45). Sera of leprosy patients, including those with tuberculoid leprosy, contain high-titer antibodies to LAM (9, 84), and a combination of PGL-I and LAM may provide the ultimate means for the serodiagnosis of incubating leprosy.

M. leprae and *M. tuberculosis* contain a second major carbohydrate-containing polymer which is readily seen on polyacrylamide gels, but which is not antigenic; this has been called LAM-A (64). However, it is virtually devoid of arabinose and is probably a lipomannan. It also contains *myo*-inositol-1-phosphate and is nonantigenic owing to the absence of arabinofuranosyl (Ara*f*) distal residues.

Arabinogalactan-Peptidoglycan Complex

Draper (28, 30) has demonstrated that the basic peptidoglycan subunit of the cell wall of *M. leprae* has features that are both typical of mycobacteria and specific for *M. leprae*. Thus, like that of other mycobacteria, it contains *N*-glycolylmuramic acid rather than *N*-acetylmuramic acid. However, unlike the peptidoglycan subunit of all other mycobacteria, it has glycine rather than L-alanine at the amino terminus of the tetrapeptide side chain (28). In addition, the detailed structure of the cell-wall mycolic acids and other fatty acids of the cell-wall mycolyl-arabinogalactan have been established (76, 86); they do not differ significantly from those of other mycobacteria. Recently, by a complex chemical process, McNeil et al (78) established that the arabinogalactan of *M. leprae* and *M. tuberculosis* contains exclusively arabinofuranosyl and arabinogalactosyl residues. Furthermore, they showed that the galactosyl residues are either 5-linked galactosylfuranosyl, 6-linked galactofuranosyl, or 5,6-linked galactofuranosyl. Thus much of the structure of the highly immunogenic arabinogalactan of *M. leprae* can now be deduced. Heteropolysaccharides containing solely furanosyl residues are rare in nature; thus the peculiar chemical properties of the copious cell wall arabinogalactan may have important biological implications. Nothing is yet known of the nature of the linkage between the arabinogalactan and the peptidoglycan unit of *M. leprae*. We expect that the dominant antigenicity of the mycolyl-arabinogalactan-peptidoglycan complex of *M. leprae* resides in the terminal 5-linked Ara*f* units. Thus the complex should elicit an antibody response indistinguishable from the response to LAM.

Relationships of Major Cell-Wall Components

Figure 3 represents the topography of the different known entities of the cell wall of *M. leprae* based on their physicochemical properties as described above. This schematic diagram incorporates ultrastructural evidence from Rastogi et al (109) that the cell wall in *M. leprae* has three layers. The innermost is an electron-dense peptidoglycan layer. Next is an electron-transparent zone, which may well equate with Minnikin's "hydrophobic interaction area" (85). This layer may be composed of acyl chains contributed partly by the mycolyl-arabinogalactan and partly by the phenolic glycolipid and dimycocerosyl phthiocerol. A major contributor to the third layer, the so-called polysaccharide outer layer, which is a feature of all mycobacteria (110), may be the triglycosyl chain of PGL-I and the distal segments of LAM. However, the interpretation of many electron micrographs of many mycobacterial cell walls (29) is apparently not in accord with the concept of a lipid permeability barrier in *M. leprae*. The model in Figure 3 is merely a first attempt to provide some conceptual order to the known cell-wall components of *M. leprae*.

THE PROTEIN ANTIGENS OF M. LEPRAE

Use of Monoclonal Antibodies

Polyacrylamide gel electrophoresis of disrupted *M. leprae* shows an array of more than 50 proteins distributed in both supernatant and pelleted fractions, ranging from ~10 to over 100 kd (15, 18). Recent interest has focused on five of these proteins, those of about 65, 36, 28, 18, and 12 kd. In the first applications of murine monoclonal antibodies to protein delineation in *M. leprae,* the 65-kd protein was selected (51). This was the first protein shown, through the use of monoclonal antibodies, to contain both cross-reactive and

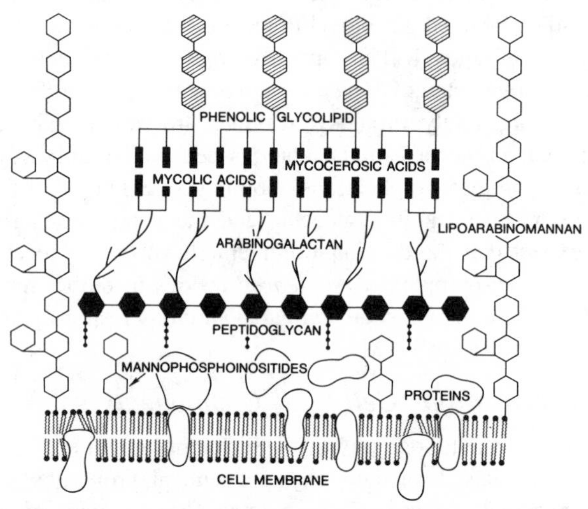

Figure 3 Schematic diagram of the interrelationships of the major cell wall components of *M. leprae*.

M. leprae–specific epitopes (51, 52). Two monoclonal antibodies (IVD8 and IIIE9) recognized an epitope within the 65-kd molecule from *M. leprae* that was selective to this species, whereas seven other antibodies reacted with regions that were shared with other mycobacterial species. Other laboratories have produced other antibodies that also recognize the *M. leprae*–specific epitope in the 65-kd protein (36). In a recent extensive interlaboratory study, 23 monoclonal antibodies produced by seven different laboratories were examined for their patterns of reactivity with the 65-kd proteins from several mycobacterial species (16). A series of cross-competition studies delineated 14 distinct epitopes; only one of these is *M. leprae*–specific, whereas the others are found to variable extents in the products from other mycobacteria. In addition, most of those antibodies reacted with the recombinant 65-kd protein obtained from the γgt11 rDNA clone Y1089 (see below).

The 12 murine monoclonal antibodies originally produced by Ivanyi and colleagues (65, 66) recognized at least four soluble antigens: a 12-kd protein, a 28-kd protein, and several other proteins in the region of 35–70 kd. One of these antibodies, ML06, showed unique specificity for the 12-kd protein from *M. leprae*. One of the monoclonal antibodies (F47-9) produced by Kolk et al (74) and Klatser et al (73) was responsible for the identification of the 36-kd protein antigen as a distinct immunogen of *M. leprae*. Presently, there are some indications that *M. leprae* also has 33-kd and 35-kd protein antigens. The 11 antibodies produced by Britton et al (15) were effective in recognizing proteins of 16 kd (which may be equivalent to the 18-kd protein) and 70 kd. Several of the antibodies described by Britton et al (15) and Ivanyi et al (66) were obviously directed to the highly immunogenic LAM. Young et al (134) discovered a 28-kd protein with the pattern of species specificity and cross-reactivity first seen in the 65-kd product, i.e. within one molecule there were *M. leprae*–specific, partially conserved, and highly cross-reactive epitopes. Incidentally, monoclonal antibodies directed to specific epitopes on several of the protein antigens are now being used in competitive inhibition assays for the serodiagnosis of leprosy (72, 119); some of these assays look promising at the tuberculoid end of the disease spectrum, and thus should complement assays based on PGL-I.

Although the 12-, 18-, 28-, 36-, and 65-kd proteins were selected through the use of murine monoclonal antibodies, they do seem to have an important role in cellular immunity in human leprosy (see succeeding section). Little is known of these proteins' physiological function or subcellular location within the bacillus; the 65-kd protein is tenaciously associated with the isolated cell wall (52). No effort has yet been made to relate proteins recognized previously in crossed immunoelectrophoresis (56) to those recently defined with monoclonal antibodies.

Molecular Cloning of M. leprae *Genes and Synthesis of Peptide Epitopes*

Recombinant DNA technology offers one means of overcoming some of the problems that arise from microbiologists' inability to cultivate *M. leprae* in vitro. Libraries of *M. leprae* DNA sequences have been constructed in *E. coli* using two somewhat different systems (26, 138). An earlier report (26) described the cloning of several mycobacterial DNAs into cosmid vectors, from which an expression vector (pYA626) was then constructed. The mycobacterial DNA was spliced into a plasmid next to an RNA promoter, which had also been placed in the plasmid by recombinant DNA procedures. While it appeared that new polypeptides were synthesized, presumably originating from *M. leprae* sequences, no proteins that were reactive with the serum of leprosy patients could be detected. More recently, Jacobs et al (68) described the complementation of a mutation in the citrate synthetase gene of *E. coli* K-12 by cloned *M. leprae* DNA that was expressed from the *asd* promoter of pYA626.

Young and his colleagues (138) obtained successful expression of protein antigens of *M. leprae* by following an approach that had been developed for *M. tuberculosis* (137). Several elements in the strategy contributed to its success. The sequences were ligated with a phage λ system (λgt11) so that the inserted *M. leprae* DNA was placed within the gene coding for β-galactosidase. Transcription was therefore placed under control of the Lac operator, which is silent until it is induced with a lactose analog. After induction, the *M. leprae* sequences were translated as a fusion protein attached to the β-galactosidase polypeptide, which helped protect the *M. leprae* sequences from degradation. The generation of the original *M. leprae* DNA sequences was randomized by mechanical shearing to assure insertion in every possible reading frame. Upon induction and expression of the fusion proteins, as many as 10^6 different phage plaques could be screened in a single experiment for *M. leprae*–derived polypeptides, using the antiprotein monoclonal antibodies described above. What is most encouraging is that sequences from all five of the major *M. leprae* protein antigens could be detected. Seven of the anti–65-kd protein monoclonal antibodies, each recognizing a different epitope, were used to isolate 12 recombinant clones, whose sequences could be aligned to demonstrate that a single gene encodes for the 65-kd protein. Recently, Mehra et al (83) sequenced the gene encoding the 65-kd protein and constructed a λgt11 gene sublibrary with fragments of the gene. They isolated recombinant DNA clones producing specific antigenic determinants by screening with the available monoclonal antibodies, and determined the sequence of their insert DNAs with a rapid primer-extension method. They then deduced the amino acid sequence of each determinant from the minimum overlap of insert DNAs from multiple antibody–positive DNA clones. In this way, the amino acid sequences for six different epitopes were elucidated. Furthermore, Mehra et al (83) synthesized the 15-residue peptide unique to *M. leprae* and showed that it is recognized only by the IIIE9 antibody. Using the predicted sequence of the 65-kd protein and solid-phase peptide synthesis, Anderson et al (2) recently demonstrated that the shortest peptide that binds the highly cross-reactive IIIC8 monoclonal antibody is DPTGGMGGMD and that the shortest that binds another cross-reactive antibody (IIC8) is EDLLKAGV. Undoubtedly, the synthetic versions of the other 14 epitopes are forthcoming.

The key to the successful molecular cloning of *M. leprae* genes is the methodology by which *M. leprae* products can be detected. To date, enzyme-linked immunoassay based on monoclonal antibodies remains our only useful handle, and these antibodies are confined to the small collection of protein antigens that have been defined. In principle it is possible to screen for mycobacterial enzymatic activities, but we are not yet sufficiently knowledgeable about what reactions to seek. The glycolipid and the lipopolysaccharide antigens may be of importance to the physiology and immunology of leprosy, but it is not likely that these complex glycoconjugates will ever be synthesized via recombinant DNA technology. It is conceivable that continued study of cultivable mycobacteria could identify specific biosynthetic enzymes (glycosyl transferases or methylases, for instance) encoded by genes that could be found through sequence homology studies of recombinant DNA techniques.

The immediate benefit of the molecular cloning of *M. leprae* genes will be to provide a large potential source of protein antigen or peptides for immunologic studies and possible vaccine manufacture. Some of the recombinant protein products have already been used as antigens in the in vitro study of T-cell mediated immunity (see section on T-cell lines and T-cell clones).

CELLULAR IMMUNITY IN HUMAN LEPROSY

Most people exposed to the leprosy bacillus develop a protective immunity to the organism. Following skin challenge with lepromin (killed *M. leprae*), such individuals develop a granulomatous reaction considered to be a delayed type hypersensitivity (DTH) response, analogous to the positive PPD reaction indicative of exposure to *M. tuberculosis*. When clinical disease does arise, the organism infects mainly the dermis. It may cause a wide range of symptoms, which can be classified according to histopathologic and immunologic criteria (111). Cases are classified on a spectrum of severity, from the least severe tuberculoid form (TT) through a borderline (BB) classification to the polar lepromatous (LL) form. There are also intermediate borderline tuberculoid (BT) and borderline lepromatous (BL) classifications.

Several recent reviews (69, 114) have discussed the histopathological findings in some detail. Briefly, the tuberculoid patient presents with a few dermal lesions, well-ordered granulomas at the site of infection, and with many epithelioid cells and multinucleate giant cells surrounded by a mantle of lymphocytes. Acid-fast bacilli are extremely rare, attesting to a healthy control of bacillary growth. Local nerve damage may result from a nonspecific bystander effect caused by the vigorous inflamatory reaction (129). Tuberculoid patients show strong positive skin-test reactions when challenged with lepromin. At the lepromatous pole, the patient may have numerous small lesions, particularly on the hands and face. The dermal infiltrates no longer appear as organized granulomas, but show large numbers of heavily parasitized macrophages. These macrophages are usually described as foamy, owing to large amounts of the *M. leprae*–derived phthiocerol-containing lipid within them. No giant cells are observed, and the relatively few lymphocytes appear to be dispersed among the macrophages rather than organized at the peripheral area. Lepromatous patients show virtually no response to lepromin challenge in the skin. The spectrum between these poles is graded, and patients may move on the spectrum, depending on drug treatment and unknown environmental factors.

In concert with the skin-test reactions, in vitro estimates of cell-mediated immunity toward leprosy show decreasing intensity from the TT pole toward the LL pole. The lepromatous donor's cells are entirely or nearly anergic. Neither the skin-test anergy nor the lack of immune response in vitro is considered to represent a generalized immune dysfunction in lepromatous patients, who will usually respond well not only to unrelated antigens, but to other highly cross-reactive mycobacteria. While not entirely paradoxical by some views of immune regulation, this phenomenon is nonetheless difficult to explain mechanistically. There is a general consensus among researchers on the observations described above and agreement that successful immune defense against *M. leprae* follows the common pattern for intracellular pathogens: The organism undergoes nonspecific uptake by phagocytic cells; antigens are presented to thymus-derived lymphocytes (T cells), which proliferate to provide specific immunologic memory; and these T cells provide feedback to the phagocytic cells, enhancing their ability to kill the intracellular bacteria. This generalized scheme of antibacterial immunology is the subject of a recent extensive review (54). Just where this scheme breaks down, giving rise to the spectrum of leprous disease, particularly at the lepromatous end of the scale, is the subject of intense current research, not a little speculation, and much disagreement.

The bulk of recent research derives not so much from the available mouse models but more from in vitro cellular experiments on cells donated by leprosy patients. These experiments typically compare normal, tuberculoid, and lepromatous cells; the assumption is that differences between the polar forms will identify key aspects of pathology. The specificity of unresponsiveness in lepromatous leprosy has been regarded as a T-cell phenomenon. The central question is whether T cells that are immune to *M. leprae* fail to expand in the LL patient, or whether they do in fact proliferate but are actively suppressed. The host cell for *M. leprae* is the macrophage, which functions in the afferent arm of the immune system as antigen presenter and in the efferent arm as microbicidal agent. Some researchers favor these cells as the site of primary immune defect. Antibody responses to *M. leprae* antigens are strong in lepromatous patients. The inverse correlation between cell-mediated immunity and humoral immunity is usually noted but rarely discussed, although it seems somewhat paradoxical. Some interesting speculations on the role of antibodies in the pathogenesis of leprosy have been published (37, 127), but virtually no compelling data have been presented. The immunology of leprosy has been reviewed quite often in recent years (6, 7, 17, 44, 50, 69, 95, 114, 127).

Studies on Uncloned Patient Cells

Early reviews (17, 114) cited considerable data indicating that lepromatous patients display a generalized defect in cell-mediated immunity. These data were derived from tests of skin sensitization to picryl chloride and in vitro blastogenic assays of reactions of patient lymphocytes toward various antigens. Bullock (17) noted that while the inability of lepromatous patients to respond to *M. leprae* was profound, the generalized immunologic anergy associated with several forms of leprosy is comparatively weak. Having pointed out that lepromatous patients do not seem highly susceptible to other infections, he favored the view that the lepromatous patient is specifically anergic, i.e. tolerant, in an operational sense, to *M. leprae*. More recent reviews continue to take this view, stressing that the lepromatous patient will give a positive PPD skin test in spite of the fact that tuberculin and lepromin are cross-reactive.

Antigen specificity in cell-mediated immunity is attributed to T cells. While many other cell types continue to be of interest, there is now nearly a consensus that the proper functioning of *M. leprae*–specific T cells is exhibited by tuberculoid but not lepromatous patients. Predictably, the question of whether T suppressor cells (46) are involved has been addressed. The question cannot be answered unequivocally as of this writing, but reflects a major thrust of current research.

SUPPRESSOR T CELLS IN PERIPHERAL BLOOD Mehra et al (79–82) published the first body of work indicating a role for suppressor T cells in lepromatous disease. More recently Bloom & Mehra (7) summarized this work. They initially described both adherent and nonadherent suppressor cell populations in peripheral blood of lepromatous patients, and proceeded to characterize the nonadherent cell cases. They postulated that a T-cell population of the expected phenotype (OKT8$^+$, or T8, the cytotoxic/suppressor phenotype) arises only in the lepromatous patients and responds to lepromin as a negative modulator of cell-mediated immunity (7). The apparent active constituent in lepromin is the phenolic glycolipid I antigen, for which the suppressor T cell apparently has exquisite specificity (79). The model explains the paradox of specific T-cell anergy: An antigen-specific T suppressor cell that can act locally to depress responses to other antigens must therefore operate through a nonspecific mechanism; blockage of synthesis or utilization of interleukin 2 (IL-2) by helper T cells was suggested (79). While pleasingly self-consistent, the model has some problems. The basic assay for suppression involves adding antigen (lepromin or the phenolic glycolipid) to a concanavalin A (ConA) mitogenic response, causing lepromatous but not tuberculoid or normal cells to show a diminished response. In some cases, cell-mixing experiments indicated that lepromatous cells could depress a normal donor's ConA response, which means that the suppression shows no genetic restriction (81, 82). In these cell-mixing experiments there was no case in which a lepromatous cell suppressed an antigen-specific response by cells from another individual. Removal of the OKT8$^+$ cells from the lepromatous cell population could restore ConA proliferation, but only occasionally made the cells responsive to lepromin (80). When these aspects of cellular immunity were addressed by a number of other laboratories, a much less coherent picture emerged. Interest in suppressor T cells has not subsided, however, as the section on cloned T cells (below) shows.

Nath et al (97–99) have repeatedly attempted to demonstrate lymphocyte-mediated suppression of either ConA or *M. leprae*–driven responses using both cell-separation and cell-mixing experiments with HLA-matched cells from differing individuals (97, 99). Nath (95) finally concluded that T cell–mediated suppression was evident only in the tuberculoid population, probably as part of the normal regulation of cell-mediated immunity. The presence of *M. leprae*–reactive lymphocytes in lepromatous patients was inferred in some of these experiments (97), but their proliferation in the lepromatous cells was tenfold less than in tuberculoid cells. These results are consistent with the view of Kaplan et al (69, 71) that some lepromatous patients have very weak positive responses to *M. leprae* because their *M. leprae*–specific T cells fail to expand. Stoner et al (120) and Touw et al (123) reached similar conclusions in experiments quite similar to those of Nath. T suppressor cells have not been convincingly demonstrated in functional assays using uncloned patient cells, but several lines of work (discussed below) make it difficult to rule them out.

The cell-mixing experiments of Nath et al and Stoner et al were designed in part to discern whether presentation of *M. leprae* antigens is impaired in lepromatous patients. Stoner et al (120) found lepromatous adherent cells capable of supporting proliferation of *M. leprae*–responsive lymphocytes from tuberculoid patients, but Nath et al (97) did not; the interpretation of the latter authors leans more toward a role for macrophages in actively suppressing the response (96, 115). Nath et al's claim that tuberculoid adherent cells can successfully present *M. leprae* antigens to lepromatous T cells depends on whether the weak response represents the activity of *M. leprae*–reactive clones or nonspecific recruitment.

Failure of antigen presentation has been suggested in several contexts, particularly in studies on macrophages, but direct evidence on the subject is lacking. Kaplan et al (71) have repeatedly demonstrated that the dendritic cell is capable of presenting antigen in the absence of macrophages. Thus it is theoretically possible to dissociate macrophage suppressive effects from antigen presentation.

INHIBITION BY ADHERENT CELLS Suppression of T-cell responses by adherent cells has been reported often in leprosy. There are ample explanations for such effects in immunology (reviewed in 1); some are of interest (production of interferons, hydrolytic enzymes, complement cleavage products) and some are not (release of thymidine). Rarely have these possible mechanisms been confirmed or ruled out in the context of leprosy. Many of the effects described are nonspecific, in the sense that they are not necessarily confined to *M. leprae*. In addition, some phenomena have not been assigned a cellular origin, but suggest macrophage effects.

Nath et al (97, 99) concluded that the lepromatous adherent cell populations actively suppress T-cell proliferation in tuberculoid lymphocytes. A nondialyzable, heat-stable factor was identified (96, 115) that can decrease human IL-2 production in the Jurkat human cell line. The factor is secreted spontaneously from bacilliferous macrophages and can be induced by giving *M. leprae* to bacteriologically negative lepromatous cells. Normal and tuberculoid cells do not produce the factor. The authors did not indicate whether the factor is made in response to other mycobacteria.

The Bombay group (112) derived a similar factor from the lysates of lepromatous macrophages exposed to *M. leprae*. The lysate suppressed mononuclear cell responsiveness to some mycobacteria (*M. leprae, M. avium, M. kansasii)* but not to others. However, induction of the factor in these species was not investigated.

The response of lepromatous macrophages to live *M. leprae* differs from that of normal or tuberculoid macrophages, according to one report (5). Of 12 mycobacteria tested, only *M. leprae* caused a decrease in Fc-receptor expression when both the bacilli and the macrophage were metabolically active. The authors suggested that low Fc-receptor response is an intrinsic property of certain individuals, and therefore a risk marker. There is some further evidence for this idea; its predictive value will be tested among a small group of patients (87).

Mehra et al's work on suppressor T cells set aside suppression of ConA responses by adherent cells (82), but in reviewing their work, Bloom & Mehra (7) reported that most mycobacteria induce a nonspecific suppression of proliferation mediated by alpha interferon (IFN-α); removal of the interferon allows the demonstration of T-cell suppression unique to *M. leprae*. They made no connection between the IFN-α and adherent cells, but the report illustrates the value of understanding (and removing) nonspecific effects before seeking specific ones.

A candidate for the molecule that causes some of the nonspecific immunodepression of mycobacteria is lipoarabinomannan (64). This lipopolysaccharide can be incorporated into liposomal and, presumably, natural membranes. Kaplan et al (70) recently showed that it decreases mitogenic and antigenic (PPD) responses in cells of normal, tuberculoid, and lepromatous patients. Decrease of ConA responses may be trivial in this case, as LAM binds ConA, but OKT-3–induced mitogenesis is also suppressed. It is not clear which cellular subpopulation is affected, but OKT8^{+} cells do not contribute to suppression of mitogenesis. Ellner (33) and Ellner & Daniel (34) have attributed a nonspecific immunosuppression of this sort to macrophages and accordingly to interference with antigen processing.

INTERLEUKINS Haregewoin et al (57, 58) presented data that show that T cell–conditioned medium, purified interleukin 2 (IL-2), or recombinant IL-2, when given with *M. leprae,* stimulates T-cell proliferation in lepromatous patient cells. The response is somewhat variable, but is taken to indicate that some *M. leprae* reactive T cells are present in the lepromatous patient, but are suppressed. Others found sporadic positive effects with IL-2 (102, 104), prompting Kaplan et al (71) to reassess *M. leprae* T-cell proliferative responses of a large group of patients. They concluded that lepromatous patients are either entirely anergic or very weakly positive. They also concluded that IL-2 enhances only the weak positive response, and only by a little. In no case could reversal of anergy be shown. Mohagheghpour et al (91) suggested that failure to respond to IL-2 may be due to a lack of IL-2 receptor (Tac antigen) in lepromatous disease, but the receptor appears to be present on many lymphocytes within lepromatous lesions (89).

Whether IL-1, which is necessary along with properly presented antigen for T-cell stimulation, is properly secreted is also in question. Some lepromatous patients' production of IL-I in response to lipopolysaccharide was poor according to one report (128), but another report showed that production in response to *M. leprae* was as good as that to BCG (58).

INTERFERON ACTIVATION OF MACROPHAGES Macrophages express nonspecific bacteriocidal activity, which is greatly enhanced by local production of immune interferon (IFN-γ) by antigen-reactive T cells (101, 116). Nogueira et al (102) observed that peripheral blood monocytes of lepromatous patients were defective in IFN-γ production, but some patients produced interferon when they were simultaneously given *M. leprae* and IL-2. The requirement for antigen argued again for the presence of *M. leprae*–reactive T cells, which must therefore be functionally suppressed in some of the lepromatous patients. Following this report, Horwitz et al (60) found that the microbicidal activity of lepromatous peripheral blood macrophages was defective, as assayed by the killing of *Legionella pneumophila,* but was reversed with factors from normal cells. This indicated that lepromatous macrophages are not intrinsically defective. Krahenbuhl and colleagues (118; R. C. Hastings & J. L. Krahenbuhl, personal communication) showed that in nude (T cell–deficient) mice infected with *M. leprae,* unparasitized peritoneal macrophages could be activated by IFN-γ and lymphokines to kill *Toxoplasma gondii*. Macrophages explanted from footpad granulomas and heavily parasitized with *M. leprae,* however, were severely compromised in ability to kill *T. gondii*. The same researchers have observed the tendency of live *M. leprae,* but not the irradiated bacillus, to inhibit the fusion of primary phagosomes with lysosomes following uptake into unactivated mouse macrophages. When activated with IFN-γ, phagolysosome fusion was improved and bacteriocidal activity appeared to occur (117). It is difficult, however, to assess the viability of *M. leprae*. Nathan et al (100) have recently begun clinical trials in which they apply low-dose recombinant IFN-γ to lepromatous patient lesions. Histological examination has revealed a tendency toward more normal granuloma formation, but data on clearance of *M. leprae* have not been impressive.

IMMUNOHISTOLOGICAL STUDIES Immunochemical staining techniques, particularly those afforded by monoclonal antibodies, have now been added to histopathological criteria for judging individual cases in leprosy. In particular, it is useful to know whether the distribution of helper and suppressor cells, expression of HLA antigens, and production of interleukins and their receptors in the dermal lesions are consistent with the results of functional studies, which are usually performed with cells derived from peripheral blood.

In general, there seems to be agreement that as the disease proceeds from the tuberculoid to the lepromatous pole, there is a decrease not only in the absolute number of lymphocytes but in the ratio of OKT-4^+ to OKT-8^+ (helper to suppressor) T cells (89, 94, 125). Two studies (89, 94) noted that T cells with the suppressor phenotype seemed confined to the lymphocytic mantle surrounding the central core, which is composed mainly of macrophages and epithelioid cells, in the tuberculoid granulomas. The OKT-8^+ cells, however, were interspersed throughout the lepromatous lesions (125).

Modlin et al (89) recently investigated cells staining for (and presumably secreting) IL-2. These cells, which costained for the T-helper phenotype, decreased in abundance by an order of magnitude as the disease progressed from the polar tuberculoid to the lepromatous form. Cells with receptors for IL-2 were present in equal numbers in all forms of the disease. The T-cell phenotype of these cells was not given.

Expression of HLA D-related (DR) antigens is important in antigen presentation and cell-cell collaboration. The antigens are expressed on macrophages and on activated T cells of helper or suppressor phenotype. van Voorhis et al (125) noted that greater than 80% of cells in granulomas from either end of the leprosy spectrum were HLA-DR positive. Collings et al (27) noted graded expression in the amount of antigen expressed, decreasing toward the lepromatous pole. Their quantitation was based on fluorescence intensity per volume of tissue, and their data could be reconciled with those of van Voorhis et al if it is assumed that an equivalent number of cells expressed less antigen per cell.

T-Cell Lines and T-Cell Clones

In vitro experimentation on cellular immune function has advanced rapidly with the ability to stimulate the continued division of cells; cultures of T or B cells can now be expanded without viral transformation or hybridization. Stimulation with antigens, in combination with growth factors (usually IL-2), can allow growth of presumably normal cells into lines of uncertain homogeneity, which can be cloned as homogeneous cultures for characterization of phenotype and function. A landmark publication (92) devoted to the subject of T-cell clones signaled the beginning of a flood of experimentation in which the interactions of T-cell subsets have been examined using cloned, normal cells. We caution that while such cells are presumed to be normal, continued division can always select for variants that may not reflect any particular cell in vivo. Even if the resulting line or clone is in some sense typical of those in vivo, it is very difficult to know with certainty whether, for instance, cell-mediated killing measured with many sister cytotoxic cells reflects a process that has relevance in the whole organism. Nevertheless, the technology is immensely useful, and it promises a large contribution to the study of leprosy.

Following directly from the cellular immunology work discussed earlier, several groups have reported the cloning of human T cells derived primarily from leprosy patients (35, 90, 93, 105, 106). Typically, lymphoid cells are grown using antigen such as an *M. leprae* sonicate. Autologous mononuclear cells are used as a source of antigen-presenting cells and lymphokines. These may be supplemented with exogenous growth factors such as IL-2. If a collection of T-cell clones can be amassed, it is hoped that these are representative of the population of antigen-specific T cells in the individual. Only after cloning are the details of antigen specificity, helper or suppressor phenotype, and function investigated. Only OKT-4$^+$ (T4, or helper) cells were cloned from the peripheral blood of vaccinated volunteers (93) and tuberculoid leprosy patients (35, 106); lines with T4, OKT8$^+$ (T8, or suppressor/cytotoxic), and mixed phenotype were established from a borderline lepromatous patient, with T8 cells predominating. T cells taken from the skin lesions of lepromatous patients (90) gave rise only to cells of the suppressor phenotype. Functional studies support the surface phenotype data. Considered together, these reports reiterate the importance of suppressor T cells in the immunology of leprosy. Each group has taken a slightly different tack in characterizing cells, and precise comparison is difficult.

Mustafa et al (93) isolated a large number of T-cell clones (all T4) from immunized healthy volunteers and restimulated the cells in culture with soluble *M. leprae* antigens. They concentrated mostly on the 11 of 42 clones that would not react with BCG or PPD of *M. tuberculosis*. These 11 were tested on a large panel of mycobacterial sonicates and against the five recombinant *M. leprae* proteins expressed in the phage *λ/E. coli* system of Young et al (137). Six of the eleven clones were unreactive with all of the recombinant proteins. The remaining five all proliferated in response to the 18-kd *M. leprae* protein, which must therefore carry at least one helper determinant (shared, it appears, between *M. leprae* and *Mycobacteria scrofulaceum*). Functional studies were not performed with these cells, nor has it been determined whether the apparent strong bias toward the 18-kd protein is a generality, since four of the five clones in question arose from the blood of one volunteer. Because the analysis of the *M. leprae* proteins using monoclonal antibodies indicates that both specific and cross-reactive epitopes exist on the same molecules, it would be worthwhile to know whether the 31 cross-reactive T-cell clones (those not characterized in the study) recognized any of the recombinant proteins. Reactivity of T cells with expressed recombinant protein, as opposed to extracted natural proteins, guarantees that the epitope recognized is not a contaminating carbohydrate or posttranslational modification.

Emmrich & Kaufmann (35) isolated many T4 clones from a tuberculoid patient. Because of their interest in development of vaccines, the authors described not only *M. leprae*–specific clones, but also those that cross-reacted with other mycobacteria. All of the cells secreted IFN-γ when stimulated. One mycobacterial strain, the Indian Cancer Research Center (ICRC) strain, stimulated T cells more potently than *M. leprae;* this observation is of possible relevance to production of vaccines.

T cells from tuberculoid or borderline lepromatous patients were established by a research group based in Leiden (53, 105, 106). They used autologous B cells transformed with Epstein-Barr virus as a source of antigen-presenting cells (32, 53), and thus circumvented the need to draw patient blood repeatedly to support the growth of clones. They have thus been able to begin to characterize the T cells.

Some 23 clones were established from the peripheral blood of a tuberculoid patient (106) following restimulation with whole *M. leprae;* about half of these showed broad cross-reactivity with other mycobacteria. The other half were either *M. leprae* specific or nearly so. All were of the helper phenotype. Six clones, representing all the categories of cross-reactivity, were then tested on the 12-kd and 36-kd proteins of *M. leprae,* prepared by immunoaffinity chromatography from whole *M. leprae*. Four of these proliferated in response to the 36-kd protein. Each of these four clones displayed a different pattern of reactivity toward other mycobacteria. This seems to confirm that the 36-kd protein has both species-specific and nonspecific determinants for T-helper cells. One of the clones that was reactive with the 36-kd protein responded even better to the 12-kd *M. leprae* product. It would be interesting to know whether these clones are all reactive with the *E. coli* recombinant proteins, to determine whether posttranslational modifications of the proteins could be involved. Proliferation of all of these T-cell clones is stimulated only when the correct antigen is presented by the properly matched (HLA-DR or HLA-DQ) antigen-presenting cell.

The same researchers were also able to establish T-cell clones with suppressive activity from the peripheral blood of a borderline lepromatous patient (105). These cells were first isolated as T-cell blasts, which were grown in IL-2 without antigen. They did not proliferate in response to *M. leprae,* but when they were combined with *M. leprae* and autologous *M. leprae*–responsive peripheral blood lymphocytes they suppressed the reactivity of the lymphocytes. Operationally, this is antigen-specific suppression. One clone suppressed peripheral blood mononuclear cells when the 36-kd *M. leprae* protein was used as the antigen; this indicates that in addition to the helper determinant defined in the earlier work (106), this protein also has a suppressor determinant as well. The mechanisms of suppression remain to be elucidated. The suppressor clones did not suppress a phytohemagglutinin (PHA) mitogenic response, but this test was done in the absence of any *M. leprae* antigen. Similarly, failure to suppress a herpes simplex virus (HSV) response was reported only in the absence of *M. leprae* antigen. The experiments demonstrate antigen-specific induction of suppression, but do not speak to the effector phase. Given the debate stimulated by the first report of T-suppressor activity (81), presented as lepromin-induced suppression of a ConA mitogenic response, it would be of interest to know whether a similar system is being defined here.

When cloned T suppressors were tested against cloned T-helper cells of the same patient, certain combinations resulted in modest suppression (50–60% of [^{3}H]thymidine at a T-suppressor: T-helper ratio of 3:1) (105). It is hoped that the cloned suppressor/cloned helper system will eventually allow clear delineation of the epitopes important in cell-mediated immunity and the mechanism of its regulation.

Modlin et al (90) have produced T-suppressor clones from lymphocytes taken from the skin biopsies of lepromatous lesions, where the distribution of T4 and T8 cells may be quite different from that in peripheral blood. These cells suppressed ConA-stimulated mitogenesis of uncloned peripheral blood mononuclear cells and cloned helper (T4) cells, but only when the T8 and T4 cells were matched with respect to HLA-DR type. The requirement for HLA-DR matching of the suppressor and helper cell populations may explain the inconsistency with which a given patient's cells were able to suppress the mitogenic response of a normal donor's cells in earlier studies (81, 82). HLA-restricted suppressive ability was also observed when lepromin-reactive T4 clones were the targets of suppression. At T4:T8 ratio of 1:1, 20–50% suppression of the proliferative response was observed (90).

These reports on cloning of human T cells from leprosy patients represent the first contributions to what will undoubtedly become a very large body of literature. They seem to argue for the existence of T-suppressor cells in the pathology of leprosy, although it is too soon to say that the relevance of these cells has been proved. The Leiden group (105, 106) has formulated its approach in accordance with a view (75, 103) of helper/suppressor regulation based on the notion that proteins encountered by T cells may have helper determinants and suppressor determinants. A T8 cell suppresses a T4 cell only if the two determinants are physically connected. If a protein lacking a suppressor determinant (because of cleavage or recombinant DNA technology, or because it is a homologous but nonidentical protein from a related species) can confer immunity, then a potential vaccine exists (88) that can expand the T-helper subset without expanding the T-suppressor subset.

What has not emerged so far is a T-cell clone of helper or suppressor function that is reactive to the phenolic glycolipid I; such a line or clone might be expected, given the historical connection of this leprosy-specific antigen with T-cell mediated suppression (7, 79). If found, such a clone might settle considerable controversy. However, there is no theory of T-cell recognition dealing specifically with carbohydrates, particularly with lipid-borne oligosaccharides. As cloned T-cell systems using particular protein antigens become well defined, it may be worthwhile to investigate the effects of the *M.*

leprae glycolipid or lipoarabinomannan in these systems. It is conceivable that these molecules affect cell-mediated immunity not by acting as antigens per se, but by interfering with processing, presentation, or recognition of other antigens. Brett et al (12) could find no such effect using the phenolic glycolipid in mice, however.

CONCLUSIONS AND SPECULATIONS

Current understanding of the basic immunology of leprosy, and particularly of the immunologic nonresponsiveness exhibited at the lepromatous end of the disease spectrum, has not yet reached a satisfactory level. While it is clear that specifically immune T lymphocytes are required to successfully amplify the bacteriocidal activity of mononuclear phagocytes, there is a spectrum of possible explanations for why T cells fail to do this in some cases. A complete lack of T-cell clones reactive toward *M. leprae* represents one possibility, but this would require further explanation. Alternatively, *M. leprae*–reactive clones may be expanded normally, but may be functionally inhibited from operating, i.e. suppressed. That some form of suppression exists seems indisputable: If other mycobacteria such as BCG share T-cell epitopes with *M. leprae,* as is undoubtedly the case, and if the lepromatous patient responds well to BCG, then either (*a*) the patient's positive response to BCG is based on a different set of determinants or (*b*) the response to these determinants is actively suppressed when the patient is challenged with *M. leprae*. While *a* has not been rigorously excluded, it seems unlikely. Accepting *b*, is suppression mediated by T cells? In the absence of an experimental system allowing adoptive cell transfer, it could be argued that identification of a T-cell clone with suppressive activity is only part of the normal regulatory network and is not pathological.

The above remarks should not obscure the fact that a search for T cell–based regulatory schemes will in fact continue, based on T-cell lines and clones and making use of the recombinant protein products of *M. leprae*. The thrust will be to determine which specific or cross-reactive determinants lead most often to effective T-cell immunity, and which determinants induce suppressive effects. This would refine both immunization and therapeutic schemes based on BCG cross-reactivity and would help in the development of anti-idiotypic probes to assess the presence of antigen-specific clones in lesions.

Experimentation with patient cells in vitro is marked by a distressing amount of variability, some of which reflects a real heterogeneity of individual cases. However, there appears to be a strong component of nonspecific suppression, some of which may be traceable to the lipoarabinomannan. Various forms of lepromin or soluble extracts of *M. leprae* may contain differing amounts of LAM, and this may contribute to variable responses in vitro. Given the number of complex and lipophilic products in mycobacteria, some of which can intercalate into membrane bilayers, much of the immunologic disturbance in leprosy may be pharmacologic. Cloned T-cell systems may allow an improved understanding of these nonspecific as well as specific effects, and use of recombinant DNA systems may be the only way to eliminate some of these products while leaving protein relatively intact.

ACKNOWLEDGMENTS

We thank Marilyn Hein for preparing the manuscript and Carol Marander for the graphics. The work cited from this laboratory was supported by funds from the National Institute of Allergy and Infectious Diseases, National Institutes of Health and the Leonard Wood Foundation.

Literature Cited

1. Allison, A. C. 1978. Mechanisms by which activated macrophages inhibit lymphocyte responses. *Immunol. Rev.* 40:3–27
2. Anderson, D. C., Young, R. A., Buchanan, T. M. 1986. Synthesis of two epitopes that react with monoclonal antibodies to the 65,000 dalton protein of *Mycobacterium leprae*. *Int. J. Lepr.* 54:735
3. Andreoli, A., Brett, S. J., Draper, P., Payne, S. N., Rook, G. A. W. 1985. Changes in circulating antibody level to the major phenolic glycolipid during erythema nodosum leprosum in leprosy patients. *Int. J. Lepr.* 53:211–17
4. Bach, M. A., Wallach, D., Flageul, B., Hoffenbach, A., Cottenot, F. 1986. Antibodies to phenolic glycolipid I and to whole *Mycobacterium leprae* in leprosy patients: evolution during therapy. *Int. J. Lepr.* 54:256–67
5. Birdi, T. J., Mistry, N. F., Mahadevan, P. R., Antia, N. H. 1983. Alterations in the membrane macrophages from leprosy patients. *Infect. Immun.* 41:121–27
6. Bloom, B. R., Godal, T. 1983. Selective primary health care: Strategies for the control of disease in the developing world. V. Leprosy. *Rev. Infect. Dis.* 5:765–80
7. Bloom, B., Mehra, V. 1984. Immunological unresponsiveness in leprosy. *Immunol. Rev.* 80:5–28
8. Brennan, P. J. 1983. The phthiocerol-containing surface lipids of *Mycobacterium leprae*—a perspective of past and present work. *Int. J. Lepr.* 51:387–96
9. Brennan, P. J. 1986. The carbohydrate-containing antigens of *Mycobacterium leprae*. *Lepr. Rev.* 57:39–51
10. Brennan, P. J., Barrow, W. W. 1980. Evidence for species-specific lipid antigens in *Mycobacterium leprae*. *Int. J. Lepr.* 48:382–87
11. Brett, S. J., Draper, P., Payne, S. N., Rees, R. J. W. 1983. Serological activity of a characteristic phenolic glycolipid from *Mycobacterium leprae* in sera from patients with leprosy and tuberculosis. *Clin. Exp. Immunol.* 52:271–79
12. Brett, S. J., Lowe, C., Payne, S. N., Draper, P. 1984. Phenolic glycolipid I of *Mycobacterium leprae* causes nonspecific inflammation but has no effect on cell-mediated responses in mice. *Infect. Immun.* 46:802–8
13. Brett, S. J., Payne, S. N., Draper, P., Gigg, R. 1984. Analysis of the major antigenic determinants of the characteristic phenolic glycolipid from *Mycobacterium leprae*. *Clin. Exp. Immunol.* 56:89–96
14. Brett, S. J., Payne, S. N., Gigg, J., Burgess, P., Gigg, R. 1986. Use of synthetic glycoconjugates containing the *Mycobacterium leprae* specific and immunodominant epitope of phenolic glycolipid I in the serology of leprosy. *Clin. Exp. Immunol.* 64:476–83
15. Britton, W. J., Hellqvist, L., Basten, A., Raison, R. L. 1985. *Mycobacterium leprae* antigens involved in human immune responses. *J. Immunol.* 135:4171–77
16. Buchanan, T. M., Mehra, V., Namaguchi, H., Young, R. A., Gillis, T. P., et al. 1987. Characterization of antigen

reactive epitopes on the 65,000 dalton protein molecules of *M. leprae*. *Infect. Immun.* 55:1000–3
17. Bullock, W. 1981. Immunobiology of leprosy. In *Immunology of Human Infection, Part I*, ed. A. J. Nahmias, R. J. O'Reilly, pp. 369–90. New York: Plenum. 651 pp.
18. Chakrabarty, A. K., Maire, M. A., Lambert, P. H. 1982. SDS PAGE analysis of *M. leprae* protein antigens reacting with antibodies from sera from lepromatous patients and infected armadillos. *Clin. Exp. Immunol.* 49:523–31
19. Chatterjee, D., Aspinall, G. O., Brennan, P. J. 1986. Chemical synthesis of the immunodeterminant trisaccharide of phenolic glycolipid I of *Mycobacterium leprae*. *Abstr. 13th Int. Carbohydr. Symp., Ithaca, NY,* p. 22. Edinburgh: Int. Carbohydr. Org.
20. Chatterjee, D., Cho, S.-N., Brennan, P. J., Aspinall, G. O. 1986. Chemical synthesis and seroreactivity in leprosy of *O*-(3,6-di-*O*-methyl-β-D-glucopyranosyl)-(1→4)-*O*-(2,3-di-*O*-methyl-α-L-rhamnopyranosyl)-(1→9)-oxynonanoyl-bovine serum albumin—the leprosy-specific natural disaccharide-octyl-neoglycoprotein. *Carbohydr. Res.* 156:39–56
21. Chatterjee, D., Douglas, J. T., Cho, S.-N., Rea, T. H., Gelber, R. H., et al. 1985. Synthesis of neoglycoconjugates containing the 3,6-di-*O*-methyl-β-D-glucopyranosyl epitope and their use in serodiagnosis of leprosy. *Glycoconjugate J.* 2:187–208
22. Cho, S.-N., Chatterjee, D., Brennan, P. J. 1986. A simplified serological test for leprosy based on a 3,6-di-*O*-methyl glucose–containing synthetic antigen. *Am. J. Trop. Med. Hyg.* 35:167–72
23. Cho, S.-N., Fujiwara, T., Hunter, S. W., Rea, T. H., Gelber, R. H., Brennan, P. J. 1984. Use of an artificial antigen containing the 3,6-di-*O*-methyl-β-D-glucopyranosyl epitope for the serodiagnosis of leprosy. *J. Infect. Dis.* 150: 311–22
24. Cho, S.-N., Hunter, S. W., Gelber, R. H., Rea, T. H., Brennan, P. J. 1986. Quantitation of the phenolic glycolipid of *Mycobacterium leprae* and relevance to glycolipid antigenemia in leprosy. *J. Infect. Dis.* 153:560–69
25. Cho, S.-N., Yanagihara, D. L., Hunter, S. W., Gelber, R. H., Brennan, P. J. 1983. Serological specificity of phenolic glycolipid I from *Mycobacterium leprae* and use in serodiagnosis of leprosy. *Infect. Immun.* 41:1077–83
26. Clark-Curtiss, J. E., Jacobs, W. R., Docherty, M. A., Ritchie, L. R., Curtiss, R. 1985. Molecular analysis of DNA and construction of genomic libraries of *Mycobacterium leprae*. *J. Bacteriol.* 161:1093–102
27. Collings, L. A., Tidman, M., Poulter, L. W. 1985. Quantitation of HLA-DR expression by cells involved in the skin lesions of tuberculoid and lepromatous leprosy. *Clin. Exp. Immunol.* 61:58–66
28. Draper, P. 1976. Cell walls of *Mycobacterium leprae*. *Int. J. Lepr.* 44:95–98
28a. Draper, P. 1982. Purification of *M. leprae*. *Rep. 5th Meet. Sci. Work. Group Immunol. Lepr., Geneva, 1980. WHO Doc. TDR/IMMLEP-SWG (5)/80.3,* Annex 4. Geneva: World Health Org.
29. Draper, P. 1982. The anatomy of mycobacteria. In *The Biology of the Mycobacteria,* ed. C. Ratledge, 1:9–52. London: Academic. 544 pp.
30. Draper, P. 1984. Wall biosynthesis: a possible site of action for new antimycobacterial drugs. *Int. J. Lepr.* 52: 527–32
31. Draper, P., Payne, S. N., Dobson, G., Minnikin, D. E. 1983. Isolation of a characteristic phthiocerol dimycocerosate from *Mycobacterium leprae*. *J. Gen. Microbiol.* 129:859–63
32. Elferink, B. G., Ottenhoff, T. H. M., deVries, R. R. P. 1985. Epstein-Barr virus transformed B-cell lines present *M. leprae* antigen to T-cells. *Scand. J. Immunol.* 22:585–89
33. Ellner, J. J. 1978. Suppressor adherent cells in human tuberculosis. *J. Immunol.* 121:2573–79
34. Ellner, J. J., Daniel, T. M. 1979. Immunosuppression by mycobacterial arabinomannan. *Clin. Exp. Immunol.* 35:250–57
35. Emmrich, F., Kaufmann, S. H. E. 1986. Human T-cell clones with reactivity to *Mycobacterium leprae* as tools for the characterization of potential vaccines against leprosy. *Infect. Immun.* 51:879–83
36. Engers, H., Abe, M., Bloom, B. R., Mehra, V., Britton, W., et al. 1985. Results of a World Health Organization–sponsored workshop on monoclonal antibodies to *Mycobacterium leprae*. *Infect. Immun.* 48:603–5
37. Ferluga, J., Colizzi, V., Ferrante, A., Colston, M. J., Holborow, E. J. 1984. Hypothesis: possible idiotypic suppression of cell mediated immunity in lepromatous leprosy. *Lepr. Rev.* 55: 221–27
38. Fujiwara, T., Aspinall, G. O., Hunter, S. W., Brennan, P. J. 1987. Chemical

synthesis of the trisaccharide unit of the species-specific phenolic glycolipid from *Mycobacterium leprae*. *Carbohydr. Res.* 155:In press

39. Fujiwara, T., Hunter, S. W., Brennan, P. J. 1986. Chemical synthesis of disaccharides of the specific phenolic glycolipid antigens from *Mycobacterium leprae* and of related sugars. *Carbohydr. Res.* 148:287–98
40. Fujiwara, T., Hunter, S. W., Cho, S.-N., Aspinall, G. O., Brennan, P. J. 1984. Chemical synthesis and serology of disaccharides and trisaccharides of phenolic glycolipid antigens from the leprosy bacillus and preparation of a disaccharide protein conjugate for serodiagnosis of leprosy. *Infect. Immun.* 43:245–52
41. Fujiwara, T., Izumi, S. 1986. The synthesis of the trisaccharide related to the phenolic glycolipid of *Mycobacterium leprae* and preparation of sugar-protein conjugate. *Int. J. Lepr.* 54:727
42. Fujiwara, T., Izumi, S., Brennan, P. J. 1985. Synthesis of 3,6-di-*O*-methylglucosyl disaccharides with methyl 3-(*p*-hydroxyphenyl)propionate as a linker arm and their use in the serodiagnosis of leprosy. *Agric. Biol. Chem.* 49:2301–8
43. Fukunishi, Y., Kearney, G. P., Whiting, J., Walsh, G. P., Meyers, W. M., et al. 1985. Isolation of characteristic glycolipids possibly included in spherical droplets around *M. leprae*. *Int. J. Lepr.* 53:447–54
44. Gaylord, H., Brennan, P. J. 1985. Antigens and host-parasite interactions. In *Parasite Antigens. Toward New Strategies for Vaccines,* ed. T. W. Pearson, pp. 49–89. New York/Basel: Dekker. 413 pp.
45. Gaylord, H., Brennan, P. J., Young, D. B., Buchanan, T. M. 1987. The *Mycobacterium leprae* carbohydrate-reactive monoclonal antibodies are directed primarily to lipoarabinomannan. *Infect. Immun.* In press
46. Gershon, R. K. 1975. A disquisition on suppressor T cells. *Transplant. Rev.* 26:170–85
47. Gigg, J., Gigg, R., Payne, S., Conant, R. 1985. The allyl group for protection in carbohydrate chemistry. Part 17. *Chem. Phys. Lipids* 38:299–307
48. Gigg, J., Gigg, R., Payne, S., Conant, R. 1985. Synthesis of propyl 4-*O*-(3,6-di-*O*-methyl-β-D-glucopyranosyl)-2,3- di-*O*-methyl-α-D-rhamnopyranoside. *Carbohydr. Res.* 141:91–97
49. Gigg, R., Payne, S. N., Conant, R. 1983. The allyl group for protection in carbohydrate chemistry. Part 14. *J. Carbohydr. Chem.* 2:207–23
50. Gill, H. K., Godal, T. 1986. Deficiency of cell mediated immunity in leprosy. *Prog. Allergy* 37:377–90
51. Gillis, T. P., Buchanan, T. M. 1982. Production and partial characterization of monoclonal antibodies to *Mycobacterium leprae*. *Infect. Immun.* 37:172–78
52. Gillis, T. P., Miller, R. A., Young, D. B., Khanolkar, S. R., Buchanan, T. M. 1985. Immunochemical characterization of a protein associated with the cell wall of *Mycobacterium leprae*. *Infect. Immun.* 49:275–81
53. Haanen, J. B. A. G., Ottenhoff, T. H. M., Voordouw, A., Elferink, B. G., Klatser, P. R., et al. 1986. HLA class-II-restricted *Mycobacterium leprae* reactive T-cell clones from leprosy patients established with minimal requirement for autologous mononuclear cells. *Scand. J. Immunol.* 23:101–08
54. Hahn, H., Kaufmann, S. 1981. The role of cell mediated immunity in bacterial infections. *Rev. Infect. Dis.* 3:1221–50
55. Hanks, J. H. 1961. Capsules in electron micrographs of *Mycobacterium leprae*. *Int. J. Lepr.* 29:84–87
56. Harboe, M. 1982. Significance of antibody studies in leprosy and experimental models of the disease. *Int. J. Lepr.* 50:342–50
57. Haregewoin, A., Godal, T., Mustafa, A. S., Belchu, A., Yemaneberhan, T. 1983. T cell conditioned media reverse T cell unresponsiveness in lepromatous leprosy. *Nature* 303:342–44
58. Haregewoin, A., Mustafa, A. S., Helle, I., Waters, M. F. R., Leiker, D. L., et al. 1984. Reversal by interleukin 2 of the T-cell unresponsiveness of lepromatous leprosy to *Mycobacterium leprae*. *Immunol. Rev.* 80:76–86
59. Hastings, R. C., ed. 1986. *Leprosy.* New York: Livingstone. 344 pp.
60. Horwitz, M. A., Levis, W. R., Cohn, Z. A. 1984. Defective production of monocyte activating lymphokines in lepromatous leprosy. *J. Exp. Med.* 159:666–78
61. Hunter, S. W., Brennan, P. J. 1981. A novel phenolic glycolipid from *Mycobacterium leprae* possibly involved in immunogenicity and pathogenicity. *J. Bacteriol.* 147:728–35
62. Hunter, S. W., Brennan, P. J. 1983. Further specific extracellular phenolic glycolipid antigens and a related diacylphthiocerol from *Mycobacterium leprae*. *J. Biol. Chem.* 258:7556–62
63. Hunter, S. W., Fujiwara, T., Brennan,

P. J. 1982. Structure and antigenicity of the major specific glycolipid antigen of *Mycobacterium leprae*. *J. Biol. Chem.* 257:15072–78

64. Hunter, S. W., Gaylord, H., Brennan, P. J. 1986. Structure and antigenicity of the phosphorylated lipopolysaccharides of the tubercle and leprosy bacilli. *J. Biol. Chem.* 261:12345–51
65. Ivanyi, J., Morris, J. A., Keen, M. 1985. Studies with monoclonal antibodies to mycobacteria. In *Monoclonal Antibodies Against Bacteria,* ed. A. J. L. Macario, E. C. Macario, 1:55–90. New York: Academic. 320 pp.
66. Ivanyi, J., Sinha, S., Aston, R., Cussell, D., Keen, M., et al. 1983. Definition of species-specific and cross-reactive antigenic determinants of *Mycobacterium leprae* using monoclonal antibodies. *Clin. Exp. Immunol.* 52:528–36
67. Izumi, S., Sugiyama, K., Fujiwara, T., Hunter, S. W., Brennan, P. J. 1985. Isolation of the *Mycobacterium leprae*–specific glycolipid antigen, phenolic glycolipid I, from formalin-fixed human lepromatous liver. *J. Clin. Microbiol.* 22:680–82
68. Jacobs, W. R., Docherty, M. A., Curtiss, R., Clark-Curtiss, J. E. 1986. Expression of *Mycobacterium leprae* genes from a *Streptococcus mutans* promoter in *Escherichia coli* K-12. *Proc. Natl. Acad. Sci. USA* 83:1926–30
69. Kaplan, G., Cohn, Z. A. 1986. The immunology of leprosy. *Int. Rev. Exp. Pathol.* 28:45–78
70. Kaplan, G., Gandhi, R. R., Weinstein, D. E., Levis, W. R., Patarroyo, M. E., et al. 1987. *Mycobacterium leprae* antigen induced suppression of T cell proliferation in vitro. *J. Immunol.* 138:3028–34
71. Kaplan, G., Weinstein, D. E., Steinman, R. M., Levis, W. R., Elvers, U., et al. 1985. An analysis of in vitro T cell responsiveness in lepromatous leprosy. *J. Exp. Med.* 162:917–29
72. Klatser, P. R., de Wit, M. Y. L., Kolk, A. H. J. 1985. An ELISA-inhibition test using monoclonal antibody for serology of leprosy. *Clin. Exp. Immunol.* 62:468–73
73. Klatser, P. R., Hartskeerl, R. A., Van Schooten, W. C. A., Kolk, A. H. J., Van Rens, M. M., et al. 1986. Characterization of the 36 K antigen of *Mycobacterium leprae*. *Lepr. Rev.* 57:77–81
74. Kolk, A. H. J., Ho, M. L., Klatser, P. R., Eggalte, T. A., Kuijper, S., et al. 1984. Production and characterization of monoclonal antibodies to *Mycobacterium tuberculosis, M. bovis* (BCG) and *M. leprae*. *Clin. Exp. Immunol.* 58:511–21
75. Krzych, V., Fowler, A. V., Sercarz, E. E. 1985. Repertoires of T-cells directed against a large protein antigen, β-galactosidase. II. Only certain T-helper or T-suppressor cells are relevant in particular regulatory interactions. *J. Exp. Med.* 162:311–23
76. Kusaka, T., Kohsaka, K., Fukunishi, Y., Akimori, H. 1981. Isolation and identification of mycolic acids in *Mycobacterium leprae* and *Mycobacterium lepraemurium*. *Int. J. Lepr.* 49:406–16
77. Levis, W. R., Meeker, H. C., Schuller-Levis, G., Sersen, E., Schwerer, B. 1986. IgM and IgG antibodies to phenolic glycolipid I from *Mycobacterium leprae* in leprosy: insight into patient monitoring, erythema nodosum leprosum, and bacillary persistence. *J. Invest. Dermatol.* 86:529–34
78. McNeil, M., Wallner, S. J., Hunter, S. W., Brennan, P. J. 1987. Demonstration that the galactosyl and arabinosyl residues in the cell wall arabinogalactan of *Mycobacterium leprae* and *Mycobacterium tuberculosis* are in the furanosyl form. *Carbohydr. Res.* In press
79. Mehra, V., Brennan, P. J., Rada, E., Convit, J., Bloom, B. R. 1984. Lymphocyte suppression in leprosy induced by unique *M. leprae* glycolipid. *Nature* 308:194–96
80. Mehra, V., Convit, J., Rubinstein, A., Bloom, B. R. 1982. Activated T suppressor cells in leprosy. *J. Immunol.* 129:1946–51
81. Mehra, V., Mason, L. H., Fields, J. P., Bloom, B. R. 1979. Lepromin induced suppressor cells in patients with leprosy. *J. Immunol.* 123:1183–88
82. Mehra, V., Mason, L. H., Rothman, W., Reinherz, E., Schlossman, S. F., Bloom, B. R. 1980. Delineation of a human T cell subset responsible for lepromin induced suppression in leprosy patients. *J. Immunol.* 125:1183–88
83. Mehra, V., Sweetser, D., Young, R. A. 1986. Efficient mapping of protein antigenic determinants. *Proc. Natl. Acad. Sci. USA* 83:7013–17
84. Miller, R. A., Dissanayake, S., Buchanan, T. M. 1983. Development of an enzyme-linked immunosorbent assay using arabinomannan from *Mycobacterium smegmatis:* a potentially useful screening test for the diagnosis of incubating leprosy. *Am. J. Trop. Med. Hyg.* 32:555–64
85. Minnikin, D. E. 1982. Lipids: complex

lipids, their chemistry, biosynthesis and roles. See Ref. 29, pp. 95–184

86. Minnikin, D. E., Dobson, G., Goodfellow, M., Draper, P., Magnusson, M. 1985. Quantitative comparison of the mycolic acid and fatty acid compositions of *Mycobacterium leprae* and *Mycobacterium gordonae*. *J. Gen. Microbiol.* 131:2013–21
87. Mistry, N. F., Birdi, T. J., Mahadevan, P. R., Antia, N. H. 1985. *Mycobacterium leprae* induced alterations in macrophage Fc receptor expression and monocyte-lymphocyte interaction in familiar contacts of leprosy patients. *Scand. J. Immunol.* 22:415–23
88. Mitchison, N. A. 1984. Rational design of vaccines. *Nature* 308:112–13
89. Modlin, R. L., Hofman, F. M., Horwitz, D. A., Husmann, L. A., Gillis, S., et al. 1984. *In situ* identification of cells in human leprosy granulomas with monoclonal antibody to IL-2 and its receptor. *J. Immunol.* 132:3085–90
90. Modlin, R. L., Kato, H., Mehra, V., Nelson, E., Fan, X. D., et al. 1986. Genetically restricted suppressor T-cell clones derived from lepromatous leprosy lesions. *Nature* 322:459–61
91. Mohagheghpour, N., Gelber, R. H., Larrick, J. W., Sasaki, D. T., Brennan, P. J., et al. 1985. Defective cell mediated immunity in leprosy: failure of T cells from lepromatous patients to respond to *Mycobacterium leprae* is associated with defective expression of interleukin 2 receptors, and is not reconstituted by interleukin 2. *J. Immunol.* 135:1443–49
92. Moller, G., ed. 1981. T-cell clones. *Immunol. Rev.* 54:1–266
93. Mustafa, A. S., Gill, H. K., Nerland, A., Britton, W. J., Mehra, V., et al. 1986. Human T-cell clones recognize a major *M. leprae* protein expressed in *E. coli*. *Nature* 319:63–66
94. Narayanan, R. B., Bhutani, L. K., Sharma, A. K., Nath, I. 1983. T-cell subsets in leprosy lesions: *in situ* characterization using monoclonal antibodies. *Clin. Exp. Immunol.* 51:421–29
95. Nath, I. 1983. Immunology of leprosy—current status. *Lepr. Rev.* Special Issue:31S–45S
96. Nath, I., Jayaraman, J., Sathish, M., Bhutani, L. K., Sharma, A. K. 1984. Inhibition of interleukin-2 production by adherent cell factors from lepromatous leprosy patients. *Clin. Exp. Immunol.* 58:531–38
97. Nath, I., Sathish, M., Tayaraman, T., Bhutani, L. K., Sharma, A. K. 1984. Evidence for the presence of *M. leprae* reactive T-lymphocytes in patients with lepromatous leprosy. *Clin. Exp. Immunol.* 58:522–30
98. Nath, I., Singh, R. 1980. The suppressive effect of *M. leprae* on the in vitro proliferative responses of lymphocytes from patients with leprosy. *Clin. Exp. Immunol.* 41:406
99. Nath, I., van Rood, R. R., Mehra, N. K., Vaidya, M. C. 1980. Natural suppressor cells in human leprosy: the role of HLA-D-identical peripheral lymphocytes and macrophages in the *in vitro* modulation of lymphoproliferative responses. *Clin. Exp. Immunol.* 42:203–10
100. Nathan, C. F., Kaplan, G., Levis, W. R., Nusrat, A., Witmer, M. D., et al. 1986. Local and systemic effects of intradermal and recombinant interferon-γ in patients with lepromatous leprosy. *N. Engl. J. Med.* 315:6–15
101. Nathan, C. F., Murray, H. W., Wiebl, M. E., Rubin, B. Y. 1983. Identification of interferon-γ as the lymphokine that activates human macrophage oxidative metabolism and antimicrobial activity. *J. Exp. Med.* 158:670–89
102. Nogueira, N., Kaplan, G., Levy, E., Sarno, E., Kushner, P., et al. 1983. Defective γ-interferon production in leprosy. Reversal with antigen and interleukin 2. *J. Exp. Med.* 158:2165–70
103. Oki, A., Sercarz, E. 1985. T-cell tolerance studied at the level of antigenic determinants. I. Latent reactivity to lysozyme peptides that lack suppressogenic epitopes can be revealed in lysozyme-tolerant mice. *J. Exp. Med.* 161:897–911
104. Ottenhoff, T. H. M., Elferink, D. G., deVries, R. R. P. 1984. Unresponsiveness to *Mycobacterium leprae* in lepromatous leprosy in vitro: reversible or not? *Int. J. Lepr.* 52:419–22
105. Ottenhoff, T. H. M., Elferink, D. G., Klatser, P. R., deVries, R. R. P. 1986. Cloned suppressor T-cells from a lepromatous leprosy patient suppress *M. leprae*-reactive helper T-cells. *Nature* 322:462–64
106. Ottenhoff, T. H. M., Klatser, P. R., Ivanyi, J., Elferink, D. G., deWitt, M. Y. L., et al. 1986. *Mycobacterium leprae*-specific protein antigens defined by cloned human helper T-cells. *Nature* 319:66–68
107. Payne, S. N., Draper, P., Rees, R. J. W. 1982. Serological activity of purified glycolipid from *Mycobacterium leprae*. *Int. J. Lepr.* 50:220–21
108. Praputpitlaya, K., Ivanyi, J. 1985. Detection of an antigen (MY4) common to

M. tuberculosis and *M. leprae* by "Tandem" immunoassay. *J. Immunol. Methods* 79:149–57
109. Rastogi, N., Frehel, C., David, H. L. 1984. Cell envelope architectures of leprosy derived corynebacteria, *Mycobacterium leprae,* and related organisms: a comparative study. *Curr. Microbiol.* 11:23–30
110. Rastogi, N., Frehel, C., David, H. L. 1986. Triple-layered structure of mycobacterial cell wall: evidence for the existence of a polysaccharide-rich outer layer in 18 mycobacterial species. *Curr. Microbiol.* 13:237–42
111. Ridley, D. S., Jopling, W. H. 1966. Classification of leprosy according to immunity: a five group system. *Int. J. Lepr.* 34:255–73
112. Salgame, P. R., Mahadevan, P. R., Antia, N. H. 1983. Mechanism of immunosuppression in leprosy: presence of suppressor factors from macrophages of leprosy patients. *Infect. Immun.* 40:1119–26
113. Sanchez, G. A., Malik, A., Tougne, C., Laubert, P. H., Engers, H. D. 1986. Simplification and standardization of serodiagnostic tests based on phenolic glycolipid I (PG-I) antigen. *Lepr. Rev.* 57(Suppl. 2):83–93
114. Sansonetti, P., Lagrange, P. H. 1981. The immunology of leprosy: speculations on the leprosy spectrum. *Rev. Infect. Dis.* 3:422–69
115. Sathish, M., Bhutani, L. K., Sharma, A. K., Nath, I. 1983. Monocyte-derived soluble suppressor factors in patients with lepromatous leprosy. *Infect. Immun.* 42:890–99
116. Schultz, R. M., Kleinschmidt, W. J. 1983. Functional identity between murine interferon and macrophage activating factor. *Nature* 305:239–40
117. Sibley, L. D., Franzblau, S. G., Krahenbuhl, J. L. 1987. Intracellular fate of *Mycobacterium leprae* in normal and activated macrophages. *Infect. Immun.* 55:680–85
118. Sibley, L. D., Krahenbuhl, J. L. 1987. *Mycobacterium leprae*–burdened macrophages are refractory to activation of gamma interferon. *Infect. Immun.* 55:446–50
119. Sinha, S., Sengupta, U., Ramu, G., Ivanyi, J. 1983. A serological test for leprosy based on competitive inhibition of monoclonal antibody binding to a MY2a determinant of *Mycobacterium leprae*. *Trans. R. Soc. Trop. Med. Hyg.* 77:869–71
120. Stoner, G. L., Mshana, R. N., Touw, T., Belehu, A. 1982. Studies on the defect in cell mediated immunity in lepromatous leprosy using HLD-D-identical siblings. Absence of circulating suppressor cells and evidence that the defect is in the T-lymphocyte rather than the monocyte population. *Scand. J. Immunol.* 15:33–48
121. Tarelli, E., Draper, P., Payne, S. N. 1984. Structure of the oligosaccharide component on a serologically active phenolic glycolipid isolated from *Mycobacterium leprae*. *Carbohydr. Res.* 131:346–52
122. Teuscher, C., Yanagihara, D., Brennan, P. J., Koster, F. T., Tung, K. S. K. 1985. Antibody response to phenolic glycolipid I in inbred mice immunized with *Mycobacterium leprae*. *Infect. Immun.* 48:474–79
123. Touw, J., Stoner, G. L., Belehu, A. 1980. Effect of *Mycobacterium leprae* on lymphocyte proliferation: suppression of mitogen and antigen responses of human peripheral blood mononuclear cells. *Clin. Exp. Immunol.* 41:397–405
124. Truman, R. W., Shannon, E. J., Hagstad, H. V., Hugh-Jones, M. E., Wolff, A., et al. 1986. Evaluation of the origin of *Mycobacterium leprae* infections in the wild armadillo, *Dasypus novemcinctus*. *Am. J. Trop. Med. Hyg.* 35:588–93
125. van Voorhis, W. C., Kaplan, G., Sarno, E. N., Horwitz, M. A., Steinman, R. M., et al. 1982. The cutaneous infiltrates of leprosy. Cellular characteristics and the predominant T-cell phenotypes. *N. Engl. J. Med.* 307:1593–97
126. Vemuri, N., Khandke, L., Mahadevan, P. R., Hunter, S. W., Brennan, P. J. 1985. Isolation of phenolic glycolipid I from human lepromatous nodules. *Int. J. Lepr.* 53:487–89
127. Watson, S. R., Bullock, W. E. 1983. Immunoregulatory defects in leprosy. *Adv. Exp. Med. Biol.* 162:203–15
128. Watson, S., Bullock, W., Nelson, K., Schauf, V., Gelber, R., et al. 1984. Interleukin 1 production by peripheral blood mononuclear cells from leprosy patients. *Infect. Immun.* 45:787–89
129. Wisniewsky, H. M., Bloom, B. R. 1975. Primary demyelination is a nonspecific consequence of cell mediated immune reactions. *J. Exp. Med.* 141:346–59
130. Young, D. B. 1981. Detection of mycobacterial lipids in skin biopsies from leprosy patients. *Int. J. Lepr.* 49:198–204
131. Young, D. B., Buchanan, T. M. 1983. A serological test for leprosy with a glycolipid specific for *Mycobacterium leprae*. *Science* 221:1057–59
132. Young, D. B., Dissanayake, S., Miller,

R. A., Khanolkar, S. R., Buchanan, T. M. 1984. Humans respond predominantly with IgM immunoglobulin to the species-specific glycolipid of *Mycobacterium leprae*. *J. Infect. Dis.* 149:870–73

133. Young, D. B., Fohn, M. J., Buchanan, T. M. 1985. Use of a polysulfone membrane support for immunochemical analysis of a glycolipid from *Mycobacterium leprae*. *J. Immunol. Methods* 79:205–11
134. Young, D. B., Fohn, M. J., Khanolkar, S. R., Buchanan, T. M. 1985. Monoclonal antibodies to a 28,000 mol. wt. protein antigen of *Mycobacterium leprae*. *Clin. Exp. Immunol.* 60:546–52
135. Young, D. B., Fohn, M. J., Khanolkar, S. R., Buchanan, T. M. 1985. A spot test for antibodies to phenolic glycolipid I. *Lepr. Rev.* 56:193–98
136. Young, D. B., Khanolkar, S. R., Barg, L. L., Buchanan, T. M. 1984. Generation and characterization of monoclonal antibodies to the phenolic glycolipid of *Mycobacterium leprae*. *Infect. Immun.* 43:183–88
137. Young, R. A., Bloom, B. R., Grossinsky, C. M., Ivanyi, J., Thomas, D., Davis, R. W. 1985. Dissection of *Mycobacterium tuberculosis* antigens using recombinant DNA. *Proc. Natl. Acad. Sci. USA* 82:2583–87
138. Young, R. A., Mehra, V., Sweetser, D., Buchanan, T., Clark-Curtiss, J., et al. 1985. Genes for the major protein antigens of the leprosy parasite, *Mycobacterium leprae*. *Nature* 316:450–52

REFERENCE ADDED IN PROOF

139. Kaldany, R.-R., Nurlign, A. 1986. Development of a dot-ELISA for detection of leprosy antigenuria under field conditions. *Lepr. Rev.* 57:95–100

Ann. Rev. Microbiol. 1987. 41:677–701

ELABORATION VERSUS SIMPLIFICATION IN REFINING MATHEMATICAL MODELS OF INFECTIOUS DISEASE

Francis L. Black and Burton Singer

Department of Epidemiology and Public Health, Yale University School of Medicine, New Haven, Connecticut 06510

CONTENTS

INTRODUCTION

Twice within the last twenty years United States public health authorities have called for the elimination of measles (65, 93), and twice our collective efforts have fallen flat. The eradication plans were based on projections of the best models available at the time (10). The failures were due to faulty data on the

0066-4227/87/1001-0677$02.00

level of immunity required for herd protection, and to the fact that the models used did not take population heterogeneity into account (30, 31). The chief direction of recent research on model building has been toward solutions to these problems, i.e. toward better definition of the parameters and development of a method for dealing with heterogeneity within populations. A model is essential to any rational plan for controlling infectious disease, but the more refined models require more and better biological data. Mathematicians have sometimes demanded unrealistic assumptions, and biologists have been slow to respond or have provided inappropriate statistics. Communication between the two scientific disciplines needs to be improved; this review is intended to facilitate one side of the exchange, offering the biological reader a better understanding of models.

It would seem logical to organize a review dealing with diverse diseases according to established microbiological classification. However, the primary purpose of a mathematical model is to generalize phenomena so that basic principles are clearly revealed and outcomes are predictable when a few parameters are specified. Microbiological taxonomy has little relevance in these models. However, different reproductive strategies can influence the choice of a model. For example, some microorganisms, such as the hepatitis A virus and *Giardia lamblia,* can give rise to substantial epidemics without replication at the source; others, such as measles and *Mycobacterium tuberculosis,* can go through many generations in a single host; and still others, such as many helminths, must find a new host for every generation of microorganism. Each of these strategies requires a different model. Furthermore, some models are designed to describe host population density or evolution of host and parasite rather than distribution of disease. Models designed for most of these circumstances had been proposed by the end of the 1960s, but there has been continuous development, in which efforts to improve precision by elaboration of the model have been countered by efforts to improve utility by simplification. Our first purpose is to examine these modifications, model by model.

Host populations are neither uniform nor uniformly distributed, and this variation affects the ability of a disease to spread. As already noted, a major emphasis in the recent work on model building has been on the problem of host heterogeneity. Efforts have been especially directed to the problem posed by spatial heterogeneity, which is more amenable to strictly mathematical treatment than qualitative heterogeneity. Unfortunately, abandonment of the assumption of homogeneous mixing can greatly complicate the models and place them beyond the reach of nonmathematicians. This not only reduces their usefulness but also separates them from the feedback control of being fitted to field data, which might otherwise reveal inconsistency with the biological world. A second purpose of this review is to examine attempts to deal with this heterogeneity. Before discussing models per se, we provide a brief guide to and commentary on the literature on infectious disease modeling.

In the most comprehensive review currently available, Bailey (9) dealt systematically with various types of models that have been developed to explain infectious disease epidemiology. He also addressed specific applications, although these are not microbially specific or disease specific, but methodologically so. Bailey provided examples of diseases to which each specific method may be applied. This review uses Bailey's book as a starting point.

Hethcote (60) set out the basic compartmentalized model of disease prevalence in a clear and concise manner. He described variations in which infectious persons become immune or susceptible again with and without intermediate vectors.

Bradley (23) has reviewed the successes and failures of models in the control of malaria and schistosomiasis. Bradley's main point is that models have often failed because of lack of communication between "determinedly non-numerate public health workers" and model builders who have a "tendency toward premature quantitative simplification to produce spurious elegance." Bradley found that models of malaria transmission have been more successful in predicting conditions for insect control than in predicting conditions for transmission of the disease, because the disease tends to be more heterogeneously distributed in space than the insects. Model builders have too often glossed over this heterogeneity. For a more recent insightful discussion of malaria modeling per se, see Molineaux (85).

To develop a model one must start with a set of data from a natural epidemic, and to test the model one must apply it to at least one other epidemic. Preferably, one should also suggest an intervention and on the basis of the model predict and verify its outcome. Unfortunately, the available data on the incidence of most diseases in most countries are below the accuracy needed for discriminating mathematical tests. Also, as Bradley emphasized, interventions in human disease are often prohibitively expensive, and many theoretically desirable interventions are ethically unacceptable. For these reasons, theoretical work in model building has outrun experimental verification. For an in-depth discussion of the tension between model complexity and utility, see Cohen's discussion of schistosomiasis (34).

Whereas Bradley decried oversimplification, Warren et al (99) have focused on the way in which a single parameter, the transmission potential (or basic reproduction rate), can "encapsulate the biological details of different transmission mechanisms." This parameter, usually designated R_0, is a measure of the infection's ability to give rise to secondary cases in a setting where there is no infection-dependent immunity. The report summarized models for a variety of transmission pathways and for both disease prevalence and infection density (number of parasites per infected host). The problems posed by spatial, temporal, and biological heterogeneity were only briefly discussed. References 14–16, 62, 77, 79, 84, and 89 provide a diverse introduction to heterogeneity modeling.

Dietz & Schenzle (43) considered comparative features of several basic prevalence models, age-specific differences in infection rates, and cyclic patterns of disease over extended periods of time. Each of these topics is considered in later sections of the present review.

Anderson & May (4, 82) reviewed disease prevalence models as a prelude to adapting them to situations in which the host population size is affected by interaction with the parasite, as in the classic predator-prey relationship. The same authors (7) also reviewed vaccination programs that might eliminate measles and rubella; they thoroughly reviewed models utilizing transmission potential.

PROGRESS IN DIVERSE TYPES OF MODELS

Common Source Models

An ostensibly simple model would be one for an epidemic resulting from infections that derive from a common source persisting for a limited period of time. Shonkwiler & Thompson (95) have published a stochastic model to predict the number of cases expected per unit time on the basis of degree and intensity of exposure and various levels of host infectability. The resultant curve is always S-shaped, but the upper asymptote approached as the epidemic wanes may be near 100% or much lower, and the rate of climb will vary. The influence of several designated variables on these characteristics of the curve can be examined with this model.

Adesina (2) has approached the analysis of common source epidemics quite differently in trying to determine the most important causes of the observed geographic distribution of cholera during a 1971 epidemic in Ibadan, Nigeria. He compared the observed distribution of the disease with travel patterns to school, work, or market. None of these patterns, alone or in conjunction, corresponded to the pattern of spread better than simple radial diffusion, but the methods used in this study should be useful elsewhere.

Models of Geographic Spread

Concentric spread of an epidemic from one or a few foci is chiefly a historic characteristic in human diseases, because we have become such a peripatetic species. There was some evidence of concentric spread in the 1947 and 1957 influenza epidemics (73), but there has been little in subsequent pandemics. However, this pattern still persists in animal diseases. Cliff & Haggett (33) have developed a mathematical model to reflect the factors that determine the rates of spread. They applied the model to the spread of measles in Iceland, where almost all introductions started in Reykjavik. Bogel et al (22) have modeled the spread of rabies across central Europe and examined the expected effectiveness of different control procedures.

Models of Disease Prevalence

The most widely known models of epidemics are those designed to predict changes in numbers of infectious and susceptible individuals over the course of an epidemic. Dietz & Schenzle have credited En'ko (or Yenko) (46) with the first published work on this kind of model. Most other writers have given primary credit to Hamer (57) and Soper (97), or to the still more recent work of Frost (52). In its simplest form, this model can be represented by two equations. For a disease that has a combined incubation-infectious period of defined duration, and death or immunization of all individuals at the end of this time,

$$I_{(t+1)} = \lambda I_t S_t$$

and

$$S_{t+1} = S_t - \lambda I_t S_t + \mu, \qquad 1.$$

where I_t is the number of infectious persons in any one period, S_t is the number of susceptible persons, λ is the frequency of effective contact between an infectious and a susceptible person, and μ is the number of new susceptible individuals entering the population, as by birth, in time t. A detailed discussion of this kind of deterministic model and its relation to a corresponding stochastic specification is given in the next section.

The above formulae also have a continuous-time analog in the differential equation system:

$$\frac{dI}{dt} = \lambda IS \text{ and } \frac{dS}{dt} = -\lambda IS + \mu. \qquad 2.$$

This model has been referred to as "compartmental," because the host population is divided into several compartments according to its relation to the disease. The number of compartments that should be defined in the model has been the subject of considerable disagreement. In the forms set out above there are only two compartments, infectious and susceptible, but one might also consider compartments for persons incubating the disease and for those who become immune after infection. Anderson & May (7) not only find it unnecessary to add these additional compartments, but also suggest that the number of infectious persons may have little relevance. Omission of this compartment would leave the number of new cases proportional to the number left susceptible by the preceding wave. This, they point out, is approximately true of an airborne infection, such as measles, in a defined space, such as a classroom. The main advantage of this simple model is that whether the infectious individuals are homogeneously mixed throughout the population is no longer a critical concern, but the model has little relevance outside its narrowly defined limits.

Briscoe (26) has also simplified his models by limiting the number of categories. He has designed four basic models, for disease with and without superinfection and with and without environmental contamination; each model has two compartments, infectives and susceptibles. He describes the four models as a system for grouping human disease, but the effects of immunity, vectors, and different levels of infectiousness are all lumped into two rate parameters for each model. From another point of view these subsumed factors seem critical, and Briscoe's four categories seem to be very heterogeneous agglomerations.

Many researchers feel that attempts to simplify in the above manner represent "spurious elegance" (23) and obscure the factors with which an epidemiologist must be most concerned if the outcome of an epidemic is to be anticipated or modified. Fukada et al (53) added a compartment for inapparent infections and found that the peak of an epidemic was sharper than that predicted solely on the basis of apparent cases. Others have gone much farther in elaborating the models defined by Equation Systems 1 and 2. Elvaback et al (45) and Longini et al (76) with an influenza model and Cvjetanovic et al (36, 37) with models for cholera, diphtheria, meningitis, pertussis, tetanus, poliomyelitis, and measles have striven for greater resolution and reliability by setting up large numbers of compartments. The poliomyelitis model of Cvjetanovic et al, for instance, has 10 compartments and 15 rate parameters for movement between compartments. Parameters include the proportion of children born with passive immunity, the duration of this immunity in those who have it, the proportion of the population receiving temporary protection from vaccination, and so on. They assigned values for each of these rates based on old and geographically restricted (to developed countries) data from the literature. This model gives a much clearer picture of an epidemic than the simpler models and makes it easier to see how a specific change applied to one compartment will affect numbers in other compartments.

The complex models, however, are also likely to lead the investigator astray, because the parameter values that are usually inserted at the beginning of an investigation are unreliable. The data of Cvjetanovic et al on durability of maternal antibody are no longer valid, because now most mothers derive their immunity from vaccine and have less specific antibody to give (74); their data on the proportion deriving immunity from vaccine are not applicable to situations in third world countries, where other enteric infections frequently interfere with the vaccine. Each of these parameters gives one value for a heterogeneous entity or process, and the distribution of this heterogeneity may be as important as its mean (36, 37). Cvjetanovic et al (described in 99) tested their model by simulating curves for age-specific antibody prevalence, but these simulations have a disturbing element of circularity because the parameters are based on data from the same sort of situation as that being modeled. Elvaback et al (45) suggested that their model could be tested by trying to stop an influenza pandemic by vaccinating school children according to their recommendations. However, there has been no major epidemic since these recommendations came out. When a pandemic does occur it will be extremely difficult to attain the recommended level of vaccine usage in time.

The complex models set out the factors that influence the course of an epidemic in a much more precise manner than the simple models, and they are useful as pedagogical devices for clarifying the nature of epidemics. To make these models effective tools for reliably assessing the consequences of planned interventions, much further development and evaluation of models remains to be carried out.

Alternatives to the above models are the differential equation systems of Cooke & Yorke (35), Hale (56), and Busenberg & Cooke (29). Integration of their equations gives rise to new constants that have biological meaning, although they are not derived directly from biological considerations. The mathematicians have produced some very promising methods in this way, but it is now up to the biologists to give meaning and values to the new constants. To do this will require both facility with the abstract and familiarity with contamination and disease. Few people since George Macdonald, who worked on malaria 30 years ago (80), have adequately combined these talents. Busenberg & Cooke made a start on relating the new constants to biologically permissible values, but they made no attempt to deal with specific diseases. Even after multitalented people have been found and these problems have been solved, the "determinedly non-numerate" (23) field persons will have to be educated to use the more complex equations. The computer simulation algorithms outlined in the next section do, however, provide a promising access to complex models for persons with minimal mathematical background but extensive knowledge of disease processes.

Dietz & Schenzle (43) have carried out an impressive analysis of the fit of six variants of Equation Systems 1 and 2 to a large data set on the spread of the common cold within families (25). This data set was particularly attractive because each family had the same number of members of approximately the same ages and thus host heterogeneity was minimized. The relatively early model of Kermack & McKendrick (68) and a recent modification developed by Schenzle (92) gave the only significant correlations. However, this study fails as a convincing test of the relative merits of the models, because the common cold is not one but many diseases. Dietz & Schenzle tried to fit the models to multimodal data with variable infectivity and variable preexisting immunity.

Algorithmic Construction of Infectious Disease Models

Before proceeding to an informal discussion of more complex models, we discuss model specifications from the perspective of computer implementation and use by persons who lack a sophisticated mathematical background.

The basic concepts and rationale behind the most intricate heterogeneous population models are likewise manifest in the classical Reed-Frost model (9) and its variants. We discuss the simplistic Reed-Frost formulations in some detail, emphasizing the manner in which modern computer technology makes model-based exploration of the character of epidemics and infectious disease transmission a routine matter that does not require elaborate analytical formulas.

REED-FROST AS A STOCHASTIC COMPARTMENT MODEL An initial population of S_0 susceptible individuals is assumed to interact with an initial population of I_0 infective individuals subject to the following rules.

1. During an infectious period each susceptible individual has probability p of an effective contact with a given infective individual. Contacts are independent, equally likely random events. Thus a given susceptible has probability $(1 - p)^{I_0}$ of not making an effective contact with the infective population. The same individual has probability $1 - (1 - p)^{I_0}$ of acquiring an infection and thereby being transferred from the susceptible to the infective population for the next time period.
2. We establish a discrete time scale, where the unit of time is the infectious period, assumed to be the same for all individuals.
3. A single attack of infection is assumed to confer permanent immunity. Thus, following the infectious period, each infective is removed from consideration in the analysis of the further development of the transmission process.

When transmission is governed by rules 1–3 and the sequential algorithm delineated below, the states "susceptible," "infective," and "removed" may be viewed as three compartments through which an initial closed population of size N moves.

The sequence of susceptible and infective populations evolving over time is defined by the pairs (S_0, I_0), (S_1, I_1), (S_2, I_2), . . ., (S_n, I_n), . . . where S_k is the number of susceptibles in time period k and I_k is the number of infectives in time period k. This sequence of random vectors is a stochastic process governed by the following algorithm.

1. For a priori given values $S_0 = s_0$, $I_0 = i_0$, the value of I_1 is determined by taking s_0 independent samples (contacts), each of which has probability $P_1 = 1 - (1 - p)^{i_0}$ of being an effective contact. Thus the probability distribution of I_1 is the specific binomial distribution

$$\text{Prob}(I_1 = i_1 \mid S_0 = s_0, I_0 = i_0) = \binom{s_0}{i_1} P_1^{i_1} (1 - P_1)^{s_0 - i_1}, \tag{3.}$$

where $0 \leq i_1 \leq s_0$. The interpretation of this sampling rule is that the S_0 initial susceptibles independently interact with the initial infective population; each has probability P_1 of becoming infective. Having determined i_1 by sampling from Equation 3, we define s_1 (size of the susceptible population in time period 1) $= s_0 - i_1$. The original i_0 infectives become immune and no longer enter into the transmission process.

2. Having determined that $S_1 = s_1$, $I_1 = i_1$, we determine the value of I_2 by taking s_1 independent samples (contacts), each of which has probability $P_2 = 1 - (1 - \text{p})^{i_1}$ of being an effective contact. Thus

$$\text{Prob}(I_2 = i_2 \mid S_1 = s_1, I_1 = i_1) = \binom{s_1}{i_2} P_2^{i_2} (1 - P_2)^{s_1 - i_2}, \tag{4.}$$

where $0 \leq i_2 \leq s_1$ and $s_2 = s_1 - i_2$.

3. The above process is repeated so that if at time k we have $S_k = s_k$, $I_\text{k} = i_\text{k}$, then the value of I_{k+1} is determined by taking s_k independent samples (contacts), each of which has probability $P_{k+1} = 1 - (1 - p)^{i_k}$ of being an effective contact. The distribution of I_{k+1} is again binomial, with

$$\text{Prob}(I_{k+1} = i_{k+1} \mid S_k = s_k, I_k = i_k) = \binom{s_k}{i_{k+1}} P_{k+1}^{i_{k+1}} (1 - P_{k+1})^{s_k - i_{k+1}}, \tag{5.}$$

where $0 \leq i_{k+1} \leq s_k$ and $s_{k+1} = s_k - i_{k+1}$.

The above algorithm can be readily implemented using a binomial random number generator. It can be used to generate multiple realizations of the sequence $(S_0, I_0), \ldots, (S_k, I_k)$ for various values of the effective contact probability, p. Each such realization is an instance of a Reed-Frost stochastic epidemic and allows an investigator to discern whether Reed-Frost dynamics mirror the observed data or not. Stochastic epidemic models, in general, can be studied via examination of multiple realizations (histories) constructed of sequences of probabilistic sampling processes analogous to the above sequence of binomial samples. Although the sampling schemes become increasingly intricate as more details of infectious disease transmission are incorporated into the model specifications, the essential logic is analogous to that of the simple Reed-Frost construction.

Within the stochastic Reed-Frost construction we can readily calculate and interpret the basic reproduction rate, R_0, for such processes. If we define R_0 as the expected number of new infectives produced in a susceptible population by a single infective individual, then (using Equation 3)

$$\begin{aligned} R_0 &= E(I_1 \mid S_0 = s_0, I_0 = 1) \text{ (expected numbers of new infectives produced from } s_0 \text{ susceptibles by 1 infective)} \\ &= s_0 P_1 = s_0[1-(1-p)] \\ &= s_0 p. \end{aligned} \tag{6}$$

Thus the usual criterion for the interruption of transmission in terms of basic reproduction rates, i.e. $R_0 < 1$, reduces in the present case to p (probability of effective contact) < 1/(number of initial susceptibles). However, from the basic stochastic process the origin of the principle that $R_0 < 1$ implies eradication is not obvious. To see this, it is useful to recall that this kind of threshold principle is usually derived within deterministic models of transmission. Thus we seek a natural deterministic counterpart to the stochastic Reed-Frost construction. If $S_n = s_n, I_n = i_n$, the expected number of infectives in period $n + 1$ is given by $E(I_{n+1} \mid S_n = s_n, I_n = i_n) = s_n P_n$ and the expected number of susceptibles in period $n + 1$ is given by $E(S_{n+1} \mid S_n = s_n, I_n = i_n) = s_n(1 - P_n)$, since $s_n = I_{n+1} + S_{n+1}$

As $P_n = 1 - (1 - p)^{i_n}$, the above formula suggests that the expected number of susceptibles satisfies a recurrence relation of the form

$$\begin{aligned} s_1 &= s_0 (1 - p)^{i_0} \\ s_2 &= s_1 (1 - p)^{i_1} \\ &\vdots \\ s_{n+1} &= s_n (1 - p)^{i_n}. \end{aligned} \tag{7}$$

Thus s_{n+1} may be represented in terms of the initial population composition (s_0, i_0) and the history of the epidemic via the formula

$$s_{n+1} = s_0 (1 - p)^{i_0 + \cdots + i_n} = s_0(1 - p)^{N - s_n}, \tag{8}$$

where $N = s_0 + i_0$ = total size of the population.

The recurrence relations of Equation System 7 represent a deterministic version of the Reed-Frost model of infectious disease transmission, where the average number of susceptibles in any time period equals the number of susceptibles in the previous time period multiplied by their probability of not becoming infected. In other words, $s_n(1 - p)^{i_n}$ represents the average number of susceptibles in period n who do not achieve an effective contact with the susceptible population. The equations $s_{n+1} = s_n(1 - p)^{I_n}$ and $i_{n+1} = s_n - s_{n+1}$ are analogous to Equation System 1. Letting the time period increase arbitrarily, i.e. $n \to \infty$, and introducing the proportions $w_n = s_n/N$, then w_n satisfies the relation

$$w_{n+1} = w_0(1 - p)^{N(1-w_n)} \tag{9}$$

and $\lim_{n \to \infty} w_n = w_\infty$ is a well-defined equilibrium proportion of susceptibles.

Some further mathematical analysis (79) established that for w_0 (the initial proportion of susceptibles) close to 1 (i.e. when the process begins with only a few infectives) there are two possible equilibria:

$$w_\infty = 1 \text{ if } R_0 < 1 \text{ and } 0 < w_\infty < 1 \text{ if } R_0 > 1.$$

Recall that in the Reed-Frost model $R_0 = s_0 p$. The condition $w_\infty = 1$ corresponds to eradication. In other words, the equilibrium condition is that the population contains only susceptibles; the infectives are eliminated. The condition $0 < w_\infty < 1$ means that transmission is sustained; there are always infectives remaining.

The following are generic aspects of this analysis for infectious disease transmission models.

(*a*) The logic behind Equation 9 is simply that the proportion of susceptibles in successive time periods is determined by the interaction (independent random contacts in this instance) between susceptibles and infectives in the previous time period. In more general models the right-hand side of an equation like 9 may involve more complex interaction terms or addition of new susceptibles and departures via mortality. However, the left-hand side of an expression always represents either a change in a proportion locally in time or a level at a subsequent time, while the right hand side has terms that determine the change or the level according to either a contact process or immigration and emigration (including mortality) of individuals.

(*b*) Equations like Equation 9 can easily be solved numerically on microcomputers; sequences $w_0, w_1, w_2, \ldots$ can be displayed for various values of p and the program can immediately indicate whether the equilibrium corresponding to eradication or maintainance of transmission is approached. Simple analytical formulas that are solutions of equations like Equation 9 are rarely producible, except with the simplest and usually most unrealistic model specifications. This was a matter of great concern in the precomputer age when many of the infectious disease transmission models were first formulated. However, now that numerical solutions even of very intricate models can be routinely generated on microcomputers, the central issues in model development are the tradeoffs between biological realism, data availability, and end use of the models. The concern with providing analytical formulas for solutions of equations like Equation 9, which has pervaded the modeling literature almost to the present time, need not be an issue when numerical solutions and graphic displays of them can be generated by computer in a matter of minutes.

(*c*) The basic reproduction rate, R_0, depends critically on the parameters in an assumed class of models (in this case, the probability of an effective contact, p). Thus conclusions about the conditions under which transmission can be interrupted depend on the extent to which a dynamic model of the propagation of infection mirrors actual processes. This dependence on dynamic models is especially interesting since the concept that basic reproduction rate equals the expected number of infections passed on to a susceptible population by a single infectious individual does not involve a formal transmission model. Unfortunately, for virtually all infectious diseases direct measurement of R_0 is a practical impossibility. Thus mathematical models and parameter estimates in them provide a practical quantitative route to R_0.

A HETEROGENEOUS EXTENSION OF THE REED-FROST MODEL If the initial susceptible population is comprised of high susceptibles, denoted by S_0^H, and low susceptibles, denoted by S_0^L, then the total number of initial susceptibles is represented as $S_0 = S_0^H + S_0^L$. We let ρ denote the probability that a susceptible is a low susceptible. We then define p_1 as the probability of effective contact between an infective and a high susceptible, and p_2 as the probability of effective contact between an infective and a low susceptible, with $p_2 < p_1$.

There are two cases to consider: (*a*) High and low susceptibles are identifiable with existing technology. (*b*) High and low susceptibles are not directly identifiable, and the probability ρ must be estimated simultaneously with the other parameters, p_1 and p_2, governing the dynamic process whose construction is outlined below.

For case *a* the heterogeneous Reed-Frost model evolves according to steps 1–5:

1. New infectives are generated from the high susceptible population via sampling from the binomial distribution, using

$$\text{Prob}(I_1^H = i_1^H \mid S_0^H = s_0^H, I_0^H = i_0^H) = \binom{s_0^H}{i_1^H} P_1^{\,i_1^H} (1 - P_1)^{s_0^H - i_1^H}, \qquad 10.$$

where I_1^H is the number of high susceptibles in the initial population who become infective in the next time period; $P_1 = 1 - (1 - p_1)^{i_0}$ is the probability that a single high susceptible makes an effective contact with the initially infective population of size i_0.

2. New infectives are generated from the low susceptible population via sampling from the binomial distribution, using

$$\text{Prob}(I_1^L = i_1^L \mid S_0^L = s_0^L, I_0^L = i_0^L) = \binom{s_0^L}{i_1^L} P_2^{i_1^L} (1 - P_2)^{s_0^L - i_1^L}, \quad 11.$$

where I_1^L is the number of low susceptibles in the initial population who become infective in the next time period; $P_2 = 1 - (1 - p_2)^{i_0}$ is the probability that a single low susceptible makes an effective contact with the initially infective population of size i_0.

3. The number of infectives in time period 1 is defined as $i_1 = i_1^L + i_1^H$. Then $s_1^L = s_0^L - i_1^L$, $s_1^H = s_0^H - i_1^H$. Then we define $s_1 = s_1^L + s_1^H$ as the number of susceptibles in time period 1.

4. Steps 1–3 are repeated using s_1^L, s_1^H, and i_1 as the initial populations of low and high susceptibles and the infectives, respectively.

5. Following time period k, steps 1–3 are repeated using s_k^L, s_k^H, and i_k as initial populations to generate i_{k+1}^L, i_{k+1}^H, and thereby $i_{k+1} = i_{k+1}^L + i_{k+1}^H$.

For case b we first generate s_0^L via binomial sampling from the full initial population of susceptibles. In particular, we set

$$\text{Prob}(S_0^L = s_0^L \mid S_0 = s_0) = \binom{s_0}{s_0^L} \rho^{s_0^L} (1 - \rho)^{s_0 - s_0^L}. \quad 12.$$

Then we define $s_0^H = s_0 - s_0^L$, and with these values at hand proceed to steps 1–4 as in the construction for case a.

The basic reproduction rate, R_0, for the above constructions is analytically cumbersome. However, R_0 can be simply evaluated for various choices of p_1, p_2, and ρ by using steps 1–3 to generate several hundred realizations of I_1 from $S_0 = s_0$, $I_0 = 1$ and then averaging the resulting counts of infectives.

Thus microsimulation of the heterogeneous Reed-Frost process allows an investigator to examine multiple histories (S_0,I_0), (S_1,I_1), . . . ,(S_n,I_n), . . . and to evaluate R_0 without concern for simple analytical formulas (which don't exist) for R_0 and such quantities as the mean and variance of S_n and I_n. This construction is the simplest prototype of the introduction of heterogeneity into infectious disease transmission models.

Models With Intermediate Vectors

The application of compartmental models to vector-borne diseases entails additional complexity, but also provides additional opportunities to identify vulnerable links in the chain of infection. During the period covered by this review a large body of new field data became available from relatively well-monitored malaria control projects in Africa (17, 86). Dietz et al (42) used these data to develop a refined model that takes immunity into account and predicts prevalence rates better than earlier models. Nedelman (87) has set the model of Dietz et al in a biologically more realistic framework. In the process he has raised questions regarding the value of mathematical goodness-of-fit tests as measures of biological relevance.

More recently, emphasis has shifted from vector control to the effects of chemotherapy and parasite resistance (27, 38). As pointed out below, the models suggest that drugs might be most effective when concentrated on the most heavily infected individuals (5). The rapid spread around the world of *Plasmodium falciparum* strains resistant to chloroquine and pyrimethamine has quite exceeded the speed that modelers might have predicted. It would still be interesting, however, to see whether the models could provide information on the extent to which this could have been spread by the progeny of a few mutants and the extent to which parallel evolution must have been involved.

Transmission Potential

It is useful to have a simple statistic that measures the capacity of a disease to spread in a population if the spread is not inhibited by immunity induced by its prior passage. Such a statistic should facilitate the prediction of whether an epidemic will persist or fade out. A parameter possessing these characteristics has been variously called "basic reproduction rate" (3, 7, 8, 41), "infectious contact number" (61), and "transmission potential" (44). Here we use the term "transmission potential," as it is both the shortest name and the clearest representation of what is meant. We use the symbol R_0, which is common to all the above treatises, for this value. The origins of the concept of R_0 go back to Macdonald's work on malaria (80), but its principles are most clearly set out by Anderson (3). The concept is applicable to diseases with vector transmission as well as to those with a direct host-to-host pattern, but it is not relevant to common source epidemics or to models for the density of infection in individual hosts. In density modeling, as for example in the case of schistosomiasis, R_0 has been defined as the average number of female offspring produced by a mature female parasite when the offspring reach reproductive maturity in the absence of immunological or other density-dependent constraints.

The concept of R_0 is focused on the number of new cases engendered by one initial case. Obviously, if the number of new cases in one cycle is less than that in the preceding cycle an epidemic will not grow, and if the number can be held at a stable level the epidemic will die out. The essential feature of the transmission potential concept, however, is that the slowing of an epidemic due to accumulation of infection-induced immunity does not affect its persistence in a large population. Except below a critical population size (a concept discussed below), any decline in an epidemic due to accumulation of immune individuals will reverse as the rate of spread decreases, and an equilibrium will be established, possibly with oscillations. Only if the rate of spread can be held at less than one secondary case per index case, in the largely hypothetical circumstance where there is no disease-induced immunity, will a disease be eliminated. The transmission rate can be reduced by sufficiently modifying factors that affect the ability of the disease to spread, e.g. by eliminating mosquito vectors to reduce malaria incidence (80), or by using condoms to reduce the infection transmission rate of acquired immune deficiency syndrome (AIDS). Alternatively, reduction of transmission rate might be achieved by reducing the susceptibility of the uninfected population through vaccination, as with measles (20). The goal of nearly all such interventions would be to hold this rate below 1.0 long enough and consistently enough across all the multiple social groupings that make up any population that the disease will disappear altogether from that population.

Guidance about the impact of particular interventions on a basic reproduction rate can be obtained from explicit formulas for R_0 that are associated with models of transmission. For example, for diseases that induce lifelong immunity in recovered individuals (e.g. measles), if homogeneous populations and uniform mixing in transmission are assumed, the basic reproduction rate is represented as

$$R_0 = \frac{N\lambda\alpha}{(\alpha + \beta)(\gamma + \beta)}, \qquad 13.$$

where N is the size of an initial disease-free population, λ is the contact rate, α is the reciprocal of the latent period of the disease, β is the reciprocal of the host's life expectancy, and γ is the reciprocal of the average duration of infectiousness.

The importance of such formulas stems from the fact that although R_0 usually cannot be measured directly, i.e. by counting secondary infections, its components (λ, α, β, and γ) can be directly ascertained. Thus the trustworthiness of estimates of R_0 based on parameters in transmission models depends on the extent to which the models mirror actual processes. R_0 is important in planning immunization programs, in that the proportion p of the population that must be immunized at or near birth to reduce the value of R_0 below unity must satisfy $p > 1 - 1/R_0$ (see Reference 8).

Furthermore, Dietz (39) has published a useful and general approximate formula for R_0 (see also Reference 8):

$$R = 1 + L/A, \qquad 14.$$

where L is the life expectancy and A is the average age at infection.

Protection provided by maternal antibody for a significant part of the time prior to the average age at infection can be simply accounted for by subtracting the mean duration of this protection from A (63). Using Formula 14, Anderson has estimated R_0 for several diseases (3). The reported values ranged from about 75 for malaria in Nigeria and 16 for measles in England to 1.5 for schistosomiasis in Brazil. These estimates must be country and time specific because the number of contacts and chance of transmission vary greatly from one cultural setting to another. Differences in the proportions of lower versus upper age groups in the population also distort R_0 (20).

Although it should be possible using Formula 14 to determine how much intervention is needed to reduce R_0 to less than 1.0, such efforts have often yielded misleading results, and diseases have proven much harder to eliminate than predicted. The transmission rates in the poorest parts of any community are commonly much higher than the average; if these poorer sections are large enough they may sustain endemicity even when the average proportion susceptible has been reduced to a level that would stop transmission. The areas with high transmission rates are likely to be the most difficult to cover adequately in an immunization program, and an ample average coverage rate is likely to be least attainable where the highest levels are needed. This seems to be why an intensive measles vaccination program in São Paulo failed to significantly change the pattern of disease distribution.

Another problem in using R_0 to plan interventions is that contact rates generally go up with increasing age. In the course of a lifetime, from the nursery or mother's back, to neighborhood play groups, to school, and perhaps to college or into large military units, the number of contacts increases progressively. The measles vaccination program in the United States that was adequate to reduce R_0 below 1.0 in 1967 when most transmission occurred in elementary schools was no longer adequate in 1984 when the proportion of susceptible young adults had increased (74). There was a spate of epidemics among young adults in colleges, who associated in groups of tens of thousands of persons (30). Dietz & Schenzle (44) have expanded the formula for the relationship between R_0 and age at infection to include integration of changing values for different ages. For this formula to be useful, however, it is necessary for the relation between age and infectiousness to be couched in an expression that can be integrated, and this has not been done.

Another practical problem that has often led estimates of R_0 astray is inaccurate measurement of the average age at infection. For many diseases the number of reported cases does not form an appropriate basis for the estimation of this value. Low overall efficiency and age-specific bias in case reporting invalidate many data sets for this purpose. Serological data are generally preferable, but it is essential that the serological test recognize all immune individuals. Tests that only transiently record some persons as positive will give an erroneously high value for R_0, whereas tests that are insufficiently sensitive may give low values.

Stochastic Versus Deterministic Models and Epidemic Periodicity

There remains considerable difference of opinion regarding the adequacy of deterministic models, in which a population is considered sufficiently large that the fraction infected in each cycle will be independent of chance. Alternatives are stochastic models, illustrated above for the Reed-Frost model, in which the number infected will be some integer that, only over an average, will correspond to the predicted value. It is obviously more important to use stochastic processes when dealing with small populations, but this method is also used to determine whether an epidemic will persist when it has been reduced to a minimal level. The method commonly requires "Monte Carlo" procedures for its solution, i.e. the selection of numbers from a random list with a predetermined distribution. Ludwig (78) has suggested approximations to avoid this, and at least in his examples, these estimates yield curves very similar to those derived by fully stochastic procedures. Severo (94) has also discussed effective analytical approximations.

A phenomenon common to many endemic diseases is cyclic variation in intensity. In the prevaccine era measles commonly attained peak frequencies at two- to three-year intervals and rubella reached peaks at approximately ten-year intervals. The basic models of Soper (97) and Frost (52) could reproduce these cycles over a period of a few years, but the waves invariably dampened with time. In 1956 Bartlett (11) developed a stochastic model for measles and showed that by introducing chance, the indefinite continuation of cycles in prevalence that are multiples of one year could be modeled. More recently, London & Yorke (75, 102) showed that if seasonal differences in transmission rate were incorporated, periodic waves could be modeled by deterministic methods. Fine & Clarkson (48) confirmed the adequacy of seasonality as an explanation for cycling of measles and pertussis in England. They showed that this seasonality was largely due to the opening and closing of primary schools. The seasonality of pertussis decreased coincident with reduced vaccine usage and a corresponding increase in the proportion of cases in the preschool age group. The measles seasonality in the United States has also become less pronounced in recent years, as coincident with extensive vaccine use, the average age has moved above the primary school years.

There is, however, a difference in the nature of periodicity shown by stochastic and seasonal models. The seasonal model predicts a more regular wave pattern than the stochastic model. The regularity of the seasonal model is appropriate for a very large population such as that of New York, which

Yorke studied reasonably well, but it is less satisfactory for smaller cities like Providence (32) and Baltimore (59), where chance events have larger roles. Clearly a more refined model would be stochastic and would incorporate seasonal variation.

Although either stochastic models or deterministic models that allow for seasonal differences can give a reasonable representation for measles periodicity, no method has been published that would explain the much longer periodicity of rubella epidemics. It seems likely that an explanation of the rubella pattern will take into account variations in infectiousness of the virus analogous to those recognized in influenza.

An important aspect of stochasticity and seasonality is the fact that in all but the very largest populations the number of new cases falls to low levels once each cycle. Yorke et al (103) have shown that this fact greatly increases the possibility of interrupting disease spread. Efforts to eliminate a disease agent, whatever the control method, will be most effective if intensified in the ebb phase.

In relatively simple modeling tasks, such as the modeling of common source epidemics, stochastic methods are both practical and desirable; Shonkwiler & Thompson used them in their study of common source epidemics (95), and Lange et al (72) used them in a study of the effect of intervention in typhoid epidemics. Haskey (58) used stochastic models in a study of cross-infection between related populations. On the other hand, when Watson (100) tried to combine stochastic methods with population heterogeneity, he obtained essentially intractable formulas and was forced to make a number of approximations. Microsimulations of his models would, however, be tractable. The model of Elvaback et al (45) for the effectiveness of vaccination against influenza is stochastic. This permits its application to relatively small populations.

Critical Population Size

An extension of the considerations of stochasticity indicates that each disease will have a critical population size for its persistence. An epidemic in smaller populations will consume all susceptible hosts, and the infecting agent will be unable to persist. If many of the hosts have been immunized and are not killed in the process of the epidemic, the host population may be able to recover and go on disease free until the agent is reintroduced from outside. Hope-Simpson (66) was the first to provide field data on this phenomenon from a human population. He showed that varicella could persist in a population of 2000 on one of the Shetland Islands, where other diseases died out.

Bartlett (12, 13) showed that measles did not persist in cities with populations of less than 300,000. Black (18) extended this observation to island populations where, because there were fewer introductions, the disease-free periods were more prolonged and more easily recognizable. In the island studies the critical population size for measles perpetuation was seen to be somewhat larger than in the city studies; however, it was dependent not only on total population, but also on population density. On a densely populated island or in a city, measles spread more rapidly, burned itself out more quickly, and required more people for its perpetuation than on a large, thinly populated island such as Iceland or New Zealand. The duration of an epidemic was shown to be proportional to the population per square kilometer.

Subsequent studies based on isolated populations in forests as well as on islands (19, 21, 54) have extended the concept of critical population size to a wide variety of diseases, with values ranging from less than 100 for herpes simplex to nearly one million for measles.

Models with Variable Host Population Size

In all the models described above it is assumed that the host population is fixed in size or that there is a constant input of new susceptibles, i.e. a constant birth rate. This is a reasonable approximation if the disease is nonfatal, but if the disease is fatal, diminution of host population will affect the spread of a disease agent. Disease agent and host survival will interact to cause genetic selection and evolutionary change in both host and parasite. An extraordinary set of field data from studies conducted on myxomatosis after its introduction to the Australian rabbit population (47) serves as a guide to modeling in this field. In this epidemic, where a nearly 100% lethal agent was introduced to an inexperienced host population, the virus mutated within a few cycles to cause reduced mortality. In addition the host gained resistance, but only slowly.

Anderson & May have presented basic mathematical methodology for modeling with variable host population size both directly (4) and indirectly transmitted diseases (83). The evolutionary implications have been clearly set out elsewhere by the same authors (6, 82). Hadeler & Dietz (55) have introduced stochasticity into the model. However, a persistent, basically biological problem with these models is caused by the complexity of the term "virulence." Virulence includes lethality, fecundity, and infectiousness of an agent. These three elements are linked but not equivalent. Greater lethality need not mean that more pathogen is produced, and increased reproduction of pathogens need not mean that infectiousness is raised. If these factors were unlinked, a pathogen would inevitably evolve to low lethality. Optimal infectiousness would be governed by whether or not immunity is induced and by the size of the host-population aggregates, i.e. by factors that determine critical population size and the probability of fade-out. Because lethality and infectiousness are linked, however, lethality may be maintained. Anderson & May (6) concluded that because of this complexity, a specific outcome for coevolution of host and pathogen cannot be predicted.

Parasite Density Models

Many protozoan and metazoan parasites do not complete a life cycle in one host. In these systems, the number of parasites per host, the severity of disease, and the infectiousness of each host organism depend on the number of parasites encountered. Recently, relatively little work has been done on models of this type of disease. Anderson & May (5) have reviewed the subject and introduced the effects of chemotherapy into their model. They suggested that treatment may be most effective if focused selectively on that part of the population with heavy parasite burden. Bundy et al (28) looked specifically at data on treatment of *Trichuris trichura* in Jamaica and reached the same conclusion.

In many instances, parasites that do not complete their life cycle in one host are sexual organisms that can only reproduce if both sexes are present. Accordingly, some workers (40, 81) have suggested that reducing the density of infection might provide better control of disease caused by such parasites than would be expected on the basis of simple models. Bradley & May (24) have reexamined this hypothesis and found that because the parasites are not uniformly distributed, the chances that an individual will be infected with too few parasites to permit pairing are not likely to be substantial. One should not expect a precipitous decline in the infection rate in the late stages of a control program.

HETEROGENEITY

Age-Based Heterogeneity

In addition to simple geographic division of a population into like groups, there may be division into dissimilar groups. The most frequently encountered dissimilar groups and the most amenable to mathematical analysis are age-based groups. Characteristically, as humans age they change their habits in ways that affect the number of contacts and change individual immune status and the immunity of the group. This problem is important in modeling (4, 20, 36, 51, 97). Recently there has been a promising start on the development of models that would take these differences into account.

Hethcote (60) designed a model that can be used for multiple age groups with different probabilities of disease transmission. However, the model only differentiates children from adults, and does not allow for movement from one group to the other. Hethcote & Yorke (64) presented more elaborate age-structured models focusing on gonorrhea.

Hoppensteadt (67) proposed that by the use of integral equations a smooth age progression could be considered. For integration of his equations the changes in infectiousness and number of contacts must be continuous and expressible in the form of mathematical functions. Hoppensteadt did not attempt to present functions that would represent these changes. Such functions would have to be quite complex to account for changes associated with discrete stages of life.

Dietz & Schenzle (41, 44) have proposed similar integral expressions leading to R_0. Their formulas can incorporate differences in contact rates and vaccination rates in different age groups, as well as differences in the duration of infection. Functions to describe the manner in which the parameters change with age are still lacking, however. If there are discontinuities in these functions, the system may become extremely complicated.

Fine & Clarkson (49) took the amount of measles vaccine given annually to each age cohort into account in estimating the number susceptible at different ages. Their comprehensive studies of pertussis (50) as well as measles (49), firmly based on field data, showed the relationship of age-specific attack rate to age at vaccination and to age at school entry. The observed changes in age of affected children can reasonably be explained by these analyses.

In models of rubella incidence, age is important not only because the virus will travel more easily through some age groups than through others and not only because some groups will have a higher proportion of immune individuals than others, but also because infection has serious consequences in only one age group. Knox (69) considered age in the outcome variable of a model for the frequency of rubella infection and used models in which vaccine is administered at discrete ages.

The greatest need for age-structured transmission models may be in the design of control methods for rubella. Flawed rubella control programs, which reduced the spread of the virus in younger groups but did not entail immunization of older groups, may lead to an increase in congenital rubella syndrome. There is also a need for further development of age-structured malaria transmission models.

Social Heterogeneity

Less systematic host-population heterogeneity than that represented by spatial aggregates and age cohorts is also important to the outcome of many epidemics. The most effective studies of such heterogeneity have dealt with gonorrhea (35, 61, 64, 71, 88, 102). The sexually inactive part of a population is at essentially no risk of contracting this disease, but qualitatively different subsets, especially prostitutes, are at very high risk. This disease also differs from most diseases considered previously in that there is no effective immunity. After a period of infectiousness, the individual returns to the susceptible pool. Lajmanovich & Yorke (71) have developed a model that takes these special circumstances into account. Reynolds & Chan (90) have applied a similar model to data from the United States. Hethcote (61) has suggested that models for vector-transmitted diseases are adaptable to gonorrhea if highly active transmitters, such as prostitutes, are considered vectors of the disease. Yorke et al (101) have examined the problems associated with R_0 for this disease. Although control conditions may be set so that most new cases give rise to fewer than one new case each, a small core group with very high R_0 values will usually remain. Thus the concept of average infection potential loses most of its significance.

AIDS is another disease in which specific population subgroups have very different transmission potential. Here the most important social groupings are the male homosexuals and the intravenous drug abusers. Control efforts, which at present mainly include attempts at behavior modification, clearly need to be focused on these groups. It might be questioned whether the transmission potential of infected females is low enough to have a negligible role in maintaining the epidemic. However, some African evidence indicates a nearly 1 : 1 sex ratio for AIDS. An appropriate model with parameter values based on good field data could answer such questions. Knox (70) has begun to develop a model for the spread of AIDS, which seems manageable and promises to be useful. Data on transmission rates, however, are inadequate as yet, particularly for heterosexual non–drug users.

In other diseases the relevant social class divisions are generally less defined and more numerous. It will be difficult to develop models that can appropriately account for the differences in measles transmission and vaccine acceptance rates in the several social strata of a large city and in the diverse social organizations found in some less-developed countries (1). No serious attempts to address these problems have been published.

Literature Cited

1. Aaby, P., Bukh, J., Lisse, I. M., Smits, A. J. 1983. Measles mortality, state of nutrition, and family structure: a community study from Guinea-Bissau. *J. Infect. Dis.* 147:693–701
2. Adesina, H. O. 1984. Identification of the cholera diffusion process in Ibadan, 1971. *Soc. Sci. Med.* 18:429–44
3. Anderson, R. M. 1982. Transmission dynamics and control of infectious disease agents. In *Population Biology of Infectious Diseases,* ed. R. M. Anderson, R. M. May, pp. 149–76. Berlin: Springer-Verlag. 315 pp.
4. Anderson, R. M., May, R. M. 1979. Population biology of infectious disease: Part I. *Nature* 280:361–67
5. Anderson, R. M., May, R. M. 1982. Population dynamics of human helminth infections: control by chemotherapy. *Nature* 297:557–63
6. Anderson, R. M., May, R. M. 1982. Coevolution of hosts and parasites. *Parasitology* 85:411–26
7. Anderson, R. M., May, R. M. 1983. Vaccination against rubella and measles: quantitative investigations of different policies. *J. Hyg.* 90:259–325
8. Anderson, R. M., May, R. M. 1985. Vaccination and herd immunity to infectious diseases. *Nature* 318:323–29
9. Bailey, N. T. J. 1975. *The Mathematical Theory of Infectious Diseases.* New York: Hafner. 413 pp.

10. Bart, K. J., Orenstein, W. A., Hinman, A. R., Amler, R. W. 1983. Measles and models. *Int. J. Epidemiol.* 12:263–66
11. Bartlett, M. S. 1956. Deterministic and stochastic models for recurrent epidemics. *Proc. Berkeley Symp. Math. Stat. Probab.* 4:31–109
12. Bartlett, M. S. 1957. Measles periodicity and community size. *J. R. Stat. Soc. Ser. A* 120:40–70
13. Bartlett, M. S. 1960. The critical community size for measles in the United States. *J. R. Stat. Soc. Ser A* 123:37–44
14. Becker, N. G. 1968. The spread of an epidemic to fixed groups within the population. *Biometrics* 24:1007–14
15. Becker, N. G. 1979. The uses of epidemic models. *Biometrics* 35:295–305
16. Becker, N. G., Hopper, J. L. 1983. Assessing the heterogeneity of disease spread through a community. *Am. J. Epidemiol.* 117:362–74
17. Bekessey, A., Molineaux, L., Storey, J. 1976. Estimation of incidence and recovery rates of *Plamodium falciparum* parasitemia from longitudinal data. *Bull. WHO* 54:685–91
18. Black, F. L. 1966. Measles endemicity in insular populations: critical community size and its evolutionary implications. *J. Theor. Biol.* 11:207–11
19. Black, F. L. 1982. Geographic and sociologic factors in the epidemiology of virus diseases. In *Viral Diseases in South-East Asia and the Western Pacific,* ed. J. S. Mackenzie, pp. 23–32. Sydney: Academic. 751 pp.
20. Black, F. L. 1982. The role of herd immunity in control of measles. *Yale J. Biol. Med.* 55:351–60
21. Black, F. L., Hierholzer, W. J., Pinheiro, F. P., Evans, A. S., Woodall, J. P., et al. 1974. Evidence for persistence of infectious agents in isolated human populations. *Am. J. Epidemiol.* 100:230–50
22. Bogel, K., Moegle, H., Knorpp, F., Arata, A., Dietz, K., Diethelm, P. 1976. Characteristics of the spread of a wildlife rabies epidemic in Europe. *Bull. WHO* 54:433–47
23. Bradley, D. J. 1982. Epidemiological models—theory and reality. In *The Population Dynamics of Infectious Diseases,* ed. R. M. Anderson, pp. 320–61. London: Chapman & Hall. 370 pp.
24. Bradley, D. J., May, R. M. 1978. Consequences of helminth aggregation for the dynamics of schistosomiasis. *Trans. R. Soc. Trop. Med. Hyg.* 72:262–73
25. Brimblecome, F. S. W., Cruickshank, R. T., Masters, P. L., Stewart, G. T. 1958. Family studies of respiratory disease infections. *Br. Med. J.* 1:119–28
26. Briscoe, J. 1980. On the use of simple analytic mathematical models of communicable diseases. *Int. J. Epidemiol.* 9:265–70
27. Bruce-Chwatt, L. J. 1981. *Chemotherapy of Malaria.* Geneva: WHO. 261 pp.
28. Bundy, D. A., Thompson, D. E., Cooper, E. S., Golden, M. H., Anderson, R. M. 1985. Population dynamics and therapeutic control of *Thrichuris trichura* infection of children in Jamaica and St. Lucia. *Trans. R. Soc. Trop. Med. Hyg.* 79:759–64
29. Busenberg, S., Cooke, K. L. 1980. The effect of integral conditions in certain equations modelling epidemics and population growth. *J. Math. Biol.* 10:13–20
30. Centers for Disease Control. 1981. Measles among children with religious exemptions to vaccination—Massachusetts, Ohio. *Morb. Mortal. Wkly. Rep.* 30:550, 555–56
31. Centers for Disease Control. 1984. Measles—United States, first 26 weeks, 1983. *Morb. Mortal. Wkly. Rep.* 32:363–64, 69
32. Chapin, C. V. 1925. Measles in Providence, R. I., 1858–1923. *Am. J. Hyg.* 5:645–55
33. Cliff, A., Haggett, P. 1982. Methods for the measurement of epidemic velocity from time-series data. *Int. J. Epidemiol.* 11:82–89
34. Cohen, J. E. 1977. Mathematical models of schistosomiasis. *Ann. Rev. Ecol. Syst.* 8:209–33
35. Cooke, K. L., Yorke, J. A. 1973. Some equations modelling growth processes and gonorrhea epidemics. *Math. Biosci.* 16:75–101
36. Cvjetanovic, B., Grab, B., Dixon, H. 1982. Epidemiological models of poliomyelitis and measles and their application in the planning of immunization programmes. *Bull. WHO* 60:405–22
37. Cvjetanovic, B., Grab, B., Uemura, K. 1978. Dynamics of acute bacterial diseases. Epidemiologic models and their application in public health. *Bull. WHO* 56:1–143 (Suppl.)
38. Davies, A., Wagner, D. H. G. 1979. Multicenter trials of praziquantel in human schistosomiasis: design and technique. *Bull. WHO* 57:761–71
39. Dietz, K. 1975. Transmission and control of arbovirus diseases. In *Epidemiology,* ed. D. Ludwig, K. L. Cooke, pp. 104–21. Philadelphia: Soc. Ind. Appl. Math. 176 pp.
40. Dietz, K. 1975. The pairing process. *Theor. Popul. Biol.* 8:81–96

41. Dietz, K. 1982. Overall population patterns in the transmission cycle of infectious disease agents. See Ref. 3, pp. 87–102
42. Dietz, K., Molineaux, L., Thomas, A. 1974. A malaria model tested in the African savannah. *Bull. WHO* 50:347–57
43. Dietz, K., Schenzle, D. 1985. Mathematical models for infectious disease statistics. In *Celebration of Statistics: The ISI Centenary Volume,* ed. A. C. Atkinson, S. E. Feinburg, pp. 167–204. New York: Springer-Verlag. 606 pp.
44. Dietz, K., Schenzle, D. 1985. Proportionate mixing models for age dependent infection transmission. *J. Math. Biol.* 22:117–20
45. Elvaback, L. R., Fox, J. P., Ackerman, E., Langworthy, A., Boyd, M., Gatewood, L. 1976. An influenza stimulation model for immunization studies. *Am. J. Epidemiol.* 103:152–65
46. En'ko, P. D. 1889. The epidemic course of some infectious diseases. *Vrach. Delo* 10:1008–10, 1039–42, 1061–63 (In Russian)
47. Fenner, F., Ratcliffe, F. N. 1965. *Myxomatosis*. Cambridge, UK: Cambridge Univ. Press. 379 pp.
48. Fine, P. E. M., Clarkson, J. A. 1982. Measles in England and Wales—I: An analysis of factors underlying seasonal patterns. *Int. J. Epidemiol.* 11:5–14
49. Fine, P. E. M., Clarkson, J. A. 1982. Measles in England and Wales—II: The impact of the measles vaccination programme on the distribution of immunity in the population. *Int. J. Epidemiol.* 11:15–25
50. Fine, P. E. M., Clarkson, J. A. 1986. Seasonal influences on pertussis. *Int. J. Epidemiol.* 15:237–47
51. Fox, J. P., Elvaback, L., Scott, W., Gatewood, L., Ackerman, E. 1971. Herd immunity: basic concept and relevance to public health immunization practices. *Am. J. Epidemiol.* 94:179–89
52. Frost, W. H. 1976. Some conceptions of epidemics in general. *Am. J. Epidemiol.* 103:141–51
53. Fukuda, K., Sugawa, K., Ishii, K. 1984. Simulation of infectious disease by Reed-Frost model with proportion of immune and inapparent infection. *Comput. Biol. Med.* 14:209–15
54. Garruto, R. M. 1981. Disease patterns of isolated groups. In *Biocultural Aspects of Disease,* ed. H. R. Rothschild, pp. 557–97. New York: Academic. 653 pp.
55. Hadeler, K. P., Dietz, K. 1984. Population dynamics of killing parasites which reproduce in the host. *J. Math. Biol.* 21:45–65
56. Hale, J. 1974. Behavior of near constant solutions of functional differential equations. *J. Differ. Equat.* 15:278–94
57. Hamer, W. H. 1906. The Milroy lectures on epidemic disease in England—the evidence of variability and persistency of type. *Lancet* 1:733–39
58. Haskey, H. W. 1957. Stochastic cross infection between two otherwise isolated groups. *Biometrika* 44:193–204
59. Hedrick, A. W. 1933. Monthly estimates of the child population "susceptible" to measles 1900–1931, Baltimore, Md. *Am. J. Hyg.* 17:613–30
60. Hethcote, H. W. 1975. Mathematical models for the spread of infectious diseases. See Ref. 39, pp. 122–31
61. Hethcote, H. W. 1976. Qualitative analysis of communicable disease models. *Math. Biosci.* 28:235–56
62. Hethcote, H. W. 1978. An immunization model for a heterogeneous population. *Theor. Popul. Biol.* 14:338–49
63. Hethcote, H. W. 1984. A vaccination model for an epidemic disease with maternal antibodies in infants. *Proc. 2nd Int. Assoc. Math. Comput. Simulation (IMACS) Int. Symp. Biomed. Syst. Model.* Amsterdam: North-Holland
64. Hethcote, H. W., Yorke, J. A. 1976. *Gonorrhea Transmission Dynamics and Control. Lect. Notes Biomath.* New York: Springer-Verlag
65. Hinman, A. R., Brandling-Bennett, D., Nieburg, I. 1979. The opportunity and the obligation to eliminate measles from the United States. *J. Am. Med. Assoc.* 242:1157–62
66. Hope-Simpson, R. E. 1954. Studies on shingles. Is the virus ordinary chickenpox? *Lancet* 2:1299–302
67. Hoppensteadt, F. 1974. An age dependent epidemic model. *J. Franklin Inst.* 297:325–33
68. Kermack, W. O., McKendrick, A. G. 1927. A contribution to the mathematical theory of epidemics. *Proc. R. Soc. London Ser. A* 115:700–21
69. Knox, E. G. 1980. Strategy for rubella vaccination. *Int. J. Epidemiol.* 9:13–23
70. Knox, E. G. 1986. A transmission model for AIDS. *Eur. J. Epidemiol.* 2:165–77
71. Lajmanovich, A., Yorke, J. A. 1976. A deterministic model for gonorrhea in a non-homogeneous population. *Math. Biosci.* 28:221–36
72. Lange, H.-J., Ulm, H., Raettig, H., Huber, H. C. 1983. The effect of various interventions during a typhoid epidemic.

Results of simulation study. *Infection* 11:97–103

73. Langmuir, A. D. 1961. Epidemiology of Asian influenza. *Am. Rev. Respir. Dis.* 83(Pt. 2):2–9
74. Lennon, J. L., Black, F. L. 1986. Maternally derived measles immunity in the era of vaccine-protected mothers. *J. Pediatr.* 108:671–76
75. London, W. P., Yorke, J. A. 1973. Recurrent outbreaks of measles, chicken pox and mumps. I. Seasonal variation in contact rates. *Am. J. Epidemiol.* 98:453–68
76. Longini, I. M., Ackerman, E., Elvaback, L. R. 1978. An optimization model for influenza A epidemics. *Math. Biosci.* 38:141–57
77. Longini, I. M., Koopman, J. S. 1892. Household and community transmission parameters from final distributions of infections in households. *Biometrics* 38:115–26
78. Ludwig, D. 1975. Approximate solutions for stochastic epidemics. See Ref. 39, pp. 151–61.
79. Ludwig, D. 1974. *Stochastic Population Theories, Lect. Notes Biomath.* New York: Springer-Verlag
80. Macdonald, G. 1955. The measurement of malaria transmission. *Proc. R. Soc. Med.* 48:295–301
81. May, R. M. 1977. Togetherness among schistosomes: its effect on the dynamics of the infection. *Math. Biosci.* 35:301–43
82. May, R. M. 1983. Parasitic infection as regulators of animal population. *Am. Sci.* 71:36–45
83. May, R. M., Anderson, R. M. 1979. Population biology of infectious diseases: Part II. *Nature* 280:455–61
84. May, R. M., Anderson, R. M. 1984. Spatial heterogeneity and the design of immunization programs. *Math. Biosci.* 72:83–111
85. Molineaux, L. 1985. The pros and cons of modelling malaria transmission. *Trans. R. Soc. Trop. Med. Hyg.* 79: 743–47
86. Molineaux, L., Gramiccia, G. 1979. *The Garki Project: Research on the Epidemiology of Human Malaria in the Sudan Savannah of West Africa.* Geneva: WHO. 311 pp.
87. Nedelman, J. 1984. Inoculation and recovery rates in the malaria model of Dietz, Molineaux, and Thomas. *Math. Biosci.* 69:209–33
88. Orenstein, W. A. 1985. Measles outbreak in São Paulo, 1984. *Consult. Rep. Pan Am. Health Org. #3892s.* Unpublished ms
89. Post, W. M., DeAngelis, D. L., Travis, C. C. 1983. Endemic disease in environments with spatially heterogeneous host populations. *Math. Biosci.* 63:289–302
90. Reynolds, G. H., Chan, Y. K. 1975. A control model for gonorrhea. *Bull. Int. Stat. Inst.* 106-2:264–79
91. Schaffer, W. M., Kot, M. 1985. Nearly one dimensional dynamics in an epidemic. *J. Theor. Biol.* 112:403–27
92. Schenzle, D. 1982. Problems in drawing epidemiological inferences by fitting epidemic chain models to lumped data. *Biometrics* 38:643–47
93. Sencer, D. J., Dull, H. B., Langmuir, A. D. 1967. Epidemiologic basis for eradication of measles in 1967. *Public Health Rep.* 82:253–56
94. Severo, N. C. 1967. Two theorems on solutions of differential-difference equations and applications to epidemic theory. *J. Appl. Probab.* 4:271–80
95. Shonkwiler, R., Thompson, M. 1982. Common source epidemics I: a stochastic model. *Bull. Math. Biol.* 44:259–79
96. Singer, B. 1984. Mathematical models of infectious diseases: seeking new tools for planning and evaluating control programs. *Child Survival: Strategies for Research. Popul. Dev. Rev.* 10:347–65 (Suppl.)
97. Soper, H. E. 1929. The interpretation of periodicity in disease prevalence. *J. R. Stat. Soc. Ser. A* 92:34–61
98. Stille, W. T., Gersten, J. C. 1978. Tautology in epidemic models. *J. Infect. Dis.* 138:99–100
99. Warren, K. S., Anderson, R. M., Capasso, V., Cliff, A. D., Dietz, K., et al. 1982. Transmission patterns and dynamics of infectious diseases. Group report. See Ref. 3, pp. 67–85
100. Watson, R. K. 1972. On an epidemic in a stratified population. *J. Appl. Probab.* 9:659–66
101. Yorke, J. A., Hethcote, H. W., Nold, A. 1978. Dynamics and control of the transmission of gonorrhea. *Sex. Transm. Dis.* 5:51–57
102. Yorke, J. A., London, W. P. 1973. Recurrent outbreaks of measles, chickenpox, and mumps. II. Systematic differences in contact rates and stochastic effects. *Am. J. Epidemiol.* 98:469–78
103. Yorke, J. A., Nathanson, N., Pianigiani, G., Martin, J. 1979. Seasonality and the requirements for perpetuation and eradication of viruses in populations. *Am. J. Epidemiol.* 109:103–23

Ann. Rev. Microbiol. 1987. 41:703–26

GENETIC RESEARCH WITH PHOTOSYNTHETIC BACTERIA

Pablo A. Scolnik and Barry L. Marrs

E. I. du Pont de Nemours & Company, Central Research and Development Department, Experimental Station, Wilmington, Delaware 19898

CONTENTS

INTRODUCTION

Coverage

The nonsulfur purple photosynthetic bacteria have been among the most studied phototrophic organisms for many years (11). The development of good genetic systems (48, 66, 73, 75) and, more recently, the determination of the crystal structures of photosynthetic reaction centers for members of

0066-4227/87/1001-0703$02.00

this group of bacteria (1a, 18a) have intensified research into not only their photosynthetic apparatus, but also their nitrogen and carbon dioxide fixation. Genetic tools for working with these metabolically most versatile organisms have been proliferating, and this is having a predictably autocatalytic effect on the pace of research in the field. This review covers genetic research done with nonsulfur purple photosynthetic bacteria during 1984–1986. Current genetic research in this field embodies two sorts of activities: tool building and tool using. The structure of this review reflects that dichotomy. As is inevitable, the review is not comprehensive; rather, we have selected topics that reflect our own perceptions of what is interesting and important. We direct readers to other recent related reviews to obtain other perspectives (22, 29, 66a).

Nomenclature Changes

In 1984 Imhoff et al (32) recommended a rearrangement of the species and genera of the phototrophic purple nonsulfur bacteria to better reflect current data on the interrelatedness of this group. Among the changes suggested was the renaming of *Rhodopseudomonas capsulata* and *Rhodopseudomonas sphaeroides* as *Rhodobacter capsulatus* and *Rhodobacter sphaeroides*. Their rationale for this proposal was sound, but the change was slow to be accepted. Thus even late in 1986 some authors were still using the older nomenclature. In general the change has now been adopted by workers in this field, and one can only hope that the confusion that accompanies such a change will be minimal.

A group of researchers attending the Fifth International Symposium on Photosynthetic Prokaryotes in Grindelwald, Switzerland (September 1985) suggested some names for the genes most studied in those bacteria. Kaplan & Marrs reported those suggestions (39), and they are used throughout this review, with one exception: The operon containing the genes for the ubiquinol–cytochrome c_2 oxidoreductase was originally designated *fbc* by Gabellini et al (24a), and that name was retained in the Grindelwald proposal. Davidson & Daldal (17) subsequently showed that the organism with which Gabellini et al had worked was incorrectly identified as *R. sphaeroides,* when in fact it was a strain of *R. capsulatus*. Davidson & Daldal have proposed that henceforward the *fbc* acronym be dropped to avoid further confusion, and that the name *pet* be used for this operon. While we are not convinced that this change lowers the level of confusion, we use the *pet* nomenclature in this review.

GENETIC TOOLS

R. capsulatus was the first phototrophic bacterium for which a useful genetic exchange system was discovered (48), and it remains the most well-developed system for genetic studies of photosynthesis. Most genetic tasks can probably also be accomplished with *R. sphaeroides,* but since it has received less attention from geneticists, fewer tools have been developed for it; consequently any given manipulation takes a little more time and effort. Two of the most important genetic developments of this review period have been the development of interposon mutagenesis (61) and the genetic mapping of the *R. capsulatus* chromosome using R plasmid–mediated conjugation (73). However, the major accomplishment of this period, overshadowing all else, is the cloning and sequencing of so many of the key genes encoding the photosynthetic apparatus (see Table 1).

Capsduction

There has been no addition to our understanding of capsduction since it was last reviewed (49), but because this transductionlike genetic exchange process is useful in interposon mutagenesis (see below), it has remained a well-used tool. This utility comes from the fact that the gene transfer agent (GTA) particle contains fragments of the donor genome, apparently cut to uniform lengths and packaged at random, that cannot replicate when deposited in the recipient's cytoplasm. Successful selection for the GTA-borne marker therefore requires chromosomal integration by homologous recombination. If an interposon can fit within the GTA head and still carry enough flanking DNA to provide regions of homology for recombination, capsduction provides a very simple means of introducing the interposon into the recipient genome. Initial studies using this technique have always included genomic Southern blot analysis of the products of capsduction (61, 78), and the only integration events observed have been those directed by homologous recombination.

Now that genetics of *R. capsulatus* is better developed it should be possible to learn more about the molecular biology of this unique genetic exchange system.

Conjugation

Willison et al (73) used pTH10, a mutant of plasmid RP4 that is temperature-sensitive for maintenance in *Escherichia coli,* to mobilize the *R. capsulatus* chromosome and develop a circular genetic map. The original strategy was to try to isolate cointegrates of pTH10 in the chromosome, creating Hfr-like strains. However, pTH10 is not temperature sensitive in *R. capsulatus.* Nonetheless, pTH10 gave as much as a 10^4 higher level of chromosomal mobilization than the parental plasmid. The authors selected against auxotrophic donor strains containing pTH10 by using minimal medium, and selected exconjugants for *nif*$^+$ or biosynthetic markers. Selections for antibiotic-resistance markers did not work. The overall pattern of transfer frequencies and linkage relationships suggested that multiple origins of transfer were involved. The mapping function of Kondorosi et al (44b) appeared to

Table 1 Genes cloned from photosynthetic bacteria

Gene	Product function	Species	Restriction mapped?	Sequenced?	Ref
bchA	Bacteriochlorophyll biosynthesis	*R. capsulatus*	Yes	No	66
bchB	Bacteriochlorophyll biosynthesis	*R. capsulatus*	Yes	No	66
bchC	Bacteriochlorophyll biosynthesis	*R. capsulatus*	Yes	No	66
bchD	Bacteriochlorophyll biosynthesis	*R. capsulatus*	Yes	No	66
bchE	Bacteriochlorophyll biosynthesis	*R. capsulatus*	Yes	No	66
bchF	Bacteriochlorophyll biosynthesis	*R. capsulatus*	Yes	No	66
bchG	Bacteriochlorophyll biosynthesis	*R. capsulatus*	Yes	No	66
bchH	Bacteriochlorophyll biosynthesis	*R. capsulatus*	Yes	No	66
bchI	Bacteriochlorophyll biosynthesis	*R. capsulatus*	Yes	No	84
bchJ	Bacteriochlorophyll biosynthesis	*R. capsulatus*	Yes	No	84
bchK	Bacteriochlorophyll biosynthesis	*R. capsulatus*	Yes	No	84
bchL	Bacteriochlorophyll biosynthesis	*R. capsulatus*	Yes	No	84
cycA	Cytochrome c_2	*R. capsulatus*	Yes	Yes	13
		R. sphaeroides	Yes	Yes	21
crtA	Carotenoid biosynthesis	*R. capsulatus*	Yes	No	66
crtB	Carotenoid biosynthesis	*R. capsulatus*	Yes	No	66
crtC	Carotenoid biosynthesis	*R. capsulatus*	Yes	No	66
crtD	Carotenoid biosynthesis	*R. capsulatus*	Yes	No	66
crtE	Carotenoid biosynthesis	*R. capsulatus*	Yes	No	66
crtF	Carotenoid biosynthesis	*R. capsulatus*	Yes	No	66
crtI	Phytoene dehydrogenase	*R. capsulatus*	Yes	No	26
crtJ	Carotenoid biosynthesis	*R. capsulatus*	Yes	No	84
glnA	Glutamine synthetase	*R. capsulatus*	Yes	No	62
hupA	Hydrogenase	*R. capsulatus*	Yes	No	12
hupB	Hydrogenase	*R. capsulatus*	Yes	No	12
nifA?	Regulatory	*R. capsulatus*	No	No	1
nifB?	Mo metabolism	*R. capsulatus*	No	No	1
nifD	Nitrogenase	*R. capsulatus*	Yes	Yes	29, 60
nifE?	Mo metabolism	*R. capsulatus*	No	No	29
nifH	Nitrogenase	*R. capsulatus*	Yes	Yes	29, 60
nifK	Nitrogenase	*R. capsulatus*	Yes	Unpublished	29
nifL?	Regulatory	*R. capsulatus*	No	No	29
nifR1	Regulatory	*R. capsulatus*	Yes	No	29
nifR2	Regulatory	*R. capsulatus*	Yes	No	29
nifR3	Regulatory	*R. capsulatus*	Yes	No	29
nifR4	Regulatory	*R. capsulatus*	Yes	No	29
nifS?	Maturation	*R. capsulatus*	No	No	29
petA	Rieske Fe-S protein	*R. capsulatus*	Yes	Part	14
		R. sphaeroides	Yes	Part	16
petB	Cytochrome *b*	*R. capsulatus*	Yes	Part	14
		R. sphaeroides	Yes	Unpublished	16
petC	Cytochrome c_1	*R. capsulatus*	Yes	Part	14
		R. sphaeroides	Yes	Unpublished	16
pucA	LH-II α	*R. capsulatus*	Yes	Yes	77
pucB	LH-II β	*R. capsulatus*	Yes	Yes	77
pufA	LH-I α	*R. capsulatus*	Yes	Yes	76
		R. rubrum	Yes	Yes	5
pufB	LH-I β	*R. capsulatus*	Yes	Yes	76
		R. rubrum	Yes	Yes	5

Table 1 (*continued*)

Gene	Product function	Species	Restriction mapped?	Sequenced?	Ref
pufL	RC-L	*R. capsulatus*	Yes	Yes	76
		R. sphaeroides	Yes	Yes	71
		R. rubrum	Yes	No	5
pufM	RC-M	*R. capsulatus*	Yes	Yes	76
		R. sphaeroides	Yes	Yes	72
		R. rubrum	No	No	5
puhA	RC-H	*R. capsulatus*	Yes	Yes	76
		R. sphaeroides	Yes	Part	20
rbcL	Rubisco, large subunit	*R. sphaeroides*	Yes	No	25
		R. rubrum	Yes	Yes	51, 64
rbcR	Rubisco, rubrum-like	*R. sphaeroides*	Yes	No	50, 58
rbcS	Rubisco, small subunit	*R. sphaeroides*	Yes	No	25
rrnA	Ribosomal RNA	*R. capsulatus*	Yes	No	79

apply to the data generated in these crosses. More than 34 mutations have been mapped in this way, and work in Willison et al's lab is progressing toward the original goal of generating Hfr-like strains.

Biel (6) has used the conjugative R' plasmid pRPS404 to introduce a mutation *(crtD223)* that blocks carotenoid synthesis at the neurosporene level into a variety of *R. capsulatus* strains. For example, a *crtD223* derivative of PAS100 *(hsd-1, str-2)* was obtained by mating pRPS404 into PAS100 from a Str^s donor, selecting for kanamycin resistance (plasmid marker) and streptomycin resistance (recipient marker), and isolating a yellow colony upon restreaking (6). The yellow derivatives arise via recombination between homologous regions on the plasmid and the chromosome. Km^s yellow colonies have lost the R-factor.

Transformation

No useful transformation system has been reported for *R. capsulatus;* in this area of genetics *R. sphaeroides* is a better tool. Fornari & Kaplan (22a) developed a method for introducing plasmid DNA into *R. sphaeroides* so that small, nonconjugative plasmids suitable for cloning could be directly transferred into that species. The system gives transformation frequencies as high as 10^{-5}. Jasper et al (33) observed transformation in *R. capsulatus,* but the frequency was approximately 10^{-9}, so other methods on genetic exchange are currently used to introduce exogenous DNA in this species. This is another area where additional genetic work might be fruitful.

Transposon Mutagenesis

Zsebo & Hearst (84) isolated and analyzed 45 independent insertions of transposon Tn*5.7* into the *R. capsulatus* DNA carried on the R' plasmid pRPS404. They made Tn*5.7* by replacing the kanamycin resistance gene of Tn*5* with the streptomycin marker from Tn*7* so that the transposon would have a selectable marker that could be distinguished from the kanamycin resistance gene already carried by pRPS404. It is ironic that Tn*5* itself carries a streptomycin resistance gene, so the construction was probably unnecessary (57). Transposition occurred in *E. coli,* and pRPS404 derivatives with inserts were selected for by the mobilization of the plasmids into streptomycin-sensitive *E. coli* (85).

Kaufmann et al (40) explored three plasmids containing Tn*5* for their usefulness in mutagenizing *R. capsulatus*. Plasmid pJB4JI was stably established in this species, and therefore is not useful for transposon mutagenesis. Plasmid pACYC184::Tn*5* was mated into *R. capsulatus,* but the frequency of transposon-induced mutations was low (3×10^{-8} per donor). The most useful plasmid found was pSUP201::Tn*5*, which was not maintained in *R. capsulatus* and which gave rise to up to 3×10^{-5} kanamycin-resistant exconjugants per donor. The authors described a system for mobilizing pSUP201::Tn*5* into *R. capsulatus* from *E. coli*.

Hüdig et al (31) used the Tn*5*-containing plasmid pSUP201::Tn*5* (63) to mutagenize the *R. capsulatus* chromosome. The plasmid was transferred to *R. capsulatus* from *E. coli* S17-1. pSUP201::Tn*5* does not replicate in *R. capsulatus,* where the Tn*5* kanamycin resistance gene is well expressed; thus selection for kanamycin-resistant exconjugants gives a good yield of mutants in which Tn*5* has transposed to the host genome. Hüdig et al found that the frequency of respiratory mutants was about 0.1% that of the kanamycin-resistant exconjugants.

Interposon Mutagenesis

It is now easy to introduce marker-bearing insertions into any cloned segment of the *R. capsulatus* genome. The insertion sites may be points or deletions, as described by Scolnik & Haselkorn (61). The technique is currently widely used, and is summarized in Figure 1. An interposon for this application is a stretch of DNA, smaller than ~4.0 kb (the smaller the better), coding for a selectable marker, flanked by useful restriction sites, and ideally containing transcription-translation terminators in each orientation. Youvan et al (78) have listed commonly used interposons. Perhaps the most useful interposon is the 2.0-kb–long Ω fragment (53), which was constructed to have the appropriate terminators in both directions and which codes for spectinomycin and streptomycin resistance.

Interposon Mapping

Although interposon mutagenesis for knock-out mutations is a valuable technique, and site-specific mutagenesis provides an enormously powerful tool

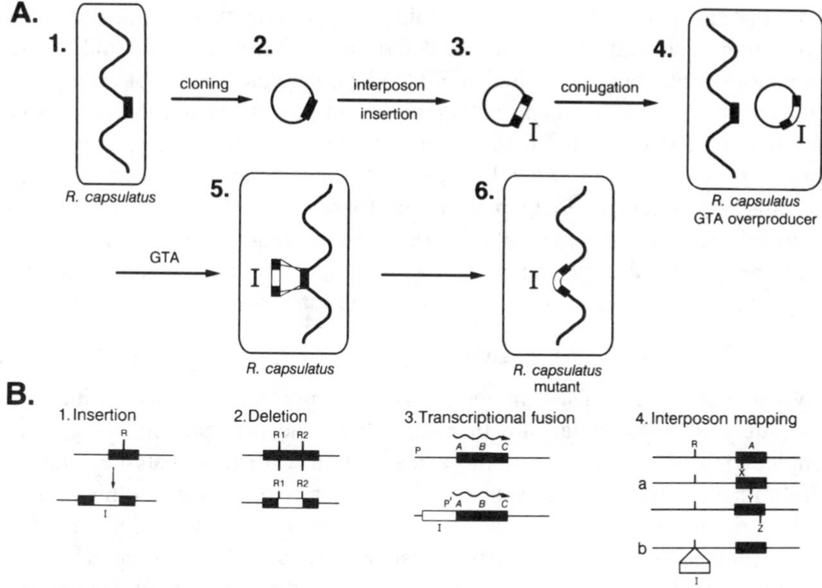

Figure 1 The principle (*A*) and applications (*B*) of interposon mutagenesis. *A:* (*1*) A segment of the *R. capsulatus* genome is cloned by standard recombinant DNA techniques into a plasmid vector that can replicate in both *R. capsulatus* and *E. coli*. (*2*) Plasmid DNA is isolated. (*3*) Using restriction enzymes and DNA ligase an interposon is inserted into the cloned sequence in such a manner that it is flanked by *R. capsulatus* sequences. (*4*) The construct is mobilized by conjugation into an *R. capsulatus* strain that overproduces gene transfer agent (GTA). (*5*) GTA is used to transduce the antibiotic resistance marker contained in the interposon into a recipient *R. capsulatus* strain. (*6*) Homologous recombination results in the insertion of the interposon into the *R. capsulatus* genome. *B:* (*1*) If a single restriction site in the target sequence is used, the result is an interposon insertion. (*2*) If at least two sites are used the result is an interposon deletion. (*3*) Example of an interposon-induced transcriptional fusion. The target sequence is the *R. capsulatus* *ABC* operon containing a promotor (*P*). An interposon deletion is constructed in such a way that both *P* and at least part of the *A* gene are deleted. The interposon contains an outward promotor (*P'*) that now controls the transcription of the remaining *ABC* operon. The *A* gene has been either partially or totally deleted and can now be reintroduced into the cell by conjugation. This approach is useful for the study of the regulation and/or the structure-function relationship of genes that are part of an operon. (*4*) Interposon mapping is used to determine the physical distance between a restriction site and one or more point mutations. In *4a* there are three point mutations (*X, Y,* and *Z*) in the *A* gene. To determine the distance between the restriction site, *R*, and the individual mutations, a construct (*4b*), which carries an interposon at *R* and a wild-type copy of the *A* gene, is used as a GTA donor. Each of the three mutants is used as a recipient, and selection for the antibiotic resistance coded for by the interposon is applied. The transductants are scored for the A phenotype. The frequency at which X, Y, and Z are rescued by the incoming fragment of DNA reflects their distance from *R*.

for testing hypotheses, randomly generated mutants obtained by selection and screening techniques are still the most efficient source of functionally altered proteins and genes. Interposon mapping is a new method for analyzing a collection of mutants genetically to determine the location of the underlying mutations. For example, Daldal et al (15) incorporated a selectable marker in the *pet* operon by interposon mutagenesis and then measured the cotransfer frequency of that marker with a series of inhibitor-resistance markers that might be expected to show linkage to the *pet* genes. A theory relating cotransfer frequencies to map distances (base pairs) has not yet been developed, because of complications arising from the length of the interposon and the predicted dependence of the map function on the length of the required homologous sequences flanking the interposon.

Site-Saturation Mutagenesis

Bylina et al (9) have used the term "site-saturation mutagenesis" to indicate the study of a particular site in a protein by the introduction via genetic engineering of all possible amino acids, one at a time, at that site. Only a preliminary report of these studies (9) had been published when this review was written, but presentations at the Seventh International Congress for Photosynthesis (Providence, Rhode Island, August, 1986) suggested that additional publications will be forthcoming. Bylina et al (9) changed the histidine residue 32 in the α-subunit of the LH-I polypeptide (thought to bind bacteriochlorophyll) to threonine, asparagine, glutamine, proline, aspartic acid, and arginine. Each of these changes resulted in the loss of LH-I antennae from the photosynthetic membranes as assayed by absorption spectra and SDS polyacrylamide gel electrophoresis. They used oligonucleotide-mediated site-specific mutagenesis of plasmid-borne genes to effect the changes, and used deletion strains (78) to examine the altered gene products in the absence of the corresponding wild-type genes. Changes of the alanine residue at position 28 of the LH-I α-subunit were more interesting. While mutations to histidine or aspartic acid each knocked out LH-I accumulation, glycine gave an antenna that absorbed near the normal absorption maximum. A valine at the same position gave a partially functional LH-I complex. Studies of this type hold great promise for analysis of roles of individual residues in complex protein structures.

Expression Vectors

Johnson et al (34) have created a set of vectors derived from the broad–host range plasmid pRK404 that use the promoter region of the *puf* operon from *R. capsulatus* to drive transcription. Either transcriptional or translational fusions can be created, and expression is regulated by oxygen. The vectors can be mobilized by conjugation among the species in the RK2 host range. The transcriptional fusion vectors are 11.2 kb long and contain appropriately positioned unique *Bam*HI and *Pst*I cloning sites.

Plasmid Curing

Magnin et al (47) have discovered that *R. capsulatus* can be cured of R-plasmids of the P1 incompatibility group by repeated subculturing in yeast extract–peptone (YP) medium. This is an alternative to introducing a second plasmid to displace the first (1b), a procedure that switches, but does not eliminate, the plasmids. The same procedure has been used to cure the endogenous plasmid from strain B10 (74). YP medium was routinely used in many laboratories for decades before it was modified to yeast extract–peptone salts (YPS) by the addition of $CaCl_2$ and $MgSO_4$ (70). Thus older laboratory strains may be significantly diverged from their wild-type ancestors. Prolonged subculturing in YP medium also resulted in the accumulation of auxotrophs, although this may also be possible in YPS (47).

Catalog of Cloned Genes

Genes beget genes. New genes can often be cloned by probing with homologs from the same or different species. Useful mutants can be created by insertional mutagenesis if the gene of interest has been cloned. The collection of genes that have already been cloned is therefore an extremely useful genetic tool. Table 1 is a catalog of cloned genes from photosynthetic bacteria.

SYSTEMS STUDIED

Photosynthesis

The nonsulfur purple photosynthetic bacteria are uniquely positioned for studying the light reactions of photosynthesis at the molecular level. Only here does one find the combination of a well-described photosynthetic apparatus and well-developed genetic tools with which to manipulate it.

REACTION CENTER AND ANTENNA PROTEIN GENES The cloning and sequencing of the reaction center and antenna genes (5, 20, 72, 76, 77), coinciding as it did with the development of X-ray crystal structures for at least two species' photochemical reaction centers (1a, 18a), marks the beginning of a significant era in the study of photosynthesis and membrane-bound electron transport proteins.

The photosynthetic apparatus of these bacteria consists of four integral membrane pigment-protein complexes plus an ATPase. Three of those complexes are specific to photosynthesis, i.e. the reaction center (RC), the light-harvesting I antenna (LH-I), and the light-harvesting II antenna (LH-II). The remaining one, the ubiquinol:cytochrome c_2 oxidoreductase (bc_1), is common to both respiration and photosynthesis. The genes for the three known reaction center polypeptides (L, M, and H) are called *pufL, pufM,* and *puhA*. The LH-I polypeptide (α, β) genes are *pufA* and *pufB,* and the LH-II polypeptide (α,β) genes are *pucA* and *pucB*. The known genes for the bc_1 complex are *petA, petB,* and *petC,* which encode the Reiske iron sulfur protein, cytochrome *b,* and cytochrome c_1, respectively. The gene order in the *puf, puc,* and *pet* operons is conserved between *R. capsulatus* and *R. sphaeroides* (14, 16, 71, 76, 78).

Maps Williams et al (72) isolated the *pufM* gene from *R. sphaeroides* using oligonucleotide probes based upon the N-terminal amino acid sequence of the protein; the *pufL* gene was found adjacent to and overlapping *pufM*. The same gene arrangement was discovered in *R. capsulatus* (75a, 76), where the *puf* and *puh* operons were located by the sequencing of restriction fragments that complemented reaction center mutants. These two operons, whose products form a stoichiometric complex, are not adjacent on the chromosome as might be expected, but instead flank a region of about 50 kb of DNA that codes for the pigment-biosynthesis genes in this species. The *puc* gene is not part of this chromosomal segment, but all the other genes needed to differentiate a respiratory membrane into a photosynthetic structure seem to be clustered.

Regulation of transcription Zhu & Hearst (81) examined the effects on message levels of various growth conditions (high and low light intensities and oxygen tensions) and one shift in light intensity (high to dark) in *R. capsulatus*. It is of interest to compare their results with those of Clark et al (10), who also examined levels of mRNA from many of the same genes. Clark et al found a striking (40×) increase in RNA that hybridized to the *pufB,-A,-L,* and *-M* genes 40 min after a shift from 20 to 2% dissolved oxygen, whereas Zhu & Hearst found only a 3–4× increase in the same RNAs when they compared cultures grown under high and low oxygen conditions. Zhu & Hearst did not discuss this discrepancy, but it might be due to either differences in the ways the cultures were grown or how the measurements were made, or both. Each group normalized their results to total RNA, which in each case was presumably mostly rRNA. The total amount of rRNA per cell increases dramatically during induction of the photosynthetic apparatus, increasing as much as seven fold over 140 min following a shift from high to low oxygen pressures (43). This effect probably caused an underestimate of the increase in mRNA levels in the experiments of Clark et al. In the steady state experiments of Zhu & Hearst, rRNA contents per cell should have been quite different at the very different growth rates associated with the different conditions tested. These factors make direct comparisons of the two reports impossible. Furthermore, Zhu & Hearst used specific probes for each of the three *puf* operon genes they examined, whereas Clark et al used one large probe for the entire region, including DNA outside the *puf* operon.

Klug et al (44) examined the kinetics of induction of RNAs from both the *puc* and *puf* regions of the *R. capsulatus* chromosome after induction by (*a*) a drop in light intensity and (*b*) a decrease in oxygen tension. In each case the extent of induction was comparable to that reported by Clark et al for the *puf* operon. They also reported that the peak of induction for the *puf* operon occurred about 30 min before that of the *puc* genes, and each RNA peak occurred about 5 min before the respective polypeptides achieved maximal rates of incorporation. These results are consistent with the notion that regulation of the pigment-binding proteins is at the transcriptional level.

Posttranscriptional regulation When Youvan et al (76) sequenced the restriction fragment of *R. capsulatus* DNA carrying the genes for LH-I and RC proteins, they suggested that the two sets of genes formed an operon, and they noted a stable hairpin that could form in the region between them. They speculated that this structure might function as a transcription terminator and thus be involved in the differential expression of these two segments of the putative operon, i.e. the 10–30-fold excess of LH-I over RC proteins observed in functional photosynthetic membranes. Belasco et al (4) confirmed that the LH-I and RC genes were cotranscribed, since they were found on a single long transcript (2.7 kb) of mRNA in *R. capsulatus*. They also found two abundant shorter fragments (0.49 and 0.50 kb), which seemed to share a common 5' terminus with the long transcript. The relative abundance of these species was 2.7:6.4:1 for the lengths 0.49:0.50:2.7 kb, which means that 10 copies of mRNA encode LH-I for each fragment encoding RC proteins. This neatly fits with the relative abundance of the corresponding protein species. These authors concluded that the difference in RNA fragment abundance accounts for the observed protein ratios; however, it must be kept in mind that physical and functional half-lives of mRNA may vary greatly (41), i.e. the fragments may or may not be translated equally well or even at all. In the present case there is a good match between the physical half-lives of the fragments (long, 5 min; short, 22 min) and the functional half-lives of the mRNAs [RC, 8 min; LH-I, 22.5 min (18)], so the coincidence is more compelling, but the relationship should remain a hypothesis until tested experimentally. The physical half-lives were not affected by the oxygen status of the cells, so this work does not suggest that oxygen regulates transcription through control of message stability (but see below).

Belasco et al (4) showed that the hairpin structure between the LH-I and RC genes does not function as a transcription terminator in *R. capsulatus* by demonstrating its lack of ability to affect transcription of a gene for β-galactosidase, which is fused to the LH-I β gene (*pufB*) and driven by a promoter that they presumed to be the normal *puf* operon promoter. They showed a precursor-product relationship between the 2.7- and 0.5-kb molecules in pulse-chase experiments, and suggested that the hairpin is more likely to stabilize the portion of the message that lies on its 5' side. Thus message processing and degradation are thought to control the relative abundance of expression of the two classes of genes identified on the *puf* operon.

Zhu & Hearst (81) revisited the question of *puf* operon message levels in *R. capsulatus* and extended the studies to include the *puc* operon, which codes for the LH-II proteins; the *puhA* gene, which encodes the H-subunit protein of the RC; *bch* genes, which encode bacteriochlorophyll biosynthetic enzymes; and *crt* genes, which code for carotenoid biosynthesis. These authors confirmed the earlier reports on *puf* operon transcript sizes (4) and showed that only one transcript (0.5 kb) hybridizes to probes for the *puc* operon, while the *puhA* probe hybridized to messages of two different sizes (1.2 and 1.4 kb).

Zhu & Kaplan (82) and Zhu et al (83) found basically the same pattern of transcripts in *R. sphaeroides* as reported for *R. capsulatus;* only two differences were reported. The two *puf* operon transcripts found in *R. sphaeroides* did not share a common 5' terminus: The 2.6-kb transcript's 5' end was found to be 75 nucleotides upstream from *pufB,* and that of the 0.5-kb segment was 104 bases upstream. The authors suggested that the difference in abundance of the short and long transcripts might be due in part to their initiation at independent promoters, since the shorter transcript could not originate from the longer, because the longer lacks a region found in the shorter. The other difference reported concerns the effect of oxygen on message half-lives. Belasco et al (4) reported that the half-lives of the long and short transcripts from *R. capsulatus* were the same in cells grown under high aeration as in cells grown under low aeration. Zhu et al (83) reported that in phototrophic cultures of *R. sphaeroides* the long and short transcripts displayed virtually the same half-lives as their *R. capsulatus* counterparts, but a shift to aerobic growth conditions significantly shortened each half-life. These authors proposed that the "accelerated degradation of the mRNA specifying photosynthetic membrane proteins would be an efficient regulatory mechanism for photosynthetic cells when they are exposed to oxygen." Zhu et al (80), working with *R. capsulatus,* reported that the half-life of LH-I mRNA was 19 min in an anaerobic photosynthetic culture and 23 min after a shift from photosynthetic to aerobic growth, but the half-life of LH-II mRNA decreased with aeration, from 24 to 17 min. It is impossible to say whether these different results stem from species differences or experimental differences.

Donohue et al (20) performed Northern blot analyses on the transcripts of the *puhA* gene and found evidence for two transcripts, one extending farther in the 5' direction than the other. The relative amounts of these two transcripts varies slightly (2:1 to 3:1) with growth conditions. The *puhA* gene is thought to be transcribed as a monocistronic operon, based on the transcript sizes measured. In a companion paper, Donohue et al (21) showed similar findings for transcripts of the *cycA* gene: Two transcripts were revealed by Northern blotting; their relative amounts varied with growth conditions (from 7:1 in chemoheterotrophic cells to 1:1 in cells grown anaerobically in the dark); and the authors presumed that the operon is monocistronic based on the size of the fragments. Since the relationship between the sizes of these fragments and the "primary transcripts" in the absence of degradation is unknown, the conclusion that these are monocistronic operons is not justified by this evidence alone.

These studies, taken together, show clearly that message processing occurs in *R. capsulatus* and *R. sphaeroides* and suggest that differential processing of segments may account for differential expression of cotranscribed genes. The area neglected in these studies was the nature of the 5' ends observed. Zhu et al (83) mentioned that the 5' ends might be the products of processing of a primary transcript in *R. sphaeroides,* but then added that "if this is the route by which these two transcripts are derived and the observed stoichiometry achieved, the mechanism(s) of post-transcriptional processing of the primary *puf* transcript of *R. sphaeroides* is far more complex than that envisioned for *R. capsulatus*." We emphasize that the site of transcription initiation of the *puf* operon has not been established in either species. Until transcripts with 5' triphosphate termini are observed or appropriate functional studies are done (e.g. *lac* fusions), it is premature to conclude that the 5' ends reported to date are indeed initiation sites. Along these lines, Beatty et al (3) have published a preliminary report indicating that they have evidence for short-lived precursors of the more stable transcripts reported previously.

Klug et al (44a) have carefully examined the effects of inhibition of bacteriochlorophyll on the synthesis of pigment-binding proteins. They used both levulinic acid, an inhibitor that blocks the formation of porphobilinogen from δ-amino levulinic acid, and a set of mutants with the biosynthetic pathway blocked at various points after addition of magnesium to the tetrapyrrole ring. Although inhibitors of bacteriochlorophyll synthesis also specifically inhibited the synthesis of pigment-binding proteins, they had no effect on the transcription of mRNA for those proteins. Klug et al therefore concluded that a form of translational control by chlorophyll precursors coordinates bacteriochlorophyll and pigment-binding protein syntheses in addition to the previously observed effects of bacteriochlorophyll on the assembly and stability of the photosynthetic membrane complexes. They did not explain the observation that mutants blocked in bacteriochlorophyll synthesis cannot derepress mRNA for the pigment-binding proteins.

PIGMENT BIOSYNTHESIS GENES Pigment biosynthesis genes were the first mapped in these organisms, because pigment biosynthesis is so easy to score in genetic crosses (75). Our current knowledge of the genetics of bacteriochlorophyll and carotenoid biosynthesis comes mostly from analyses of point mutations (66) and transposon-induced mutations (85).

Maps The block of genes coding for pigment biosynthesis in *R. capsulatus* has been shown to map between the genes for the pigment-binding proteins (66). The whole cluster for photosynthesis maps between genes for histidine synthesis and nitrogen fixation on the circular map of Willison et al (73). Zsebo & Hearst's analysis of polar effects of transposon mutations (84) suggest functional arrangements in operons, but very little is known with confidence about the existence of regulons or operons for these genes.

The *crtI* gene was first described by Zsebo & Hearst (84) as a locus between *crtA* and *crtB*, which when interrupted by transposon Tn*5.7* mutagenesis gave the blue-green phenotype. The blue-green phenotype is characteristic of mutants that accumulate no colored carotenoids but are still able to synthesize bacteriochlorophyll and grow photosynthetically. Giuliano et al (26) have subsequently shown that the *crtI* gene most likely codes for the structural gene for phytoene dehydrogenase, an enzyme early in the carotenoid biosynthetic pathway. CrtI mutants accumulate phytoene, and Guiliano et al localized the lesions in two mutants of this type by genetic complementation and deletion mapping. They also developed a simple in vitro complementation assay to show that membranes from CrtI mutants can complement those from CrtB mutants, producing a mixture of colored carotenoids in cell-free extracts.

Regulation of transcription Zhu & Hearst (81) and Zhu et al (80) examined the effects of light intensity and oxygen on transcripts from *crt* genes in *R. capsulatus* by hybridizations of dot blots or electrophoretically separated RNAs using labeled restriction fragments as probes. Both sets of data lack sufficient resolution because the probes carry DNA from more than one gene, including undefined genetic regions. Although that caveat should be kept in mind, Zhu & Hearst (81) showed that the levels of transcripts of *crtE, -C, -D* and *-F* in cells grown under low light, high oxygen, or low oxygen were slightly lower than in cells grown under strong light. Levels of a transcript(s) from the *Bam*HI-H fragment increased slightly in response to high oxygen; the authors argued that this transcript must have been the *crtA* mRNA, since that gene codes for an oxygenase. This hypothesis is supported by an observation that a transposon insertion in *crtA* reversed the twofold increase in levels of RNA hybridizing to *Bam*HI-H that had been seen upon shifting a culture from photosynthetic to aerobic conditions (80). Both *crtE* and *crtF* RNA levels, measured with another restriction fragment, dropped after the same shift. Do these small shifts in RNA content represent transcriptional or posttranscriptional controls, or are they unregulated changes that are not related to the genetic content of the molecules, simply reflecting differential degradation of different mRNA fragments?

Biel & Marrs (7) presented evidence that oxygen does not directly regulate carotenoid synthesis in *R. capsulatus*. They showed that bacteriochlorophyll synthesis was required for normal regulation of carotenoid synthesis in response to a decrease in oxygen tension by examining various mutants defective in bacteriochlorophyll synthesis. These experiments may not rule out the possibility that carotenoids turn over faster in the absence of carotenoid-binding proteins (which are themselves unstable in the absence of bacteriochlorophyll); however, there is no evidence for turnover of carotenoids.

CYTOCHROME GENES The ubiquinol–cytochrome *c* oxidoreductase (bc_1 complex) appears to contain the only cytochromes essential for photosynthesis. An additional cytochrome, cytochrome c_2, is part of the wild-type photosynthetic electron transport system, but Daldal et al (13) clearly demonstrated that it is dispensable in *R. capsulatus*. They deleted the central portion of the *cycA* gene, verified by immunological and spectroscopic as well as genetic criteria that no cytochrome c_2 was produced in the resulting strains, and then observed that those strains could grow photosynthetically with a 120-min doubling time under high light intensity. (The wild-type control doubled in 108 min under the same conditions.) Only in dim light was the mutant phenotype clearly impaired, requiring 600 min in contrast to the wild type's 300-min doubling time. Donohue et al (21) claimed that this cytochrome is indispensable for photosynthetic growth of *R. sphaeroides*, but we must await publication of their data to evaluate the nature of this discrepancy.

Using the cytochrome c_2 deletion mutant described above, Prince et al (54, 55) have gathered spectral, thermodynamic, and kinetic evidence that the cytochrome c_1 of the bc_1 complex is able to donate electrons to the reaction center directly and rapidly ($t_{1/2} < 100$ μsec). This pioneering work points the way to exciting new studies in bioenergetics.

As discussed in the introduction, study of the genetics of the bc_1 complex has gotten off to a rocky start owing to an unfortunate misidentification of the species of *Rhodobacter* under investigation (17, 24a), but even this mix-up has its brighter side. It seems unlikely that investigators would knowingly have repeated the arduous task of sequencing the bc_1 operon on two different strains of the same species, but we now have such data to provide evidence on the relatedness of the strains and, perhaps more significantly, on which areas of the proteins are most or least conserved.

Daldal et al (15) have carried out an interesting application of interposon mutagenesis and interposon mapping on the *pet* operon. They introduced an interposon carrying a kanamycin resistance marker at the carboxy-terminal end of the *petC* gene, which lies at the end of the *pet* operon. This mutant is photosynthetically competant. Daldal et al transferred the insertion construction to a strain of *R. capsulatus* that overproduces GTA, and then performed GTA-mediated mapping of a series of spontaneous inhibitor-resistant mutants. The inhibitors, stigmatellin and myxothiazol, were known to inhibit photosynthesis by interfering with bc_1-mediated electron transport. The mutant collection was split into subcategories as a result of the mapping, since some mutants showed no linkage to the *pet* operon, while most of the mutations clearly mapped there and thus were probably structural changes in the bc_1 protein subunits.

Nitrogen Fixation

In terms of physiological diversity, photosynthetic bacteria are among the most successful microorganisms on Earth. They are capable of growing and fixing nitrogen under photoheterotrophic, photoautotrophic, microaerobic organotrophic, microaerobic chemoheterotrophic, and fermentative conditions. Many prokaryotes can enzymatically reduce atmospheric nitrogen to ammonia; this ability to fix nitrogen has been extensively studied in the coliform *Klebsiella pneumoniae*. However, the photosynthetic bacteria have unique ecological and physiological characteristics. How the process of nitrogen fixation integrates into the overall metabolic picture of the group is by and large unknown. Most of the information currently available about the molecular genetics of nitrogen fixation and assimilation in photosynthetic bacteria was obtained with *R. capsulatus;* therefore we center our report on this organism. A more extensive review which includes other photosynthetic prokaryotes has recently been published (27). Hallenbeck et al (28) purified the *R. capsulatus* nitrogenase. This task was considerably facilitated by the fact that under optimal conditions nitrogenase can make up as much as 30% of the *R. capsulatus* protein. As in other diazotrophs, the enzyme consists of two proteins: an iron- and molybdenum-containing protein (MoFe protein, dinitrogenase, component I) and an iron-containing protein (Fe protein, dinitrogenase reductase, component II). The MoFe protein is an $\alpha_2\beta_2$ tetramer. The estimated molecular mass of the α- and β-subunits is 59.5 and 55 kd, respectively. The estimated mass of the Fe protein is 33.5 kd (27). These values are in good agreement with those predicted from the DNA sequences of the genes coding for the Fe protein and the α-subunit of the MoFe protein (60).

NITROGENASE GENES Ruvkun & Ausubel (59) showed that the *K. pneumoniae* genes for nitrogenase are homologous to the equivalent genes of many nitrogen-fixing organisms. Scolnik & Haselkorn (61) used this information to clone the genes for nitrogenase in *R. capsulatus*.

Maps Scolnik & Haselkorn (61) used *K. pneumoniae* probes in Southern hybridizations of *R. capsulatus* DNA digested with *Hin*dIII and detected several bands of homology. The two most prominent bands were 11.8 and 4.8 kb long, and at least eight more bands were observed (61). Both the 11.8-kb and 4.8-kb *Hin*dIII fragments were cloned from a cosmid library. The 11.8-kb fragment could complement *nif* point mutants. Using a combination of Tn*5* mutagenesis and Southern hybridization, Avtges et al (2) determined that the *nifH*, *-D*, and *-K* genes are organized in an operon, transcribed in the direction of the *nifK* gene. In the first use of the interposon mutagenesis technique, Scolnik & Haselkorn (61) constructed insertions and deletions within this region of the genome by inserting a Km^r gene from Tn*5*. The phenotype of the resulting mutants was Nif^-, which indicates that, at least under the conditions used, the only active *nif* structural genes are contained within the 11.8-kb *Hin*dIII fragment. However, the investigators were able to select Nif^+ pseudorevertants from the interposon mutants. These pseudorevertants had approximately 10% of the nitrogenase activity of the wild-type cells, as judged by the acetylene reduction assay (61). Southern blot analysis indicated that no rearrangements had taken place within the *nifHDK* region; thus the appearance of nitrogenase activity must have been due to the activation of cryptic *nif* genes somewhere else in the genome (61). The authors suggested that the cryptic *nif* genes might be contained within some of the *nif*-homologous DNA fragments detected in the Southern blots. However, a recent characterization of the 4.8 kb *Hin*dIII fragment showed that no cryptic *nif* gene is present in this region (60). Instead, a copy of a 16S rRNA gene in the 4.8 kb *Hin*dIII fragment hybridized unspecifically to *nif* probes (60). However, not all the hybridization bands detected with *nif* probes can be explained in terms of spurious homology to rRNA genes; it is still possible that the cryptic genes activated in the pseudorevertants are homologous to the characterized *nifHDK* set. Interestingly, an alternate nitrogenase system was described in *Azotobacter vinelandii* (8). Also, multiple regions of homology to *nif* genes were observed in several species (19, 23, 35, 37).

Hearst et al (30) identified an open reading frame within the *R. capsulatus* photosynthesis region that could code for a protein with homology to nitrogenase reductase. The fact that this homology is clustered around the cysteine residues suggests that this protein, if it is indeed made in the cell, also binds Fe. The role of this polypeptide in nitrogen metabolism, if any, is not presently known. The location of the open reading frame among other genes that affect photosynthesis and the nature of the protein suggest that this polypeptide may be part of the electron transport system.

In summary, when cells are grown photoheterotrophically in the absence of fixed nitrogen, a classical nitrogenase is produced by the *nifHDK* operon contained in the 11.8-kb *Hin*dIII fragment. When this operon is inactivated by interposon mutagenesis the resulting phenotype is Nif^-, but Nif^+ pseudorevertants can be isolated. The genes responsible for the nitrogenase activity in the pseudorevertants have not been identified and may or may not be related to the multiple areas of *nif* homology in the *R. capsulatus* genome. These genes are silent under photoheterotrophic conditions. However, they may be active under at least one of the remaining growth conditions listed above. If this is the case, the pseudorevertants may arise from mutations that would allow these genes to be activated under different physiological conditions. Alternatively, the genes for the alternate nitrogenase system may be silent under all conditions and may be activated by mutation in the pseudorevertants. Clearly, solution of the alternate nitrogenase puzzle requires the identification of the enzyme that reduces dinitrogen in the pseudorevertants. With recent advances in micro methods for protein isolation and sequencing, it should be possible to identify the genes that code for the alternate nitrogenase.

Transcriptional and posttranscriptional regulation For each molecule of N_2 fixed, 18 mol of ATP are consumed. As a consequence of this high energy requirement, an array of regulatory systems controls nitrogenase activity according to environmental conditions. Fixed nitrogen, oxygen, light, and perhaps Mo regulate nitrogenase activity. Regulation can take place at the mRNA (46, 68) or protein level. NH_4 repression of nitrogenase is indirect, as demonstrated by the fact that a glutamine-deficient mutant of *R. capsulatus* is constitutive for nitrogenase expression (68). However, ammonia control can be restored by complementing this mutant with a copy of the wild-type gene for glutamine synthetase, which indicates that the effector molecule is either glutamine or a metabolite thereof (46). Covalent modification of nitrogenase as a way of regulating its activity has been extensively documented in photosynthetic bacteria. The modifying group in *Rhodospirillum rubrum* is an adenosine diphosphoribose that becomes bound to the Fe protein at the Arg residue in the sequence Gly-Arg-Gly-Val-Ile-Thr (52). This modification inactivates the enzyme. Interestingly, the same amino acid sequence is found in the *R. capsulatus* Fe protein (60), which shows a similar regulatory mechanism (36). However, it is not known whether the structure of the modifying groups is the same in both cases. In *R. rubrum* a membrane-bound enzyme can reactivate nitrogenase by removing the modifying group. In vivo there seems to be a balance between activation and inactivation (38) that provides fine-tuning of regulation at the protein level. The genetics of this regulatory mechanism is virtually unexplored. As previously mentioned, both oxygen and fixed nitrogen regulate transcription of the genes for nitrogenase. Recently, evidence was obtained that DNA topology has a role in the transcriptional regulation of these genes (46). A hitherto unexplored possibility is that mRNA processing regulates *nif* gene expression. There is evidence for this in many bacteria, including *R. capsulatus,* in which processing of a polycistronic message affects the expression of genes for components of the photosynthetic apparatus (4).

OTHER GENES THAT AFFECT NITROGEN FIXATION Wall and coworkers (67, 69) have isolated several point mutations that affect nitrogen fixation. GTA mapping resulted in the identification of six linkage groups. More recently several groups have studied the organization of *nif* genes in *R. capsulatus*. A cosmid library was constructed in the wide–host range vector pLAFR1 (24). Mobilization of this library by conjugation into *nif* mutants resulted in the cloning of several regions of the genome that contain *nif* genes (1b). Avtges et al (1b) also extended the results by constructing new mutants from recombination of Tn*5* transposons with the *R. capsulatus* chromosome. Puhler et al (56) isolated a 20-kb fragment of DNA containing *nif* genes. Vignais's group mapped *nif* mutants with the R plasmid pTH10 (73) and also conducted biochemical and immunological assays on the mutants.

It is possible to correlate the original GTA linkage groups with more recent information. Three mutants in group IV (J56, J57, and J60) were complemented by a plasmid carrying the *nif* structural genes (2). Biochemical tests showed that these mutants lacked nitrogenase (45). Mutant LJ1 (group VI) was shown by GTA mapping to be unlinked to other *nif* mutants (69). However, it was complemented by an *Eco*RI fragment that is adjacent to the fragment containing the *nifHDK* operon (1b). Thus, groups IV and VI are linked and contain the *nifHDK* operon and a regulatory gene, which may be equivalent to the *K. pneumoniae nifA* gene (P. Vignais, personal communication). A cluster of up to eight *nif* genes was found to correspond to Wall et al's (69) groups I and II (1b). Puhler et al (56) also cloned this region. These genes are not regulatory (62), and mutants in the region show a decreased nitrogenase content (45). Therefore, it is likely that this region contains genes that code for structural proteins that affect synthesis or stability of nitrogenase. The group III mutants were localized in a separate cluster (1b), which contains regulatory genes (62). Finally, group V mutants seem to be regulatory (45). In conclusion, *nif* genes in *R. capsulatus,* unlike those in *K. pneumoniae,* seem to be arranged in three or four clusters that are not in the same region of the genome. The next challenge is to assign roles to these genes and to study the regulation of their expression in the context of the overall metabolism of the cell.

Other Systems

Johnson et al (34) have achieved expression of three different cellulase genes from the gram-positive bacterium *Cellulomonas fimi* in *R. capsulatus* by cloning them in the appropriate expression vectors described above. The carboxymethyl cellulase 2 gene was expressed at high levels in a strain carrying an expression plasmid with the cellulase gene fused in the same frame as the *pufB* gene. However, the resulting cells were unable to use carboxymethyl cellulose as a sole carbon source, presumably because of a lack of excretion of the gene product. Similarly, a *R. capsulatus* strain producing lower levels of carboxymethyl cellulase 1 from a transcriptional fusion was unable to grow on the same carbon source.

PROSPECTUS

Where is the frontier in genetic research with photosynthetic bacteria? Several important tools are currently being built: methods for quick and sure site-directed mutagenesis, Hfr-like genetic exchange systems, and techniques for cloning and ordering the entire genome, to name just a few. Other important tools (transformation for *R. capsulatus,* transduction for *R. sphaeroides*) are not yet being developed, to our knowledge. Basic genetic systems for other genera are in very early stages of development. In general, however, workers in this field will not be frustrated by a lack of tools. What then are the biological problems to which these tools will be applied? The major impact of this area of research will most likely be in protein chemistry (the structure/function rules for electron transport proteins) and molecular genetics (the mechanisms for control of gene expression by light and oxygen).

Literature Cited

1. Ahombo, G., Willison, J. C., Vignais, P. M. 1986. The *nif-hdk* genes are contiguous with a *nif-a*-like regulatory gene in *Rhodobacter capsulatus. Mol. Gen. Genet.* 205:442–45

1a. Allen, J. P., Feher, G., Yeates, T. O., Rees, D. C., Diesenhofer, J., et al. 1986. Structural homology of reaction centers from *Rhodopseudomonas sphaeroides* and *Rhodopseudomonas viridis* as determined by X-ray diffraction. *Proc. Natl. Acad. Sci. USA* 83:8589–93

1b. Avtges, P., Kranz, R., Haselkorn, R. 1985. Isolation and organization of genes for nitrogen fixation in *Rhodopseudomonas capsulata. Mol. Gen. Genet.* 201:363–69

2. Avtges, P., Scolnik, P., Haselkorn, R. 1983. Genetic and physical map of the structural genes *(nifHDK)* coding for the nitrogenase complex of *Rhodopseudomonas capsulata. J. Bacteriol.* 156:251–56

3. Beatty, J. T., Adams, C. W., Cohen, S. N. 1986. Regulation of expression of the

rxcA operon of *Rhodopseudomonas capsulata*. In *Microbial Energy Transduction*, ed. D. C. Youvan, F. Daldal, pp. 27–29. New York: Cold Spring Harbor Lab.

4. Belasco, J. G., Beatty, J. T., Adams, C. W., von Gabain, A., Cohen, S. N. 1985. Differential expression of photosynthesis genes in *R. capsulata* results from segmental differences in stability within the polycistronic *rxcA* transcript. *Cell* 40:171–81
5. Bérard, J., Bélanger, G., Corriveau, P., Gingras, G. 1986. Molecular cloning and sequence of the B880 holochrome gene from *Rhodospirillum rubrum*. *J. Biol. Chem.* 261:82–87
6. Biel, A. J. 1986. Control of bacteriochlorophyll synthesis by light in *Rhodobacter capsulatus*. *J. Bacteriol.* 168:655–59
7. Biel, A. J., Marrs, B. L. 1985. Oxygen does not directly regulate carotenoid biosynthesis in *Rhodopseudomonas capsulata*. *J. Bacteriol.* 162:1320–21
8. Bishop, P. 1986. A second nitrogen fixation system in *Azotobacter vinelandii*. *Trends Biol. Sci.* 11:225–27
9. Bylina, E. J., Ismail, S., Youvan, D. C. 1986. Site-specific mutagenesis of bacteriochlorophyll-binding sites affects biogenesis of the photosynthetic apparatus. See Ref. 3, pp. 63–70
10. Clark, W. G., Davidson, E., Marrs, B. L. 1984. Variation of levels of mRNA coding for antenna and reaction center polypeptides in *Rhodopseudomonas capsulata* in response to changes in oxygen concentration. *J. Bacteriol.* 157:945–48
11. Clayton, R. C., Sistrom, W. R., eds. 1978. *The Photosynthetic Bacteria*. New York: Plenum. 946 pp.
12. Colbeau, A., Godfroy, A., Vignais, P. M. 1986. Cloning of DNA fragments carrying hydrogenase genes of *Rhodopseudomonas capsulata*. *Biochimie* 68:147–55
13. Daldal, F., Cheng, S., Applebaum, J., Davidson, E., Prince, R. C. 1986. Cytochrome c_2 is not essential for photosynthetic growth of *Rhodopseudomonas capsulata*. *Proc. Natl. Acad. Sci. USA* 83:2012–16
14. Daldal, F., Davidson, E., Cheng, S. 1987. Isolation of the structural genes for the Rieske Fe-S Protein, cytochrome *b* and cytochrome c_1, all components of the ubiquinol: cytochrome c_2 oxidoreductase complex of *Rhodopseudomonas capsulata*. *J. Mol. Biol.* In press
15. Daldal, F., Davidson, E., Cheng, S., Naiman, B., Rook, S. 1986. Genetic analysis of the structure and function of the ubiquinol-cytochrome c_2 oxidoreductase of *Rhodopseudomonas capsulata*. See Ref. 3, pp. 113–19
16. Davidson, E., Daldal, F. 1987. Primary structure of the bc_1 complex of *Rhodopseudomonas capsulata:* nucleotide sequence of the *pet* operon encoding the Reiske, cytochrome *b*, and cytochrome c_1 apoproteins. *J. Mol. Biol.* In press
17. Davidson, E., Daldal, F. 1987. The *fbc* operon, encoding the Rieske FeS protein, cytochrome *b* and cytochrome c_1 apoproteins, previously described from *Rhodopseudomonas sphaeroides*, is from *Rhodopseudomonas capsulata*. *J. Mol. Biol.* In press
18. Dierstein, R. 1984. Synthesis of pigment-binding protein in toluene-treated *Rhodopseudomonas capsulata* and in cell-free systems. *Eur. J. Biochem.* 138:509–18

18a. Diesenhofer, J., Epp, O., Miki, R., Huber, R., Michel, H. 1984. X-ray structure analysis of a membrane protein complex: electron density map at 3 Å resolution and partial model of the photosynthetic reaction center of *Rhodopseudomonas viridis*. *J. Mol. Biol.* 180:385–98

19. Donald, R., Nees, D., Raymond, C., Loroch, A., Ludwig, R. 1986. Characterization of three genomic loci encoding *Rhizobium* sp. strain ORS571 nitrogen fixation genes. *J. Bacteriol.* 165:72–81
20. Donohue, T. J., Hoger, J. H., Kaplan, S. 1986. Cloning and expression of the *Rhodobacter sphaeroides* reaction center H gene. *J. Bacteriol.* 168:953–61
21. Donohue, T. J., McEwan, A. G., Kaplan, S. 1986. Cloning, DNA sequence, and expression of the *Rhodobacter sphaeroides* cytochrome c_2 gene. *J. Bacteriol.* 168:962–72
22. Drews, G., 1985. Structure and functional organization of light-harvesting complexes and photochemical reaction centers in membranes of phototrophic bacteria. *Microbiol. Rev.* 49:59–70

22a. Fornari, C. S., Kaplan, S. 1982. Genetic transformation of *Rhodopseudomonas sphaeroides* by plasmid DNA. *J. Bacteriol.* 152:89–97

23. Fornari, C., Kaplan, S. 1983. Identification of nitrogenase and carboxylase genes in the photosynthetic bacteria and cloning of a carboxylase gene from *Rhodopseudomonas spheroides*. *Gene* 25: 291–99
24. Friedman, A., Long, S., Brown, S., Buikema, W., Ausubel, F. 1982. Construction of a broad host range cosmid

cloning vector and its use in the genetic analysis of *Rhizobium* mutants. *Gene* 18:289–96

24a. Gabellini, N., Harnisch, U., McCarthy, J. E. G., Hauska, G., Sebald, W. 1985. Cloning and expression of the *fbc* operon encoding the FeS protein, cytochrome *b* and cytochrome c_1 from *Rhodopseudomonas sphaeroides* b/c_1 complex. *EMBO J.* 4:549–53

25. Gibson, J., Tabita, F. R. 1986. Isolation of the *Rhodopseudomonas sphaeroides* form I ribulose 1,5-bisphosphate carboxylase/oxygenase large and small subunit genes and expression of the active hexadecameric enzyme in *Escherichia coli*. *Gene* 44:271–78

26. Giuliano, G., Pollock, D., Scolnik, P. A. 1986. The gene *crtI* mediates the conversion of phytoene into colored carotenoids in *Rhodopseudomonas capsulata*. *J. Biol. Chem.* 261:12925–29

27. Hallenbeck, P. 1987. Molecular aspects of nitrogen fixation by photosynthetic bacteria. *CRC Crit. Rev. Microbiol.* 14:1–48

28. Hallenbeck, P., Meyer, C. M., Vignais, P. M. 1982. Nitrogenase from the photosynthetic bacterium *Rhodopseudomonas capsulata:* purification and molecular properties. *J. Bacteriol.* 149:708–17

29. Haselkorn, R. 1986. Organization of the genes for nitrogen fixation in photosynthetic bacteria and cyanobacteria. *Ann. Rev. Microbiol.* 40:525–47

30. Hearst, J., Alberti, M., Doolittle, R. 1985. A putative nitrogenase reductase gene found in the nucleotide sequences from the photosynthetic gene cluster of *R. capsulata*. *Cell* 40:219–20

31. Hüdig, H., Kaufmann, N., Drews, G. 1986. Respiratory deficient mutants of *Rhodopseudomonas capsulata*. *Arch. Microbiol.* 145:378–85

32. Imhoff, J. F., Truper, H. G., Pfennig, N. 1984. Rearrangement of the species and genera of the phototrophic "purple nonsulfur bacteria." *Int. J. Syst. Bacteriol.* 34:340–43

33. Jasper, P., Hu, N. T., Marrs, B. 1978. Transfer of plasmid-borne kanamycin-resistance genes to *Rhodopseudomonas capsulata* by transformation and conjugation. *Abstr. Annu. Meet. Am. Soc. Microbiol.* 1978:114

34. Johnson, J. A., Wong, W. K. R., Beatty, J. T. 1986. Expression of cellulase genes in *Rhodobacter capsulatus* by use of plasmid expression vectors. *J. Bacteriol.* 167:604–10

35. Jones, R., Woodley, P., Robson, R. 1984. Cloning and organization of some genes for nitrogen fixation from *Azotobacter chroococcum* and their expression in *Klebsiella pneumoniae*. *Mol. Gen. Genet.* 197:318–27

36. Jouanneau, Y., Meyer, C., Vignais, P. 1983. Regulation of nitrogenase activity through iron protein interconversion into an active and an inactive form in *Rhodopseudomonas capsulata*. *Biochim. Biophys. Acta* 749:318–28

37. Kallas, T., Rebiere, M., Rippka, R., Tandeau de Marsac, N. 1983. The structural *nif* genes of the cyanobacteria *Gloeothece* sp. and *Calothrix* sp. share homology with those of *Anabaena* sp. but the *Gloeothece* genes have a different arrangement. *J. Bacteriol.* 155:427–31

38. Kanemoto, R., Ludden, P. 1984. Effect of ammonium, darkness and phenazide methosulfate on whole-cell nitrogenase activity and Fe protein modification in *Rhodospirillum rubrum*. *J. Bacteriol.* 158:713–18

39. Kaplan, S., Marrs, B. L. 1986. Proposed nomenclature for photosynthetic procaryotes. *ASM News* 52:242

40. Kaufmann, N., Hüdig, H., Drews, G. 1984. Transposon Tn*5* mutagenesis of genes for the photosynthetic apparatus in *Rhodopseudomona capsulata*. *Mol. Gen. Genet.* 198:153–58

41. Kennell, D., Riezman, H. 1977. Transcription and translation initiation frequencies of the *Escherichia coli lac* operon. *J. Mol. Biol.* 114:1–21

42. Klug, G., Drews, G. 1984. Construction of a gene bank of *Rhodopseudomonas capsulata* using a broad host range DNA cloning system. *Arch. Microbiol.* 139:319–25

43. Klug, G., Kaufmann, N., Drews, G. 1984. The expression of genes encoding proteins of B800–850 antenna pigment complex and ribosomal RNA of *Rhodopseudomonas capsulata*. *FEBS Lett.* 177:61–65

44. Klug, G., Kaufmann, N., Drews, G. 1985. Gene expression of pigment-binding proteins of the bacterial photosynthetic apparatus: Transcription and assembly in the membrane of *Rhodopseudomonas capsulata*. *Proc. Natl. Acad. Sci. USA* 82:6485–89

44a. Klug, G., Liebetanz, R., Drews, G. 1986. The influence of bacteriochlorophyll biosynthesis on the formation of pigment-proteins and asembly of pigment protein complexes in *Rhodopseudomonas capsulata*. *Arch. Microbiol.* 146:284–91

44b. Kondorosi, A., Kiss, G. B., Forrai, T., Vincze, E., Banfalvi, Z. 1977. Circular

linkage map of *Rhizobium meliloti*. *Nature* 268:525–27
45. Kranz, R., Haselkorn, R. 1985. Characterization of *nif* regulatory genes in *Rhodopseudomonas capsulata* using *lac* gene fusions. *Gene* 40:203–15
46. Kranz, R., Haselkorn, R. 1986. Anaerobic regulation of nitrogen fixation genes in *Rhodopseudomonas capsulata*. *Proc. Natl. Acad. Sci. USA* 83:6805–9
47. Magnin, J. P., Willison, J. C., Vignais, P. M. 1987. Elimination of R plasmids from the photosynthetic bacterium *Rhodobacter capsulatus*. *FEMS Lett.* 41:157–61
48. Marrs, B. 1974. Genetic recombination in *Rhodopseudomonas capsulata*. *Proc. Natl. Acad. Sci. USA* 71:971–73
49. Marrs, B. L. 1983. Genetics and molecular biology. In *Studies in Microbiology,* Vol. 4, *The Phototrophic Bacteria: Anaerobic Life in the Light,* ed. J. G. Ormerod, pp. 186–214. Oxford / London / Edinburgh / Boston/Melbourne: Blackwell
50. Muller, E. D., Chory, J., Kaplan, S. 1985. Cloning and characterization of the form II ribulose-1,5-bisphosphate carboxylase gene of *Rhodopseudomonas sphaeroides*. *J. Bacteriol.* 161:469–72
51. Nargang, F., McIntosh, L., Somerville, C. 1984. Nucleotide sequence of the ribulose bisphosphate carboxylase gene from *Rhodospirillum rubrum*. *Mol. Gen. Genet.* 193:220–24
52. Pope, M., Murrell, S., Ludden, P. 1985. Covalent modification of the iron protein of nitrogenase from *Rhodospirillum rubrum* by adenosine diphosphoribosylation of a specific arginine residue. *Proc. Natl. Acad. Sci. USA* 82:3173–72
53. Prentki, P., Krisch, H. M. 1984. In vitro insertional mutagenesis with a selectable DNA fragment. *Gene* 29:303–13
54. Prince, R. C., Davidson, E., Daldal, F. 1986. Genetic and biophysical approaches to elucidating the mechanism of the cytochrome bc_1 complex. See Ref. 3, pp. 87–92
55. Prince, R. C., Davidson, E., Haith, C. E., Daldal, F. 1986. Photosynthetic electron transfer in the absence of cytochrome c_2 is not essential for electron flow from the cytochrome bc_1 complex to the photochemical reaction center. *Biochemistry* 25:5208–14
56. Puhler, A., Aguilar, M., Hynes, M., Muller, P., Klipp, W., et al. 1984. Advances in the genetics of free-living and symbiotic nitrogen fixing bacteria. In *Advances in Nitrogen Fixation Research,* ed. C. Veeger, W. Newton, pp. 609–19. The Hague: Nijhoff/Junk
57. Putnoky, P., Kiss, G. B., Ott, J., Kondorosi, A. 1983. Tn*5* carries a streptomycin resistance determinant downstream from the kanamycin resistance gene. *Mol. Gen. Genet.* 191:288–94
58. Quivey, R. G., Tabita, F. R. 1984. Cloning and expression in *Escherichia coli* of the form II ribulose 1,5-bisphosphate carboxylase/oxygenase gene from *Rhodopseudomonas sphaeroides*. *Gene* 31:91–101
59. Ruvkun, G., Ausubel, F. 1980. Interspecies homology of nitrogenase genes. *Proc. Natl. Acad. Sci. USA* 77:191–95
60. Schumann, J., Waitches, G., Scolnik, P. 1986. A DNA fragment hybridizing to a *nif* probe in *Rhodobacter capsulatus* is homologous to a 16S rRNA gene. *Gene* 48:79–90
61. Scolnik, P., Haselkorn, R. 1984. Activation of extra copies of genes coding for nitrogenase in *Rhodopseudomonas capsulata*. *Nature* 307:289–92
62. Scolnik, P., Virosco, J., Haselkorn, R. 1983. The wild-type gene for glutamine synthetase restores ammonia control of nitrogen fixation to Gln^- *(glnA)* mutants of *Rhodopseudomonas capsulata*. *J. Bacteriol.* 155:180–85
63. Simon, R., Priefer, U., Peuhler, A. 1983. Vector plasmids for in-vivo and in-vitro manipulations of gram-negative bacteria. In *Molecular Biology of Bacteria-Plant Interaction,* ed. A. Peuhler, p. 99. Berlin/Heidelberg/New York: Springer
63a. Sistrom, W. R. 1977. Transfer of chromosomal genes mediated by plasmid R68.45 in *Rhodopseudomonas sphaeroides*. *J. Bacteriol.* 131:526–32
64. Somerville, C., Somerville, S. C. 1984. Cloning and expression of the *Rhodospirillum rubrum* ribulose bisphosphate carboxylase gene in *E. coli*. *Mol. Gen. Genet.* 193:214–19
65. Deleted in proof
66. Taylor, D. P., Cohen, S. N., Clark, W. G., Marrs, B. L. 1983. Alignment of the genetic and restriction maps of the photosynthesis region of the *Rhodopseudomonas capsulata* chromosome by a conjugation-mediated marker rescue technique. *J. Bacteriol.* 154:580–90
66a. Vignais, P. M., Colbeau, A., Willison, J. C., Jouanneau, Y. 1985. Hydrogenase, nitrogenase, and hydrogen metabolism in the photosynthetic bacteria. *Adv. Microb. Physiol.* 26:156–234
67. Wall, J., Braddock, K. 1984. Mapping

of *Rhodopseudomonas capsulata nif* genes. *J. Bacteriol.* 158:404–2

68. Wall, J., Gest, H. 1979. Derepression of nitrogenase activity in glutamine auxotrophs of *Rhodopseudomonas capsulata*. *J. Bacteriol.* 137:1459–63
69. Wall, J., Love, J., Quinn, S. 1984. Spontaneous Nif⁻ mutants of *Rhodopseudomonas capsulata*. *J. Bacteriol.* 159:652–57
70. Weaver, P. F., Wall, J. D., Gest, H. 1975. Characterization of *Rhodopseudomonas capsulata*. *Arch. Microbiol.* 105:207–16
71. Williams, J. C., Steiner, L. A., Feher, G., Simon, M. I. 1984. Primary structure of the L subunit of the reaction center of *Rhodopseudomonas sphaeroides*. *Proc. Natl. Acad. Sci. USA* 81:7303–8
72. Williams, J. C., Steiner, L. A., Ogden, R. C., Simon, M. I., Feher, G. 1983. Primary structure of the M subunit of the reaction center from *Rhodopseudomonas sphaeroides*. *Proc. Natl. Acad. Sci. USA* 80:6505–9
73. Willison, J., Ahombo, G., Chabert, J., Magnin, J.-P., Vignais, P. 1985. Genetic mapping of the *Rhodopseudomonas capsulata* chromosome shows nonclustering of genes involved in nitrogen fixation. *J. Gen. Microbiol.* 131:3001–15
74. Willison, J. C., Magnin, J. P., Vignais, P. M. 1987. Isolation and characterization of *Rhodobacter capsulatus* strains lacking endogenous plasmids. *Arch. Microbiol.* 147:134–42
75. Yen, H. C., Marrs, B. 1976. Map of genes for carotenoid and bacteriochlorophyll biosynthesis in *Rhodopseudomonas capsulata*. *J. Bacteriol.* 126:619–29

75a. Youvan, D. C., Alberti, M., Begusch, H., Bylina, E. J., Hearst, J. E. 1984. Reaction center and light-harvesting I genes from *Rhodopseudomonas capsulata*. *Proc. Natl. Acad. Sci. USA* 81:189–92

76. Youvan, D. C., Bylina, E. J., Alberti, M., Begusch, H., Hearst, J. E. 1984. Nucleotide and deduced polypeptide sequences of the photosynthetic reaction center, B870 antenna, and flanking polypeptides from *R. capsulata*. *Cell* 37:949–57
77. Youvan, D. C., Ismail, S. 1985. Light-harvesting II (B800–B850 complex) from *Rhodopseudomonas capsulata*. *Proc. Natl. Acad. Sci. USA* 82:58–62
78. Youvan, D. C., Ismail, S., Bylina, E. J. 1985. Chromosomal deletion and plasmid complementation of the photosynthetic reaction center and light-harvesting genes from *Rhodopseudomonas capsulata*. *Gene* 38:19–30
79. Yu, P.-L., Hohn, B., Falk, H., Drews, G. 1982. Molecular cloning of the ribosomal RNA genes of the photosynthetic bacterium *Rhodopseudomonas capsulata*. *Mol. Gen. Genet.* 188:392–98
80. Zhu, Y. S., Cook, D. N., Leach, F., Armstrong, G. A., Alberti, M., et al. 1986. Oxygen-regulated mRNAs for light-harvesting and reaction center complexes and for bacteriochlorophyll and carotenoid biosynthesis in *Rhodobacter capsulatus* during the shift from anaerobic to aerobic growth. *J. Bacteriol.* 168:1180–88
81. Zhu, Y. S., Hearst, J. E. 1986. Regulation of expression of genes for light-harvesting antenna proteins LH-I and LH-II; reaction center polypeptides RC-L, RC-M, and RC-H; and enzymes of bacteriochlorophyll and carotenoid biosynthesis in *Rhodobacter capsulatus* by light and oxygen. *Proc. Natl. Acad. Sci. USA* 83:7613–17
82. Zhu, Y. S., Kaplan, S. 1985. Effects of light, oxygen, and substrates on steady-state levels of mRNA coding for ribulose-1,5-bisphosphate carboxylase and light-harvesting and reaction center polypeptides in *Rhodopseudomonas sphaeroides*. *J. Bacteriol.* 162:925–32
83. Zhu, Y. S., Kiley, P. J., Donohue, T. J., Kaplan, S. 1986. Origin of the mRNA stoichiometry of the *puf* operon in *Rhodobacter spheroides*. *J. Biol. Chem.* 261:10366–74
84. Zsebo, K. M., Hearst, J. E. 1984. Genetic-physical mapping of a photosynthetic gene cluster from R. capsulata. *Cell* 37:937–47
85. Zsebo, K. M., Wu, F., Hearst, J. E. 1984. Physical mapping and Tn*5.7* mutagenesis of pRPS404 containing photosynthesis genes from *Rhodopseudomonas capsulata*. *Plasmid* 11:182–84

SUBJECT INDEX

Q

R

CUMULATIVE INDEXES

CONTRIBUTING AUTHORS, VOLUMES 37–41

CHAPTER TITLES, VOLUMES 37–41

nual Reviews Inc.

)NPROFIT SCIENTIFIC PUBLISHER

4139 El Camino Way
P.O. Box 10139
Palo Alto, CA 94303-0897 • USA

ORDER FORM

Now you can order
TOLL FREE
1-800-523-8635
(except California)

Reviews Inc. publications may be ordered directly from our office by mail or use our Toll Free one line (for orders paid by credit card or purchase order, and customer service calls only); n booksellers and subscription agents, worldwide; and through participating professional es. Prices subject to change without notice. ARI Federal I.D. #94-1156476

ividuals: Prepayment required on new accounts by check or money order (in U.S. dollars, check drawn on . bank) or charge to credit card — American Express, VISA, MasterCard.
titutional buyers: Please include purchase order number.
dents: $10.00 discount from retail price, per volume. Prepayment required. Proof of student status must provided (photocopy of student I.D. or signature of department secretary is acceptable). Students must d orders direct to Annual Reviews. Orders received through bookstores and institutions requesting student s will be returned. You may order at the Student Rate for a maximum of 3 years.
fessional Society Members: Members of professional societies that have a contractual arrangement Annual Reviews may order books through their society at a reduced rate. Check with your society for information.
Free Telephone orders: Call 1-800-523-8635 (except from California) for orders paid by credit card or chase order and customer service calls only. California customers and all other business calls use -493-4400 (not toll free). Hours: 8:00 AM to 4:00 PM, Monday-Friday, Pacific Time.

r orders: Please list the volumes you wish to order by volume number.
ng orders: New volume in the series will be sent to you automatically each year upon publication. Cancel-ay be made at any time. Please indicate volume number to begin standing order.
lication orders: Volumes not yet published will be shipped in month and year indicated.
nia orders: Add applicable sales tax.
e paid (4th class bookrate/surface mail) **by Annual Reviews Inc.** Airmail postage or UPS, extra.

UAL REVIEWS SERIES		Prices Postpaid per volume USA & Canada/elsewhere	Regular Order Please send:	Standing Order Begin with:
			Vol. number	Vol. number
l Review of ANTHROPOLOGY				
. 1-14	(1972-1985)	**$27.00/$30.00**		
. 15-16	(1986-1987)	**$31.00/$34.00**		
17	(avail. Oct. 1988)	**$35.00/$39.00**	Vol(s). ________	Vol. ________
l Review of ASTRONOMY AND ASTROPHYSICS				
. 1-2, 4-20	(1963-1964; 1966-1982)	**$27.00/$30.00**		
. 21-25	(1983-1987)	**$44.00/$47.00**		
26	(avail. Sept. 1988)	**$47.00/$51.00**	Vol(s). ________	Vol. ________
l Review of BIOCHEMISTRY				
. 30-34, 36-54	(1961-1965; 1967-1985)	**$29.00/$32.00**		
. 55-56	(1986-1987)	**$33.00/$36.00**		
57	(avail. July 1988)	**$35.00/$39.00**	Vol(s). ________	Vol. ________
l Review of BIOPHYSICS AND BIOPHYSICAL CHEMISTRY				
. 1-11	(1972-1982)	**$27.00/$30.00**		
. 12-16	(1983-1987)	**$47.00/$50.00**		
17	(avail. June 1988)	**$49.00/$53.00**	Vol(s). ________	Vol. ________
l Review of CELL BIOLOGY				
1	(1985)	**$27.00/$30.00**		
. 2-3	(1986-1987)	**$31.00/$34.00**		
4	(avail. Nov. 1988)	**$35.00/$39.00**	Vol(s). ________	Vol. ________

ANNUAL REVIEWS SERIES		Prices Postpaid per volume USA & Canada/elsewhere	Regular Order Please send:	Standi Begi
			Vol. number	Vol.
Annual Review of COMPUTER SCIENCE				
Vols. 1-2	(1986-1987)	**$39.00/$42.00**		
Vol. 3	(avail. Nov. 1988)	**$45.00/$49.00**	Vol(s). ________	Vol. ___
Annual Review of EARTH AND PLANETARY SCIENCES				
Vols. 1-10	(1973-1982)	**$27.00/$30.00**		
Vols. 11-15	(1983-1987)	**$44.00/$47.00**		
Vol. 16	(avail. May 1988)	**$49.00/$53.00**	Vol(s). ________	Vol. ___
Annual Review of ECOLOGY AND SYSTEMATICS				
Vols. 2-16	(1971-1985)	**$27.00/$30.00**		
Vols. 17-18	(1986-1987)	**$31.00/$34.00**		
Vol. 19	(avail. Nov. 1988)	**$34.00/$38.00**	Vol(s). ________	Vol. ___
Annual Review of ENERGY				
Vols. 1-7	(1976-1982)	**$27.00/$30.00**		
Vols. 8-12	(1983-1987)	**$56.00/$59.00**		
Vol. 13	(avail. Oct. 1988)	**$58.00/$62.00**	Vol(s). ________	Vol. ___
Annual Review of ENTOMOLOGY				
Vols. 10-16, 18-30	(1965-1971; 1973-1985)	**$27.00/$30.00**		
Vols. 31-32	(1986-1987)	**$31.00/$34.00**		
Vol. 33	(avail. Jan. 1988)	**$34.00/$38.00**	Vol(s). ________	Vol. ___
Annual Review of FLUID MECHANICS				
Vols. 1-4, 7-17	(1969-1972, 1975-1985)	**$28.00/$31.00**		
Vols. 18-19	(1986-1987)	**$32.00/$35.00**		
Vol. 20	(avail. Jan. 1988)	**$34.00/$38.00**	Vol(s). ________	Vol. ___
Annual Review of GENETICS				
Vols. 1-19	(1967-1985)	**$27.00/$30.00**		
Vols. 20-21	(1986-1987)	**$31.00/$34.00**		
Vol. 22	(avail. Dec. 1988)	**$34.00/$38.00**	Vol(s). ________	Vol. ___
Annual Review of IMMUNOLOGY				
Vols. 1-3	(1983-1985)	**$27.00/$30.00**		
Vols. 4-5	(1986-1987)	**$31.00/$34.00**		
Vol. 6	(avail. April 1988)	**$34.00/$38.00**	Vol(s). ________	Vol. ___
Annual Review of MATERIALS SCIENCE				
Vols. 1, 3-12	(1971, 1973-1982)	**$27.00/$30.00**		
Vols. 13-17	(1983-1987)	**$64.00/$67.00**		
Vol. 18	(avail. August 1988)	**$66.00/$70.00**	Vol(s). ________	Vol. ___
Annual Review of MEDICINE				
Vols. 1-3, 6, 8-9	(1950-1952, 1955, 1957-1958)			
11-15, 17-36	(1960-1964, 1966-1985)	**$27.00/$30.00**		
Vols. 37-38	(1986-1987)	**$31.00/$34.00**		
Vol. 39	(avail. April 1988)	**$34.00/$38.00**	Vol(s). ________	Vol. ___
Annual Review of MICROBIOLOGY				
Vols. 18-39	(1964-1985)	**$27.00/$30.00**		
Vols. 40-41	(1986-1987)	**$31.00/$34.00**		
Vol. 42	(avail. Oct. 1988)	**$34.00/$38.00**	Vol(s). ________	Vol. ___